CROP PROTECTION AGENTS FROM NATURE: NATURAL PRODUCTS AND ANALOGUES

The editor would like to express his sincere thanks to Ms Helen Cottrell for her support throughout the preparation of this book and her encouragement, advice and editorial expertise as the volume went to press. Without her wisdom and experience the project would not have been completed.

Crop Protection Agents from Nature: Natural Products and Analogues

Edited by

Leonard G. Copping

LGConsultants, Saffron Walden, Essex, UK

THE ROYAL
SOCIETY OF
CHEMISTRY
Information
Services

Critical Reports on Applied Chemistry Volume 35

ISBN 0–85404–414–0

A catalogue record of this book is available from the British Library.

Published by The Royal Society of Chemistry,
Thomas Graham House, Science Park, Milton Road, Cambridge CB4 4WF, UK

Typeset by Paston Press Ltd, Loddon, Norfolk, NR14 6JD
Printed in Great Britain by Hartnolls Ltd, Bodmin, UK

Contents

Introduction

It has often been argued that the agrochemical industry is a mature industry that is unlikely to return to the days of the 1960s and 1970s, when most major chemical companies were investing in crop protection discovery research. In those times, the returns that could be almost guaranteed from the introduction of a new agrochemical were considerable; the market was large and expanding, the competion was limited, and regulation and environmental controls were of little concern.

Things have changed. The subsidies of crops in Europe and the Far East are being withdrawn, there is increasing concern about the environmental fate and effects on non-target organisms of crop protection agents, and the standards of weed, pathogen, and insect control are exceedingly high. This is a reflection of the excellence of the science directed towards agricultural practice, and the outstanding biological effects that are achieved routinely by today's agrochemicals, without adverse effects on the environment, are a testament to the scientific endeavours of the agrochemical industry. Nevertheless, a study of the 10th Edition of *The Pesticide Manual* shows 725 different active ingredients which are available commercially. It is true that this number will have changed a little since the date of publication (1994), new compounds being introduced and some having been lost or withdrawn. Nevertheless, there is a large number of compounds with a diversity of chemical complexity and biological efficacy.

However, some of these are related to naturally occurring chemicals or represent biological control systems. The pyrethroid insecticides represent the largest single group that owes its origins to chemicals that occur naturally. *The Pesticide Manual* lists 33 synthetic compounds within this chemical class, plus the pyrethrins, chrysanthemates and pyrethrates. They are all represented in Table 1.

If the synthetic pyrethroids are omitted from this list there are only 28 (plus three natural pyrethrins) chemical and 28 biological compounds included. The 'natural' chemicals represent only 3.9% of the chemical entries in *The Pesticide Manual* (8.4% if synthetic pyrethroids are included) and 3.9% of the entries are biological systems. A number of the entries are closely related and often the products of the same company. Pyrethroids plus natural pyrethrins represent 36 of the total entries, derivatives of nereistoxin represent three, biologically active compounds from higher plants have three entries, strobilurin analogues have two entries, as do pyrrolnitrin derivatives, pheromones and pheromone analogues have three entries, and fermentation product fungicides used in rice culture are represented by four compounds.

Table 1 *Compounds and biological control systems of agrochemical significance [data from* The Pesticide Manual, *10th Edn., C. Tomlin (ed)]*

Compound	*Activity*	*Company*
Abamectin	Microbial insecticide/ acaricide	Merck Sharp and Dohme
Milbemectin	Microbial insecticide/ acaricide	Sankyo
Acrinathrin	Pyrethroid insecticide	Roussel-Uclaf
Allethrin	Pyrethroid insecticide	Sumitomo
Bifenthrin	Pyrethroid insecticide	FMC
Bioallethrin	Pyrethroid insecticide	Roussel-Uclaf
Bioallethrin S-cyclopentyl isomer	Pyrethroid insecticide	Roussel-Uclaf
Bioresmethrin	Pyrethroid insecticide	FMC, Roussel-Uclaf, Sumitomo, Wellcome
Cycloprothrin	Pyrethroid insecticide	Nippon Kayaku
Cyfluthrin	Pyrethroid insecticide	Bayer
β-Cyfluthrin	Pyrethroid insecticide	Bayer
Cyhalothrin	Pyrethroid insecticide	Zeneca
λ-Cyhalothrin	Pyrethroid insecticide	Zeneca
Cypermethrin	Pyrethroid insecticide	Ciba, Zeneca, Cyanamid
α-Cypermethrin	Pyrethroid insecticide	Cyanamid
β-Cypermethrin	Pyrethroid insecticide	Chinoin Pharmaceutical and Chemical Works
ζ-Cypermethrin	Pyrethroid insecticide	FMC
Cyphenothrin	Pyrethroid insecticide	Sumitomo
Delthamethrin	Pyrethroid insecticide	Roussel-Uclaf
Empenthrin	Pyrethroid insecticide	Sumitomo
Esfenvalerate	Pyrethroid insecticide	Sumitomo
Fenpropathrin	Pyrethroid insecticide	Sumitomo
Fenvalerate	Pyrethroid insecticide	Sumitomo
Flucythrinate	Pyrethroid insecticide	Cyanamid
τ-Fluvalinate	Pyrethroid insecticide	Sandoz
Permethrin	Pyrethroid insecticide	FMC, Zeneca, Cyanamid, Sumitomo, Penick, Wellcome
Phenothrin	Pyrethroid insecticide	Sumitomo
Prallethrin	Pyrethroid insecticide	Sumitomo
Resmethrin	Pyrethroid insecticide	FMC, Penick, Sumitomo
RU 15525	Pyrethroid insecticide	Rousell-Uclaf
Tefluthrin	Pyrethroid insecticide	Zeneca
Tetramethrin	Pyrethroid insecticide	Sumitomo
Tetramethrin [(1*R*)-isomers]	Pyrethroid insecticide	Sumitomo
Tralomethrin	Pyrethroid insecticide	Roussel-Uclaf
Transfluthrin	Pyrethroid insecticide	Bayer
Natural pyrethrins, chrysanthemates, and pyrethrates	Botanical insecticides	Various
Adoxyphyes orana granulosis virus	Baculovirus – insecticide	Andermatt

Continued

Table 1 *Continued*

Compound	*Activity*	*Company*
Cydia pomonella granulosis virus	Baculovirus – insecticide	Andermatt, Natural Plant Protection
Mamestra brassicae nuclear polyhedrosis virus	Baculovirus – insecticide	Natural Plant Protection
Spodoptera exigua nuclear polyhedrosis virus	Baculovirus – insecticide	DuPont, Brinkmann
Neodiprion sertifer nuclear polyhedrosis virus	Baculovirus – insecticide	USDA Forestry Service, UK Forestry Service
Metaseiulus occidentalis	Predatory mite – acaricide	Koppert
Phytoseiulus persimilis	Predatory mite – acaricide	Ciba Bunting, Koppert
Ampelomyces quisqualis	Fungus – fungicide	Ecogen
Beauveria bassiana	Fungus – insecticide	Natural Plant Protection
Beauveria brongniartii	Fungus – insecticide	Andermatt, Natural Plant Protection
Phlebiopsis gigantea	Fungus – fungicide	Kemira
Verticillium lecanii	Fungus – insecticide	Koppert, Hansen
Aphidius colemani	Predatory wasp – insecticide	Ciba Bunting, Koppert
Dacnusa sibirica	Predatory wasp – insecticide	Ciba Bunting, Koppert
Diglyphus isaea	Predatory wasp – insecticide	Ciba Bunting, Koppert
Encarsia formosa	Predatory wasp – insecticide	Ciba Bunting, Koppert
Leptomastix dactylopii	Predatory wasp – insecticide	Koppert
Metaphycus helvolus	Predatory wasp – insecticide	Ciba Bunting
Trichogamma evanescens	Predatory wasp – insecticide	Koppert
Bacillus subtilis	Bacterium – fungicide	Gustafson
Bacillus thuringiensis	Bacterium – insecticide	Various
Bacillus thuringiensis delta endotoxin	Bacterial toxin – insecticide	Mycogen
Streptomyces griseoviridis	Bacterium – fungicide	Kemira
Azadirachtin	Botanical insecticide	Agridyne
Nicotine	Botanical insecticide	Various
Rotenone	Botanical insecticide	Various
BAS 490F	Strobilurin analogue – fungicide	BASF
ICI A5504	Strobilurin analogue – fungicide	Zeneca
Bensultap	Nereistoxin analogue – insecticide	Takeda
Cartap	Nereistoxin analogue – insecticide	Takeda
Thiocyclam	Nereistoxin analogue – insecticide	Sandoz
Bilanafos	Microbial herbicide	Meiji Seika Kaisha
Glufosinate	Bilanafos analogue – herbicide	AgrEvo
Blasticidin-S	Microbial fungicide	Kaken, Kumiai, Nihon Nohyaku

Continued

Table 1 *Continued*

Compound	*Activity*	*Company*
Kasugamycin	Microbial fungicide	Hokko, Kaken, Kumiai
Polyoxins	Microbial fungicide	Nihon Nohyaku
Validomycin	Microbial fungicide	Takeda
Natamycin	Microbial fungicide	Gist-Brocades
Streptomycin	Microbial bactericide	Various
Chrysoperla carnea	Predatory lacewing – insecticide	Biolab, Ciba Bunting, Rincon Vitova
Cryptolaemus montrouzieri	Predatory beetle – insecticide	Koppert
Ergocalciferol	Vitamin D_2 – rodenticide	Rentokil, Sorex
Vitamin D_3	Vitamin-based rodenticide	Bayer
Fenpiclonil	Fungicide based on pyrrolenitrin	Ciba
Fludioxonil	Fungicide based on pyrrolenitrin	Ciba
Gibberellins	Plant growth regulator	Zeneca
Indol-3-ylacetic acid	Plant growth regulator	Various
Heterorhabditis bacteriophora and *H. megidis*	Nematode – insecticide	Andermatt, AGC MicroBio, Koppert
Steinernema feltiae	Nematode – insecticide	AGC Microbio, Andermatt, Koppert
Steinernema scapterisci	Nematode – insecticide	Ecogen
Hippodamia convergens	Ladybird – insecticide	Koppert
Hydroprene	Pheromone analogue	Sandoz
Methoprene	Pheromone analogue	Sandoz
Muscalure	Insect pheromone	Sandoz
Macrolophus caliginosus	Predatory bug – insecticide	Koppert
Orius spp.	Predatory bug – insecticide	Koppert
Strychnine	Mammalian pest control	Various

Why should the agrochemical industry be interested in natural products as a potential source of new products, either new chemical structures that are amenable to synthetic modification or as models for new biochemical modes of action? Traditionally, the crop protection industry lags behind the pharmaceutical industry in introducing new concepts into discovery research. It is true that the pharmaceutical groups are more interested in finding new compounds with known and approved modes of action, as the registration process for these compounds is easier and less costly than the introduction of a new compound with a completely new mode of action. Hence, there is a tendency to stay with the traditional sources of compound evaluated in traditional biochemical assays. This is not the case with the crop protection business. Normally, the company that discovers a compound with a brand new mode of action (and, possibly, use) will be the winner. There is no relaxation of regulatory requirements because a compound has the same mode of action as an existing product.

It is no less costly to register an analogue of a market leader than a brand new molecule of new or unknown mode of action. So the search is on for new classes of chemistry that will yield tomorrow's new pesticides. All too often, however, the problems experienced with natural product chemistry, be it clean-up of the biological extract or synthesis of the active compound with the correct chirality, deter the synthesis chemist from initiating a search for natural based toxins.

What can be learned from the use of natural products in medicine? At present, 25% of all prescription drugs in the USA include natural products as the active ingredient. It has been estimated that the value of plant products to developed world medicine may be well over $6.25 billion a year. The sales of the anti-cancer drugs vincristine and vinblastine, obtained from the Madagascan rosy periwinkle, are estimated to be in excess of $100 million annually. The National Cancer Institute in Bethseda, Maryland, has evaluated over 28 000 plant products from more than 20 countries as potential, marketable anti-cancer products. Their most significant discovery is taxol, a diterpenoid from the bark of *Taxus brevifolia* Nutt. and the leaves of various other *Taxus* species. Kew Gardens have isolated a polyhydroxy alkaloid from *Castanospermum australe* (castanospermine) that is showing potential as a treatment for human immundeficiency virus (HIV) infections.

Micro-organisms are a rich source of biologically active pharmaceuticals. The discovery of penicillin led to the development of other naturally occurring β-lactam antibiotics, endectocides, immunosuppressives and cholesterol-lowering agents. Indeed, currently Xenova of Slough, UK, are working closely with a number of companies in the search for biologically active secondary metabolites derived from micro-organisms that occupy unusual ecological niches with potential as pharmaceuticals. Algae are also a useful source of medicines. *Caulerpa* spp. produce a range of anti-microbial and cytotoxic metabolites, *Ulva* spp. have been shown to produce anthelmintics, anti-microbials, antiviral agents and cardio-inhibitory agents. *Chara globularis* produces the compounds charatoxin and charamin which have been shown to possess insecticidal, herbicidal and antibiotic effects. Red algae are a rich source of anti-microbial, cytotoxic, anti-inflammatory and antiviral compounds, brown algae produce predominately antimicrobial and cytotoxic compounds, while blue–green algae produce large numbers of different biologically active compounds, especially cyclic peptides and macrolides.

Local custom and folklore contributes to the discovery of pharmaceutically useful medicines. It is estimated that 74% of plant-derived major medicines were discovered by following up empirical folklore leads. The plant-derived artemisinin, from *Artemisia annua*, well known as an insecticidally active compound, has also been shown to inhibit the protozoa *Plasmodium falciparum* and *P. vivax*, which cause malaria. Certain genera of plants are well known to possess particular properties:

- Hallucinogenic compounds – Rubiaceae, Loganiaceae, Apocynaceae, Solanaceae, Convolvulaceae.

- Biodynamically active alkaloids – Rubiaceae, Solanaceae, Apocynaceae, Leguminosae, Ranunculaceae, Berberidaceae, Menispermaceae, Papaveraceae.
- Ethnobotanically active – Bignoniaceae.

This has led to a major input from naturally derived products into commercially successful medicine (Table 2).

Using knowledge of folklore medicinal preparations as a starting point for the search for sources of compounds with potential for the treatment of new diseases was demonstrated by the search of higher plants with anti-HIV activity (Table 3). Of 776 plants examined, 18% of those with a history of medicinal use showed activity while only 10% of those with no history of medicinal use were active.

Local knowledge is abundant in tropical areas. For example, 75 out of 200 (38%) species of the genus *Bignoniaceae* have specific non-horticultural, mainly medicinal, uses.

Table 2 *The world's best selling pharmaceuticals – 1991*

Position in 1991	*Product*	*Therapeutic class*	*Sales ($m)*
1	Ranitidine	H_2 antagonist	3,032
2	Enalapril*	ACE inhibitor	1,745
3	Captopril*	ACE inhibitor	1,580
4	Diclofenac*	NSAID	1,185
5	Atenolol	β antagonist	1,180
6	Nifedipine	Ca^{2+} antagonist	1,120
7	Cimetidine	H_2 antagonist	1,097
8	Mevinolin*	HMGCo-R inhibitor	1,090
9	Naproxen*	NSAID	954
10	Cefaclor*	β-lactam antibiotic	935
11	Diltiazem	Ca^{2+} antagonist	912
12	Fluoxetine	5HT reuptake inhibitor	910
13	Ciprofloxacin	Quinolone	904
14	Amlodipine	Ca^{2+} antagonist	896
15	Amoxycillin/clavulanic acid*	β-lactam antibiotic	892
16	Acyclovir	Anti-herpetic	887
17	Ceftriaxone*	β-lactam antibiotic	870
18	Omeprazole	H^+ pump inhibitor	775
19	Terfenadine	Anti-histamine	768
20	Salbutamol*	β_2-agonist	757
21	Cyclosporin*	Immunosuppressive	695
22	Piroxicam*	NSAID	680
23	Famotidine	H_2 antagonist	595
24	Alprazolam	Benzodiazepine	595
25	Oestrogens*	HRT	569

*Indicates that the product is derived from natural sources.

Table 3 *Results of anti-HIV tests on higher plants from southeast Asia*

Test results	*Species with a history of medicinal use*	*Species with no history of medicinal use*	*Subtotal*
Active	62	44	106
Inactive	289	371	660
Subtotal	351	415	766
Percentage active	17.7%	10.6%	13.8%

The pharmaceutical industry demonstrates well the advantages that exist in examining naturally occurring compounds for new products. However, it may be that the opportunities to explore living organisms for useful activity are contracting. The Earth's species are disproportionately concentrated in the 7–8% of the planet's land surface which is tropical forest. It is clear that we still do not know how many species exist in the world, as shown by the various estimates of insect numbers, which vary from 5 million to 80 million. Nevertheless, either figure far exceeds the known number of 750,000 species. There are about 250,000 known plant species, but it has been estimated that over 10,000 additional, undescribed species exist, mainly in the tropics. It is believed that between one-half and two-thirds of the plant species in the world occur only in the tropics. These figures pale into insignificance when compared with the estimates of fungal numbers, for which the 69,000 known species of fungus contrasts significantly to the various estimates of 1.5 million to 13.5 million actual species. However, the unexplored areas of the Earth are disappearing rapidly. Only 4% of the original forested area of West Ecuador remains today, a story repeated world-wide; and with the loss of these unexplored areas go the, as yet, unexamined and unrecognized species and any useful biological activity their secondary metabolites possess.

This volume examines the compounds that have been found to show useful biological effects and describes their modes of action and the attempts to modify their chemical structure to render them more suitable for application as agrochemicals. Naturally occurring defence systems are considered and the use of biological control techniques, in both niche markets or in more widely used agricultural systems, are discussed.

Micro-organisms have been widely exploited in medicinal chemistry and the Japanese have introduced fermentation products for agrochemical use. In general, these compounds have shown potential in the culture of rice, but their application in Western agriculture has been somewhat limited. In addition, the micro-organisms examined are from a few taxonomic groups that have been recognized as being producers of biologically active secondary metabolites. Chapter 1, by Lange and Lopez, re-examines the world of micro-organisms and lists biologically active compounds by taxonomic class of the producer, by biochemical mode of action and by chemistry. Yamaguchi (Chapter 2) describes the compounds discovered and used in agriculture from the fermentation of micro-organisms and discusses the possibility of transferring their traits to crop plants by the application of molecular biological techniques. Sauter, Ammermann, and Roehl (Chapter 3) take the theme of opportunities from micro-organisms much further by revealing the discovery of useful biological activity from wood-invading fungi and the application of targeted chemical synthesis to ensure that this biological effect was stabilized and, consequently, had application to crop protection. An excellent example of the value of integrated research.

Rhizosphere bacteria have attracted the attention of researchers for many years and in Chapter 4 by Duke *et al.*, the potential that exists in the micro-organisms that inhabit the root zones of crops and weeds is demonstrated. It is felt unlikely that higher plants will be a rich source of herbicidally active compounds, but bacteria have already produced bilanafos and there are indications of other possibly useful compounds from other micro-organisms.

The algae are well known to produce a wide range of biologically active compounds. However, the difficulties associated with their establishment in pure axenic culture have restricted their use as a source of new compounds. It is anticipated that advances in culture and isolation techniques will allow the exploitation of algae in the not too distant future. They have the advantage of being photosynthetically active and, thereby, have no requirement for the addition of a carbon source. In Chapter 5, Dixon examines, in depth, the chemical diversity and biological activity of algal-derived compounds. Many show potential in medicine, as food additives and in crop protection. In addition, many algae have been shown to produce a range of potent toxins, some of which are possible starting points for useful biologically active compounds. Others, however, are so toxic that the use of certain species of algae as producers of secondary metabolites is precluded.

Higher plants are a potent source of biologically active compounds that have been exploited for crop protection purposes. Contributions from Benner, Shang Zhi Zheng, and Miana *et al.* (Chapter 6) describe a wide range of compounds from higher plants which show potential and discuss their application in developing countries. In addition, the pitfalls of pursuing natural plant defence systems and characterizing the active components are reviewed.

It is well known that higher plants were the starting point for the development of the synthetic pyrethroids. Chapter 7 by Michael Elliott is a wonderful description of the work undertaken by scientists at Rothamsted Experimental Station that led to the commercialization of these compounds. In addition, it reveals the single-mindedness of key scientists and the importance of communication within the scientific world such that discoveries are discussed, reviewed, and progressed to the benefit of the entire scientific community. The importance of biological screening systems to support the work of synthesis chemists is also emphasized; this is a further indication of the importance of integrating all disciplines of discovery research.

It should also be remembered that plants gain advantage if they can deter predators, including mammals, as well as insects and pathogens. Hence, Chapter 8 by Parr and Rhodes examines the role of natural plant defence mechanisms and warns that natural compounds may not necessarily be any better in environmental or toxicological terms than synthetic crop protection chemicals!

Arthropods and other animals have developed chemical systems to protect themselves from predation and to assist in the gathering of their prey. It would not be surprising if these chemicals showed some potential as sources of new agrochemicals. Blagbrough and Moya (Chapter 9) describe the diversity of chemistry and biochemical modes of action that can be found in toxins from the animal kingdom and give examples of how these compounds have been examined with a view to the discovery of new insecticides. It should be remembered that compounds that were discarded some years ago as being too complex, too unstable, or too bulky to act as contact insecticides, or sometimes because they were polypeptides or proteins, are once again potential areas of investigation. The development of the molecular biological techniques of gene isolation and plant transformation has meant that the potential of many of these compounds can be exploited by the transfer of the gene coding for the protein into the target crop. In addition, developments in computerized molecular modelling have meant that molecular structures can be elucidated quickly and their biological activity reproduced by the synthesis of simpler, more stable compounds.

Biological systems have already shown their potential in contained situations, such as for glasshouse crops, and in niche markets where chemical control is banned, such as forestry. It is expected that the use of biological systems will increase and several chapters in this book review the progress that has been achieved and identify areas where significant impact will be made. *Bacillus thuringiensis* endotoxins currently represent over 90% of the total usage of biological compounds globally. Chapter 10 by Adams *et al.* describes the structure and biological efficacy of these toxins and outlines, in simple terms, the classification of the different toxin structures and the potential for their use.

Baculoviruses have found opportunities in the control of insects in forestry, as the forest insects are often communal in nature and thus render themselves prone to viral epizootics. Furthermore, specific trees are often attacked by a single insect species, hence offering a clear target for a biological control agent, most of which are highly specific in their action. The often slow action of these biological agents is only a minor problem in forestry, as trees are usually able to withstand the moderate levels of damage that will occur before the pest succumbs to the virus. These very advantages of viruses in forestry are distinct disadvantages in agriculture. Leisy and Fuxa (Chapter 11) describe the biology of baculoviruses and the commercial products that are available today. Progress in the production of viruses and their formulation and application are discussed along with the opportunities that exist for reducing the cost of application. Advances made in the use and genetic manipulation of viruses, which offer potential to overcome the disadvantages of slow action and high specificity of baculoviruses and enhance the potential for their use in agriculture, are explored.

The use of living bacteria to control fungal attack has been described by several workers and in Chapter 12, Pusey reviews research to date and identifies specific opportunities for living systems to control plant pathogens in agriculture. The biological spectrum of these disease-control micro-organisms is discussed and examples given of their possible commercial exploitation. In many cases it is believed that disease control is brought about by the production of antifungal secondary metabolites.

In addition, there are some plant pathogens that have been commercialized for the control of particular hard-to-control weeds. Greaves (Chapter 13) identifies the possibilities that exist for the extension of the use of plant pathogenic organisms into agriculture as weed control systems and affirms that opportunities for mycoherbicides do exist and will be exploited in the foreseeable future.

The very high cost of registration requirements for conventional pesticides restricts their development to meet the needs of the major economic markets. This has resulted in a lack of good crop protection agents for the so-called minor crops. In the case of naturally derived chemicals or biological control systems there is a move towards a reduced registration requirement. This may give wider market opportunities and, consequently, a more rapid development of such crop protection agents. Plimmer (Chapter 14) reviews the regulation situation that exists with particular reference to the USA.

Overall, this book examines all aspects of the natural world as a source of crop protection agents, be they chemical, biochemical, or biological. There is a huge variety of options available, but the crop protection industry is only just beginning to investigate these resources. The future, however, looks rosy.

Leonard G. Copping

Contributors

Hamed K. Abbas, *USDA, ARS, Southern Weed Science Laboratory, PO Box 350, Stoneville, Mississippi 38776, USA*

Lee F. Adams, *Novo Nordisk Entotech Inc., Davis, California 95616, USA*

Tadashi Amagasa, *Sankyo Co Ltd, Shiga-ken, Japan*

E. Ammermann, *BASF AG, Landwirtschaftliche Versuchsstation, D-67114 Limburgerhof, Germany*

Jill P. Benner, *Zeneca Agrochemicals, Jealott's Hill Research Station, Bracknell, Berks RG12 6EY, UK*

Hafsa Bibi, *Department of Chemistry, Gomal University, D I Khan, Pakistan*

Ian S. Blagbrough, *School of Pharmacy and Pharmacology, University of Bath, Claverton Down, Bath BA2 7AY, UK*

M. Iqbal Choudhary, *HEJ Research Institute of Chemistry, University of Karachi, Karachi-75270, Pakistan*

Graham K. Dixon, *Zeneca Pharcaceuticals, Mereside, Alderley Park, Macclesfield, Cheshire SK10 4TG, UK*

Stephen O. Duke, *USDA-ARS, Southern Weed Science Laboratory, PO Box 350, Stoneville, MS 38776, USA*

Michael Elliott, *Biological and Ecological Chemistry Department, Rothamsted Experimental Station, Harpenden, Herts AL5 2JQ, UK*

J. R. Fuxa *Department of Entomology, Louisiana Agricultural Experimental Station, Louisiana State University Agricultural Center, Baton Rouge, Louisiana, USA*

M. P. Greaves, *Department of Agricultural Sciences, University of Bristol, BBSRC Institute of Arable Crop Research, Long Ashton Research Station, Bristol BS18 9AF, UK*

G. Jilani, *Pakistan Agricultural Research Council, Grains Storage Research Laboratory, Karachi, Karachi-75270, Pakistan*

Lene Lange, *Novo Nordisk A/S, Novo Alle 1, 2880 Bagsvaerd, Denmark*

D. J. Leisy, *Department of Agricultural Chemistry, Oregon State University, Corvallis, Oregon 97331, USA*

Chi-Li Liu, *Novo Nordisk Entotech Inc., Davis, California 95616, USA*

Carmen Sanchez Lopez, *Novo Nordisk Entotech Inc., Davis, California 95616, USA*

Susan C. MacIntosh, *Novo Nordisk Entotech Inc., Davis, California 95616, USA*

G. A. Miana, *Department of Chemistry, Gomal University, D I Khan, Pakistan*

Eduoardo Moya, *School of Pharmacy and Pharmacology, University of Bath, Claverton Down, Bath BA2 7AY, UK*

A. J. Parr, *Institute of Food Research, Norwich Science Park, Colney, Norwich NR4 7UA, UK*

Jack R. Plimmer, *ABC Laboratories, PO Box 197, Columbia, Missouri 65205, USA*

P. Lawrence Pusey, *USDA-ARS, Pacific West Area, Tree Fruit Research Lab., 1104 N. Western Avenue, Wenatchee, Washington 98801, USA*

Atta-ur-Rahman, *HEJ Research Institute of Chemistry, University of Karachi, Karachi-75270, Pakistan*

M. J. C. Rhodes, *Institute of Food Research, Norwich Science Park, Colney, Norwich NR4 7UA, UK*

F. Roehl, *BASF AG, Hauptlaboratorium, D-67056 Ludwigshafen, Germany*

H. Sauter, *BASF AG, Hauptlaboratorium, D-67056 Ludwigshafen, Germany*

Robert L. Starnes, *Novo Nordisk Entotech Inc., Davis, California 95616, USA*

Tatsumi Tanaka, *Ube Industries Ltd., Yamaguchi, Japan*

Isamu Yamaguchi, *The Institute of Physical and Chemical Research (RIKEN), Wako, Saitama 351-01, Japan*

Shang Zhi Zhen, *Institute of Elemento-Organic Chemistry, Nankai University, Tianjin 300071, China*

CHAPTER 1

Micro-organisms as a Source of Biologically Active Secondary Metabolites

LENE LANGE and CARMEN SANCHEZ LOPEZ

1.1 Introduction

Secondary metabolites have been found in all major groups of micro-organisms. They are responsible for interactions within populations and between the organism and the environment, in contrast to the primary metabolites, which serve to produce and reproduce the cellular structures. The microbial secondary metabolism, as such, represents the 'lateral activity' of the micro-organism, while the primary metabolites represent the linear activity.[1]

By studying these 'lateral activities' we learn more about the complexity of microbial interaction, and obtain building blocks for understanding their ecology; we are at the same time able to detect compounds which can be of potential applied interest.

Micro-organisms as such serve as an active source of new discoveries for agriculture. Pillmoor *et al.*[2] estimated that approximately 100 patents are filed annually, claiming novel and interesting discoveries of potential use in agriculture, horticulture and silviculture. The discoveries have three objectives: for use directly as fermented product in, *e.g.* plant protection; for use as fermentable products which require subsequent chemical modifications; or for use as lead compounds for chemical synthesis. Also, control of certain pests by unexpected or novel modes of action can lead to the design of structurally unrelated molecules for synthesis and testing.

The application of microbial products has been in the past and is still very broad: as antifungal, antibacterial, antiviral, and antitumour compounds; as growth regulators in plants and animals; and as insecticidal, acaricidal, or herbicidal agents for agriculture, horticulture and forestry.[3]

In total, somewhere between 6,000–10,000 bioactive compounds have been found.[4,5] The vast majority of these have been isolated as metabolites from

Actinomycetes (approx. 65%). The remaining balance is approx. 20% from fungi, 7% from *Bacillus*, and 1–2% from *Pseudomonas*. Surprisingly few secondary metabolites have been found so far from groups like yeasts and water moulds (Oomycetes).

The metabolites also seem to be of importance to the producing organism. Interesting reviews have been published, elucidating the possible role in nature of secondary metabolites.[6,9] Some suggested roles for secondary metabolites include weapons against competitors, part of symbiotic partnerships, metal-transporting agents, sexual hormones, reserve pools of new pathways, and excretion of unwanted products. A full understanding of this concept is far from being achieved.[6,7]

Micro-organisms which have the ability to degrade cellulose seem, in general, to be rich in biologically active metabolites. An explanation for this could be that it is of more importance for cellulose degraders to be fighting their competitors for access to the substrate. A similar example could be that of dung-inhabiting micro-organisms.

An example of the specific activity of a compound for the producing organisms is the siderophore activity (interfering with membrane activity) of myxochelin A from Myxobacteria.[10]

The genetic evidence for the positive selection value of biologically active metabolites for the producing organism has been provided by Hopwood.[11] It has been demonstrated that the genes that code for the enzymes involved in the biosynthesis of the biologically active compound cluster alongside the resistance gene. Whether this picture, mostly elucidated for prokaryotic producers (actinomycetes), also has bearing for the eukaryotes, as exemplified by the fungal producers of secondary metabolites, still has to be demonstrated.

A picture of the proliferating richness and generosity of the micro-organisms as producers of secondary metabolites is slowly emerging. In their two new review chapters on Myxobacteria as producers of secondary metabolites, Reichenbach and Höfle[5,7] highlight some interesting figures – as many as 18 new basic structures have been found from just one species (*Sorangium cellulosum*); and compounds may be produced in cascades of variety, *e.g.*, as many as 50 soraphen derivatives have been isolated from just one strain of *S. cellulosum*, and 35 different myxothiazoles have been isolated from just one isolate of *Myxococcus fulvus*. This can be seen as Nature's own way of making derivatives.

Previously, it was the general understanding that, as opposed to primary metabolites, secondary metabolites were produced in the lag phase, following exponential growth. However, we now know that this is much too simplistic. The secondary metabolites may be produced in all phases of the growth curve,[5] with highest probabilities of their production when the growth of the organism is limited by, *e.g.*, oxygen, ammonia, *etc*. Exciting new research on the correlation between physiology and morphology of the producing organism and the onset of secondary metabolite production has become feasible due to the development of novel high-resolution light microscopes – used in combination with molecular probes.

Table 1.1 *Secondary metabolites, active against plant pathogenic fungi (F), insects (I), and weeds (H), with potential as plant growth regulators (R) or produced by biocontrol agent (B/*). (?) indicates that the role – if any – of secondary metabolites is unknown. The table includes the name of the compound, the name and the taxonomic grouping of the producing organisms, and one or more key references*

Activity	*Compound name*	*Producing organisms*	*Key refs*
		Gram negative bacteria	
		Pseudomonadales	
F	Pyrrolnitrin	*Pseudomonas pyrrocinia*	70
		Gram positive bacteria	
		Eubacteriales (endospore forming)	
B/F	Iturin A	*Bacillus subtilis*	71,72
		Actinomycetales	
I	Avermectins	*Streptomyces avermitilis*	19
F	Blasticidin S	*Streptomyces griseochromogenes*	73
F	Tautomycetin	*Streptomyces griseochromogenes*	74
B/F	(?)	*Streptomyces griseoviridis*	75
F	Polyoxin	*Streptomyces cacaoi* spp. asoensis	76
F	Kasugamycin	*Streptomyces kasugaensis*	77
I	Tetranactin	*Streptomyces aureus*	39
I	Alanosine	*Streptomyces alanosinicus*	43
H	Bilanafos	*Streptomyces hygroscopicus*	54
F	Validamycin	*Streptomyces hygroscopicus* spp. *limoneus*	20
I	Milbemycin	*Streptomyces hygroscopicus* spp. *aureolacvimosus*	78
F	Fumaramidmycin	*Streptomyces kurssanovii*	79,80
I	Altemicidin	*Streptomyces sioyansis*	81
I	Bafilomycins	*Streptomyces* sp.	82
I	Allosamidin	*Streptomyces* sp. (1713)	40,41
H	Phthoxazolin	*Streptomyces* sp.	22
H	Oxetin	*Streptomyces* sp	83
I	AB3217-A	*Streptomyces platensis*	84
F	Mildiomycin	*Streptoverticillium rimofaciens*	85
F	Rustmicin	*Micromonospora narashinoensis*	47–49
I	Trehazolin	*Micromonospora* sp.	42
I	Trehalostatin	*Amycolatopsis trehalostatica*	86
		Myxobacteria	
F	Soraphen	*Sorangium cellulosum*	87
F	Myxothiazole	*Myxococcus fulvus*	88
		Oomycetes	
B/F	(?)	*Pythium oligandrum*	89
B/I	(?)	*Entomophthora* spp.	90
		Fungi	
		Basidiomycota	
		Hymenomycetes	
		Agaricales	
F	Strobilurin	*Strobiluria tenacellus*	31,92

Continued

Table 1.1 *Continued*

Activity	*Compound name*	*Producing organisms*	*Key refs*
I	*trans*-2-Dec-2-enedoic acid	*Pleurotus oestreatus*	24
B/F	(?)	Red sterile fungus	93,94
		Ascomycota	
		Plectomycetes	
		Euroriales	
F	Asperfuran	*Aspergillus oryzae*	52
I	Nominine	*Aspergillus nomius*	95
		Pyrenomycetes	
		Hypocreales	
R	Moniliformin	*Fusarium moniliforme*	96
R	Gibberellins	*Gibberella fujikoroi*	91
B/F	Several metabolites	*Glicladium virens*	97
B/F	Several metabolites	*Trichoderma harzianum & T. viride*	98,99
B/I	Several metabolites	*Verticillium lecanii*	103
		Phyllachorales	
B/H	(?)	*Colletotrichum gloeosporoides*	105
		Mitosporic Ascomycota	
B/I	Several metabolites	*Metarrhizium anisopliae*	100,101
B/I	Several metabolites	*Beauveria bassiana*	102
F	Xanthofusin	*Fusicoccum solieri*	27,104
I	Bursaphelocide	*Mycelia sterilia*	106

The rest of the problem leaves no room for generalization. How to cultivate an organism to make it produce to the utmost of its capabilities regarding secondary metabolites varies between the taxonomic groups, between species – even between isolates.

Recently, excellent reviews of bioactive microbial secondary metabolites have been published.[3,5,6,7,12–18] In this chapter on microbial secondary metabolites we have chosen to illustrate the state-of-the-art by selecting 32 secondary metabolites, which together demonstrate the potential for isolating much-needed novel microbial products for the agricultural market.

We have tried to do this by examining these 32 compounds using the following questions:

- What is the taxonomic position of the producing organism (Table 1.1)?
- To which chemical class do they belong (Table 1.2)?
- What are their modes of action (Table 1.3)?

To make the text more reader-friendly we include comments about biodegradability, potentials regarding biotechnological solutions, biosynthesis, practical application and toxicity – and finish with concluding remarks about future potential and research.

Table 1.2 *The table indicates the chemical group to which each of the listed bioactive secondary metabolites belong. Further, the molecular weight of the compound is included. The key for activity (I, F, H, R, B/*) follows the definitions in Table 1.1*

Activity	*Compound name*	*Chemical class*	*Mol. weight*
F	Pyrrolnitrin	Pyrrole derivative	257
I	Ivermectins	Macrolide, milbemycin type	873
F	Blasticidin S	Cytosine glycoside	422
F	Tautomycetin	Lactone	607
F	Nikkomycin	Uracil glycoside	495
F	Kasugamycin	Aminoglycoside-like	379
I	Tetranactin	Cyclopolylactone, nonactin-type	793
I	Alanosine	Amino acid	149
H	Bilanafos	Oligopeptide (phosphorous)	323
F	Validamycin A	Aminoglycoside-like	497
I	Milbemycin	Macrolide	557
F	Fumaramidmycin	Benzene derivative	232
I	Altemicidin	Monoterpene-alkaloid	376
I	Allosamidin	Oxazoline aminosugar derivative	622
H	Phthoxazolin	Oxazole	290
H	Oxetin	Amino acid	117
I	Bafilomycins	Macrolide-like antibiotic, humidin type	721
I	AB3217-A,B,C	Pyrrolidine derivative	434
F	Mildiomycin	Cytosine glycoside, gougerotin type	514
F	Rustmicin	Macrolide	380
I	Trehalostatin/ trehazolin	Aminosugar derivative	366
F	Soraphen	Macrolide	521
F	Myxothiazole	Thiazole derivative	488
R	Gibberellin A3	Diterpene	346
F	Strobilurin	Benzene derivative	342
I	*trans*-2-Decenedioic acid	Fatty acid	200
F	Asperfuran	Benzofuran derivative	218
I	Nominine	Indole diterpene	175
R	Moniliformin	Cyclobutene derivative	98
F	Xanthofusin	Tetronic acid derivative	168
I	Bursaphelocide A,B	Peptolide	609,595

Table 1.3 *The primary target for the secondary metabolite is given, along with the (suggested or proven) mode of action of the molecule; one or more key references are given*

Compound	*Targets*	*Mode of action*	*Key refs*
Pyrrolnitrin	Seed-borne and leaf-spot fungi	Inhibition of respiration and electron transport	107,108
Avermectin	Parasitic worms, mites, aphids	Enhancing binding of GABA to receptor	109,110
Blasticidin S	Rice blast	Inhibition of protein synthesis	21
Polyoxin	Rice blast	Inhibition of cell wall synthesis	111
Kasugamycin	Rice blast	Inhibition of protein synthesis	112
Tetranactin	Spider mites	Ionophore; interference with membrane function	39
Alanosine	Lepidoptera (army worm and silk worm)	Inhibition of insect ecdysis	43
Bilanafos	Dicotyledon herbicide	Inhibition of glutamine synthetase	55
Validamycin	Sheath blight	Inhibition of trehalase	56
Milbemycin	Mites; dog heart worm	Inhibition of neurotransmission at GABA receptor	113
Fumaramidmycin	Vine downy mildew	Unknown	–
Altemicidin	Two-spotted spider mite	Unknown	–
Allosamidin	Silk worm	Inhibition of chitinase	41
Phthoxazolin	Dicotyledon herbicide	Inhibition of cellulose synthesis	114
Oxetin	Dicotyledon herbicide	Inhibition of glutamine synthetase	115
Bafilomycins	Lepidoptera, coleoptera, diptera, heteroptera	Unknown	–
AB3217-A,B,C	Mites	Unknown	–
Mildiomycins	Powdery mildews	Inhibition of protein synthesis	116
Rustmicin	Rusts	Inhibition of cell wall (germ tube elongation)	47
Trehalostatin/ trehazolin	Tobacco budworm and silk worm	Inhibition of trehalase	86
Soraphen	Oomycetes	Blocking fungal acetylcoenzyme carboxylase	117
Myxothiazole	(Biochemical probe)	Inhibition of electron transport (complex III)	5
Gibberellins	Plant growth regulator	Growth hormone	91
Strobilurin	Broad fungicidal	Inhibition of mitochondrial respiration	92
trans-2-Decenedoic acid	*Panagrellus redivivus*	Nematicidal (m.o.a. unknown)	–
Asperfuran	Broad fungicidal	Inhibition of chitin synthesis	52
Nominine	*Heliothis zeae*	Deterrent of feeding	95
Moniliformin	Herbicide	Destruction of apical dominance	96
Xanthofusin	Oomycetes	Condense nucleophilic groups	27,118
Bursaphelocide A/B	Pine wood nematode	Nematicidal (m.o.a. unknown)	–

1.2 Discovery of New Bioactive Molecules – Selected Examples

1.2.1 The Discovery of an Insecticide: Avermectin

Avermectin (**1**, Figure 1.1) was discovered in an anthelmintic screening programme in which microbial fermentation broths were tested in mice against

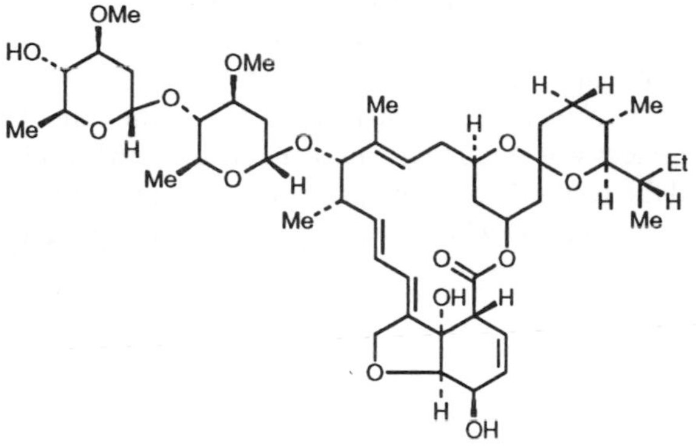

1 Avermectin A_{1a}

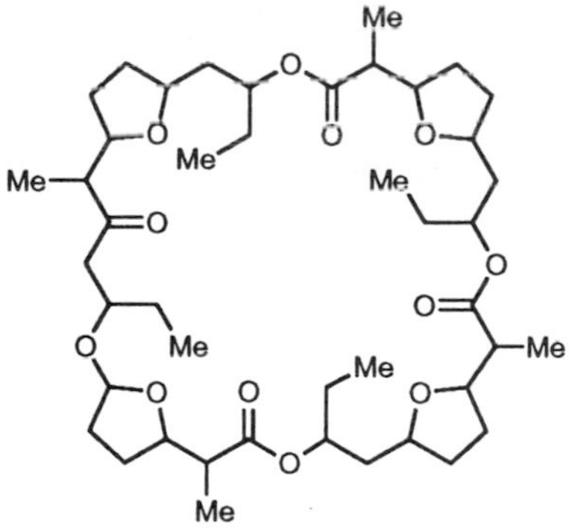

2 Tetranactin

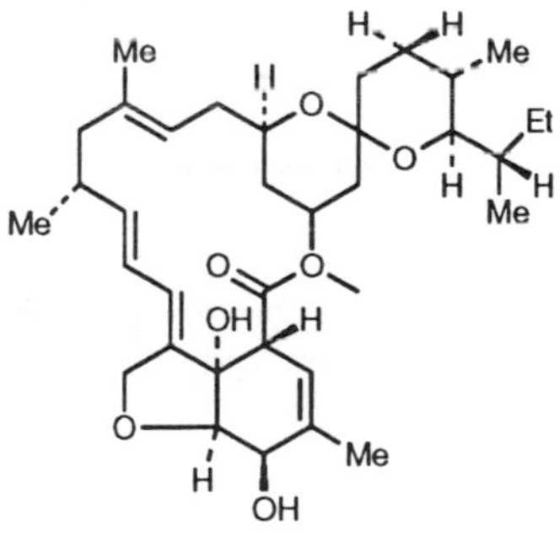

3 Milbemycin D

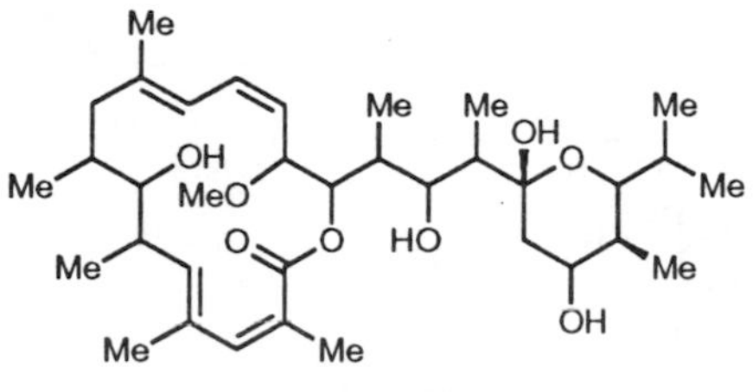

4 Bafilomycin A_1

Figure 1.1 *Insecticidal compounds with macrocyclic structures*

the nematode *Nematospiriodes dubious* (*i.e.*, a real *in vivo* test), testing the crude broth directly in a dual mice–nematode system. A deliberate choice of testing culture broths only from micro-organisms with unusual morphology predisposed the discovery.[19]

1.2.2 The Discovery of a Fungicide: Validamycin

Validamycin (**8**, Figure 1.2) was discovered by an *in vitro* test called the spot inoculation method.[20,21] In this test, agar containing a plant extract was sandwiched between agar holding the active metabolites (from preincubation of the test organism on plain agar) on the lower side and the fungal pathogen on the upper side. The test was scored simply by measuring the inhibition zones.

The idea behind the test is that the plant extract agar simulates – *in vitro* – the fungicide treated plant leaf on which *Rhizoctonia solani* hyphae proliferate.

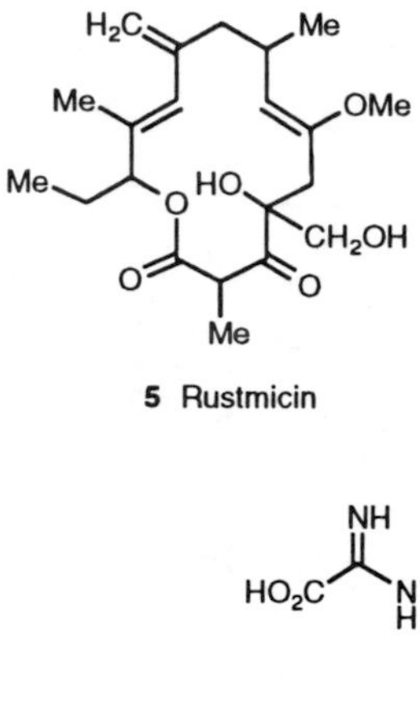

5 Rustmicin

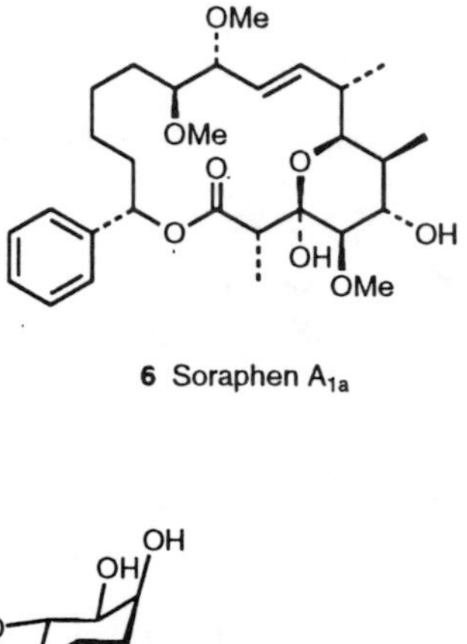

6 Soraphen A_{1a}

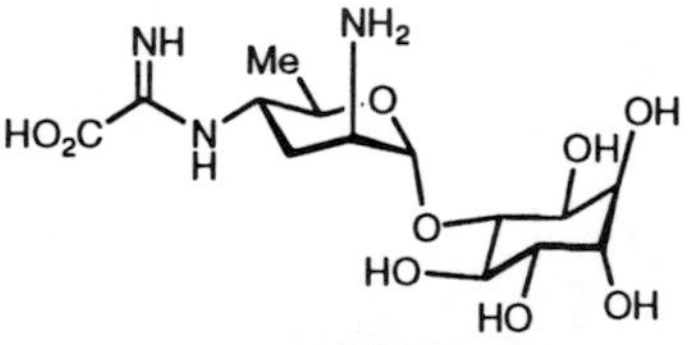

7 Kasugamycin

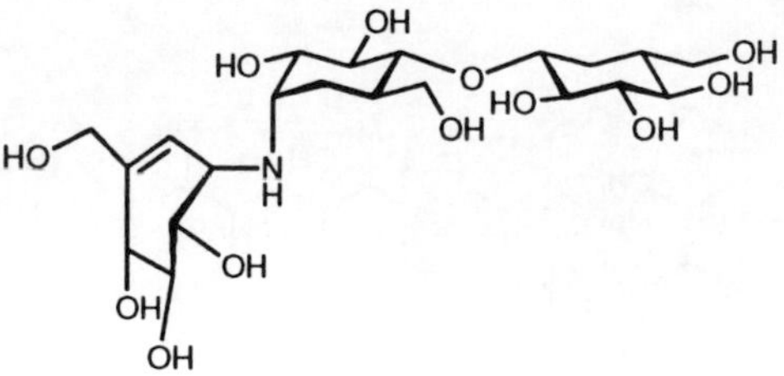

8 Validamycin A

Figure 1.2 *Structures of fungicidal compounds*

1.2.3 The Discovery of a Herbicide: Phthoxazolin

In the discovery of phthoxazolin (**10**, Figure 1.3), a screen was constructed to target selectively the molecules which showed inhibition of cellulose synthesis. This was achieved by performing a set of fungal and bacterial *in vitro* tests, covering most of the major taxonomic groups, only one of which had cellulose as a main constituent in its walls (*i.e.*, *Phytophthora parasitica*).[22]

The test was scored by selecting only the test samples which gave inhibition towards *P. parasitica* and had no effect against any of the other organisms.

The screen was successful in that a novel molecule with a new mode of action, cellulose synthesis inhibition, was discovered. The molecule has shown some potential as a herbicide, but only weak activity as fungicide against Oomycetous species.

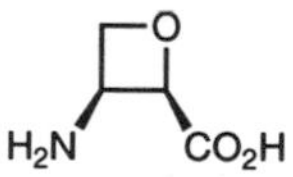

9 Oxetin

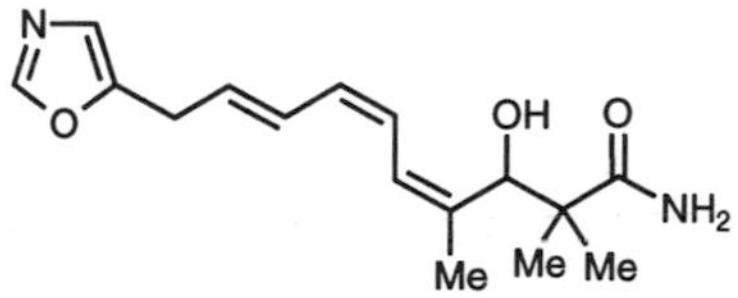

10 Phthoxazolin

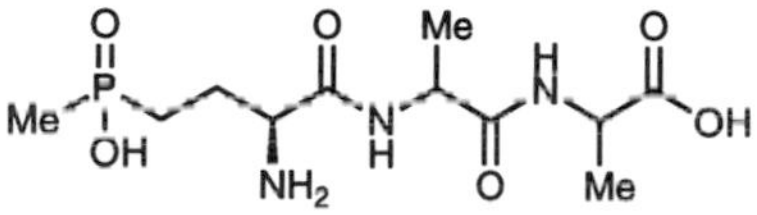

11 Bilanafos

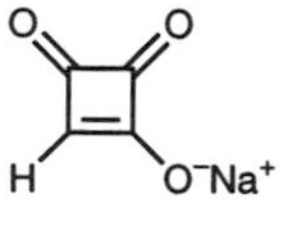

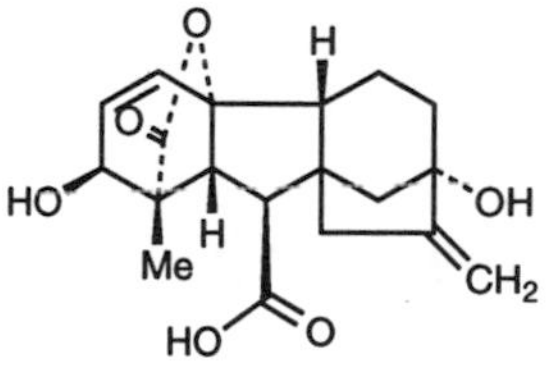

12 Moniliformin

13 Gibberellin-A3

Figure 1.3 *Structures of microbial herbicides – oxetin, phthoxazolin and bilanafos – and structures of microbial plant-growth regulators – moniliformin and gibberellin-A3*

1.2.4 The Discovery of a Nematicide: *trans*-2-Decenedoic Acid

Wood-decaying Basidiomycetes, *e.g.*, *Pleurotus* spp., have been shown to destroy (capture and consume) nematodes.[23] This inspired tests for a possible nematicidal effect of the secondary metabolites from isolates of *Pleurotus* (the oyster fungus),[24] which led to the discovery of *trans*-2-decenedoic acid (**17**,

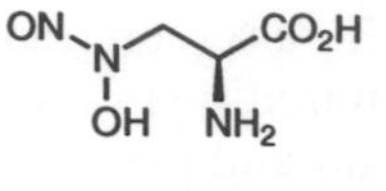

14 Alanosine

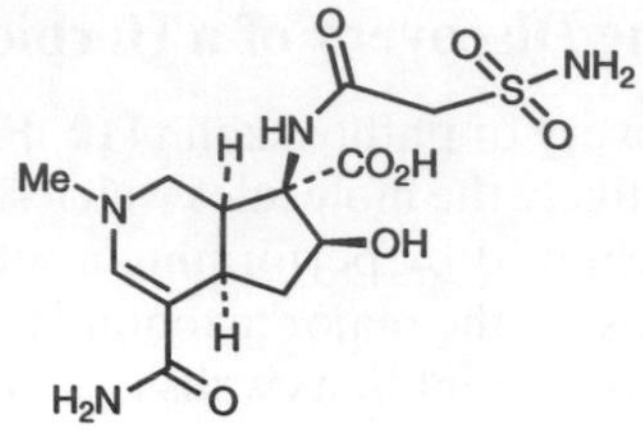

15 Altemicidin

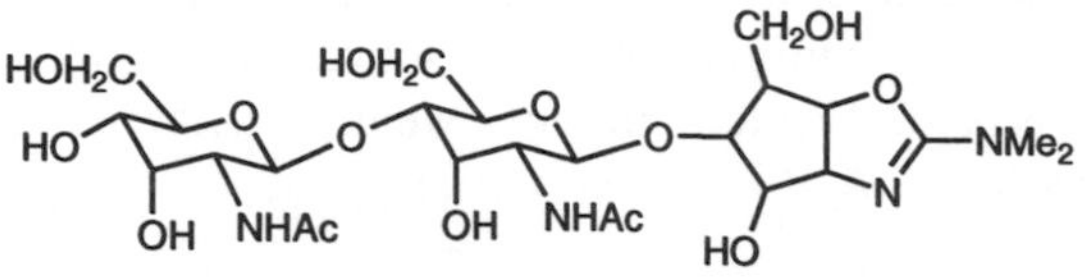

16 Allosamidin

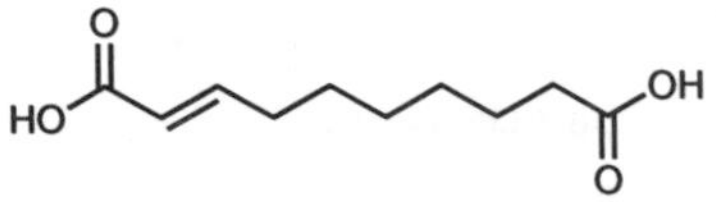

17 *trans*-2-Decenedioic acid

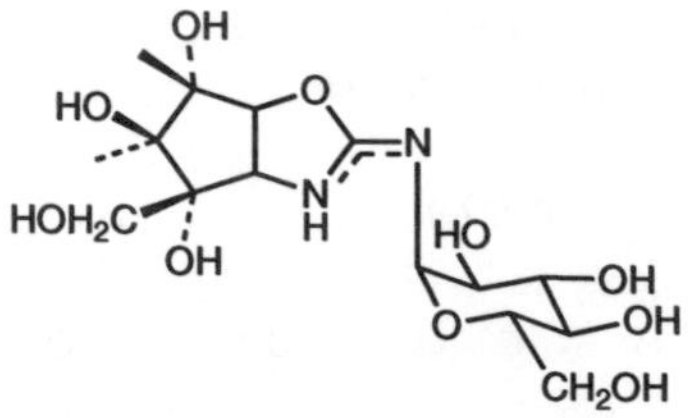

18 Trehazolin/Trehalostatin

Figure 1.4 *Structures of insecticidal compounds*

Figure 1.4), a simple fatty acid derivative. The potential and mode of action of this new molecule are still not clarified.

1.3 Where in the Taxonomic System Do We Find the Producers of Important Secondary Metabolites?

Studies in certain areas are by now so numerous that it makes sense to analyse the correlation between taxonomic positioning of the micro-organism and its potential as a producer of secondary metabolites.

The material in Table 1.1 gives some indication of apparent hot spots (*Streptomyces*, Myxobacteria, *Tricholomatacea* (Agaricales), *Aspergillus/ Penicillium,* and *Fusarium*). It lists 32 bioactive metabolites together with 11 produced by biological control agents. The importance of secondary metabolites in the activity of the biocontrol agents is, in general, not yet fully understood. However, we have found it useful to include the biocontrol agents when the objective is to obtain an overview on where in the microbial taxonomic system interesting bioactivity can be found.

As is obvious from Table 1.1, the majority of agriculturally interesting compounds (21 out of 32) has been isolated from Actinomycetes. As many as

17 of the leading compounds are from *Streptomyces* alone, distributed on 10 (identified) species.

Streptomyces and *Streptoverticillium* belong to the same subgroup (*sensu* Iwai and Takahashi[25]) of Actinomycetes (no sporangia; aerial mycelium; cell wall type I). *Micromonospora* and *Amycolatopsis* belong to a related group (neither sporangia nor aerial mycelium), distinguished from each other by their difference in cell wall type (*Micromonospora*, type II; *Amycolatopsis*, type IV).

In the period during which these discoveries were made, the resources, methodology, and expertise were very much focused on these groups. Hardly any screening activities were performed on other taxonomic groups. Only time will show whether the many findings in these groups reflect unique and outstanding characteristics or merely the effort invested.

The second largest group in Table 1.1 is the *Ascomycetes* of the imperfect fungi. Many more examples could have been included here as frequently new discoveries are being reported from this group of fungi.[26,27] It is also this group of fungi that is responsible for production of most of the mycotoxins described so far (*e.g.*, *Penicillium, Aspergillus, Fusarium* spp.)

Pseudomonads have for long been in focus as potential biocontrol agents against soil-borne diseases. Now it also seems likely that the Pseudomonad product pyrrolnitrin (**19**, Figure 1.5) may serve as the lead compound for a new group of fungicides, produced by chemical synthesis (fenpiclonil).[14,28,29]

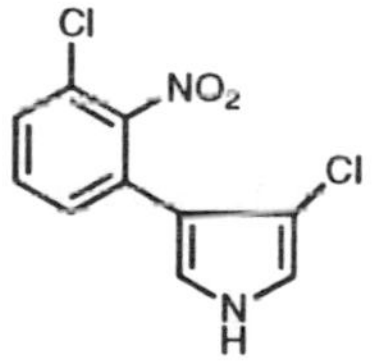

19 Pyrrolnitrin

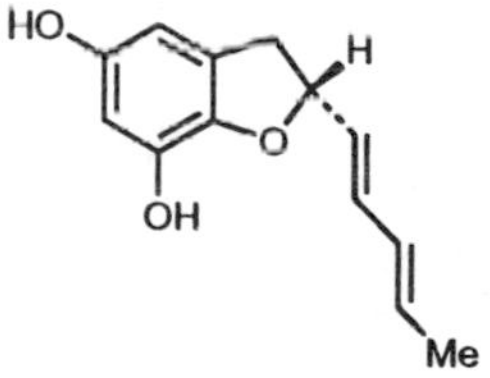

20 Asperfuran

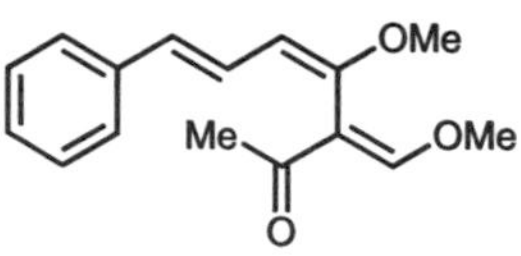

21 Strobilurin A

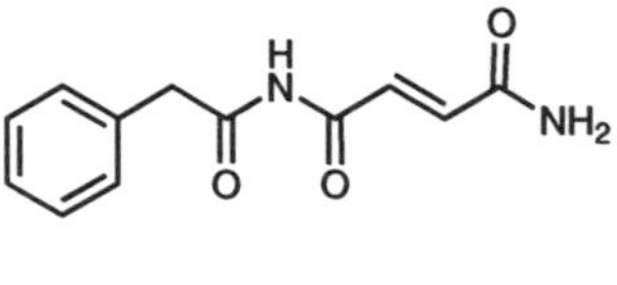

22 Fumaramidmycin

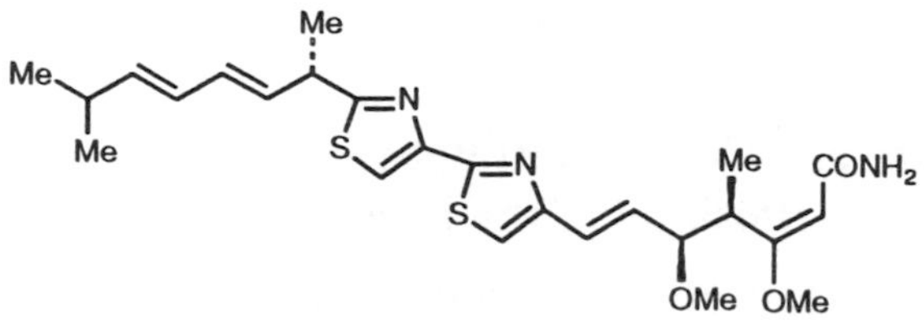

23 Myxothiazole

Figure 1.5 *Structures of fungicidal compounds*

Two of the smallest groups in Table 1.1, the Myxobacteria and the Basidiomycetes have been studied extensively over the past decade.[5,30,31] Extremely interesting findings have been published and we consider it highly probable that in the coming few years discoveries from these groups will be developed successfully to practical large scale application, for agricultural and other purposes.

Myxobacteria are, in general, found to be very rich producers of secondary metabolites.[5] Within the Myxobacteria certain species and genera also seem to be more active (have larger genomes?) than others. The two most potent molecules against plant protection relevant targets, soraphen (**6**) and myxothiazole (**23**) are produced by what Reichenbach and Höfle characterize as the two, in general, richest sources within Myxobacteria: *Sorangium cellulosum* and the genus *Myxococcus*. Maybe the most interesting and most active molecules are to be harvested at the top of a pile of metabolites?

The interesting Basidiomycete class of compounds, the strobilurins (**21**) (see Chapter 3) or the oudemansins, were discovered primarily from the toadstool genera, belonging to the complex around the genus *Collybia sensulato* (*Tricholomataceae*-Agaricales). However, related compounds have been found to be produced by species belonging to the dark-spored Agaric genera, *Crepidotus* and *Cyphellopsis*.[32,33] Also, myxothiazole (**23**) produced by *Myxococcus*, belongs to this group of molecules. Maybe the common denominator between these microbial producers is not a taxonomic relationship, but a convergent development towards a biological weapon, useful for microbes invading the cellulose substrates in a competitive way?

From such a hypothesis the logical next step is to determine if there is any interesting pattern in regard to specific ecological niches that seem to harbour the richest secondary metabolite producers.

Information from comprehensive surveys of *Penicillium, Aspergillus,* and *Fusarium* secondary metabolites, isolated from all over the world, enabled Frisvad *et al.*[34] to interpret ecological parameters in relation to the production of bioactive metabolites.

In an ecological analysis, building on more than 15 years of experience, Iwai and Takahashi[25] from the Kitasato group in Japan have provided a solid basis for selecting the optimal samples and microbial sources for the search for bioactive compounds. Whether the chances for new and interesting compounds are highest if the search is continued among the *Strepomyces* or if it is made selectively from the rarer genera is very difficult to deduce, balancing risk for re-discoveries with richness in metabolites.

1.3.1 Is There any Correlation Between Activity and Taxonomic Position of the Producing Organism?

By correlating the type of activity, be it fungicidal, insecticidal, herbicidal, or plant-growth regulatory, it seems apparent (Table 1.1) that no taxonomic group has specialized in one type of activity. All types can be found equally throughout the system.

1.4 Structural Diversity of Pesticidal Compounds from Micro-organisms

The remarkable potential of micro-organisms to produce structurally diverse secondary metabolites can once again be confirmed by taking a quick look at Table 1.2 and Figures 1.1–1.7 (compounds **1–34**), which illustrate the structures of the herbicidal, insecticidal, and fungicidal compounds discussed in this review. These compounds vary enormously in their structures, although they possess a similar spectrum of activities. For examples, avermectin shows activity against mites (Table 1.3) as does altemicidin, but both compounds belong to completely different chemical classes. Examples of this can also be found with the fungicidal and herbicidal compounds.

On the other hand, compounds belonging to similar chemical classes are found to possess different activities, as shown by milbemycin (**3**) and rustmicin (**5**), both of which are macrolides. The first is reported insecticidal and the second is reported fungicidal. This phenomenon is not unusual and many reported cases of multiple activities for one chemical class have been found. Perhaps different organisms have similar receptors to which these compounds bind, producing some sort of biological response, or perhaps it is because

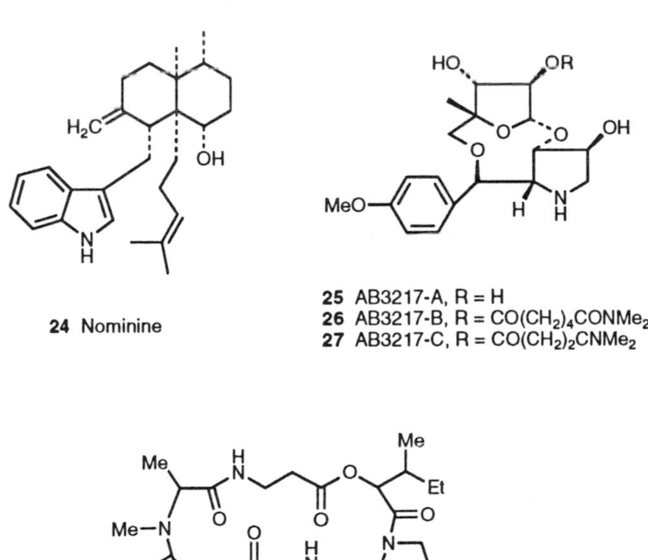

Figure 1.6 *Structures of insecticidal compounds*

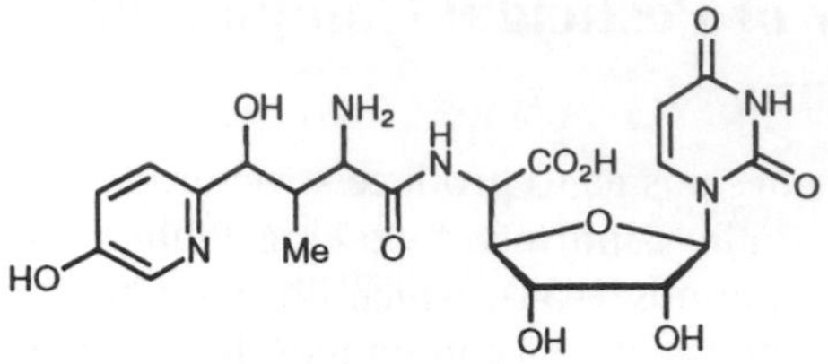

30 Nikkomycin

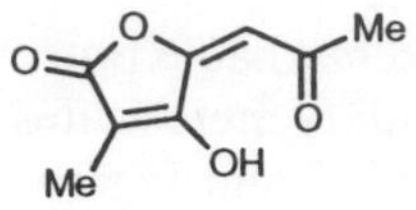

31 Xanthofusin

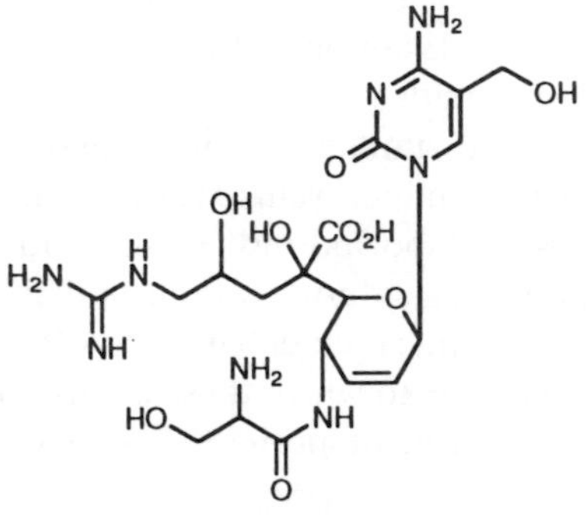

32 Mildiomycin

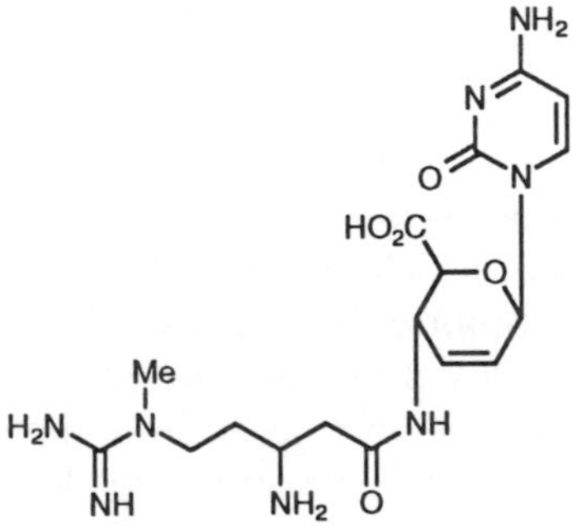

33 Blasticidin S

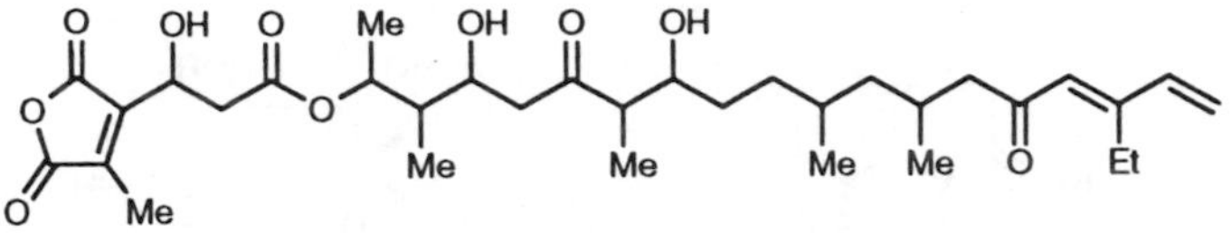

34 Tautomycetin

Figure 1.7 *Structures of fungicidal compounds*

microbial products are so complex in their chemical structure that one compound may have two or more totally different chemical moieties which can interact with different receptors. It is this complexity in their structures that makes natural products chemically and biologically diverse.

As Tables 1.1 and 1.2 show, no strict correlation between chemical class of the secondary metabolites and the taxonomic position of the producing organisms exists. Both bacteria and fungi have been found to produce similar classes of compounds, both in complexity and size, as is well illustrated in Tables 1.1 and 1.2.

Most of the compounds in this review are very complex in their structures, which makes it difficult to deduce their biosynthesis. The majority of biosynthetic studies have been made on the commercially important compounds such as penicillin, erythromycin, gibberellins, and avermectins. Improvements to our understanding of the biosynthesis of a broader range of secondary metabolites are now being published, including for metabolites from Myxobacteria[5] and Basidiomycetes.[31]

Another aspect in which important progress still has to be made is in the understanding of structure–activity relationships.[16] The large number of derivatives made (*e.g.*, in the struggle for achieving the optimal derivative of avermectins, strobilurins and pyrrolnitrins) provide interesting material for such analysis. However, only a limited part of such information is published in a comprehensive form.

1.5 Mode of Action Studies

1.5.1 Insecticidal, Acaricidal, and Anthelmintic

Five different modes of action have been elucidated for the metabolites selected for this review: inhibition of neurotransmission, interference with membrane function, chitinase inhibition, trehalase inhibition and feeding deterrence. The target for the two major discoveries within microbial products for insect and helminth control (avermectins and milbemycins) is the γ-aminobutyric acid (GABA) receptor in the peripheral nervous system.

Both avermectins and milbemycins (Figure 1.1) stimulate the release of GABA from nerve endings and enhance the binding of GABA to receptor sites on the post-synaptic membrane of inhibitory motor neurons of nematodes, and on the post-junction membrane of muscle cells of insects and arthropods. The enhanced GABA binding results in an increased flow of chloride ions into the cell, with consequent hyperpolarization and elimination of signal transduction, resulting in an inhibition of the neurotransmission.[35–38]

Tetranactin (**2**) has been shown to act by forming a metal complex which presumably interferes with the membrane function of mites. It is only active in the presence of water, which suggests that tetranactin acts as an alkali ion carrier across the membrane to cause leakage of important ions (*e.g.*, K^+).[39]

Two enzyme inhibitors have proved to have interesting insecticidal effects – allosamidin (**16**), an insect chitinase inhibitor, and trehalostatin/trehazolin (**18**), a trehalase inhibitor. The former interferes with the growth and ecdysis of, especially, lepidopteran insects,[40,41] while the latter blocks the utilization of the stored energy resource (in the form of trehalose).[42]

Alanosine (**14**) is an insect growth regulator, shown to interfere in an unknown way with larval ecdysis.[43] Nominine (**24**), which is isolated by organic solvent extraction of sclerotia of *Aspergillus nomius*, has antifeeding effects, causing up to 40% mortality in *Heliothis zea*.[44]

Metabolites may also play an important role in the activity and efficacy of entomopathogenic fungi (see also Chapter 2).

1.5.2 Fungicide

Six different modes of action – inhibition of protein synthesis, respiration, cell wall synthesis, chitin synthesis and trehalase and acetyl Co-A activity – are represented among the antifungal metabolites listed in Table 1.3.

The commercially most successful group is found within the protein synthesis inhibitors (blasticidin S (**33**) – kasugamycin (**7**), and mildiomycin (**32**) – all actinomycete metabolites). Also, the widely found but rather toxic molecule, cycloheximide, acts as a potent protein synthesis inhibitor. The modes of action of blasticidin S and mildiomycin have been studied in the refined *Escherichia coli* cell free protein synthesis system.[45] Results from these studies, as well as biological observations (as, *e.g.*, kasugamycin does not inhibit spore germination, but mildiomycin does), suggest that these three fungicides have different modes of action, even though they all are protein synthesis inhibitors.

The new fungicide lead molecules for chemical synthesis, pyrrolnitrin (**19**), from *Pseudomonas*, and the strobilurins [from toadstools (agarics) like *Strobiluria, Oudemansiella, Xerula, etc.*], are inhibitors of (mitochondrial) respiration and electron transport (see Chapter 3). Myxothiazole (**23**) also has this mode of action. However, pyrrolnitrin (**19**) which is produced by several genera of myxobacteria, may have other activities as well. Mitochondria are observed to swell up after treatment and reports vary in their description of the activities.[7]

Nikkomycins (**30**) and maybe also the rustmicins (including neorustmicins) are cell wall synthesis inhibitors.[46–49] Likewise are the commercially widely used rice sheath blight fungicidal polyoxins. These three types of molecule are all produced by Actinomycetes.

Polyoxins inhibit chitin synthesis by competitive binding of *N*-acetylglucosamine.[50] The mode of action of polyoxins has been clarified by studies in *Alternaria kikuchiana*.[51]

The very first example of a fungal producer of a chitin synthesis inhibitor was the discovery of asperfuran ((**20**) produced by *Aspergillus*).[52] The activity may, however, prove to be too low to validate its development as a commercial fungicide.

For the fungicide validamycin (**8**), it has been shown that the active mode molecule is the aglycone of validamycin A, called validoxylamine A.[53] This compound acts as a specific inhibitor of the enzyme trehalase, thereby diminishing the utilization of the stored energy resource in the form of trehalose. Trehalose is a common carbohydrate for both sclerotial fungi and some insects.

Soraphen (**6**), produced by the myxobacterium *Sorangium cellulosum* (and found to be produced by many other organisms within or outside the myxobacteria), is an efficient blocker of fungal acetyl Co-A carboxylase. It shows no activity against the corresponding enzyme systems in chloroplasts in plants. It does, however, exhibit activity against acetyl Co-A carboxylase from rat liver, indicating a risk for adverse non-target activity.[5]

1.5.3 Herbicides/Plant Growth Regulators

Two different modes of action are included among the few examples of herbicidally active metabolites in Table 1.3: inhibition of glutamine synthetase and inhibition of cellulose synthesis (see Chapter 4).

Both bilanafos (**11**) and oxetin (**9**) have been shown to have an inhibitory effect on glutamine synthetase. For bilanafos (**11**) it is not the intact molecule which is directly responsible for the activity; it is the metabolic breakdown product from bilanafos, phosphinothricin, which is responsible. This allowed an opening in the patenting of this interesting molecule (discovery[54] and patenting of phosphinothricin[55]).

The plant growth regulatory activity of gibberellin (**13**) is already well described. However, moniliformin (**12**) deserves a comment. The activity of this unique molecule is a specific destruction of the apical dominance in dicotyledons, as well as necrosis and necrotic spots. It has been used as the lead molecule for the development of non-selective desiccant herbicides (the squarates).[14]

The cellulose synthetase activity of phthoxazolin (**10**) is described in section 1.2.3 above.

1.5.4 Fungicides as Insecticides? Fungicides as Herbicides?

Interesting reports have been published on the insecticidal effect of validoxylamine A, the aglycone of validamycin A (**8**). Validoxylamine A has been shown to act as a trehalase inhibitor in tobacco cutworm. Also, nikkomycin (**30**) and mildiomycin (**32**) have been reported as miticides.[56]

Further, the *Pseudomonas* product pyrrolnitrin (**19**) may, in future, be used as a lead molecule for new insecticides[57] and not only as a lead molecule for fungicides.[28]

It is also noteworthy that the herbicidal molecule bilanafos (**11**) was first discovered as a fungicide – with some problematic phytotoxicity. Maybe fungi in general are good indicators for interesting bioactivities?

All in all there seems to be a trend that compounds shown to be fungicidal or fungistatic may be of interest as a group to investigate for the presence of other biological activities. However, this may also result in a higher number of disappointments in the discovery process of fungicides – as more will fail due to unwanted non-target activities.

Very recently it has been shown[58] that chitosan, manufactured from crustacean chitin, has interesting potential as a biocompatible biopolymer used to dress surgical and traumatic wounds. Maybe chitosan prepared directly from microbial wall materials could have equal or better characteristics for this purpose?

1.6 Resistance

One of the parameters which plays a significant role in the dynamics of the global pesticide market is the development of resistance.

Serious resistance problems are increasing as a consequence of repeated and continued use of many classes of commercialized plant protection agents. This creates a need for new, alternative plant protection agents, with new modes of action, which do not exhibit cross-resistance to the already commercialized ones. Owing to their diversity, microbial products may provide such opportunities.

However, microbial metabolites may also run into resistance problems, as with the fungicides polyoxins and kasugamycin (**7**), for which resistance has been recorded.[59,60] Similarly, resistance has been observed against avermectin (**1**) and tetranactin (**2**).[61,62]

In Japan, it was observed that resistance to polyoxins disappeared from the fungal population after use of the fungicide had been stopped for 2–3 years.[21]

Sensible use of valuable compounds, including avoiding repeated over-use and alternating with other mode of action agents, is the only feasible way to keep these weapons effective for a long period in the struggle to protect crops.

1.7 The Need for Standardization

The risk of rediscoveries is the single most time-consuming pitfall in the process of screening for new and interesting biologically active secondary metabolites. Rediscoveries cannot be avoided totally, but what can be achieved is to minimize the time of study before lack of novelty is realized.

Advanced but simple high performance liquid chromatography (HPLC) techniques, combined with PDA (photo diode array), and matched with computerized libraries of biologically active secondary metabolites, are powerful tools to cut down the time of recognizing known compounds.

A number of data bases on biologically active microbial products are available commercially. Four of these are listed and characterized in Table 1.4: Berdy Natural Product Data Base from Hungary; Kitasato Microbial Chemistry Data Base from Kitasato Institute, Tokyo, Japan; Actfund (Microbial

Table 1.4 *Overview giving the characteristics of four of the commercially available data bases, including information about biologically active, microbially produced secondary metabolites*

Data base name	*Number of compounds*	*Language*	*Compatibility*	*Chemical structure*
Berdy Natural Product Data Base	23,000	English	DOS	No
Kitasato Microbial Chemistry Data Base	16,000	Japanese	NEC	No
New Kitasota Chemistry Data Base	16,000	English	DOS	No
Chapman & Hall	15,000	English	DOS	Yes
Actfund	10,800	English	DOS & NEC	No

products from Actinomycetes and Fungi, compiled and updated by Okuda and Takeda), Tokyo, Japan; and Chapman & Hall, USA.

It is encouraging that it is becoming possible to access all these data from most computers (regarding language and software compatibilities). However, a need still exists to standardize the conditions by which the data have been obtained, *e.g.*, exact specification on how the ultraviolet spectra have been generated and by which bioassay protocol the toxicity and activity data have been assessed.

Another field where better standarization and unanimous application is much needed is in the naming of the producing organisms. Multiple attempts are being made to advance this field, both from academia and industry.[63,64]

Also, the naming of chemical structures is an area where full international standarization is required. This is, however, within closer reach.

Only if standarization in all these areas is obtained can all the data compiled in the natural product data bases be utilized fully, to the mutual benefit of all laboratories world-wide.

1.8 Conclusion

It has been shown beyond doubt that microbial secondary metabolites have interesting potential:

- To discover leads for chemical synthesis of new chemistry.
- As sources of new, fermented products for plant protection.
- As a way to monitor the production of biocontrol agents.

The enormous, still undiscovered, microbial diversity existing on this planet serves in this way as an indefinite pool for innovative biotechnology.[65,66] Novel structures of compounds from micro-organisms have proved to hold interesting new modes of action (Table 1.3). The diversity and complexity of biological products reflects more creativity than mankind can produce.

So far, the Actinomycetes have been by far the richest source of secondary metabolites. However, fungal metabolites have contributed a surprising proportion of novel fungicidal and insecticidal modes of action (see Table 1.3). Investigation of fungal, algal, and protozoan metabolites has only just begun.

The increased understanding of secondary metabolites from all different taxonomic groups may pave the way to even more interesting modes of action, to intelligent chemical modifications of natural products and to the synthesis of new products.

Further, advancement of biosynthesis studies will forward this whole area tremendously. Knowledge of the biosynthesis of interesting biologically active compounds could help improve the efficacy of the production process and perhaps mediate a pathway engineering approach. Further, biosynthesis data would be of importance for a better chemical grouping of the secondary metabolites and to add new light to both taxonomical, phylogenetic and ecological studies.

It is generally accepted that naturally produced microbial secondary metabolites would be inherently biodegradable. However, a more scientific investigation of this parameter, at least for some model studies, would be very interesting. The work of Yamaguchi[67] is pioneering in this respect.

The majority of the compounds discovered so far have been discovered in Japan. Also, by far the most reported use of these compounds in practical agriculture has taken place in Japan.[68] Further, new reports[69] suggest that the use of microbial preparations for plant protection in the People's Republic of China may exceed in volume that world-wide.

The potential for the rest of the world of the Japanese and Chinese compounds and concepts still remains to be seen. However, it will hopefully be evaluated without any bias through the banning of microbial secondary metabolites due to their 'relationship' to antibiotics used as human and veterinary drugs. Cross-resistance develops due to similarities in chemical structure and mode of action and has no relevance with respect to origin, source, or production method. In this way, microbial products do not possess greater risk for triggering cross-resistance than do any synthetic products.

Microbial secondary metabolites hold the potential to provide solutions for many problems in modern agri-, horti-, and silvi-culture, alone or in a clever, balanced use with real biocontrol agents as well as synthetic chemicals. Modern plant protection requires skilful play with many different cards. Let us take good care of the full hand; and support more biological solutions for the benefit of present and future generations.

1.9 References

1. J. Davies, Introduction, in 'Secondary Metabolites: Their Function and Evolution', ed. J. Chadwick and J. Whelan, Ciba Foundation Symposium 171, John Wiley and Sons, Chichester, 1992, p. 1.
2. J. B. Pillmoor, K. Wright and A. D. Terry, Natural products as a source of agrochemicals and leads for chemical synthesis, *Pestic. Sci.*, 1993, **39**(2), 131.
3. S. Omura, Trends in the search for bioactive microbial metabolites, *J. Indust. Microbiol.*, 1992, **10**, 135.
4. S. Omura, in 'The Search for Bioactive Compounds from Microorganisms', ed. S. Omura, Springer Verlag, New York, 1989, p. vii.
5. H. Reichenbach and G. Höfle, Production of bioactive secondary metabolites, in 'Myxobacteria II', Vol. 16, ed. M. Dworkin and D. Kaiser, Washington DC (in press).
6. L. C. Vining, Roles of Secondary Metabolites from Microbes, in 'Secondary Metabolites: Their Function and Evolution', ed. J. Chadwick and J. Whelan, Ciba Foundation Symposium 171, John Wiley and Sons, Chichester, 1992, p. 184.
7. H. Reichenbach and G. Höfle, Biologically active secondary metabolites from myxobacteria, *Biotechnol. Adv.*, 1993, **11**, 219.
8. C. Christophersen, Evolution in molecular structure and adaptive variance in metabolism, *Comp. Biochem. Physiol., B: Comp. Biochem.* 1991, **98**, 427.
9. D. H. Williams, M. J. Stone, P. R. Hauck, and S. K. Rahman, Why are secondary metabolites (natural products) biosynthesized? *J. Nat. Prod.*, 1989, **52**, 1189.

10. W. Rabsch and G. Winkelmann, The specificity of bacterial siderophore receptors probed by bioassays, *Bio. Metals*, 1991, **4**, 244.
11. D. A. Hopwood, Towards an understanding of gene switching in Streptomyces, the basis of sporulation and antibiotic production, *Proc. Roy. Soc. Lond.*, 1998, **B23**, 121.
12. S. Omura, The expanded horizon for microbial metabolites – a review, *Gene,* 1992, **115**, 141.
13. J. Chadwick and J. Whelan (ed.), 'Secondary Metabolites: Their Function and Evolution', Ciba Foundation Symposium 171, John Wiley and Sons, Chichester, 1992, 318 pp.
14. H. P. Fischer, R. Nyfeler, and J. P. Pachlatko, New agrochemicals based on microbial metabolites, in 'New Biopesticides', The Research Center for New Biomaterials in Agriculture, Seoul, 1992, p. 17.
15 C. M. M. Franco and L. E. L. Coutinho, Detection of novel secondary metabolites, *Crit. Rev. Biotechnol.* 1991, **11**, 193.
16. R. A. Maplestone, M. J. Stone, and D. H. Williams, The evolutionary role of secondary metabolites – a review, *Gene*, 1992, **115**, 151.
17. P. A. Worthington, Antibiotics with antifungal and antibacterial activity against plant diseases, *Nat. Prod. Rep.*, 1988, **5**, 47.
18. I. Yamaguchi, Antibiotics as antifungal agents, in 'Modern Selective Fungicide', (in press).
19. J. R. Babu, Avermectins: biological and pesticidal activities, in 'Biologically Active Natural Products Potential Use in Agriculture', H. Cutler, American Chemical Society, Washington DC, 1988, p. 92.
20. T. Iwasa, E. Higashide, H. Yamamoto, and M. Shibata, Studies on validamycins, new antibiotics. II. Production and biological properties of validamycins A and B, *J. Antibiot.*, 1970, **23**, 595.
21. S. Okuda and Y. Tanaka, Fungicides and antibacterial agents, in 'The Search for Bioactive Compounds from Microorganisms', ed. S. Omura, Springer Verlag, New York, 1992, p. 213.
22. S. Omura, Y. Tanaka, I. Kanaya, M. Shinose, and Y. Takahashi, Phthoxazolin, a specific inhibitor of cellulose biosynthesis, produced by a strain of *Streptomyces* sp., *J. Antibiot.*, 1990, **43**, 738.
23. G. L. Barron and R. G. Thorn, Destruction of nematodes by species of Pleurotus, *Can. J. Bot.*, 1987, **65**, 774.
24. O. C. H. Kwok, R. Plattner, D. Weisleder, and D. T. Wicklow, A nematocidal toxin from *Pleurotus ostreatus* NRRL 3526, *J. Chem. Ecol.*, 1992, **18**, 127.
25. Y. Iwai and Y. Takahashi, Selection of microbial sources of bioactive compounds, in 'The Search for Bioactive Compounds from Microorganisms', ed. S. Omura, Springer Verlag, New York, 1989, p. 281.
26. L. Lange, Microbes and microbial products in plant protection, *Prog. Bot.*, 1992, **53**, 252.
27. L. Lange, J. Breinholdt, F. W. Rasmussen, and R. I. Nielsen, Microbial fungicides – the natural choice, *Pestic. Sci.*, 1993, **39**(2), 155.
28. D. Devill, R. Nyfeler, and D. Sozzi, CGA 142705: A novel fungicide for seed treatment, *Brighton Crop Protection Confer. – Pests Dis.*, 1988, 65.
29. K. Gehmann, R. Nyfeler, A. J. Leadbeater, D. Nevill, and D. Sozzi, CGA 173506: A new phenylpyrrol fungicide for broad spectrum disease control, *Brighton Crop Protection Conf. – Pests Dis.*, 1990, 339.

30. A. J. S. Whalley, The Xylariaceaae, some ecological considerations, *Sydowia Ann. Mycol.*, 1986, **38**, 369.
31. T. Anke, G. Schramm, B. Schwalge, B. Steffan, and W. Steglich, Antibiotics from Basidiomycetes 20. Synthesis of Strobilurin A and revision of stereochemistry of natural strobilurins, *Liebigs Ann. Chem.*, 1984, **9**, 1616.
32. W. Weber, T. Anke, M. Bross, and W. Steglich, Strobilurin D and Strobilurin F, two new cytostatic and antifungal E-beta methoxyacrylate antibiotics from *Cyphellopsis anomala, Planta Med.*, 1990, **56**, 446.
33. W. Weber, T. Anke, B. Steffan, and W. Steglich, Antibiotics from Basidiomycetes 32. Strobilurin E, a new cystatic and antifungal E-beta methoxyacrylate from *Crepidotus fulvotomentosus, J. Antibiot.*, 1990, **43**, 207.
34. J. C. Frisvad, O. Filtenborg, U. Thrane, and S. B. Mathur, Analysis and screening for mycotoxins and other secondary metabolites in fungal cultures by thin layer chromatography and high performance liquid chromatography, *Arch. Environ. Contam. Toxicol.*, 1989, **18**, 331.
35. M. H. Fisher, Novel avermectin insecticides and miticides, in 'Recent Advances in the Chemistry of Insect Control II', ed. L. Crombie, The Royal Society of Chemistry, 1990, p. 52.
36. J. R. Babu, Avermectins: Biological and pesticidal activities, in 'Biologically Active Natural Products Potential Use in Agriculture', ed. H. Cutler, American Chemical Society, Washington DC, 1988, p. 95.
37. W. C. Campbell, M. H. Fisher, E. O. Stapley, G. Albers-Schonberg, and T. A. Jacob, Invermectin, a potent new antiparasitic agent, *Science,* 1983, p. 823.
38. C. C. Wang and S. S. Pong, Actions of avermectin Bla on GABA nerves, in 'Membranes and Genetic Disease', A. R. Liss, New York, 1982, p. 373.
39. K. Ando, H. Oishi, D. Hirabi, T. Okutomi, K. Suzuki, H. Osaki, M. Sawada, and T. Sagawa, Tetranactin, a new miticidal antibiotic, *J. Antibiot.,* 1971, **24**, 347.
40. S. Sakuda, A. Isogai, S. Matsumoto, and A. Suzuki, The structure of allosamidin, a novel insect chitinase inhibitor, produced by *Streptomyces* sp., *Tetrahedron Lett.,* 1986, **27**, 2475.
41. D. Koga, A. Isogai, S. Sakuda, S. Matsumoto, A. Suzuki, S. Imura, and A. Ide, Specific inhibition of *Bombyx mori* chitinase by allosamidin, *Agric, Biol. Chem.,* 1987, **51**, 471.
42. S. Takeuchi, K. Hirayama, K. Ueda, H. Sakai, and H. Yonehara, Blasticidin S, a new antibiotic, *J. Antibiot.*, 1958, **11**, 1.
43. S. Matsumoto, S. Sakuda, A. Isogai, and A. Suzuki, Search for microbial insect growth regulators; L-alanosine as an ecdysis inhibitor, *Agric. Biol. Chem.*, 1984, **48**, 827.
44. J. B. Gloer, R. M. TePaske, J. S. Sima, D. T. Wicklow, and P. F. Dowd, Antiinsectan aflavinine derivatives from the sclerotia of *Aspergillus flavus, J. Org. Chem.*, 1988, **53**, 5457.
45. Y. Om, I. Yamaguchi, and T. Misato, Inhibition of protein synthesis by mildiomycin, an anti-mildew substance, *Nippon Noyaku Gakkaishi,* 1984, **9**, 317.
46. U. Dähnm, U. Hagenmaier, H. Höhne, W. A. König, G. Wolf, and H. Zähner, Metabolic products of microorganisms, part 154: Nikkomycin, a new inhibitor of fungal chitin synthesis, *Arch. Microbiol.*, 1976, **107**, 143.
47. T. Takatsu, H. Nakayama, A. Shimazu, K. Furihata, K. Ikeda, H. Seto, and N. Otake, Rustmicin, a new macrolide antibiotic active against wheat stem rust fungus, *J. Antibiot.*, 1985, **38**, 1806.
48. Y. Abe, H. Nakayama, A. Shimazu, K. Furihata, K. Akea, H. Seto and N. Otake,

Neorustmin A, a new macrolide antibiotic active against wheat stem rust fungus, *J. Antibiot.*, 1985, **38**, 1810.

49. H. Nakayama. T. Hanamura, Y. Abe, A. Shimazu, K. Furihata, H. Seto, and N. Otake, Structures of neorustmicins B, C and D, new congeners of rustmicin and neorustmicin A, *J. Antibiot.*, 1986, **39**, 77.
50. K. Isono and S. Suzuki, The polyoxins: pyrimidin nucleoside peptide antibiotics inhibiting fungal cell wall biosynthesis, *Heterocycles,* 1979, **13**, 333.
51. M. Hori, J. Eguchi, D. Kakiki, and T. Misato, Mode of action of polyoxins, VI. Effect of polyoxin B on chitin synthesis in polyoxin-sensitive and resistant strains of *Alternaria kikuchiana, J. Antibiot.*, 1974, **27**, 260.
52. W. Pfeferle, H. Anke, M. Bross, B. Steffan, R. Vianden, and W. Steglich, Asperfuran, a novel antifungal metabolite from *Aspergillus oryzae,* 1990, **43**, 648.
53. N. Asano, M. Takeuchi, Y. Kameda, K. Matsui, and Y. Kono, Trehalase inhibitors, validoxylamine A and related compounds as insecticides, *J. Antibiot.*, 1990, **43**, 722.
54. Y. Kondo, T. Shomura, Y. Ogawa, T. Tsuruoka, H. Watanabe, K. Totsukawa, T. Suzuki, C. Moriyama, J. Yoshida, S. Inouye, and T. Niida, Studies on a new antibiotic SF-1293. I. Isolation and physicochemical and biological characterization of SF-1293 substance, *Sci. Rep. Meiji Seika Kaisha,* 1973, **13**, 34.
55. W. Rupp, M. Finke, H. Bieringer, and P. Langelcuddeke, Ger. Offen. 2,717,440, 1977. Herbicidal composition, *Chem. Abs.*, 1978, **88**, 70494e.
56. T. Ando, B. Tecle, R. F. Toia, and J. E. Casida, Titritionikkomycin Z radiolabeling of a potent inhibitor of fungal and insect chitin synthetase, *J. Agric. Food Chem.*, 1990, **38**, 1712.
57. D. G. Kuhn, R. W. Addor, R. E. Diehl, J. A. Furch, K. E. Henegar, G. T. Kamhi, B. C. Lowen, B. C. Black, T. P. Miller, and M. F. Treacy, in 'Abstracts from 203rd ACS International Meeting', American Chemical Society, Washington DC, 1992, p. 161.
58. R. A. A. Muzzarelli, Role and fate of exogenous chitosans in human wound tissues, in 'International Symposium on Chitin Enzymology', Senigelli, Italy, 1993, p. 35 (abstr).
59. R. Maria, S. B. Sullia, Acquired resistance of *Altenaria solani* and *Sclerotium rolfsii* to polyoxin D, *J. Phytopathol.*, 1986, **116**, 60.
60. E. Coppin-Raynal, Identification of two genes controlling kasugamycin resistance in the filametous fungus *Podospora anserina, Genet. Res.*, 1988, **51**, 179.
61. J. M. Clark and J. A. Argentine, Mechanisms of avermectin resistance in the colorado potato beetle, in 'Fourth Chemical Congress of North America', New York, Aug. 25–30, 1991.
62. R. Plater and J. A. Robinson, Cloning and sequence of a gene encoding macrotetralide antibiotic resistance from *Streptomyces griseus, Gene,* 1992, **112**, 117.
63. Y. Tanaka and O. Shigenobu, Insecticides, acaricides, and anticoccidial agents, in 'The Search for Bioactive Compounds from Microorganisms', ed. S. Omura, Springer Verlag, New York, 1992, p. 237.
64. H. Gürtler and L. Anker, Industries requirements with regard to identification of bacteria, in 'Proceedings FEMS Meeting on Identification of Bacteria, Current Status and Future Prospects', ed. F. G. Priest, A. Ramos-Cormenza and B. Tindal (in press).
65. A. T. Bull, M. Goodfellow, and J. H. Slater, Biodiversity as a source of innovation in biotechnology, *Ann. Rev. Microbiol.*, 1992, **46**, 219.

66. D. L. Hawksworth, The fungal dimension of biodiversity: magnitude, significance and conservation, *Mycol. Res.*, 1991, **95**, 641.
67. I. Yamaguchi, Future prospects for plant protection by pesticides of microbial origin, *Jpn Pestic. Information*, **50**, 17.
68. B. T. Misato and I. Yamaguchi, Pesticides of microbial origin, *Outlook Agric.,* 1984, **13**, 136.
69. Y. Chen, R. Mei, L. Liu, and J. W. Kloepper, The use of yield increasing bacteria as plant growth promoting rhizobacteria in Chinese agriculture (in press).
70. K. Arima, H. Imanaka, M. Kousaka, A. Fukuda, and G. Tamura, Studies on pyrrolnitrin, a new antibiotic, isolation and properties of pyrrolnitrin, *J. Antibiot.*, 1965, **18**, 201.
71. L. R. Schreiber, G. F. Gregory, C. R. Krause, and J. M. Ichida, Production, partial purification and antimicrobial activity of a novel antibiotic produced by a *Bacillus subtilis* isolate from *Ulmus americana, Can. J. Bot.* 1988, **66**, 2338.
72. C. Sandrin, F. Peypoux, and G. Michel, Coproduction of surfactin and iturin A, lipopeptides with surfactant and antifungal properties, by *Bacillus subtillis, Biotechnol. Appl. Biochem.*, 1990, **12**, 370.
73. S. Takeuchi, K. Hirayama, K. Ueda, H. Sakai, and H. Yonehara, Blasticidin S, a new antibiotic, *J. Antibiot. Ser. A,* 1958, 1.
74. X. C. Cheng, M. Ubukata, and K. Isono, The structure of tautomycetin, a dialkylmaleic anhydride antibiotic. *J. Antibiot.*, 1990, **43**, 890.
75. M. L. Lahdenperä, E. Simon, and J. Uoti, Mycostop, a novel biofungicide based on Streptomyces bacteria, in 'Biotic Interactions and Soilborne Diseases', ed. A. B. R. Beemster, Elsevier, Amsterdam, 1991, p. 258.
76. K. Isono, J. Nagatsu, K. Kobinata, K. Sasaki, and S. Suzuki, Studies on polyoxins, antifungal antibiotics. Part I. Isolation and characterization of polyoxins A and B, *Agric. Biol. Chem.*, 1965, **29**, 848.
77. H. Umezava, Y. Okami, T. Hashimoto, Y. Suhara, M. Hamada, and T. Takeuchi, A new antibiotic, kasugamycin, *J. Antibiot.*, 1965, 101.
78. H. Mishima, Milbemycins, a family of macrolide antibiotics with insecticidal activity, in 'Pesticide chemistry: Human Welfare and the Environment', ed. J. Miyamoto, Pergamon Press, Oxford, Vol. 2, 1983, 129.
79. H. B. Marauyama, *J. Antibiot.*, 1975, **28**, 636.
80. B. Loubinoux and P. Gerardin, Activity of fumaridmycin mimics against Oomycetes, *Pestic. Sci.*, 1991, **33**, 263.
81. A. Takahashi, S. Kurasawa, D. Ikeda, Y. Okami, and T. Takeuchi, Altemicidin, a new acaricidal and antitumor substance I. Taxonomy, fermentation, isolation and physicochemical and biological properties, *J. Antibiot.*, 1989, **42**, 1556.
82. O. D. Hensens, R. Monaghan, L. Huang, and G. Albers-Schonberg, Structure of the sodium and potassium ion activated ATPase ec-3.61.3 inhibitor L-681110, *J. Am. Soc. Chem. Soc.*, 1983, **105**, 3672.
83. S. Omura, M. Murata, N. Imamura, Y. Iwai, and H. Tanaka, Oxetin, a new antimetabolite from actinomycete. Fermentation, isolation structure and biological activity, *J. Antibiot.*, 1984, **37**, 1234.
84. K. Kanabe, Y. Mimura, T. Tamamura, S. Yatagai, Y. Sato, A. Takahashi, K. Sato, H. Naganawa, T. Takeuchi, and Y. Itaka, AB3217-A, a novel anti-mite substance produced by a strain of *Streptomyces platensis, J. Antibiot.*, 1992, **45**, 458.
85. S. Harada and T. Kishi, Isolation and characterization of mildiomycin, a new nucleoside antibiotic, *J. Antibiot.*, 1978, **31**, 519.

86. S. Murao, T. Sakai, T. Gibo, T. Nakayama, and T. Shin, A novel trehalase inhibitor, trehalostatin produced by *Amycolatopsis trehalostaticus, Agric. Biol. Chem.* 1991, **55**, 895.
87. M. Sutter, B. Boehlendorf, N. Bedorf, and G. Hoefle, European Patent, EP 359706, 1990.
88. K. Beautiment, J. M. Clough, P. J. Fraine, and C. R. A. Godfrey, Fungicidal beta methoxyacrylates from natural products to novel synthetic agricultural fungicides, *Pestic. Sci.*, 1991, **31**, 499.
89. K. Lewis, J. M. Whipps, and R. C. Cooke, Mechanisms of biological disease control with special reference to the case of study of *Pythium ologandrum* as an antagonist, in 'Biotechnology of Fungi for Improving Plant Growth', ed. J. M. Whipps and R. D. Lumsden, Cambridge University Press, Cambridge, 1989, p. 191.
90. N. Claydon and J. F. Grove, Metabolic products of *Enthomophthora virulenta, J. Chem. Soc., Perkin Trans. 1*, 1978, 171.
91. N. Takahashi (ed.) 'Gibberellins', Springer Verlag, Berlin, 480 pp.
92. E. Ammermann, G. Lorenz, K. Schelberger, B. Wenderoth, H. Sauter, and C. Rentzea, BAS490F – a broad-spectrum fungicide with a new mode of action, *Brighton Crop Protection Conf. – Pests Dis.*, 1992, 403.
93. M. M. Dewan and K. Sivasithamparam, A plant growth promoting sterile fungus from wheat and rye grass roots with potential for suppressing take-all, *Trans. Br. Mycol. Soc.*, 1988, **91**, 687.
94. M. M. Dewan and K. Sivasithamparam, Effects of colonization by a sterile red fungus on viability of seed and growth and anatomy of wheat roots, *Mycol. Res.*, 1990, **94**, 553.
95. J. B. Gloer and B. Rinderknect, Nominine: a new insecticidal indole diterpene from the sclerotia of *Aspergillus nomius. J. Org. Chem.* 1989, **54**, 2530.
96. H. P. Fischer and D. Bellus, Phytotoxicants from microorganisms and related compounds, *Pestic. Sci.*, 1983, **14**, 334.
97. R. D. Lumsden, J. C. Locke, S. T. Adkins, J. F. Walter, and C. F. Ridout, Isolation and localization of the antibiotic gliotoxin produced by *Gliocladium cirens* from alginate prill in soil and soiless media, *Phytopath.*, 1992, **82**, 230.
98. J. M. Lynch, Biological control within microbial communities of the rhizosphere, in 'Ecology of Microbial Communities', ed. M. Fletcher and T. R. G. Gray, Cambridge University Press, Cambridge, 1987, p. 55.
99. J. M. Lynch, Fungi as antagonists, in 'New directions in Biological Control', ed. R. R. Baker and P. E. Dunn, Alternatives for Suppressing Agricultural Pests and Diseases, Liss, New York, 1990, p. 243.
100. S. Gupta, S. B. Krasnoff, J. A. A. Renwick, and D. W. Roberts, Viridoxins A and B: Novel toxins from the fungus *Metarrhizium flavoviride, J. Org. Chem.*, 1993, **58**, 1062.
101. D. W. Roberts, Toxins from Entomopathogenic fungi, in 'Microbial Control of Pests and Plant Disease', ed. H. D. Burges, Academic Press, London, 1981, p. 441.
102. R. C. Prince, A. R. Crofts, and L. K. Steinrauf, A comparison of Beauveracin enniatin, valinomycin as calcium transporting agents in liposomes and chromatophores, *Biochem. Biophys. Res. Commun.*, 1974, **59**, 697.
103. B. A. Federici, Bright horizons for invertebrate pathology, in 'Int. Coloq. Invertebrate Pathology and Microbial Control, Adelaide', Australia, 1990. p. 138.
104. J. Breinholt, H. Demuth, L. Lange, A. Kjær, and C. Pedersen, Xanthofusin, an

antifungal tetronic acid from *Fusicoccum* sp.: production, isolation and structure, *J. Antibiot.*, 1993, **46**, 1013.

105. R. M. D. Makovski and K. Mortensen, The first mycoherbicide in Canada: *Colletotrichum gloeosporioides* f. sp. *malvae* for round-leaved mallow control, *Weed Sci. Soc. Victoria,* 1992, **2**, 298.
106. K. Kawazu, *et al.*, Isolation and characterization of two novel nematicidal depsipeptides from an imperfect fungus, strain D1084, *Biosci. Biotech. Biochem.,* 1993, **57**, 98.
107. H. Lyr, Effects of fungicides on energy production and intermediary metabolism, in 'Antifungal Compounds', ed. M. R. Seigel and H. D. Sisler, New York, 1977, Vol. 2, p. 301.
108. P. Leroux, Similarity between the antifungal activities of fenpiclonil iprodione and tolclofos-methyl, *Agronomie,* 1991, **11**, 115.
109. R. A. Dybas, N. J. Hilton, J. R. Babu, F. A. Presier, and G. R. Dolce, Novel second generation avermectin insecticides and miticides for crop protection, in 'Novel Microbial Products for Medicine and Agriculture', ed. A. L. Demain *et al.*, Society for Industrial Microbiology, Washington DC, 1989, p. 203.
110. R. A. Dybas, Abamectin use in crop protection, in 'Invermectin, and Abamectin', ed. W. C. Campbell, Springer Verlag, New York, 1989. p. 288.
111. K. Isono, New Aspects in research and development of agricultural antibiotics, *Nippon nogei Kagaku kaishi,* 1989, **63**, 1351.
112. K. Sato, Biological properties of kasugamycin, in 'Pesticide Chemistry, Human Welfare and the Environment, Vol. 2, Natural Products', Pergamon Press, Oxford, 1983, p. 293.
113. M. H. Fischer and H. Mrozig, The avermectin family of macrolide-like antibiotics, in 'Macrolide Antibiotics: Chemistry, Biology, and Practice', ed. S. Omura, Academic Press, Orlando, 1984, p. 553.
114. S. Omura, Y. Tanaka, K. Kanaya, M. Shinose, and Y. Takahashio, Phthoxazolin, a specific inhibitor of cellulose biosynthesis, produced by a strain of *Streptomyces* sp., *J. Antibiot.*, 1990, **43**, 738.
115. S. Omura, M. Murata, H. Hanaki, K. Hinotozawa, R. Oiwa, and H. Tanaka, Oxetin, a new antimetabolite from Actinomycete, fermentation, isolation, structure, and biological activity, *J. Antibiot.*, 1984, **37**, 1234.
116. Y. Om, I. Yamaguchi, and T. Misato, Inhibition of protein synthesis by mildiomycin, an anti-mildew substance, *J. Pestic. Sci.*, 1984, **9**, 317.
117. L. Pridzun, PhD Thesis, Technical University, Braunschweig, Germany, 1992.
118. J. Larsen, L. Lange, and L. W. Olson, Mastigomycotina developmental sensitivity to alfa-, beta-unsaturated carbonyl compounds, *J. Phytopath.*, 1992, **134**, 336.

CHAPTER 2

Pesticides of Microbial Origin and Applications of Molecular Biology

ISAMU YAMAGUCHI

2.1 Introduction

Agrochemicals have played an important role in preventing crop losses, caused by pests and diseases, thus meeting the ever-growing needs of the population in the world. However, there is much concern about the side effects of pesticides on non-target organisms and their environmental impact. These concerns highlight the need for selective agrochemicals with higher degradability in nature. Since pesticides of microbial origin (biochemical pesticides) are synthesized biologically, they are generally specific for target organisms and are expected to be inherently biodegradable. Thus, world-wide interest in them has been renewed. However, the use of microbial products as agrochemicals also has limitations. One disadvantage is that they are apt to suffer from the emergence of natural resistance due to their highly specific mechanisms of action. Another is the concern that their wide use may create natural resistance that will hinder the medical treatment of humans by drugs with similar modes of action. In this chapter, the use of microbial products as agrochemicals is reviewed, focusing on their action mechanisms, mostly on the basis of more than 30 years' experience of the use of pesticides of microbial origin in Japan. Recent advances in the application of molecular biology to biochemical pesticides are also discussed.

2.2 Present Status of Pesticides of Microbial Origin

One of the main obstacles in crop production is yield loss and quality damage caused by plant diseases and insect pests. While pesticides are indispensable to ensure high agricultural productivity, there is growing concern about pesti-

Table 2.1 *Microbial products used in agriculture in Japan*

Substances	*Effective against*
Antifungal	
Blasticidin S*	Rice blast
Kasugamycin*	Rice blast
Polyoxins*	Rice sheath blight, Fungal diseases of fruit trees
Validamycin A*	Rice sheath blight, Fungal diseases of vegetables
Mildiomycin*	Powdery mildew of rose and crape myrtle
Antibacterial	
Streptomycin	Bacterial diseases of fruit trees and vegetables
Dihydrostreptomycin	Bacterial diseases of fruit trees and vegetables
Oxytetracyclin	Bacterial diseases of fruit trees and vegetables
Insecticidal	
Polynactins*	Mites of fruit trees and tea plants
Milbemectins*	Mites of fruit trees and tea plants
Herbicidal	
Bilanafos*	Weeds in orchards and mulberry fields

*Substances discovered in Japan.

cides, in terms of their side effects on mammals and wildlife and as a possible source of environmental pollution.

Among crop protection agents, pesticides of microbial origin, so-called agricultural antibiotics, are generally quite selective to target organisms and are considered to be easily decomposable in the environment. The first pesticides of microbial origin introduced in agriculture were those originally developed for medicinal purposes. The first success was the use of streptomycin for the control of pear fire blight in the USA. The use of most medicinal antibiotics, however, is limited in the agricultural field due to different requirements, such as the need for less phytotoxicity, higher durability to sunlight and lower manufacturing cost. Therefore, many attempts were made to find microbial products for the prime purpose of plant disease control, and eventually an epoch-making substance, named blasticidin S, was discovered.[1] It exhibited significant efficacy in controlling rice blast disease, the most serious and damaging of all the diseases of rice plants in temperate and humid climates, and it has been widely used for rice blast control by farmers in Japan and other countries. Since then many antibiotics have been screened and some put into practical use in agriculture. Table 2.1 shows the pesticides of microbial origin used commercially in Japan. Below, the characteristics of major pesticides of microbial origin, and some promising microbial products for future developments as pesticides, are outlined.

2.2.1 Blasticidin S

Blasticidin S (**1**) is produced by *Streptomyces griseochromogenes,* and is used for its preventive and curative effects on rice blast, caused by *Pyricularia oryzae* (perfect stage: *Magnaporthe grisea*).[2]

Blasticidin S (**1**) is obtained as a white needle crystal, highly soluble in water, but mostly insoluble in organic solvents, stable in solution at pH 5–7, but unstable in alkaline conditions.

The chemical structure of blasticidin S (**1**) consists of a novel nucleoside, designated cytosinine, and a β-amino acid, named blasticidic acid (ε-*N*-methyl-β-arginine)[3] as shown in Figure 2.1. The pyrimidine ring of blasticidin S is biosynthetically derived from cytosine, the sugar moiety from glucose, the *N*-methyl group of blasticidic acid from methionine, and arginine is used as the precursor of the blasticidic acid skeleton.[4] A metabolic intermediate in its biosynthetic pathway is leucylblasticidin S.[5]

Blasticidin S (**1**) exhibits a wide range of inhibitory activity on the growth of bacterial and fungal cells, as well as antiviral and antitumour activities.[1,6] It inhibits spore germination and mycelial growth of *P. oryzae* at less than 1 μg ml^{-1}, and its excellent curative effect on rice blast is attributed to this potent inhibitory activity on the growth of the pathogen.

The antibiotic markedly interferes with the incorporation of ^{14}C amino acids into protein in the intact cells, as well as in cell-free systems of *P. oryzae*.[7,8] While the action mechanism of blasticidin S (**1**) is still not clarified at the molecular level, its mode of inhibition is suggested to occur in the processes of peptidyl transfer in protein synthesis.[9]

Blasticidin S (**1**) is rather toxic to mammals, with an oral LD_{50} of 53.3 mg kg^{-1} for mice, but its toxicity is low to fish. It causes irritation and inflammation if it comes into contact with eyes and mucous membranes. A simple method to alleviate eye irritation, without impeding the antiblast effect, is to add calcium acetate to the blasticidin S formulation. This improved dust formulation is now

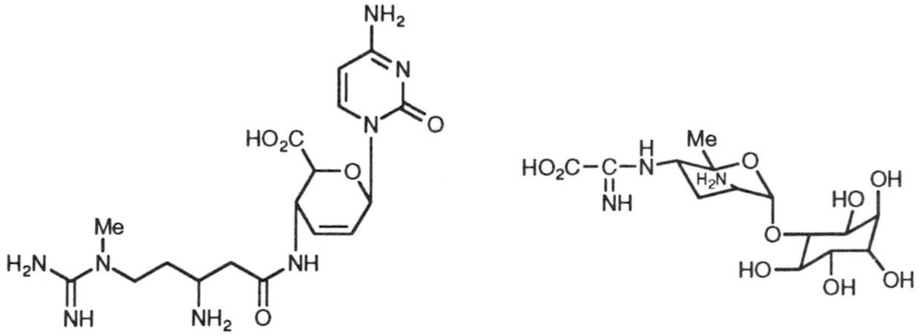

Figure 2.1 *Chemical structures of blasticidin S* (**1**) *and kasugamycin* (**2**)

used in practice for agriculture. Detoxin isolated from the culture broth of *Streptomyces caespitosus* also reduces the adverse effect without reducing antifungal activity,[10] but its use is too expensive.

In field application to control rice blast, the effective amount of blasticidin S (**1**) is as low as 1–3 g active ingredient per 1000 m^2. It occasionally causes chemical injury to rice leaves if sprayed at high concentrations. Blasticidin S benzylaminobenzene sulphonate (the applied formulation) was found to be the least phytotoxic to rice plants for equal control efficacy on blast disease. Among other crops, the tobacco plant is most susceptible to blasticidin S, followed by aubergine, tomato, and potato; the mulberry plant is rather susceptible; grape, pear, and peach plants are resistant; water melon and cucumber are as resistant as the rice plants.

The behaviour and fate of blasticidin S (**1**) in the environment were investigated using radioactive compounds prepared biosynthetically from [^{14}C]-cytosine and [^{14}C]-L-methionine.[11] The sprayed antibiotic is found mostly on the surface of the rice plant, and little is taken up by the tissue. From a wound or infected part, however, the substance is incorporated and translocated to the apices. Blasticidin S (**1**) located on the plant surface is decomposed by sunlight and gives rise to cytosine as the main degradation product. A considerable quantity of blasticidin S spray falls to the ground and is bound onto the soil surface. Significant generation of [^{14}C]-carbon dioxide from soil treated with [^{14}C]-blasticidin S occurs; however, several microbes that usually inhabit the paddy fields reduce the biological activity of blasticidin S. From the results obtained, it is suggested that, after application of blasticidin S to a crop field at a low concentration, the antibiotic is easily degraded in the environment, so that there is no danger of environmental pollution and food contamination. In fact, the residual amount of blasticidin S in an unpolished rice cropped in the field is estimated as less than 0.05 p.p.m. by biological assay.

2.2.2 Kasugamycin

Kasugamycin (**2**) is an aminoglycoside antibiotic produced by *Streptomyces kasugaensis*.[12] It has been widely used for rice blast control without any phytotoxicity and with very low toxicity to mammals and fish.

Kasugamycin (**2**) is a water-soluble and basic substance, its hydrochloride (applied form) is also water-soluble, but insoluble in most organic solvents. The molecule consists of three moieties, namely, D-inositol, kasugamine (2,3,4,6-tetradeoxy-2,4-diaminohexopyranose) and an iminoacetic acid side chain[13,14] (Figure 2.1). In the biosynthesis of kasugamycin, the kasugamine moiety is derived from glucose or mannose, and the other parts from myo-inositol and glycine.[15,16]

Kasugamycin (**2**) inhibits the growth of *P. oryzae* specifically under acidic conditions (pH 5.0), but hardly at all in neutral media.[17] Its effect on *P. oryzae* is possibly expressed *in planta* under acidic conditions. Kasugamycin also interferes with the growth of some bacteria, including *Pseudomonas* species, but unlike the fungicidal effect it provides stronger inhibition against the bacteria at pH 7 than at pH 5.

Kasugamycin inhibits protein synthesis in a cell-free system of *Escherichia coli*, by interfering with the binding of aminoacyl-tRNA to the mRNA-30S ribosomal subunit complex without causing miscoding.[18] Kasugamycin sensitivity involves the 16S RNA of the 30S ribosomal subunit, and the resistant mutant of *E. coli* cannot methylate two adjacent adenine residues near the 3′ end of the 16S RNA.[19]

Kasugamycin has quite low acute or chronic toxicity to mice, rats, rabbits, dogs and monkeys. The oral LD_{50} for mice is 2 g kg^{-1}, and the TL_m to carp is 1 mg ml^{-1}.

Kasugamycin (**2**) controls rice blast disease at concentrations as low as 20 p.p.m.[20] It is mainly applied as a dust containing 0.3% of the active ingredient. When rice seeds are coated with 2% kasugamycin wettable powder, the plants are protected from rice blast for a month in the field.

The emergence of resistance to kasugamycin was recognized a few years after its first application in 1965, and the development of resistant strains of *P. oryzae* in the field became a serious problem in 1972. Since then, mixtures of kasugamycin and chemicals with different modes of action have been used in practice. In laboratory experiments, kasugamycin-resistant strains of *P. oryzae* proved less infective than the sensitive parent strains.[21] In fact, once application of kasugamycin had been discontinued in the field, the population of the resistant strains declined quite rapidly.

2.2.3 Polyoxins

The polyoxins, a group of peptidylpyrimidine nucleoside antibiotics, are produced by *Streptomyces cacaoi* var. *asoensis*.[22,23] Polyoxins are composed of fourteen components (A–N) of closely related peptidic nucleosides. They are safely used for the control of many fungal diseases with no toxicity to livestock, fish or plants. Such excellent characteristics are due to polyoxins selectively inhibiting the synthesis of cell-wall chitin in sensitive fungi.[24,25] Polyoxins have been widely used since 1967 for crop protection against such pathogenic fungi as *Rhizoctonia solani, Cochliobolus miyabeanus,* and *Alternaria alternata*.

The isolation and chemical elucidation of polyoxins are due to the excellent work of Isono *et al.*[23,26–28] (see Figure 2.2). The C polyoxin lacks antifungal activity, but it is a key compound in elucidating the structure of polyoxins, since hydrolytic degradation of all the polyoxins yielded polyoxin C or its analogues. The identification of a nucleoside with 5-aminohexuronic acid as its sugar group is the first example in Nature. The biosynthesis of polyoxins is also attractive since they contain a unique amino acid (polyoximic acid) and 5-substituted pyrimidine ring in their molecules. The distribution of ^{14}C in this unique cyclic amino acid proved that the intact carbon skeleton of L-isoleucine was utilized directly in its biosynthesis.[29] The 5-substituted uracil in polyoxins was reported to be biosynthesized from uracil and C-3 of serine by a new enzyme which differs from thymidylate synthetase.[30]

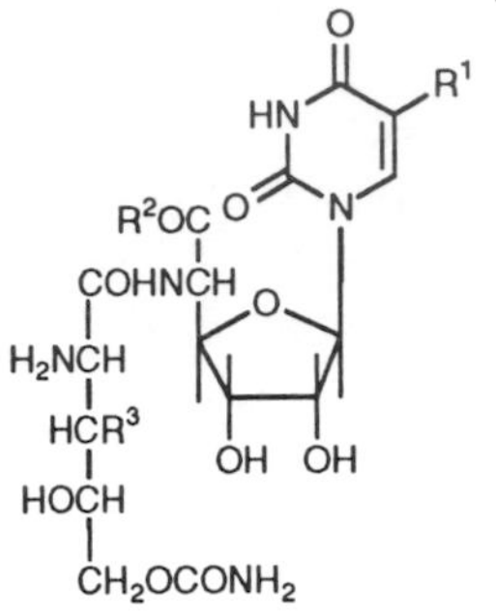

Polyoxin	R^1	R^2	R^3
A	CH_2OH	*	OH
B	CH_2OH	HO	OH
D	CO_2H	HO	OH
E	CO_2H	HO	H
F	CO_2H	*	OH
G	CH_2OH	HO	H
H	CH_3	*	OH
J	CH_3	HO	OH
K	H	*	OH
L	H	HO	OH
M	H	HO	H

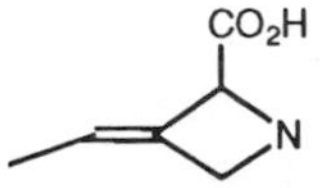

3

* Polyoximic acid

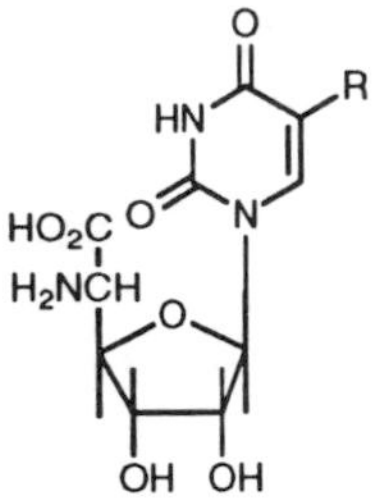

Polyoxin	R
C	HO
I	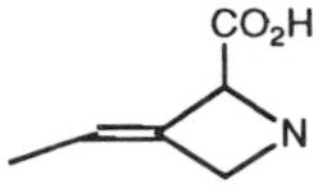 (polyoximic acid residue)

4

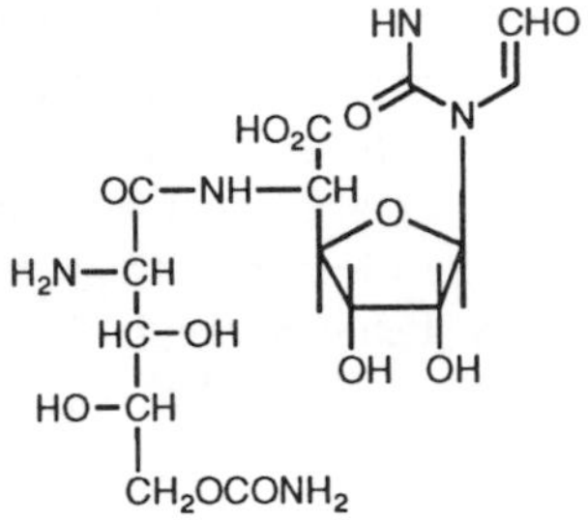

5 Polyoxin N

Figure 2.2 *Chemical structures of polyoxins*

Polyoxins inhibit the growth of some fungi, but are inactive against bacteria and yeasts. Among polyoxins, polyoxin D is most effective for the control of rice sheath-blight caused by *R. solani*, whereas polyoxin B is effective for pear black-spot fungus and apple cork-spot fungus caused by *Alternaria* sp. As for its toxicity, oral administration at 15 g kg^{-1} and injection at 800 mg kg^{-1} for mice did not cause any adverse effects, nor is it toxic to fish during 72 hours of exposure at 10 p.p.m. Foliar sprays of 200 p.p.m. polyoxin have produced no phytotoxicity on fruit trees, and no injury to the rice plant was observed even at 800 p.p.m. application. Trials to control the diseases confirmed the efficacy of polyoxins and their persistence in the field. Polyoxin complexes have been used in practice in two forms; a polyoxin D-rich fraction for sheath-blight control, and a B-rich fraction for diseases caused by *Alternaria* spp.

With regards to the mechanism of fungicidal action of polyoxins, a specific physiological phenomenon was observed in *Alternaria* spp.:[31] polyoxins caused a marked abnormal swelling on germ tubes of spores and hyphal tips of the pathogen, which made the pathogen non-infectious. Further, the incorporation of [^{14}C]-glucosamine into cell-wall chitin of *C. miyabeanus* was shown to be markedly inhibited by polyoxin D.[24] In cell-free systems of *Neurospora crassa*, polyoxin D was proved to inhibit the incorporation of *N*-acetylglucosamine (GlcNAc) into chitin in a competitive manner between UDP-GlcNAc and polyoxin D. The relation between polyoxin structure and inhibitory activity on chitin synthase was clarified by kinetic analysis;[25] the pyrimidine nucleoside moiety of the antibiotics was shown to fit into the binding site of the enzyme protein, and the carbamoylpolyoxamic acid moiety of polyoxins was shown to stabilize the polyoxin–enzyme complex. Therefore, the excellent characteristics of polyoxins can be explained by the fact that the antibiotics inhibit cell-wall synthesis of sensitive fungi, but have no adverse effects on organisms which have no chitinous cell walls.

Polyoxin-resistant strains of *A. alternata* were recognized in some orchards in Japan after intensive use of the antibiotics. The resistance was caused by a lowered permeability of the antibiotic through the cell membrane into the site of chitin synthesis, and the inhibition in mycelial growth of *R. solani* by polyoxins was antagonized by glycyl-L-alanine, glycyl-DL-valine, and DL-alanylglycine.[32] Thus, polyoxins are thought to be incorporated into fungal cells through the channels of dipeptide uptake.

2.2.4 Validamycins

Validamycin A (**6**) was isolated from the culture filtrate of *Streptomyces hygroscopicus* var. *limoneus*, which produced five additional components designated as validamycins B–F, together with validoxylamine A.[33] It has been used for the control of *Rhizoctonia* disease, without phytoxicity and with very low toxicity to mammals, birds, fish, and insects.[34]

The chemical structure of validamycin A (**6**) was determined as having two kinds of new hydroxymethyl-branched cyclitols in its molecule,[35] giving a unique example in the field of pseudosugar chemistry (Figure 2.3). Validamycins A,C,D,E, and F contain validoxylamine A as a common moiety in their molecules, but they differ from each other in either the configuration of the anomeric centre of the glucoside, the position of the glucosidic linkage, or the number of D-glucose molecules. On the other hand, validamycin B contains validoxylamine B in its molecule, which yields hydroxyvalidamine, instead of validamine derived from other validamycins. Studies on the microbial transformation of validamycins showed the conversion of validamycin C into A by *Endomycosis* spp. or *Candida* spp.,[36] which is of importance because validamycin C is about 1000 times less active than validamycin A against *R. solani*.

Validamycin A is specifically effective against certain plant diseases caused by *Rhizoctonia* spp., such as sheath blight of rice, web blight, bud rot, damping-off, seed decay, root rot, black scurf of several crops and southern blight of vegetables.[37] While the antibiotic showed no fungicidal action to *Rhizoctonia* spp., it caused abnormal branching at the tips of hyphae of the pathogen, followed by cessation of further development. When it was applied to rice, sufficient control of sheath blight was achieved by a spray of 30 p.p.m. Thus,

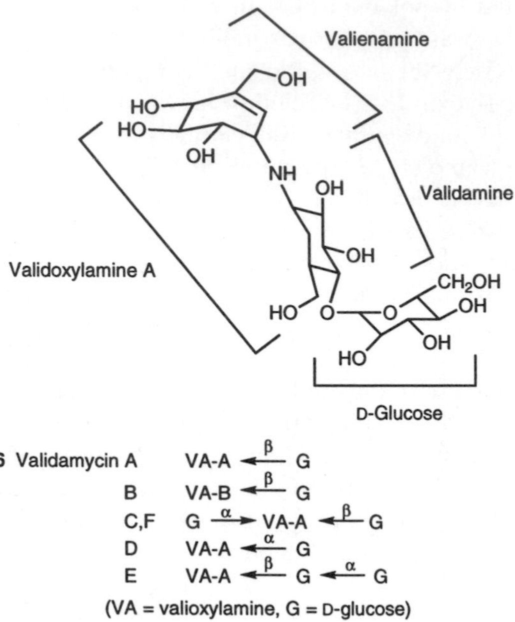

Figure 2.3 *Chemical structure of validamycin A* (**6**) *and components of the validamycins*

validamycin A has been used commercially for the control of the disease since 1973.

Although validamycin A (**6**) did not suppress significantly the growth of *R. solani* on a nutritionally rich medium, it caused extensive branching of hyphae and the cessation of colony development on a water agar, i.e. under nutrientless conditions. It was also observed that validamycin A had no need of continual contact with the pathogen for effective control of fungal growth.[37] This is important in actual disease control, since *R. solani* is one of the typical fungi that grows rapidly by transporting nutrients from the basal part to hyphal tips through long stretches of hyphae; this type of growth provides the tip part with rather nutritionally poor conditions. As for the mode of action of validamycin A, it was shown that validamycin A had potent inhibitory activity against trehalase in *R. solani* AG-1, without any significant inhibitory effect on other glycohydrolytic enzymes tested.[38] Trehalose is well-known as a storage carbohydrate in the fungus and trehalase is thought to play an essential role to digest trehalose and transport D-glucose to the hyphal tips.

Antimicrobial activity of validamycin A was not detected on many fungi and bacteria by ordinary methods, nor was disturbance of the microflora on rice and other field crops examined. No phytotoxicity was observed for over 150 species of plants sprayed with validamycin A, even at concentrations of 1000 p.p.m. Furthermore, toxicities to mammals were markedly low; in the oral administration of validamycin A at a dose of 10 g kg^{-1} to mice and rats, or in subcutaneous and intravenous administration at a dose of 2 g kg^{-1} to mice, all the animals tested underwent no change for 7 days, and no irritating effects were observed on the skin and the cornea in rabbits. Oral subacute toxicity for 4 months in beagle dogs indicated no significant abnormalities of the morphological and biochemical parameters accompanying the daily administration of 200 mg kg^{-1} of validamycin A. LD_{50} for killifish was found to be greater than 1000 p.p.m.

Validamycins are susceptible to microbial degradation and their addition to the soil resulted in quick loss of biological activity by soil microbes. Microbial degradation of validoxylamine A by *Pseudomonas denitrificans* gave rise to D-glucose and validoxylamine A, which was further decomposed into valienamine and validamine (see Figure 2.3). The metabolic fate of validamycins in animals is that orally administered validamycin A is easily decomposed to CO_2 by enteric bacteria, and validamycin A is rapidly excreted in the urine when injected intravenously. So far, cross-resistance with medicinal aminoglycoside antibiotics, such as dihydrostreptomycin or kanamycin, has not been detected in the human pathogens tested.[34]

Validamycin A has been used to protect sheath blight of rice, mainly in formulations of 3–5% solution or 0.3% dust. Residues in rice grain and straw were less than the detectable limit by gas chromatography. Thus, validamycin A is considered to be one of the ideal chemicals with respect to safety and environmental pollution.

2.2.5 Mildiomycin

Mildiomycin (**7**, Figure 2.4) was isolated from the culture filtrate of *Streptoverticillium rimofaciens* B-98891.[39] It is a water soluble and basic antibiotic which belongs to nucleoside antibiotics. Mildiomycin (**7**) is specifically active against the pathogens that cause powdery mildews, but less active on bacteria, and most fungi and yeasts. It shows excellent curative activity on powdery mildew of various plants *in vivo*, and thus it has been used practically for the control of disease on rose, spindle tree and Indian lilac.

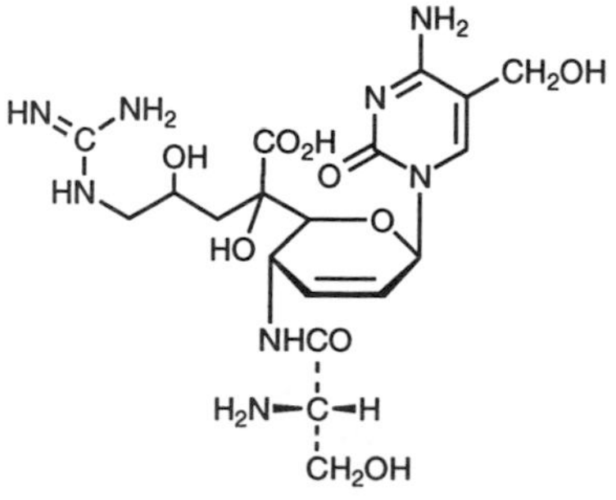

7 Mildiomycin

Figure 2.4 *Chemical structure of mildiomycin* (**7**)

The toxicity of mildiomycin is very low; LD_{50} for acute toxicity in rats and mice is 500–1000 mg kg^{-1} by intravenous and subcutaneous injections and 2.5–5.0 g kg^{-1} by oral administration. At a concentration of 1000 μg ml^{-1} no irritation has been observed to the cornea and skin of rabbits over 10 days. Toxicity to killifish is not observed at a concentration of 20 p.p.m. for 7 days. Protein synthesis is remarkably inhibited by mildiomycin at concentrations down to 0.02 mmol dm^{-3}; it markedly interfered with the incorporation of [^{14}C]-phenylalanine into polypeptides in the cell free system of *E. coli*.[40] The mammalian cell-free system from rabbit reticulocytes, however, proved to be much less sensitive to mildiomycin in the synthesis of polypeptides than did the system from *E. coli*.

2.2.6 Streptomycins

Streptomycin and dihydrostreptomycin are well-known aminoglycoside antibiotics isolated from *Streptomyces griseus*.[41] With their bactericidal activity against a wide range of bacterial species, streptomycin and dihydrostreptomycin have become the first antibiotics introduced into agriculture for plant disease control. They have been used for the control of apple and pear fire blights by foliar application at a concentration of 200 p.p.m. They are also effective against wild fire of tobacco and bacterial leaf blight of rice. A mixture of streptomycin and oxytetracycline is highly effective for the control of bacterial canker of peach, citrus canker, soft rot of vegetables and various other bacterial diseases.

Streptomycin-resistant strains are distributed in a wide range of plant pathogenic bacteria, such as *Xanthomonas oryzae, X. citri, Pseudomonas tabaci* and *P. lachrymans*. In agricultural use, the alternative or combined applications of streptomycin and other chemicals with different action mechanisms is recommended in order to reduce the development of streptomycin-resistant strains in the field. With regard to the phytotoxicity of streptomycin, it occasionally causes chemical injuries to vegetables and rice if it is applied at high concentrations. A mixture of streptomycin sulfate and iron chloride or citrate is effective for reducing the phytotoxicity of the antibiotic.

Streptomycin inhibits protein synthesis in bacterial cells by binding to the 30S ribosomal subunit, and causes misreading of the genetic codes in protein synthesis.[42] Mutants of *E. coli* highly resistant to streptomycin are known to involve modification of the P10 protein of the bacterial ribosome 30S subunit.

2.2.7 Oxytetracycline

Oxytetracycline (**8**, Figure 2.5) is produced by *Streptomyces rimosus*;[43] in the group are tetracycline produced by *S. viridifaciens*, chlortetracycline (aureomycin) by *S. aureofaciens* and demethylchlortetracycline by a variant of *S. aureofaciens*. Tetracyclines are active against a wide range of microorganisms, including bacteria, rickettsiae and protozoa, and they are effective

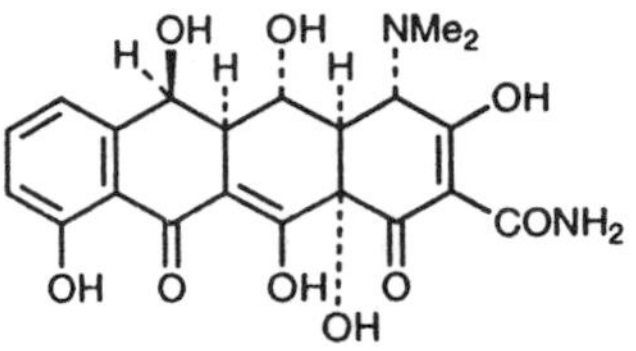

8 Oxytetracycline

Figure 2.5 *Chemical structure of oxytetracycline* (**8**)

in controlling some plant bacterial diseases caused by *Pseudomonas, Xanthomonas,* and *Erwinia* species. They also show protective effects for plants against diseases caused by mycoplasma-like organisms.[44] Oxytetracycline is easily taken up by plant leaves, expecially through stomata, and is rapidly translocated to plant tissues. In the agricultural field, oxytetracycline has been used to control bacterial diseases by mixing it with streptomycin to avoid the development of streptomycin-resistant strains. A mixture of oxytetracycline and streptomycin occasionally causes chlorosis on the leaves of some crops when sprayed under conditions of high temperature and humidity.

Tetracycline is a potent inhibitor of bacterial protein biosynthesis, with less activity on mammalian cells. It binds to the 30S and 50S bacterial ribosomal subunits, and inhibits the binding of aminoacyl-tRNA and the termination factors RF1 and RF2 to the A site of bacterial ribosome.[45]

2.2.8 Polynactins

Polynactins are macrotetrolide antibiotics produced by *Streptomyces aureus* strain S-3466.[46] They are soluble in most organic solvents, but insoluble in water, and fairly stable at pH 2–13 and on exposure to sunlight. Polynactins are effective against the carmine spider mite, two-spotted spider mite and European red mite. Adzuki bean weevils and larvae of mosquito are moderately sensitive to the antibiotics, but house flies and cockroaches are insensitive. Polynactins were the first antibiotics developed as a miticide for plants. While polynactins show least activity by direct contact with mites in a dry state, they exert marked miticidal activity under wet conditions, which means water is an essential factor for their miticidal activity. As for the mode of action of polynactins, it is suggested they cause leakage of a basic cation, such as K^+, through the lipid layer of the membrane in mitochondria.[47]

Acute toxicity of polynactins for mice and rats is very low, with an oral LD_{50} of more than 25 g kg^{-1} and 2.5 g kg^{-1}, respectively; but their toxicity to fish is high, with the TL_m to carp being 0.003 p.p.m.

When tetranactin suspensions are applied to an apple tree, proliferation of both the Kanzawa spider and the European red mites under field conditions is completely retarded for 32 days. In order to avoid the development of resistant mites, mixtures of tetranactin with other chemicals, such as fenobucarb and chlorfenson are used in the field.

2.2.9 Avermectins and Milbemectins

Avermectins, a novel class of macrocyclic lactones, were isolated from the culture of *Streptomyces avermetilis*.[48] They have extremely high potencies against agricultural and household insect pests, phytophagous mites and plant parasitic nematodes. None of the components of avermectins possess antibacterial or antifungal activity, and a study of the mode of insecticidal action indicates that avermectin B_{1a} has an agonistic activity to a γ-aminobutyric acid (GABA), so abnormally stimulating the GABA-mediated inhibitory postsynaptic potential at the neuromuscular junction.

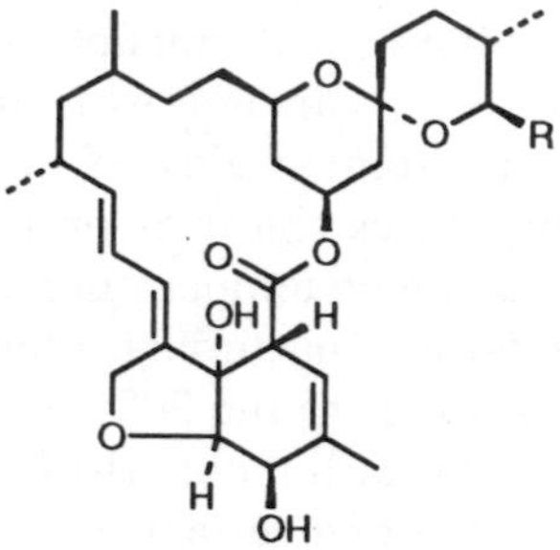

9 A_3, R = Me
10 A_4, R = Et

Figure 2.6 *Chemical structures of milbemycin A_3* (**9**) *and A_4* (**10**)

Milbemycins (Figure 2.6), which belong to the same class of 16-membered macrolide as avermectins, were isolated from *Streptomyces hygroscopicus* f. sp. *aureolacriosus*.[49] They have been shown to have high insecticidal and acaricidal, but no antimicrobial activities.[50] A mixture of milbemycin A_3 (M.A_3, **9**) and A_4 (M.A_4, **10**) has been developed as an acaricide (M.A_3/M.A_4 = 3/7). It was shown to stimulate Cl^- uptake by the leg muscles of the American cockroach at concentrations as low as 10^{-7} mol dm^{-1}, suggesting that its action mechanism is to open the chlorine channel in the plasma membrane of the nervous and neuromusclar systems.[51] However, its irreversible agonistic activity to GABA appears not to be directly mediated through binding to a GABA receptor, since bicuculline, a typical antagonist of GABA, has no effect on the activity of milbemycins. The mixture (milbemectin) has been practically used for the control of *Tetranychidae* on tea trees and egg plants since 1990. Toxicity to mammals is low, but TL_m to fish is high (1.7 p.p.m. for carp).

2.2.10 Bilanafos

Several microbial products are also known to possess herbicidal activity (see also Chapter 4). These are cycloheximide, anisomycin, toyocamyin, herbicidin A and B, and herbimycin. In practice, a synthetic analogue of anisomycin, 3,3′-dimethyl-4-methoxybenzophenone, was developed commercially as a herbicide.[52] Bilanafos (bialaphos), another microbial product, was shown to have strong herbicidal activity against a wide range of weeds when foliage is treated.[53] The application rates are 1–3 kg ha^{-1}, depending upon the growth stage of the weeds and whether they are annuals or perennials. It exerts potent herbicidal activity by foliage application, but shows very weak phytotoxic activity towards vegetables when it is present in soils. This is believed to be because of its remarkable biodegradability by soil micro-organisms. It indicates the feasibility of bilanafos for presowing treatments and in applications to orchard and mulberry fields. Interestingly, the combined application of bilanafos with fertilizers, such as urea or ammonium sulfate, enhances the herbicidal activity. This may be because bilanafos inhibits the glutamine synthetase system in weeds, resulting in a deficiency of glutamine in the plant cells, as well as abnormal accumulation of ammonia,[54] known to be a strong uncoupler of photophosphorylation.

2.2.11 Promising Microbial Products

Among the microbial products there are still other promising fungicides, such as miharamycins,[55] neopeptins,[56] and irumamycin.[57] In the area of livestock disease, salinomycin[58] and ivermectin[59] are in commercial use, respectively as a coccidiostat in poultry and as an antiparasitic agent in cattle, horses, and sheep. Carriomycin[60] and cationomycin[61] have been shown to have coccicidal activity without mammalian toxicity. If gibberellins, well-known plant-growth regulators, are also added to the group, microbial products are being used in every sphere of pesticides.

2.3 Applications of Molecular Biology to Pesticides of Microbial Origin

2.3.1 Fungicide Tolerance and a New Selective Marker for Transgenic Plants

Blasticidin S (**1**) is a fungicide with a potent control efficacy on rice blast disease, as described earlier.[1] It occasionally exhibits adverse phytotoxic effects on some sensitive plants, such as tobacco and aubergine, due to its inhibitory activity on their protein synthesis. During study of its metabolic fate in the environment, an inactivating enzyme, named blasticidin S deaminase (aminohydrolase, EC 3.5.4.23), was discovered in *Aspergillus terreus*,[62] and

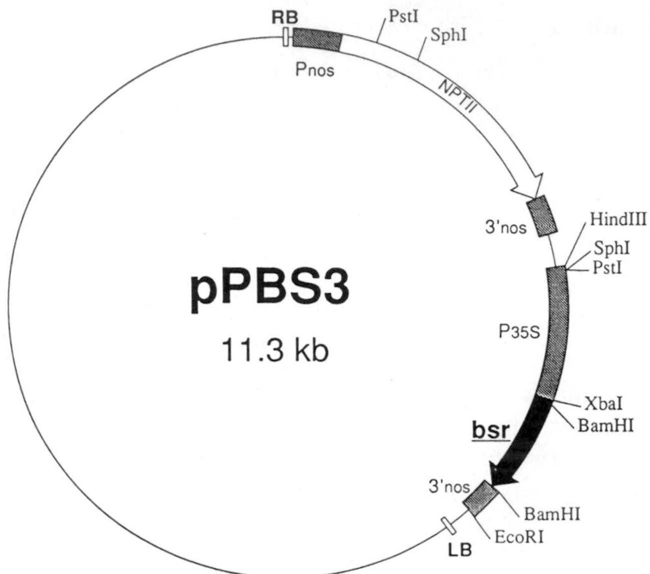

Figure 2.7 *Construction of plasmid pPBS3. The solid box indicates the fragment containing the coding region of* bsr *(420 bp). RB indicates right border sequence. 'P' means promoter and '3' terminator. The fragment from the* Hin*fI site to a* Nde*I site, which exists 10 bp downstream of the termination codon of* bsr, *was introduced into the* Bam*HI site of pPCV701 by using the* Bam*HI linker*

later in a resistant strain, K55-S1, of *Bacillus cereus*.[63] This enzyme converts blasticidin S into an inactive form by catalysing the hydrolytic deamination of cytosine nucleus in the molecule. From the bacterium, a 10.5 kb plasmid was isolated and correlated with the enzyme production,[64] and a structural gene, *bsr*, of the enzyme was successfully cloned in *B. subtilis* and *Escherichia coli*.[65] Analysis of the gene resulted in its application as a selectable marker for microorganisms such as *E. coli*, as well as for mammalian cells.[66] Further, introduction of the gene into tobacco plants was performed by using a Ti plasmid vector to afford resistance against blasticidin S, as described below.

The DNA sequencing of *bsr* revealed an open reading frame comprising 420 nucleotides coded for proteins of 140 amino acids.[67] There existed a *Hin*fI site at 36 bp upstream of the initiation codon and there were no other ATG sequences between the *Hin*fI site and the initiation codon. This minimum fragment containing the coding region of the *bsr* gene was ligated to the 35S promoter of cauliflower mosaic virus. The resulting plasmid, pPBS3 (Figure 2.7) was used for *Agrobacterium*-mediated transformation into tobacco *Nicotiana tabacum* cv. Havana SR1 by the leaf disk infection method. pPBS3 has both kanamycin resistant gene and a *bsr* gene as expected plant-selective markers. Subsequently, the leaf disks were cultured on the shooting media containing 70 μg ml^{-1} of kanamycin and 2.5 μg ml^{-1} of blasticidin S, respectively. Transformed shoots were successfully selected on both media, and the existence of the *bsr* gene was confirmed by Southern hybridization in the transformants resistant to blasticidin S. The resistant phenotype was easily detected on the whole plants by spraying with 50 $\mu g\ ml^{-1}$ of blasticidin S.[68] Thus, *bsr* was proved to be usable as a direct selection marker for the plant transformation system.

Owing to the broad antimicrobial activity of blasticidin S, it is also interesting to introduce the resistant gene into sensitive plants to control microbial attack; this may not only avoid the phytotoxicity of the chemical, but it may also widen the applicability of the chemical to the control of other diseases on the originally sensitive plants. Some advantages of the *bsr* gene are as follows:

- The *bsr* gene is small (420 bp), and easy to replace as a *Hin*fI–*Nde*I fragment is an almost minimum component of coding region.
- Blasticidin S is a stable substance under usual conditions, but is quite unstable in alkaline solution; therefore, its disposal after treatment is environmentally safe.
- The activity of blasticidin S deaminase is easily detectable by measuring ultraviolet absorption at 282 nm using blasticidin S as a substrate.
- The phytotoxic effect of blasticidin S on some plants is significant in the whole plant as well as in the tissue culture.
- The expression of the gene can be safely tested in the environment, since the safety of blasticidin S has been confirmed.

2.3.2 Disease Resistance in Transgenic Tobacco Plants Engineered for Detoxification of Tabtoxin

Advances in the genetic engineering of plants have allowed the development of agronomically important traits, such as herbicide, insect, and viral resistances. These have occurred by the transfer of single dominant genes into plants. Such strategies could be applied to achieve plant protection against bacterial and fungal diseases. In particular, detoxification of pathogenesis-related toxins has the advantage that a greater resistance can be exerted with lower levels of detoxifying enzyme. Tabtoxin, a phytotoxic dipeptide produced by *Pseudomonas syringae* pv. *tabaci*, is composed of tabtoxinine β-lactam [2-amino-4-(3-hydroxy-2-oxoazacyclobutan-3-yl)butanoic acid] and either serine or threonine.[69] Several lines of evidence suggest that this toxin causes the chlorotic symptom associated with the wildfire disease caused by the pathogen. It is suggested that, *in planta*, some peptidases convert tabtoxin into tabtoxinine β-lactam, which inhibits the target enzyme glutamine synthetase.[70] This inhibition results in an abnormal accumulation of ammonium in tobacco cells, causing the characteristic chlorosis.

The herbicide bilanafos is produced by *Streptomyces hygroscopicus* SF1293, which contains a resistance gene to protect itself from the toxic metabolites.[71] The bilanafos resistance gene, which encodes an acetyltransferase, inherently functions in the biosynthesis of bilanafos in *S. hygroscopicus*.[72] This example of 'self-resistance' suggests the existence of the tabtoxin resistance genes in the wildfire pathogen *P. syringae* pv. *tabaci*.

*Bcl*I-digested genomic DNA of *P. syringae* pv. *tabaci* MAFF 03-01075 was ligated to the *Bam*HI site of pUC13 and transformed into *Escherichia coli* DH1 sensitive to tabtoxin. After selection on ampicillin-supplemented medium, the resultant clones were reselected on minimum medium containing tabtoxin. Consequently, the recombinant plasmid pARK10, including a 2 kb insert of genomic DNA, was obtained.[73] Subcloning experiments showed that a 700 bp *Xba*I–*Sph*I fragment was enough to confer resistance to tabtoxin (Figure 2.8).

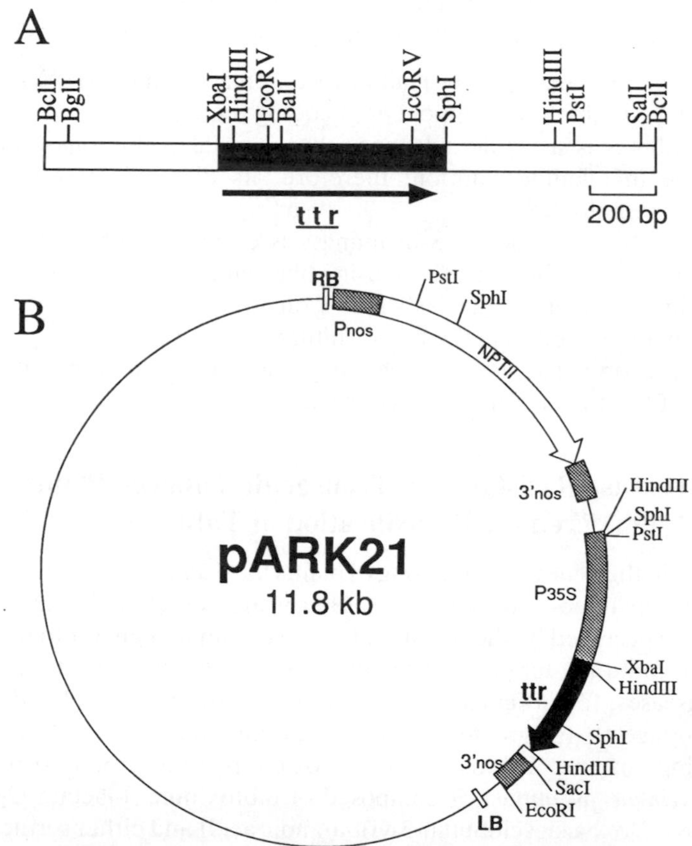

Figure 2.8 *Construction of plasmid pARK21. (a) The restriction sites determined in the 2 kb* Bcl*I fragment containing the* ttr *gene. The solid box indicates the 700 bp of the minimum region conferring resistance to tabtoxin and the direction of transcription is shown by an arrow. (b) Diagram of pARK21. The 700 bp* Xba*I*–Sph*I fragment carrying the* ttr *gene was cloned into the same restriction sites of pTZ18R. From the resulting plasmid, an 800 bp* Xba*I*–Pvu*II fragment containing the 700 bp insert was re-inserted between the* Xba*I and* Sma*I sites of pUC19 to obtain a fragment with* Xba*I*–Sac*I restriction sites. The* Xba*I*–Sac*I fragment replaced the fragment containing the β-glucuronidase gene with the same restriction sites on the plant vector pBI121, locating the* ttr *gene between the 35S promoter of CaMV and the nopaline synthase (nos) terminator of pTiC58*

The gene, *ttr*, was specific for the inactivation of tabtoxin, and not for other inhibitors of glutamine synthetase, such as bilanafos and methionine sulfoximine.

The direction of transcription was determined by the expression of the *ttr* gene linked to the *lac*Z promoter on the vector, as shown in Figure 2.8. To examine the function of the *ttr* gene, tabtoxin was incubated with [^{14}C]-acetyl-CoA in crude bacterial extracts. Radioactive acetylated products were detected only in extracts of *E. coli* transformed with pARK10, indicating that the *ttr* gene encodes an enzyme that acetylates tabtoxin and/or tabtoxinine β-lactam. The acetylation probably occurs at a tabtoxinine β-lactam amino group, in a mechanism similar to that for the inactivation of bilanafos.

To test whether this *ttr* gene could confer resistance to wildfire disease in tobacco, the coding sequence was inserted between the cauliflower mosaic virus 35S promoter and the nopaline synthase polyadenylation signal on the binary vector pBI121. The resulting plasmid pARK21 was used for the *Agrobacterium*-mediated transformation of the chimeric *ttr* gene into tobacco (*Nicotiana tabacum* cv. Havana SR1) by the leaf disk infection method.[74] Transformed shoots were first selected on a medium containing kanamycin. Leaf disks from kanamycin-resistant plants were checked by tabtoxin resistance by callus induction on a medium containing tabtoxin. Calli from independent transformed plants were resistant, demonstrating that the *ttr* gene expressed resistance to tabtoxin in tobacco cells. The integration of the *ttr* gene into the genomic DNA of transgenic tobacco plants was confirmed by Southern analysis, and tested the expression levels of the *ttr* gene by Northern analysis. When the transgenic tobacco plants were inoculated with *P. syringae* pv. *tabaci,* none of the transformed plants produced any chlorotic halo typical of wildfire disease. This strongly indicates that the transgenic tobacco plants expressing the *ttr* gene have become resistant not only to tabtoxin, but also to infection by *P. syringae* pv. *tabaci*.[74] These results demonstrate a successful approach to obtain disease-resistant plants by detoxification of the pathogenic toxins, which play an important role in pathogenesis.

The system described here represents a useful model for plant protection from disease by introducing a toxin detoxifying gene derived from the pathogen itself. Such strategy could be applied to other diseases caused by pathogenesis-related toxins. If a pathogen does not have any detoxifying enzyme, it might be possible to screen such enzymes or genes from other organisms.

2.4 Summary and Conclusion

In recent years, standards for the disposal of industrial wastes have become more and more strict, and consequently the manufacture of synthetic chemicals has become quite expensive, since waste disposal facilities, as well as production plants, must be set up for each new product. In this respect, pesticides of microbial origin have an economic advantage, since a variety of substances can be manufactured using one set of equipment and facilities. Additionally, they are produced not from limited fossil resources, but from renewable agricultural products through fermentation by micro-organisms. Therefore, pesticides of microbial origin are more advantageous, especially for developing countries where better results can be expected in the future.

As is true for every scientific technique, the use of these pesticides also has limitations. One disadvantage is the difficulty in microanalysis, especially as they consist of many components. The second difficulty is that, because of their highly specific mechanism of action, they are apt to suffer from the emergence of resistant strains. To cope with this problem, it is sometimes a requisite to combine the agents with other chemicals that have different action mechanisms, or to use these in rotation. Further, the severest limitation to the use of microbial products in agriculture is a concern that their wide use might create resistant strains which could hinder the medical treatment of humans. This concern seems particularly important in advanced Western countries, where the use of microbial products began by an application of medicinal antibiotics, as mentioned earlier. Fortunately, the use of pesticides of microbial origin in Japan for more than 30 years has not met any problem involving cross-resistance with medicinal antibiotics. Naturally, sufficient precautions must be taken to ensure no resistance in human pathogens in the future. However, this problem is not limited to pesticides of microbial origin. There is no difference between microbial products and synthetic chemicals in this respect, so the important point is whether or not any pesticides may induce resistance to medicines. At present, the use of microbial products in agriculture is subjected to rigid regulations and restrictions, apparently because they are called 'antibiotics'. This should be reconsidered from the scientific point of view; it can be argued that such products should be treated in the same way as synthetic chemicals.

Among the pesticides listed in Table 2.1, polyoxins and validamycins are quite safe – non-phytotoxic, and non-toxic to humans, livestock, and wildlife. Such excellent characteristics seem to be due to their modes of action. Polyoxins selectively inhibit the synthesis of fungal cell-wall chitin, which does not exist in mammalian cells.[24,25] Validamycin A has an exceptional character, *i.e.*, it is not fungicidal, but only deteriorates the normal mycelial growth of the pathogen on plants.[34] Probably because of such moderate activity, no resistant strains have been reported in validamycin treatment, despite it being the most used among pesticides of microbial origin. Further, components of avermectins,[48] insecticidal microbial products developed in the USA, have been successfully transformed into significantly more active derivatives by simple chemical modifications of the original structures. These findings demonstrate considerable scope for the future development of microbial products. In addition, new microbial products with novel characteristics may be found against pests and diseases, such as soil-borne diseases, which are at present difficult to control by synthetic chemicals. In the near future, crop and pest biochemistry and molecular biology will make rapid further advances, and a new biotechnology will be an important factor for plant protection research and strategy. For example, more efficient production of microbial products is becoming possible by using modern gene engineering. Biorational approaches will also become feasible in the design of new pesticides from microbial products, through the extensive use of computer and data processing procedures.

The future shape of agriculture in the world will depend largely on the availability of the appropriate kinds of pesticides in adequate quantities. The following considerations might be important for the future development of microbial products as fungicidal agents:

- Establishment of screening directions. The development of some new antibiotics occurred shortly after establishing a new screening project, *e.g.*, polyoxins against rice-sheath blight.
- Establishment of new screening test methods. The introduction of a novel screening test method carries with it the possibility for development of new, effective antibiotics. Kasugamycin and validamycins were discovered by adopting new assay systems.
- Avoidance of pathogens' resistance to antibiotics. Kasugamycin and polyoxins have high selective toxicity to pathogenic fungi. When pathogens resistant to these antibiotics emerged, the alternate or combined application of chemicals with different action mechanisms was effective. Such applications are essential in some cases to avoid the development of resistant pathogens.
- The use of antibiotics as leads for new pesticides. Modification of existing antibiotics provides new potent chemicals. The best examples of these are semi-synthetic penicillin and avermectin. More attention must be focused on finding the molecular relationship between chemical structure and biological activity.

2.5 References

1. S. Takeuchi, K. Hirayama, K. Ueda, H. Sasaki, and H. Yonehara, Blasticidin S, a new antibiotic, *J. Antibiot. Ser. A.*, 1958, **11**, 1.
2. T. Misato, I. Ishii, M. Asakawa, Y. Okimoto, and K. Fukunaga, Antibiotics as protectant fungicides against rice blast. II. The therapeutic action of blasticidin S, *Ann. Phytopathol. Soc. Jpn*, 1959, **24**, 302.
3. N. Otake, S. Takeuchi, T. Endo, and H. Yonehara, Chemical studies on blasticidin S. III. The structure of blasticidin S, *Agric. Biol. Chem.*, 1966, **30**, 132.
4. H. Seto, I. Yamaguchi, N. Otake, and H. Yonehara, Biogenesis of blasticidin S, *Tetrahedron Lett.*, 1966, 3793.
5. H. Seto, N. Otake, and H. Yonehara, Studies on the biosynthesis of blasticidin S. Part II. Leucylblasticidin S, a metabolic intermediate of blasticidin S biosynthesis, *Agric. Biol. Chem.*, 1968, **32**, 1299.
6. N. Tanaka, Y. Sakagami, H. Yamaki, and H. Umezawa, Activity of cytomycin and blasticidin S against transplantable animal tumors, *J. Antibiot. Ser. A.*, 1961, **14**, 123.
7. T. Misato, Y. Okimoto, I. Ishii, M. Asakawa, and K. Fukunaga, Antibiotics as protectant fungicides against rice blast. IV. Effect of blasticidin S on the metabolism of *Pyricularia oryzae*, *Ann. Phytopathol. Soc. Jpn*, 1961, **26**, 25.

8. K. T. Huang, T. Misato, and H. Suyama, Effect of blasticidin S on protein synthesis of *Pyricularia oryzae, J. Antibiot. Ser. A.*, 1964, **17**, 65.
9. M. Yukioka, T. Hatayama, and S. Morisawa, Affinity labelling of the ribonucleic acid component adjacent to the peptidyl recognition center of peptidyl transferase in *Escherichia coli* ribosomes, *Bioch. Biophys. Acta.*, 1975, **390**, 192.
10. H. Yonehara, H. Seto, S. Aizawa, T. Hidaka, A. Shimazu, and N. Otake, The detoxin complex, selective antagonists of blasticidin S, *J. Antibiot.*, 1968, **21**, 369.
11. I. Yamaguchi, K. Takagi, and T. Misato, The sites for degradation of blasticidin S, *Agric. Biol. Chem.*, 1972, **37**, 1719.
12. H. Umezawa, Y. Okami, T. Hashimoto, Y. Suhara, M. Hamada, and T. Takeuchi, A new antibiotic, kasugamycin, *J. Antibiot. Ser. A.*, 1965, **18**, 101.
13. Y. Suhara, K. Maeda, and H. Umezawa, Chemical studies on kasugamycin. V. The structure of kasugamycin, *Tetrahedron* Lett., 1966, **12**, 1239.
14. Y. Suhara, F. Sasaki, K. Maeda, H. Umezawa, and M. Ohno, The total synthesis of kasugamycin, *J. Am. Chem. Soc.*, 1968, **90**, 6559.
15. Y. Fukagawa, T. Sawa, T. Takeuchi, and H. Umezawa, Studies on biosynthesis of kasugamycin. I. Biosynthesis of kasugamycin and the kasugamine moiety, *J. Antibiot.*, 1968, **21**, 50.
16. Y. Fukagawa, T. Sawa, T. Takeuchi, and H. Umezawa, Studies on biosynthesis of kasugamycin. V. Biosynthesis of the amidine group, *J. Antibiot.*, 1968, **21**, 410.
17. M. Hamada, T. Hashimoto, S. Takahashi, M. Yoneyama, T. Miyake, Y. Takeuchi, Y. Okami, and H. Umezawa, Antimicrobial activity of kasugamycin, *J. Antibiot. Ser. A.*, 1965, **18**, 104.
18. N. Tanaka, H. Yamaguchi, and H. Umezawa, Mechanism of kasugamycin action on polypeptide synthesis, *J. Biochem.*, 1966, **60**, 429.
19. T. L. Helser, J. E. Davies, and J. E. Dahlberg, Change in methylation of 16S×165 ribosomal RNA associated with mutation to kasugamycin resistance in *Escherichia coli, Nature*, 1971, **233**, 12.
20. T. Ishiyama, I. Hara, M. Matsuoka, K. Sato, S. Shimada, R. Izawa, T. Hashimoto, M. Hamada, Y. Okami, T. Takeuchi, and H. Umezawa, Studies on the preventive effect of kasugamycin on rice blast, *J. Antibiot. Ser. A.*, 1965, **18**, 115.
21. K. Ohmori, Studies on characters of *Pyriculara oryzae* made resistant to kasugamycin, *J. Antibiot. Ser. A.*, 1967, **20**, 109.
22. S. Suzuki, K. Isono, J. Nagatsu, T. Mizutani, Y. Kawashima, and T. Mizuno, A new antibiotic, polyoxin A, *J. Antibiot. Ser. A.*, 1965, **18**, 131.
23. K. Isono, J. Nagatsu, Y. Kawashima, and S. Suzuki, Studies on polyoxins, antifungal antibiotics. Part I. Isolation and characterization of polyoxins A and B, *Agric. Biol. Chem.*, 1965, **29**, 848.
24. S. Sasaki, N. Ohta, I. Yamaguchi, S. Kuroda, and T. Misato, Studies on polyoxin action. Part I. Effect on respiration and synthesis of protein, nucleic acids and cell-wall of fungi, *Nippon Nogei Kagaku Kaishi*, 1968, **42**, 633.
25. M. Hori, K. Kakiki, and T. Misato, Studies on the mode of action of polyoxin. Part IV. Further study on the relation of polyoxin structure to chitin synthetase inhibition, *Agric. Biol. Chem.*, 1974, **38**, 691.
26. K. Isono, J. Nagatsu, K. Kobinata, K. Sasaki, and S. Suzuki, Studies on polyoxins, antifungal antibiotics. Part V. Isolation and characterization of polyoxins, C,D,E,F,G,H and I, *Agric. Biol. Chem.*, 1967, **31**, 190.
27. K. Isono and S. Suzuki, The structures of polyoxins D,E,F,G.H,I,J,K and L, *Agric. Biol. Chem.*, 1968, **32**, 1193.

28. K. Isono, K. Asahi, and S. Suzuki, Studies on polyoxins, antifungal antibiotics. XIII. The structures of polyoxins. *J. Am. Chem. Soc.*, 1969, **91**, 7490.
29. K. Isono, S. Funayama, and R. H. Suhadolnik, Biosynthesis of the polyoxins, nucleoside peptide antibiotics. A new metabolic role for L-isoleucine as a precursor for 3-ethylidene-L-azetidine-2-carboxylic acid (polyoximic acid), *Biochemistry*, 1975, **14**, 2992.
30. K. Isono and R. H. Suhadolnik, The biosynthesis of natural and unnatural polyoxins by *Streptomyces cacaoi, Arch. Biochem. Biophys.*, 1976, **173**, 141.
31. J. Eguchi, S. Sasaki, N. Ohta, T. Akashiba, T. Tsuchiyama, and S. Suzuki, Studies on polyoxins, antifungal antibiotics. Mechanism of action on the diseases caused by *Alternaria* spp, *Ann. Phytopathol. Soc. Jpn*, 1968, **34**, 280.
32. M. Hori, K. Kakiki, and T. Misato, Antagonistic effect of dipeptides on the uptake of polyoxin A by *Alternaria kikuchiana, J. Pestic. Sci.*, 1977, **2**, 139.
33. S. Horii, Y. Kameda, and K. Kawahara, Studies on validamycins, new antibiotics. VIII. Validamycins C, D, E and F. *J. Antibiot.*, 1972, **25**, 48.
34. K. Matsuura, Characteristics of validamycin A in controlling *Rhizoctonia* diseases, in 'IUPAC Pesticide Chemistry', ed. J. Miyamoto and P.C. Kearney, Pergamon Press, Oxford, 1983, Vol. 2, pp. 301.
35. T. Suami, S. Ogawa, and N. Chida, The revised structure of validamycin A, *J. Antibiot.*, 1980, **33**, 98.
36. Y. Kameda, S. Horii, and T. Yamoto, Microbial transformation of validamycins, *J. Antibiot.*, 1975, **28**, 298.
37. O. Wakae and K. Matsuura, Characteristics of validamycin A as a fungicide for *Rhizoctonia* disease control, *Rev. Plant Protect. Res.*, 1975, **8**, 81.
38. R. Shigemoto, T. Okuno, and K. Matsuura, Effect of validamycin A on the activity of trehalase of *Rhizoctonia solani* and several sclerotial fungi, *Ann. Phytopathol. Soc. Jpn*, 1989, **55**, 238.
39. S. Harada and T. Kishi, Isolation and characterization of mildiomycin, a new nucleoside antibiotic, *J. Antibiot.*, 1978, **31**, 519.
40. Y. Om, I. Yamaguchi, and T. Misato, Inhibition of protein synthesis by mildiomycin, an anti-mildew substance, *J. Pestic. Sci.*, 1984, **9**, 317.
41. A. Schatz, E. Bugie, and S. A. Waksman, Streptomycin, a substance exhibiting antibiotic activity against Gram-positive and Gram-negative bacteria, *Proc. Soc. Exp. Biol. Med.*, 1994, **55**, 66.
42. T. E. Likover and C. G. Kurland, The contribution of DNA to translation errors induced by streptomycin *in vitro, Proc. Natl. Acad. Sci. U.S.A.*, 1967, **58**, 2385.
43. A. C. Finlay, G. L. Hobby, S. Y. Pan, P. P. Regna, J. B. Routier, D. B. Seeley, G. M. Shull, B. A. Sobin, I. A. Solomons, J. W. Vinson, and J. H. Kane, Terramycin, a new antibiotic, *Science*, 1950, **111**, 85.
44. T. Ishii, Y. Doi, K. Yora, and H. Asuyama, Suppressive effects of antibiotics of tetracycline group on symptom development of mulberry dwarf disease. *Ann. Phytopathol. Soc. Jpn*, 1967, **33**, 267.
45. C. T. Caskey, Inhibitors of protein synthesis, in 'Metabolic Inhibitors', ed. R. M. Hochster, M. Kates and J. H. Quastel, Academic Press, New York, 1973, Vol. IV, p. 131.
46. K. Ando, H. Oishi, S. Hirano, T. Okutomi, K. Suzuki, H. Okazaki, M. Sawada, and T. Sagawa, Tetranactin, a new miticidal antibiotic. I. Isolation, characterization and properties of tetranactin. *J. Antibiot.*, 1971, **24**, 347.
47. K. Ando, T. Sagawa, H. Oishi, K. Suzuki, and T. Nawata, Tetranactin, a pesticidal antibiotic, *Proc. 1st Intersect. Congr. IAMS (Sci. Counc. Jpn)*, 1974, **3**, 630.

48. R. A. Dybas, Avermectins: their chemistry and pesticidal activities, in 'IUPAC Pesticide Chemistry', ed. J. Miyamoto, and P. C. Kearney, Pergamon Press, Oxford, 1983, Vol. 1, p. 83.
49. Y. Takiguchi, H. Mishima, M. Okuda, M. Terao, A. Aoki, and R. Fukuda, Milbemycins, a new family of macrolide antibiotics: Fermentation, isolation and physicochemical properties, *J. Antibiot.*, 1980, **33**, 1120.
50. M. Mishima, Milbemycin: A family of macrolide antibiotics with insecticidal activity, in 'IUPAC Pesticide Chemistry', ed. J. Miyamoto and P. C. Kearney, Pergamon Press, Oxford, 1983, Vol. 2, p. 129.
51. K. Tanaka and F. Matsumura, Action of avermectin B_{1a} on the leg muscles and the nervous system of the American cockroach, *Pestic. Biochem. Physiol.*, 1985, **24**, 124.
52. K. Munakata, O. Yamada, S. Ishida, F. Futatsuya, K. Ito, and H. Yamamoto, NK-049: From natural products to new herbicides, in *Proc. 4th Asian-Pacific Weed Sci. Soc. Conf.*, 1975, p. 215.
53. K. Tachibana, T. Watanabe, Y. Sekizawa, M. Konnal, and T. Takematsu, Finding of the herbicidal activity and the mode of action of bialaphos, L-2-amino-4-[(hydroxy)(methyl)phosphinoyl]butyrylalanylalanine, in 'Abstr. 5th Inst. Congr. Pestic. Chem', 1982, IVa-19.
54. K. Tachibana, T. Watanabe, Y. Sekizawa, and T. Takematsu, Accumulation of ammonia plants treated with bialaphos, *J. Pesticide Sci.*, 1986, **11**, 33.
55. H. Seto, N. Otake, M. Koyama, H. Ogino, Y, Kodama, N. Nishizawa, T. Tsuruoka, and S. Inoue, The structures of novel nucleoside antibiotics, miharamycin A and miharamycin B, *Tetrahedron Lett.*, 1983, **24**, 495.
56. T. Satomi, H. Kusakabe, G. Nakamura, T. Nishio, M. Uramoto, and K. Isono, Neopeptins A and B, new antifungal antibiotics, *Agric. Biol. Chem.*, 1982, **46**, 2621.
57. S. Omura, Y. Tanaka, A. Nakagawa, Y. Iwai, M. Inoue, and H. Tanaka, Irumamycin, a new antibiotic active against phytopathogenic fungi, *J. Antibiot.,* 1982, **35**, 256.
58. Y. Miyazaki, M. Shibuya, H. Sugawara, O. Kawaguchi, C. Hirose, J. Nagatsu, and S. Esumi, Salinomycin, a new polyether antibiotic, *J. Antibiot.*, 1974, **27**, 814.
59. W. C. Campbell, M. H. Fisher, E. O. Stapley, G. Albers-Schonberg, and T. A. Jacob, Ivermectin: A potent new antiparasitic agent, *Science*, 1983, **221**, 823.
60. N. Otake, H. Nakayama, H. Miyamae, S. Sato, and Y. Saito, X-ray crystal structure of thallium salt of carriomycin, a new polyether antibiotic, *J. Chem. Soc., Chem. Commun.*, 1977, 590.
61. G. Nakamura, K. Kobayashi, T. Sakurai, and K. Isono, Cationomycin, a new polyether ionophore antibiotic produced by *Actinomadura* nov. sp, *J. Antibiot.,* 1981, **34**, 1513.
62. I. Yamaguchi, H. Shibata, H. Seto, and T. Misato, Isolation and purification of blasticidin S deaminase from *Aspergillus terreus, J. Antibiot.*, 1973, **28**, 7.
63. T. Endo, K. Furuta, A. Kaneko, T. Katsuki, K. Kobayashi, A. Azuma, A. Watanabe and A. Shimazu, Inactivation of blasticidin S by *Bacillus cereus.* I. Inactivation mechanism, *J. Antibiot.*, 1987, **40**, 1791.
64. T. Endo, K. Kobayashi, N. Nakayama, T. Tanaka, T. Kamakura, and I. Yamaguchi, Inactivation of blasticidin S by *Bacillus cereus.* II. Isolation and characterization of a plasmid, pBSR8, from *Bacillus cereus, J. Antibiot.*, 1988, **41**, 271.
65. T. Kamakura, K. Kobayashi, T. Tanaka, I. Yamaguchi, and T. Endo. Cloning and

expression of a new structural gene for blasticidin S deaminase, a nucleoside aminohydrolase, *Agric. Biol. Chem.*, 1987, **51**, 3165.

66. M. Izumi, H. Miyazawa, T. Kamakura, I. Yamaguchi, T. Endo, and F. Hanaoka, Blasticidin S – resistance gene (*bsr*): A novel selectable marker for mammalian cells, *Exp. Cell Res.*, 1991, **197**, 229.
67. K. Kobayashi, T. Endo, T. Kamakura, I. Yamaguchi, and T. Tanaka, Nucleotide sequence of the *bsr* gene and *N*-terminal amino acid sequence of blasticidin S resistant *Escherichia coli* TK121, *Agric. Biol. Chem.*, 1991, **55**, 3155.
68. T. Kamakura, K. Yoneyama, and I. Yamaguchi, Expression of the blasticidin S deaminase gene (*bsr*) in tobacco: fungicide tolerance and a new selective marker for transgenic plants, *Mol. Gen. Genet.*, 1990, **223**, 332.
69. W. Stewart, Isolation and proof of structure of wild-fire toxin, *Nature*, 1971, **229**, 174.
70. S. L. Sinden and R.D. Durbin, Glutamine synthetase inhibition: possible mode of action of wildfire toxin from *Pseudomonas tabaci, Nature*, 1968, **219**, 378.
71. T. Murakami, H. Anzai, S. Imai, A. Satoh, K. Nagaoka, and C. J. Thompson, The bialaphos biosynthetic genes of *Streptomyces hygroscopicus:* molecular cloning and characterization of the genc cluster, *Mol. Gen. Genet.*, 1986, **205**, 42.
72. Y. Kumada, H. Anzai, E. Takano, T. Murakami, O. Hara, R. Itoh, S. Imai, A. Satoh and K. Nagaoka, The bialaphos resistance gene (*bsr*) plays a role in both self-defense and bialaphos production in *Streptomyces hygroscopicus, J. Antibiot.*, 1988, **41**, 1838.
73. H. Anzai, K. Yoneyama, and I. Yamaguchi, Transgenic tobacco resistant to bacterial disease by detoxification of a pathogenic toxin, *Mol. Gen. Genet*, 1989, **219**, 492.
74. R. B. Horsch, J. E. Fry, N. L. Hoffmann, D. Eicholtz, S. F. Rogers, and R. T. Fraley, A simple and general method for transferring genes into plants, *Science*, 1985, **227**, 1229.

CHAPTER 3

Strobilurins – From Natural Products to a New Class of Fungicides

H. SAUTER, E. AMMERMANN, and F. ROEHL

3.1 Introduction

In the 1970s, important innovations in fungicide chemistry were developed from synthetic chemistry. New classes, such as the morpholines, triazoles, dicarboximides, and acylalanines, were established. In the 1980s, after the peak of intensive optimization for the known classes and with the threats of the increasing incidence of resistance against some of these fungicides, synthetic chemists looked for new fungicidal classes. The approach taken, already established in pharmaceutical and pesticide chemistry, used natural products as potential leads and generated two new major classes of fungicides, which have already reached development status – the phenylpyrroles, derived from pyrrolnitrin,[1,2] and the strobilurins. The evolution of the latter class is described here.

3.2 Historical Background

In 1969, antifungal activity in extracts from cultures of the basidiomycete *Oudemansiella mucida* (Schrader ex Fries) Hoehnel, which grows on decaying wood, was observed by Musilek *et al.*[3–5]. The use of these extracts was recommended for topical application against dermatomycosis. Extracts were later marketed as 'mucidermin' by Spofa, CSFR. In 1974 and 1979, the putative structure **1** of the antifungal compound from *O. mucida* and its synthesis were published by the Czechoslovakian group of Vondracek *et al.*[6,7]. They named the new antibiotic 'mucidin'.

Since 1977, Anke and Steglich[8–10] have reported their efforts to isolate antifungal compounds from basidiomycetes. In 1978, they proposed structure **1** – identical with the putative structure of 'mucidin' – for the antibiotic constituent isolated from *Strobilurus tenacellus* (Pers. ex Fries) Singer, which they named 'strobilurin'. On the basis of spectroscopic evidence and by the stereospecific synthesis of both isomers, **1** and **2**, its structure was revised by Anke *et al.* in 1984[11] as **2**, with the correct *E,Z,E*-configuration of the natural strobilurins. Different synthetic routes to **1** and **2** were also reported later by Beautement and Clough[12] and by Sutter.[13]

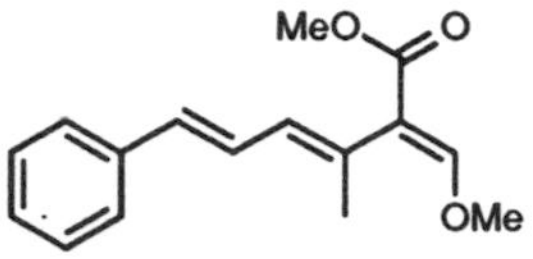

1 Mucidin

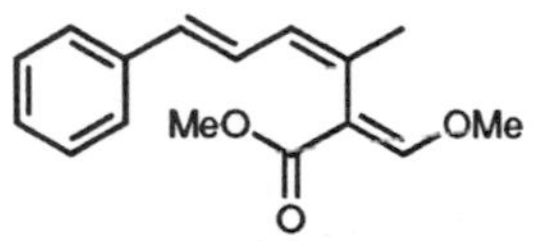

2 Strobilurin A

Figure 3.1 shows a computer model of the calculated minimum energy conformation of strobilurin A, with the typical perpendicular arrangement between the β-methoxy acrylate pharmacophore and the rest of the molecule. As shown here, the molecular shape of strobilurins is reminiscent of a rake.

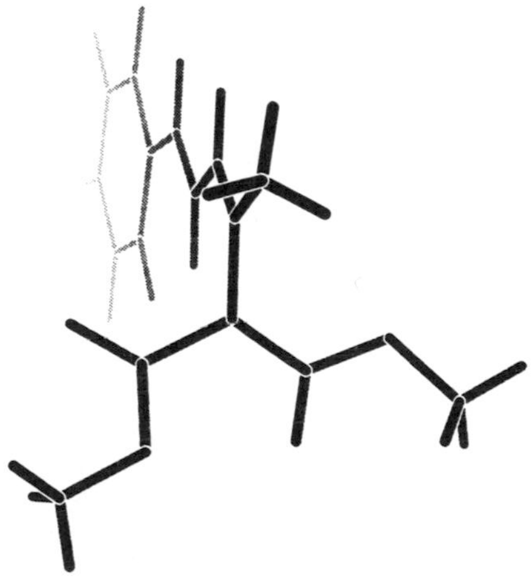

Figure 3.1 *Computer model of the calculated minimum energy conformation of strobilurin A*

In 1979, Anke *et al.*[14] reported another new structure, **3**, formally related to strobilurin A by the addition of methanol, which had been isolated from culture filtrates of *O. mucida* and named 'Oudemansin A'. Subsequently, about 20 species of basidiomycetes belonging to the genera *Crepidotus*, *Cyphellopsis*, *Hydropus*, *Mycena*, *Oudemansiella*, *Strobilurus*, and *Xerula* have been reported to produce related compounds.[15,16] Three more strobilurins, F, G and H, have been isolated from the ascomycete *Bolinea lutea* Sacc.[17,18] A comprehensive review of strobilurins, oudemansins, and myxothiazoles, with their isolation, chemistry, biochemistry, and synthesis, has recently been given by Clough.[19]

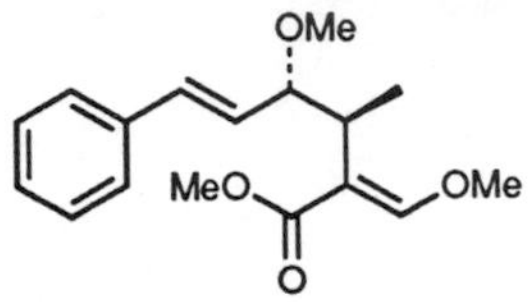

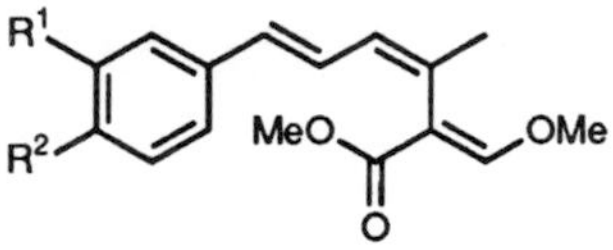

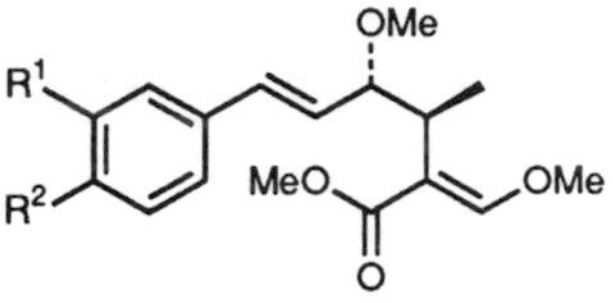

OMe
MeO OMe
O

3 Oudemansin A

Strobilurin	R^1	R^2	Oudemansin
A	H	H	A
B	MeO	Cl	B
C	$Me_2C{=}CHCH_2O$	H	–
X	H	MeO	X
F-1	HO	H	–
F-2	HO	$Me_2C{=}CHCH_2O$	–
H	MeO	H	–

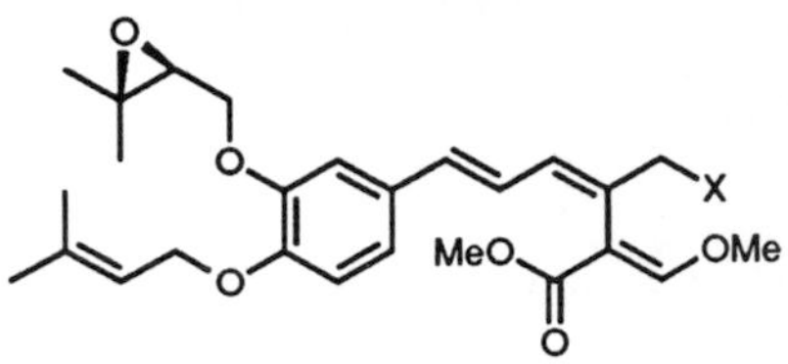

4 X = H Strobilurin D
X = OH Hydroxystrobilurin D

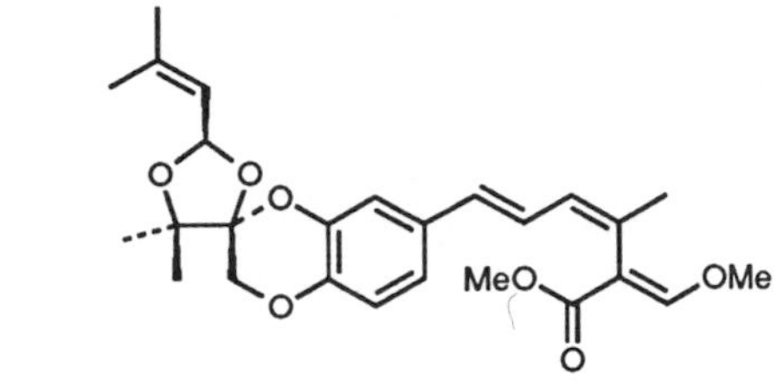

5 Strobilurin E

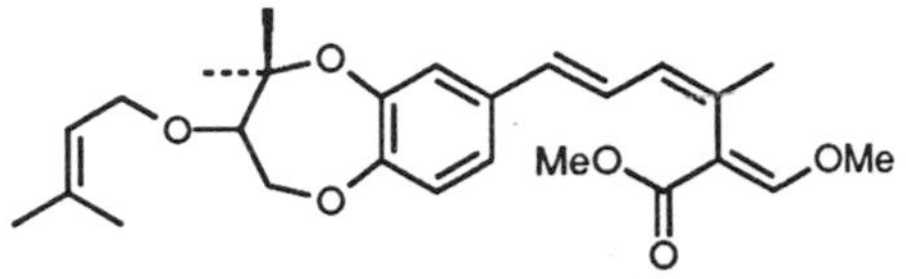

6 Strobilurin G

All these structures have in common the α-substituted β-methoxyacrylic acid-methyl ester moiety. A different type,**7**, of β-methoxyacrylic-derivative was isolated in 1980 from the gliding bacterium *Myxococcus fulvus* (Cohn) Jahn by Gerth *et al.*[20] and Trowitzsch *et al.*[21] and given the name 'myxothiazole'. Its mode of action was identified a year later as an inhibition of mitochondrial respiration.[22,23]

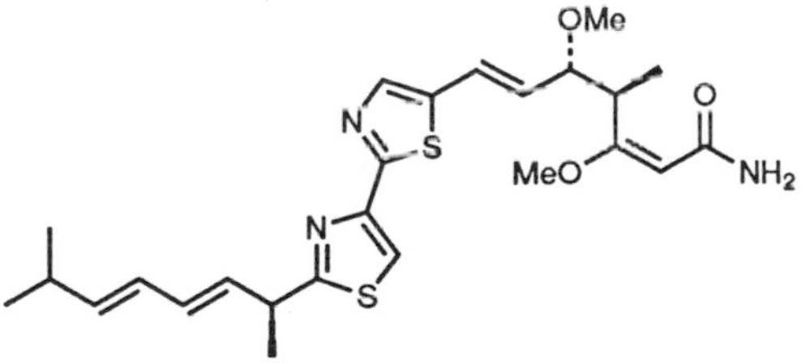

7 Myxothiazol A

A more general relationship between these natural products and their common mode of action was established by the landmark publication of Becker *et al.*[24] in 1981. They reported that the four chemically related natural products oudemansin, strobilurin A, strobilurin B, and myxothiazol A share the same biochemical mode of action: they inhibit the mitochondrial electron transport at the site of complex III (bc_1-complex) of the respiratory chain. The mechanisms of the different steps in mitochondrial respiration and the interaction of these inhibitors have been explored during the past 10 years through their intensive use in university research.[25]

3.3 Proposed Nomenclature

Owing to their common structural feature, this group of compounds was originally named β-methoxyacrylates or MOAs.[26] However, many different types of molecules with the same mode of biochemical action at the bc_1-complex have since been found which are structurally related to strobilurin, but do not possess the β-methoxyacrylate moiety, as is shown below (*cf.*, compounds **31–62**, **66–85**). Therefore, we propose to call them strobilurins, on the basis of the most simple, natural lead molecule on which the structural variations are formed and which thus led to the development of this new class of fungicides.

3.4 From the Natural Product to a New Synthetic Lead

In the early 1980s, the strobilurins began to receive attention from the agrochemical industry. Increasing numbers of reports of resistance against some recently introduced fungicides, such as dicarboximides, triazoles, and acylalanines, generated a great demand for new leads in synthetic fungicide chemistry. Several companies started programmes to test natural products as potential new leads for synthesis. They were selected on the basis of literature studies, company projects to isolate and elucidate the structure of potent candidates from natural products, and from outside sources.

BASF obtained strobilurin A, **2**, from Professor Anke, University of Kaiserslautern, Germany, in July 1983. His work had already documented its remarkable antifungal activity *in vitro*.[9] His group[27,28] had also demonstrated the broad spectrum antifungal activity of hydroxystrobilurin D, **4** (X = OH), in agar plate tests. Glasshouse tests at BASF with strobilurin A showed a better than 90% control of late blight [*Phytophthora infestans* (Montagne) de Bary] on tomatoes and of powdery mildew on wheat (*Erysiphe graminis* D.C. ex Merit) at the relatively high rate of 250 p.p.m. used as a foliar spray. At lower concentrations, or with other pathogens, its activity rapidly fell below the level of interest. Despite this unusual fungicidal profile of simultaneous control of a downy and a powdery mildew, it was felt that the activity did not fulfil the potential expected from the good antifungal *in vitro* activity of strobilurins. Discussions between our colleague Dr Schirmer and Professor Steglich led to the hypothesis that the inherent lability of the triene system might allow rapid chemical, photolytic, or metabolic degradation under glasshouse test conditions, and thus be responsible for the unsatisfactory activity *in vivo*. The question thus arose as to whether there was a simple way to *use* this hypothesis directly, without *proving* it first. The aim was to obtain analogues which combined antifungal activity with higher stability, with the hope that they would retain sufficient activity *in vivo*.

One of the most favoured molecules resulting from these reflections was **8**, the enol ether stilbene,[26,29] which is also referred to as MOA-stilbene in the literature. In this molecule the former (*E,Z,E*)-triene system is stabilized by bridging the *Z*-double bond and the methyl group to form an aromatic ring.

Supporting this concept was that G. Schramm,[30] in Steglich's group, had already synthesized the first simple derivatives of the α-phenyl-β-methoxy-

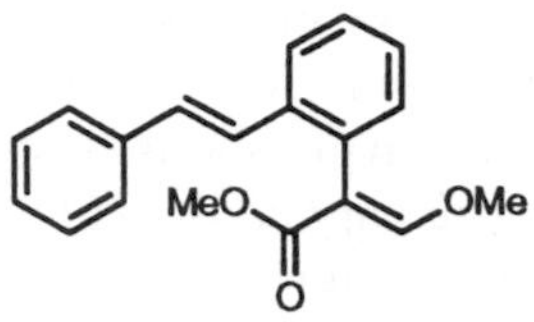

8 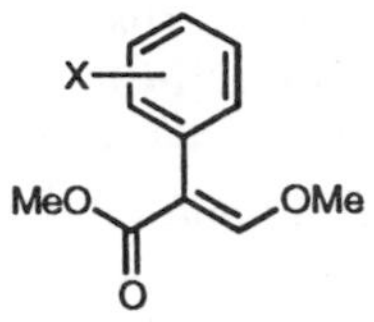 **9**

acrylate type **9**, which also showed some weak fungicidal activity *in vitro* and in glasshouse tests; this activity paralleled respiration inhibition in *Penicillium notatum* Westling.

The preparatory synthetic work for the reinvestigation of the stereochemistry of strobilurin A and for the simple derivatives **9** quickly made **8** available.[31] What we had scarcely dared to hope for came true: **8** exhibited strong and broad fungicidal activity in glasshouse tests, far superior to that of the natural lead molecule. Following a foliar application of 60 p.p.m., **8** showed excellent control (≥95%) of many important plant diseases, such as powdery mildew on wheat (*E. graminis*), net blotch on barley [*Drechslera teres* (Sacc.) Shoemaker], glume blotch on wheat [*Septoria nodorum* (Berkeley) Berkeley], rice blast (*Pyricularia oryzae* Cavara), downy mildew on grape vine [*Plasmopara viticola* (Berkeley & Curtis) Berlese et de Toni], late blight on tomatoes (*P. infestans*), and apple scab [*Venturia inaequalis* (Cooke) Aderhold]. As a seed treatment, 100 p.p.m. of **8** reduced powdery mildew on wheat to 20%. This observation gave the first hint of the systemic potential in this class. Crop tolerance in some sensitive crops was improved in comparison with strobilurin A. **8** did not show any signs of rapid degradation or loss of activity under glasshouse conditions.

In contrast to the positive glasshouse results, small-plot field tests of the stilbene **8** in 1985 were disappointing. They showed that rates of 1 kg (a.i.) ha^{-1} (a.i. = active ingredient) could reduce some plant diseases only in the very early phase of trials against powdery mildew in wheat, late blight in potatoes, and apple scab. Trials in grape vines had to be stopped after two applications due to severe phytotoxicity problems.

As expected generally for stilbene-type compounds,[32,33] photostability tests of **8** explained the loss of activity under field conditions. The half-life of the stilbene as a thin film on the surface of glassplates exposed to artificial irradiation, which simulated the intensity and spectrum of sunlight, was found to be only minutes. A major aim resulting from this was to increase the photostability by new structural modifications.

Independently, parallel work was performed at ICI, as comprehensively described by Beautement *et al.*[26] Their starting point, somewhat earlier in the summer of 1982, was the biological activity of oudemansin and myxothiazol. Both compounds showed good activity against several plant pathogenic fungi in *in vivo* models in the glasshouse when applied as a foliar spray at the low concentration of 33 p.p.m. Strobilurin A was not available and was synthesized at ICI. They also encountered the problem of stereochemistry, which was still unclear at that time (1983). This was solved by synthesizing both the wrong *E,E,E*- and the correct *E,Z,E*-geometrical isomers, **1** and **2**. Only the latter showed inhibition of mitochondrial respiration. In glasshouse tests the *E,Z,E*-isomer showed only poor fungicidal activity when applied as a foliar spray or as a root drench at 25 p.p.m. The reason for this poor fungicidal performance was the very poor photostability of strobilurin A, resulting in a half-life of only seconds[26]; attempts to overcome the problems associated with photoinstability led to the design of amide analogues, of the type **10**, in which the sensitive *Z*-double bond is replaced by an amide moiety, and to the enol ether stilbene **8**, previously mentioned.

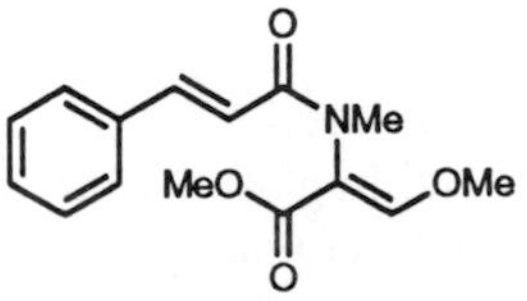

O
NMe
MeO
OMe
O

10

10 and related compounds showed good activity against a range of plant pathogenic fungi in *in vivo* models in glasshouse tests.[34] However, the limited scope for possible structural modifications[26] resulted in a preference for **8** as the basis for further synthesis.[35]

This parallel design of the enol ether stilbene **8** and its function as a central new lead for further structural variations at BASF, in cooperation with Steglich, and at ICI, is certainly one of the most striking examples of how independently generated ideas finally result in similar, if not identical, developments in an interesting and highly competitive field such as the strobilurins.

3.5 Optimization to Development Products

However, the poor level of disease control by the stilbene **8** under field conditions, due to its photoinstability, made further optimization efforts necessary.

At ICI, the expectations of better fungicidal activity and, parallel to this, an increased photostability proved to be the case for several structural modifications of the side chain, as described by Beautement *et al.*,[26] such as the highly active and very photostable diphenyl ether system **11**.[35] This cluster showed very interesting fungicidal activity. Furthermore, it was sufficiently distinct from other interesting side-chain variations that were the subject of competing research at BASF which finally led to partially overlapping patents.[29,36–38] The aim to increase systemicity within this class stimulated further structural modifications, particularly through the introduction of more hydrophilic heterocycles into the side chain.[39] This strategy led to the development product ICI A5504, **12**, which was announced for the first time at the Brighton Conference in November 1992.[40]

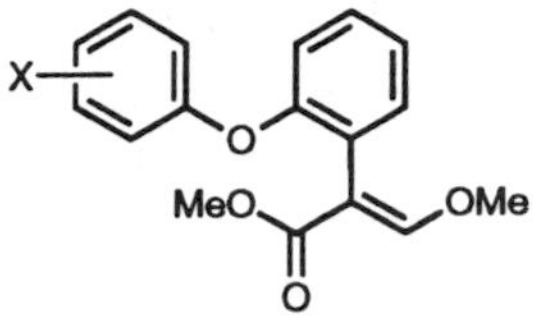

11

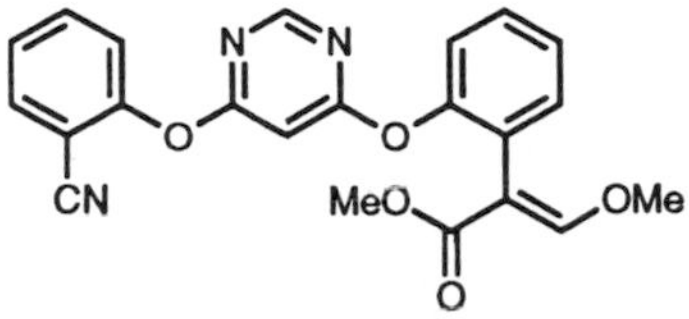

12 ICI A5504

According to the Brighton Proceedings,[40] the water solubility of ICI A5504 is 10 mg l^{-1} at 25°C, its *n*-octanol–water partition coefficient is 440 (log P = 2.64), and its vapour pressure is $\ll 10^{-5}$ Pa at 20°C. The product is photostable: its half-life in a simulated sunlight test is 24 h versus 12 s for strobilurin A.

ICI A5504 is toxicologically a very safe product. Its acute oral toxicity LD_{50} in rats is >5000 mg kg^{-1}, the acute dermal toxicity is LD_{50} >2000 mg kg^{-1} (rat), skin and eye irritation in rabbits is only slight, while no skin sensitization was observed in guinea pigs. The Ames test is negative. ICI A5504 is characterized as a systemic fungicide with excellent protectant, curative, eradicative, and translaminar properties against a broad spectrum of plant pathogenic fungi.

In cereal seed treatment, ICI A5504 has shown good crop safety, with 100 mg (a.i.) kg^{-1} seed, and has given good control of powdery mildew (*E. graminis*), equivalent to commercial standards. As a foliar spray the product controls important cereal diseases, such as *Puccinia recondita* Roberge et Desmazieres and *P. hordei* Otth., *S. nodorum* and *S. tritici* Roberge et Desmazieres, and *D. teres* at a rate of 250 g (a.i.) ha^{-1} to an extent similar to modern triazole standards.

In rice, ICI A5504 controls the two most important diseases, blast (*P. oryzae*) and sheath blight [*Corticium sasakii* = *Thanatephorus cucumeris* (Frank) Donk], simultaneously when applied either as granules directly to paddy water with 1600–1800 g (a.i.) ha^{-1} or as a foliar spray at appropriately lower rates.

In vines, downy mildew (*P. viticola*) on the leaves and grape bunches is controlled excellently by 250 g (a.i.) ha^{-1} in a prophylactic schedule, superior to dithiocarbamate standards. Powdery mildew [*Uncinula necator* (Schwein.) Burrill] on the leaves and grape bunches is controlled at a rate of 125 g (a.i.) ha^{-1}, superior to sulfur, but slightly inferior to triazole standards.

In potatoes, ICI A5504 gives control of late blight (*P. infestans*) at a rate of 200 g (a.i.) ha^{-1} in a prophylactic schedule equivalent to dithiocarbamate standards. In apples, ICI A5504 controls scab (*V. inaequalis*) at a rate of 120 mg (a.i.) l^{-1} and *Alternaria mali* Roberts at a rate of 200 mg (a.i.) l^{-1} equivalent to standard products. Problems of phytotoxicity have been observed on younger vine leaves where transient chlorotic symptoms were caused by ICI A5504. In a limited number of apple varieties, necrosis on young leaves and buds has also been observed.

At BASF, variations of stilbene **8** followed similar patterns to those described for ICI. We considered variations for three parts of the molecule: the pharmacophore, the bridging aromatic ring, and the side-chain, as illustrated in Figure 3.2. Side-chain variations were the starting point. The intermediate

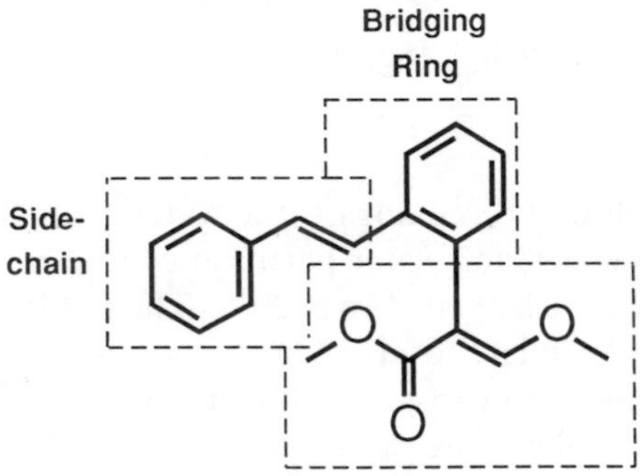

Bridging Ring

Side-chain

Pharmacophore

Figure 3.2 *Regions for structural variation*

chain between the two aromatic ring systems was hydrogenated to give the dihydrostilbene and heteroatoms were introduced into the intermediate C_2-unit. The synthetically easily accessible benzyloxy derivatives **13**[36] and phenoxymethyl derivatives **14**[37] were thoroughly investigated and yielded very promising fungicidal results.

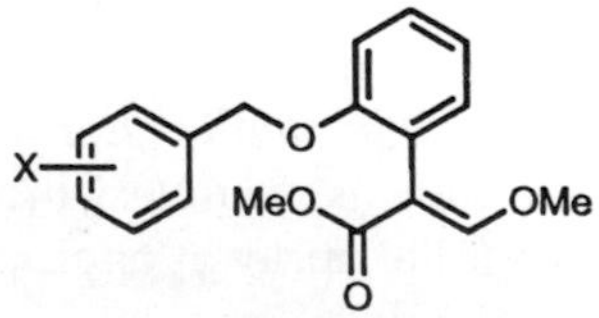

X MeO OMe O **13**

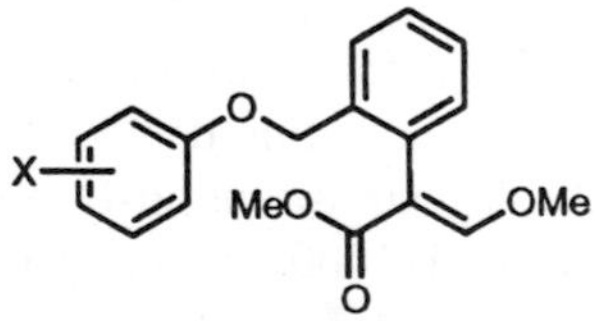

X O MeO OMe O **14**

Publication of the ICI European patent application 178 826[35] at the end of April 1986 changed the optimization situation at BASF dramatically and blocked many routes that had already been started. It stimulated intensive searches to exploit different structural types. Great emphasis was placed on possible modifications of the β-methoxyacrylate pharmacophore, which had been considered as essential up to that time. An early success was obtained with compounds of the oximether type, *e.g.*, **15** and **16**.[41]

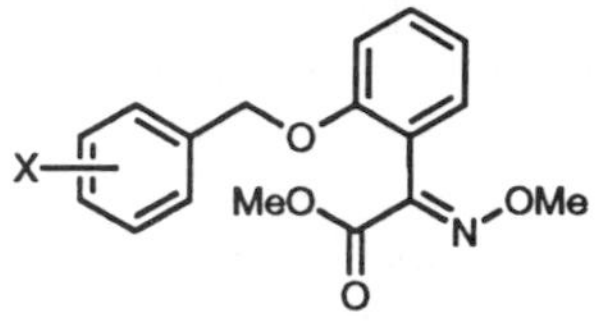

15

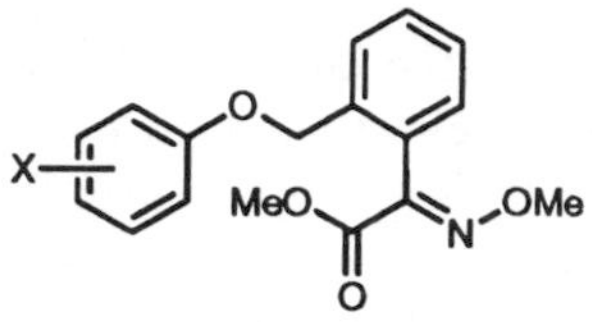

16

Again, independent work at ICI resulted in the same type of compounds.[42] However, in this case, BASF gained a priority advantage of only 2 days and therefore, good opportunities for further structure optimization in this particular field.

In this situation, important hints came from an *in vitro* test at the mitochondrial target level. In this test, inhibition of yeast mitochondrial respiration was measured in parallel with the usual glasshouse screening. In contrast to the glasshouse experiments, in which the results were, at times, ambiguous, the mitochondrial test offered a very clear and early idea of the trends. First, compounds containing the oximether pharmacophore possess the same high target activity as those with the enol–ether pharmacophore. Second, derivatives with the phenoxymethyl side-chain, **16**, are, by a factor of approximately 10, more active than the corresponding compounds with a benzyloxy side-chain, **15**. These findings stimulated not only a concentration of synthetic effort on cluster **16** by varying the phenoxy ring substitution, but also encouraged us to investigate preferentially a larger number of compounds of this type in a field screening programme, where, in several cases, their exceptionally good performance could be confirmed.

Field selection combined with concern of the possible costs of production finally led to the selection of BAS 490 F, **17**, as the most promising development product; this was presented at the Brighton Conference 1992.[43] As described in the Brighton Proceedings,[43] the water solubility of **17** is 2 mg l^{-1} at 20°C, log P of the n-octanol–water partition coefficient is 3.4 at 25°C, and its vapour pressure at 20°C is 2.3×10^{-6} Pa.

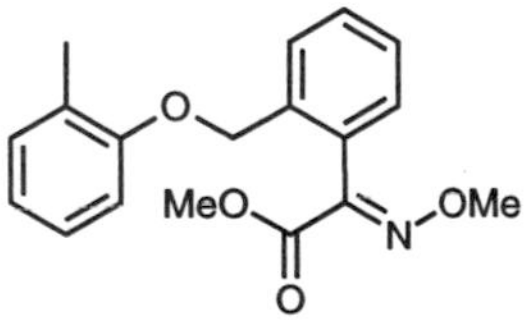

17 BAS 490 F

BAS 490 F is toxicologically a very safe product. Its acute oral toxicity LD_{50} in rats is >5000 mg kg^{-1}, the acute dermal toxicity LD_{50} is >2000 mg kg^{-1} (rat), and skin and eye irritation in rabbits is not observed. The Ames test is negative and to date no teratogenicity has been observed.

BAS 490 F can be characterized as a fungicide with excellent protectant properties against a broad spectrum of plant pathogenic fungi. Against some of them, particularly against powdery mildew and apple scab, it has a strong eradicative potential. BAS 490 F shows translaminar movement and quasi-systemic properties *via* diffusion in the gas phase.

In cereals, foliar sprays of BAS 490 F give excellent control of some important diseases, such as powdery mildew (*E. graminis*), and good control of net blotch (*D. teres*) at rates of 125 g (a.i.) ha^{-1}, equivalent to triazole and superior to morpholine standards. *S. nodorum* and *S. tritici*, scald [*Rhynchosporium secalis* (Oudemans) Davis], and rust diseases, such as *P. recondita* and *P. striiformis* Westend., are well controlled in a prophylactic spray schedule, equivalent to triazole standards.

In rice, the two important diseases, blast (*P. oryzae*) and sheath blight (*C. sasakii*), are well controlled by foliar treatments with rates of 200 g (a.i.) ha^{-1} of BAS 490 F.

In vines, 100 g (a.i.) ha^{-1} give good control of powdery mildew (*U. necator*) on the leaves and grapes in prophylactic as well as in eradicative situations, equivalent to standard triazoles. Control of downy mildew (*P. viticola*), comparable to dithiocarbamate standards, is achieved with 375 g (a.i.) ha^{-1} of BAS 490 F.

In apples, excellent control of scab (*V. inaequalis*) on the leaves and fruits and of powdery mildew [*Podosphaera leucotricha* (Ellis et Ev.) Salmon] is achieved by rates of 100 g (a.i.) ha^{-1}, superior to triazole standards.

In potatoes, early and late blight [*Alternaria solani* (Ellis et Martin) Sorauer and *P. infestans*, respectively] are significantly reduced by 400 g (a.i.) ha^{-1}, equivalent to dithiocarbamate standards.

In sugar beet, powdery mildew (*Erysiphe betae* D.C.) is excellently controlled by 200 g (a.i.) ha^{-1}, equivalent to triazole standards. Leaf spot (*Cercospora beticola* Saccardo) is significantly reduced by 300 g (a.i.) ha^{-1}, comparable with the control given by fentin acetate.

In cucurbits, powdery mildew [*Sphaerotheca fuliginea* (Schlecht.) Pollacci] is well controlled by 100–150 g (a.i.) ha^{-1}, equivalent to triazole standards.

Slight phytotoxicity was sometimes observed, but only on the young expanding leaves of grape vines at rates $>$200 g (a.i.) ha^{-1}. This was dependent upon the variety and on the prevailing climatic conditions.

The Shionogi Company, as the first of the other companies not originally involved in this area of chemistry, published their strobilurin development product in August 1993. On the occasion of the 6th International Congress of Plant Pathology in Montreal they presented SSF-126, **18**.[44,45] According to the Congress poster,[46] the water solubility of SSF-126 is 128 mg l^{-1} at 20°C, and log *P* of the *n*-octanol–water partition coefficient is 2.3 at 20°C.

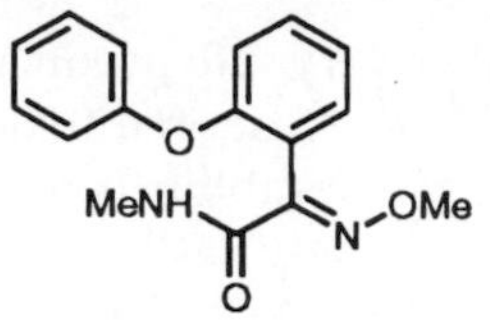

MeNH OMe N O

18 SSF-126

Toxicological studies of SSF-126 were described as being under examination. The Ames test and the micronucleus test are negative. Information about the fish toxicity are given as *Oryzias latipes* LD_{50} (48 h) 20.7 p.p.m., *Cyprinus carpio* LD_{50} (48 h) 16.8 p.p.m., and *Daphnia pulex* EC_{50} (3 h) >100 p.p.m.

SSF-126 is characterized as a systemic fungicide with excellent activity as a foliar spray against rice blast, powdery mildew of vegetables and cereals, scab and rust of apple and pear, and grey mould and Sclerotinia rot of vegetables in glasshouse trials. In field tests, excellent simultaneous control of rice blast and rice sheath blight following drip application is reported.

As discussed, the development products from the class of the strobilurins possess an unusually broad fungicidal profile which has not been observed for fungicides before. From this class, products are being developed which control diseases caused by fungi belonging to different classes. Although the simultaneous control of diseases, such as powdery mildews (ascomycetes) and rusts (basidiomycetes), is already known for triazole fungicides, the strobilurins for the first time offer tools by which the control of powdery and downy mildews (ascomycetes and phycomycetes) can be achieved with one and the same fungicide. This will lead to the development of one product for several diseases caused by fungi of different classes in one crop. A good example of this is the simultaneous control of the rice diseases blast (*P. oryzae*, ascomycete) and sheath blight (*C. sasakii*, basidiomycete).

The mode of action of strobilurins probably offers a good explanation for their broad-spectrum fungicidal activity. However, the three development products described above differ in their systemic properties and in their fungicidal profile. Selectivity and, thus, the individual fungicidal activities of each product are determined by their biokinetic properties.

3.6 Synthesis of the Oxime ethers

The vicinal positions of the pharmacophore and side-chain at the bridging aromatic ring in strobilurins are fascinating features of synthesis, because the selection of suitable *ortho*-substituted starting materials may be critical with regard to costs in industrial production. For the synthesis of the oxime ethers several approaches have been realized, three of which are reported here.

One possibility is to start with *ortho*-substituted bromobenzenes **19** and to introduce the side-chain R first to give **20**. Then the pharmacophore is constructed. Metallation of **20** followed by reaction with a suitable oxalic ester derivative (e.g. Y = 1-imidazolyl) yields the ketoesters **21**, which are then converted into the final oxime ethers **22** by treatment with *O*-methylhydroxylamine, as exemplified in the patent.[41] Depending on R, yields in the step **20** → **21** in this sequence are sometimes only moderate.

In the final step, generally mixtures containing the *E*/*Z*-isomers of the oxime ethers are obtained; the ratio depends on the reaction conditions. Interestingly, only the thermodynamically more stable *E*-isomer has biological activity

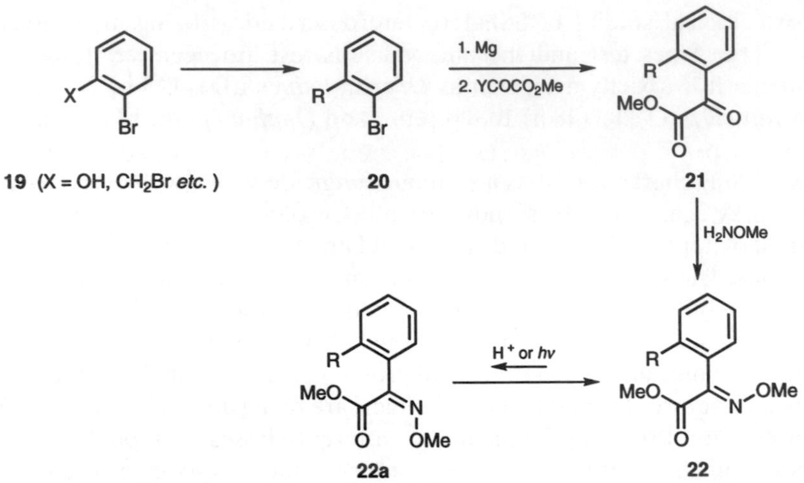

19 (X = OH, CH_2Br *etc.*) 20 21 22a 22

1. Mg
2. $YCOCO_2Me$
H_2NOMe
H^+ or $h\nu$

at the mitochondrial target. In order to direct the equilibrium in its favour, the presence of excess acid is useful. The *E*/*Z* equilibrium ratio can be achieved starting with **22** or **22a**. With R = CH_3 and for BAS 490 F it is approximately 95:5 in methylene chloride/HCl. Similar ratios are obtained after photoequilibration in ether with Pyrex-filtered UV-light. This photoequilibration explains the fact that, despite its lack of target activity, the *Z*-isomer of BAS 490 F shows similar fungicidal performance to BAS 490 F itself under field conditions.

An alternative route for the subsequent synthesis of the side-chain and pharmacophore leads to phenoxymethyl derivatives **16** with excellent overall yields.[47] Here, the starting material is phthalide, **23**, which is a cheap product available by hydrogenation of phthalic anhydride. The side-chain is introduced by opening the γ-lactone ring with phenolates. The resulting carboxylic acids,

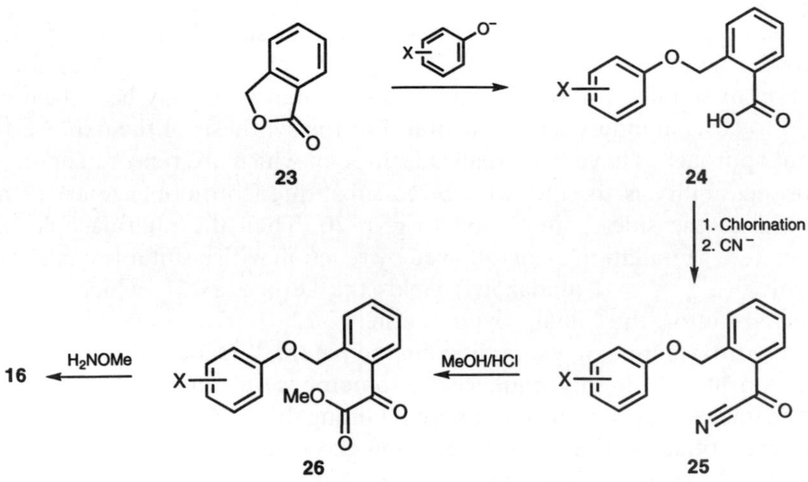

23 24 25 26 16

1. Chlorination
2. CN^-
MeOH/HCl
H_2NOMe

24, are first converted into the corresponding acid chlorides and then into the α-ketonitriles, **25**. Subsequent Pinner reactions yield the ketoesters, **26**, which, again, can easily be converted into the oxime ethers, **16**.

Alternatively, starting with *o*-bromotoluene, **27**, the oximether pharmacophore can be introduced first, using synthetic steps as described above, leading to **28**.[41] Subsequent bromination with *N*-bromosuccinimide (NBS) gives **29** in good yield as a versatile synthon, in which different side-chains can be introduced by reaction with nucleophiles, e.g., with phenolates to yield compounds to type **16**.

Me Br → → Me MeO N OMe O —NBS→ BrCH$_2$ MeO N OMe O → **16**

27 **28** **29**

3.7 Structures and Activities Around BAS 490 F

Further variations of the lead compound at BASF resulted in many different structures which have also shown a more or less strobilurin-type activity. This is best demonstrated when BAS 490 F is used as a central focus, because it is the reference structure around which we have the most systematic collection of data available.

A preparation of submitochondrial particles prepared from yeast cells was used as a model system for testing target activity. The activity of compounds was measured by monitoring the succinate-dependent cytochrome C reduction photometrically at 546 nm using an automatic enzyme analyser. Each compound was measured at six concentrations. The concentration at which 50% inhibition occurred – the I_{50} value – was then calculated by fitting a sigmoidal curve to the inhibition data using a computer program. The average I_{50} value for the enol ether stilbene **8** from the latest 50 tests is 2.2×10^{-8} mol l^{-1}, with a standard deviation of $\pm 1.4 \times 10^{-8}$ mol l^{-1}. As a consequence of variation in the activity of different enzyme preparations, larger deviations have been obtained in individual tests; these ranged from a minimum I_{50} of 0.61×10^{-8} mol l^{-1} to a maximum value of 5.8×10^{-8} mol l^{-1}. To allow for this, the enol ether stilbene **8** is always included as standard in each test and the relative activity, expressed as the ratio F, given by the I_{50} of the test compound divided by the I_{50} of the standard enolether stilbene, is calculated, equation (3.1). Consequently, by definition F is equal to 1 for the standard and activity increases with decreasing F and vice versa. So, a smaller F means a greater activity.

$$F = \frac{I_{50}\text{ (test compound)}}{I_{50}\text{ (enol ether stilbene } \mathbf{8})} \tag{3.1}$$

During the investigation of structural variation, some efforts were applied to variations of the bridging ring[48,49] (*cf.* Figure 3.2). The results in terms of activity did not justify the considerable expenditure necessary for their synthesis; thus, we concentrated on variations of the pharmacophore and side-chain.

3.7.1 Variations of the Pharmacophore

In the following structural formulae, the bridging aromatic ring and the side-chain of BAS 490 F are kept constant, as in **30**. Its relative activity, expressed as the ratio *F* defined above, is 0.67, which means that it is slightly more active than the standard enol ether stilbene **8**.

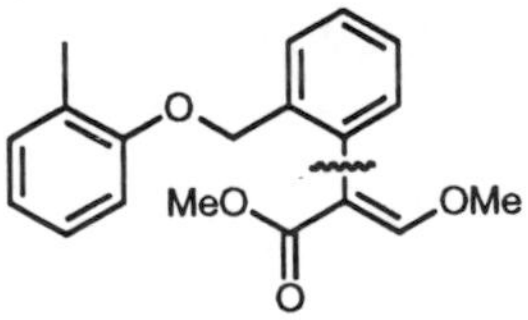

MeO OMe O — *F* = 0.67 **30** 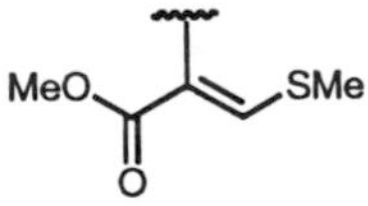
MeO SMe O — 1.1 **31** 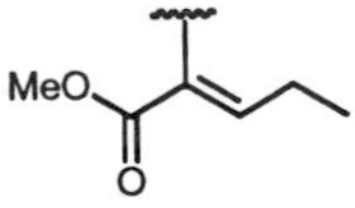
MeO O — 9.0 **32**
MeO NHMe O — 15 **33** 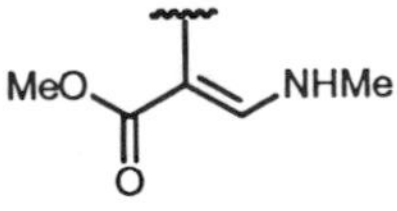

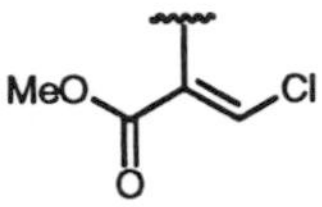

MeO Cl O — *F* = 0.58 **34**
MeO OH O — >100 **35** 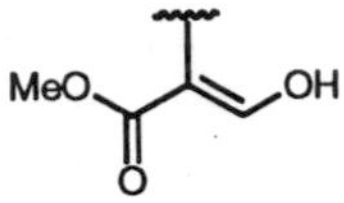
MeO CN O — 180 **36** 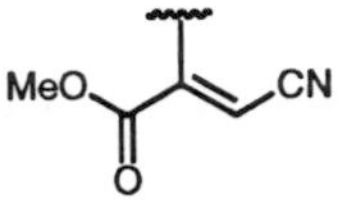

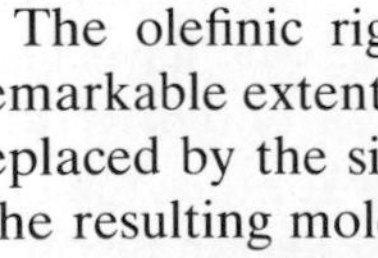

MeO N OMe O — *F* = 0.6 **17**
MeO N NHMe O — 11 **37** 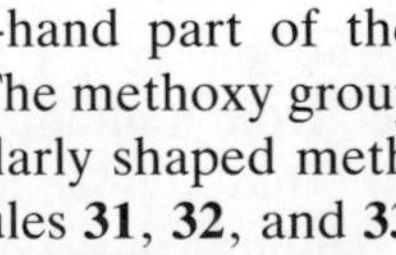
MeO N NH_2 O — 95 **38** 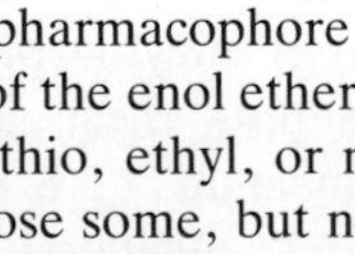
MeO N NMe_2 O — 910 **39** 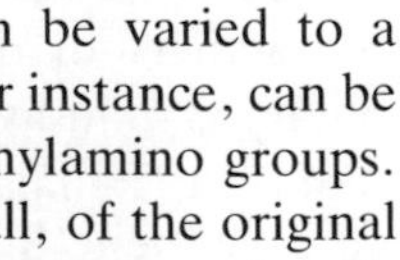

The olefinic right-hand part of the pharmacophore can be varied to a remarkable extent. The methoxy group of the enol ether, for instance, can be replaced by the similarly shaped methylthio, ethyl, or methylamino groups. The resulting molecules **31**, **32**, and **33** lose some, but not all, of the original activity of **30**. The result is better if methoxy is replaced by chlorine to give **34**, whereas substitution with more polar and hydrophilic groups, such as hydroxy or cyano as in **35** and **36**, respectively, leads to more drastic losses of activity.

Replacement of the olefinic carbon–carbon double bond by a carbon–nitrogen double bond yields the oximether-type strobilurine **17** (BAS 490 F). As a consequence of that change one might intuitively expect a considerable increase in polarity and hydrophilicity. This is not the case; the logarithm of the *n*-octanol–water partition coefficient, log *P*, decreases by only 0.2 units. Nor, interestingly, is there any loss of activity of the oximether-type **17** versus the enol ether type **30**. This has been confirmed in general by hundreds of analogues with other side-chains.

Exchange of the methoxy group in **17** for the more polar methylamino group to give **37** parallels the analogous step from **30** to **33**. In both cases, activity is diminished considerably by factors greater than 10. When polarity increases further, as in **38**, the decrease in activity is also enhanced. Branching at the γ-position of the right-hand side with a methyl group, as in **39**, leads to an almost complete loss of activity.

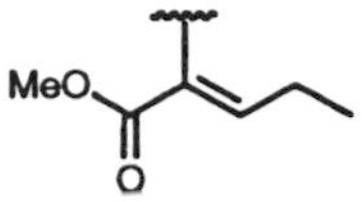

MeO O MeO O MeO O MeO O

F = 9.0 **32** | 1.2 **40** | 3.2 **41** | 530 **42**

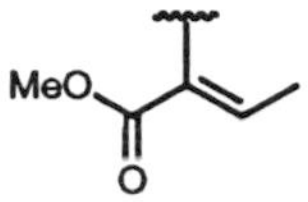

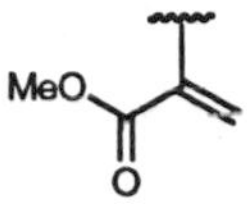

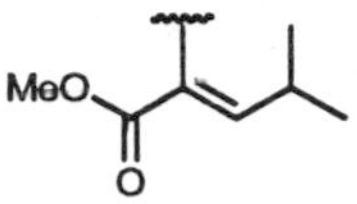

MeO O Cl | MeO O Br | MeO O O

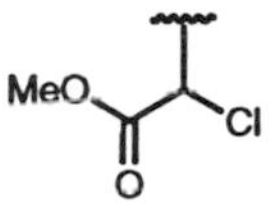

F = 48 **43** | 120 **44** | 540 **45**

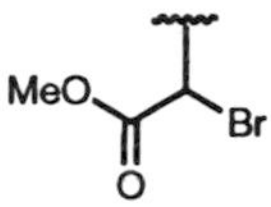

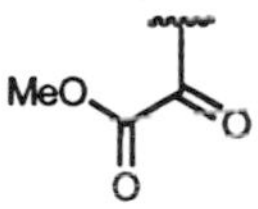

Interestingly, the activity variation in the series of hydrogen- or alkyl-substituted acrylic esters[50–52] depends on the chain length. Removing one methyl group at the right-hand side by going from **32** to **40** leads to a remarkable increase in activity. The optimum chain length thus seems to be that of **40** and not that of **32**. Further shortening, as in **41**, again gives decreased target binding. With compound **42**, branching in the γ-position, comparable to the dimethylhydrazone **39**, took place. The resulting disappearance of activity was also comparable.

More strongly modified derivatives with tetrahedral geometry at the central carbon atom, as in **43** and **44**, retain some weaker activity. Comparison of these two compounds with **41** shows a strong preference for sp^2-hybridizations at the central carbon atom. In spite of this, the activity of the very polar ketoester **45** is of no interest.

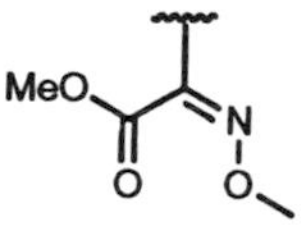

MeO O N O | MeO O

F >100 **46** | >1000 **47**

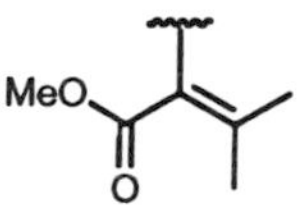

In the presence of a *Z*-substituent at the double bond, as in **46** (which is the *Z*-isomer of BAS 490 F) or in **47**, good target binding is obviously completely prevented. The very small residual activity of **46** is probably the result of a partial isomerization during sample preparation and measurement to give the highly active BAS 490 F.

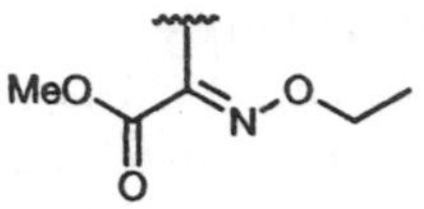

F = 35 **48**; 300 **49**; 280 **50**

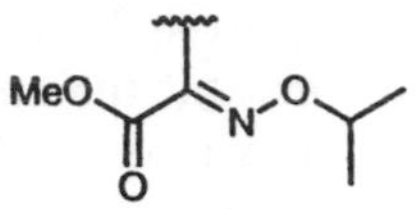

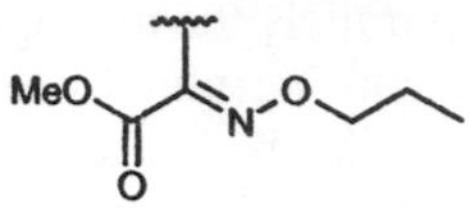

Extension of the right-hand side above the dimensions given by **17** very rapidly leads to decreasing activities, as illustrated with **48–50**. In the case of the left-hand side, lengthening and/or branching also results in a loss of activity, as can be seen with compounds **51–54**. This loss, however, is not as severe as in the cases discussed above for the right-hand side.

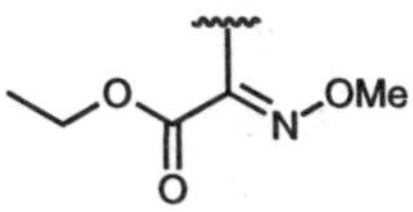

F = 2.2 **51**; 34 **52**; 28 **53**; 86 **54**

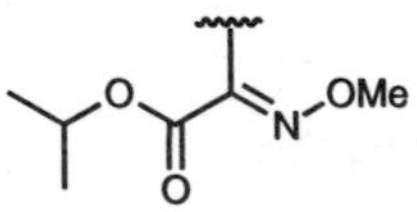

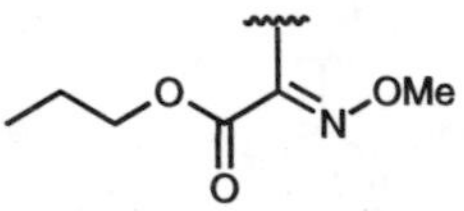

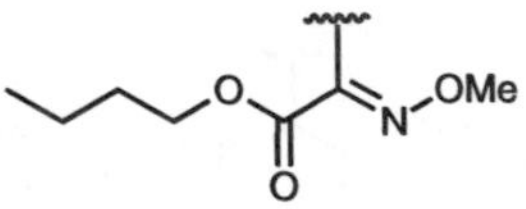

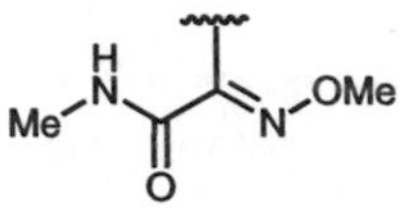

F = 10 **55**; 52 **56**; 230 **57**; >1000 **58**

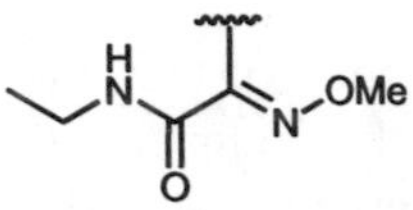

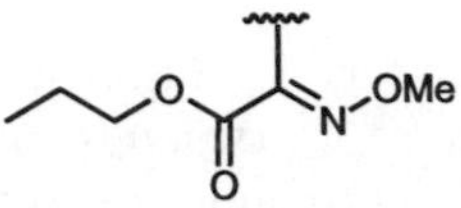

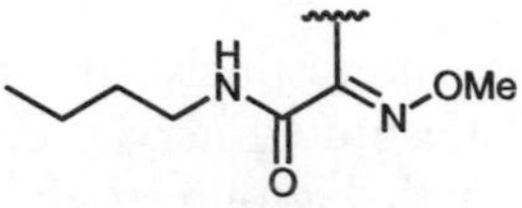

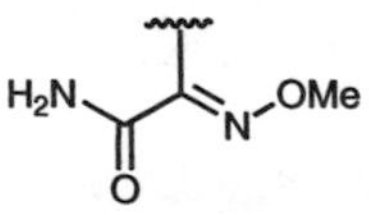

F = 63 **59**; 370 **60**; 33 **61**; 300 **62**

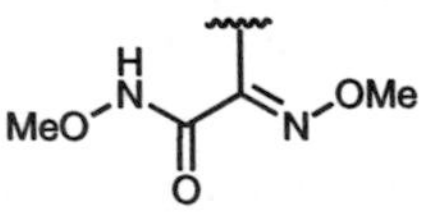

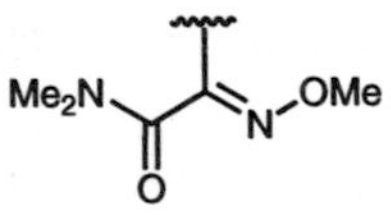

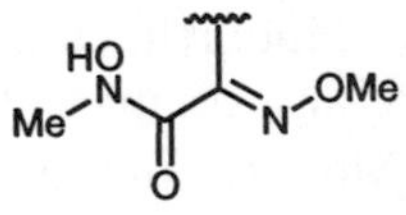

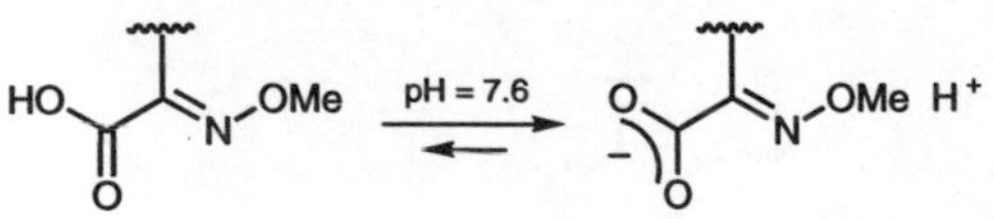

pH = 7.6

F >1000 **63**

Introduction of a methylamino group gives the *N*-methylamide **55**. This type of pharmacophore has been independently synthesized by Shionogi and BASF.[45,53,54] Due to its higher hydrophilicity, **55** loses much of its target activity compared with **17**. Higher lipophilicity through the introduction of longer alkyl groups at the nitrogen atom cannot compensate for this; in contrast, what occurred with the esters also happened here – enlarging the alkyl group always resulted in less active compounds (*cf.* **56–58**). On the other hand, further increasing hydrophilicity, as in **59**, decreases activity further. This effect can also be seen by comparing **56** with **60** and **61** with **62**. The two pairs have a similar shape, but hydrophilicity is considerably higher in **60** and **62** (due to the replacement of one carbon atom by oxygen), and thus activity is lower.

The carboxylic acid **63** is completely inactive. This is best explained by the extremely high hydrophilicity of the corresponding anion, which is by far the predominant species under test conditions at pH 7.6. Most remarkably, and in contrast to the oximether amide **55**, the corresponding crotonic amide **64** and the β-methoxyacrylic amide **65** are completely inactive.

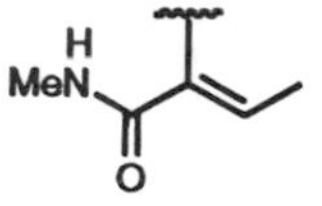

F >1000 **64** >1000 **65**

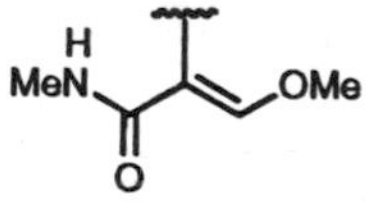

Ketones[55] represent another group of highly active derivatives. Ethyl ketones such as **67**, bind particularly well to the target, this being affected only slightly by shortening to give the methyl ketone **66** or by branching in the α-position, as in **68**. The activity lost on lengthening the chain to give **69** is also similar to that in the ester or monoalkylamide series.

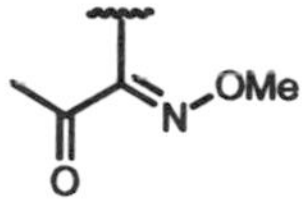

F = 1.3 **66** 0.7 **67** 1.7 **68** 15 **69**

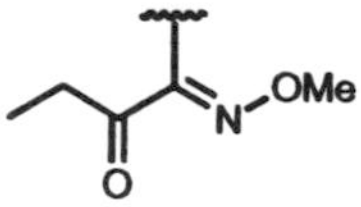

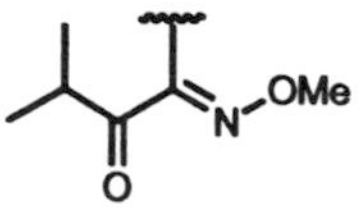

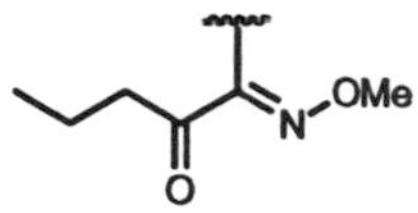

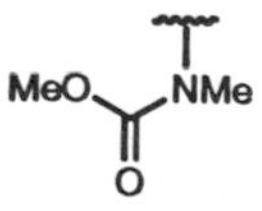

F = 17 **70** 31 **71** 56 **72** 30 **73** 1.4 **74**

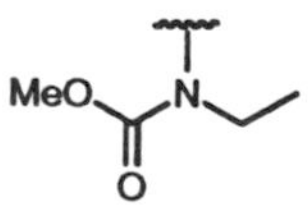

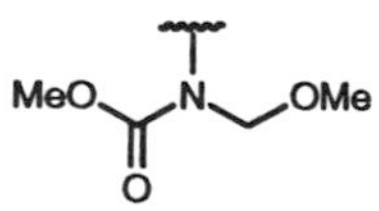

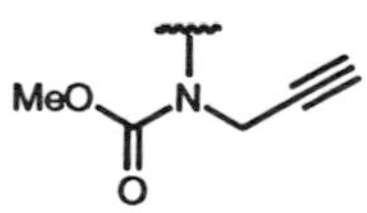

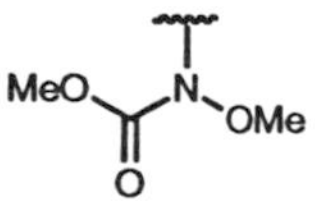

Even changing the centre of the pharmacophore, as in the carbamates **70–73**, is possible. Some of these were first patented by Ishihara,[56] but they were also made independently by BASF[57] – yet another interesting example of coincidence of ideas in the strobilurin field. For the *N*-methoxycarbamates **74**, surprisingly strong activity is observed. In this case patent applications have been filed first by BASF[57] and also independently by Nihon Nohyaky.[68] Considering the pronounced structural change which occurs with the introduction of the central nitrogen atom, this type of strobilurin retains a remarkable degree of activity.

With so many examples of the large extent to which the strobilurin pharmacophore may be varied, without basically losing its target activity, the question arises: is there any structural feature which needs to be kept constant? Only the carbonyl group is retained in all the active derivatives discussed above. One of the most essential properties of this group is the ability of its oxygen atom to function as a hydrogen bond acceptor. In considering structure–activity relationships, we came to the conclusion that this hydrogen bond – as symbolized with the ester **17** (Figure 3.3) – plays the most important role in pharmacophore binding.

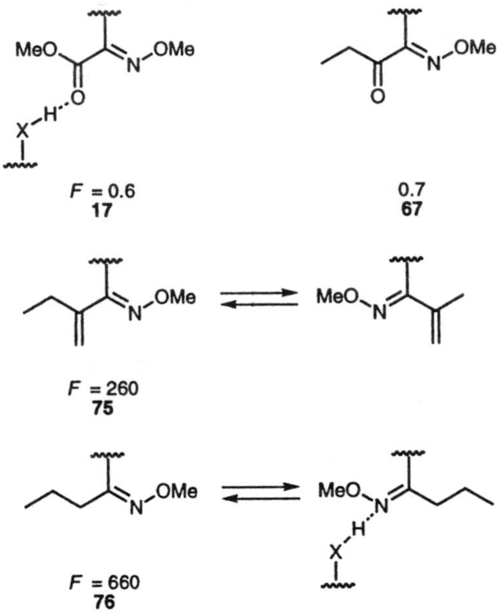

Figure 3.3 *Hydrogen bond interaction*

The same kind of interaction can also occur in the ketone **67**, but not in the derivatives **75** and **76** where the carbonyl group is absent. Consequently, their binding ability is almost completely lost. It cannot be excluded that their very weak residual activity may be due to a different and less effective docking mode, in which one of the heteroatoms of the oximether moiety takes over the role as a hydrogen acceptor, as illustrated for **76** (Figure 3.3).

Sulfur derivatives are also interesting. In contrast to the *S*-ester **77**, which contains a carbonyl group, the isomeric *O*-ester **78** shows very weak binding. This corresponds to the extremely poor hydrogen bonding potency of thiocarbonyl groups.

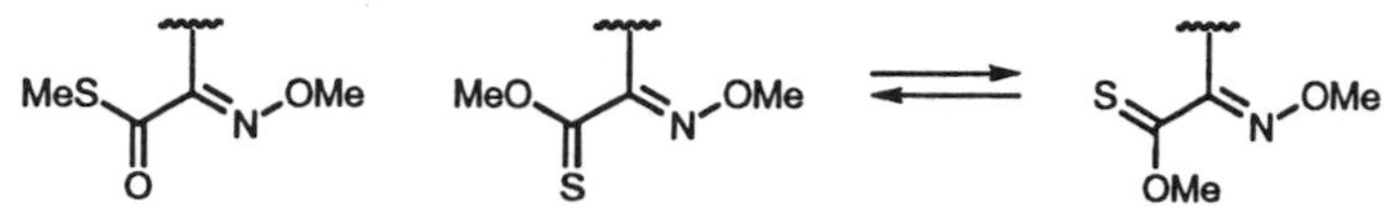

F = 0.34 **77**   190 **78**

The structures **79–85** have one feature in common: they all contain groups which are, to a certain extent, also able to form hydrogen bonds as acceptors in place of the carbonyl group. They are thus expected to have at least some activity. The best of them, **83** with its small hydroxy group, still retains medium activity. In the remaining compounds, however, it appears that docking is more or less sterically hindered by substituents such as methyl, amino, or hydroxy groups.

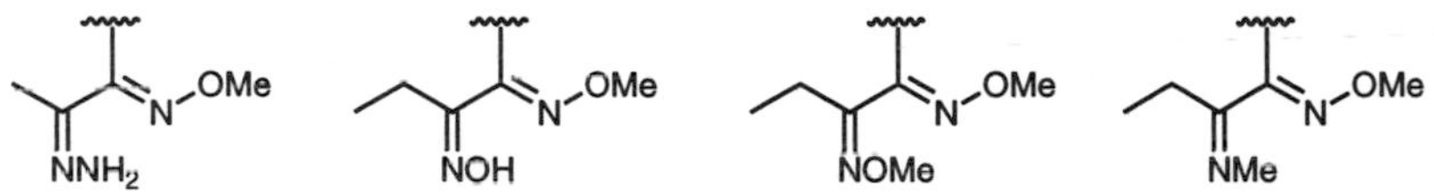

F = 100 **79**   82 **80**   118 **81**   13 **82**

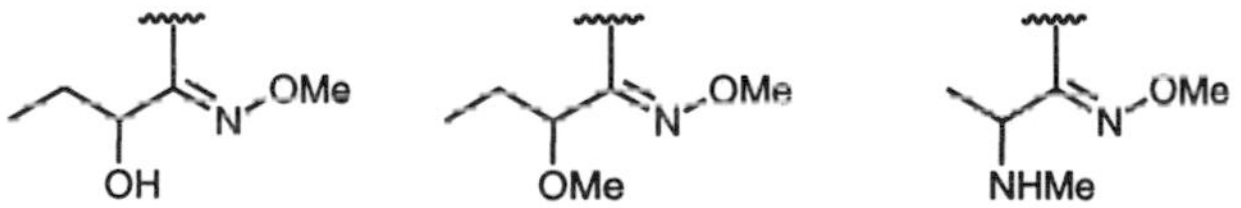

F = 7.2 **83**   213 **84**   72 **85**

The case of the completely inactive nitrile **86** is very interesting. On the one hand its sp-hybridized nitrogen atom is located in a position quite similar – relative to the rest of the molecule – to that of the carbonyl oxygen in the conformation in Figure 3.4. On the other hand, the nitrogen of this molecule is obviously not able to meet the correct geometrical arrangement for binding with the target hydrogen donor when the rest of the molecule is superimposed

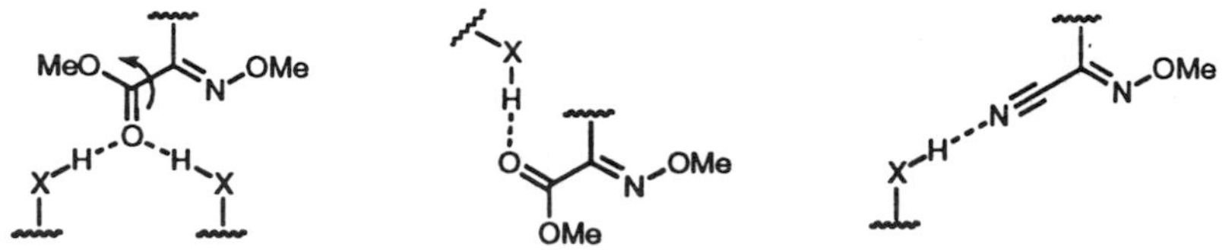

17 (*s-cis*)   **17** (*s-trans*)   *F* >1000 **86**

Figure 3.4 *Effect of conformation on binding*

to **17** (*s-cis*). In other words, the target hydrogen donor is not located in a position where the nitrile group of **86** can interact with it. However, it must be located where it can interact with the carbonyl group of **17**, which is, of course, able to meet it in quite different orientations; for instance, when the hydrogen appoaches from the other side of the oxygen or from completely different directions within the target pocket, as illustrated with the *s-trans* conformation shown in Figure 3.4.

The following points summarize the conclusions which can be drawn from the structure–activity relationships of the pharmacophore variations obtained mainly with oximether and crotonic ester derivatives (Figure 3.5):

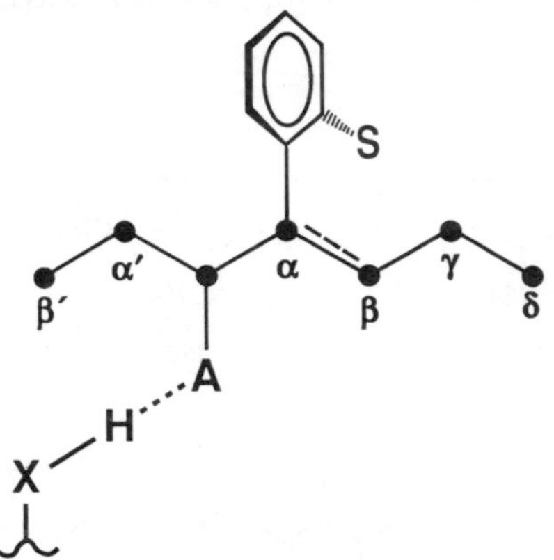

Figure 3.5 *Model of target binding for strobilurins*

- The strongest binding contribution comes from a hydrogen bond between the target X–H donor and A, which is preferentially a carbonyl oxygen in the pharmacophore.
- There are several conformations of the pharmacophore in which hydrogen bond interactions with the target can occur. The actual one remains unclear. The direction of hydrogen bonding relative to the pharmacophore is also unclear.
- It is clear that no heteroatoms, apart from A, are essential for maximum activity, as is shown by the ketones and crotonic esters. Thus, there is also no need for a second hydrogen bond interaction, as has been proposed.[58,59]
- The atom in the α-position is preferentially an sp^2-hybridized carbon with trigonal geometry. It can be replaced by an sp^2-like nitrogen, but then activity seems to be reduced.
- Excessive hydration of the pharmacophore tends to diminish target binding, as is shown by the decreasing activities with increasing pharmacophore hydrophilicity.
- The pocket receiving the pharmacophore appears to be quite narrow. This

follows from the observed loss of activity in the presence of *Z*-substituents in the β-position, branching in the γ-position, or elongation beyond δ or β' in the derivatives investigated to date.

Some related observations and general conclusions have also been made for the enol ether derivatives.[59]

3.7.2 Side-Chain Variations

In addition to the binding of the pharmacophore, target–side-chain interactions play a major role in activity. The presence of the side-chain in the *ortho*-position of the bridging ring, as in **87**, clearly gives the highest activity. Connection to the *meta*-position, as in **88**, is less favourable, and linking to the *para*-position (**89**) is completely unfavourable. The side-chain should also not be too small, as demonstrated by the lack of satisfactory activity with **90**.

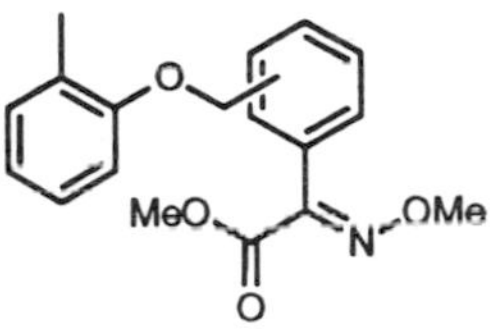

MeO OMe N O

87 *ortho*: *F* = 0.67
88 *meta*: *F* = 5.4
89 *para*: *F* = 270

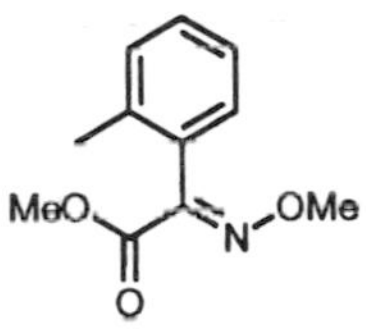

MeO OMe N O

90 *F* = 210

Further side-chain variations are shown in the following structures, in which the pharmacophore and the bridging aromatic ring of BAS 490 F are kept constant. For instance, the chain length between the two aromatic rings can be varied to a large extent, as demonstrated with **91** and **92**. Structures with a four-membered intermediate chain, such as **93–95** show high activity. However, too

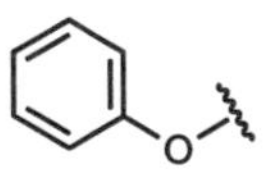

F = 2.6
91

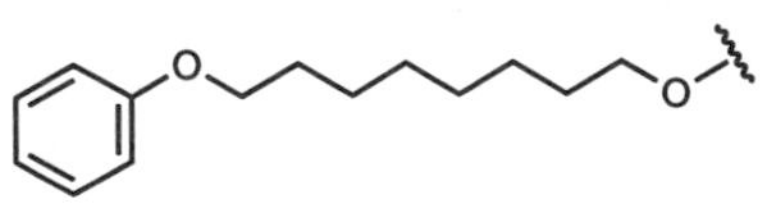

1.3
92

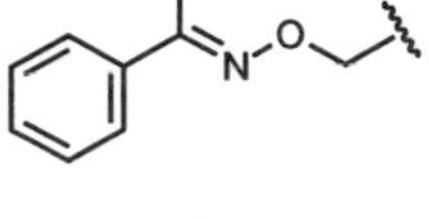

F = 0.83
93

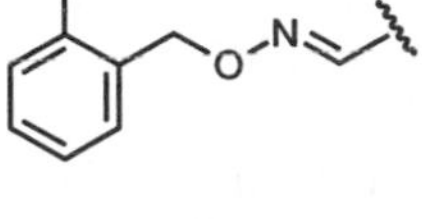

2.3
94

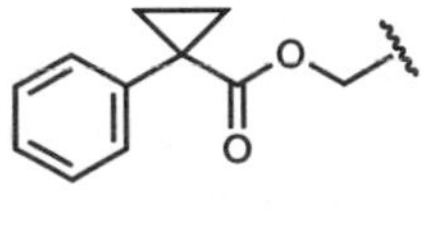

0.33
95

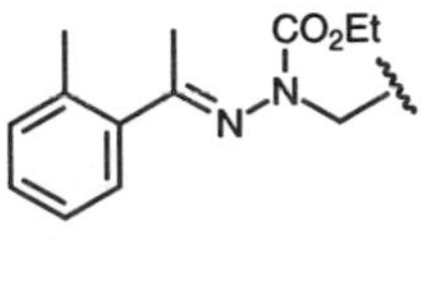

CO_2Et

>1000
96

high hydrophilicity, as in **96**, seems to be disastrous for activity. Derivatives of type **93** are obviously of particular interest, as can be seen from a series of more or less simultaneous patent applications of this type[54,60–65] from as many as five different companies.

In the case of the two-membered intermediate chain, as in BAS 490 F, the exchange of heteroatoms does not lead to any remarkable changes in activity, with the exception of compound **99**, in which the heteroatom is directly connected to the bridging ring, and of **101**, **102**, and **103**, in which the intermediate chain is branched and/or very hydrophilic.

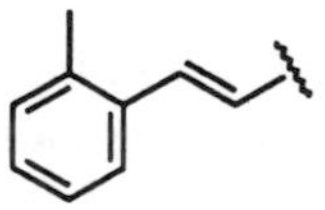

F = 0.4
97

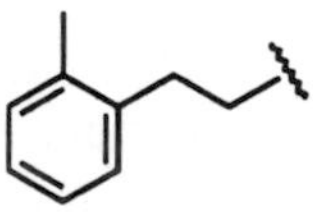

1.1
98

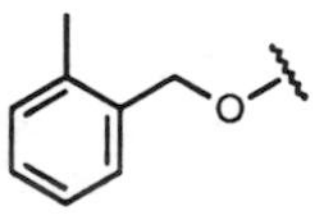

6.5
99

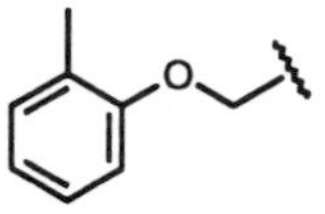

F = 0.6
17

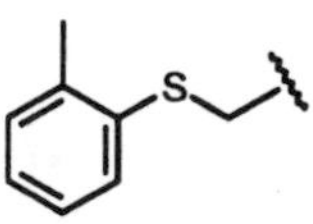

1.0
100

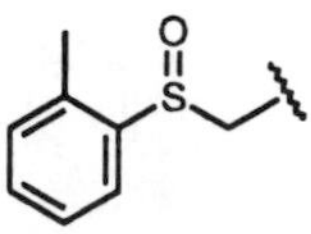

>500
101

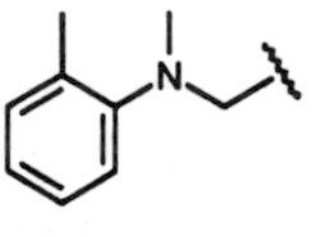

F = 150
102

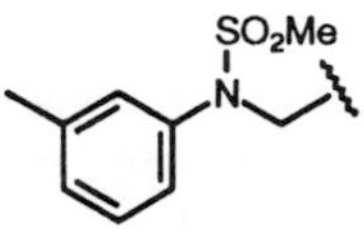

>1000
103

After synthesizing so many side-chain derivatives at BASF, we have gained the impression that by far the most important side-chain property determining target activity is simply lipophilicity.

This has been confirmed with a QSAR study (quantitative structure activity relationship) for a series of 72 derivatives of **16** with different X substituents in their phenoxy ring. Their pI_{50} values from the target test on yeast mitochondria are plotted versus the calculated lipophilicity, log P_{ow}, of the corresponding compounds (Figure 3.6). The pI_{50} is the negative decadic logarithm of the I_{50} value and thus increases with activity. The plotted derivatives contain, for example, the substituents shown in key A.

At the beginning of the plot there is a sharp increase in activity as lipophilicity increases. Then comes a broad optimum, followed by a slowly declining area. Electronic effects of the substituents do not contribute significantly to

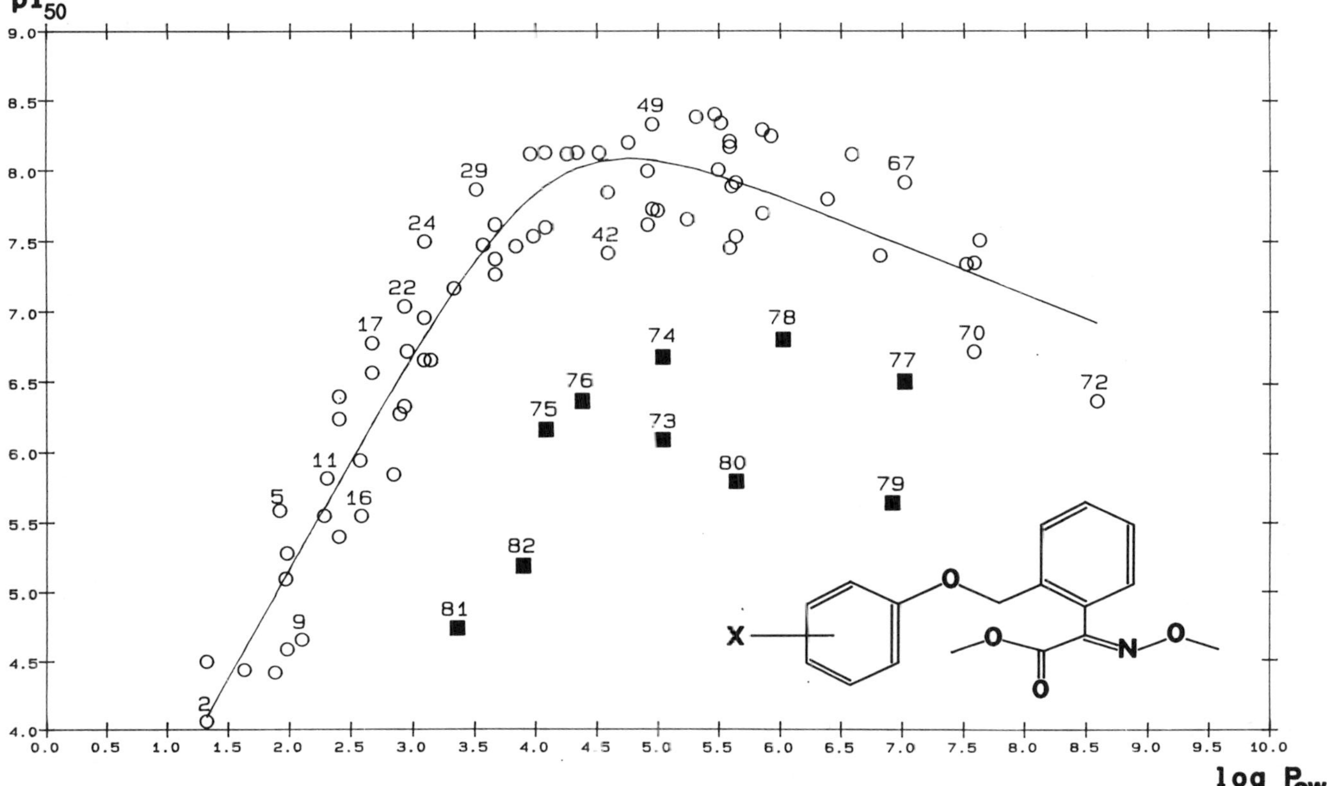

Figure 3.6 *Target activity versus lipophilicity for a series of strobilurins*

Key A

Compound	*X*
2	4-SO_2CH_3
5	4-CH_2OH
9	3-$CH{=}NNHCONH_2$
11	3-CHO
16	4-$NHCOOCH_3$
17	3-NO_2
22	3-OCH_3
24	2-F
29	2-CH_3
42	4-$O(CH_2)_4CH_3$
49	4-$C(CH_3){=}N{-}O{-}(CH_2)_3CH_3$
67	2-CH_3, 4-$C(C_2H_5){=}NO(CH_2)_5CH_3$
70	4-$O(CH_2)_9CH_3$
72	3-$O(CH_2)_{11}CH_3$

activity. Regression analysis of the data fits both parabolic (3.2) and bilinear (3.3) equations with high significance.

$$pI_{50} = -0.228(\pm 0.03)\cdot[\log P]^2 + 2.530(\pm 0.24)\cdot\log P + 1.119(\pm 0.51)$$
$$(n = 72;\ r = 0.951;\ s = 0.367;\ F = 323.40) \quad (3.2)$$

$$pI_{50} = +1.584(\pm 0.13)\cdot\log P - 1.933(\pm 0.21)\cdot\log(\beta\cdot P + 1) + 2.005(\pm 0.43)$$
$$\log\beta = -4.094\ (n = 72;\ r = 0.953;\ s = 0.359;\ F = 225.54) \quad (3.3)$$

However, several compounds exist which do not achieve the activity calculated on this basis, as illustrated by the squares in Figure 3.6. They contain the substituents shown in key B. These compounds are typical underperformers due to the special arrangements of bulky and/or rigid substituents in certain positions, which hinder effective docking of the side-chain at the target. This is true particularly of larger X substituents in the *ortho*-position, for phenoxy groups in any position of the ring, for bulky groups, such as *t*-butyl in the 3,5-disubstitution, and even for small groups, such as methyl or chloro, in the case of 2,6-disubstitution. The underperformance of such compounds together with the good activity of long-chain derivatives indicates that the docking area of the side-chain is more limited with respect to its breadth and depth than its length. These data give the impression that the docking area should be something like a distinctly limited furrow or groove, rather than a fluid, ill-defined region of membrane lipids.

To conclude, as a result of these structure–activity studies, an idea of the possible structures of the docking sites for the strobilurin pharmacophore and

Key B	
Compound	*X*
73	3-Phenoxy
74	4-Phenoxy
75	2,6-$(CH_3)_2$
76	2,6-Cl_2
77	4-(3-*t*-Butylphenoxy)
78	3,5-Diisopropyl
79	3,5-Di-*t*-butyl
80	2-*E*-β-Styryl
81	2-CH=$NOCH_3$
82	2-CH=$NOCH_2CH_3$

side-chain is obtained. Their relative arrangement at the target surface may well resemble that of perpendicular (rake-like), low energy conformations of strobilurins.

3.8 Future Perspectives

In spite of these findings, the full picture is poorly resolved and remains hypothetical and diffuse. In the future there may be chances to synthesize better interacting and thus more active inhibitors as details of the three-dimensional structure of the target emerge. However, this may provide more opportunities for the design of completely new structural types of inhibitors than the strobilurins, for which target activity has been optimized already to a fairly high level by empirical approaches.

It must also be acknowledged that, while target binding is an important prerequisite, it is only one of several factors which contribute to the *in vivo* activity and overall performance of an agricultural fungicide. Biokinetic properties, such as uptake, translocation and metabolic degradation in the fungus and, eventually, the host plant also play major roles. This is quite trivial to say, but very difficult to approach experimentally in detail.

Figure 3.7 illustrates what can happen by moving just one step away from the mitochondrial target to the next highest level of increased complexity, the whole fungal cell. The parameter measured is inhibition of spore germination in *Botrytis cinerea*. As expected, activities given as pI_{50} values are generally lower in this case, simply because fewer fungicide molecules are available for the target inside the cell. More important is the observation that, within this series of homologous compounds, there are quite different optima for the target and the whole cell level. As everyone concerned with fungicide research

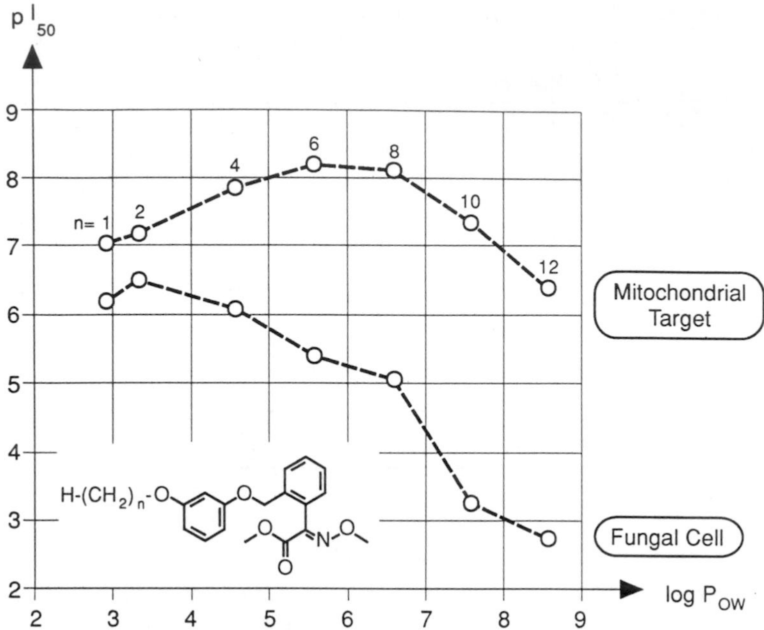

Figure 3.7 *Activities of a series of strobilurins at the target and the cell level*

knows, this sometimes erratic shifting of optima frequently occurs when examining different pathogens or when moving to higher levels of complexity, such as plant-pathogen systems. However, knowledge of the target activity contributes much to a better understanding of *in vivo* results.

Activity, however, is not the only aspect which determines the final usefulness and practical value of a fungicide. Selective toxicity against fungal pathogens combined with safety for plant and animal species (and, of course, humans) is also an indispensible prerequisite. Since strobilurins are inhibitors of the bc_1-complex, a target which is present in all eukaryotes, they could, in principle, be toxic to all of them. Statistical analysis reveals little species selectivity at the target level. This has been shown by comparing the activity of 14 different strobilurins and myxothiazole A in mitochondrial preparations from the following five species: yeast, Botrytis, house fly, rat, and maize.[66] With the exception of the less sensitive maize, there is little variation in the level of sensitivity of the enzymes irrespective of their origin. Additionally, when the different compounds are ranked according to their activity the trends are parallel for all species, as demonstrated with the Spearman rank correlation coefficients $\rho \geqslant 0.8$ (see Table 3.1). This is also illustrated with the paired pI_{50} plots (Figure 3.8) for the examples fly versus yeast ($\rho = 0.80$) and fly versus rat ($\rho = 0.97$). This implies that a structural change which improves activity towards a fungal enzyme will, in general, also improve activity towards the enzymes of other species.

Table 3.1 *Spearman rank correlation table of the* pI_{50} *values for* bc_1 *complex inhibition from different species for 14 strobilurins and myxothiazole A*

	Rat	*Fly*	*Maize*	*Yeast*
Fly	0.97			
Maize	0.90	0.93		
Yeast	0.80	0.82	0.92	
Botrytis	0.83	0.90	0.88	0.90

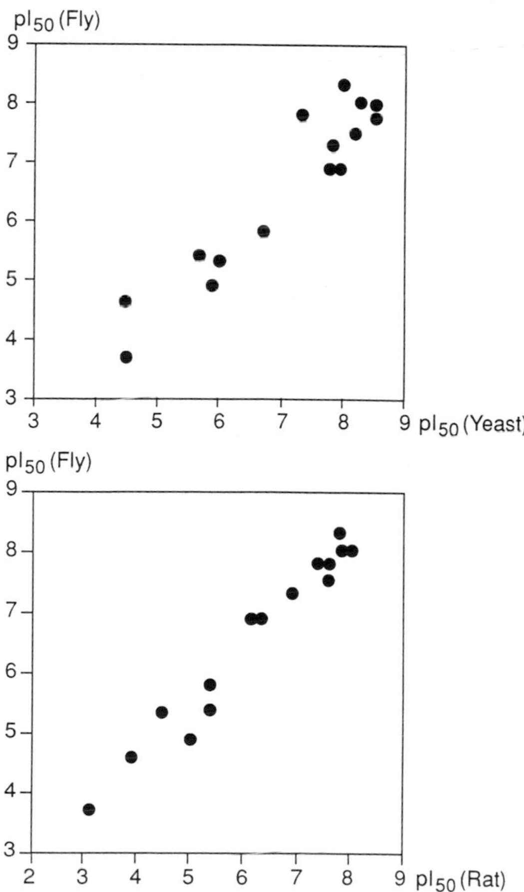

Figure 3.8 *Plots of* pI_{50} *values for* bc_1 *complex inhibition from different species for 14 strobilurins and myxothiazole A*

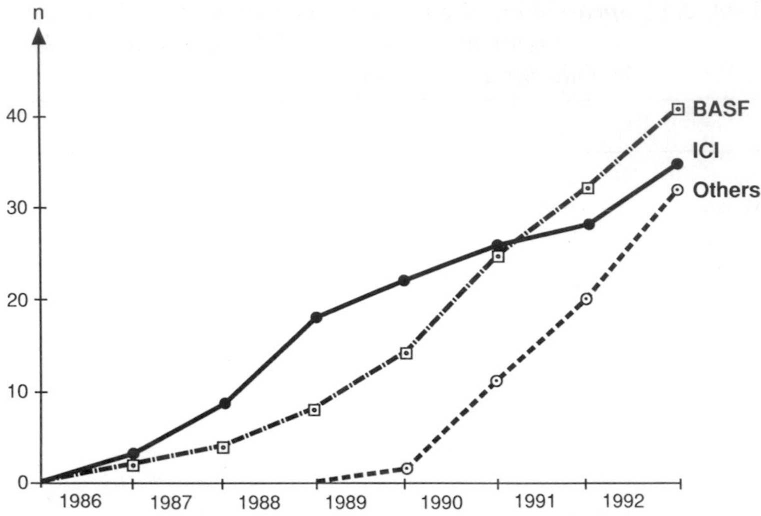

Figure 3.9 *Cumulated number* (n) *of European and World Patent applications for synthetic strobilurins from different companies 1986–1992 (Key date: Dec. 31 each year; process patents not included)*

These results strongly emphasize the importance of the biokinetic properties of strobilurins, not only for the overall *in vivo* performance towards fungal pathogens,[67] but also for selectivity towards other species.

That strobilurins have great potential as a new type of agrochemical has been recognized, not only by the academic and industrial pioneers in this area (Professors Steglich and Anke, ICI and BASF), but subsequently also by many others. This is illustrated by the cumulative number of European and World Patent applications for new strobilurins (Figure 3.9), which suggests further growth of interesting and useful inventions in this area.

3.9 Acknowledgements

We thank all those colleagues from chemistry and biology at BASF who contributed to this article, in particular H. Bayer, R. Benoit, S. Brand, W.v. Deyn, R. Doetzer, W. Grammenos, B. Hellandahl, R. Kirstgen, B. Mueller, K. Oberdorf, C. Rentzea, U. Schirmer, F. Schuetz, H. Theobald, B. Wenderoth, H. Wingert and B. Wolf for the design and synthesis of many strobilurins; R. E. Gold, H. Guggolz, H. Koehle, G. Lorenz, R. Saur, K. Schelberger and M. Scherer for biological investigations; and G. Klebe and H. Kubinyi for QSAR and computer modeling studies. Finally, we thank T. Anke and W. Steglich for the long years of fruitful cooperation.

3.10 References

1. D. Nevill, R. Nyfeler, and D. Sozzi, 'Proceedings of the Brighton Crop Protection Conference – Pests and Diseases', 1988, vol. 1, p. 65.
2. K. Gehmann, R. Nyfeler, A. J. Leadbeater, D. Nevill, and D. Sozzi, 'Proceedings of the Brighton Crop Protection Conference – Pests and Diseases', 1990, vol. 2, p. 399.
3. V. Musilek, J. Cerna, V. Sasek, M. Semerdzieva, and M. Vondracek, *Folia Microbiol. (Prague)*, 1969, **14**, 377.
4. V. Musilek, Czech. Patent 136 492, 1965.
5. M. Vondracek, J. Capkova, J. Slechta, A. Benda, V. Musilek, and J. Cudlin, Czech. Patent 136 495, 1970.
6. M. Vondracek, J. Capkova, and K. Culik, Czech. Patent 180 775, 1974.
7. M. Vondracek, J. Capkova, V. Musilek, and J. Machova, 'International Symposium on Antibiotics', Weimar, 1979, Paper B-12.
8. T. Anke, F. Oberwinkler, W. Steglich, and G. Hoefle, *J. Antibiot.*, 1977, **30**, 221.
9. T. Anke, F. Oberwinkler, W. Steglich, and G. Schramm, *J. Antibiot.*, 1977, **30**, 806.
10. G. Schramm, W. Steglich, T. Anke, and F. Oberwinkler, *Chem. Ber.*, 1978, **111**, 2779.
11. T. Anke, G. Schramm, B. Schwalge, B. Steffan, and W. Steglich, *Liebigs Ann. Chem.*, 1984, 1616.
12. K. Beautement and J. M. Clough, *Tetrahedron Lett.*, 1987, **28**, 475.
13. M. Sutter, *Tetrahedron Lett.*, 1989, **30**, 5417.
14. T. Anke, H.-J. Hecht, G. Schramm, and W. Steglich, *J. Antibiot.*, 1979, **32**, 1112.
15. T. Anke and W. Steglich, in 'Biologically Active Molecules', ed. U. P. Schlunegger, Springer-Verlag, Berlin and Heidelberg, 1989, p. 9.
16. W. Weber, T. Anke, B. Steffan, and W. Steglich, *J. Antibiot.*, 1990, **43**, 207.
17. A. Fredenhagen, A. Kuhn, H. H. Peter, V. Cuomo, and U. Giuliano, *J. Antibiot.*, 1990, **43**, 655.
18. A. Fredenhagen, P. Hug, and H. H. Peter, *J. Antibiot.*, 1990, **43**, 661.
19. J. M. Clough, *Nat. Prod. Rep.*, 1993, **10**, 565.
20. K. Gerth, H. Irschik, H. Reichenbach, and W. Trowitzsch, *J. Antibiot.*, 1980, **33**, 1474.
21. W. Trowitzsch, G. Reifenstahl, V. Wray, and K. Gerth, *J. Antibiot.*, 1980, **33**, 1480.
22. G. Thierbach and H. Reichenbach, *Biochim. Biophys. Acta*, 1981, **638**, 282.
23. G. von Jagow and W. D. Engel, *FEBS Lett.*, 1981, **136**, 19.
24. W. F. Becker, G. von Jagow, T. Anke, and W. Steglich, *FEBS Lett.*, 1981, **132**, 329.
25. G. von Jagow and U. Brandt, Electron and Proton transfer in chemistry and biology, in 'Studies in Physical and Theoretical Chemistry', ed. A. Mueller, H. Ratajczak, W. Junge, and E. Diemann, Elsevier Science Publisher, Amsterdam, London, New York, Tokyo, 1992, vol. 78, p. 42.
26. K. Beautement, J. M. Clough, P. de Fraine, and C. R. A. Godfrey, *Pestic. Sci.*, 1991, **31**, 499.
27. J. Bäuerle, PhD Thesis, University of Tübingen, FRG, 1981.
28. S. Backens, W. Steglich, J. Bäuerle, and T. Anke, *Liebigs Ann. Chem.*, 1988, 405.
29. U. Schirmer, S. Karbach, E. H. Pommer, E. Ammermann, W. Steglich, B. A. M. Schwalge, and T. Anke, EP Appl. 203 606, BASF, Priority 30.05.1985.

30. G. Schramm, PhD Thesis, University of Bonn, FRG, 1980.
31. B. A. M. Schwalge, PhD Thesis, University of Bonn, FRG, 1986.
32. E. V. Blackburn and C. J. Timmons, *Quart. Rev.*, 1969, **23**, 482.
33. F. R. Stermitz, in 'Organic Photochemistry', ed. O. L. Chapman, Dekker, New York, 1967, vol. I, p. 247.
34. J. M. Clough and I. T. Kay, EP Appl. 178 808, ICI, Priority 19.10.1984.
35. M. J. Bushell, K. Beautement, J. M. Clough, P. de Fraine, V. M. Anthony, and C. R. A. Godfrey, EP Appl. 178 826, ICI, Priority 19.10.1984.
36. U. Schirmer, S. Karbach, E. H. Pommer, E. Ammermann, W. Steglich, B. A. M. Schwalge, and T. Anke, EP Appl. 203 608, BASF, Priority 30.05.1985.
37. U. Schirmer, S. Karbach, E. H. Pommer, E. Ammermann, W. Steglich, B. A. M. Schwalge, and T. Anke, EP Appl. 226 917, BASF, Priority 20.12.1985.
38. U. Schirmer, S. Karbach, E. H. Pommer, E. Ammermann, W. Steglich, B. A. M. Schwalge, and T. Anke, EP Appl. 229 974, BASF, Priority 20.12.1985.
39. J. M. Clough, P. J. de Fraine, and C. R. A. Godfrey, lecture given at the 'Symposium toward the Rational Design of Pesticides' organized by the Physicochemical and Biophysical Panel of the SCI Pesticides group at the University of Surrey, Guildford, UK, Sept. 2–4, 1991.
40. J. R. Godwin, V. M. Anthony, J. M. Clough, and C. R. A. Godfrey, 'Proceedings of the Brighton Crop Protection Conference – Pests and Diseases', 1992, vol. 1, p. 435.
41. B. Wenderoth, C. Rentzea, E. Ammermann, E. H. Pommer, W. Steglich, and T. Anke, EP Appl. 253 213, BASF, Priority 16.07.1986.
42. V. M. Anthony, J. M. Clough, C. R. A. Godfrey, and T. E. Wiggins, EP Appl. 254 426, ICI, Priority 18.07.1986.
43. E. Ammermann, G. Lorenz, K. Schelberger, B. Wenderoth, H. Sauter, and C. Rentzea, *Brighton Crop Prot. Conf. – Pests Dis.*, 1992, vol. 1, p. 403.
44. M. Masuko, M. Niikawa, T. Kataoka, M. Ichinari, H. Takenaka, Y. Hayase, Y. Hayashi, and R. Takeda, Abstracts of the 6th International Congress of Plant Pathology, Montreal, Canada, July 28–August 6, 1993, p. 91.
45. Y. Hayase, T. Kataoka, H. Takenaka, M. Ichinari, M. Masuko, T. Takahashi, and N. Tanimoto, EP Appl. 398 692, Shionogi, Priorities 17.05.1989, 29.12.1989.
46. M. Masuko, M. Niikawa, T. Kataoka, M. Ichinari, H. Takenaka, Y. Hayase, Y. Hayashi, and R. Takeda, Poster at the 6th International Congress of Plant Pathology, Montreal, Canada, July 28–August 6, 1993.
47. H. Wingert, B. Wolf, R. Benoit, H. Sauter, M. Hepp, W. Grammenos, and T. Kükenhöner, EP Appl. 493 711, BASF, Priority 31.12.1990.
48. B. Müller, H. Sauter, E. Ammermann, and G. Lorenz, EP Appl. 438 726, BASF, Priority 20.01.1990.
49. B. Müller, F. Röhl, S. Brand, H. Sauter, E. Ammermann, and G. Lorenz, EP Appl. 534 216, BASF, Priority 20.09.1991.
50. B. Wenderoth, H. Sauter, E. Ammermann, and E. H. Pommer, EP Appl. 280 185, BASF, Priority 20.02.1987.
51. S. Brand, B. Wenderoth, F. Schütz, H. Sauter, E. Ammermann, and G. Lorenz, EP Appl. 348 766, BASF, Priority 29.06.1988.
52. W. Grammenos, R. Kirstgen, K. Oberdorf, H. Sauter, F. Röhl, R. Otter, E. Ammermann, G. Lorenz, U. Kardorff, and C. Künast, EP Appl. 513 580, BASF, Priority 17.05.1991.
53. S. Brand, E. Ammermann, G. Lorenz, H. Sauter, K. Oberdorf, U. Kardorff, and C. Künast, EP Appl. 477 631, BASF, Priority 22.09.1990.

54. S. Brand, U. Kardorff, R. Kirstgen, B. Müller, K. Oberdorf, H. Sauter, G. Lorenz, E. Ammermann, C. Künast, and A. Harreus, EP Appl. 463 488, BASF, Priority 27.06.1990.
55. R. Benoit, H. Sauter, R. Kirstgen, G. Lorenz, and E. Ammermann, EP Appl. 498 188, BASF, Priority 07.02.1991.
56. T. Komyoji, I. Shigehava, N. Matsuo, H. Shimoharada, T. Ohshima, T. Akagi, and S. Mitani, EP Appl. 498 396, Ishihara, Priorities 07.02.1991, 23.10.1991.
57. B. Müller, H. Sauter, F. Röhl, R. Dötzer, G. Lorenz, and E. Ammermann, Internat. PCT Appl. WO 93/15046, BASF, Priorities 29.01.1992, 26.06.1992, 05.10.1992.
58. T. E. Wiggins, *Biochem. Soc. Trans.*, 1992, **21**, 29.
59. T. E. Wiggins, *Biochem. Soc. Trans.*, 1992, **21**, 19.
60. P. J. de Fraine and A. Martin, EP Appl. 370 629, ICI, Priorities 21.11.1988, 09.03.1989.
61. H. P. Isenring, S. Trah, and B. Weiss, Internat. PCT Appl. WO 90/07493, Hoffmann La Roche, Priorities 29.12.1988, 03.11.1989.
62. H. P. Isenring and B. Weiss, EP Appl. 460575, Ciba Geigy, Priorities 05.06.1990, Dr. Maag AG, 23.04.1991.
63. K. Tsubata, N. Niino, K. Endo, Y. Yamamoto, and H. Kanno, EP Appl. 414 153, Nihon Nohaku, Priority 22.08.1989.
64. M. Watanabe, T. Tanaka, H. Kobayashi, and S. Yokoyama, EP Appl. 426 460, Ube, Priorities 02.11.1989, 27.02.1990, 02.03.1990.
65. J. M. Clough, C. R. A. Godfrey, and J. P. de Fraine, EP Appl. 472 300, ICI, Priority 22.08.1990.
66. F. Röhl and H. Sauter, *Biochem. Soc. Trans.*, 1994, **22**, 63S.
67. H. Köhle, R. E. Gold, E. Ammermann, H. Sauter, and F. Röhl, *Biochem. Soc. Trans.*, 1994, **22**, 65S.
68. H. Ohnishi, S. Tajma, Y. Yamamoto and H. Kanno, EP Appl. 619 301, Nihon Nohyaka, Priority 04.04.1993.

CHAPTER 4

Phytotoxins of Microbial Origin with Potential for Use as Herbicides

STEPHEN O. DUKE, HAMED K. ABBAS, TADASHI AMAGASA and TATSUMI TANAKA

4.1 Introduction

Natural products of microbes offer a vast array of secondary compounds that are phytotoxic. Many of these compounds have potential to be used directly as herbicides or as structural clues for new synthetic herbicides. The most notable success of this strategy is glufosinate, a microbial phytotoxin that is synthesized and used throughout the world. Bilanafos, a precursor of glufosinate, is produced by fermentation and sold as a herbicide in Japan. Use of microbial compounds as herbicides has been hindered by the often complex structures of the molecules, rediscovery of known compounds after costly research efforts, and the limited environmental stability of the compounds. However, advances in chemistry and biotechnology are reducing the time and cost in using microbial products as leads for new herbicides. Thus far, all successful microbial compounds have come from non-plant-pathogenic organisms. However, plant pathogens produce many potent phytotoxins with varying degrees of selectivity.

Microbes produce a myriad of secondary products with biological activity. The pesticide industry has recognized this and has included this source of novel compounds in their pesticide discovery strategies. Larger amounts of herbicides than any other class of pesticides are used in agriculture,[1,2] so the pesticide industry is especially interested in this source of new herbicides based on natural products. Both plants and microbes are excellent sources of bioactive secondary products, but plants have not been a good source of phytotoxins with potential for herbicide development.[3–5] Plants, however, have been an excellent source for discovery of other pesticide classes, particularly insecticides.[3,5] They may not be particularly good sources of phytotoxins because the evolution of phytotoxins may be more limited than that of other toxins, due to autotoxicity. Thus, microbes are the most promising source of natural compounds with the potential for use as herbicides.

The topic of microbial compounds as herbicides has been reviewed previously in articles wholly devoted to the topic,[6–16] or as part of broader reviews on microbial products as pesticides[17] or natural products as herbicides.[18–22] This review covers some of the same material as these earlier reviews; however, it emphasizes more recent findings from both the authors' laboratories and others. In addition to the main theme of this review, how microbial products have been useful tools in probing the mode of action of synthetic herbicides, basic plant biochemistry and physiology research is described. For the purposes of this review, bacteria, fungi, and unicellular algae are included as microbial sources of phytotoxins.

4.2 Microbial Products in Herbicide Discovery Strategies

4.2.1 Advantages over other Strategies

Herbicide discovery strategies can be grouped into four categories. The traditional strategy that has been used to discover virtually all commercial herbicides is that of screening synthesized compounds for activity as phytotoxins. To some herbicide chemists' objections, this process has been termed 'random screening'. With this strategy, promising groups of compounds are subjected to structure–activity analyses in order to optimize herbicidal activity at the whole plant level.

Another strategy is to design a phytotoxin based on knowledge of the targeted molecular site of action or of structure–activity relationships of the phytotoxin at the site of action. This strategy has been called the 'biorational approach'. Few successes, one being metribuzin, have resulted from this process.[23] Designing herbicides around loopholes in the herbicide patents of competitors is a third strategy, an approach that may require more legal than scientific expertise. However, it has obviously been successful in some cases. The fourth strategy is to use natural products with known or suspected phytotoxic activity as herbicides or as templates for new herbicides. Only two successful herbicides, phosphinothricin (called glufosinate when chemically synthesized) and bilanafos are known to have come from microbial sources.[24,25]

Why not continue to use random screening, the strategy that has been the most rewarding? Millions of compounds have been screened by this approach, so this strategy has reached diminishing returns.[23] The return on investment has reached the marginal point and, thus, other strategies are beginning to become more attractive. The traditional screening method of herbicide discovery has resulted in several hundred commercial herbicides, yet these compounds target only about a dozen molecular sites of action (Table 4.1).[19,26]

Table 4.1 *Summary of the known molecular modes of action of herbicides and microbial phytotoxins (from Duke* et al.[7] *and Devine* et al.[19])

Physiological site product	*Molecular site*	*Herbicide or microbial*
Amino acid synthesis		
Aromatic amino acids	EPSP synthase	Glyphosate
Branched chain amino acids	Acetolactate synthase	Imidazolinones, sulfonylureas
Glutamine synthesis	Glutamine synthetase	*Phosphinothricin, oxetin, bilanafos, phosalacine, tabtoxin*
Glutamate synthesis	Aspartate amino transferase	*Gostatin*
General amino acid synthesis	Many transaminases	*Gabaculin*
Ornithine synthesis	Ornithine carbamyol transferase	*Phaseolotoxin*
Methionine synthesis	β-Cystathionase	*Rhizobitoxin*
Photosynthesis		
Electron transport	D-1, quinone-binding protein	Many, including triazines, substituted ureas, *etc.*
Photophosphorylation	CF_1 ATPase	*Tentoxin*
Electron transport diverters	Photosystem 1	Bipyridiliums (*e.g.*, paraquat)
Pigment synthesis		
Porphyrins	Protoporphyrinogen oxidase	*p*-Nitrodiphenyl ethers, oxadiazoles, *etc.*
	ALA synthase	*Gabaculin*
Carotenoids	Phytoene desaturase	Many, including pyridazinones, *etc.*
	IPP isomerase and/or prenyl transferase	Isoxazolidinones
Cell division		
Mitotic disruptors	β-Tubulin	Dinitroanilines, phosphoric amides
Vitamin synthesis		
Folate synthesis	Dihydropteroate synthase	Asulam
Lipid synthesis		
	Acetyl-CoA carboxylase	Aryloxyphenoxypropanoates, cyclohexanediones
	Acetyl-CoA transacylase	*Thiolactomycin*
	3-Oxoacyl-ACP synthase	*Cerulenin*
Nucleic acid synthesis		
Plastid nucleic acid synthesis	RNA polymerase	*Tagetitoxin*
Plasma membrane function		
	H^+-ATPase	*Syringomycin*
Cell wall synthesis		
	Cellulose synthesis?	Dichlobenil

*Compounds in italics are microbial products.

The molecular sites of action for most commercial herbicides are now known. Herbicides with new sites of action are needed to combat the rapid increase in the evolution of herbicide resistance by weeds. One would suspect that, having screened so many compounds, virtually all viable herbicide sites of action would now be exploited. Yet, the only microbial phytotoxins used as herbicides, glufosinate and bilanafos, attack a molecular site that was apparently not discovered in the random screening approach.

Could the 'synthesize-and-screen' method favour certain sites of action? This question cannot be answered with certainty; however, there is a rationale that would suggest that this is true. First, there is very little overlap between the known molecular sites of action for commercial herbicides and microbial phytotoxins (Table 4.1). These microbial phytotoxins obviously kill plants by attacking such sites. Thus, one must conclude that these are good sites of action that have been missed in herbicide discovery efforts. Why have they been missed?

Most microbial phytotoxins are water-soluble, non-halogenated compounds of the type in which traditional pesticide chemists have had little interest. Also, most microbial compounds tend to be so structurally complex that, for practical reasons, a pesticide synthesis chemist would not produce them. Another reason that compounds which attack only certain sites of action have been found is that many of these currently exploited sites have the potential for many herbicidal analogues. For example, the PS II D1 protein,[27] protoporphyrinogen oxidase,[28] and acetolactate synthetase[29] each have the potential for thousands of different effective herbicidal inhibitors. The odds of stumbling onto one of these is relatively great, compared to the odds of discovering an inhibitor for a site of action for which there are relatively few effective inhibitors. This aspect is discussed further below.

Another impetus for the exploitation of microbial products as herbicides is the impression of many people that natural products, or compounds based on them, are more toxicologically and environmentally benign than synthetic compounds. This is a popular misconception, yet it may influence research activities of some companies. This aspect is discussed further in Section 4.4.2.

Phytotoxins from plant pathogens and some non-pathogenic microbes often have built-in species selectivity, a highly desirable property for which those who design synthetic herbicides work hard to produce in their molecules. Many examples of this property are discussed below. Another impetus for exploiting microbes as sources of herbicides is that of advances in technology. Rapid improvements in chemical methods and biotechnology have made this strategy less costly and time-consuming. Finally, the successes of glufosinate and bilanafos have stimulated interest in this approach.

4.2.2 Disadvantages, Pitfalls, and Cautionary Notes

There are, however, vexing problems associated with microbes as sources of new herbicides. This discovery process involves many more potential steps than does the more traditional process, which becomes an iterative cycle once a

A. Traditional herbicide discovery strategy

synthesis → bioassay
↑ ↓
QSAR ← evaluation

B. Microbial product herbicide discovery strategy

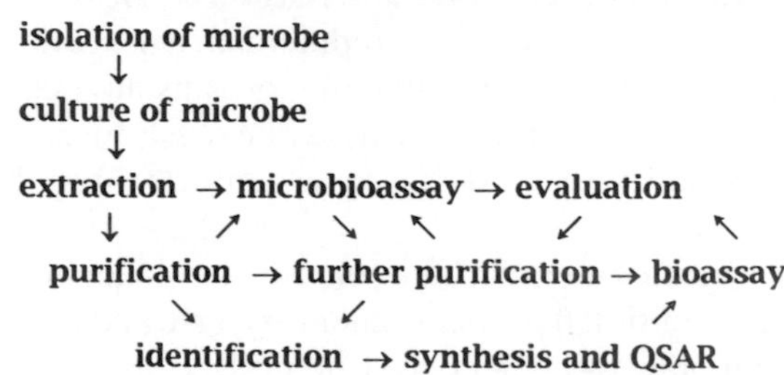

Figure 4.1 *Schematics of the process of herbicide discovery for the synthesis and screen strategy (A) and the use of microbial compounds in herbicide discovery (B).*

promising lead compound has been discovered (Figure 4.1). With microbes, one must find a method of growing the microbe to levels that can produce enough matcrial to evaluate. The microbial culture may not be stable, *i.e.*, the titre of phytotoxin may vary widely for unknown reasons. If the microbe broth has some activity, is this due to a small amount of a very potent phytotoxin or a large amount of a weak and uninteresting phytotoxin? This question can only be answered by purification. Purification may be conducted to varying degrees in order to eliminate uninteresting compounds early. After purification, there may be such a small amount of the compound of interest that bioassay is difficult.

Microbioassays have been developed for screening natural compounds as herbicides. Examples of microbioassays include the use of extremely small whole plants [*e.g.*, duckweed (*Lemna* spp.)], radicle growth of small-seed plants, and leaf disc assays. Microbioassays may not reflect the true potential of a phytotoxin under field conditions as adequately as do the greenhouse screening methods used with the synthetic chemical approach to herbicide discovery. Many of the bioassays for phytotoxicity used by plant pathologists or natural product chemists who are not familiar with herbicide discovery methods are almost useless in evaluating the herbicidal potential of phytotoxins.

After a promising compound has been purified, its structure must be determined. This can be extremely difficult for some compounds. After all of this effort, there is a strong possibility that a known phytotoxin will be rediscovered in a new organism. Ayer *et al.*[30] reported that 72% of compounds that reached the structure determination stage in their company's discovery programme were known compounds. They would have had a higher level of rediscovery if they had not used a phytotoxin profile data base to eliminate known compounds. Similarly, most of the phytotoxic compounds found in fermentation broths of soil microbes by another group were previously discovered phytotoxins, such as cycloheximide.[31]

The level of patent protection for known phytotoxins as herbicides is unclear. One might argue that use of a known phytotoxin as a herbicide is an 'obvious' use for the compound. In the USA, this could be grounds for a patent challenge. If a company produces the compound by synthesis, a synthesis patent may protect their investment. An alternative is to patent a synthesized analogue without reference to the known phytotoxin. There is at least one example of a synthetic analogue of a natural phytotoxin that was under development by a company that would not publicly acknowledge the origin of their compound. Nevertheless, there are numerous patents for microbial compounds as herbicides. Table 4.2 provides a sample of these patents.

The structural complexities of many microbial products make their economical synthesis impossible. Many of these compounds have more than one chiral centre. Furthermore, economic production for agrochemical use by biosynthesis is generally not possible. The cost of many microbially-derived drugs well illustrates this difficulty. In cases of daunting structural complexity, structure–activity efforts to simplify the structure while retaining or improving herbicide properties may be productive. An example of this is methoxyphenone, the synthesized analogue of anisomycin.[32] However, in many cases the phytotoxin structure may have been optimized by æons of co-evolution of the producing microbe with its host(s) or enemies. In such cases, structure–activity efforts may be fruitless.

Table 4.2 *Examples of microbial compounds patented for use as herbicides arranged by chemical type*[11,37]

Amino acids	**Nucleosides**
Rhizobitoxin	Herbicidins
Gabaculine	Tubercidin
Homoalasinone	Formycin A
Glutarimides	**Macrocycles**
Cycloheximides	Herbimycin
Streptimidone	Ascotoxin
Sugars	**Miscellaneous**
Nojirimycin	Oxetin
Phosphinates	Irpexil
Phosphinotricin (glufosinate)	Moniliformin
Bilanafos	
Phosalacine	

Another potential problem with many microbial products is that their physicochemical properties and environmental half-life may be unsuited for use as a herbicide. The chemical characteristics of many of these molecules may make them unsuited for uptake by plant roots or through leaf cuticles. Furthermore, if taken up, they may not exist as a weak acid suitable for adequate translocation. Good activity at the molecular site of action is seldom a good predictor of the efficacy of a herbicide on whole plants (*e.g.*, Nandihalli *et al.*[33]). Little is known of the environmental half-life of microbial secondary products. However, these compounds have been in the environment for æons, and one would suspect that they have relatively short half-lives compared with synthetic, halogenated chemical structures.

Finally, optimization of this strategy for herbicide development requires personnel with training other than that commonly found in the synthetic pesticide industry. Plant pathologists, natural product chemists, and microbiologists with expertise in fermentation methods are needed. For a company to hire a critical mass of these specialists requires a strong commitment to this strategy.

4.2.3 Relationships in the Use of Plant Pathogens as Bioherbicides

Why not use the phytotoxin-producing organism as a biological control agent? This approach would appear to offer several advantages over using the phytotoxin or related compound as a herbicide. With this approach, synthesis of the compound and its uptake are no longer problems because the microorganism manufactures the compound in the host plant. However, there are currently extreme limitations to this method. Of the scores of microbial agents that have been patented for herbicide use,[34] less than five have reached the market, and they have had very limited success[35] (see also Chapter 13).

The first limitation is that only plant pathogens lend themselves to this approach. This eliminates a large number of phytotoxin-producing microorganisms. Biotechnology could introduce into plant pathogens the capacity to produce toxins from non-pathogenic organisms. This approach, however, could be quite dangerous from an environmental standpoint. Of plant pathogens, only those that attack weeds can be used, although some effort has been expended to overcome this limitation through biotechnology.[36] Very often, those pathogens that only attack weeds are too host-specific; that is, they affect only one weed species. If the weed is not an extremely important one in an important crop, this factor alone precludes development. An example of a microbial herbicide that satisfies this criterion is that of *Drechslera* spp. for control of barnyardgrass and related weed species (*Echinochloa* spp.) in rice.[37] Host range extension by formulation may increase the commercial viability of some microbials. For example, the host range of *Alternaria crassa* (Sacc.), which normally only infects jimsonweed (*Datura stramonium* L.), can be extended to include hemp sesbania [*Sesbania exaltata* (Raf.) Rydb. ex A.W.

Hill], a completely unrelated species, by formulating with pectin and an extract of hemp sesbania.[38]

Living organisms are difficult to store, handle, and apply, compared with chemicals. They generally have a relatively short shelf life, and so may be sufficiently viable for one season after production. The economic repercussions of this are obvious. Storage conditions may have to be managed to avoid loss of viability. Formulation with agrochemicals may be severely limited because of incompatibility of the organism with certain fungicides, insecticides, and/or herbicides. Formulation requirements of microbes may be unusual and costly.[39] Exotic formulations may require specialized application technology. For example, invert emulsion formulations are so viscous that standard spray equipment does not work. Such products require more sophisticated air-assist nozzles fed with metering pumps, such as those used for the ultra-low-volume application of oils.[40]

The environmental window for success with microbials is usually very narrow compared with that for chemical herbicides. Many fungal pathogens require a prolonged period of dew at the proper temperature in order for substantial infection to occur. There are often microenvironments within a field that are not suitable for adequate infection, resulting in the more than 90% weed control expected by farmers not being achieved.

4.3 Sources and Examples of Microbial Phytotoxins

4.3.1 Phytotoxins from Non-Pathogenic Microbes

Many non-pathogenic microbes produce highly potent phytotoxins. Most studies of these microbes have been with soil-borne microbes, such as the actinomycetes (*e.g.*, Heisey *et al.*[31]). However, there are also many microbial saprophytes and aquatic microbes that are potential sources of useful phytotoxins. The only successful herbicidal compounds from microbes, glufosinate and bilanafos, are from soil-borne microbes, and the herbicide industry continues to focus on this source. Soil samples from remote locations are examined for new organisms that may yield new, promising compounds. The greater interest in these organisms than in plant pathogens may be due to the relative ease with which they can be cultured.

Bilanafos (**3**) and phosphinothricin (**1**, Figure 4.2) are the most successful microbial products for use as herbicides. Bilanafos is a tripeptide with two alanyl groups that is metabolically converted into phosphinothricin within the plant by hydrolytic cleavage of the two alanyl groups.[41] Phosphinothricin is a potent, irreversible inhibitor of glutamine synthetase (GS),[42–44] whereas bilanafos has no *in vitro* activity on this enzyme.[45,46] Thus, bilanafos is a proherbicide. It is produced by both *Streptomyces viridochromeogenes* (Krainsky) Waksman & Henrici and *S. hygroscopicus* (Jensen) Waksman & Henrici. Phosphinothricin was first isolated from *S. hygroscopicus*.[47]

Two properties of phosphinothricin (**1**) that make it an ideal non-selective

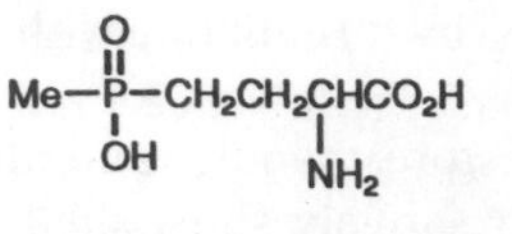

1 Phosphinothricin (Glufosinate)

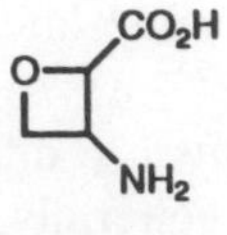

2 Oxetin

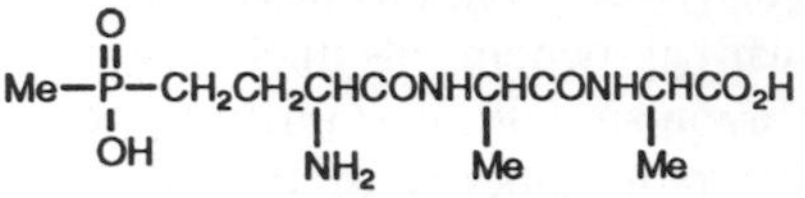

3 Bilanafos

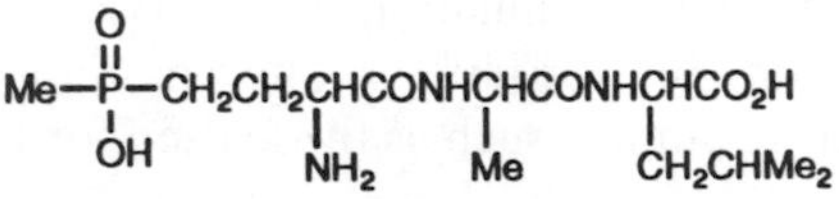

4 Phosalacine

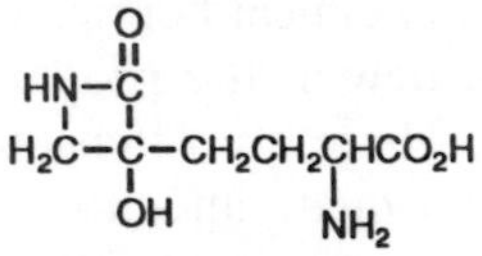

5 Tabtoxinine-β-lactam

Figure 4.2 *Structure of microbial glutamine synthetase inhibitors.*

herbicide are that it is not significantly metabolically degraded in higher plants[48] and that it is translocated readily both in phloem and xylem.[49] The chemically synthesized version of phosphinothricin, glufosinate, is marketed as a non-selective herbicide in the form of an ammonium salt. A significant amount of structure–activity work has been conducted to improve on the molecular structure of glufosinate, but without success.[50]

The phenomenology of the herbicidal action of glufosinate on whole plants is too rapid to be the result of starvation for glutamine and other amino acids derived from glutamine. Initially, it was thought that the high ammonium ion accumulation caused by glufosinate was responsible for most of the herbicidal effects. However, glufosinate effects can be reversed by supplying the plant with glutamine, a treatment that does not lower the ammonium ion levels.[51] Most of the phytotoxicity of inhibiting GS in C_3 plants is due to rapid cessation of photorespiration, which is due to glutamine depletion, resulting in accumulation of glyoxylate in the chloroplast and fast inhibition of RuBP carboxylase.[52,53] Inhibition of carbon fixation in the light leads to a series of events ending in severe photodynamic damage.[19]

The tripeptide phosalacine (**4**), produced by *Kitasatosporia phosalacinea*, is structurally related to bilanafos, differing only by a terminal leucine rather than an alanine residue.[54] Like bilanafos, it is metabolized to phosphinothricin within plants.[41] Oxetin (**2**) is another *Streptomyces* product that inhibits GS.[55] Structurally, it is only marginally related to the other microbial GS inhibitors and is a relatively poor GS inhibitor. Tabtoxinine-β-lactam (T-β-L) (**5**) is a related GS inhibitor from the plant pathogen *Pseudomonas syringae* van Hall, which is responsible for 'wildfire' disease in a number of crops.[56] Like phosphinothricin, it is the hydrolysis product of a larger molecule, tabtoxin. It is discussed in this section because it is a close analogue of the GS inhibitors from non-pathogens. T-β-L causes symptomology very similar to that of glufosinate. Like glufosinate, T-β-L is readily taken up by plant cells and translocated.[57,58] However, it is less potent than glufosinate as a herbicide.

A large number of herbicidal compounds, other than GS inhibitors, have been discovered from non-pathogenic microbes. For example, highly phytotoxic blastocidins have been found in culture broths of soil *Streptomyces* strains.[59] These nucleoside compounds also have activity against the plant pathogenic fungus *Pyricularia oryzae* Cavara and have been commercialized for this purpose. Other nucleoside phytotoxins from non-pathogenic microbes have been reported. SF 2494, a tubercidin produced by *S. mirabilis* is a non-selective herbicide with activity comparable to bilanafos (**3**).[60] Tubercidin had earlier been reported to be herbicidal.[8] Toyocamycin (**11**) is a pyrrolopyridine nucleoside from a *Streptomyces* species.[61] It is also a non-selective herbicide. Other herbicidal nucleosides from non-pathogenic microbes include gougerotin,[62] rodaplutin,[63] sangivamycin,[8] herbiplanin,[8] and the herbicidins.[64] The herbicidins, a group of adenine nucleosides from *S. saganonenis*, provide good dicotyledonous weed control in rice. In an extensive study, five herbicidal nucleosides were discovered in five different actinomycetes.[65] Three were new compounds (5′-deoxyguanosine, coaristeromycin, and 5″-deoxytoyocamycin) and two were previously discovered compounds not known to be phytotoxic (coformyin and adenine 9-β-D-arabinofuranoside). The mechanisms of action of these compounds are unknown; however, their structures suggest interference in nucleic acid metabolism, a site of action currently unexploited by the herbicide industry. Blasticidins inhibit protein synthesis, but a direct effect has not been demonstrated.[59] All herbicides affect protein synthesis at some time after treatment.[19]

Anisomycin (**14**), another *Streptomyces* sp. product, is a good selective herbicide for rice, with excellent activity against barnyardgrass and crabgrass [*Digitaria sanguinalis* (L.) Scop.][61] Methoxyphenone (**15**), a synthetic rice herbicide, was developed from the structure of anisomycin.[32,66] It has been sold on the Japanese herbicide market, and its mode of action is inhibition of photosynthesis.

Herboxidiene (**13**) from a *Streptomyces* sp. A7847 has recently been reported as a potent soil-applied herbicide against many weed species.[67,68] However, it is inactive against wheat at rates as high as 5.6 kg ha^{-1}. No structural modifications to the molecular structure improved herbicidal activity and all significant reductions in structural complexity resulted in a loss of activity.[68] Its mechanism of action is unknown.

Vulgamycin (**12**) was isolated from a new *Streptomyces* isolate.[69] It controls dicotyledonous weeds at doses between 125 and 500 g ha^{-1}. Furthermore, cotton, barley, and maize are unaffected or only slightly affected at these

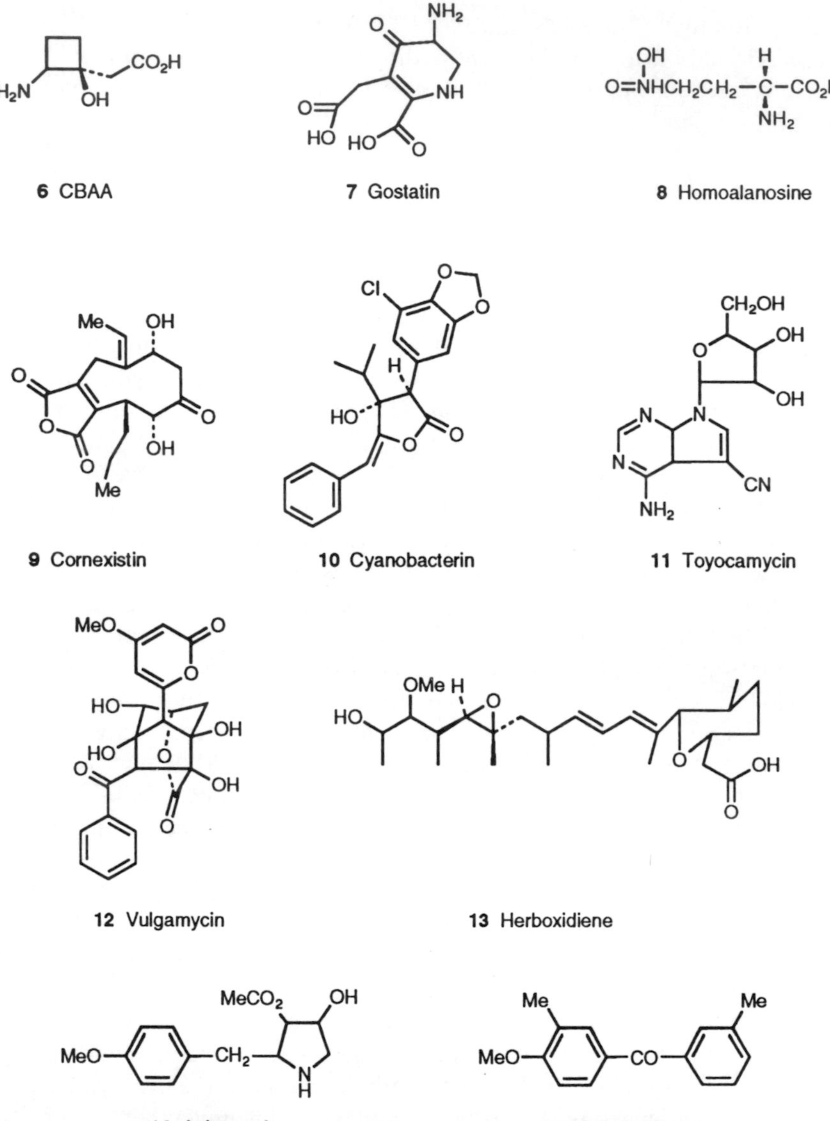

Figure 4.3 *Phytotoxins from non-pathogenic microbes.*

doses. It appears to interfere with an isoleucine-dependent step in the cell cycle, resulting in the arrest of cell division in the G1 stage.

Streptomyces rocheii A13018 was found to produce *cis*-2-amino-1-hydroxycyclobutane-1-acetic acid (CBAA) (**6**), a component of a larger, previously described dipeptide.[70] Its structural similarity to the microbial GS inhibitors suggests that it could be a GS inhibitor; however, chlorosis in *Arabidopsis thaliana* Heynh. caused by CBAA could be reversed by supplying

L-cysteine or L-methionine. It has activity against members of the genus *Ipomoea* and *Brassica juncea* Coss.

Cornexistin (**9**) is a recently discovered nonadride phytotoxin isolated from a culture filtrate of the saprophytic, coprophyllic basidiomycete *Paecilomyces variotii* Bainer (SANK 21086).[71] It has excellent herbicidal activity against both monocotyledonous and dicotyledonous weeds, but maize (*Zea mays* L.) is tolerant to it. Rubratoxin B, a related fungal metabolite, is only weakly phytotoxic.[71] Little is known of the mechanism of action of cornexistin. Its herbicidal activity against duckweed (*Lemma pausicostata* L.) can be prevented by supplying the plants with L-aspartate, L-glutamate, and/or tricarboxylic acid cycle intermediates.[72] This phenomenon and the general herbicidal effects are much like that of amino-oxyacetic acid, a well-known aspartate amino transferase (AAT) inhibitor. However, no *in vitro* effect on AAT activity has been found. It is structurally related, but not closely, to the AAT inhibitor gostatin (**7**) produced by *S. sumanensis*.[73] The herbicidal effects of gostatin have not been reported. The growth-retarding effects of homoalanosine (**8**), an amino acid from *S. hygroscopicus*, on bacteria can be reduced by supplying L-aspartate and L-glutamate.[74] Homoalanine is an effective herbicide for dicotyledonous weeds in rice. Like cornexistin, the development of symptoms is slow.

Isoxazole-4-carboxylic acid from *Streptomyces* sp. NK-489 has been reported to be phytotoxic.[75] Arabenoic acid from an unidentified soil microbe was phytotoxic enough to be isolated and identified in a screen for natural products with herbicidal activity.[76] In the same laboratory, α-methylene-β-alanine from *Streptomyces* sp. A12701 was found to have weak herbicidal activity at 5.6 kg ha^{-1} on velvetleaf (*Abutilon theophrastii* Medic.).[77] This compound is also produced by at least two sponge genera.

Ansamitocins from *S. hygroscopicus*, such as the herbimycins, have been patented as herbicides for use in rice. Some herbimycins have mammalian activity as a protein tyrosine kinase inhibitor and inhibitor of signal transduction by the T cell antigen receptor.[78] The structurally related geldanamycin, also from *S. hygroscopicus*, is a good plant growth inhibitor.[79,80] Nigericin, another *S. hygroscopicus* product, is a good plant growth inhibitor at doses of 300 g ha^{-1}. It is both a photophosphorylation inhibitor[81] and a potassium ionophore.[82]

SF 701 is a streptomycin-like product of a *Streptomyces* sp. which is herbicidally active against barnyardgrass [*Echinochloa crus-galli* (L.) Beauv.], but with little phytotoxicity to rice.[83] It has no direct effect on photosynthesis; however, it inhibits starch synthesis. This may be a secondary effect of inhibited plastid protein synthesis, the known mode of action of streptomycin.

Two antibiotics (thiolactomycin and cerulinin) from non-pathogenic microbes (*Cephalosporium caerulens* and *Nocardia* sp., respectively) are phytotoxic by inhibiting lipid synthesis.[84] They are not good candidates for new herbicides, as their efficacy is much lower than that of synthetic lipid-synthesis-inhibiting herbicides (*e.g.*, cyclohexanediones and aryloxyphenoxy propanoates). Nevertheless, these compounds inhibit different lipid synthesis sites than the commercial herbicides that target this pathway (Table 4.1), indicating that their enzymatic target sites are worthy of further examination in biorational herbicide design studies.

The myxobacterium *Stigmatella aurantiaca* Berkeley & Curtis produces stigmatellin and aurachins. These compounds inhibit PS II by inhibition of electron transport through the cytochrome b_6/f complex.[85,86] At least two potent photosynthesis inhibitors, cyanobacterin and fischerellin, are produced by cyanobacteria.[87,88] Cyanobacterin (**10**) is a halogenated compound that inhibits PS II at a site between Q_A(D-2) and Q_B(D-1).[89] Its activity is comparable to that of DCMU (diuron). Attempts to produce more effective synthetic analogues have been unsuccessful.[90] Since DCMU is about as active as any PS II inhibitor, it may be unlikely that structural manipulation of a compound with similar activity could produce a more active molecule. The existence of several good PS II-inhibiting natural products with molecular sites of action different from that of commercial PS II-inhibiting herbicides is highly significant. Why have no compounds with these sites of action been discovered by conventional efforts? The availability of compounds with these new sites of action could be extremely valuable in combating herbicide resistance, either as rotational herbicides or by combining conventional PS II-inhibiting herbicides with one of these compounds.

4.3.2 Phytotoxins from Plant Pathogens

Compared with non-phytopathogens, relatively little research and development have been conducted with plant pathogens as sources of new herbicides. This is partly due to the relative difficulty in culturing many plant pathogens, compared with most non-pathogens. Also, some may have the false impression that plant phytotoxins have been well-studied or that such toxins tend to have a very limited range of susceptible species. Furthermore, plant pathologists who study phytotoxins have had little interest in the potential of these compounds as herbicides. Very little has been done with pathogens that infect weeds, and the susceptibility of weed species to most known phytotoxins is unknown. Extrapolation from the literature can be misleading, as illustrated in the case of AAL-toxin discussed below.

4.3.2.1 Non-Host-Specific Phytotoxins

Phytotoxins produced by plant pathogens (Figure 4.4) generally have a wider range of susceptible plant species than do the producing microbes, although the phytotoxins are often important virulence factors for the pathogen. Furthermore, it is common for a plant pathogen to produce an array of phytotoxins and for a particular phytotoxin to be produced by several microbial species. There are many structurally described non-host-specific phytotoxins from plant pathogens, and the known compounds probably represent only a small fraction

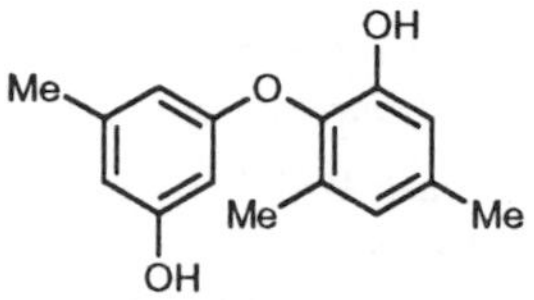

16 Cyperine

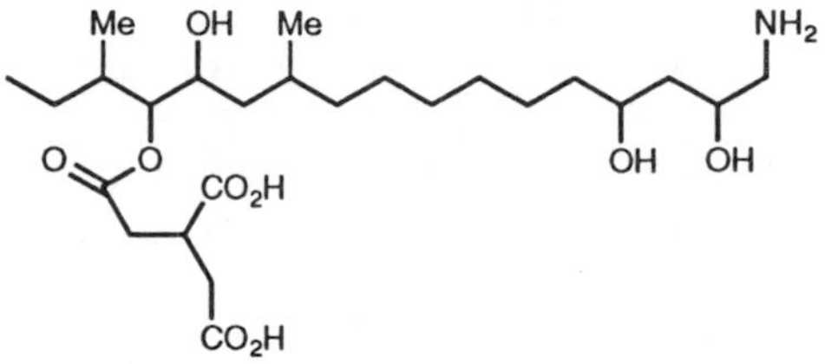

17 AAL-Toxin

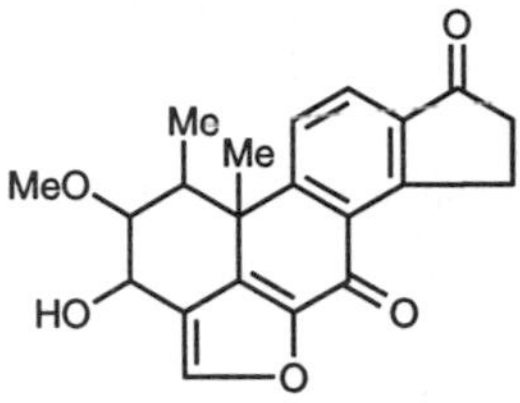

18 Viridiol

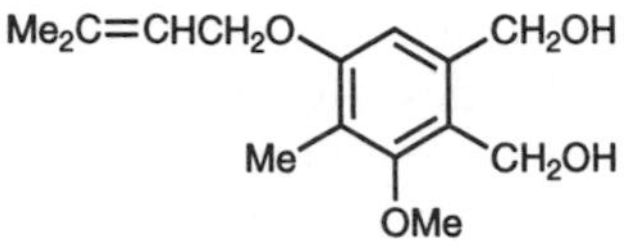

19 Zinniol

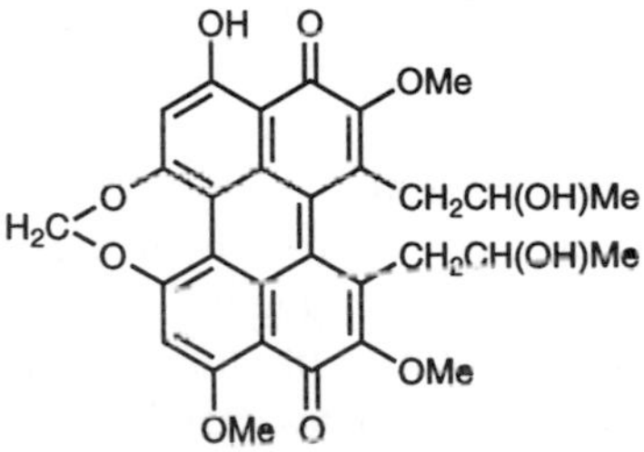

20 Cercosporin

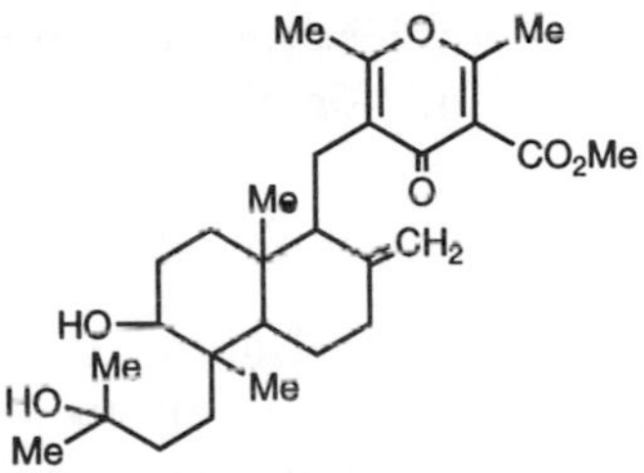

21 Colletotrichin

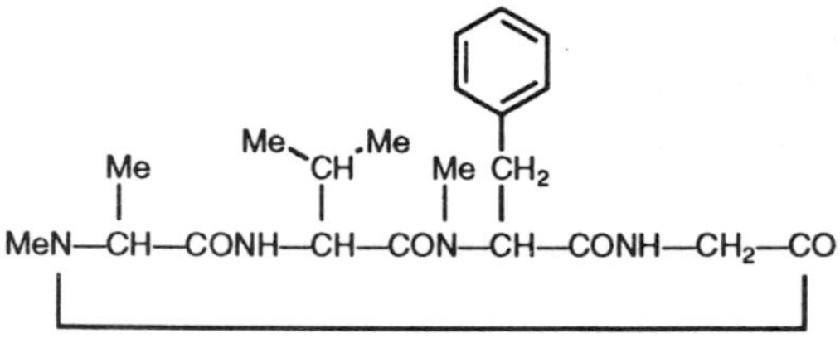

22 Tentoxin

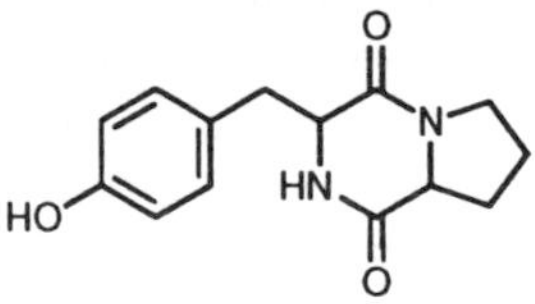

23 Maculosin

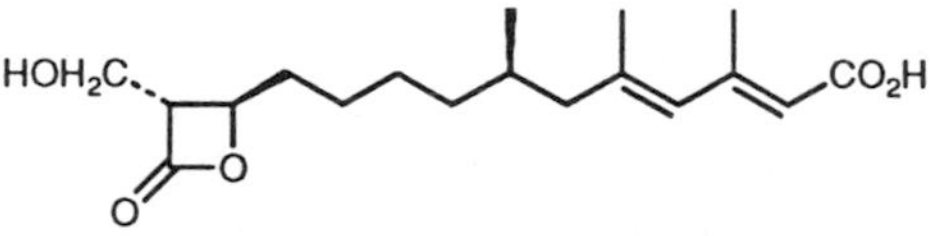

Absolute configuration

24 Antibiotic 1233A

Figure 4.4 *Phytotoxins from plant pathogens.*

of those produced by plant pathogens. Only a small fraction of these are mentioned here.

Bacterial plant pathogens produce a number of interesting phytotoxins (tabtoxin is discussed above). *Pseudomonas syringae* subsp. *syringae* van Hall produces an arsenal of phytotoxins, including syringomycins,[91] syringostatins,[92] syringotoxin,[93] and syringopeptins.[94] The first three of these phytotoxins are all closely related cyclic compounds containing common amino acids and non-protein amino acids with an *N*-terminal serine *N*-acylated by a long-chain unbranched 3-hydroxy fatty acid. All of these compounds inhibit plasma membrane H^+-ATPase activity by both direct effects on the enzyme and, partly, by a detergent-like action.[95] The syringopeptins are of such high molecular weight and complexity that they would be of no interest in development as herbicides. However, they may be proherbicides with smaller phytotoxic fragments. Syringopeptin 25A is considerably more phytotoxic than syringomycins E and G in a variety of tests.[96]

The agent of halo blight of beans, *Pseudomonoas syringae* van Hall pv. *phaseolicola*, produces the tripeptide toxin phaseolotoxin.[97] Like bilanafos and tabtoxin, phaseolotoxin must be metabolically modified by the plant to produce the actual phytotoxin *in vivo*.[98] The active compound is octicidine, an inhibitor of ornithine carbamoyltransferase.[99] There are no known attempts to exploit this rather simple plant and microbe-specific toxin as a herbicide.

Coronatine is produced by numerous pathovars of *P. syringae*.[100] This non-host-specific toxin causes growth inhibition and hypertrophy in some tissues; however, its principle symptom appears to be the induction of chlorosis in green leaves.[101] The chlorosis is not a consequence of photobleaching,[101] but no determinations of the effects on chlorophyll synthesis have been made. Its precise mode of action is unknown.

Tagetitoxin, another chlorosis-causing phytotoxin, is produced by yet another pathovar of *P. syringae*.[102] This eight-membered, sulfur-containing ring structure causes chlorosis by inhibiting chloroplast RNA polymerase, thereby preventing chloroplast development.[103] However, its inhibiting activity towards RNA polymerase III promoter-directed transcription extends to vertebrates, insects, and yeast.[104]

Many more non-host-specific phytotoxins have been described from fungal than from bacterial pathogens, only a few of which are described here.

Pyrichalasin H, an analogue of the cytochalains, is produced by *Pyricularia grisea* (Cooke) Sacc. (IFO 7287), the causative agent of a blast disease in crabgrass [*Digitaria sanguinalis* (L.) Scop.].[105] Although not produced by the *Pyricularia* that infests rice, pyrichalasin H is highly phytotoxic to rice.[105,106] *Pyricularia* isolates from a number of other monocotyledonous species also produce this compound, which suggests that it is herbicidally active against a number of monocotyledonous species. Like the cytochalasins, this compound is too structurally complex to be of great interest for herbicide development. Whether its mechanism of phytotoxicity is through interaction with cellular microfilaments, like the cytochalasins, is unknown.

Cercosporin (**20**) is a red, photodynamic compound produced from several species of *Cercospora*.[107] Isocercosporin, a closely related compound, is produced by *Scolecotrichum graminis* Fuckel, the agent of leaf streak in orchardgrass (*Dactylis glomerata* L.)[108] These compounds are photodynamic, producing singlet oxygen in the presence of sunlight and molecular oxygen.[109] Similar compounds, such as hypericin,[110] have been proposed as herbicides; however, such compounds are toxic to all life that exists in sunlight and an oxygen atmosphere.

Viridiol (**18**) is a broad-spectrum phytotoxin produced by the fungus *Gliocladium virens* Miller, Giddens & Foster.[111,112] It is particularly effective against composite and other dicotyledonous species, with less effect on monocotyledonous species. However, these studies are difficult to interpret with respect to the actual phytotoxicity of the toxin because the results are extrapolated from studies in which the phytotoxicity of viridiol, estimated to have been generated by the fungus in the soil, is determined. In other words, the fungus is used as a biocontrol agent to deliver the toxin.

Phytotoxins from the fungus *Ascochyta hyalospora* (Cooke & Ell.) Boerema, Mathur, & Neergaard, including ascochytine, pyrenolide A, and hyalopyrone, are active against several weed species, including lambsquarters (*Chenopodium album* L.), prickly sida (*Sida spinosa* L.), and johnsongrass [*Sorghum halepense* (L.) Pers.].[113] The compounds caused electrolyte leakage from leaf tissue of lambsquarters and inhibition of johnsongrass root growth. These metabolites are also active as growth inhibitors of bacteria and fungi.

Stagonospora apocyni, the fungal agent of leaf spot disease in hemp dogbane (*Apocynum cannabinum* L.), produces the phytotoxins citrinin, mellein, tyrosol, and α-acetylorcinol.[114] All of these non-host-specific phytotoxins cause necrotic lesions in dog hempbane, prickly sida, morningglory (*Ipomoea* sp.), and several other weed species.

Colletotrichin (**21**), a product of several *Colletotrichum* species, is active against cucumber, tobacco, and solanaceous weed species, including nightshades and horsenettle (*Solanum* spp.).[7,115,116] The first ultrastructural symptom of phytotoxicity is the loss of structural integrity of the plasma membrane. Lipid peroxidation is associated with the membrane damage; however, radical-quenching agents do not protect plant cells from the toxin. Colletotrichin reduces or prevents the phytotoxicity of the synthetic herbicide, acifluorfen.[116] Acifluorfen requires the activity of a plasma membrane-associated redox enzyme for its activity.[117] Thus, colletotrichin may interfere with this plasma membrane function, and others.

Tentoxin (**22**), a cyclic tetrapeptide from *Alternaria alternata* f. sp. *tenuis* (Fr.) Keissler and other *Alternaria* species, is a potent, chlorosis-producing phytotoxin.[118] Susceptibility is characterized by causing achlorophyllous (yellow) plastids in developing leaf or cotyledonary tissues. The range of susceptible species shows no taxonomic pattern, with both susceptible and resistant monocotyledonous and dicotyledonous species.[119] Even within a single genus, both susceptible and resistant species can be found. For example, several species of *Nicotiana* are resistant and several are susceptible,[120–122] the differences being marked. Typical tentoxin-caused chlorosis cannot be induced in resistant species, even at millimolar levels of tentoxin (Duke, unpublished).

Numerous important weed species, including ivyleaf morningglory [*Ipomoea hederacea* (L.) var *Hederacea*],[123] johnsongrass,[6] sicklepod (*Cassia obtusifolia* L.), and cocklebur (*Xanthium stramonium* L.), are susceptible to tentoxin (Duke, unpublished data). Tentoxin (**22**) is a good pre-emergence, soil-incorporated herbicide. The desirable pattern of susceptibility, the extreme sensitivity of susceptible weed species, and a unique mode of action have made this compound the focus of considerable effort in herbicide development, in both the private and public sectors.

Tentoxin (**22**) has several physiological effects, including inhibition of coupling factor 1 (CF_1) ATPase activity of the photosynthetic coupling factor complex,[121,124,125] inhibition of the uptake of at least one nuclear-coded protein by plastids,[126–128] and effects on membrane properties.[129] From the literature it is not clear as to which of these effects causes the chlorosis by which susceptibility is judged. The membrane effects are clearly unrelated to susceptibility in that they are seen in resistant species at concentrations that do not cause chlorosis in susceptible species. The other two primary effects (inhibition of CF_1 ATPase and nuclear-coded protein uptake by plastids) are not so easily separated.

There are two proofs that the chlorosis caused by tentoxin (**22**) is not the result of inhibition of photophosphorylation, the secondary effect of inhibition of CF_1 ATPase. First, the etioplast of tentoxin-treated tissues of susceptible species is clearly abnormal, with complete inhibition of the uptake and processing of nuclear-coded polyphenol oxidases, ultrastructural abnormalities, and reduced capacity to synthesize protochlorophyllide.[130,131] None of these effects can be attributed to inhibition of photophosphorylation. During chloroplast development, no electron transport capacity develops in tentoxin-treated plants, whereas it can be easily measured within 15 minutes in control plants, so it is not likely to be the result of inhibited photophosphorylation. The second proof is that in seedlings of a lethal mutant of *Oenothera*, in which CF_1 ATPase is inactive because of a genetic lesion, the plastids become light green before the seedlings die, whereas tentoxin causes yellow seedlings and complete loss of polyphenol oxidase (PPO).[132] There is not always a good correlation between tentoxin effects on photophosphorylation and susceptibility.[132] In fact, the photophosphorylation effect was discovered using maize, a resistant species.[124]

The gene coding for a tentoxin-susceptible CF_1 ATPase was transferred to the alga *Chlamydomonas reinhardtii* Dang., an organism that is resistant to tentoxin (**22**).[121] The resulting transformant was susceptible to tentoxin. Recent work in our laboratory indicates that there is an absolute relationship between the susceptibility of CF_1 ATPase and the susceptibility of the whole plant.[133] However, the physiological results (discussed above) demonstrate that there is no causal relationship between inhibition of photophosphorylation and arrested chloroplast development. The data can be reconciled by only two theories. One is that CF_1 ATPase has a role in chloroplast development unrelated to photophosphorylation. The other is that the subunit of CF_1 ATPase that confers resistance to the CF_1 is also a component of a plastid envelope ATPase involved in movement of certain nuclear-coded proteins into the plastid. There appear to be several mechanisms for the movement of nuclear-coded proteins into plastids.[134] Support for the latter theory is provided by findings that certain plastid envelope ATPases have been found to be inhibited by tentoxin in susceptible plant species.[118,135]

Tentoxin (**22**) has been a valuable tool in probing the mechanisms of action of synthetic herbicides and in studying fundamental questions in plant physiology. It has been used to eliminate chlorophylls in tissues, in order to measure the effects of light on phytochrome synthesis *in vivo*,[136] to determine the role of photosynthesis in the mechanism of action of diphenyl ether herbicides,[137] to determine the role of PPO in phenolic compound biosynthesis,[138] and to measure effects of herbicides on porphyrin synthesis *in vivo*.[139,140]

Cyperine (**16**), a diphenyl ether, is produced by *Ascochyta cypericola*, a fungal pathogen of purple nutsedge (*Cyperus rotundus* L.).[141] It has some selectivity for weeds in the genus *Cyperus*. The mechanism of action of this compound is unknown, and its structure does not suggest that its activity would be like that of synthetic *p*-nitrodiphenyl ethers (protoporphyrinogen oxidase inhibitors) or diphenyl ether aryloxyphenoxyphenoxy propanoates (acetyl CoA carboxylase inhibitors).[19]

An unnamed antibiotic (1233A) (**24**) from *Scopulariopsis candida* (Guég.) Vuill., *Cephalosporium* spp., and *Fusarium* spp. is a potent phytotoxin.[142] This compound is an inhibitor of 3-hydroxy-3-methylglutaryl coenzyme A (HMG-CoA) synthase,[143] an early enzyme of the terpenoid pathway that is not known to be affected by any commercial herbicides. This enzymic site has been studied as a possible site of action of the commercial herbicide clomazone.

Zinniol (**19**) is produced by a number of *Alternaria* species and in *Phoma macdonaldii* Boerema.[144,145] It causes necrosis in affected tissues by an unknown mechanism. An analogue, porritoxin, produced by *A. porri* (Ellis) Cif. causes similar symptoms.[146] Zinniol binds to specific binding sites in carrot cells.[147] This binding is associated with stimulation of Ca^{2+} uptake by the cells at only 0.1–1 μmol dm^{-3} levels of zinniol. Calcium-regulated cell processes may be disrupted by zinniol. Zinniol is a relatively simple compound that could be the basis for herbicide-discovery QSAR studies.

The destruxins are produced by *Alternaria brassicae* (Berk.) Sacc. and *Metarrhizium anisopliae* (Metch.) Sorok.[148,149] These compounds have a much greater range of susceptible plant species than do the producing pathogens. Destruxin B causes necrotic and chlorotic symptoms in susceptible species.[149]

Some plant pathogens produce large peptides that are very phytotoxic to the infected plant. For example, *Trichoderma viride* Pers.:Fr. produces a 22 kD protein that causes severe phytotoxicity to certain tobacco varieties.[150] Only a small portion of the three-dimensional structure of the protein may be involved in its phytotoxicity. Modelling of the necessary steric and electrical charge distribution features of active portions of such large phytotoxins could result in synthesis of smaller chemical mimics that might have utility as herbicides.

These are but a small fraction of the known non-host-specific phytotoxins from plant pathogens. Very few of these compounds have been the focus of intensive study for their potential as herbicides or as leads for new herbicides. Relatively little is known of the molecular mechanism of action of these compounds. However, what little is known suggests that these compounds generally attack sites of action that are currently unexploited by the herbicide industry.

4.3.2.2 Host-Specific Phytotoxins

Host-specific phytotoxins are those that significantly affect only one plant species, the species to which the producing pathogen is a pathogen. Some host-specific phytotoxins are only toxic to certain cultivars of the host plant species. For example, *Alternaria alternata* (Fr.:Fr.) Keissl. f. sp. *lycopersici* is pathogenic to certain tomato cultivars with the recessive genotype *asc/asc–Asc/Asc* or *Asc/asc* genotypes are resistant.[151] AAL-toxin (**17**) produced by the fungus is responsible for the symptoms on susceptible tomato cultivars and does not affect resistant cultivars. All known host-specific phytotoxins are from fungal pathogens.

Since more than 99% of plant pathology studies have been with crop or ornamental plants as the host, the only reported host-specific phytotoxins until recently were with these species. Less than twenty host-specific phytotoxins that affect crops have been reported.[152] These phytotoxins have been reported to have exactly the same host range as the pathogens producing them. To our knowledge, the only host-specific toxin that affects crops and has been tested on weed species is AAL-toxin (Figure 4.4; **17**). These studies demonstrated that it is not a host-specific toxin after all and that it is highly phytotoxic to a wide range of weed species.[7,153–155] Destruxin B (discussed above) was reported to be host-specific;[156] however, it was later demonstrated to be non-host-specific.[149] Part of the confusion may be the result of inappropriate bioassays.[149] These findings indicate that other host-specific phytotoxins should be tested against a broad range of plant species with more than one bioassay.

AAL-toxin is a hydroxylated long-chain alkylamine with one tricarboxylic acid moiety attached. In fact, it is a close analogue of the fumonisins discussed below in Section 4.3.3. It is produced by *Alternaria alternata* f. sp. *lycopersici*. In susceptible varieties of tomatoes, it causes rapid wilting and necrosis. Its physiological effects appear to be identical to those of the fumonisins and, as such, are discussed below in Section 4.3.3.

The small amount of effort that has been expended to discover a host-specific toxin for a weed has yielded only one, maculosin (**23**).[157,158] Maculosin is a diketopiperazine with a phenolic moiety, apparently produced by the cyclization of a dipeptide of L-proline and L-tyrosine. It is produced by a strain of *Alternaria alternata* and is host-specific for spotted knapweed (*Centaurea* *maculosa* L.), a serious rangeland weed of the northwest states of the USA. It causes no phytotoxicity to a wide range of monocotyledonous and dicotyledonous crop and weed species at concentrations up to 1 μmol dm^{-3}. Its mechanism of action is unknown.

Bipolaroxin, from *Bipolaris cynodontis* (Marignoni) Shoemaker, a fungal pathogen of bermudagrass [*Cynodon dactylon* (L.) Pers.], is host-selective at low concentrations.[159] At concentrations 20 times larger than required to affect bermudagrass, it causes phytotoxicity symptoms to wild oats (*Avena fatua* L.), sugarcane (*Saccharum officinarum* L.), and maize.

Host-specific phytotoxins will only be of interest as potential products if their host range can be extended by manipulation of their chemical structure or if they are host-specific for extremely important weeds. The economics of production of a chemical herbicide that targets only one weed species is considered prohibitive by the pesticide industry.

4.3.3 Phytotoxins Produced by both Plant Pathogens and Non-Pathogenic Microbes

Several compounds of interest are produced by both plant pathogens and saprophytes. That some compounds are generated by such a wide range of organisms has only recently been recognized. An example of such a compound is fumonisin B_1, a phytotoxin produced by both saprophytic *Fusarium* species,[160,161] as well as by an *Alternaria alternata* strain known to cause stem canker disease on certain tomato varieties.[162]

Fumonisin B_1 (FB_1), first isolated from *F. moniliforme* J. Scheld. MRC 826 in 1988 by Bezuidenhout *et al.*,[163] is a hydroxylated, long-chain alkylamine with two tricarboxylic acid moieties attached. It is highly phytotoxic to most weed and crop species tested.[7,153,154,164,165] Ultrastructural and physiological studies of jimsonweed leaves with FB_1 revealed that it causes light-dependent plasma membrane and tonoplast disruption through an unknown mechanism.[153,154] We have demonstrated that structurally related fumonisins and AAL-toxin (**17**) have similar activity on plants, although AAL-toxin is generally more potent than FB_1.[154] The hydrolysis product, amino alcohols, of the fumonisins have very little phytotoxic effect.

FB_1 is also a potent mammalian toxin,[166,167] and in animal systems the mechanism of action is apparently altered sphingolipid synthesis.[168] The effect of FB_1 on sphingolipid synthesis is also likely to cause the observed membrane deterioration and dysfunction that we have described in plant cells. We have found sphingolipids themselves, at relatively high concentrations, to cause phytotoxicity symptoms similar to those of fumonisins and AAL-toxin (**17**).[154,169] For example, sphingosine and phytosphingosine required micromolar concentrations to cause effects similar to those of nanomolar levels of FB_1 and AAL-toxin. The mammalian toxicity of this compound would preclude its use as a herbicide. However, this research clearly demonstrates that the site of action of FB_1 is an excellent site of action for herbicides with less mammalian toxicity, which might be produced by biorational approaches.

Fusaric acid is produced by both the virulent plant pathogen *Fusarium oxysporum* Schlechtend.:Fr., the cause of wilt in several plant species, and by non-pathogenic *Fusarium* species that grow saprophytically on maize kernels.[170] Recently, fusaric acid has been demonstrated to have herbicidal activity against several weed species, including jimsonweed[169,171] and duckweed.[172]

Moniliformin can be obtained from isolates of saprophytic *Fusarium* species, as well as from pathogenic isolates of *Fusarium oxysporum*.[170] Moniliformin causes growth inhibition, necrosis, and chlorosis to many weed species.[159,171,172] However, it has no selectivity and is highly toxic to mammals.[10]

4.4 Development Considerations

4.4.1 Production Considerations

The cost of production of a compound is a critical factor in deciding whether to commercialize it and, if commercialized, in determining the profit margin. Many excellent herbicides have been abandoned because of the costs of synthesis. Microbial products offer two options – fermentation or synthesis. As mentioned above, the herbicide bilanafos is produced by fermentation in Japan, and a herbicidal derivative of bilanafos, glufosinate, is synthesized and sold throughout much of the world.

Biosynthesis by fermentation might be the only viable option for overly complicated compounds. However, one might argue that this method of production is only likely to be economical for specialized markets. The one successful case of biosynthesis, bilanafos, is for the specialized rice herbicide market of Japan, where farmers can afford to spend up to US\$200 ha^{-1} on weed management chemicals. Furthermore, there are many technical problems in fermentation, such as strain stability.

The methodology of secondary product biosynthesis by fermentation is a highly sophisticated technology, thanks to the pharmaceutical industry. Advances in biotechnology are improving the yield and quality of fermentation processes. For example, improved knowledge of the genetics of toxin production[100] will allow genetic manipulation of producing strains to increase greatly the toxin yield. Such advances will improve the prospects for production of microbial compounds for use as herbicides.

At present, chemical synthesis is likely to be the only viable option for many compounds. If the natural compound does not lend itself to economical synthesis, an active analogue may be more amenable. For example, the commercial herbicide methoxyphenone was the result of QSAR studies with anisomycin.[32,66]

4.4.2 Toxicological, Environmental, and Regulatory Aspects

Some of the most potent mammalian toxins are microbial products. This fact seems to be unknown or ignored by those who assume that natural products are intrinsically safer than synthetic compounds. Compounds that are handled in bulk and are to be sprayed into the environment should be subject to the same regulations, regardless of source. Nevertheless, microbial products do generally offer at least one toxicological and environmental advantage – a relatively short half-life.

The majority of synthetic herbicides are halogenated compounds. Although there are some halogenated natural products, these are relatively rare. Halogenated compounds generally are thought to persist longer *in vivo* and in the environment than are non-halogenated compounds. Of course, herbicides must persist long enough to kill the target weeds, and some microbial products may not fulfil this requirement. Nevertheless, one might generalize that significant contamination of food products, the soil, and water would be less likely with microbial compounds than with most synthetic herbicides used at the same rates.

4.5 Conclusions and Predictions

Microbial products offer, perhaps, the best readily accessible source of novel compounds with biological activity towards plants. Tremendous effort has been expended in chemically characterizing thousands of these compounds, yet comparatively little effort has been made to determine their herbicidal potential. Most of the information available on the phytotoxicity of microbial products is not useful in evaluating their potential as herbicides. In only a few cases in which such compounds have been found to be phytotoxic has a mechanism of action been determined. In the few cases in which a molecular target site has been established, it has generally been one that has not yet been exploited by the herbicide industry. The potential for future discovery is enormous.

Some microbial products will be useful without modification. The examples of bilanafos and phosphinothricin (glufosinate) provide ample evidence of the potential of already discovered compounds for development. The success of these products will probably further increase the interest in microbial products as herbicides. In other cases, modification of the structure of microbial phytotoxins will improve their utility as herbicides. Those compounds with prohibitively complex structures may still provide valuable clues to unexploited molecular sites of action for those involved in the biorational design of herbicides. Furthermore, structure–activity efforts might elucidate from the molecular complexity of microbial compounds more simple molecules that retain good herbicidal activity and can be produced effectively. A major advantage of these approaches is that they promise to produce new herbicides with previously unexploited molecular sites of action. The chances of finding new sites of action by exploitation of older chemical classes and by empirical screening are relatively low. With the rapidly increasing incidence of weeds becoming resistant to herbicides, new mechanisms of action will be a highly desirable trait in herbicides of the future.

There are, however, some formidable obstacles to success with such a strategy, including the structural complexity of many microbial toxins, the difficulties in working with very small amounts of materials, and the cost of rediscovering known compounds. The last two of these difficulties can be minimized with advances in technology.

4.6 References

1. Anonymous, *Agrow World Crop Protection News*, 1991, **135**, 18.
2. H. M. LeBaron and J. McFarland, Herbicide resistance in weeds and crops. An overview and prognosis, *Amer. Chem. Soc. Symp. Ser.*, 1990, **421**, 336.
3. J. Lydon and S. O. Duke, The potential of pesticides from plants, in 'Herbs, Spices, and Medicinal Plants – Recent Advances in Botany, Horticulture, and Pharmacology', ed. L. E. Craker and J. E. Simon, Oryx Press, Phoenix, AZ, 1989, vol. 4, p. 1.
4. S. O. Duke, Natural pesticides from plants, in 'Advances in New Crops', ed. J. Janick and J. E. Simon, Timber Press, Portland, OR, 1990, p. 511.
5. S. O. Duke, Plant terpenoids as pesticides, in 'Handbook of Natural Toxins, vol. 6, Toxicology of Plant and Fungal Compounds', ed. R. F. Keeler and A. T. Tu, Marcel Dekker, New York, 1991, p. 269.
6. S. O. Duke, Microbial phytotoxins as herbicides – a perspective in 'The Science of Allelopathy', ed. A. R. Putnam and C. S. Tang, John Wiley, New York, 1986, p. 287.
7. S. O. Duke, H. K. Abbas, C. D. Boyette, and M. Gohbara, Microbial compounds with the potential of herbicidal use, *Brighton Crop Prot. Conf. – Weeds*, 1991, vol. 1, 155.
8. H. Fischer and D. Bellus, Phytotoxicants from microorganisms and related compounds, *Pestic. Sci.*, 1983, **14**, 334.
9. H. G. Cutler, Perspectives on discovery of microbial phytotoxins with herbicidal activity, *Weed Technol.*, 1988, **2**, 525.
10. R. E. Hoagland, Microbes and microbial products as herbicides, *Amer. Chem. Soc. Symp. Ser.*, 1990, **439**, 1.
11. N. J. Poole and E. J. T. Chrystal, Microbial phytotoxins, *Brighton Crop Prot. Conf. – Weeds*, 1985, vol. 2, 591.
12. G. Strobel, D. Kenfield, G. Bunker, F. Sugawara, and J. Clardy, Phytotoxins as potential herbicides, *Experientia*, 1991, **47**, 819.
13. G. Strobel, F. Sugawara, and J. Hershenhorn, Pathogens and their products affecting weedy plants, *Phytoparasitica*, 1992, **20**, 307.
14. H. G. Cutler, Phytotoxins of microbial origin, in Handbook of Natural Toxins, vol. 6, 'Toxicology of Plant and Fungal Compounds', ed. R. F. Keeler and A. T. Tu, Marcel Dekker, New York, 1991, p. 411.
15. G. A. Strobel, D. Kenfield, G. Bunkers, and F. Sugawara, in Handbook of Natural Toxins, vol. 6, 'Toxicology of Plant and Fungal Compounds', ed. R. F. Keeler and A. T. Tu, Marcel Dekker, New York, 1991, p. 397.

16. Y. Sekizawa and T. Takematsu, How to discover new antibiotics for herbicida use, in 'Pesticide Chemistry, Human Welfare and the Environment', ed. N. Takahashi, H. Yoshioka, T. Misato and S. Matsunaka, Pergamon, Oxford, 1983, vol. 2, p. 261.
17. T. Misato and I. Yamaguchi, Pesticides of microbial origin, *Outlook Agric.*, 1984, **3**, 136.
18. S. O. Duke, Natural products as herbicides, *Proc. First International Weed Control Conference*, 1992, **1**, 302.
19. M. D. Devine, S. O. Duke, and C. Fedtke, 'Physiology of Herbicide Action', Prentice Hall, Englewood Cliffs, NJ, 1993, 441 pp.
20. S. O. Duke and J. Lydon, Natural phytotoxins as herbicides, *Amer. Chem. Soc. Symp. Ser.*, 1993, **524**, 110.
21. S. O. Duke and J. Lydon, Herbicides from natural compounds, *Weed Technol.*, 1987, **1**, 122.
22. S. O. Duke, Naturally-occurring chemical compounds as herbicides, *Rev. Weed Sci.*, 1986, **2**, 15.
23. K. Wright, J. B. Pillmoor, and G. G. Briggs, Rationality in herbicide design, in 'Topics in Photosynthesis – vol. 10, Herbicides', ed. N. R. Baker and M. P. Percival, Elsevier, Amsterdam, 1991, p. 337.
24. H. Müllner, P. Eckes, and G. Donn, Engineering crop resistance to the naturally occurring glutamine synthetase inhibitor phosphinothricin (glufosinate), *Amer. Chem. Soc. Symp. Ser.*, 1993, **524**, 38.
25. T. Tachibana and K. Kaneko, Development of a new herbicide, bialaphos, *J. Pestic. Sci.*, **11**, 297.
26. S. O. Duke, Overview of herbicide mechanisms of action, *Environ. Health Perspect.*, 1990, **87**, 263.
27. J. R. Bowyer, P. Camilleri, and W. F. J. Vermaas, Photosystem II and its interactions with herbicides, in 'Topics in Photosynthesis – vol. 10, Herbicides', ed. N. R. Baker and M. P. Percival, Elsevier, Amsterdam, 1991, p. 27.
28. U. B. Nandihalli and S. O. Duke, The porphyrin pathway as a herbicide target site, *Amer. Chem. Soc. Symp. Ser.*, 1993, **524**, 62.
29. M. Stidham, Herbicidal inhibitors of branched chain amino acid biosynthesis, in 'Topics in Photosynthesis – vol. 10, Herbicides', ed. N. R. Baker and M. P. Percival, Elsevier, Amsterdam, 1991, p. 247.
30. S. W. Ayer, B. G. Isaac, D. M. Krupa, K. E. Crosby, L. J. Letendre, and R. J. Stonard, Herbicidal compounds from microorganisms, *Pestic. Sci.*, 1989, **27**, 221.
31. R. M. Heisey, S. K. Mishra, A. R. Putnam, J. R. Miller, C. J. Whitenack, J. E. Keller, and J. Huang, Production of herbicidal and insecticidal metabolites by soil microorganisms, *Amer. Chem. Soc. Symp. Ser.*, 1988, **380**, 68.
32. K. Ito, F. Futatsuya, K. Hibi, S. Ishida, O. Yamada, and K. Munakata, Herbicidal activity of 3,3′-dimethyl-4-methoxybenzophenone (NK-049) in paddy field. 1. Herbicidal characteristics of NK-049 on weeds, *Weed Sci. (Jpn)*, 1974, **18**, 10.
33. U. B. Nandihalli, M. V. Duke, and S. O. Duke, Relationships between molecular properties and biological activities of *O*-phenyl pyrrolidino- and piperidino-carbamate herbicides, *J. Agric. Food Chem.*, 1992, **40**, 1993.
34. M. H. Julien (ed.), 'Biological Control of Weeds', CAB International, Oxon, 1992, 186 pp.
35. P. S. Zorner, S. L. Evans, and S. D. Savage, Perspectives on providing a realistic technical foundation for the commercialization of bioherbicides, *Amer. Chem. Soc. Symp. Ser.*, 1993, **524**, 79.

36. D. C. Sands, R. V. Miller, E. J. Ford, and K. A. Glass, Altering the host range of mycoherbicides by genetic manipulation, *Amer. Chem. Soc. Symp. Ser.*, 1993, **524**, 101.
37. G. Gohbara and K. Yamaguchi, Biological control agents for rice paddy weed management in Japan, *Proc. Internat. Symp. Biol. Cont. Integ. Manage. Paddy Aquatic Weeds in Asia*, in press.
38. C. D. Boyette, P. C. Quimby, W. J. Connick, D. J. Daigle, and F. E. Fulghum, Progress in the production, formulation, and application of mycoherbicides, in 'Microbial Control of Weeds', ed. D. O. TeBeest, Chapman and Hall, New York, 1991, p. 209.
39. D. J. Daigle and W. J. Connick, Formulation and application technology for microbial weed control, *Amer. Chem. Soc. Symp. Ser.*, 1990, **439**, 288.
40. C. G. McWhorter and J. E. Hanks, Spray equipment for applying herbicides in ultra-low volume, *Proc. First Internal, Weed Control. Cong. (Melbourne)*, 1992, **1**, 165.
41. S. Omura, M. Murata, H. Hanaki, K. Hinotozawa, R. Oiwa, and H. Tanaka, Phosalacine, a new herbicidal antibiotic containing phosphinothricin. Fermentation, isolation, biological activity and mechanism of action, *J. Antibiot.*, 1984, **37**, 829.
42. E. Bayer, K.K. Gugel, K. Kaegel, H. Hagenmaier, S. Jessipov, W. A. Konig, and H. Zähner, Stoffwechselprodukte von Mikroorganismen. Phosphinothricin und Phosphinothricyl-alanyl-alanin, *Helv. Chim. Acta*, 1972, **55**, 224.
43. M. Leason, D. Cunliffe, D. Parkin, P. J. Lea, and B. J. Miflin, Inhibition of pea leaf glutamine-synthetase by methionine sulfoximine, phosphinothricin, and other glutamate analogs. *Phytochemistry*, 1982, **21**, 855.
44. R. Manderscheid and A. Wild, Studies on the mechanism on inhibition by phosphinothricin of gluamine synthetase isolated from *Triticum aestivum* L., *J. Plant Physiol.*, 1986, **123**, 135.
45. K. Tachibana, T. Watanabe, Y. Sekizawa, and T. Takematsu, Action mechanism of bialaphos. 1. Inhibition of glutamine-synthetase and quantitative changes of free amino acids in shoots of bialaphos-treated Japanese barnyard millet, *J. Pestic. Sci.*, 1986, **11**, 27.
46. A. Wild and C. Ziegler, The effects of bialaphos on ammonium-assimilation and photosynthesis. I. Effect on the enzymes of ammonium assimilation, *Z. Naturforsch.*, 1989, **44c**, 97.
47. Y. Ogawa, H. Yoshida, S. Inouye, and T. Niida, Studies on a new antibiotic SF-1293. III. Synthesis of a new phosphorus containing amino acid, a component of antibiotic SF-1293, *Meiji Seika Kenkyu Nempo*, 1973, **13**, 49.
48. D. Komoßa and H. Sandermann, Jr, Plant metabolism of herbicides with C–P bonds: phosphinothricin, *Pestic. Biochem. Physiol.*, 1992, **43**, 95.
49. B. J. Shelp, C. J. Swanton, and J. C. Hall, Glufosinate (phosphinothricin) mobility in young soyabean shoots, *J. Plant Physiol.*, 1992, **139**, 626.
50. P. J. Lea, The inhibition of ammonia assimilation: a mechanism of herbicide action, in 'Topics in Photosynthesis – vol. 10, Herbicides', ed. N. R. Baker and M.P. Percival, Elsevier, Amsterdam, 1991, p. 267.
51. H. Sauer, A. Wild, and W. Rühle, The effect of phosphinothricin (glufosinate) on photosynthesis. I. Inhibition of photosynthesis and accumulation of ammonia, *Z. Naturforsch.*, 1987, **42c**, 270.
52. C. Ziegler and A. Wild, The effect of bialaphos on ammonium-assimilation and

photosynthesis. II. Effect on photosynthesis and photorespiration, *Z. Naturforsch.*, 1989, **44c**, 103.
53. C. Wendler, A. Putzer, and A. Wild, Effect of glufosinate (phosphinothricin) and inhibitors of photorespiration on photosynthesis and ribulose-1,5-bisphosphate carboxylase activity, *J. Plant Physiol.* 1992, **139**, 666.
54. S. Omura, K. Hinotozawa, N. Imamura, and M. Murata, The structure of phosalacine, a new herbicidal antibiotic containing phosphinothricin, *J. Antibiot.*, 1984, **37**, 939.
55. S. Omura, M. Murata, N. Imamura, Y. Iwai, H. Tanaka, A. Furusaki, and T. Matsumoto, Oxetin, a new antimetabolite from an actinomycete. Fermentation, isolation, structure and biological activity, *J. Antibiot.*, 1984, **37**, 1324.
56. D. G. Gilchrist, Molecular modes of action, in 'Toxins and Plant Pathogenesis', ed. J. M. Daly and B. J. Deverill, Academic Press, New York, 1983, p. 81.
57. D. R. Bush and P. J. Langston-Unkefer, Tabtoxinine-β-lactam transport into cultured corn cells: uptake via an amino acid transport system, *Plant Physiol.*, 1987, **85**, 845.
58. T. J. Knight, D. R. Bush, and P. J. Langston-Unkefer, Oats tolerant of *Pseudomonas syringae* pv. *tabaci* contain tabtoxinine-β-lactam-insensitive leaf glutamine synthetases, *Plant Physiol.*, 1988, **88**, 333.
59. A. Scacchi, R. Bortolo, G. Cassani, G. Pirali, and E. Nielson, Detection, characterization, and phytotoxic activity of the nucleoside antibiotics, blasticidin S and 5-hydroxymethylblasticidin S, *J. Plant Growth Regul.*, 1992, **11**, 39.
60. M. Iwata, T. Sasaki, H. Iwamatsu, S. Miyadoh, K. Tachibana, K. Matsumoto, T. Shomura, M. Sezaki, and T. Watanabe, A new antibiotic, SF 2494 (5′-*O*-sulfamoyltubercidin) produced by *Streptomyces mirabilis*, *Sci. Rep. Meija Seika Kaisha*, 1987, **26**, 17.
61. O. Yamada, A. Kurozumi, F. Futatsuya, K. Ito, S. Ishida, and K. Munakata, Studies on plant growth-regulating activities of anisomycin and toyocamycin, *Agric. Biol. Chem.*, 1974, **36**, 2013.
62. S. Murao and H. Hayahi, Gougerotin as a plant growth inhibitor from *Streptomyces* sp. No. 179, *Agric. Biol. Chem.*, 1983, **47**, 1135.
63. H. Dellweg, A. Henssen, J. Kurz, W. Pflueger, M. Schedel, and G. Vobis, Production of rodaplutin, a peptidynucleoside from *Nocardioides albus*, and its use as a microbicide, pesticide and herbicide. German Patent DE 3,529,733 (1987) CA 106:212589f, 1987.
64. M. Arai, T. Haneishi, A. Terahara, N. Kitahara, H. Kayamori, Y. Kondo, and K. Kawakubo, Antimicrobial and herbicidal compositions containing herbicidins. Japanese Patent 74,101,523 (1973) CA 83:2315u, 1973.
65. B. G. Isaac, S. W. Ayer, L. J. Letendre, and R. J. Stonard, Herbicidal nucleosides from microbial sources, *J. Antibiot.*, 1991, **44**, 729.
66. O. Yamada, A. Kurozumi, F. Futatsuya, K. Ito, S. Ishida, and K. Munakata, Studies on chlorosis-inducing activities and plant growth inhibition of benzophenone derivatives, *Agric. Biol. Chem.*, 1979, **43**, 1467.
67. M. Miller-Wideman, N. Makkar, M. Tran, B. Isaac, N. Biest, and R. Stonard, Herboxidiene, a new herbicidal substance from *Streptomyces chromofuscus* A7847: taxonomy, fermentation, isolation, physico-chemical and biological properties, *J. Antibiot.*, 1992, **45**, 914.
68. M. B. Isaac, S. W. Ayer, R. C. Elliott, and R. J. Stonard, Herboxidine: a potent phytotoxic polyketide from *Streptomyces* sp. A7847, *J. Org. Chem.*, 1992, **57**, 7220.

69. P. Babczinski, M. Dorgerloh, A. Löbberding, H-J. Santel, R. R. Schmidt, P. Schmitt, and C. Wünsche, Herbicidal activity and mode of action of vulgamycin, *Pestic. Sci.*, 1991, **33**, 439.
70. S. W. Ayer, B. G. Isaac, K. Luchsinger, N. Makkar, M. Tran, and R. J. Stonard, *cis*-2-amino-1-hydroxycyclobutane-1-acetic acid, a herbicidal antimetabolite produced by *Streptomyces rochei* A13018, *J. Antibiot.*, 1991, **44**, 1460.
71. M. Nakajima, K. Itoi, Y. Takamatsu, S. Sato, Y. Furukawa, K. Furuya, T. Honma, J. Kodotani, M. Kozasa, and T. Haneishi, Cornexistin: a new fungal metabolite with herbicidal activity, *J. Antibiot.*, 1989, **44**, 1065.
72. T. Amagasa, R. N. Paul, J. J. Heitholt, and S. O. Duke, Physiological effects of cornexistin on *Lemna pansicostata*, *Pestic. Biochem. Physiol.*, 1994, **49**, 37.
73. T. Nishino and S. Murao, Isolation and some properties of an aspartate amino transferase inhibitor, gostatin, *Agric. Biol. Chem.*, 1983, **47**, 1961.
74. S. Fushimi, S. Nishikawa, N. Mito, M. Ikemoto, M. Sasaki, and H. Seto, Studies on a new herbicidal antibiotic, homoalanosine, *J. Antibiot.*, 1989, **42**, 1370.
75. K. Kobinata, S. Sekido, M. Uramoto, M. Ubukata, H. Osada, I. Yamaguchi, and K. Isono, Isoxazole-4-carboxylic acid as a metabolite of *Streptomyces* sp. and its herbicidal activity, *Agric. Biol. Chem.*, 1991, **55**, 1415.
76. B. G. Isaac, S. W. Ayer, and R. J. Stonard, Arabenoic acid, a natural product herbicide of fungal origin, *J. Antibiot.*, 1991, **44**, 793.
77. B. G. Isaac, W. W. Ayer, and R. J. Stonard, The isolation of α-methylene-β-alanine, a herbicidal microbial metabolite, *J. Antibiot.*, 1991, **44**, 795.
78. M. Graber, C. H. June, L. W. Samuelson, and A. Weiss, The protein tyrosine kinase inhibitor herbimycin A, but not genistein, specifically inhibits signal transduction by the T cell antigen receptor, *Int. Immunol.*, 1992, **11**, 1201.
79. R. M. Heisey and A. R. Putnam, Herbicidal effects of geldanamycin and nigericin antibiotics from *Streptomyces hygroscopicus*, *J. Nat. Prod.*, 1986, **49**, 859.
80. R. M. Heisey and A. R. Putnam, Herbicidal activity of the geldanamycin and nigericin, *J. Plant Growth Regul.*, 1990, **9**, 19.
81. N. Shavit and A. San Pietro, K^+-dependent uncoupling of photophosphorylation by nigericin, *Biochem. Biophys. Res. Commun.*, 1967, **28**, 277.
82. H. Sze, Nigericin-stimulated ATPase activity in microsomal vesicles of tobacco callus, *Proc. Natl. Acad. Sci. USA*, 1989, **77**, 5904.
83. T. Kida, T. Kishikawa, and H. Shibai, Isolation of two streptomycin-like antibiotics, Nos. 6241-A and B, as inhibitors of *de novo* starch synthesis and their herbicidal activity, *Agric. Biol. Chem.*, 1985, **49**, 1839.
84. A. Feld, K. Kobek, and H. K. Lichtenthaler, Inhibition of *de novo* fatty-acid biosynthesis in isolated chloroplasts by different antibiotics and herbicides, *Z. Naturforsch.*, 1989, **44c**, 976.
85. W. Oettmeier, D. Godde, B. Kunze, and G. Höfle, Stigmatellin, a dual type inhibitor of photosynthetic electron transport, *Biochim, Biophys. Acta*, 1985, **807**, 216.
86. W. Oettmeier, R. Dostantni, C. Majewski, G. Höflc, T. Fecker, B. Kunze, and H. Reichenbach, The aurachins, naturally occurring inhibitors of photosynthetic electron flow through photosystem II and the cytochrome b_6/*f*-complex, *Z. Naturforsch.*, 1990, **45c**, 322.
87. C. P. Mason, K. R. Edwards, R. E. Carlson, J. Pignatell, F.K. Gleason, and J.M. Wood, Isolation of chlorine-containing antibiotic from the fresh water cyanobacterium, *Scytonema hofmani*, *Science*, 1982, **215**, 400.

88. E. M. Gross, C. P. Wolk, and F. Jüttner, Fischerellin, a new allelochemical from the cyanobacterium *Fischerella muscicola*, *J. Phycol.*, 1991, **27**, 686.
89. F. K. Gleason, D. E. Case, K. D. Sipprell, and T. S. Magnuson, Effects of the natural algacide, cyanobacterin, on a herbicide-resistant mutant of *Anacystic nidulans* R2, *Plant. Sci.*, 1986, **46**, 5.
90. J. L. Carlson, T. A. Leaf, and F. K. Gleason, Synthesis and activity of analogs of the natural herbicide cyanobacterin, *Amer. Chem. Soc. Symp. Ser.*, 1987, **355**, 141.
91. A. Ballio, D. Barra, F. Bossa, J. E. DeVay, I. Grgurina, N. S. Iacobellis, G. Marino, P. Pucci, M. Simmaco, and G. Surico, Multiple forms of syringomycin, *Physiol. Mol. Plant Pathol.*, 1988, **33**, 493.
92. A. Isogai, N. Fukuchi, S. Yamashita, K. Suyama, and A. Suzuki, Structures of syringostatin A and syringostatin B, novel phytotoxins produced by *Pseudomonas syringae* pv. *syringae* isolated from lilac blights, *Tetrahedron Lett.*, 1990, **31**, 695.
93. A. Ballio, F. Bossa, A. Collina, M. Gallo, N. S. Iacobellis, M. Paci, P. Pucci, A. Scaloni, A. Segre, and M. Simmaco, Structure of syringotoxin, a bioactive metabolite of *Pseudomonas syringae* pv. *syringae*, *FEBS Lett.*, 1990, **269**, 377.
94. A. Ballio, D. Barra, A. Collina, I. Grgurina, G. Marino, G. Moneti, M. Paci, P. Pucci, A. Segre, and M. Simmaco, Syringopeptins, new phytotoxic lipodepsipeptides of *Pseudomonas syringae* pv. *syringae*, *FEBS Lett.*, 1991, **291**, 109.
95. F-S. Che, K. Kasamo, N. Fukuchi, A. Isogai, and A. Suzuki, Bacterial phytotoxins, syringomycin, syringostatin and syringotoxin, exert their effect on the plasma membrane H^+-ATPase partly by a detergent-like action and partly by inhibition of the enzyme, *Physiol. Plant*, 1992, **86**, 518.
96. N. S. Iacobellis, P. Lavermicocca, I. Grgurina, M. Simmaco, and A. Ballio, Phytotoxic properties of *Pseudomonas syringae* pv. *syringae* toxins, *Physiol. Mol. Plant Pathol.*, 1992, **40**, 107.
97. R. E. Mitchell, Isolation and structure of a chlorosis inducing toxin of *Pseudomonas phaseolicaola*, *Phytochemistry*, 1976, **15**, 1941.
98. R. E. Mitchell, and R. L. Beilski, Involvement of phaseolotoxin in halo blight of beans: transport and conversion to functional toxin, *Plant Physiol.*, 1977, **60**, 723.
99. R. E. Moore, W. P. Niemczura, O. C. H. Kwok, and S. S. Patil, Inhibitors of ornithine carbamoyltransferase from *P. syringae* pv. *phaseolicola*. Revised structure of phaseolotoxin, *Tetrahedron Lett.*, 1984, **25**, 3931.
100. D. K. Willis, T. M. Barta, and T. G. Kinscherf, Genetics of toxin production and resistance to phytopathogenic bacteria, *Experientia*, 1991, **47**, 765.
101. J. Kenyon and J. G. Turner, Physiological changes in *Nicotiana tabacum* leaves during development of chlorosis caused by coronatine, *Physiol. Molec. Plant Pathol.*, 1990, **37**, 463.
102. J. H. Lukens and R. D. Durbin, Tagetitoxin affects plastid development in seedling leaves of wheat, *Planta*, 1985, **165**, 311.
103. D. E. Mathews and R. D. Durbin, Tagetitoxin inhibits RNA synthesis directed by RNA polymerases from chloroplasts and *Escherichia coli.*, *J. Biol. Chem.*, 1990, **265**, 493.
104. T. H. Steinberg, D. E. Mathews, R. D. Durbin, and R. R. Burgess, Tagetitoxin: a new inhibitor of eukaryotic transcription by RNA polymerase III, *J. Biol. Chem.*, 1990, **265**, 499.
105. M. Nukina and T. Namai, Productivity of pyrichalasin H, a phytotoxic metabolite, from different isolates of *Pyricularia grisea* and from isolates of *Pyricularia* spp., *Agric. Biol. Chem.*, 1991, **55**, 1899.

106. M. Nukina, Pyrichalsasin H, a new phytotoxic metabolite belonging to the cytochalains from *Pyricularia grisea* (Cooke) Saccardo, *Agric. Biol. Chem.*, 1987, **51**, 2625.
107. G. Nasini, R. Locci, L. Camarda, L. Merlini, and G. Nasini, Screening of the genus *Cercospora* for secondary metabolites, *Phytochemistry*, 1977, **16**, 243.
108. H. Tabuchi, A. Tajimi, and A. Ichihara, (+)-Isocercosporin, a phytotoxic compound isolated from *Scolecotrichum graminis* Fuckel, *Agric. Biol. Chem.*, 1991, **55**, 2675.
109. P. E. Hartman, W. J. Dixon, T. A. Dahl, and M. E. Daub, Multiple modes of photodynamic action by cercosporin, *Photochem. Photobiol.*, 1988, **47**, 699.
110. J. P. Knox and A. D. Dodge, Isolation and activity of the photodynamic pigment hypercin, *Plant Cell Environ.*, 1985, **8**, 19.
111. R. W. Jones, W. T. Lanini, and J. G. Hancock, Plant growth response to the phytotoxin viridiol produced by the fungus *Gliocladium virens*, *Weed Sci.*, 1988, **36**, 683.
112. C. R. Howell and R. D. Stipanovic, Phytotoxicity to crop plants and herbicidal effects on weeds of viridiol produced by *Gliocladium virens*, *Phytopathology*, 1984, **74**, 1346.
113. P. Venkatasubbaiah and W. S. Chilton, Phytotoxins of *Ascochyta hyalosopora*, causal agent of lambsquarters leaf spot, *J. Nat. Prod.*, 1992, **55**, 461.
114. P. Venkatasubbaiah, A. B. A. M. Baudoin, and W. S. Chilton, Leaf spot of hemp dogbane caused by *Stagonospora apocyni*, and its phytotoxins, *J. Phytopath.*, 1992, **135**, 309.
115. S. O. Duke, M. Gohbara, R. N. Paul, and M. V. Duke, Colletotrichin causes rapid membrane damage to plants, *J. Phytopath.*, 1992, **134**, 289.
116. M. Gohbara, Y. Kosuge, S. Yamaaki, Y. Kimura, A. Suzuki, and S. Tamura, Isolation, structures and biological activities of colletotrichins, phytotoxic substances from *colletotrichum nicotianae*, *Agric. Biol. Chem.*, 1978, **42**, 1037.
117. J. M. Jacobs, N. J. Jacobs, T. D. Sherman, and S. O. Duke, Effect of diphenyl ether herbicides on oxidation of protoporphyrinogen to protoporphyrin in organellar and plasma membrane-enriched fractions of barley, *Plant Physiol.*, 1991, **97**, 197.
118. A. R. Lax and H. S. Shepherd, Tentoxin: a cyclic tetrapeptide having potential herbicidal usage, *Amer. Chem. Soc. Symp. Ser.*, 1988, **380**, 24.
119. R. D. Durbin and T. F. Uchytil, A survey of plant insensitivity to tentoxin, *Phytopathology*, 1977, **67**, 602.
120. L. G. Burk and R. D. Durbin, The reaction of *Nicotiana* species to tentoxin, *J. Hered.*, 1978, **69**, 117.
121. A. Avni, J. D. Anderson, N. Holland, J-D. Rochaix, A Gromet-Elhanan, and M. Edelman, Tentoxin sensitivity of chloroplasts determined by codon 83 of β subunit of proton-ATPase, *Science*, 1992, **257**, 1245.
122. A. R. Lax, K. C. Vaughn, and G. E. Templeton, Nuclear inheritance of polyphenol oxidase in *Nicotiana*, *J. Hered.*, 1984, **75**, 285.
123. S. O. Duke, R. N. Paul, and J. L. Wickliff, Tentoxin effects on ultrastructure and greening of ivyleaf morningglory [*Ipomoea hederacea* (L.) var *Hederacea*] cotyledons, *Physiol. Plant*, 1980, **49**, 27.
124. C. J. Arntzen, Inhibition of photophosphorylation by tentoxin, a cyclic tetrapeptide, *Biochim. Biophys. Acta*, 1972, **283**, 539.
125. J. A. Steele, T. F. Uchytil, R. D. Durbin, P. Bhatnager, and D. H. Rich,

Chloroplast coupling factor 1: A species-specific receptor for tentoxin, *Proc. Natl. Acad. Sci. USA*, 1978, **73**, 2245.

126. K. C. Vaughn and S. O. Duke, Tentoxin-induced loss of plastidic polyphenol oxidase, *Physiol. Plant*, 1981, **53**, 421.
127. K. C. Vaughn and S. O. Duke, Tentoxin effects on *Sorghum*: the role of polyphenol oxidase, *Protoplasma*, 1982, **110**, 48.
128. K. C. Vaughn and S. O. Duke, Tentoxin stops the processing of polyphenol oxidase into an active enzyme, *Physiol. Plant*, 1984, **60**, 257.
129. M. G. Klotz, The action of tentoxin on membrane processes in plants, *Physiol. Plant*, 1988, **74**, 575.
130. J. L. Wickliff, S. O. Duke, and K. C. Vaughn, Involvement of photobleaching and inhibition of protochlorophyll(ide) accumulation in tentoxin effects on greening mung bean seedlings, *Physiol. Plant*, 1982, **56**, 399.
131. S. O. Duke, J. L. Wickliff, K. C. Vaughn, and R. N. Paul, Tentoxin does not cause chlorosis in greening mung bean leaves by inhibiting photophosphorylation, *Physiol. Plant*, 1982, **56**, 387.
132. A. R. Lax and K. C. Vaughn, Lack of correlation between effects of tentoxin on chloroplast coupling factor and chloroplast ultrastructure, *Physiol. Plant*, 1986, **66**, 384.
133. S. O. Duke, Tentoxin effects on variable fluorescence and P515 electrochromic absorbance changes in tentoxin-sensitive and -resistant plant species, *Plant Sci.*, 1993, **90**, 119.
134. K. C. Vaughn and S. O. Duke, Transport of proteins into the chloroplast, in 'Models in Plant Physiology and Biochemistry', ed. D. W. Newman and K. G. Wilson, CRC Press, Boca Raton, Fl, 1987, p. 69.
135. A. R. Lax, H. S. Shepherd, and J. V. Edwards, Tentoxin, a chlorosis-inducing toxin from *Alternaria* as a potential herbicide, *Weed Technol.*, 1988, **2**, 540.
136. S. O. Duke and A. D. Lane, Phytochrome control of its own accumulation and leaf expansion in tentoxin- and norflurazon-treated mung bean seedlings, *Physiol. Plant*, 1984, **60**, 341.
137. S. O. Duke, K. C. Vaughn, and R. L. Meeusen, Mitochondrial involvement in the mode of action of acifluorfen, *Pestic. Biochem. Physiol.*, 1984, **21**, 368.
138. S. O. Duke and K. C. Vaughn, Lack of involvement of polyphenol oxidase in ortho-hydroxylation of phenolic compounds in mung bean seedlings, *Physiol. Plant*, 1982, **54**, 381.
139. L. P. Lehnen, T. D. Sherman, J. M. Becerril, and S. O. Duke, Tissue and cellular localization of acifluorfen-induced porphyrins in cucumber cotyledons, *Pestic. Biochem. Physiol.*, 1990, **37**, 239.
140. S. O. Duke, M. V. Duke, T. D. Sherman, and U. B. Nandihalli, Spectrophotometric and spectrofluorometric methods in weed science, *Weed Sci.*, 1991, **39**, 505.
141. A. Stierle, R. Upadhyay, and G. Strobel, Cyperine, a phytotoxin produced by *Ascochyta cypericola*, a fungal pathogen of *Cyprus rotundus*, *Phytochemistry*, 1991, **30**, 2191.
142. J. M. Jacyno, H. G. Cutler, R. G. Roberts, and R. M. Waters, Effects on plant growth of the HMG-CoA synthase inhibitor, 1233A/F244/L-659,699, isolated from *Scopulariopsis candidus*, *Agric. Biol. Chem.*, 1991, **55**, 3129.
143. M. D. Greenspan, J. B. Yudkovitz, C-Y. L. Lo, J. S. Chen, A. W. Alberts, V. M. Hunt, M. N. Chang, S. S. Yang, K. L. Thompson, Y-C. P. Chiang, J. C. Chabala, R. L. Monaghan and R. L. Schwartz, Inhibition of hydroxymethylglutaryl-coenzyme A synthase by L-659,699, *Proc. Natl. Acad. Sci. USA*, 1987, **84**, 7488.

144. K. G. Tietjen, E. Schaller, and U. Matern, Phytotoxins from *Alternaria carthami* Chowdhury: structural identification and physiological significance, *Physiol. Plant Pathol.*, 1983, **23**, 387.
145. F. Sugawara and G. Strobel, Zinniol, a phytotoxin, is produced by *Phoma macdonaldii*, *Plant Sci.*, 1986, **43**, 19.
146. R. Suemitsu, K. Ohnishi, M. Horiuchi, A. Kitaguchi, and K. Odamura, Porritoxin, a phytotoxin of *Alternaria porri*, *Phytochemistry*, 1992, **31**, 2325.
147. P. Thuleau, A. Graziana, M. Rossignol, H. Kauss, P. Auriol, and R. Ranjeva, Binding of the phytotoxin zinniol stimulates the entry of calcium into plant protoplasts, *Proc. Natl. Acad. Sci. USA*, 1988, **85**, 5932.
148. L. Buchwaldt and J. S. Jenson, HPLC purification of destructins produced by *Alternaria brassicae* in culture and in leaves of *Brassica napus*. Assignment of the ^{1}H- and ^{13}C-NMR spectra by 1D- and 2D-techniques, *Phytochemistry*, 1991, **30**, 2311.
149. L. Buchwaldt and H. Green, Phytotoxicity of destructin B and its possible role in the pathogenesis of *Alternaria brassicae*, *Plant Pathol.*, 1992, **41**, 55.
150. B. A. Bailey, J. F. D. Dean, and J. D. Anderson, An ethylene biosynthesis-inducing endoxylanase elecits electrolyte leakage and necrosis in *Nicotiana tabacum* cv. Xanthi leaves, *Plant Physiol.*, 1990, **94**, 1849.
151. H. K. Abbas, T. Tanaka, and S. O. Duke, Pathogenicity of *Alternaria alternata* and *Fusarium moniliforme* and phytotoxicity of AAL-toxin and a fumonisin B_1 on tomato varieties, *J. Phytopath.*, 1995, **143**, 329.
152. R. P. Scheffer and R. S. Livingston, Host-selective toxins and their role in plant diseases, *Science*, 1984, **112**, 17.
153. H. K. Abbas, R. N. Paul, C. D. Boyette, S. O. Duke, and R. F. Vesonder, Physiological and ultrastructural effects of fumonisin on jimsonweed leaves, *Can. J. Bot.*, 1992, **70**, 1824.
154. T. Tanaka, H. K. Abbas, and S. O. Duke, Structure-dependent phytotoxicity of fumonisins and related compounds in a duckweed (*Lemna pausicostata* L.) bioassay, *Phytochemistry*, 1993, **33**, 779.
155. H. K. Abbas, R. F. Vesonder, C. D. Boyette, and S. W. Peterson, Phytotoxicity of AAL-toxin and other compounds produced by *Alternaria alternata* to jimsonweed (*Datura stramonium*), *Can. J. Bot.*, 1993, **71**, 155.
156. P. S. Bains and J. P. Tewari, Purification, chemical characterization and host-specificity of the toxin produced by *Alternaria brassicae*, *Physiol. Molec. Plant Pathol.*, 1987, **30**, 259.
157. G. Strobel, A. Stierle, S. H. Park, and J. Cardellina, Maculosin: a host-specific phytotoxin from *Alternaria alternata* on spotted knapweed, *Amer. Chem. Soc. Symp. Ser.*, 1990, **439**, 53.
158. A. Stierle, J. H. Cardellina, and G. Strobel, Maculosin, a host-specific phytotoxin for spotted knapweed from *Alternaria alternata*, *Proc. Natl. Acad. Sci. USA*, 1989, **85**, 8008.
159. R. Sugawara, G. Strobel, L. E. Fisher, G. D. Van Duyne, and J. Clardy, Bipolaroxin, a selective phytotoxin produced by *Bipolaris cynodontis*, *Proc. Natl. Acad. Sci. USA*, 1985, **82**, 829.
160. H. K. Abbas, C. D. Boyette, R. E. Hoagland, and R. F. Vesonder, Bioherbicidal potential of *Fusarium moniliforme* (Sheldon) and its phytotoxin fumonisin, *Weed Sci.*, 1991, **39**, 673.
161. H. K. Abbas, R. F. Vesonder, C. D. Boyette, R. E. Hoagland, and T. Krick,

Production of fumonisins by *Fusarium moniliforme* cultures isolated from Jimsonweed in Mississippi, *J. Phytopathol.*, 1992, **136**, 199.

162. J. Chen, C. J. Miroscha, W. Xie, L. Hogge, and D. Olson, Production of the mycotoxin fumonisin B_1 by *Alternaria alternata* f. sp. *lycopersici*, *Appl. Environ. Microbiol.*, 1992, **58**, 3928.
163. S. C. Bezuidenhout, W. C. A. Gelderblom, C. P. Gorst-Allman, R. M. Horak, W. F. O. Marasas, G. Spiteller, and R. Vleggaar, Structure elucidation of the fumonisins, mycotoxins from *Fusarium moniliforme*, *J. Chem. Soc., Chem. Commun.*, 1988, 743.
164. M. A. J. Van Asch, F. H. J. Rijkenberg, and T. A. Coutinho, Phytotoxicity of fumonisin B_1, moniliformin, and T-2 toxin to corn callus cultures, *Phytopathology*, 1992, **82**, 1330.
165. H. K. Abbas and C. D. Boyette, Phytotoxicity of fumonisin on weed and crop species, *Weed Technol.*, 1992, **6**, 548.
166. W. T. Shier, H. K. Abbas, and C. J. Mirocha, Toxicity of the mycotoxins fumonisins B_1 and B_2 and *Alternaria alternata* f. sp. *lycopersici* toxin (AAL) in cultured mammalian cells, *Mycopathologia*, 1991, **116**, 97.
167. H. K. Abbas, W. C. A. Gelderblom, M. F. Cawood, and W. T. Shier, Biological activities of fumonisins, mycotoxins from *Fusarium moniliforme*, in jimsonweed (*Datura stramonium* L.) and mammalian cell cultures, *Toxicon*, 1993, **31**, 345.
168. E. Wang, W. P. Norred, C. W. Bacon, R. T. Riley, and A. H. Merrill, Inhibition of sphingolipid biosynthesis by fumonisins, *J. Biol. Chem.*, 1991, **266**, 14486.
169. R. F. Vesonder, R. E. Peterson, D. Labeda, and H. K. Abbas, Comparative phytotoxicity of the fumonisins, AAL-toxin and yeast sphingolipids in *Lemna minor* L. (duckweed), *Arch. Environ. Contam. Toxicol.*, 1992, **23**, 464.
170. H. K. Abbas, C. J. Mirocha, T. Kommedahl, R. F. Vesonder, and P. Golinski, Production of trichothecene and non-trichothecene mycotoxins by *Fusarium* species isolated from maize in Minnesota, *Mycopathologia*, 1989, **108**, 55.
171. H. K. Abbas, C. D. Boyette, and R. E. Hoagland, Phytotoxicity of *Fusarium*, other fungal isolates, and of their phytotoxins fumonisin, fusaric acid and moniliformin to jimsonweed, *Phytoprotection*, 1995, **76**, 17.
172. R. F. Vesonder, D. P. Labeda, and R. E. Peterson, Phytotoxic activity of selected water-soluble metabolites of *Fusarium* against *Lemna minor* L. (duckweed), *Mycopathologia*, 1992, **118**, 185.

CHAPTER 5

Biologically Active Compounds from Algae

GRAHAM K. DIXON

And Moses and Aaron did so, as the Lord commanded; and he lifted up the rod, and smote the waters that were in the river, in the sight of Pharaoh, and in the sight of his servants; and all the waters that were in the river were turned to blood. And the fish that was in the river died; and the river stank, and the Egyptians could not drink of the water of the river; and there was blood throughout all the land of Egypt.

Exodus 7; 20,21

5.1 Introduction

This excerpt from the bible is thought to be the earliest documentation of a bloom of 'toxic' algae, a so-called 'red tide', associated with the death of aquatic organisms. Examples of this phenomenon have been reported on numerous occasions throughout the world.[1–4] The majority of toxin-producing algae are found in three of the 15 divisions as classified by Lee.[5] These are the Prymnesiophyta, Cyanophyta, and Dinophyta, the last division producing toxins which are among the most potent non-proteinaceous lethal materials known.[2]

Toxins are not the only extracellular products produced by algae. Other substances, including carbohydrates, lipids, peptides, organic phosphates, volatile substances, vitamins, and antibiotics have been identified.[6,7] The nature of algal extracellular products and their release has been discussed in detail by Fogg,[8–11] Hellebust,[12] Aaronson *et al.*,[13] Jensen,[14] and Jones.[15]

Some of these extracellular products, including toxins often referred to as 'external metabolites' or 'ectocrines' (exocrines) are known to possess antibacterial, antiviral, or growth-inhibiting activity[16–18] which may affect other organisms at the same or higher trophic levels.[19] Jones[20] concluded that interactions between mixtures of algae and bacteria in the laboratory could be stimulatory, antagonistic, and competitive. These relationships appear to be influenced by the algal extracellular products in a zone around the algae termed the 'phycosphere'. Other active substances, called 'endotoxins', are synthesized and retained within the algal cells, and exert their effects when consumed and passed through the food chain or when released into the water by cell breakage. Biologically active algal products have been known for a long time, but only relatively recently has their potential importance in trophic relationships and algal ecology been discussed.[15,19]

The variety of compounds produced by algae and the diversity which exists in the algal division, consisting of some 30,000 known species, gives tremendous potential for commercial exploitation. Developing areas of commercial interest for algal products are shown in Table 5.1. However, the algal products industry of today has concentrated on two main areas: first, farming of edible seaweeds, for example *Porphyra* (Nori), *Laminaria* (Kombo) and *Undaria* (Wakame), which has an annual turnover of approximately $1.5 billion,[21] and, second, the fine chemicals and polysaccharide phycocolloids industry. For the past 60 years, the production of alginates, carageenan, and agar (used as emulsifying and gelling agents, stabilizers, and thickeners for the food and pharmaceutical industries) from members of the Phaeophyta and Rhodophyta have dominated this latter category, with an annual turnover in the region of $250 million.[22]

To date, there has been little commercial development of algal secondary metabolites as pharmaceutical and/or agricultural compounds, even though these two areas constitute large potential markets (>$200 billion and approximately $21 billion, respectively[23]).

Knowledge of the range of biological activities ascribed to algal compounds has increased tremendously in recent years, and in this chapter the chemical nature, structure, and mode of action of these compounds is reviewed. However, basic classification and isolation and culturing techniques are also

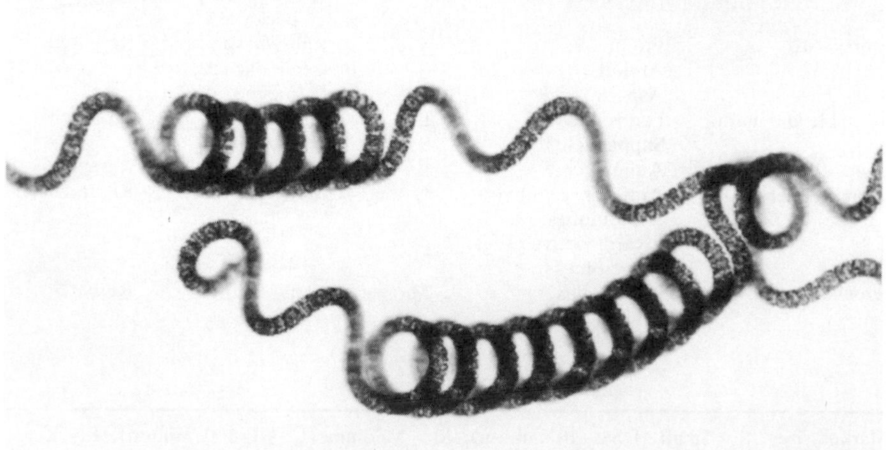

Figure 5.1 Spirulina *sp. (×190)*

Table 5.1 *Commercial algal products and areas under research (adapted from Benneman* et al.[24]*)*

Products	*Uses*	*Approx. Market**	*Algal genus or Type*	*Current status*
Radiolabelled compounds	Medicinal research	S	Many	Commercial
Vitamins	Dietary supplements	M	Green algae	Research
Health foods	Supplements	M–L	*Chlorella* sp., *Spirulina* sp. (Figure 5.1)	Commercial
Restriction enzymes	Research	S	Cyanobacteria	Commercial
Lipids				
1. Eicosapentaenoid acid	Health foods, dietary supplements, substrate for synthesis in pharmaceutical industry	M–L	*Chlorella* sp.	Research
2. γ-Linolenic acid			*Spirulina* sp.†	Research
2. Arachadonic acid			*Porphyridium cruentum*	Research
Pigments				
1. β-Carotene	Food supplement	S	*Dunaliella* sp.	Commercial
	Food colouring	M		
2. Xanthophylls	Chicken feed	M	Green algae diatoms	Research
3. Phycobiliproteins	Research: fluorescent dyes, immuno-chemical reagents, clinical diagnostics	U	Cyanobacteria, cryptomonads, and red algae	Research
	Food colouring	S	Cyanobacteria, *Spirulina* sp.,† red algae, and cryptomonads	Commercial
Polysaccharides	Viscosifiers, gums	M–L	*Porphyridium* sp.	Research
Soil inoculum	Conditioner, fertilizers	U	*Chlamydomonas* sp., N-fixing species	Commercial, research
Amino acids	Proline	S	*Chorella* sp.	Research
	Arginine	S	Cyanobacteria	
	Aspartic acid	L	Cyanobacteria	
Vegetable and marine oils	Foods, feeds	L	Green algae	Research
	Supplements	S	Diatoms	
Single cell protein	Animal feeds	L	Green algae	Research
Pharmaceuticals	Anticancer, antiviral antibiotics, cardioactive drugs, and others	L	Many	Research
Agrochemicals	Insecticides, fungicides, herbicides, and plant growth regulators	M	Many	Research

*Market sizes: S = Small (US$1–10 million); M = Medium (US$10–100 million); L = Large (US$ > 100 million); U = Unknown.
†See Figure 5.1

introduced to give the reader a broader understanding of algal biology. The methods used in screening biologically active compounds are also described briefly.

5.2 Basic Characteristics of Algae

Phycology (from the Greek word *Phykos* meaning 'seaweed'), or algology, is the study of algae. The algae are thallophytes (plants lacking roots, stems, and leaves) that have chlorophyll *a* as their primary photosynthetic pigment.

Most algae are aquatic, either freshwater or marine. However, they are ubiquitous and can be found in almost every other environment on Earth. The aquatic forms are principally free-living, although some occur growing in symbiotic associations with invertebrate animals (*e.g.*, sponges and corals) and fungi (*e.g.*, lichens). Photoautotrophic algae are of fundamental importance environmentally, first through production of oxygen in photosynthesis (which is essential for all aerobic forms of life) and, second, by functioning as primary producers in the food chain, producing organic material from sunlight, carbon dioxide, and water. Other types of nutrition, however, are also found among the algae (Table 5.2).

Table 5.2 *Types of nutrition found in algae*

Type of nutrition	*Principal energy source for growth*	*Principal carbon source for growth*
Autotrophic		
Photoautotrophic	Light	Carbon dioxide
Chemoautotrophic	Oxidation of organic compounds	Carbon dioxide
Heterotrophic		
Photoheterotrophic	Light	Organic compounds
Chemoheterotrophic	Oxidation of organic compounds	Organic compounds

Marine algae are of particular importance, since their total mass (and consequently gross photosynthetic activity) is at least equal to that of all land plants. This is by no means evident, because the most conspicuous of marine algae, the seaweeds or macroalgae, occupy a very limited area of the oceans, being attached to rocks in the intertidal zone and the shallow coastal waters of the continental shelves. The great bulk of the algae, however, are unicellular floating organisms which are distributed through the surface water of lakes and oceans. These algae are termed phytoplankton or microalgae and can be subdivided further according to their size (Table 5.3).

It is only recently that sampling, isolation, and culturing techniques have been developed to allow detailed study of the picoplankton, which are by far the most abundant of all phytoplankton (usually in the region 10^4 cells ml^{-1} of water).[25]

Table 5.3 *Size grading of the phytoplankton*

Size range (μm)	*Category*
>200	Macroplankton
20–200	Microplankton
10–20	Nanoplankton
2–10	Ultraplankton
0.2–2	Picoplankton

Although phytoplankton are present throughout the year, the seasonal successions which occur are not always evident. However, during periods when suitable nutrient (*e.g.*, nitrogen, phosphate) and physical (*e.g.*, turbulence, thermal stratification) conditions exist, massive local growths or blooms of certain algae can occur. These blooms can result in discolouration of the water, for example the red tides mentioned earlier produced by some marine dinoflagellates. The physicochemical factors affecting the growth of phytoplankton have been discussed in detail.[3,26–29] As mentioned previously, the production of biologically active substances by phytoplanktonic algae has also been considered an important factor in interspecific interactions, and hence in the seasonal succession and in bloom formation.[19,30–32] Interactions of this nature are not confined to the phytoplankton, as seaweeds also appear to control the numbers of microbial epiphytes colonizing their surface by the production of such compounds.[15]

5.3 Classification

The primary classification of algae is based on both biochemical and structural properties of the cell. Details of some of these characteristics are given in Table 5.4. On the basis of these characteristics the algae can be divided into 15 divisions within four distinct evolutionary groups. A more complete coverage of algal taxonomy is given by Lee.[5]

The first group contains the prokaryotic algae and includes two divisions; the Cyanophyta (blue–green algae) and the Prochlorophyta (prochlorophytes). The prokaryotic algae were the first to evolve and are characterized by the prokaryotic cellular organization (see Section 5.3.1 for details). They are closely related to the bacteria.[33]

The second evolutionary group contains those algae which possess a chloroplast, surrounded only by the two membranes of the chloroplast envelope. These algae are thought to have evolved through an evolutionary event that involved the capture and retention of a prokaryotic algal cell in a food vesicle of a phagocytic non-photosynthetic protozoan. Eventually in the process of evolution, the plasma membrane of this endosymbiotic prokaryotic alga became the inner membrane of the chloroplast envelope and the food vesicle membrane of the host protozoan became the outer membrane of the chloro-

Table 5.4 *Some basic characteristics of the different taxonomic groups of algae*

	Major pigments				
Algal class	*Chorophyll*	*Others*	*Cell wall*	*Storage product*	*Flagella**
Cyanophyta	a	Phycocyanin, phycoerythrin	Peptidoglycan	Glycogen	0
Prochlorophyta	a,b	Zeaxanthin	Peptidoglycan	?	0
Euglenophyta	a,b	Neoxanthin	Proteinaceous	Paramylon, Peculiar starch	+ or 0
Dinophyta	a,c_2	Peridinin, neoperidinin	Cellulosic plates	Starch oils	+
Cryptophyta	a,c_2	Phycoerythrin, phycocyanin	Proteinaceous	Starch, lipids	+
Raphidophyta	a,c	Fucoxanthin	Mucous secretion	Oils	
Chrysophyta	a,c_1,c_2	Fucoxanthin	Cellulosic and siliceous scales	Chrysolaminarin	+
Prymnesiophyta	a,c_1,c_2,c_3	Fucoxanthin	Cellulosic and/or calcareous scales	Chrysolaminarin, lipids, sterols	+
Bacillariophyta	a,c_1,c_2	Fucoxanthin	Silica	Chrysolaminarin	0
Xanthophyta	a,c_1,c_2	Diadinoxanthin, heteroxanthin, vaucheriaxanthin	Cellulose, pectin	β-1,3-glucan similar to paramylon	+ or 0
Eustigmatophyta	a	Violaxanthin, vaucheriaxanthin	Microfibrillar material	Lamellate material	0
Phaeophyta	a,c_1,c_2	Fucoxanthin	Cellulose	Laminarin	0
Rhodophyta	a,d	Phycocyanin, phycoerythrin, zeaxanthin	Cellulose, some calcareous	Floridean starch	0
Chlorophyta	a,b	Lutein	Cellulose, pectin	Starch (found inside chloroplast)	+ or 0

*Gametes and spores omitted; +, present; 0, absent.

plast envelope. The Glaucophyta (not included in Table 5.4) represent an intermediate evolutionary stage in this group with the Rhodophyta (red algae) and Chlorophyta (green algae) completing the group and representing the completion of this evolutionary pathway.

The third group comprises algae in which the chloroplast is surrounded by a single membrane of the chloroplast endoplasmic reticulum. This evolutionary pathway is believed to have resulted when a chloroplast from a eukaryotic algae was taken up into a food vesicle of a phagocytic protozoan. Eventually the food vesicle membrane of the host protozoan became the single membrane of chloroplast endoplasmic reticulum surrounding the chloroplast in this group of algae. This group consists of the Euglenophyta (euglenoids) and the Dinophyta (dinoflagellates).

The final evolutionary group contains those algae with two membranes of chloroplast endoplasmic reticulum surrounding the chloroplast. In the evolutionary event that led to this group of algae, a phagocytic protozoan took up a red alga into a food vesicle. Subsequently, the food vesicle membrane of the protozoan was lost along with the nucleus of the red alga and the functions of the red algal symbiont became controlled by the nucleus of the host protozoan. This protozoan, along with its algal symbiont, was then taken up by a second phagocytic protozoan into a food vesicle. The initial protozoan plus algal symbiont were retained in the cytoplasm of the second protozoan and the functioning of the combined cellular apparatus taken over by the nucleus from the first protozoan. The second protozoan lost its nucleus and the food vesicle membrane, while the outer nuclear envelope of the first protozoan was also lost. These events resulted in the characteristic two membranes of chloroplast endoplasmic reticulum found in this group, the outer membrane being derived from the plasma membrane of the first protozoan, and the inner membrane derived from the food vesicle membrane of the first protozoan. This group contains most of the algal divisions including the Cryptophyta (cryptophytes), Chrysophyta (golden-brown algae), Prymnesiophyta (prymnesiophytes or haptophytes), Bacillariophyta (diatoms), Raphidophyta (chloromonads), Xanthophyta (yellow–green algae), Eustigmatophyta, and Phaeophyta (brown algae).

Of the 15 divisions of algae mentioned above, eight are of primary importance in terms of abundance, species diversity, and known production of biologically active substances. Therefore, these eight divisions are focussed on in this chapter.

5.3.1 Cyanophyta

The Cyanophyceae or blue–green algae are prokaryotic organisms that have an outer plasma membrane enclosing protoplasm containing photosynthetic thylakoids, 70S ribosomes, and DNA fibrils. They contain chlorophyll *a* and phycobiliproteins as the major photosynthetic pigments, produce glycogen as a storage product, and possess cell walls consisting of mainly peptidoglycan, which may constitute up to 50% of the cell's dry weight.[34] All orders of the

Cyanophyceae except the Chamaesiphonales[35] possess gas vacuoles which are thought to function in light shading and/or buoyancy.

Cyanophyceae occur in terrestrial, freshwater, and marine environments. Terrestrial species include some examples which are capable of nitrogen fixation, which have been shown to contribute significantly to the level of soil fertility.[36,37] The freshwater species occur throughout the world as phytoplankton (unicellular, colonial, or filamentous) or as attached algae in standing or running water. In the marine environment, the picoplankton comprise mostly small coccoid Cyanophyceae. They are also found in the intertidal zone in tidal pools or as blackish zones on rocks at the high tide mark. Planktonic blue–green algae cause extensive blooms under certain conditions in both marine and freshwater environments. Many of these blooms have been associated with toxin production and the death of aquatic organisms.[3,4,38]

5.3.2 Rhodophyta

This division consists of a single class of algae, the Rhodophyceae, which is probably one of the oldest groups. They lack motile cells at any stage in their life cycle, have chlorophylls *a* and *d*, phycobiliproteins, produce floridean starch as a storage product, and possess thylakoids which occur singly within the chloroplast.

There are approximately 4000 species in this class, a major proportion of which are multicellular macroalgae (seaweeds); this is more species than all of the other major seaweed groups combined. Few species are found in polar and subpolar regions, where brown and green algae predominate, but in temperate and tropical regions they are predominant. Their possession of phycobiliproteins ($\lambda_{max} = 565$ nm) as accessory photosynthetic pigments allows Rhodophytes, through utilization of the shorter (green and blue) wavelengths of light which penetrate deepest in the ocean, to survive at depths of up to 200 m. Approximately 200 species of red algae are found in freshwater.

5.3.3 Chlorophyta

The Chlorophyta, or green algae, are made up of four classes: Micromonadophyceae, Charaphyceae, Ulvophyceae, and Chlorophyceae. They all possess chlorophylls *a* and *b* as the major photosynthetic pigments and form starch as the storage polysaccharide within the chloroplast. The Chlorophyta thus differ from other eukaryotic algae in forming the storage product in the chloroplast rather than in the cytoplasm.

This group is mainly freshwater in habitat, only about 10% being marine.[39] The freshwater species have a cosmopolitan distribution, with few species being endemic to a particular area. In the marine environment the algae in the warmer tropical and semitropical waters are similar species, while those in the colder waters of the Northern and Southern Hemispheres differ markedly from each other in their species distribution.

5.3.4 Dinophyta

The Dinophyta, sometimes referred to as the Pyrrophyta, contains only one class, the Dinophyceae or dinoflagellates. These are important members of the plankton in marine, estuarine, and freshwater environments, particularly in tropical regions. A greater variety of forms is, however, found among the marine members.

The Dinophyceae are mostly unicellular and autotrophic, although some colourless heterotrophic forms are known. A typical motile dinoflagellate is a biflagellate organism with a highly elaborate cellular structure. In photosynthetic forms chlorophylls *a* and c_2 are present, with peridinin[40] and, occasionally, 19′-hexanoyloxyfucoxanthin[41] being the main carotenoids. The storage product is starch, similar to that found in higher plants[42] which is found in the cytoplasm. The dinoflagellates differ from all other algae in respect of their nuclei. The nucleus of a dinoflagellate possesses permanently condensed chromosomes, and is referred to as a dinokaryotic or mesokaryotic nucleus.[43] This characteristic is neither prokaryotic nor eukaryotic, but falls somewhere between the two.

Dinoflagellates, particularly marine species, can cause extensive blooms or red tides which colour the water red (as mentioned earlier). Some of these blooms are associated with toxin production and the death of other marine life.[2,3]

Animal and human poisoning that results from eating contaminated shellfish are known by various names: mussel poisoning, paralytic shellfish poisoning, and saxitoxin poisoning.[44]

5.3.5 Chrysophyta

The Chrysophyta, or gold–brown algae, contains two classes, the Chrysophyceae and the Synurophyceae. They have a small number of chloroplasts, each containing chlorophylls *a* and c_1/c_2, fucoxanthin, and β-carotene. The storage product is chrysolaminarin (leucosin), a β-1,3-linked glucan. Species are mainly biflagellate and are primarily freshwater in habit, being more abundant in soft waters (*i.e.*, low in calcium). Most species are sensitive to environmental changes, but can survive unfavourable periods as resting spores called statospores.

5.3.6 Prymnesiophyta

This group was until recently considered part of the division Chrysophyta. There is one class, originally named the Haptophyceae (after the presence of the haptonema), but now referred to by many authors under the typified name, the Prymnesiophyceae.[45] There are unicellular flagellates characterized by the possession of a haptonema between two smooth flagella. The haptonema is a filamentous appendage of unclear function, although it can serve as a temporary organ of attachment to a surface and in some species as a prey-capture organelle. The chloroplasts contain chlorophylls a, c_1, c_2, and c_3, with fucoxanthin as the major carotenoid. Diadinoxanthin and the fucoxanthin derivative, 19′-hexonoylfucoxanthin, also occur in this group.[41] The storage product is chrysolaminarin contained within vesicles in the cytoplasm. The cells are commonly covered with scales and in some cases are calcified, as in the Coccolithophorids.

The Prymnesiophyceae are primarily marine organisms that constitute a major part of the nanoplankton, although there are some freshwater representatives within the class. Several of the representatives within this group cause blooms in temperate waters,[46,47] some of which have been associated with production of potent toxins.[48–50]

5.3.7 Bacillariophyta

The Bacillariophyceae (sometimes referred to as the Diatomaphyceae) or diatoms are unicellular, sometimes colonial algae found in almost every aquatic habitat as free-living photoautotrophs, colourless heterotrophs, or photosynthetic symbionts.[51] A characteristic feature of this group is their ability to secrete an external wall composed of silica, the provision of silica being essential for cell division.[52] The photosynthetic pigments are chlorophylls a, c_1, and c_2, with the major carotenoid being fucoxanthin. The storage product is chrysolaminarin.

Diatom blooms occur often in spring and autumn in the freshwater[53,54] and marine habitat.[55,56] The production of biologically active compounds by diatoms was not confirmed until relatively recently and has been reviewed by Lincoln *et al.*[57]

5.3.8 Phaeophyta

The Phaeophyceae, or brown algae, derive their characteristic colour from the large amounts of fucoxanthin (a carotenoid) present in the chloroplast, as well as from any phaeophycean tannins which might be present. They also possess chlorophylls a, c_1, and c_2. The long-term storage product is laminarin, although the sugar alcohol, D-mannitol, is the accumulation product of photosynthesis. There are no unicellular or colonial representatives within this division, all members being macroalgae of a filamentous, pseudoparenchymatous, or parenchymatous nature. They are almost exclusively found in the marine environment growing near the intertidal belt and upper littoral region, although a number of marine forms penetrate into brackish water.

5.4 Laboratory Culture of Microalgae

A great deal of time and effort is often involved in the isolation and culture of algae. For this reason, an overview of the isolation and cultivation techniques currently used for algae is given, but for a greater understanding of this branch of phycology see the publications of Lewin,[58] Droop,[59,60] and Guillard.[62]

Although it is possible to obtain samples of planktonic algae in relatively large amounts from natural waters, particularly in bloom situations, it is usually preferable to grow the organisms of interest in laboratory culture. The culture conditions can then be manipulated easily to obtain optimal growth.

5.4.1 Isolation and Purification

Laboratory cultivation of micro-organisms dates back to the nineteenth century. Early methods for isolating algae from associated organisms, including bacteria, by plating on gelatin or agar are still useful, although have largely been superseded. Cultures can be initiated from single cells or spores isolated using a specifically designed micropipette.[63] This technique, when combined with the removal of contaminants by serial transfer of the cell from one sterile solution to another, has proved successful for initiating unialgal and, in some cases, axenic (*i.e.*, free from bacteria) and/or clonal (single genotype) cultures.[60,63,64] Motile autotrophic algae, *e.g.* dinoflagellates, can also be isolated by exploring their phototactic behaviour[60,63] (tendency to move towards or away from light). Once a unialgal culture has been obtained, several methods have been used to make these axenic, including the use of gamma irradiation,[65] the use of osmotic shock designed to take advantage of the salt tolerance of certain algae,[66] or by using mixtures of antibiotics.[63,67–69] The last technique has proved very successful, but it is worth keeping in mind that cells exposed to strong antibiotic solutions may be altered in ways not readily recognizable and thus give rise to cultures no longer representative of the species. Antibiotics are best used when initial efforts at purification by plating, serial washings, and/or phototaxis have proved unsuccessful. The ultimate goal of these treatments is an axenic, unialgal, clonal culture – only one species, one genotype.

5.4.2 Culture Media

Early culture media utilized soil and peat extracts to improve growth of algae in laboratory culture. Pringsheim[70] suggested that soil extracts were important in supplying soluble iron compounds and that peat extracts supplied hydrogen ions and various salts, both helping to regulate the pH of the media. Although culture media have improved over the years, some still contain soil extract which seems essential for the growth of certain algae. Many media have now been developed based on natural seawater or freshwater, which are considered 'enriched media'. However, the most advanced media available are totally artificial and termed 'chemically defined' media. Tables 5.5 and 5.6 give examples of media used to grow autotrophic microalgae. Similar media are used to grow algae heterotrophically, but with the addition of one or more organic carbon sources, for example glucose, glutamate, acetate, or succinate. Terrestrial algae are frequently grown on simple inorganic media similar to those used for freshwater species.[71,72]

Table 5.5 *Some freshwater culture media used for autotrophic microalgae (molar concentration)*

Compound	*Chu 10*	*ASM*[74]	*WC*[62]	*Dy III*[75]	*Ch*[76]
$CaCl_2$		10^{-4}	2.5×10^{-4}	5×10^{-4}	4.5×10^{-4}
$MgSO_4$	10^{-4}	2×10^{-4}	1.5×10^{-4}	2×10^{-4}	4.06×10^{-4}
$MgCl_2$		2×10^{-4}			
KCl				4×10^{-5}	6.71×10^{-5}
Na_2CO_3	1.9×10^{-4}				
$NaHCO_3$			1.5×10^{-4}		
$NaNO_3$		10^{-3}	10^{-3}	2.35×10^{-4}	1.18×10^{-3}
NH_4NO_3				6.25×10^{-5}	
$Ca(NO_3)_2$	2.4×10^{-4}				
K_2HPO_4	6.5×10^{-5}	10^{-4}	5×10^{-5}		5.74×10^{-5}
Na_2 glycerophosphate				4.6×10^{-5}	
Na_2SiO_3	2×10^{-4}		1×10^{-4}	5.3×10^{-5}	
EDTA*			1.2×10^{-5}		
Na_2EDTA*		2×10^{-5}		2.15×10^{-5}	2.69×10^{-5}
TRIS†					1.27×10^{-3}
MES‡				1.03×10^{-3}	
FeEDTA*			1.17×10^{-5}		
$FeCl_3$	4.9×10^{-6}	2×10^{-6}		1.25×10^{-5}	1.79×10^{-6}
$FeSO_4$					
$ZnSO_4$			8×10^{-8}		
$ZnCl_2$		8×10^{-7}		6.15×10^{-7}	7.65×10^{-7}
$MnCl_2$		7×10^{-6}	9×10^{-7}	3.6×10^{-6}	7.28×10^{-6}
$CoCl_2$		2×10^{-8}	5×10^{-8}	1.35×10^{-7}	1.7×10^{-7}
$CuSO_4$			4×10^{-8}		
$CuCl_2$		2×10^{-9}			
Na_2MoO_4			3×10^{-9}	2×10^{-7}	
H_3BO_3		1×10^{-5}	1.16×10^{-4}	7.3×10^{-5}	1.85×10^{-4}
Vitamin B_{12}			3.7×10^{-10}	3.7×10^{-10}	8.12×10^{-10}
Thiamine			3×10^{-7}	6×10^{-7}	1.19×10^{-6}
Biotin			2×10^{-9}	2×10^{-9}	4.09×10^{-9}
Niacin					1.63×10^{-6}
Deionized water		1 l	1 l	1 l	1 l
pH		Usually in the range 6.5–7.2			

*EDTA, ethylenediaminetetraacetic acid.
†TRIS, tris(hydroxymethyl)methylamine.
‡MES, 2-morpholinoethanesulfonic acid.

5.4.3 Measurement of Growth Rates

The growth rates of algae in culture can be determined from the change of biomass with time. Biomass can be estimated in many different ways. These include measurement of optical density, cell number, chlorophyll, protein or particulate carbon content, and dry weight.

The effects of physical (light, temperature, salinity, pH, *etc.*) and chemical

Table 5.6 *Some marine culture media used for autotrophic microalgae (molar concentration)*

	Artificial seawater media		*Enriched seawater media*	
Compound	*Aquil*[77]	*ESAW*[78]	*F/2*[62]	*GPM*[68]
NaCl	4.2×10^{-1}	3.63×10^{-1}	8.83×10^{-4}	
$CaCl_2$	1.05×10^{-2}	9.14×10^{-3}		
KBr	8.4×10^{-4}	7.3×10^{-4}		
NaF	7.14×10^{-5}	6.57×10^{-5}		
KCl	9.39×10^{-3}	8.04×10^{-3}		
H_3BO_3	4.85×10^{-4}	3.72×10^{-4}		5.53×10^{-4}
Na_2SO_4	2.88×10^{-2}	2.5×10^{-2}		
$NaHCO_3$	2.38×10^{-3}	2.07×10^{-3}		
$SrCl_2$	6.38×10^{-5}	8.2×10^{-5}		
$MgCl_2$	5.46×10^{-2}	4.72×10^{-2}		
$NaNO_3$	1.0×10^{-4}	5.49×10^{-4}		
KNO_3				2×10^{3}
NaH_2PO_4	1.0×10^{-5}			
K_2HPO_4			3.63×10^{-5}	
Na_2 glycerophosphate		2.18×10^{-5}		2.01×10^{-4}
Na_2SiO_3	1.25×10^{-5}	1.06×10^{-4}		
Na_2EDTA*	5.0×10^{-6}	1.49×10^{-5}		
$FeCl_3$	4.51×10^{-7}	5.92×10^{-7}	1.17×10^{-5}	5.37×10^{-6}
$Fe(NH_4)_2(SO_4)_2$		5.97×10^{-6}		
$ZnSO_4$	4.0×10^{-9}	2.54×10^{-7}	8.0×10^{-8}	
$MnCl_2$	2.3×10^{-8}		9.0×10^{-7}	2.18×10^{-5}
$MnSO_4$		2.42×10^{-6}		
$CoCl_2$	2.5×10^{-9}		5.0×10^{-8}	9.24×10^{-7}
$CoSO_4$		5.69×10^{-8}		
$CuSO_4$	9.97×10^{-10}		4.0×10^{-8}	
Na_2MoO_4			3.0×10^{-8}	
$(NH_4)_6Mo_7O_{24}$	1.5×10^{-9}			
Vitamin B_{12}	4.06×10^{-10}	1.47×10^{-9}	3.7×10^{-10}	7.4×10^{-10}
Biotin	2.05×10^{-10}	4.09×10^{-9}	2.1×10^{-9}	8.2×10^{-9}
Thiamine HCl	2.97×10^{-10}	2.97×10^{-7}	3.0×10^{-7}	3.0×10^{-6}
Soil extract				15 ml
	Deionized water to 1 l		Filtered seawater to 1 l	
pH	Usually in the range 7.6–8.3			

*EDTA, ethylenediaminetetraacetic acid.

(nutrient status, autotrophic *versus* heterotrophic growth, *etc.*) factors on growth and the production of secondary metabolite(s) can be studied and optimized relatively easily.

5.5 Collection and Cultivation of Macroalgae

The culture of macroalgae has not been studied in the same detail as that of microalgae, probably because collection from the natural habit, and removal of epiphytes and washing with sterile water, is far easier. Certainly, for laboratory culture the amount of space required for these large plant-like algae is a restricting factor in view of the biomass required to extract biologically active metabolites.

5.5.1 Collection

The harvesting of seaweed on a commercial scale dates back to the end of the nineteenth century with the extraction of saltwort, iodine, and, later, phycocolloids.[79,80] Seaweed gathering remained traditional for a long time, the seaweeds being collected as drift, either by hand or using rudimentary tools. Changes in this activity are only very recent, with the development of seaweed harvesting (reviewed by Briand[81]) remaining closely linked to the colloid industry.

5.5.2 Cultivation

The cultivation of seaweeds in large quantities, particularly in Europe, is still in its infancy. One reason for this is the considerable natural resources of seaweeds,[82] which were adequate for our needs. There are a number of reasons why cultivation may be considered advantageous over harvesting natural populations of seaweeds, which include:

- Demand on natural algal resources exceeding supply.
- Harvesting itself may be difficult in natural algal beds due to the terrain and exposure to the elements.
- There is a poor quality of the harvested material; it is likely to be a mixture both of species and of ages.
- Improvement of the quality through genetic selection is only feasible with cultivated stocks.

In principle, several technical approaches to mariculture are practised. These include:

- Management of natural populations or cultivation in land-locked sea areas, such as lagoons, fjords, and embayments.
- Cage culture, rope culture, and floating maring farms.
- Onshore cultivation in land-based tanks, ponds, or raceways.
- Spray culture (occasional misting of algae).

Seaweed cultivation has been comprehensively reviewed by Mathieson,[83,84] Kain,[85] and Schramm.[86]

5.6 Bioassay Methods

Perhaps the most fundamental aspect of natural product research is the discovery and development of rapid, sensitive, and preferably specific assays.

The systematic screening of algal extracts for the detection of biologically active substances was initiated in the 1950s with screens for antimicrobial activity (see references 355, 356, and 359–362). However, more recently the screens have been extended to detect a wide spectrum of additional activities.[22]

5.6.1 *In Vivo* Methods

Many pharmacological *in vivo* bioassays have been developed, including as test animals cats, mongooses, chickens, mice, and fish.[87–89] Traditionally, the standardized mouse bioassay has been the method of choice[90] to detect bioactive compounds from algae. Various minor alterations in the assessment of potency have been utilized,[91] but most current procedures involve intraperitoneal injection. Such a standardized assay is of value for several reasons. First, inbred strains of genetically similar animals are available to minimize variations in response. Second, maintenance is easy, dosing is conveniently and simply effected, and response (for example, death) is easily judged. However, *in vivo* bioassays have been criticized as being inefficient routine testing procedures. Some of the reasons for this are that the tests often involve time-consuming procedures, they are not particularly accurate or precise, and are costly to perform; also, a factor which is becoming increasingly important is the public outcry at the use of live animals in experimentation. Other *in vivo* tests which have been developed include housefly,[92] mosquito,[93] and brine shrimp[94] bioassays.

Methods to measure the effects of bioactive compounds on plant systems have been reviewed by Leather and Einhellig.[95] The most widely used bioassays being the inhibition of seed germination[96–100] and/or plant growth and development, for example radicle and hypocotyl/coleoptile elongation.[101–103] Assays have also been developed using fungal[104,105] and fern[106] spore germination and development, and growth of marine microalgae.[107]

The need to develop high throughput, specific assays which are easy to perform and more closely define the desired activity has prompted much effort in the search for alternative *in vitro* techniques.

5.6.2 *In Vitro* Methods

In vitro methods to detect bioactive compounds from algae have included using freshly prepared rat hepatocytes,[108,109] use of isolated organ preparations,[110] isolated guinea-pig ileum,[111] haemaglutinin assays,[112] tissue culture assays,[113,114] detection of growth-promoting agents in cultured malignant cell lines,[115] antineoplastic agents in potato discs infected with *Agrobacterium tumifaciens*,[116] cytochemical staining techniques,[117] and agar plate based assays for the detection of antimicrobial compounds.[118–124]

Enzyme inhibitor assays in which culture filtrates and cell extracts are assayed against enzyme preparations have also been developed for pharmacological screening.[125–128] A majority of these assays are colorimetic and based on the use of microtitre plates, which allows a high throughput of samples. The enzymes chosen are either known to be involved in or be implicated in human or animal disease conditions.

Other methods which have been developed particularly for the study of algal toxins, with varying degrees of success, are based on chromatographic separations using acidic exchange resins,[129] thin layer chromatography,[130] liquid chromatography,[131,132] electrophoretic,[133] immunometric,[134] and radioimmunometric[135–138] methods and receptor binding assays[139–142] have also been developed.

5.7 Biologically Active Compounds from Algae

In this section the range of biological activities attributed to compounds from algae is discussed, focusing primarily on their physiological and/or biochemical modes of action. This is to give the reader an indication of the potential for commercial development that exists within the algae. These compounds could be of value, first, as products *per se* produced by mass culture or fermentation technology; second, they could act as leads for total or partial chemical synthesis; third, they may also be valuable as research tools in the search for novel modes of action; and, fourth, if they are proteinaceous compounds the gene (or genes) may be expressed in transgenic species, to confer resistance.

The compounds are split for convenience into three major sections: toxins, compounds of pharmaceutical interest, and those of interest for agrochemicals. There are obvious overlaps between these broad categories, but cross-references are made where appropriate. Biologically active compounds from algae have been extensively reviewed.[6,7,15,17,19,20,22,57,143–155]

5.7.1 Toxins

Toxins are defined as substances poisonous to multicellular organisms, and are probably the most notorious of algal secondary metabolites. They have been reported to be produced mainly by freshwater Cyanophyta and marine Dinophyta, Cyanophyta, and Prymnesiophyta (Table 5.7), although other classes do have representatives that produce toxins. These toxins are a diverse group of compounds and are grouped for discussion in this section according to their chemical nature or biological activity. Many of the toxins have been shown to be important ecologically and economically, causing mass mortality of fish and other aquatic animals, and intoxication of man and terrestrial animals. Toxins produced by algae have been the subject of several reviews.[1–4,19,22,38,46,88,156,157]

5.7.1.1 *Guanidine-Derived Toxins*

These toxins are neurotoxins responsible for paralytic shellfish poisoning

Table 5.7 *Algal toxins**

Organism	*Toxin*	*Structure group*	*Structure*
Prymnesiophyta			
Prymnesium parvum (Figure 5.2)	Ichthyotoxin, cytotoxin, haemolysin	Acid polar lipids	**131**
Cyanophyta			
Marine & brackish water			
Lyngbya majuscula	Lyngbyatoxin A	Indole alkaloid	**18**
	Aplysiatoxin	Phenolic bislactone	**109**
	Debroimoaplysiatoxin	Phenolic bislactone	**17**
Schizothrix calcicola	Debromoaplysiatoxin	Phenolic bislactone	**17**
Oscillatoria nigroviridis	Oscillatoxin A	Phenolic bislactone	**19**
Nodularia spumigena	Nodularia toxin	Peptide	ND
Freshwater			
Microcystis aeruginosa (Figure 5.5)	Fast Death Factor	Peptide	
	Microcystin	Peptide	**13**
	Cyanogenosin	Peptide	**14**
Aphanizomenon flos-aquae (Figure 5.3)	Aphantoxin (neosaxitoxin)	Alkaloids	**1**
	Aphantoxin II (saxitoxin)		**1**
Anabaena flos-aquae (different strains) (Figur 5.11)	Anatoxin *a*	Alkaloid	**12**
	Anatoxin *b*	Unknown	ND
	Anatoxin *c*	Peptide	ND
	Anatoxin *d*	Unknown	ND
	Anatoxin *a*(s)	Unknown	ND
	Anatoxin *b*(s)	Unknown	ND
Oscillatoria agardhii (Figure 5.4)	Microcystins	Peptides	
Microcystis viridis (Figure 5.5)	Cyanoviridin	Peptide	**15**
Gloeotrichia echinulata (Figure 5.5)	Gloeotrichia toxin	Unknown	ND
Synechocystis sp.	Synechocystis toxin	Unknown	ND
Cylindrospermopsis raciborskii	Cylindrospermopsis toxin	Unknown	ND
Nodularia spumigena	Nodularin	Peptide	
Dinophyta			
Marine			
Gonyaulax spp.	Various PSP toxins	Alkaloid	
catenella	Saxitoxin		**1**
tamarensis	Gonyautoxins, 1,2,3,4		**1**
	Neosaxitoxin		**1**
	Saxitoxin		**1**
acatenella	Gonyautoxins, 1,2,3,4		**1**
phoneus	Cryptic (B_{1-2}, C_{1-4})		

Continued

Table 5.7 *Continued*

Organism	*Toxin*	*Structure group*	*Structure*
Dinophyta *cont.*			
Pyrodinium bahamense var. Compressa	Decarbamolysaxitoxin Gonyautoxins IV and VI		I
Ptychodiscus brevis	Neurotoxin	Polyether	
	Haemolysins	Unknown	**130**
Dinophysis fortii *D.acuminator* *D.acuta* *D.norvegica* *Prorocentrum lima*	Diarrhoetic shellfish poisons	Polyether	**4–11**
Gambierdiscus toxicus	Ciguatoxin	Lipid soluble	**3**
	Maitotoxin	Water soluble	ND
	Scaritoxin	Lipid soluble	
Protogonyaulax chorticula	Gonyautoxins 1,2,3,4	Alkaloids	**1**
	Saxitoxin	Alkaloid	**1**
Noctiluca miliaris	Ammonia	Amine	
Goniodoma pseudogoniaulax	Goniodomin A	Polyether	ND
Freshwater			
Peridinium polonicum (Figure 5.6: *Peridinium* sp.)	Glenodinine	Unknown	ND
Rhodophyta			
Chondria armata	Domoic acid	Amino acid	**21**
Chondria baileyana	Domoic acid	Amino acid	**21**
Alsidium corallitinum	Domoic acid	Amino acid	**21**
Jania sp.	Gonyautoxins, 1,2,3	Alkaloids	**1**
Chlorophyta			
Caulerpa spp.	Caulerpin	Mixture of hydroxyamides	**41**
	Caulerpicin	Pirazine derivative	**40**
Ulva pertusa	Palmitic acid	Fatty acid	**42**
Chaetomorpha minima	Unnamed	Fatty acids	
Bacillariophyta			
Amphora coffeaeformis (Figure 5.7: Amphoro. sp.)	Domoic acid	Amino acid	**21**
Nitzchia pungens	Domoic acid	Amino acid	**21**

*ND, structure not determined.

(PSP). They are water-soluble toxins (**1**) produced mainly by marine members of the Dinophyta, and by *Aphanizomenon flos-aquae* (Cyanophyceae) and *Jania* sp. (Rhodophyceae) (Table 5.7). This group of toxins (**1**, Figure 5.8) is made up of saxitoxin and its naturally occurring derivatives neosaxitoxin and

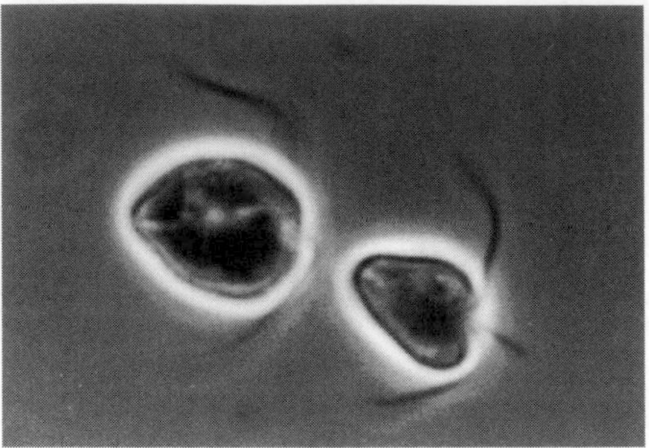

Figure 5.2 Prymnesium parvum, phase contrast (×2000)

the gonyautoxins. The purification and characterization of saxitoxin was first accomplished in 1957[158] from Alaskan butter clams. The 12 currently known PSP toxins can be placed readily in four groups based on the type and location of derivatization. Neosaxitoxin, which was first described by Shimizu *et al.*,[159] and saxitoxin comprise Class I PSPs. The remaining ten toxins are grouped into three classes based on sulfated 11-hydroxy substitution (Class II), *N*-sulfo-conjugation on the carbamoyl position (Class III), or dual sulfo-conjugation at the 11-hydroxy and carbamoyl positions (Class IV).

The relative potencies are known to differ according to the source and extraction method used. Hall[160] reported the toxins to be in the order (decreasing potency) STX and NEO > GTX3 > C2 > B2 > B1 > C1. Vari-

Figure 5.3 Aphanizomenon flos-aquae (×60)

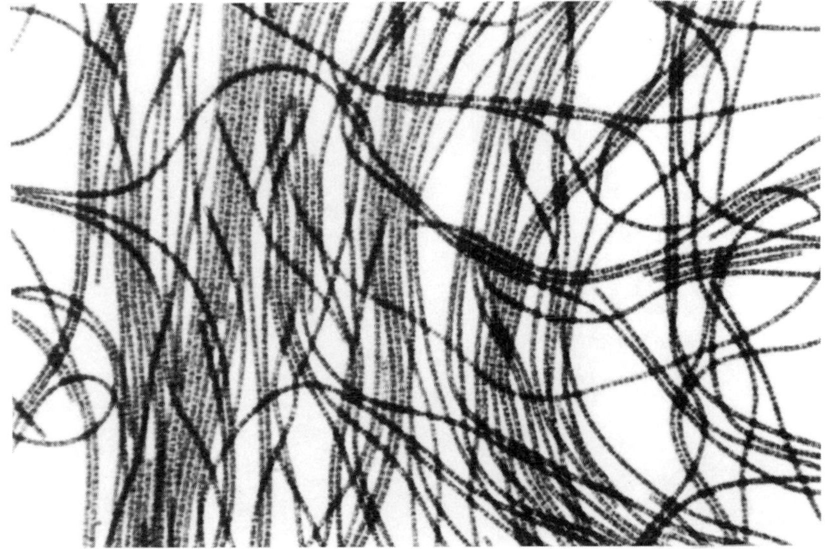

Figure 5.4 Oscillatoria agardhii (×130)

ability in the observed potency may also result from hydrolysis of the toxins[160,161] or from metabolic transformation of the toxins within the shellfish vector.[162–164] The characteristic neurological symptoms resulting from consumption of contaminated vectors include paraesthia of lips, mouth, face, and extremities, nausea, vomiting, weakness, incoherence, respiratory distress, and death in as little as 12 hours.

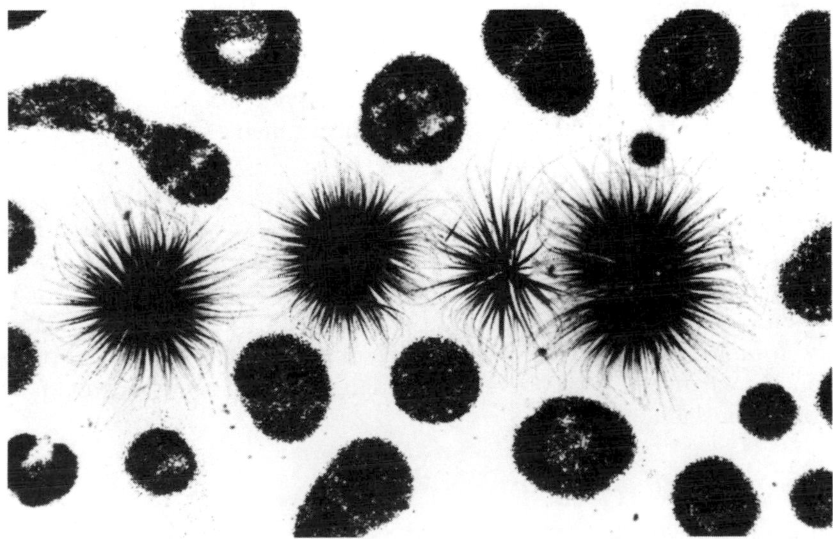

Figure 5.5 *Smooth colonies* Microcystis aeruginosa *and spiny colonies* Gloeotrichia echinulata (×40)

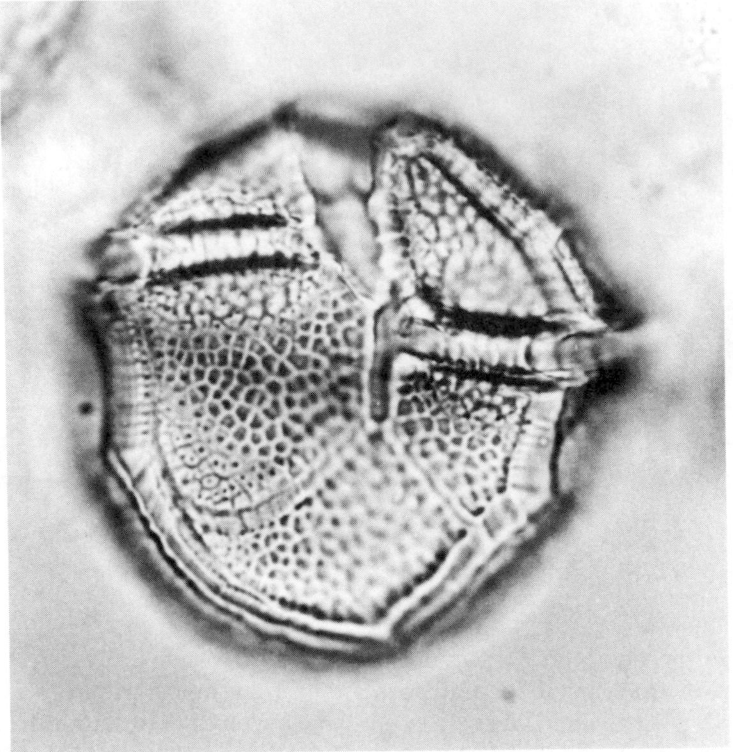

Figure 5.6 Peridinium *sp. (×1385)*

The pharmacological mechanism of action of PSP toxins has been described in detail.[2] They act predominantly by altering transmembrane fluxes of cations, primarily sodium. Saxitoxin selectively inhibits the inward flux of sodium without affecting the resting membrane potential or potassium channels. This prevents the propagation of nerve–muscle action potentials.[165–167] STX binds to specific receptors on the exterior of the nerve membrane in a 1:1 stoichiometry.[168,169]

The mechanism by which STX interfered with sodium channels was thought to be a physical plugging of the channel by the toxin.[167] However, it was subsequently shown that the active groups were the C12–OH groups, 7,8,9 guanidinium, and the carbonyl group,[170] and that electrostatic interaction between the guanadinium group and the anionic surface charge of the membrane around the sodium channel, which results in the STX molecule forming a lid on the channel.[170,171] The STX binding site has been shown to have glycoprotein characteristics.[172]

5.7.1.2 *Polyether Toxins*

5.7.1.2.1 *Brevetoxins* The brevetoxins represent a class of lipid-soluble neurotoxins produced by the dinoflagellate *Ptychodiscus brevis* (formerly

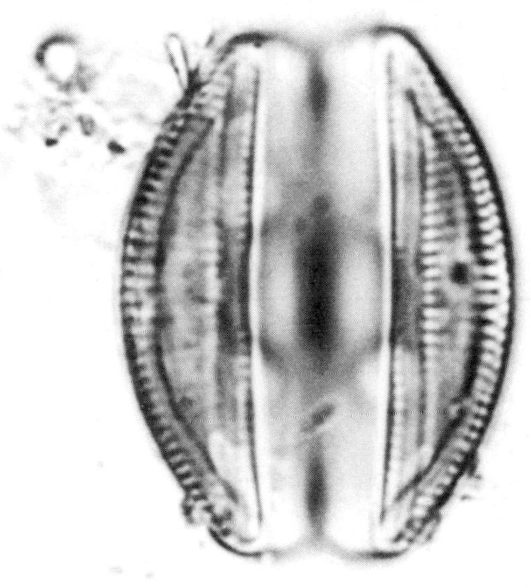

Figure 5.7 Amphora *sp. (×2200)*

Gymnodinium breve) which have been isolated and characterized in laboratory cultures and from toxic shellfish.[173–180] Two structurally related families of polyether brevetoxins have been identified. The first group contains type 1 backbones consisting of a skeleton of 11 contiguous transfused ether rings (**2**), which includes PbTx-2, PbTx-3, PbTx-5, PbTx-6, PbTx-8, and PbTx-9. The second family contains type 2 backbones of 10 ether rings (**2**) and includes PbTx-1, PbTx-7 and PbTx-10. Both families possess an α,β-unsaturated lactone functionality which is essential for biological activity. They differ, however, in their flexibility – type 1 backbones being planar and rigid, whereas type 2 backbones can assume two stereochemical conformations relative to ring G (**2**). The toxins within each family differ in their side-chain structure at C_{37} and C_{38} respectively.

The brevetoxins have been implicated in massive fish kills and human intoxications[89] resulting from ingestion of contaminated shellfish and/or inhalation of toxic aerosols, leading to asthma-like attacks[181] and skin irritation.

Brevetoxins have been reported to stimulate excitable tissues,[182,183] resulting in repeated neuronal discharge.[184–186] Muscle twitching, agitated behaviour, lumbar contractions, and convulsions have been observed in mice subjected to these toxins. Sodium channels are involved in the observed brevetoxin-induced membrane depolarization.[185–187] The toxin is suggested to bind to a novel neurotoxin binding site on voltage-sensitive sodium channels,[180,188,189] a specificity shared with ciguatoxin,[189,190] which is described

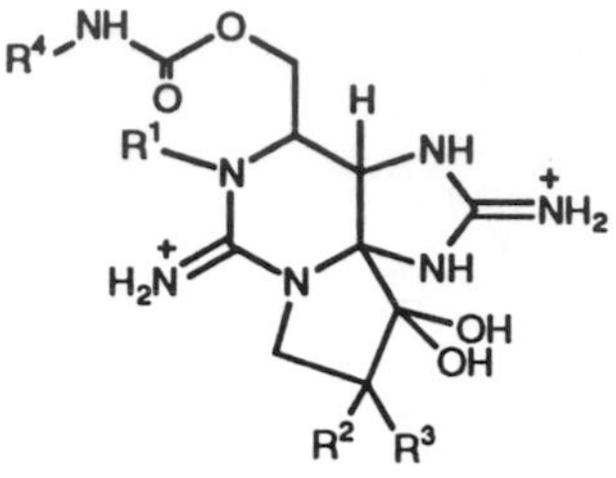

Compound	R_1	R_2	R_3	R_4	*Nomenclature*[162]
Class I					
Saxotoxin	H	H	H	H	STX
Neosaxitoxin	OH	H	H	H	NEO
Class II					
Gonyautoxin 3	H	OSO_3^-	H	H	GTX3
Gonyautoxin 2	H	H	OSO_3^-	H	GTX2
Gonyautoxin 4	OH	OSO_3^-	H	H	GTX4
Gonyautoxin 1	OH	H	OSO_3^-	H	GTX1
Class III					
Gonyautoxin 5	H	H	H	SO_3^-	B1
Gonyautoxin 6	OH	H	H	SO_3^-	B2
Class IV					
Epigonyautoxin 8	H	H	OSO_3^-	SO_3^-	C1
Gonyautoxin 8	H	OSO_3^-	H	SO_3^-	C2
C3	OH	H	OSO_3^-	SO_3^-	C3
C4	OH	OSO_3^-	H	SO_3^-	C4

Figure 5.8 *Structure of saxitoxin and its naturally occurring derivatives* (adapted from Steidinger and Baden[2])

more fully later (Section 5.7.1.2.2). Brevetoxin-mediated opening of these sodium channels occurs more slowly than in the absence of the toxin;[188,191,192] however, inactivation of these channels is delayed, resulting in prolonged channel opening. The observed nerve excitabilities caused by brevetoxins are, indeed, thought to be the result of sodium channel inactivation via membrane depolarization. Alternatively, it is suggested that the toxin may cause a direct modification of the channel, which results in the channel adopting a non-conducting state.[188,192] Other possible mechanisms of action include enhanced release of acetylcholine, anti-acetylcholinesterase activity, or an enhanced postsynaptic effect.[2] Altered calcium fluxes in muscle have also been demonstrated.[193]

5.7.1.2.2 *Ciguatoxin* The sporadic toxicity of tropical reef fish is a major human health problem referred to as ciguatera. Ciguatoxin (CTX) (**3**), a lipid-soluble neurotoxin, is the major active component of ciguatera poisoning and was first isolated from the flesh of the red snapper (*Lutjanus bohar*)[194] and

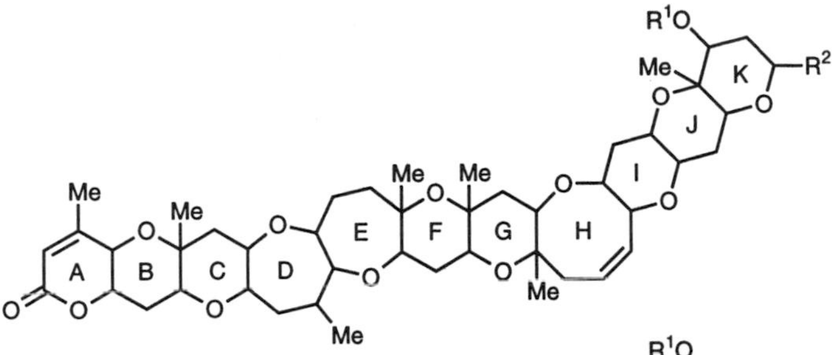

2 Type 1

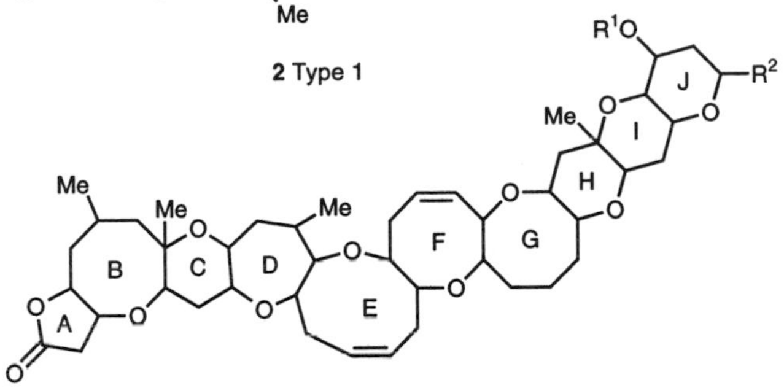

2 Type 2

Toxin	*Type*	R^1	R^2	
PbTx-1	2	H	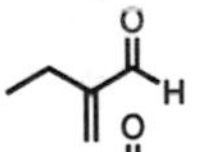 O, H	
PbTx-2	1	H	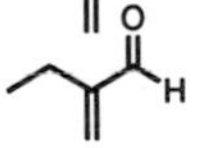 O, H	
PbTx-3	1	H	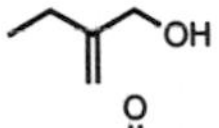 OH	
PbTx-5	1	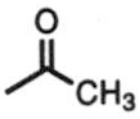 O, CH_3	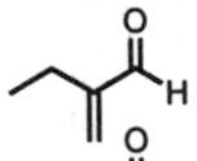 O, H	
PbTx-6	1	H	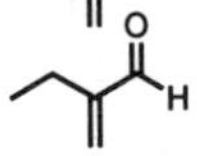 O, H	27, 28 epoxide
PbTx-7	2	H	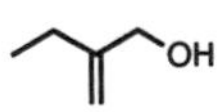 OH	
PbTx-8	1	H	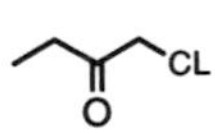 O, CL	
PbTx-9	1	H	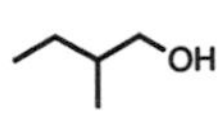 OH	
PbTx-10	2	H	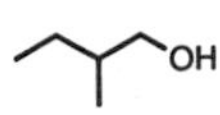 OH	

Figure 5.9 *The brevetoxins*

subsequently from the Moray eel (*Gymnothorax javanicus*).[195] Various molecular formulae have been postulated, but in 1987[196] the molecular mass was confirmed as 1111.7 and a molecular formula of $C_{60}H_{80}O_{19}$ was proposed.[197] Finally, after many years of research, the structure of CTX (**3**), which is

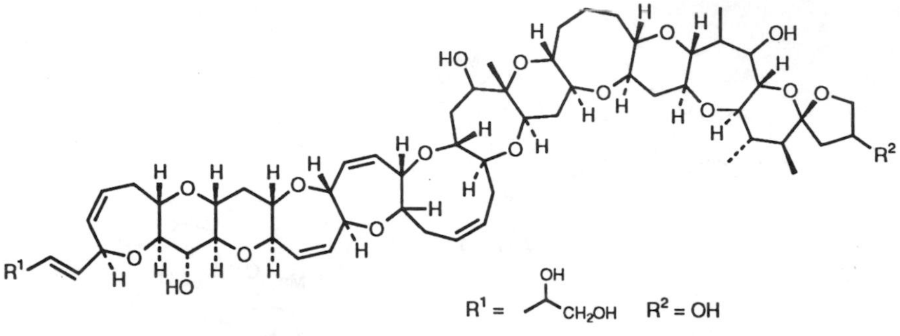

3 Ciguatoxin

reported to exist in two interchangeable forms that differ in polarity, has been elucidated by spectroscopic methods.[198]

The onset of the symptoms of ciguatera poisoning can occur within minutes of ingestion of contaminated reef fish or at any time up to 30 hours later. Symptoms include neurological, gastrointestinal, respiratory, and cardiovascular abnormalities.[91,199] An LC_{50} of 0.45 μg kg^{-1} has been reported upon intraperitoneal adminstration of CTX to mice.

Significant progress on elucidating the mechanism of action has been hampered by the unavailability of purified material.[200] Like brevetoxins, CTX is known to bind directly to the sodium channel,[190] but the receptors are different from those to which other neurotoxins bind.[201] Binding to the sodium channel results in membrane depolarization.[202] The effects of CTX on sodium channels and sodium currents have been reported in several experimental systems.[201,203–209] Indirect noradrenaline release from presynaptic sites in the neuromuscular junction in response to CTX has been reported in *in vitro* studies with guinea-pig taenia caecum,[210] vas deferens,[211] and atria[203,211,212] and *in vivo* during cardiovascular and respiratory effects. From these studies a role for adrenoreceptors has been implicated in the actions of CTX.[213] The toxins responsible for ciguatera are not unequivocally demonstrable in laboratory cultures of any one suspect dinoflagellate, although three lipid-soluble toxins (GT1, GT2, and GT3) and the water soluble toxin, maitotoxin (Section 5.7.1.2.3), which cause ciguatera-type symptoms, have been isolated from cultures of the benthic dinoflagellate, *Gambierdiscus toxicus*.[214]

5.7.1.2.3 *Maitotoxin* Maitotoxin (MTX) is another toxin thought to contribute to ciguatera poisoning. MTX is a water-soluble substance that is often found in the liver and gut of affected fish.[215] It has also been identified in laboratory cultures of *G. toxicus*,[216] and has been found to be larger than CTX, non-dialysable, and polyhydroxylated.[217] The existence of amino acids, sugars, and sphingosines in this toxin has been disputed,[217] although its positive reaction with Dragendorf's reagent indicates the presence of nitrogen. More recent analytical evidence suggests that MTX is disulfated and contains no known repeating units, carbonyl chains, or side-chains other than methyl groups, the disodium salt having an estimated molecular weight of 3424.5.[218] MTX has been suggested to be a precursor of CTX.[3]

The physiological effects of MTX are similar to those of CTX, namely neurological, gastrointestinal, respiratory, and cardiovascular abnormalities. However, MTX is reported to be more potent than CTX,[219] with a minimum lethal dose of $0.2\,\mu g\,kg^{-1}$, which makes it one of the most potent of marine toxins.[220] MTX has been reported to activate directly voltage-sensitive calcium channels.[218,221,222] This results in calcium-dependent smooth[223–226] and skeletal[227] muscle contraction, a calcium-mediated release of hormone[228] or the neurotransmitters acetylcholine[229] and noradrenaline[220,229,230] and stimulation of calcium uptake into neuronal cells, but not into excitable fibroblasts that lack voltage-sensitive calcium channels.[231] All these processes can be modified in the presence of known calcium channel antagonists. It has also been suggested that MTX may exert its effects by creating a pore similar in properties to a calcium channel in the membrane, which in turn induces the measured membrane current. Recently, it has been observed that MTX stimulates phosphoinositide breakdown in several cell systems. This breakdown appears to be dependent upon extracellular calcium, and results in the generation of secondary messengers.[232–234]

5.7.1.2.4 *Diarrhoetic Shellfish Poisons* These are all polyether, lipid-soluble toxins and are the causative agents of DSP, a condition characterized by gasterointestinal symptoms. This group of toxins can be divided into three groups based on their structure (Figure 5.10).

- *Okadaic acid and okadaic acid derivatives.* This group includes okadaic acid (**4**), (a C_{38} polyether derivative), 35(*S*)-methylokadaic acid (**5**) [Dinophysistoxin 1 (DTX-1)] and two 7-*O*-acyl derivatives of 35(*S*)-methylokadaic acid (**6**) [Dinophysistoxins 2 and 3 (DTX-2 and DTX-3)].
- *Pectenotoxins.* Pectenotoxins 1 (PTX-1) (**7**), 2 (PTX-2) (**8**), 3 (PTX-3) (**9**), and 6 (PTX-6) (**10**) share a common polyether lactone structure, but differ in their substituent at C_{43}. Although pectenotoxins 4 and 5 have been isolated, the quantities in which they were available were insufficient for structural determination.
- *Yessotoxins.* Yessotoxin (YT) (**11**) and 45-hydroxyyessotoxin comprise a series of tranfused ether rings, similar to the brevetoxins, but also contain a terminal side-chain with nine carbon atoms and two sulfate esters.

The DSPs are known to be produced by dinoflagellates, including *Dinophysis fortii*, *D. acuminata*, *D. acuta*, *D. norvegica*, and *Prorocentrum lima*.[235,236]

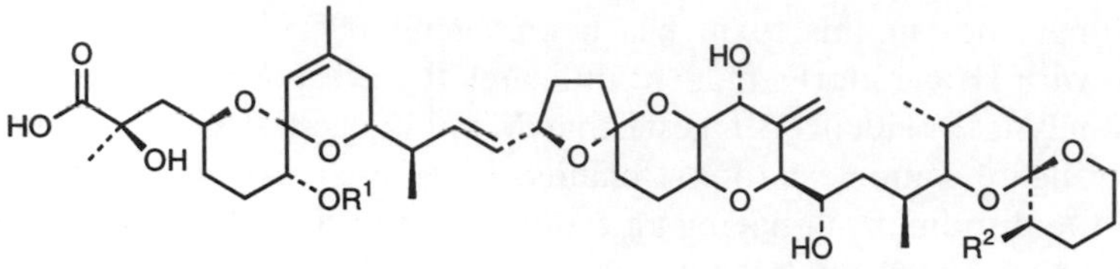

4 Okadaic Acid	$R^1 = H$	$R^2 = H$
5 Dinophysistoxin	$R^1 = H$	$R^2 = Me$
6 Dinophysistoxin	R^1 = palmitoyl	$R^2 = Me$

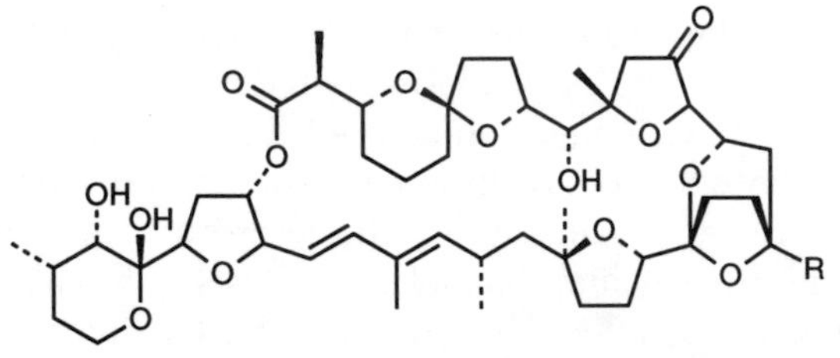

7 Pectenotoxin-1	R = CH_2OH
8 Pectenotoxin-2	R = Me
9 Pectenotoxin-3	R = CHO
10 Pectenotoxin-6	R = COOH

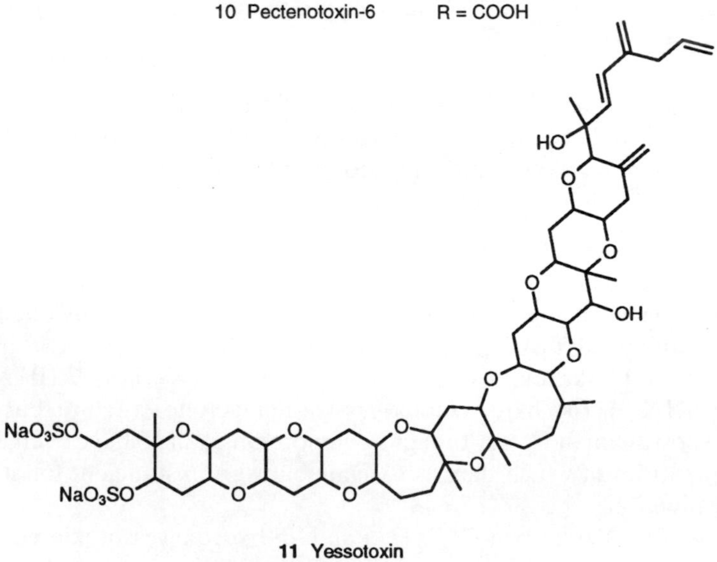

11 Yessotoxin

Figure 5.10 *Diarrhoetic shellfish poisons*

Human intoxication results from ingestion of shellfish vectors. Okadaic acid has been detected in European mussels, DTX-1 in Japanese and Norwegian mussels, and pectenotoxins only in Japanese mussels.

The mechanisms of action of these toxins remain undetermined. However, okadaic acid has been found to cause a long-lasting tonic contraction of umbilical arteries[237] and tonic contraction of both the guinea-pig ileum and rabbit aorta; it is suggested these are independent of any action of the normal stimulatory mechanisms, for example, adrenoreceptor, histamine, or seratonin. The contraction also appeared independent of any involvement of sodium and calcium channels, since treatment with sodium channel blockers and calcium-free medium [containing ethylene glycol *O,O'*-bis(β-aminoethyl)-*N,N,N',N'*-tetraacetic acid (EGTA)] had no effect.[22]

Other effects of this group of toxins have been observed. DTX-1 has been reported to affect the intestinal mucosa, causing leakage of blood from villi vessels, degeneration of the absorptive epithelium, and removal of the outer layer of this epithelium.[238] PTX-1 has been observed to cause liver damage.[238] Both okadaic acid and DTX-1 have been reported to be potent tumour-promoting agents[239] (see Section 5.7.2.3).

5.7.1.3 *Anatoxins*

Anatoxins *a*, *b*, *c* and *d* and anatoxins *a*(s) and *b*(s) are water-soluble alkaloid neurotoxins produced by strains of the freshwater cyanophyte *Anabaena flos-aquae*[240] (Figure 5.11). The anatoxin *a* (**12**) of *A. flos-aquae* strain (NRC-44-1) was the first toxin from a freshwater cyanobacteria to be defined chemically and as a result has been the most extensively studied. It is a bicyclic secondary amine, 2-acetyl-9-azabicyclo[4.2.1]non-2-ene,[241,242] with a molecular weight of 165. It can be synthesized through ring expansion of cocaine,[243,244] from iminium salts,[245,246] from nitrone,[247,248] and from 4-cycloheptenone or tetrabromotricyclooctane.[249] Anatoxin *a* is a potent, postsynaptic, depolarizing, neuromuscular blocking agent that affects both nicotinic and muscarinic acetylcholine receptors.[250–257] A related toxin, anatoxin *a*(s) produced by *A. flos-aquae* NRC-525-17 is thought to differ structurally from anatoxin *a*, because of its higher toxicity in a mouse bioassay (LD_{50}s of 50 μg kg^{-1} and 250 μg kg^{-1}, respectively), and because it apparently acts as an anticholinesterase.[258,259] The other anatoxins can be physiologically distinguished, but have not yet been fully characterized,[4] although all have been implicated in animal and human poisonings. Anatoxin *c* is reported to be a hepatotoxin[260] (Section 5.7.1.4).

5.7.1.4 *Peptide Toxins*

Low-molecular weight peptide toxins (Figure 5.12) that affect the liver, hepatotoxins, are the predominant toxins involved in cases of animal poisonings due to cyanobacteria.[261–263] They are primarily found in various strains of *Microcystis aeruginosa*, although they have also been isolated from other

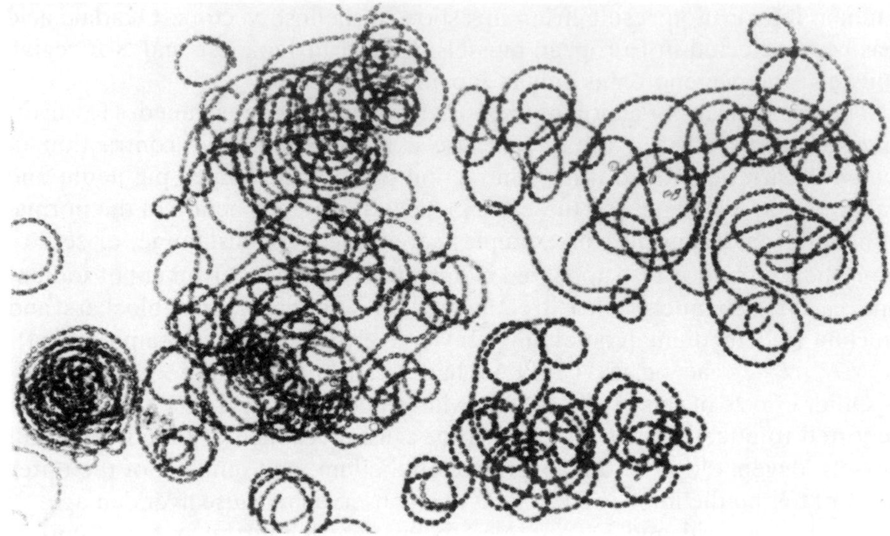

Figure 5.11 Anabaena flos-aquae *(×140)*

cyanobacteria, including *Anabaena*, *Aphanizomenon*, and *Oscillatoria* spp. (Table 5.8);[264,265] they have been extensively studied. Although only one or two toxic peptides are usually found in individual *M. aeruginosa* isolates, up to six different peptides have been found in some cases.[267,268] However, all of these peptides appear to conform to a basic pattern, which is that of a cyclic

Table 5.8 *Hepatotoxin-producing cyanobacteria and their toxins (adapted from Lincoln* et al.[22]*)*

Alga	*Toxin*	*Structure**	*Reference*
Anabaena flos-aquae	Anatoxin *c*	ND	260
Cylindrospermopsis raciborskii	Cylindrospermopsis toxin	ND	282
Microcystis aeruginosa	Fast Death Factor	†	271, 497
	Microcystin	**13**†	267, 272, 498–502
	Cyanogenosin-LA	**14**	269, 275
	Cyanogenosin-LR, -YR, -YA, -YM		270
Microcystis viridis	Cyanoviridin-RR	**15**	274
Nodularia spumigena mertens	Nodularia toxin	ND	503–505
Oscillatoria agardhii	Oscillatoria toxin	**16**	266, 280

*ND, not determined.
†Varying amino acid constituents have been reported.

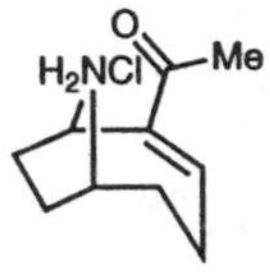

12 Anatoxin *a*

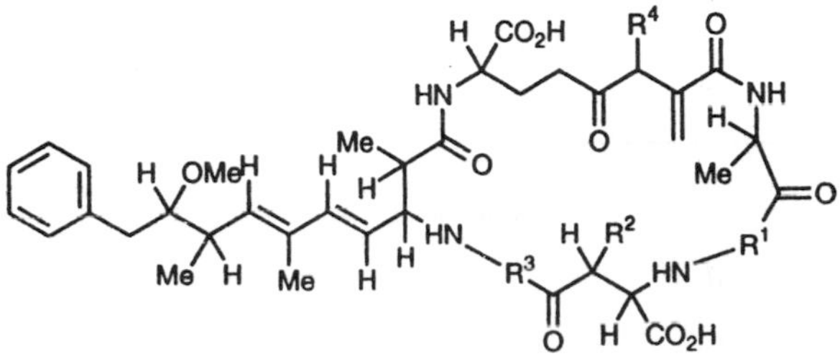

13 Microcystin

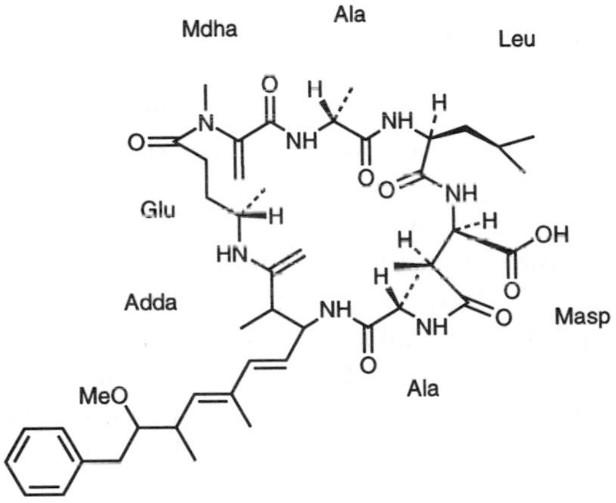

14 Cyanogenosin-LA

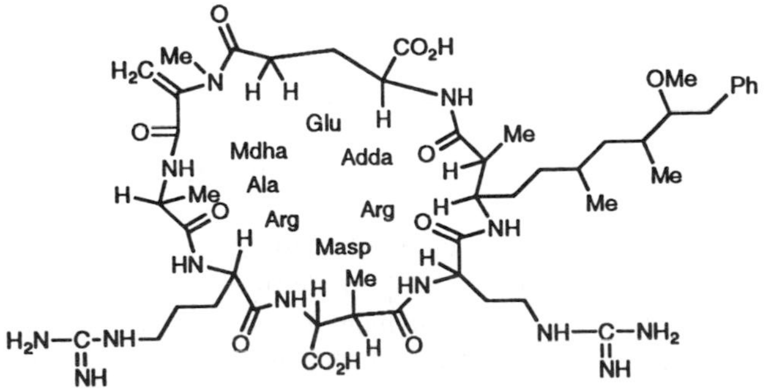

15 Cyanoviridin-RR

Figure 5.12 *Peptide toxins*

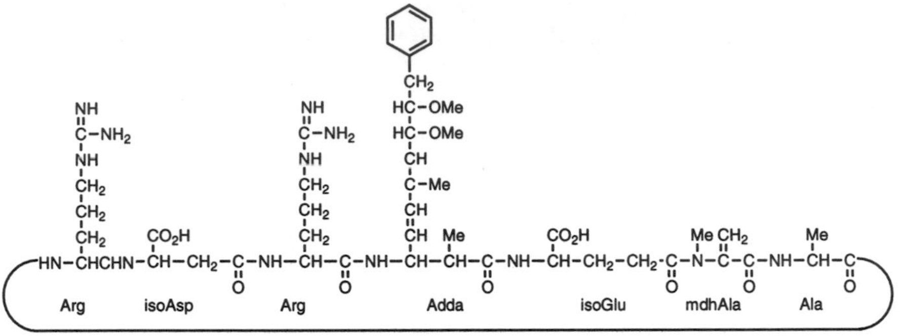

16 Oscillatoria toxin

Figure 5.12 *Continued.*

heptapeptide with a molecular weight of about 1000. The general structure is:

$$\text{cyclo}(\text{D-Ala-}R_1\text{-Masp-}R_2\text{-Adda-D-Glu-Mdha})$$

(where R_1 and R_2 are various amino acids, including leucine, alanine, arginine, tyrosine, and methionine; Masp is β-methylaspartate; Adda is a novel β-amino acid residue of 3-amino-9-methoxy-2,6,8-trimethyl-10-phenyldeca-4,6-dienoic acid; and Mdha is methyldehydroalanine[267,269,270]). Since the members of this family of toxins show similar toxicities in the standard intraperitoneal mouse bioassay (LD_{50} of 50–100 $\mu g\ kg^{-1}$), it is inferred that the toxicity of the molecule resides in the invariable components. The hepatotoxins have been given various names, including Fast Death Factor,[271] microcystin (**13**),[272] cyanoginosin,[273] cyanoviridin (**15**),[274] and cyanogenosin (**14**).[275] This multiple naming system for the hepatotoxins has been criticized by many workers in this field and a system of nomenclature based on the original term microcystin (MCYST)[276] has been proposed.

The cyanobacterial hepatotoxins show similar symptoms of poisoning, which include lethargy, diarrhoea, heavy breathing, paralysis of limbs, and death. The most obvious internal symptom of poisoning is a dark mottled liver, swollen with blood to about twice its normal weight.[156,277,278] The toxicity of the hepatotoxic peptides isolated from strains of *Oscillatoria agardhii* (XVI) and *Anabaena flos-aquae*[260,279–281] is generally lower than that of the microcystins. The principle toxins from the severely hepatotoxic *Cylindrospermopsis raciborskii* has not been fully characterized. Although it may be similar in structure to the microcystins, *i.e.*, a peptide, it differs in the pattern and onset of hepatocyte necrosis and haemorrhage *in vivo*.[282]

In vivo evidence characteristic of this liver damage includes jaundice, haemorrhaging,[283] elevated serum levels of the enzymes aspartate aminotransferase, lactate dehydrogenase, glutamate, and alkaline phosphatase,[283–287] massive hepatic necrosis, and cytological hepatocyte cell changes.[283,287] Increases in hepatic microsomal cytochrome levels have also been reported.[288] The cytological changes have been confirmed using isolated hepatocyte preparations *in vitro*,[289–291] perfusion studies with isolated liver,[292] and *in vivo* studies.[293] Radiolabelled microcystin has recently been used to confirm that the liver is the main target organ for accumulation and excretion of these toxins.[4,288,294,295] Mdha has been suggested to be of importance in target-cell specificity and cellular uptake, which may occur via the bile acid transporter.[296]

Death in animals receiving a lethal dose of toxin has been attributed to the decrease in volume of circulating blood resulting from pooling of blood in the liver following sinusoid destruction.[296–298] It has been suggested that blood pooling in the liver may result from heart failure due to pulmonary congestion;[299] however, this has been disputed recently.[297,298]

5.7.1.5 *Dermatotoxins*

Episodes of dermatitis and/or irritation from contact with dermatotoxins (Figure 5.13) from freshwater and marine cyanobacteria are occurring with increasing frequency.[3,38] The most common and best documented toxic reaction is a severe contact dermatitis known as 'swimmer's itch'. This is a cutaneous inflammation characterized by red patchiness, followed by blisters and scaling of the skin within 12 hours of exposure to the causative algae.[162]

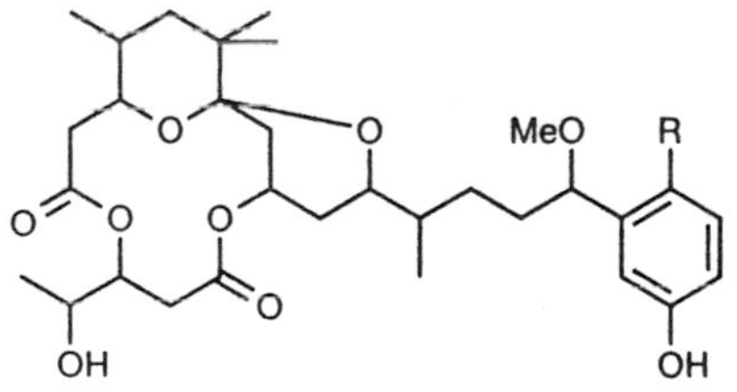

17 Debromoaplysiatoxin

18 Lyngbyatoxin A

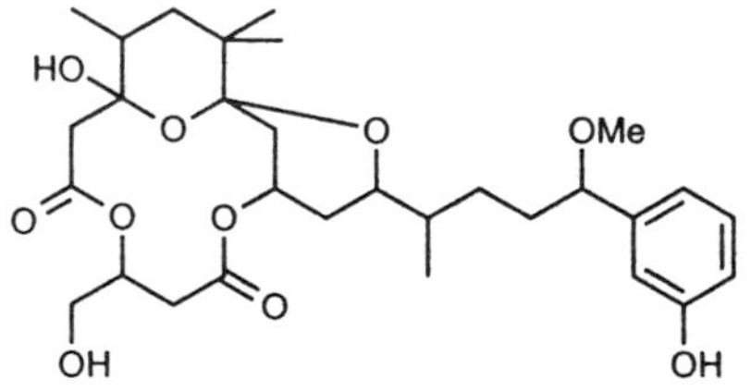

19 Oscillatoxin A

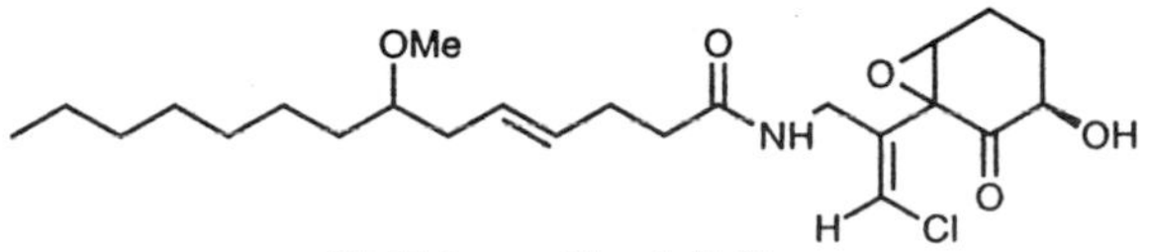

20 Malyngamides A, B, C

Figure 5.13 *Dermatotoxins*

This condition is induced by exposure to two compounds, debromoaplysiatoxin (**17**), an acetogenic bislactone, and lyngbyatoxin A (**18**), an indole alkaloid isolated from the marine cyanobacteria *Lyngbya majuscula*.[3,300–304] Similar irritant properties have also been assigned to oscillatoxin A (**19**), isolated with debromoaplysiatoxin from a mixture of the cyanobacteria *Oscillatoria nigroviridis* and *Schizothrix calcicola*.[305,306] In freshwater environments dermatotoxic blooms have been shown to be dominated by *Anabaena*, *Aphanizomenn*, *Gloeotrichia*, and *Oscillatoria* spp., although the toxins are not yet characterized.[307] Lesser irritants, malyngamides A, B, and C (**20**)[308,309] have been isolated from shallow water varieties of *Lyngbya majuscula*, and malyngamides D and E from deep water varieties.[310] Both debromoaplysiatoxin and lyngbyatoxin have also been observed to be potent tumour promoters (Section 5.7.2.3) in mice.[311] These compounds, when topically applied to mouse skin, induce epidermal ornithine decarboxylase activity.[3] How, and if, this is related to their dermatotoxic mechanism of action remains unclear.

5.7.1.6 *Other Toxins*

5.7.1.6.1 *Domoic Acid* The nitrogen-containing heterocyclic amino acid, domoic acid (**21**), has been isolated from three Rhodophytes, *Alsidium corallinum*,[312] *Chondria armata*,[313,314] and *Chondria baileyana*,[315] and two diatoms, *Amphora coffeaeformis*[316] and *Nitzchia pungens*.[317,318] Domoic acid is a neurotoxin and has been shown to cause amnestic shellfish poisoning, intoxication resulting from eating contaminated cultivated blue mussels (*Mytilis edulis*), which accumulate the toxin. The symptoms include abdominal cramp and neurological responses involving disorientation and memory loss.[317,318] Domoic acid is reported to be a potent neuronal depolarizing and excitory substance in invertebrates,[319,320] and to activate neurons of the spinal cord in the frog and the rat.[321,322]

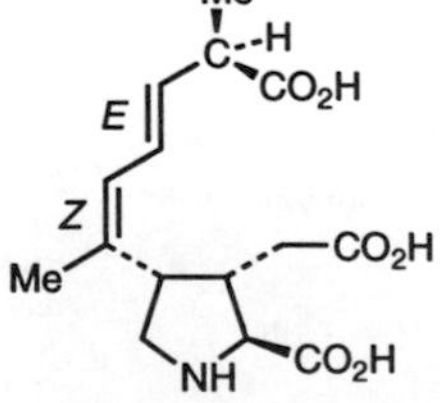

21 Domoic acid

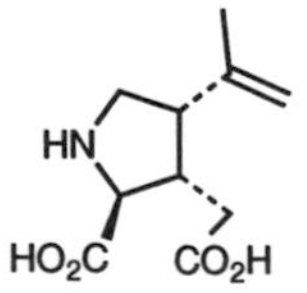

22 Kainic acid

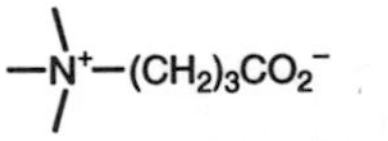

23 γ-Aminobutyric acid

Domoic acid is structurally related to the excitory amino acid α-kainic acid (**22**), and both compounds antagonize the effects of the central nervous system neurotransmitter glutamate. Glutamate is found in particularly high concentrations in the brain, where it is involved in the regulation of nerve-cell toxicity and as a precursor to the inhibitory neurotransmitter γ-aminobutyric acid (GABA) (**23**).[323,324]

5.7.1.6.2 *Ichthyotoxins* Many of the toxins already mentioned (Table 5.7) have been associated with the death of fish and are, therefore, termed ichthyotoxins, shown in Table 5.9 and Figure 5.14. The Prymnesiophyte *Prymnesium parvum* (Figure 5.2) is known to produce a cation activated ichthyotoxin[325–327] which is responsible for reversible gill damage in aquatic animals, such as fish or tadpoles. The name 'prymnesin' was first used by Yariv and Helstrin to describe the extracellular toxic material collected from water samples rich in *P. parvum*.[325] However, it now seems that there are a family of toxins that are acidic, polar phospho-proteolipids,[49] showing a broad spectrum of different biological activities, including haemolytic, cytotoxic, and bacteriolytic activities. The mode of action of these *Prymnesium* toxins seems to be based on a reversible increase in the permeability of biological membranes.[329,330]

5.7.1.6.3 *Endotoxins* Cyanobacterial lipopolysaccharide (LPS) endotoxins, also produced by other Gram-negative prokaryotes,[331] have been

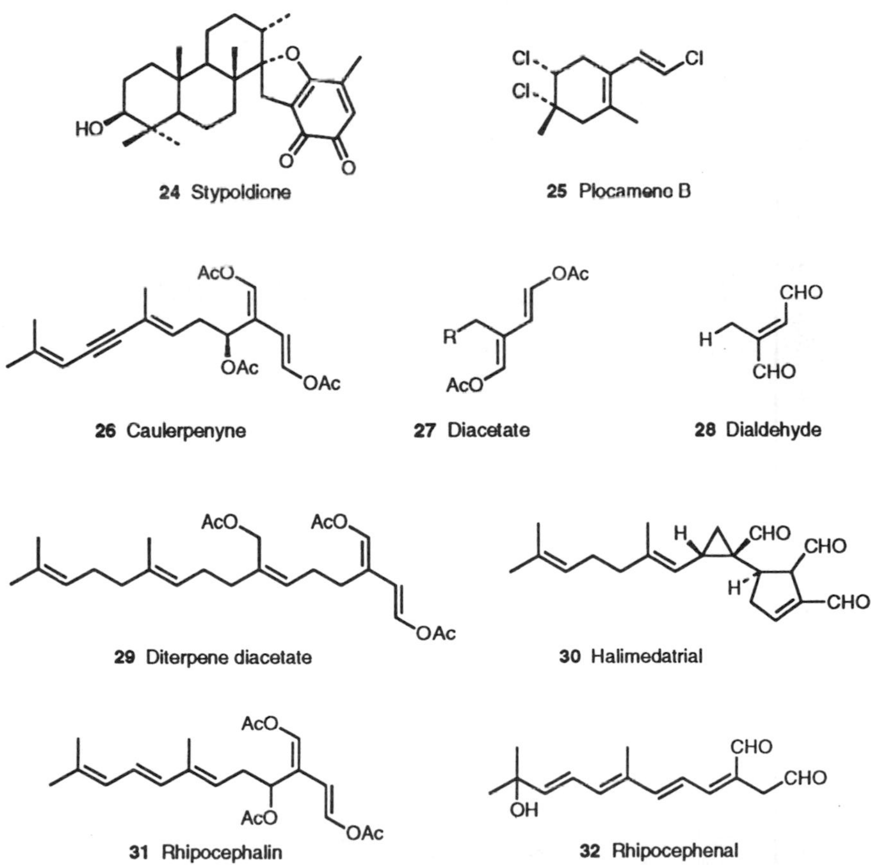

Figure 5.14 *Ichthyotoxins (continued on p. 150)*

Table 5.9 *Ichthyotoxic compounds from algae (adapted from Lincoln* et al.[22]*)*

Chemical Nature	*Toxin*	*Structure*	*Source*	*Ichthyotoxic or natural fish kills*	*References*
Alkaloid	Glenodinine	ND	*Peridinium polonicum*	FK	506
Amides	Choline esters		*Amphidinium carterae*	I	507
	Saxitoxin, neosaxitoxin, gonyautoxins		*Gonyaulax monilata*		
			Gonyaulax tamarensis	FK	2
			Pyrodinium bahamense	I/FK	2
Fatty acids	Unnamed	ND	*Chaetomorpha minima*	I	341
Peptides	Hormothamnin A	ND	*Hormothamnon enteromorphoides*	I	508
Polyether toxins	Brevetoxins		*Ptychodiscus brevis*	I/FK	89, 328
Proteolipid–sugar complex	Prymnesium toxin	**131**	*Prymnesium parvum*	I/FK	325–327
Quinone	Stypoldione	**24**	*Stypopodium zonale*	I/FK	509
Terpenes					
(i) Monoterpene	Plocamene B	**25**	*Plocamium cartilagineum*	I	510
(ii) Diterpenoid	Caulerpenyne	**26**	*Caulerpa ashmeadii*	I	511
	Diacetate	**27**	*Caulerpa bikinensis*	I	512
	Dialdehyde	**28**	*Caulerpa bikinensis*	I	512
	Diterpene diacetate	**29**	*Penicillus diemetosus*	I	370
	Halimedatrial	**30**	*Halimeda* sp.	I	367
(iii) Sesquiterpenoids	Rhipocephalin	**31**	*Rhipocephalus phoenix*	I	513
	Rhipocephenal	**32**	*Rhipocephalus phoenix*	I	513

(iv) Cyclic diterpenoids	Acetoxycruenulide	**33**	*Dictyota crenulata*	I	514
	5,6-Diacetoxy-10,18-dihydroxy-2,7-dolabelladiene	**34**	*Dilophus fasciola*	I	515
	5,6,10-Triacetoxy-18-hydroxy-2,7-dolabelladiene	**35**	*Dilophus fasciola*	I	515
	5,6,10,18-Tetraacetoxy-2,7-dolabelladiene	**36**	*Dilophus fasciola*	I	515
	Dilophic acid	**37**	*Dilophus guineensis*	I	516
	Laurinterol	**38**	*Laurencia* sp.	I	517
(v) Cyclic sequiterpenoids	Chamigrene alcohol	**39**	*Laurencia* sp.	I	517
	Prepacifenol		*Laurencia* sp.	I	517
Unknown	Unnamed	ND	*Amphidinium klebsii*	I	518
	Unnamed	ND	*Amphidinium rhynchocephalum*	I	518
	Unnamed	ND	*Ochromonas dancia*	I	519
	Unnamed	ND	*Ochromonas malhamensis*	I	519

*I, Ichthyotoxic; FK, natural fish kills; ND, not determined.

33 Acetoxycruenulide

34 5,6-Diacetoxy-10,18-dihydroxy-2,7-dolabelladiene

35 5,6,10-Triacetoxy-18-hydroxy-2,7-dolabelladiene

36 5,6,10,18-Tetracetoxy-2,7-dolabelladiene

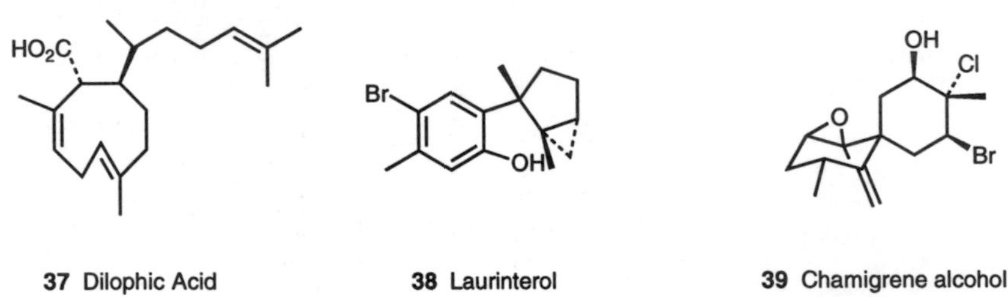

37 Dilophic Acid

38 Laurinterol

39 Chamigrene alcohol

Figure 5.14 *Continued.*

implicated as causative agents of various ailments, including toxaemias and gastroenteritis.[332,333] These conditions have occurred among many people at freshwater sites containing *Anabaena flos-aquae*, other *Anabaena* spp., *Microcystis aeruginosa*, *Oscillatoria* spp., and *Schizothrix calcicola*.[3,332–336] Chemical studies have established a detailed structure of many bacterial LPS endotoxins, which are amphiphilic compounds composed of lipid A and polysaccharide moieties, with lipid A being the toxic principle.[337] Cyanobacterial LPS differs from bacterial LPS because of the variable presence of 2-keto-3-deoxyoctonate, heptose, galactose, and glucosamine, and the usual absence of phosphates from lipid A.[335] Cyanobacterial LPS and lipid A hydrolysate of LPS preparations are active in mouse lethality tests, but are not as potent as bacterial LPS preparations.[335,336] The precise mechanism of action of these toxins remains undetermined.

5.7.1.6.4 *Miscellaneous Toxins from Chlorophytes* *Caulerpa racemosa* and certain other *Caulerpa* species produce two substances known as caulerpicin (**40**) and the unusual orange–red pigment caulerpin (**41**).[338–340] These are thought to be toxins which, after being transferred through the food chain, are responsible for several effects, including a mild numbness of the tongue, dizziness coupled with a cold sensation in the feet and hands, difficulty in

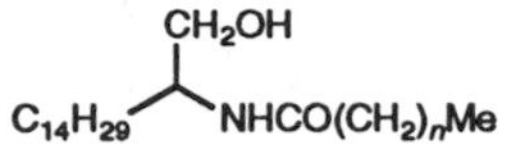

40 Caulerpicin 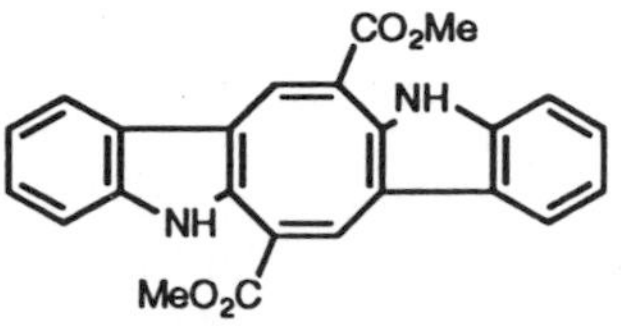 **41** Caulerpin **42** Palmitic acid

Miscellaneous toxins from chlorophytes

breathing, and loss of balance, although their involvement is questionable. The green alga *Chaetomorpha minima* is toxic to fish and has haemolytic activity, although little is known of its definitive biological properties; the ichthyotoxicity and haemolytic activity have been attributed to free fatty acids.[341] Another green alga, *Ulva pertusa*, also has several haemolytic fractions,[342] two being water-soluble and one fat-soluble. The fat-soluble haemolysin is palmitic acid (**42**), one of the water-soluble haemolysins is thought to be a galactolipid ($C_{31}H_{58}O_{14}$), while the other is thought to be a sulfolipid ($C_{25}H_{47}O_{11}SK$).[88]

5.7.2 Compounds with Pharmaceutical Potential

The pharmaceutical uses of algal products are varied, but to date have primarily involved the phycocolloids isolated from selected members of the Phaeophyceae and Rhodophyceae. These uses include the use of carrageenan in cough syrup emulsions, of fucoidan and agar as binding agents for medical tablets, and of agar as a blood anticoagulant, laxatives, or dental moulds. The uses of algae for pharmaceutical purposes have been reviewed previously.[145,147,149] Although the use of such algal products in these and other ways will continue, considerable efforts are now being made to exploit novel 'biologically-active' compounds produced by algae, which is the subject of this section. However, two groups of compounds are described which have no direct relevance to the development of pharmaceuticals. These are neoplastic agents, which are of interest as research tools to help understand the biochemical and molecular processes involved with the growth of tumours, and haemolysins, which are used in the chemical treatment of blood cells to allow the analysis of haemoglobin or enzymes present within these cells.

5.7.2.1 *Antimicrobial Compounds*

Algal broths and extracts have been screened extensively for the presence of antimicrobial compounds, including those showing antialgal, antibacterial, antifungal, and antiviral activity. Examples of such compounds have been reported from every algal class and can conveniently be divided into those from microalgae[18,117,119,121,123,343–349] and those from macroalgae.[120,350–354] Although the algae are not being used presently as a commercial source of antibiotics, the potential exists.

5.7.2.1.1 *Antimicrobials from Microalgae* Active compounds from microalgae include indole alkaloids, lactones, halogenated lipids, nucleosides, peptides, fatty acids, gallotannins, glycolipids, lipoproteins, terpenoids, and chlorophyll precursors or degradation products.

The first partly identified antimicrobial substance isolated from algae was chlorellin,[355] a mixture of peroxides of unsaturated fatty acids[356,357] obtained from unicellular green algae, particularly *Chlorella* (Figure 5.15). Chlorellin

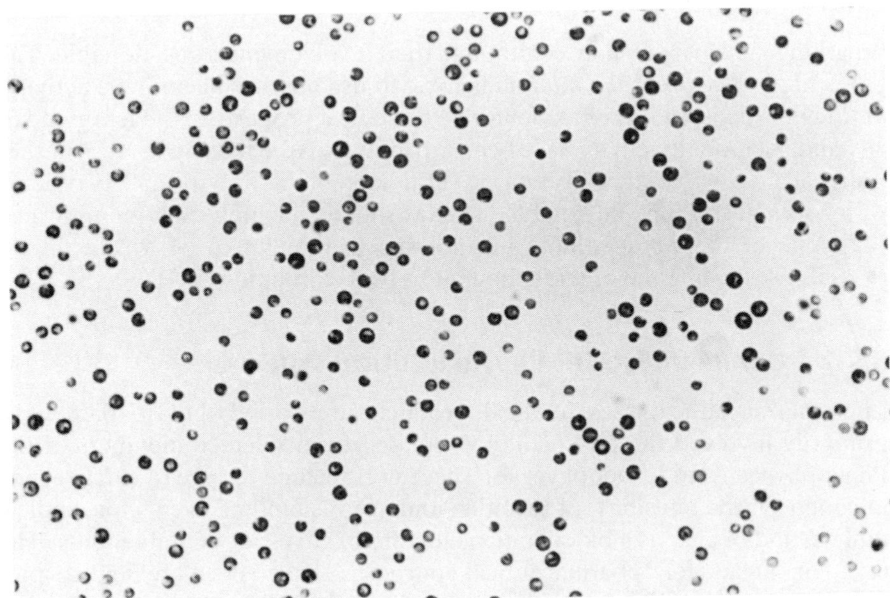

Figure 5.15 Chlorella sp. *(×580)*

exhibited inhibitory activity against both Gram-positive and Gram-negative bacteria. Since this discovery many substances exhibiting antimicrobial activity have been identified from members of the Chlorophyta, Bacillariophyta, Cyanophyta, Chrysophyta, and Dinophyta. Many algae may produce at least two different antibiotic substances.[344] The antimicrobials produced by microalgae are too numerous to mention, but examples of active compounds for which partial or complete purification has been achieved are shown in Table 5.10 and Figure 5.16.

More recently, compounds showing anti-human immunodeficiency virus (HIV) activity have been identified. Sulfonic acid containing glycolipids (**58**) isolated from unialgal laboratory cultures of the cyanobacteria *Lyngbya lagerheimii* and *Phormidium tenue* have been found to suppress the cytopathic effects of HIV-1 virus in cultured human lymphoblastoid CEM, MT-2, LDV-7, and C3-44 cell lines. Inhibition was found to be dependent on both the cell line and the mode of HIV infection. These compounds were inhibitory over a relatively wide (1–100 μg ml^{-1}) concentration range and, importantly, at non-cytotoxic concentrations.[358] Other cyanobacterial extracts known to contain phospholipids (*Anabaena variabilis*, *Calothrix elenkinii, Oscillatoria raciborskii*, *Phormidium cebennse*, and *Scytonema burmanicum*) have also been observed to be inhibitory to HIV-1.[359]

5.7.2.1.2 *Antimicrobials from Macroalgae* More is known of the chemical nature of antimicrobials in macroalgae than for those in microalgae, and an indication of their variety is shown in Table 5.11 and Figure 5.22. Earliest reports on compounds exhibiting such activity date from 1951[117,345,360–363] and have been reviewed in detail by Sieburth,[364,365] Glombitza,[300] and Berdy *et al.*[146]

The *in vitro* antimicrobial activity associated with Rhodophyceae has been attributed to the production of a variety of halogenated compounds including, haloforms, halogenated alkanes, alkenes, alcohols, aldehydes, and ketones. These halogenated compounds act primarily as alkylating agents and are generally considered non-specific toxins, but have, however, commercial potential, particularly as antiseptic agents. Haloacetones have been shown to exert their inhibitory effect on enzymes via cross-linking of serine and histidine residues.[22]

Other compounds that exhibit antimicrobial activity and are produced by macroalgae include acrylic acid, terpenes, sulfur-containing heterocyclic compounds, phenolic compounds, and isoprenelated and brominated hydroquinones.

The bacteriostatic action of the phenolic compounds, which are thought to act as non-specific enzyme inhibitors,[15] is the basis for these compounds' use as antiseptics.

There are relatively fewer reports of antimicrobials from green seaweeds, although antimicrobial and cytotoxic terpenoids from some tropical Udotaceae have been reported,[366] along with other reports of related sesquiterpenoids and diterpenoids.[366–370]

The range of antimicrobial activities identified from algae suggests a great deal of commercial potential as therapeutic or antiseptic and cleansing agents in soaps, lotions, and shampoos. However, it has been commented that algae seem a rather disappointing source of commercial compounds, in so far as only a very small mumber show activity *in vivo* at toxicologically safe concentrations.[348] Of the compounds which were safe, many became bound to serum proteins and were either inactivated, not absorbed, or not circulated to the site of infection, or were metabolized to inactive forms or excreted.[15] The most promising potential application for these algal compounds is as leads for the synthesis and design of antimicrobial drugs.

A recent discovery, as previously mentioned (Section 5.7.2.1.1), is the inhibition of HIV-1 by a sulfated carrageenan (polysaccharide) from the red alga *Schizymenia pacifica*. This has been reported to cause an inhibition of viral replication *in vitro*, as a result of inhibition of the enzyme reverse transcriptase, with minimal effect on the human DNA polymerase and RNA polymerase III enzymes.[371,372]

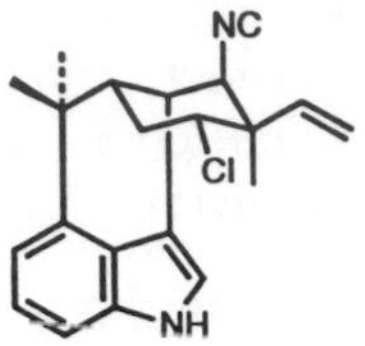

43 Hapalindole A

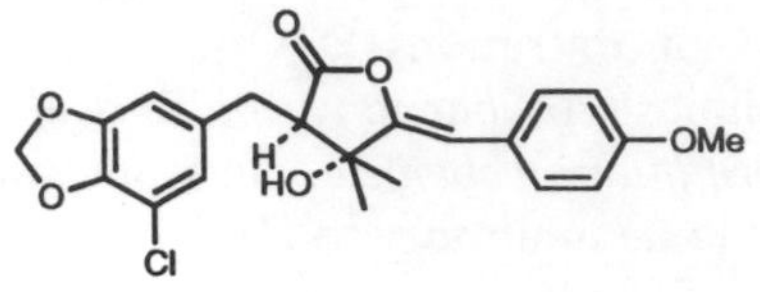

44 Cyanobacterin

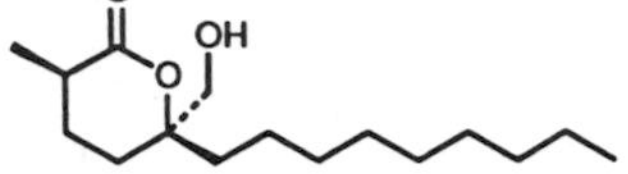

45 Malyngolide

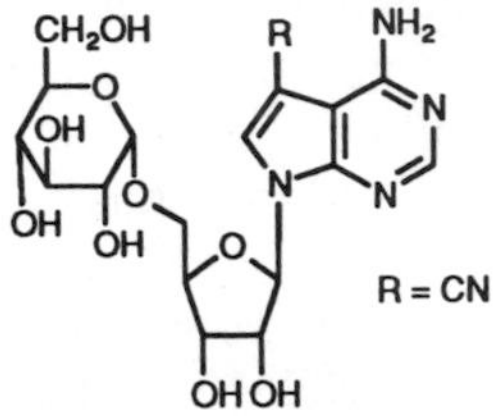

46 Toyocamycin 5-α-D-glucopyranose

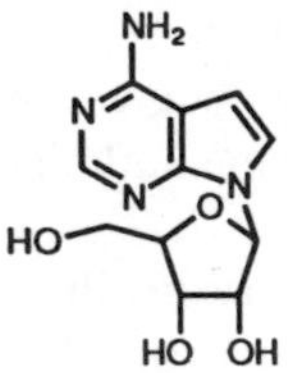

47 Tubercidin

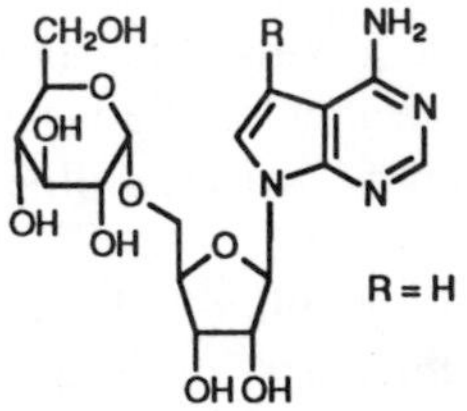

48 Tubercidin 5-α-D-glucopyranose

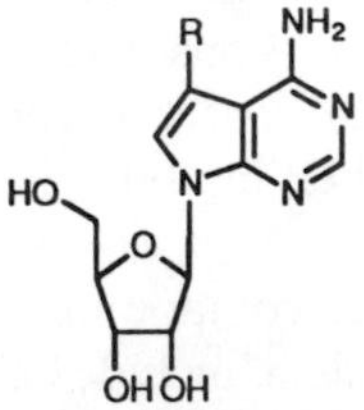

49 Toyocamycin

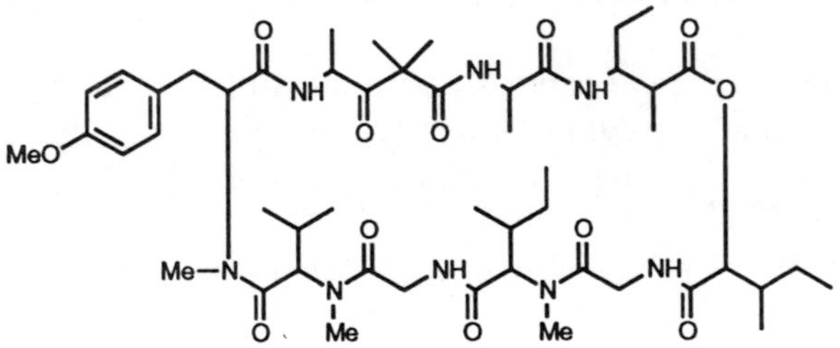

50 Majusculamide C

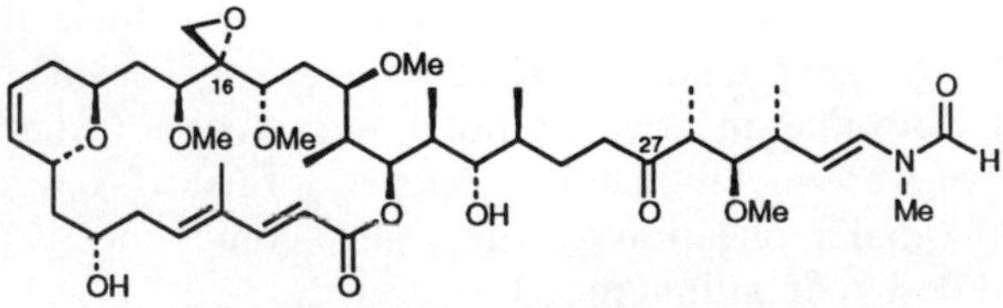

51 Scytophycin B

Scytophycin A: same as B except at C-27

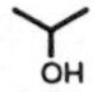

Scytophycin C: same as B except at C-16

Scytophycin D: same as B except at C-16

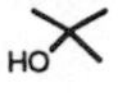

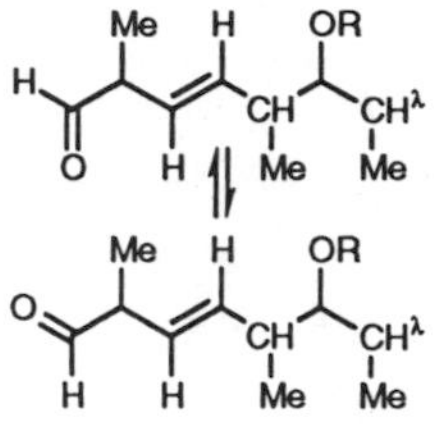

52 Tolytoxin

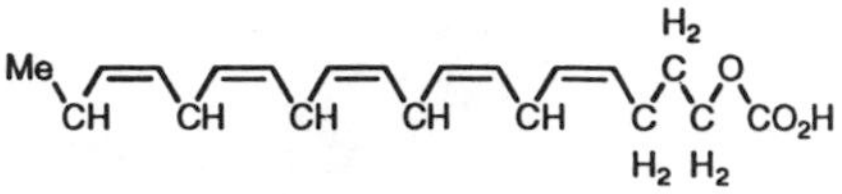

53 *Cis*-Eicosapentaeonic acid

R = (*E*)-phytyl
54 Phytyl ester

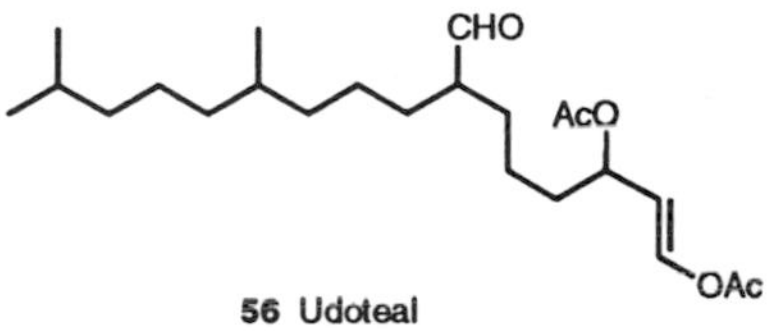

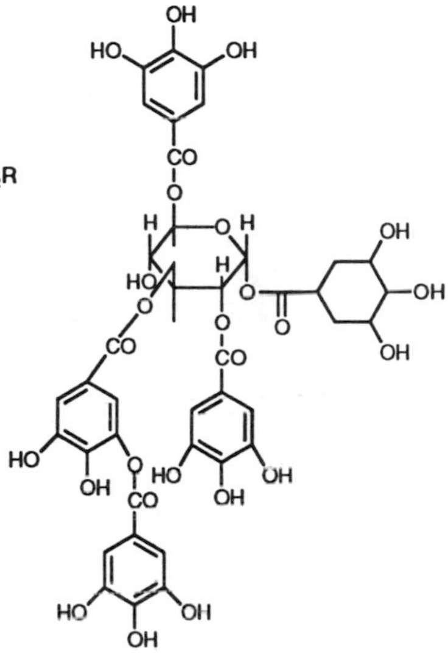

55 Pentagalloyl glucose

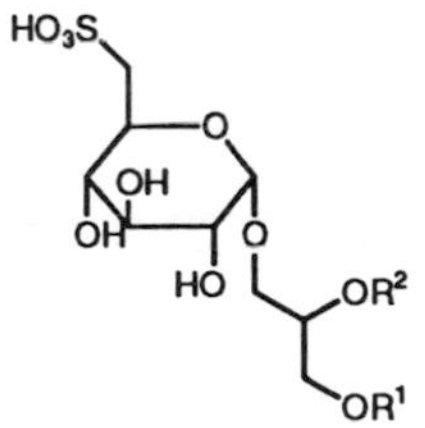

56 Udoteal

$H_2C{=}CHCO_2H$
57 Acrylic acid

R^1	R^2
18.3	16.0
18.2	16.0
18.1	16.0
16.1	16.0

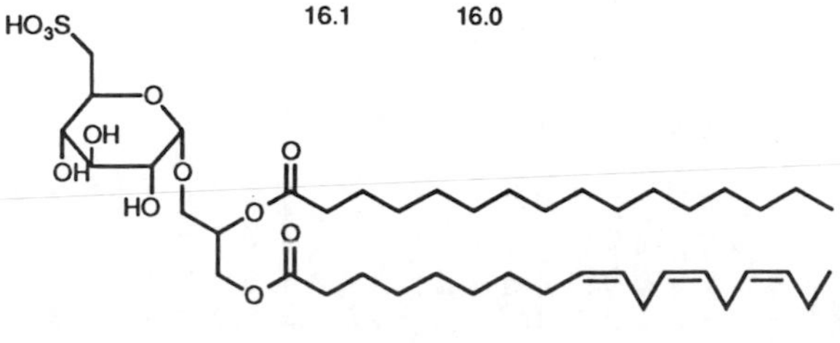

58 Sulfonic acid containing glycolipids

Figure 5.16 *Antimicrobials from microalgae*

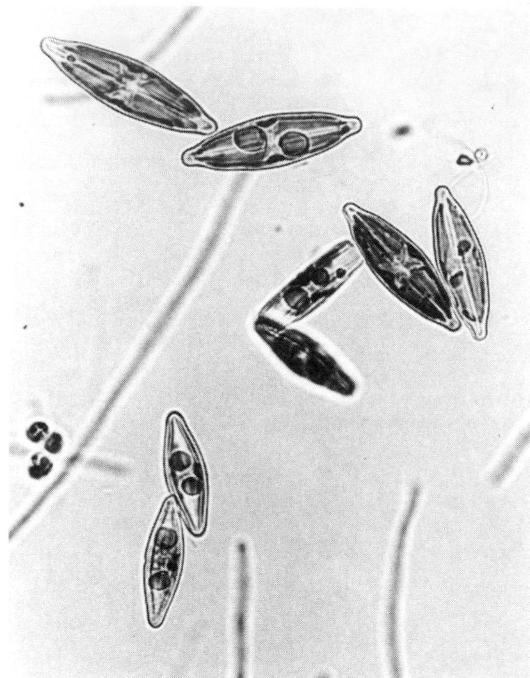

Figure 5.17 Navicula *sp. (×550)*

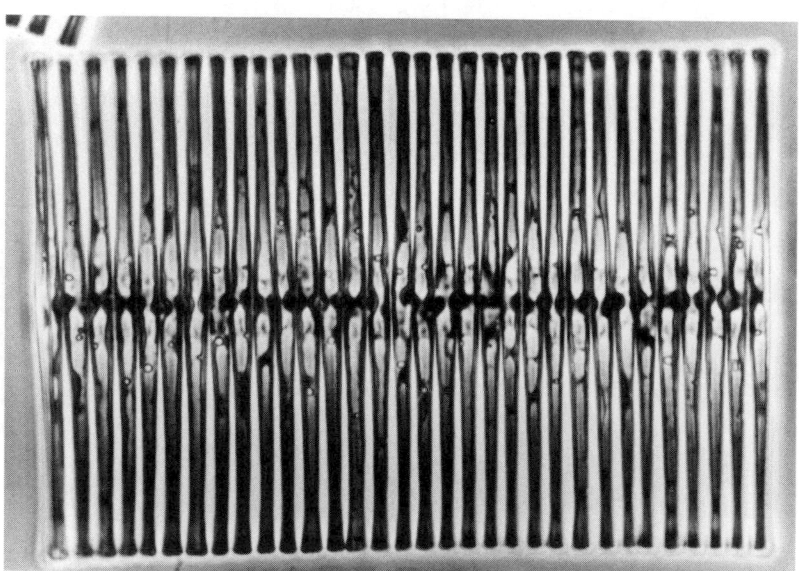

Figure 5.18 Fragilaria crotonensis *phase contrast (×780)*

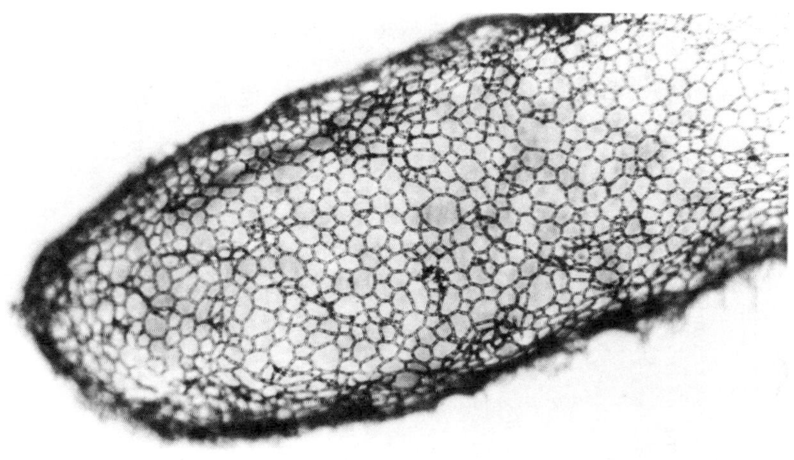

Figure 5.19 *Part of a young net of* Hydrodictyon reticulatum (×40)

Figure 5.20 Spirogyra *sp. (×130)*

Table 5.10 *Antimicrobials from microalgae (adapted from Lincoln* et al.[22]*)*

Chemical nature	*Compound name*	*Structure**	*Antibiotic spectrum*	*Source*	*Reference*
Cyanobacteria					
Indole alkaloid	Debromoaplysiatoxin	**17**	Antibacterial	*Lyngbya majuscula*	308
	Hapalindole A	**43**	Algicidal, Antifungal	*Hapalosiphon fontinalis*	520
Diacyl substituted lactone	Cyanobacterin	**44**	Algicidal, Antibacterial	*Scytonema hofmanni*	521–523
Lactone	Malyngolide	**45**	Antifungal, Antibacterial	*Lyngbya majuscula*	306, 349
Nucleosides	Toyocamycin 5-α-D-glucopyranose	**46**	Antifungal	*Tolypothrix tenuis*	524
	Tubercidin	**47**	Antifungal	*Tolypothrix byssoidea*	525
				Scytonema saleyeriense	524
	Tubercidin 5-α-D-glucopyranose	**48**	Antifungal	*Plectonema radiosum*	524
				Tolypothrix distorta	524
	Toyocamycin	**49**	Antifungal	*Tolypothrix tenuis*	524
Peptides	Hormothamion A	ND	Antibacterial	*Hormothamnion enteromorphoides*	508
	Majusculamide C	**50**	Antifungal	*Lyngbya majuscula*	495
	Scytophycins	**51**	Antifungal	*Scytonema pseudohofmani*	526
	Scytonema A		Antifungal		
	Tolytoxin	**52**	Antifungal	*Scytonema ocellatum*	306
				Scytonema mirabile	524
Terpene/carbohydrate	Unnamed	ND	Antibacterial	*Trichodesmium erythraeum*	527
Phospholipids/glycolipids	Unnamed		Antiviral	Various	358
Bacillariophyta					
Aqueous phases and lipophilic substances	Mixed fatty acids: $C_{16:4}$, $C_{16:3}$, $C_{16:2}$, $C_{18:4}$, $C_{20:5}$	Main lipid not determined	Antibacterial	*Skeletonema* sp.	528
	Unnamed	ND	Antibacterial	*Phaeodactylum tricornutum*	528
Nucleosides	Unnamed	ND	Antibacterial	*Asterionella japonica*	529

Photoactivated lipid	*cis*-Eicosapentaeonic acid	**53**	Antibacterial	*Asterionella japonica*	529–532
			Antibacterial	*Chaetoceros lauderi*	532
	Phytlester	**54**	Antibacterial	*Navicula delognei* (Figure 5.17: *Navicula* sp.)	151, 531, 532
Acid polysaccharide	Unnamed	ND	Antifungal	*Chaetoceros lauderi*	347
Fatty acids	Unnamed	ND	Antibacterial	*Bacteriastrum elegans*	530
				Chaetoceros lauderi	533
				Chaetoceros socialis	344
				Chaetoceros peruvianus	344
				Skeletonema costatum	344
				Thalassiosira spp.	344
Peptide	Unnamed	ND	Antifungal	*Fragilaria pinnata* (Figure 5.8: *F. crotonensis*)	*347*
Chlorophyta					
Unsaturated fatty acids	Chlorellin	ND	Antibacterial	*Chlorella* sp. (Figure 5.15)	355, 356
Fatty acids	Unnamed	ND	Antibacterial	*Chlamydomonas reinhardtii*	534
	Unnamed	ND	Antibacterial	*Hydrodictyon reticulatum* (Figure 5.19)	535
	Unnamed	ND	Antibacterial	*Protosiphon botryoides*	536
	Unnamed	ND	Antibacterial	*Stichococcus mirabilis*	536
Acrylic, fatty acid terpenoid fractions	Unnamed	ND	Antibacterial	*Stichococcus mirabilis*	117, 362, 363
Gallotannins	Gallotannin	ND	Antiviral	*Spirogyra* sp. (Figure 5.20)	537
	Pentagalloyl glucose	**55**	Antibacterial	*Spirogyra varians*	538
Precursor/chlorophyll degradation product	Chlorophyllides	ND	Antibacterial	*Chlorella vulgaris*	539
				Chlorella spp.	540
				Scendesmus quadricauda (Figure 5.21)	541
				Chlamydomonas reinhardtii	542
Udoteal	Unnamed	**56**	Antibacterial	*Udoteya floabellum*	543, 544

Continued

Table 5.10 *Continued*

Chemical nature	*Compound name*	*Structure**	*Antibiotic spectrum*	*Source*	*Reference*
Chrysophyta					
Fatty acids			Algicidal	*Ochromonas danica*	545, 546
Peptide			Antibacterial	*Stichochrysis immobilis*	547
Terpene/carbohydrate	Unnamed	ND	Antibacterial	*Isochrysis* spp., *Coccolithus* sp., *Monochrysis* sp.	346
Glycolipid/lipoprotein	Unnamed	ND	Antibacterial Antiprotozoan	*Prymnesium parvum* (Figure 5.2)	330 49
Dinophyta					
Terpenoid lactone	Unnamed	ND	Antifungal	Symbiotic dinoflagellates	548
Prymnesiophyta					
Acrylic acid	Acrylic acid	**57**	Antibacterial	*Phaeocystis pouchetii*	549

*ND, not determined.

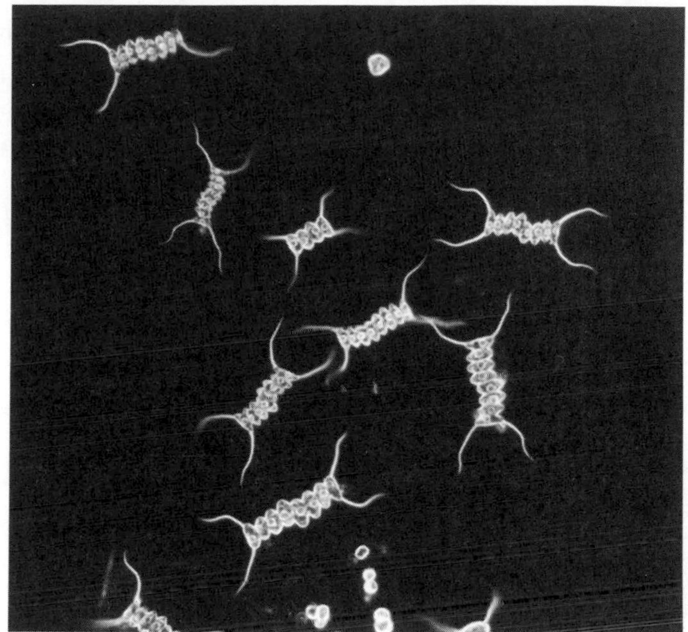

Figure 5.21 Scenedesmus quadricauda, *dark field illumination (×230)*

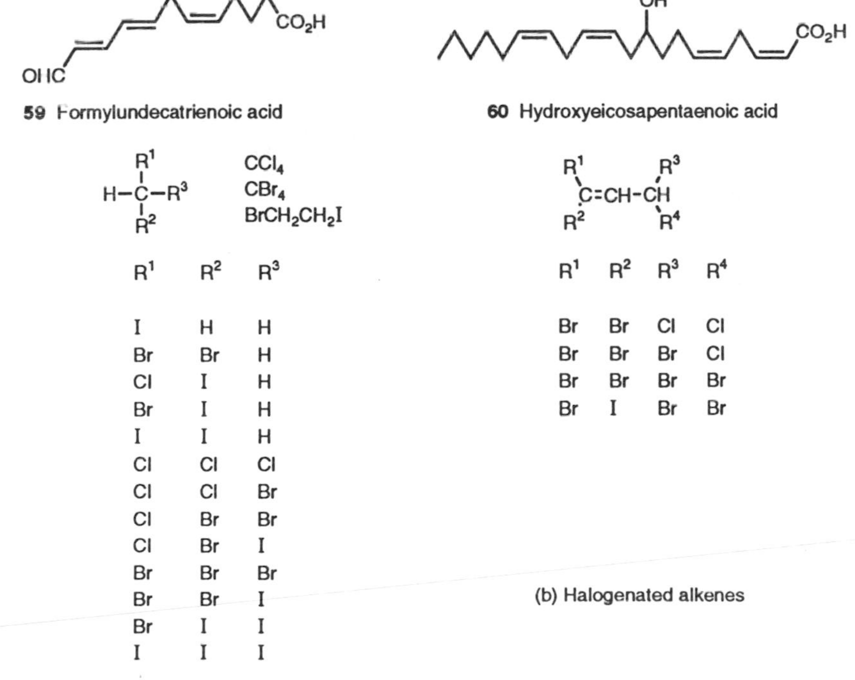

Figure 5.22 *Antimicrobials from macroalgae*

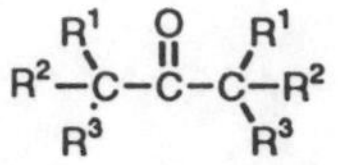

$R^2-C(R^1)(R^3)-C(=O)-C(R^1)(R^2)(R^3)$

R[1]	R[2]	R[3]	R[4]	R[5]	R[6]
Br	H	H	H	H	H
I	H	H	H	H	H
Cl	H	H	Cl	H	H
Cl	H	H	Br	H	H
Cl	H	H	I	H	H
Br	H	H	Br	H	H
Br	H	H	I	H	H
Br	Br	H	H	H	H
Cl	Cl	H	Cl	H	H
Cl	Cl	H	Br	H	H
Cl	Br	H	Cl	H	H
Cl	Br	H	Br	H	H
Br	Br	H	Cl	H	H
Br	Br	H	Br	H	H
Br	Br	H	I	H	H
Cl	Cl	H	Cl	Cl	H
Cl	Br	H	Cl	Cl	H
Cl	Br	H	Cl	Br	H
Br	Br	H	Cl	Cl	H
Br	Br	H	Cl	Br	H

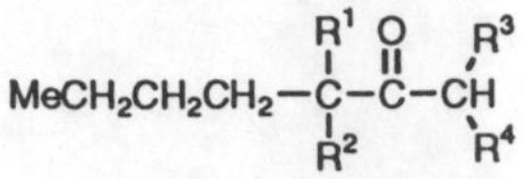

$MeCH_2CH_2CH_2-C(R^1)(R^2)-C(=O)-CH(R^3)(R^4)$

R[1]	R[2]	R[3]	R[4]
Br	H	Br	H
Br	H	Br	Br
Br	Br	Br	H
Br	Br	I	H
Br	Br	Br	Br

62 Saturated ketones

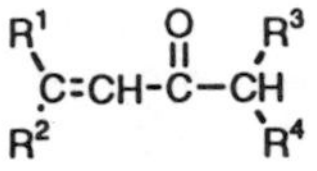

$R^1R^2C{=}CH-C(=O)-CH(R^3)(R^4)$

R[1]	R[2]	R[3]	R[4]
Br	Br	H	H
Br	Br	Br	H
Br	Br	I	H
Br	Br	Br	Br
Br	Br	Br	Cl
Br	Cl	Br	Br
Cl	Cl	Br	Br
Br	Cl	Br	Cl

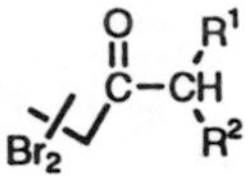

Br_2 (cyclopropyl) $-C(=O)-CH(R^1)(R^2)$

R[1]	R[2]
H	H
Cl	H
Br	H
I	H
Cl	Br
Br	Br

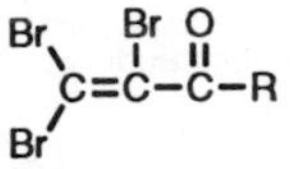

$Br_2C{=}C(Br)-C(=O)-R$

R
CH_2Br
$CH{=}CHCBr_2CH_2Me$
$CHBrCH_2CH_2CH_2Me$
$CBr_2CH_2CH_2CH_2Me$

63 Unsatuarated ketones

$Br_2CH-CHO$

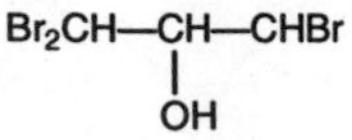

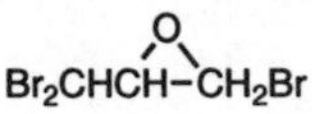

64 3,3-Dibromacrolein

$Br_2CH-CH(OH)-CHBr$

65 1,1,3-Tribromo-2-propanol

$Br_2CHCH(-O-)CH-CH_2Br$

66 1,3,3-Tribromoepoxypropane

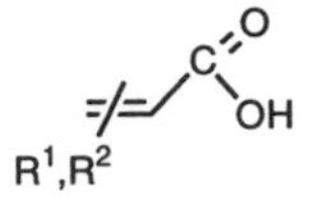

R^1	R^2
I	H
Br	Cl
Br	Br
Br	I
I	I

67 Halogenated derivatives of acrylic acid

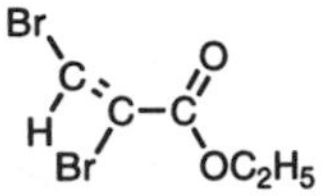

68 Halogenated derivatives of acetic acid

$Cl_2CHCONH_2$

69 Dichloroacetamide

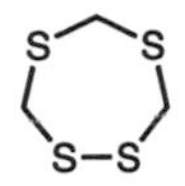

70 Tetrathiepane

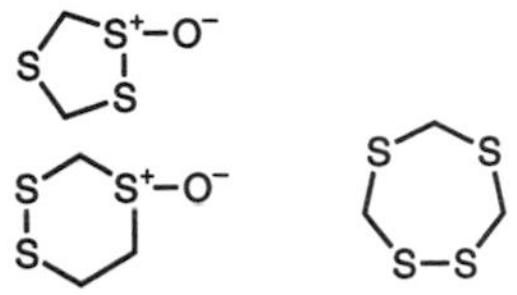

71 Theolanes, e.g. Lenthionin

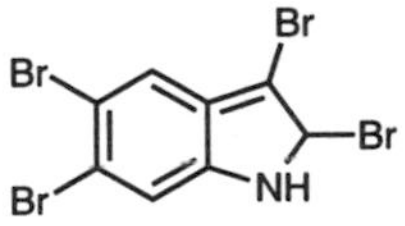

72 2,3,5,6-Tetrabromoindole

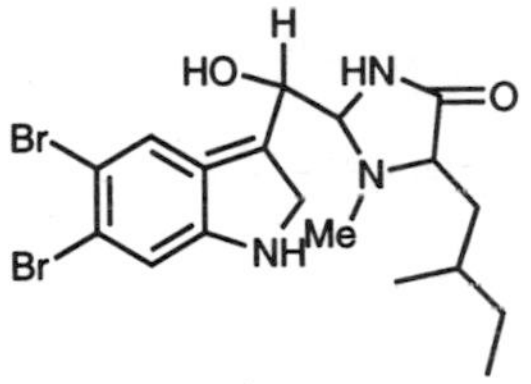

73 Martensine A

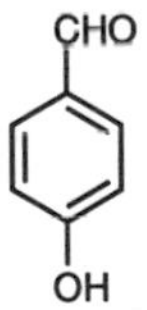

74 *p*-Hydroxybenzaldehyde

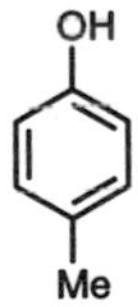

75 *p*-Cresol

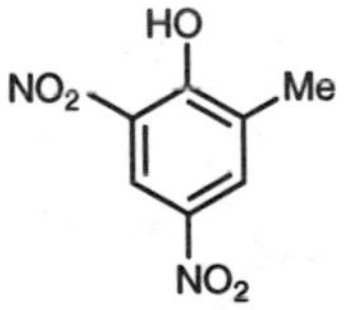

76 3,5-Dinitroguiacol

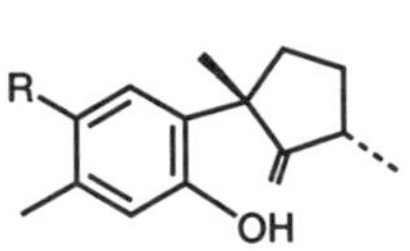

R = Br
77 Isolaurinterol

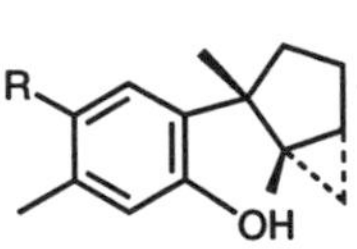

R = H
78 Debromolaurinterol

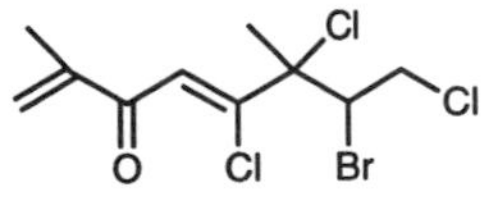

79 Plocamenone

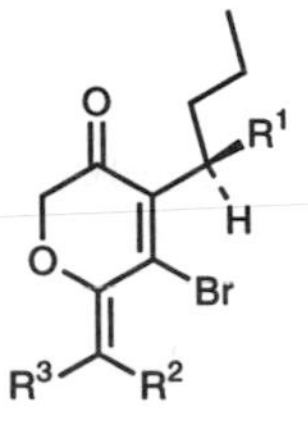

R^1 = H
R^2, R^3 = H, Cl, I
R^1 = CO_2CH_3
R^2, R^3 = H, Cl Br, I

80 Fimbrolides

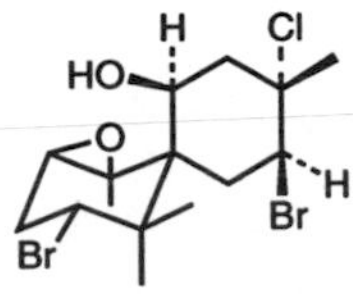

81 Prepacifenol

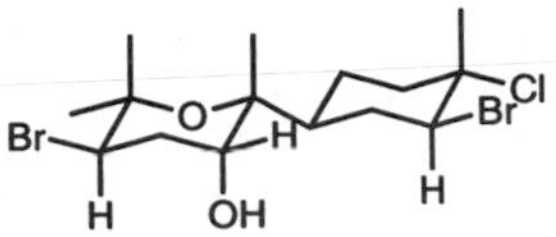

82 Caespitol

Figure 5.22 *Continued*

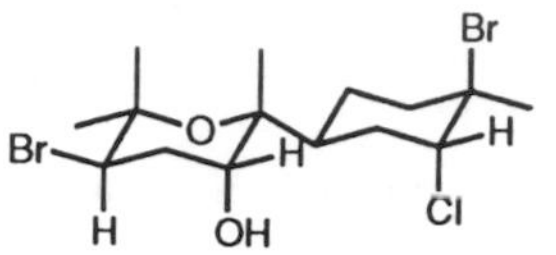

83 R = H Isocaespitol
84 R = Ac Isocaespitol acetate

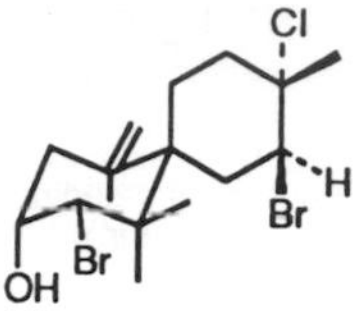

85 Obtusol

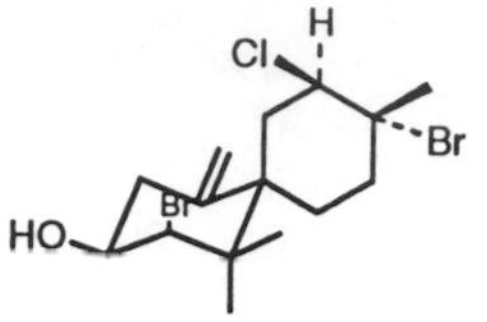

86 Isoobtusol

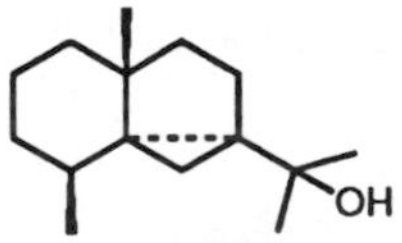

87 Cycloeudesmol

88 Phytol

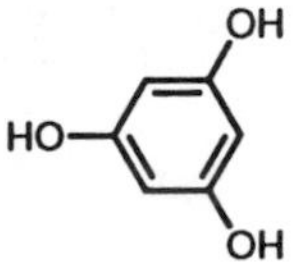

89 Phloroglucinol

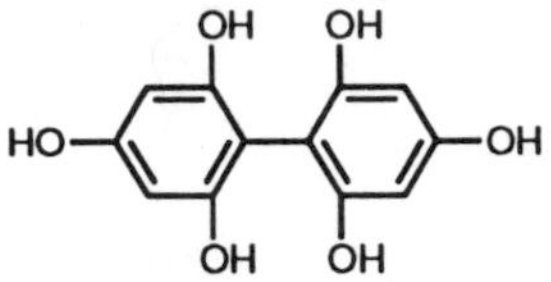

90 Difucol

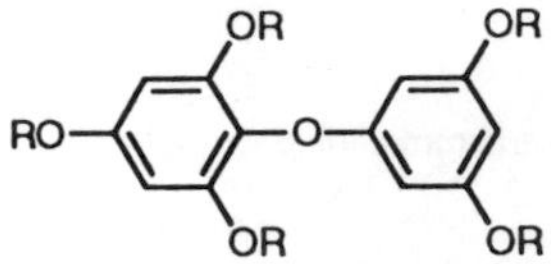

91 Diphlorethal

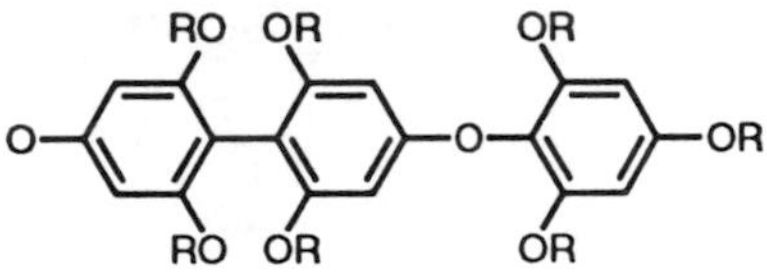

R = Ac

92 Fucophloretol

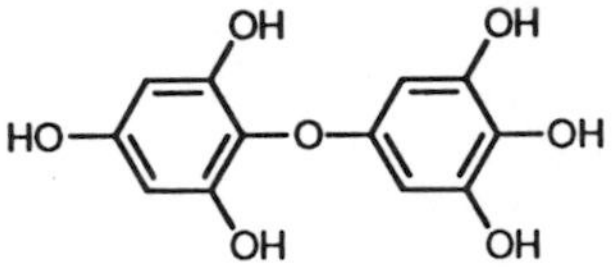

93 Bifuhaol

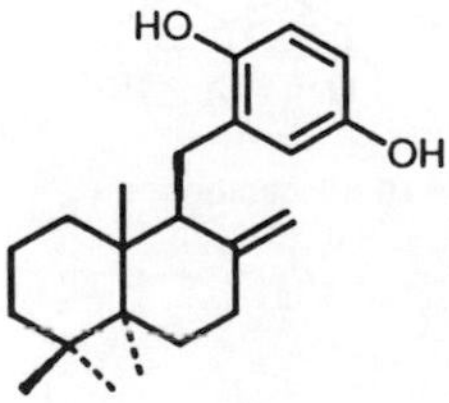

94 Zonarol

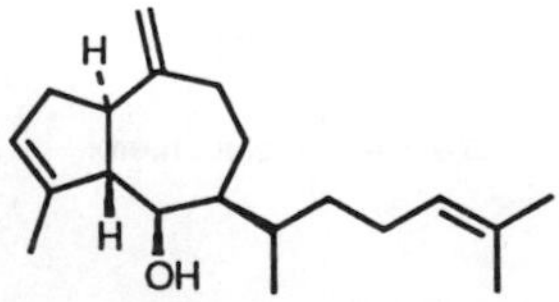

95 Pachydictyol-A-epoxide

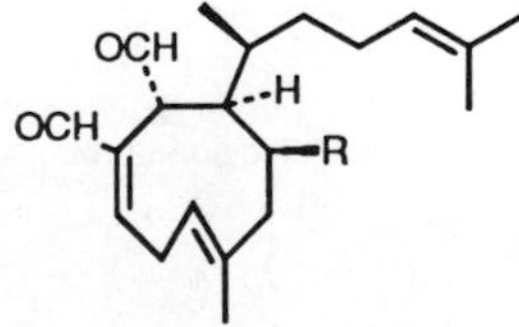

96 R = OH Hydroxydictyodial
97 R = H Dictyodial

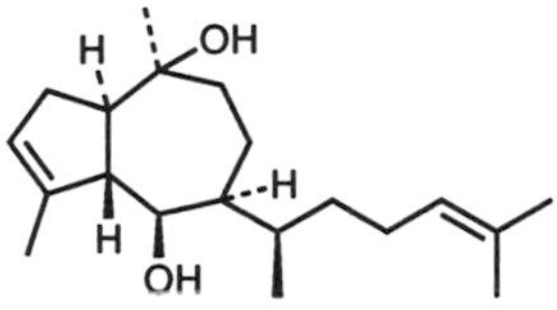

Figure 5.22 *Continued*

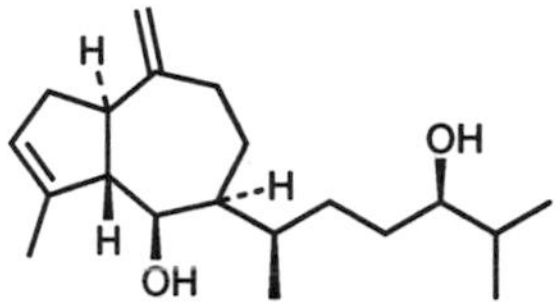

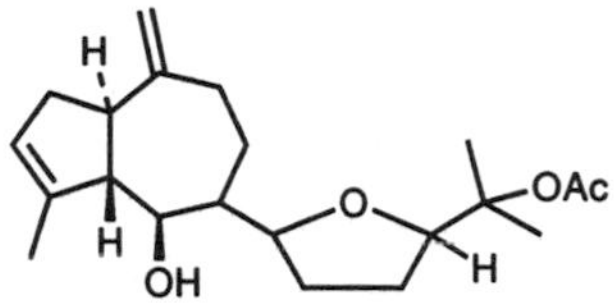

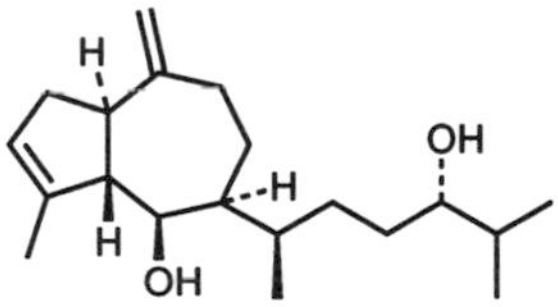

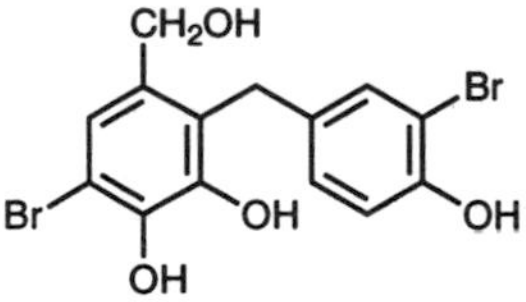

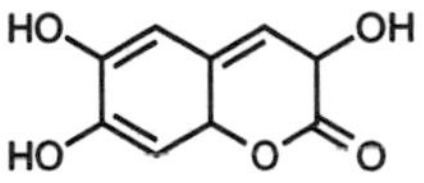

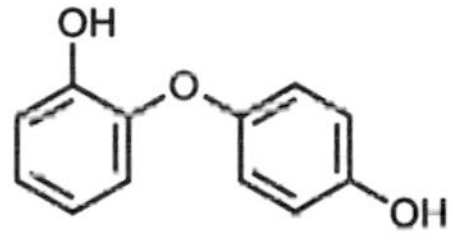

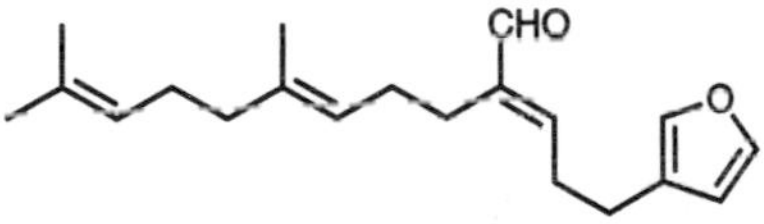

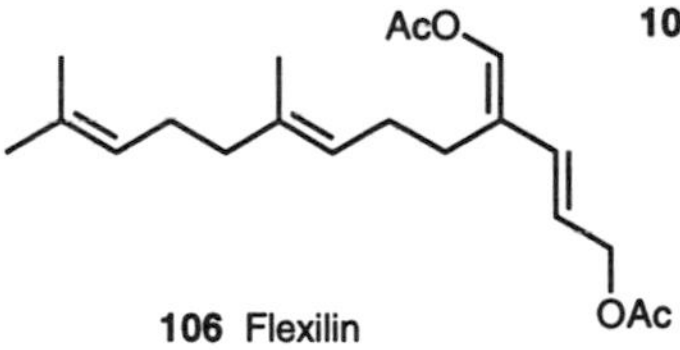

Table 5.11 *Antimicrobials from macroalgae (adapted from Lincoln* et al.[22]*)*

Chemical nature	*Compound name*	*Structure**	*Antibiotic spectrum*	*Source*	*Reference*
Rhodophyta					
Unsaturated aliphatic carboxylic acids	Acrylic acid	**57**	Antibacterial	*Asparagopsis armata*	361
				Chondrus chrispus	550
				Laurencia spp.	300
				Polysiphonia spp.	300
				Porphyra tenera	300
				Rhodomela	300
	F_1 (predominantly $C_{16:1}$, $C_{16:3}$)		Antibacterial	*Cystoclonium purpureum*	301
	F_2 (predominantly $C_{20:4}$, $C_{16:2}$)		Antibacterial	*Cystoclonium purpureum*	301
	F_3 (predominantly $C_{16:2}$, $C_{20:4}$, $C_{18:3}$)		Antibacterial	*Cystoclonium purpureum*	301
	11-Formylundeca-5,8,10-trienoic acid	**59**	Antibacterial, Antifungal	*Laurencia hybrida*	302
	9-Hydroxyeicosa-2,5,7,11,14-pentaenoic acid	**60**	Antibacterial, Antifungal	*Laurencia hybrida*	302
Halogenated aliphatic compounds					
(i) Halogenated alkanes	Various	**61**	Antibacterial	*Asparagopsis armata*	303, 304
	Various	**61**	Antibacterial	*Asparagopsis taxiformis*	303, 304, 551, 552
(ii) Saturated ketones	Unnamed	**62**	Antibacterial, Antifungal	*Asperagopsis armata*	304
				Asparagopsis taxiformis	304, 552, 553
				Bonnemaisonia hamifera	304, 554
				Falkenbergia rufolanosa	303
(iii) Unsaturated ketones	Various	**63**	Antibacterial	*Asparagopsis armata*	304
				Asparagopsis taxiformis	551, 552
				Ptilonia australasica	304, 555
	Polyhalo-1-octen-3-ones		Antibacterial, Antifungal	*Bonnemaisonia asparagoides*	304

	1,1,3,3-Tetrabromo-2-heptanone		Antibacterial	*Bonnemaisonia hamifera*	304
	1,1,3-Tribromo-2-heptanone		Antibacterial		
(iv) Aldehydes	3,3-Dibromacrolein	**69**	Antibacterial	*Asparagopsis taxiformis*	551
(v) Alcohols	1,1,3-tribromo-2-propanol	**65**	Antibacterial	*Asparagopsis taxiformis*	552
(vi) Epoxides	*trans*-1,3,3-Tribromo-epoxypropane	**66**	Antibacterial	*Asparagopsis taxiformis*	552
(vii) Halogenated derivatives of acrylic acid and acetic acid	Unnamed	**67**	Antibacterial	*Asparagopsis armata*	304
	Unnamed	**68**	Antibacterial	*Asparagopsis armata,* *Asparagopsis taxiformis*	304 304
	Dichloroacetamide	**69**	Antifungal	*Marginisporum aberrans*	556
(viii) Halogenated amide S-containing heterocyclic compounds	Tetrathiepane	**70**	Antibacterial, Antifungal	*Chondria californica*	557
	Theolanes, *e.g.*, lenthionin	**71**	Antibacterial	*Chondria californica*	557
Halogenated indoles	2,3,5,6-Tetrabromoindole	**72**	Antibacterial, Antifungal	*Laurencia brongniartii*	558
	Martensine A	**73**	Antibacterial	*Martensia fragilis*	559
Free phenols	*p*-Hydroxybenzaldehyde	**74**	Antibacterial	*Dasya pedicellata*	560
(i) Brominated and unbrominated derivatives of 3,4-dihydroxybenzyl alcohols			Antibacterial	*Marginisporum aberrans*	556 561
	p-Cresol	**75**	Antibacterial	Many Rhodophyceae	562
				Phaeophyceae	562
				Chlorophyceae	562
	3,5-Dinitroguiacol	**76**	Antibacterial	*Marginisporum aberrans*	561
(ii) Isopyrenated, brominated hydroxyquinones	Unnamed		Antibacterial, Antialgal		300

Continued

Table 5.11 *Continued*

Chemical nature	*Compound name*	*Structure**	*Antibiotic spectrum*	*Source*	*Reference*
Halogenated/aromatic terpenes					
(i) Phenolic halogenated terpenes	Laurinterol	**38**	Antibacterial	*Laurencia intermedia*	563
			Antifungal	*Laurencia okamurai*	556, 564, 565
	Isolaurinterol	**77**	Antibacterial	*Laurencia intermedia*	563
(ii) Phenolic unhalogenated terpene	Debromolaurinterol	**78**	Antibacterial	*Laurencia okamurai*	300
			Antifungal		564, 565
(iii) Halogenated nonaromatics terpines					
(a) Aliphatic monoterpenes	Plocamenone	**79**	Antibacterial	*Plocamium* spp.	566
				Chondrococcus spp.	566
	Fimbrolides	**80**	Antifungal	*Delisea fimbriata*	304
				Microcladia spp.	567
(b) Sesquiterpenes	Prepacifenol	**81**	Antibacterial	*Laurencia* spp.	300
	Caespitol	**82**	Antibacterial	*Laurencia caespitosa*	568
	Isocaespitol	**83**	Antibacterial	*Laurencia caespitosa*	568
	Isocaespitol acetate	**84**	Antibacterial	*Laurencia caespitosa*	568
	Obtusol	**85**	Antibacterial	*Laurencia obtusa*	568
	Isoobtusol	**86**	Antibacterial	*Laurencia obtusa*	568
	Cycloeudesmol	**87**	Antibacterial	*Chondria oppositiclada*	565
			Antifungal		
(b) Diterpenes	Phytol	**88**	Antibacterial	*Hizikia fusiformis*	452
Sulfated carageenan	Unnamed		Antiviral	*Schizymenia pacifica*	371, 372
Phaeophyta					
Carboxylic acids	Acrylic acid	**57**	Antibacterial	Various spp.	362
Free phenols (phloroglucinol polymers)					
(i) Phloroglucinol (PG)	Phloroglucinol	**89**	Antibacterial	*Fucus vesiculosis*	570
(ii) Fucols (1–4 PG units)	Difucol	**90**	Antibacterial	*Fucus* spp.	300
(iii) Phlorethols (aryl linkages)	Diphlorethal	**91**	Antibacterial	Various spp.	572

(iv) Fucophlorethiols (diaryl and diacryl ether linkages)	Fucophloretol	**92**	Antibacterial	Various spp.	300
(v) Fuhaols (hydroxylated PG)	Bifuhaol	**93**	Antibacterial	Various spp.	300
Terpenes					
(i) Diterpenoidomono acetyl spatane skeleton	Zonarol	**94**	Antibacterial	*Dictyopteris undulata*	343
	Pachydictyol-A epoxide	**95**	Antibacterial	*Pachydictyon coriaceum*	343
(ii) Cyclic diterpenes	Hydroxydictyodial	**96**	Antibacterial	*Dictyota spinulosa*	573
	Dictyodial	**97**	Antibacterial	*Dictyota crenulata*	574
			Antifungal	*Dictyota flabellata*	574
	Dictyol C	**98**	Antibacterial, Antifungal	*Dictyota dichotoma*	575
	Dictyol F	**99**	Antibacterial	*Dictyota dichotoma*	575
	Dictyol H	**100**	Antibacterial, Antifungal	*Dictyota dichotoma*	575
	Epidictyol F	**101**	Antibacterial, Antifungal	*Dictoyta dichotoma*	576
Chlorophyceae					
Carboxylic acids	Acrylic acid	**57**	Antibacterial	Various spp.	500, 577
Phenols	Avrainvilleol	**102**	Antibacterial	*Avrainvillea longicaulis*	578
	3,6,7-Trihydroxycoumarine	**103**	Antibacterial	*Cymopolia barbata*	579
	Diphenyl ether	**104**	Antibacterial	*Cladophora fascicularis*	580
Sesqui- and di-terpenoids	Udoteafuran	**105**	Antibacterial, Antifungal	*Udotea flabellum*	544, 366
	Halimedatrial	**30**	Antibacterial, Antifungal	*Halimeda* sp.	367
	Unnamed	ND	Antibacterial	*Halimeda* sp.	366
			Antifungal		368
	Flexilin	**106**	Antibacterial	*Caulerpa flexilis*	366, 369, 370

*ND, not determined.

5.7.2.2 *Anthelmintics*

Anthelmintics or vermifuges are compounds, such as the drug dichlorophen, which are used to destroy parasitic worms (helminths) and/or remove them from the body. Compounds that possess such activity are mainly present in marine macroalgae from the divisions Rhodophyta, Chlorophyta, and Phaeophyta, although one freshwater species, *Rhizoclonium rivulare*, has also been reported to be associated with anthelmintic activity.[150,314]

In the Phaeophyta the active compounds produced by *Sargassum vulgare*, *S. confusum*, and *Sympocladia fragilis* are phenols, which are particularly active against *Ascularis* sp. of worms. Members of the Laminariales and other brown algae have been demonstrated to produce laminine (**107**), a general anthelmintic.[150] In contrast, the anthelmintics produced by members of Rhodophyta are domoic acid (**21**), the amino acid neurotoxin (Section 5.7.1.6.1) produced by *Alsidium helminthochorton*, *A. corallinum*, and *Chondria armata*, and the related proline derivative α-kainic acid (**22**) from *Digenea simplex*.[314] This is a broad-spectrum anthelmintic and has been shown to kill adult *Ascaris* worms by causing a neuromuscular block with no apparent adverse side effects in humans at a dose between 5 and 10 mg. This was commercially marketed as a general anthelmintic, but is no longer available in western countries.[314]

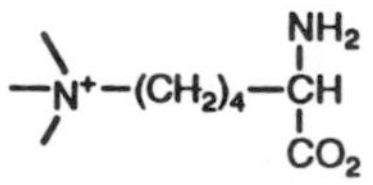

$$-N^{+}(CH_3)_2-(CH_2)_4-CH(NH_2)-CO_2^{-}$$

107 Laminine

5.7.2.3 *Neoplastic Agents*

Neoplastic agents (Figure 5.23) or tumour promoters are compounds which increase the yield of tumours in initiated cells. Tumour promoters of algal origin have been classified into two groups, the teleocidin and aplysiatoxin classes. Both groups are similar to the phorbol ester tumour-promoting constituent of croton oil, 12-*O*-tetradecanoylphorbol-13-acetate (TPA). The teleocidin class includes the indole alkaloid lyngbyatoxin A (**18**) (Section 5.7.1.5), isolated from shallow water varieties of *Lyngbya majuscula*, and structurally and biochemically similar to teleocidin B (**108**) from *Streptomyces mediocidicus*.[373,374] Lyngbyatoxin has been reported to be a tumour promoter in mice,[311] being of equal potency to TPA in two-stage carcinogenesis studies. It was found, like TPA, to induce inflammatory effects in mice, induce ornithine decarboxylase activity (associated with cell proliferation), cause cell surface changes, bind to phorbol ester receptors, and activate protein kinase C *in vitro*. The aplysiatoxin class of algal tumour promoters includes the cyanobacterial toxins aplysiatoxin (**109**), debromoaplysiatoxin (**17**), oscillatoxin A

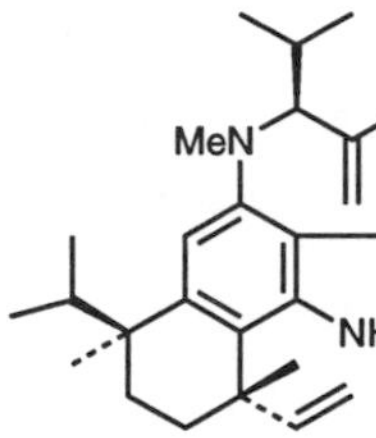

108 Teleocidin B

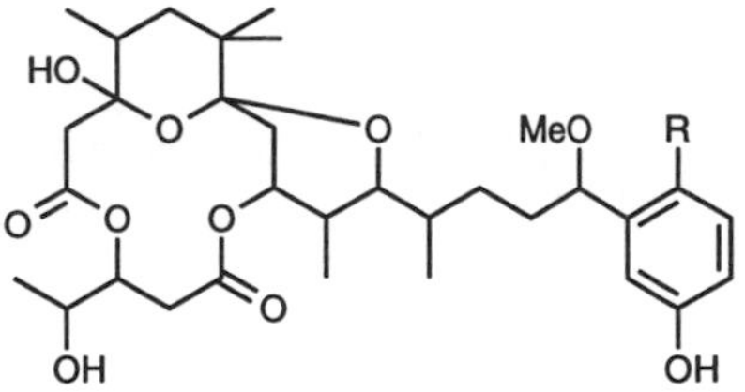

109 Aplysiatoxin

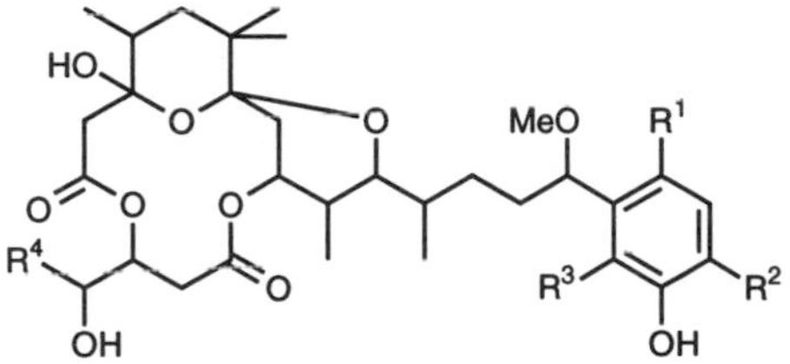

R^1 = Br
R^2 = Br
R^3 = H
R^4 = Me

110 Bromoaplysiatoxin

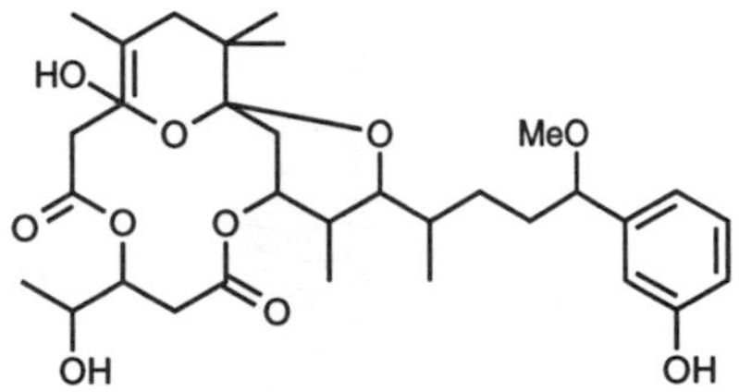

111 Anhydrobromoaplysiatoxin

Figure 5.23 *Neoplastic agents*

(**19**), bromoaplysiatoxin (**110**), and anhydrobromoaplysiatoxin (**111**) (Section 5.7.1.5). Like the teleocidin class tumour-promoters, these act via phorbol ester receptors on cell membranes and activate protein kinase C *in vitro*. However, the potency of these compounds differs: aplysiatoxin shows equal potency to TPA, teleocidin B, and lyngbyatoxin A, while debromoaplysiatoxin and bromoaplysiatoxin are weaker tumour promoters.

A non-TPA-type promotor, okadaic acid (**4**) (Section 5.7.1.2.4.1), isolated from various dinoflagellates, has been reported to cause dermal irritation in mice and to stimulate prostaglandin metabolism. It has been suggested that these effects are mediated via an okadaic acid receptor rather than a phorbol ester receptor.[375] An unidentified tumour-promoter from extracts of the cyanophyte *Microcystis* sp. has been reported also.[376]

5.7.2.4 *Antineoplastic and Cytotoxic Compounds*

Antineoplastic compounds which reduce the growth of tumours are generally also cytotoxic; however, cytotoxins do not always exhibit antineoplastic activity *in vivo*. *In vivo* screens to detect antineoplastic activity[377–380] and *in vitro* cytotoxicity assays[343,377,381,382] have been developed to identify compounds that exhibit these activities. Compounds identified and purified to date are shown in Tables 5.12 and 5.13 and Figures 5.24 and 5.25, and are thought to

Table 5.12 *Antineoplastic compounds from algae (adapted from Lincoln* et al.[22]*)*

Alga	*Active compound*	*Structure**	*Dose required*	*Reference*
Cyanophyta				
Lyngbya gracilis	Debromoaplysiatoxin	**17**	0.01 mg twice daily	378
Lyngbya gracilis	Debromoaplysiatoxin	**17**	1.8 μg kg^{-1}	305
Lyngbya majuscula (deep water)	Debromoaplysiatoxin	**17**	0.6 mg kg^{-1}	581
Lyngbya majuscula (shallow water)	Debromoaplysiatoxin	**17**		581
	Aplysiatoxin	**109**		581
Oscillatoria acutissima	Acutiphycin	**112**	50 g kg^{-1}	582
Mix of *Schizothrix calcicola* and *Oscillatoria nigroviridis*	Oscillatoxin A	**19**	0.2 μg kg^{-1}	581
	31-(Nordebromoaplysiatoxin)			
	Debromoaplysiatoxin	**17**		
Mix of *Cyanobacteria*, *Spirulina*, and *Dunaliella*	β-Carotene and other carotenoids	**113**	350 μg kg^{-1}	383
Phaeophyta				
Sargassum kjellmanianum	Polysaccharide	ND	100 mg kg^{-1}	583
Laminaria angustata	Polysaccharide	ND		584
Laminaria angustata	Mainly polysaccharide, some peptides, amino acid, nucleic acid also present	ND	100 mg kg^{-1} every 2 days for 10 days	384
Laminaria angustata var. *longissima*				
Laminaria angustata	Glycolipid	ND	40 mg kg^{-1} daily for 7 days	382
Kjellman	Phospholipid	ND	6.4 mg kg^{-1} daily for 7 days	382
Sargassum fulvellum	Sulfated polysaccharide, either *S*-peptidogluronoglycan or glycuron	ND	100 mg kg^{-1} daily for 10 days	384, 585

Sargassum ringgoldianum	Fucoidan	ND	40 mg kg^{-1} daily for 7 days	382
	Neutral lipid	ND	40 mg kg^{-1} daily for 7 days	382
	Glycolipid	ND	40 mg kg^{-1} daily for 7 days	382
	Phospholipid	ND	40 mg kg^{-1} daily for 7 days	382
Undaria pinnatifida	Polysaccharide	ND	12.5–400 mg kg^{-1}	385
	Fucoidan	ND	40 mg kg^{-1} for 7 days	382
Dinophyta				
Amphidinium sp.	Amphidinolide A	**114**		586
	Amphidinolide B	**115**		587
	Amphidinolide C	**116**		463, 588
	Amphidinolide D	**117**		463
Rhodophyta				
Porphyra yezoensis	Porphyran	ND		382
	Phospholipid	ND	6.7 mg kg^{-1} for 7 days	382
Chlorophyta				
Monostroma nitidum	Sulfated polysaccharide	ND	400 mg kg^{-1} for 28 days	382

*ND, not determined.

Table 5.13 *Cytotoxic compounds from algae (adapted from Lincoln* et al.[22]*)*

Alga	*Chemical nature*	*Active compound*	*Structure*	*In vitro cell culture*	*Cytotoxicity*	*Reference*
Cyanophyta						
Halimeda spp.	Diterpenoid trialdehyde	Halimedatrial	**30**	Sea urchin sperm	1 μg ml^{-1}	367
Hormothamnion enteromorphoides	Peptide	Hormothamnin A	ND	Human lung Carcinoma SW 1271	IC_{50} 0.2 μg ml^{-1}	508
				Carcinoma A529	IC_{50} 0.16 μg ml^{-1}	508
				Murine melanoma *B*16-*F*10	IC_{50} 0.13 μg ml^{-1}	508
				Human colon HCT-116	IC_{50} 0.72 μg ml^{-1}	508
Hormothamnion enteromorphoides	Strichromone	Hormothamnione	**120**	P-388 lymphocytic leukaemia	IC_{50} 4.6 μg ml^{-1}	589
				HL-60	IC_{50} 0.1 μg ml^{-1}	589
Oscilatoria acutissima	Peptide	Acutiphycin	**112**	KB	$ED_{50} < 1$ μg ml^{-1}	582
		20,21-Didehydroacutiphycin	**121**	N1H/3T3	$ED_{50} < 1$ μg ml^{-1}	582
Scytonema pseudohofmanni	Peptide	Scytophycins A,B,C,D,E	**51**	KB cells	0.2 μg ml^{-1}	582
Tolypothrix conglutinata	Peptide	Tolytoxin	**52**	P388 cells	ND	306

Scytonema mirabile	Peptide	Tolytoxin	**52**	KB cells	$<18\ \mu g\ ml^{-1}$†	524
Scytonema ocellatum	Peptide	Tolytoxin	**52**	KB cells	$0.5\ \mu g\ ml^{-1}$†	524
Tolypothrix byssoidea	Nucleoside	Tubercidin	**47**	KB	Complete kill	525
				N1H/3T3 cells	Complete kill	525
				KB	$<20\ \mu g\ ml^{-1}$†	524
Scytonema saleyeriense	Nucleoside	Tubercidin	**47**	KB	$2\ \mu g\ ml^{-1}$†	524
					$25\ \mu g\ ml^{-1}$†	524
Plectonema radiosum	Nucleoside	Tubercidin-5-D-glucopyranose	**48**	KB	$14, 3\ \mu g\ ml^{-1}$†	524
Tolypothrix distorta	Nucleoside	Tubercidin-5-D-glucopyranose	**48**	KB	$18, 49\ \mu g\ ml^{-1}$†	524
Tolypothrix tenuis	Nucleoside	Toyocamycin-5-D-glucopyranose	**46**	KB	$12\ \mu g\ ml^{-1}$†	524
				AL-60	$6\ \mu g\ ml^{-1}$	524
Symbiodinium sp.	Sphingosine derivative	Sphingosine derivative	**151**	LH20	$9.5\ \mu g\ ml^{-1}$	448
Phaeophyceae						
Colpomenia peregrina	Fatty acids	Fucoxanthin and other saturated and unsaturated fatty acids	ND		ND	590
Turbinaria ornata	Secasqualene carboxylic acid	Turbinaric acid	**119**	Murine melanoma	$26.6\ \mu g\ ml^{-1}$	591
				Human colon	$12.5\ \mu g\ ml^{-1}$	591

*ND, not determined.
†MIC values (minimal inhibitory concentration).

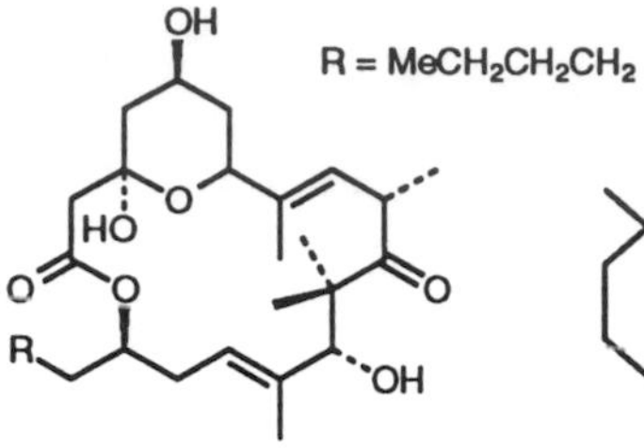

112 Acutiphycin

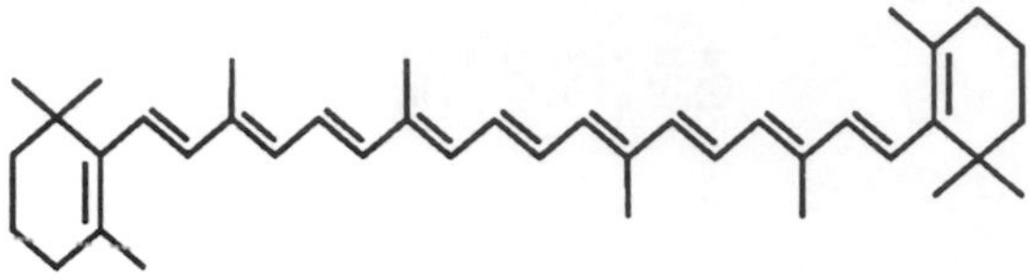

113 β-Carotene

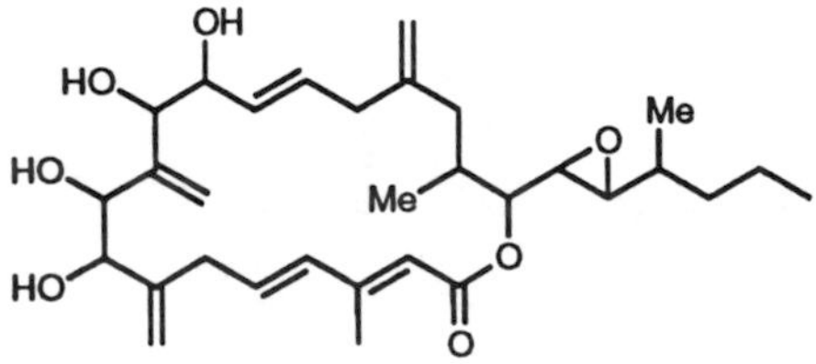

114 Amphidinolide A

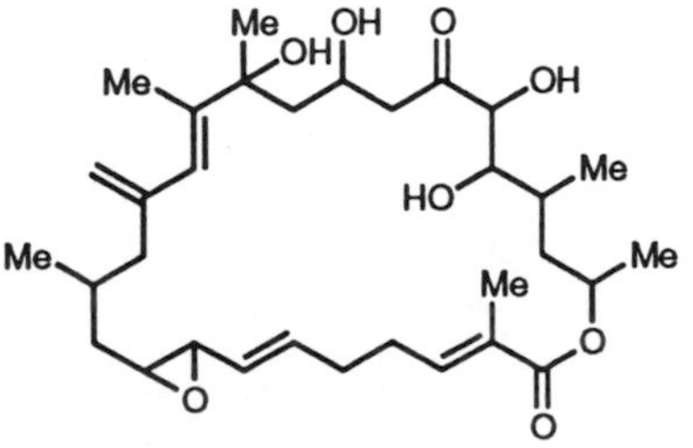

115 Amphidinolide B

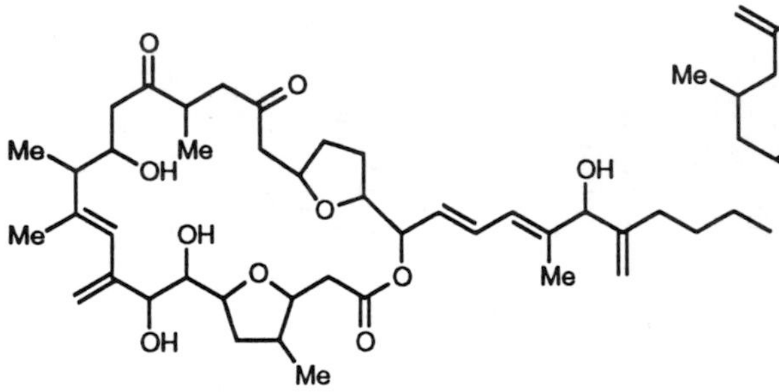

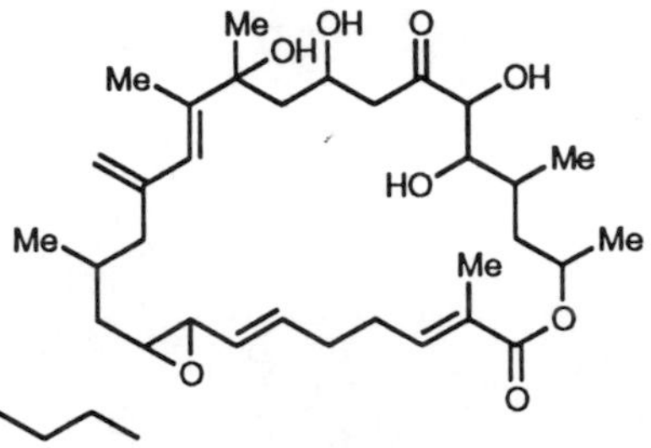

117 Amphidinolide D

116 Amphidinolide C

Figure 5.24 *Antineoplastic agents*

exert their effects via stimulation–activation of the immune system.[383–385] The natural product, named Viva-natural, isolated from the brown alga *Undaria pinnatifida* is thought to be a polysaccharide; it has been shown to be therapeutically active against Lewis lung carcinoma, to enhance the natural cytolytic activity of peritoneal macrophages *in vitro* against human carcinoma cells, and to activate splenic lymphoid tissues (T and B cells).[385] Similar increases in the cytolytic activity of peritoneal macrophages have been attributed to *β*-carotene in cyanobacterial extracts, which has also been reported to increase the number of tumour necrosis factor positive cells,[383,386,387] which are

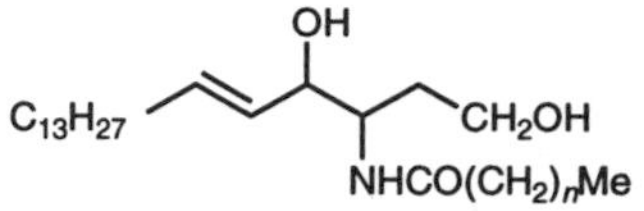

118 *N*-Acylsphingosines

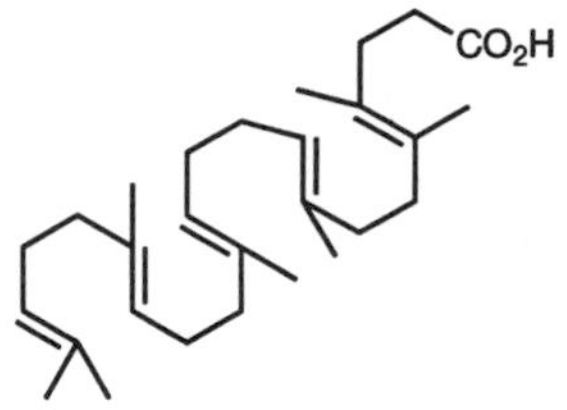

$n = 12,14,20,22$

119 Turbinaric acid

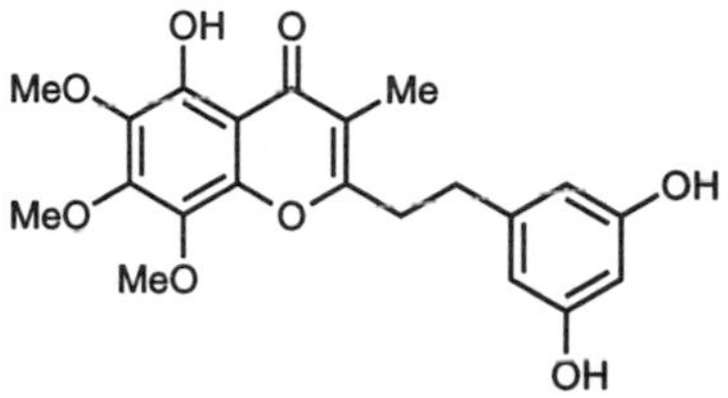

120 Hormothamnione

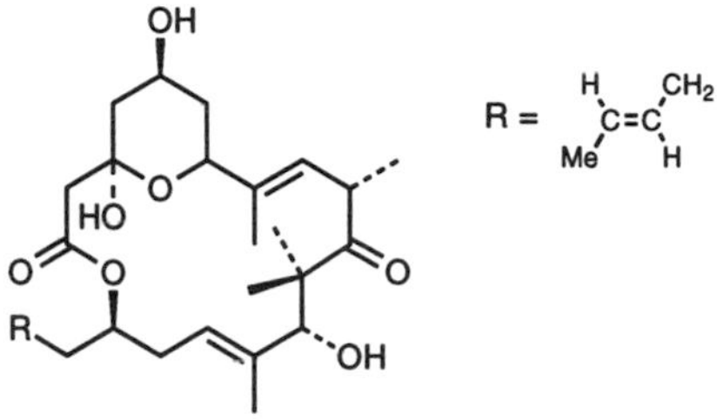

121 20, 21-Didehydroacutiphycin

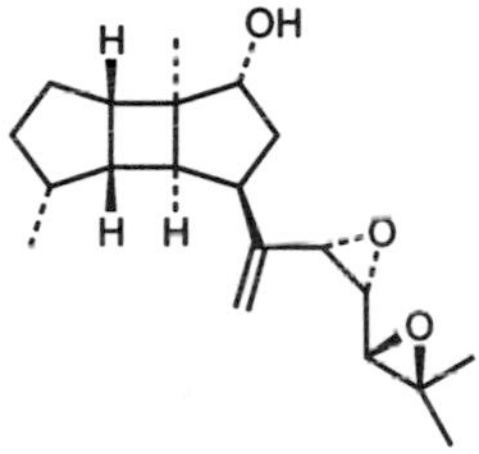

122 Spatol

Figure 5.25 *Cytotoxic compounds from algae*

considered to be endogenous antineoplastic agents.[388] Other carotenoids in these extracts have been found to be mitogenic[389] and to enhance the cytotoxic action of thymus derived cells.[390]

Included in this class of compounds are inhibitors of microtubulin polymerization, for which the inhibition of synchronous cell division in fertilized sea urchin eggs has proved a good selective screen.[391] Fat-soluble extracts of several algae, including *Anabaena* sp., *Oscillatoria* sp., *Micractinium* sp., *Scenedesmus* sp., *Pediastrum* sp., and *Goniodoma pseudogoniaulax* have been shown to contain inhibitors of microtubulin polymerization, some active at concentrations $<1\,\mu g\,ml^{-1}$.[392] Spatol (**122**), a novel tricyclic diterpenoid isolated from the brown alga *Spatoglossum schmittii*, has been shown to be cytotoxic via disruption of mitotic spindle formation as a result of an inhibition of microtubulin polymerization (ED_{50} $1.2\,\mu g\,ml^{-1}$ in the sea urchin egg assay[393]). This mechanism of action is characteristic of several potent clinical antineoplastic agents, including colchicine, vincristine, and vinblastine.[394] Stypoldione (**24**), an orthoquinone oxidation product of stypotriol, isolated from the brown alga *Stypopodium zonale*, has also been observed to inhibit microtubule polymerization, either by binding to a low affinity site on the tubulin dimer or acting via an unknown mechanism prior to the onset of mitosis.[395] Microcystin (**13**), from the cyanobacteria *Microcystis aeruginosa*, has also been suggested to affect microtubule polymerization with a resultant disruption of the cytoskeletal structure in isolated rat hepatocytes.[395].

5.7.2.5 *Growth Stimulants*

Culture filtrates of the thermophilic cyanobacteria *Phormidium* sp. and extracts of the thermophiles *Syneococcus elongatus* (Cyanophyta) and *Spirulina subsala* (Cyanophyta) have been shown to contain growth-promoting substances for several lines of hybridoma, lymphocyte and tumour cells, bacteria, and plant cultures.[396–398]

5.7.2.6 *Anti-Inflammatory Compounds*

The principle anti-inflammatory of the brown alga *Caulocystis cephalornithos* has been identified as 6-tridecylsalicylic acid (**123**) (Figure 5.26).[399] A series of

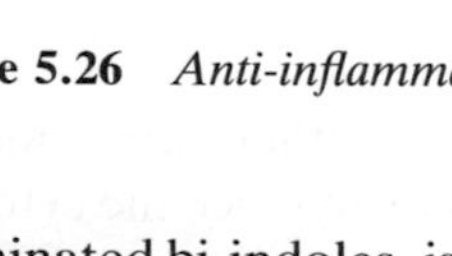

123 6-Tridecylsalicylic acid

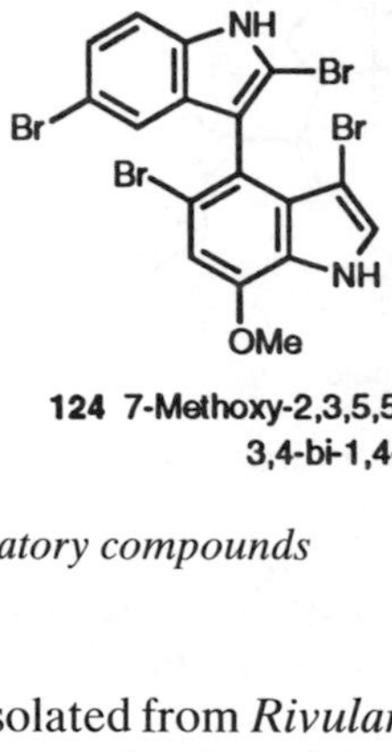

124 7-Methoxy-2,3,5,5-tetrabromo-3,4-bi-1,4-indole

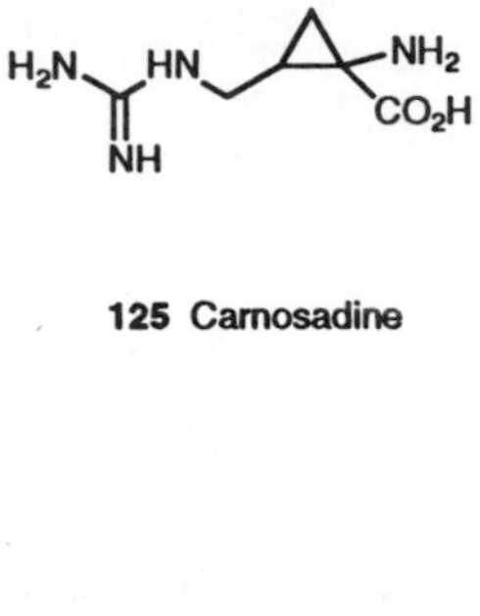

125 Carnosadine

Figure 5.26 *Anti-inflammatory compounds*

brominated bi-indoles, isolated from *Rivularia firma* (Cyanophyta), have been found to exhibit significant analgesic activity, reduce swelling due to the build-up of fluid, and inhibit allergic reactions, the last suggesting potential as an anti-allergenic. The most active compound from this series was (+)-7-methoxy-2,3,5,5-tetrabromo-3,4-bi-1,4-indole (MTBI) (**124**), which has also been shown to exhibit CNS anti-amphetamine activity.[400] Finally, carnosadine (**125**) (1-amino-2-guanidinomethylcyclopropane-1-carboxylic acid), from the red alga *Grateloupia carnosa*, has been patented as an anti-inflammatory compound with additional carcinostatic and immunological effects.[401]

5.7.2.7 *Anticoagulant and Haemostatic Compounds*

Anticoagulants are agents that prevent the clotting of blood and are used in the treatment of thrombosis and embolism. Haemostatic agents are those which stop or prevent haemorrhage. Compounds identified to date from algae exhibiting this type of activity are all polysaccharides and are shown in Table 5.14 and Figure 5.27.

The most important naturally occurring polysaccharide anticoagulants are the heparins, a family of sulfated glycosaminoglycans, which are reported to act via an interaction with the plasma protein antithrombin III. On binding, heparin induces a conformational change in the antithrombin III molecule which promotes the inactivation of numerous proteases, *e.g.*, factor Xa and

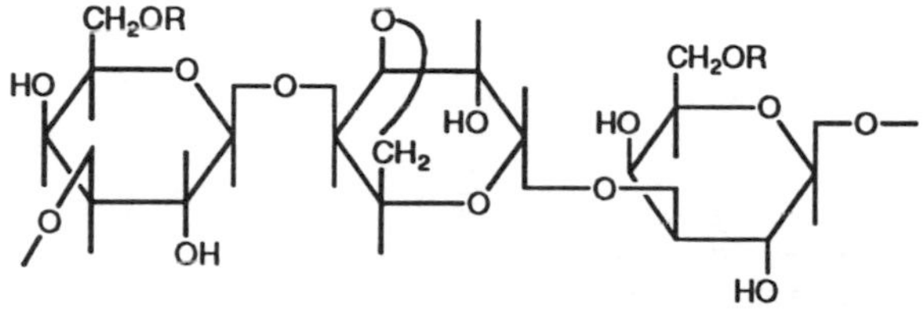

R = H or Me

126 Agar

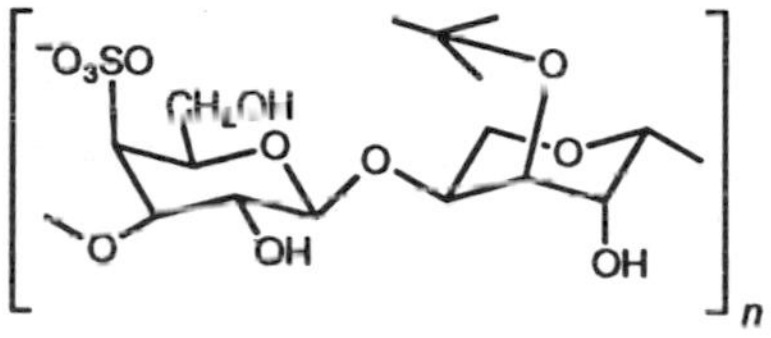

127 κ-Carrageenan

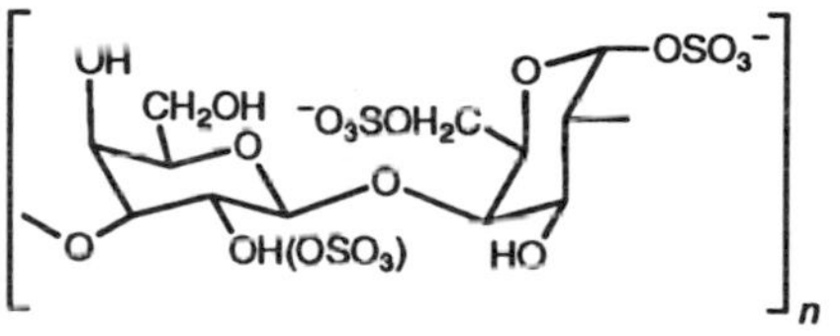

128 λ-Carrageenan

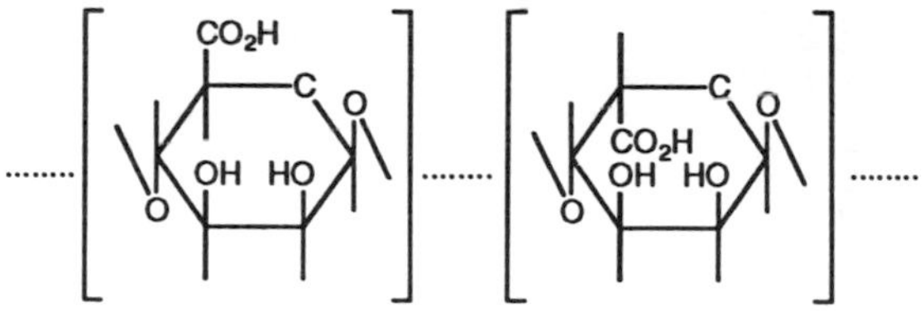

D-Mannuronic acid

L-Guluronic acid

129 Alginic acid

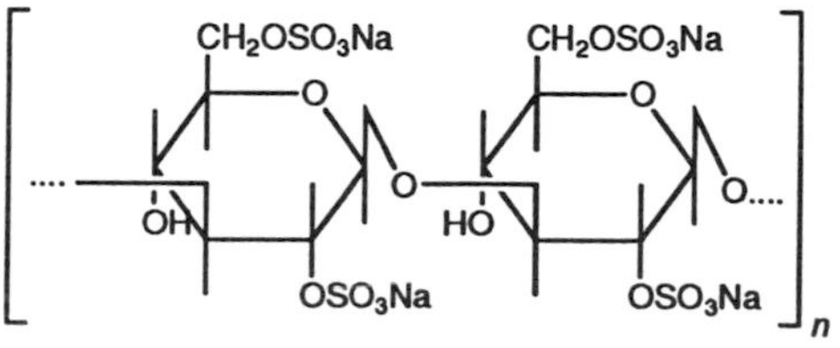

130 Laminarine

Figure 5.27 *Anticoagulant compounds from algae*

Table 5.14 *Anticoagulant compounds from algae (adapted from Lincoln* et al.[22]*)*

Chemical nature	*Name*	*Constituent sub-units*	*Source*	*Anticoagulant activity compared with heparin**	*Reference*
Natural sulfated polysaccharides	Agar, **126**	Agarose = alternate units 1,3-linked β-D-galactose and 1,4-linked 3,6-anhydro-L-galactose	*Gelidium* sp. *Gracilaria* sp. *Ahnfeltia* *Phyllophora* sp. *Hetrocladia*	ND	592
	κ-Carrageenan, **127**	1,3-Linked D-galactose-4-sulfate 1,4-linked 3,6-anhydro-D-galactose-2-sulfate	*Chondrus* sp. *Eucheuma* sp. *Gigartina* sp. *Irideae*	+	402 593 403
	λ-Carrageenan, **128**	1,3-Linked D-galactose-2-sulfate 1,4-Linked D-galactose-2,6-disulfate	*Chondrus crispus*	–	402 403
	Fucans	10% Sulfur, 65% L-fucose, 1% fast sugar	*Fucus vesiculosus*	+	594
	Fucans	35–40% Sulfur, 35–40% fucose, sugars (% dependent on alga)	*Pelvetia canuculata*	–	404
			Fucus vesiculosus	–	404
			Laminaria digitata	–	404
			Sargassum muticum	–	404
			Ascophyllum nodosum	–	404
	Fucoidan	Fucose, 2.3% sulfate, neutral sugar, 0.3% uronic acid	*Eisenia bicyclis*	–	595
	Fraction A	27.2% Fucose, 22.9% galactose, 10.5% glucuronic acid, 5.1% ester sulfate, 2.2% protein	*Undaria pinnatifida*	–	405
	Fraction B	24.2% Fucose, 23.1% galactose, 17% ester sulfate, 1.3% protein	*Undaria pinnatifida*	–	405

	Fraction C	12.5% Fucose, 24.3% galactose, 2.5% glucuronic acid, 24.7% ester sulfate	*Undaria pinnatifida*	+	405
	pD-I	3.3% Nitrogen, 44.9% fucose, 2.3% uronic acid, 26.2% sulfate, 67% fucose, xylose (trace), 2% mannose, 31% galactose	*Hizikia fustiforme*	–	596
	pD-II	2.1% Nitrogen, 29.1% fructose, 1.0% uronic acid	*Hizikia fustiforme*	–	596
	pD-III	31.6% Sulfate, 77% fucose, xylose (trace), mannose (trace), galactose (trace)	*Hizikia fustiforme*	–	596
	Sargassan	Backbone = glucuronic acid, side chains = partially mannose, galactose, sulfated galactose, xylose, fucose residues	*Sargassum linifolium*	+	597
	Unnamed	55% Carbohydrate, 8.2% protein, 4.9% sulfate	*Padina tetrastromatica*	+	598
	Unnamed	D-Glucuronic acid, L-fucose, D-xylose, D-mannose, D-glucose, D-galactose	*Padina pavonia*	+	599
	Unnamed	D-Glucuronic acid, D-galactose, fucose	*Dictyota dichotoma*	+	600
	Unnamed	Polysaccharide	*Codium fragile*	ND	601
Man-made sulfated polysaccharides	Laminarine, **130**	B-1,3 Glucan units with occasional branch points at C-6, sulfated at reducing end	*Laminaria cloustoni*	–	406 602

*+ = More potent than heparin; – = less potent than heparin; ND, not determined.

thrombin, which results in the disruption of the coagulation cascade. Mediation of the anticoagulant activity of sulfated and unsulfated brown algal polysaccharides was initially attributed to an ability, like heparin, to inhibit the conversion of fibrinogen into fibrin.[402] It was suggested that carrageenans react with fibrinogen *in vitro* to form an insoluble precipitate which interferes non-specifically with the clotting factors.[403] However, anticoagulant activity was then shown to be predominantly the result of direct polysaccharide–thrombin interaction, the polysaccharide affecting the fibrinogen binding site on thrombin.[404] Additionally, polysaccharides were also found to exhibit catalytic effects similar to heparin on the intrinsic and extrinsic coagulation pathways, inferring the involvement of a heparin-like mechanism, although the abilities of the polysaccharides and heparin to bind antithrombin III probably differ. Polysaccharides, however, unlike heparin, were not found to affect factor Xa.[404]

The anticoagulant activity of the algal polysaccharides has been shown to be dependent on the degree of sulfation of the polysaccharide – the higher the sulfation the higher the anticoagulant activity.[402,404,405] There has been much debate as to whether anticoagulant activity is dependent[404] or not[406] on molecular weight.

Calcium alginate (**129**), a hemicellulose mannuronoglucan from the brown seaweed *Laminaria digitata*, has been reported to stimulate blood clotting before being absorbed by the tissues.[407]

5.7.2.8 *Haemolysins*

Haemolysins are compounds which have a lytic effect on red blood cells and other cell types. The toxin from *Prymnesium parvum* (Figure 5.2; Section 5.7.1.6.2) has been separated into six compounds, all of which possess haemolytic activity.[408] The major component of this mixture is haemolysin I, which has been found to be a mixture of 1′-*O*-octadecatetraenoyl-3′-*O*-(6-*O*-*β*-D-galactopyranosyl-*β*-D-galactopyranosyl)glycerol (**131a**) and 1′-*O*-octadecapentaenoyl-3′-*O*-(6-*O*-*β*-D-galactopyranosyl-*β*-D-galactopyranosyl)-glycerol (**131b**). Several dinoflagellates are also known to produce haemoly-

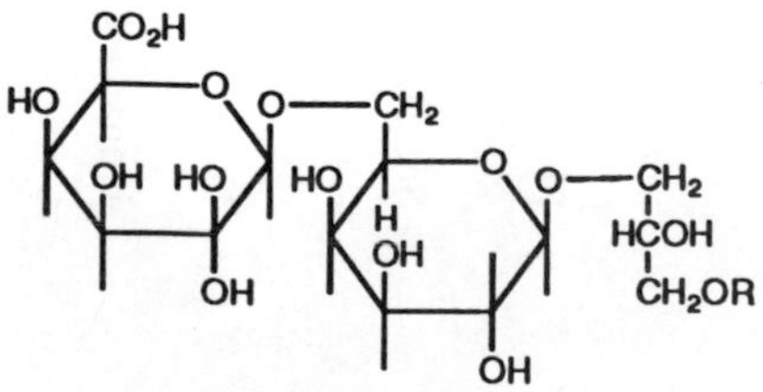

a R = $COCH_2CH_2CH_2(CH_2CH{=}CH)_4CH_2Me$

b R = $CO(CH_2CH{=}CH)_5CH_2Me$

131 Digalactosyldiglycerides
Haemolysins

sins, including an unsaturated lipid, two galactoglycerolipids, and several polyether haemolysins, isolated from *Gonyaulax monilata*,[409] and galactoglycerolipid haemolysins, from *Amphidinium carterae*[215] and *Ptychodiscus brevis*. Other dinoflagellates, including *Amphidinium klebsii*, *Gambierdiscus toxicus*, *Ostreopsis ovata*, *O. streopsis siamensis*, *P. concavum*, *P. lima*, and *Prorocentrum mexicanum*, are reported to produce haemolysins,[2] as are the cyanophyte *Microcystis aeruginosa*[410] and the chlorophyte algae *Chaetomorpha minima*[341] and *Ulva pertusa*.[342]

5.7.2.9 *Anticonvulsants*

Anticonvulsants (Figure 5.28) are compounds, such as the synthetic drugs sodium valproate and phenytoin, that prevent or reduce the severity of convulsions in various types of epilepsy.

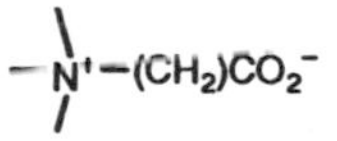

$$-\overset{|}{\underset{|}{N^+}}-(CH_2)CO_2^-$$

132 Glycinebetaine

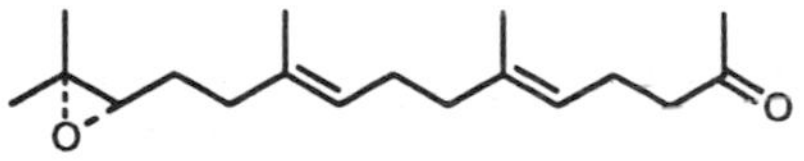

133 Farnesylacetone epoxide

Figure 5.28 *Anticonvulsants*

Glyncinebetaine (**132**) possesses anticonvulsant properties.[411–413] It has been found in many algal families, including Bacillariophyceae, Prymnesiophyceae, Prasinophyceae,[414] Chlorophyceae,[415–419] and Phaeophyceae,[417–419] where it is postulated to play a role in cytoplasmic osmoregulation.[420] Glycinebetaine also occurs naturally in mammals, where it is known to function as a methyl donor for the remethylation of homocysteine to methionine during sulfur conservation.[421] It is not surprising, therefore, that it is of benefit in the treatment of homocystinuria,[422–424] the failure to form cystathionine from homocysteine and serine in patients deficient in the enzyme cystathionine β-synthase.

Farnesylacetone epoxide (**133**), isolated from the brown seaweed *Cystophora moniliformis*, has also shown anticonvulsant activity in mice following intraperitoneal administration. Although less potent than the anticonvulsant drug phenytoin, it is less toxic.[400]

5.7.2.10 *Antiulcer Activity*

All algal-derived compounds reported to possess antiulcer activity are polysaccharides, some of which have been previously described as possessing anticoagulant or haemostatic properties (Section 5.7.2.7). The sulfated polysaccharides, κ-carrageenan (**127**) and its low viscosity degradation product have been found to inhibit gastroduodenal ulceration via an effect on histamine-stimulated gastric secretion.[425–428] The algal polysaccharide appears to protect the intestinal lining by combining with mucoproteins that line the stomach to form a resistant membraneous structure.[427,429,430]

5.7.2.11 *Cardiotonic and Cardiovascular Agents*

Cardiotonic and cardiovascular agents (Figure 5.29) are compounds which affect the functioning of the heart or cardiovascular system, bringing about an alteration in the circulation of blood around the body. Some such compounds from algae are known toxins, as described previously (Section 5.7.1). However, novel cardiotonic and/or cardiovascular agents, such as the chlorinated cyclic decapeptide puwainaphycin C (**134**) isolated from the cyanophyte *Anabaena* sp.[431] (ED_{50} on mice atria of 0.2 μg ml^{-1}), have been reported and are shown in Table 5.15. Ciguatoxin is reported to be effective at concen-

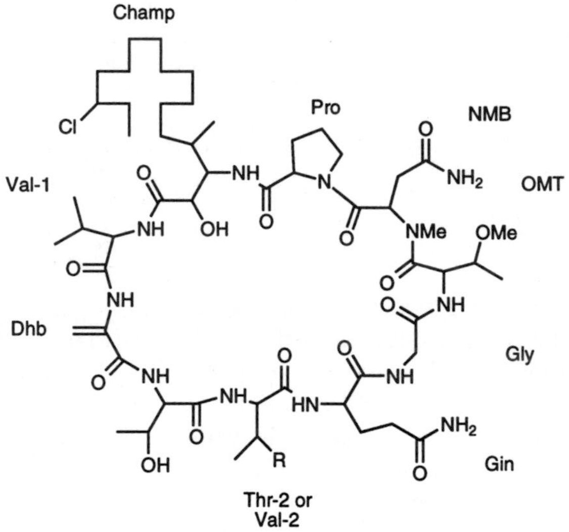

Champ, Pro, NMB, Val-1, OMT, Dhb, Gly, Gin, Thr-2 or Val-2

a R = OH
b R = Me

134 Puwainaphycin C

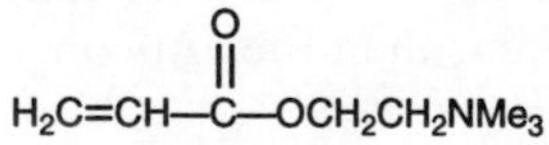

$H_2C{=}CH{-}C({=}O){-}OCH_2CH_2NMe_3$

135 Acrylcholine

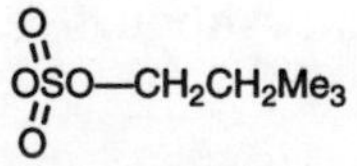

$O\overset{O}{\underset{O}{S}}O{-}CH_2CH_2Me_3$

136 Choline *O*-sulfate

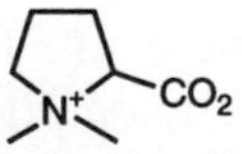

137 Prolinebetanine

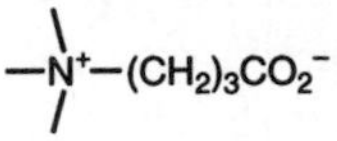

$-N^+{-}(CH_2)_3CO_2^-$

138 γ-Aminobutyric acid

Figure 5.29 *Cardiotonic and cardiovascular agents*

Table 5.15 *Cardiotonic and cardiovascular agents from algae (adapted from Lincoln* et al.[22]*)*

Chemical nature of compound	*Compound name*	*Structure**	*Source*	*Reference*
Amines	Saxitoxin	**1**	*Aphanizomenon flos-aquae* and other PSP toxin producers (Figure 5.3)	603
	Choline esters:		*Amphidinium carterae*	
	Acrylcholine	**135**		507
	Choline-*O*-sulfate	**136**		604
Betanines	Prolinebetanine	**137**		433
	γ-Aminobutyric acid	**138**		434
	Laminine	**107**	*Laminaria angustata*	605
Peptides	Oscillatoria toxin	**16**	*Oscillatoria agardhii* (Figure 5.4)	280
	Puwainaphycin C	**134**	*Anabaena* sp.	431
Polyether toxins	Brevetoxins	**2**	*Ptychodiscus brevis*	182, 605
	Ciguatoxin	**3**	*Gambierdiscus toxicus*	203, 204, 223
	Maitotoxin	ND	*Gambierdiscus toxicus*	607
Proteolipid-sugar complex	Prymnesium toxin	**131**	*Prymnesium parvum* (Figure 5.2)	130
Unknown	Tolyphycins A & B	ND	*Tolypothrix byssoidea*	608

*ND, not determined.

trations $>2.8 \times 10^{-2}\,\mu g\,ml^{-1}$ in rat atria[212] and maitotoxin to be effective on guinea-pig heart at 10^{-3}–$10^{-4}\,\mu g\,ml^{-1}$,[221,432] prymnesium toxin at 5–10 mg kg^{-1} on frog heart,[330] prolinebetanine at 5–100 mg kg^{-1},[433] and λ-aminobutyric acid betanine at 40 μg kg^{-1} on mouse atria.[434]

5.7.2.12 *Hypocholesterolaemic and Hypotensive Agents*

The occurrence of heart disease has been shown to be closely linked to elevated levels of plasma cholesterol (hypercholesterolaemia) and elevated blood pressure (hypertension).

The unsaponifiable sterols fucosterol (**139**), isolated from the brown algae *Fucus evanescens* and *Fucus gardeneri* and the red alga *Porphyra wrightii*, and sargasterol (**140**), isolated from the brown alga *Sargassum muticum*, have been reported to possess hypocholesterolaemic activity (Figure 5.30). Reductions of 83% and 59% in blood cholesterol levels in chickens have been reported on the

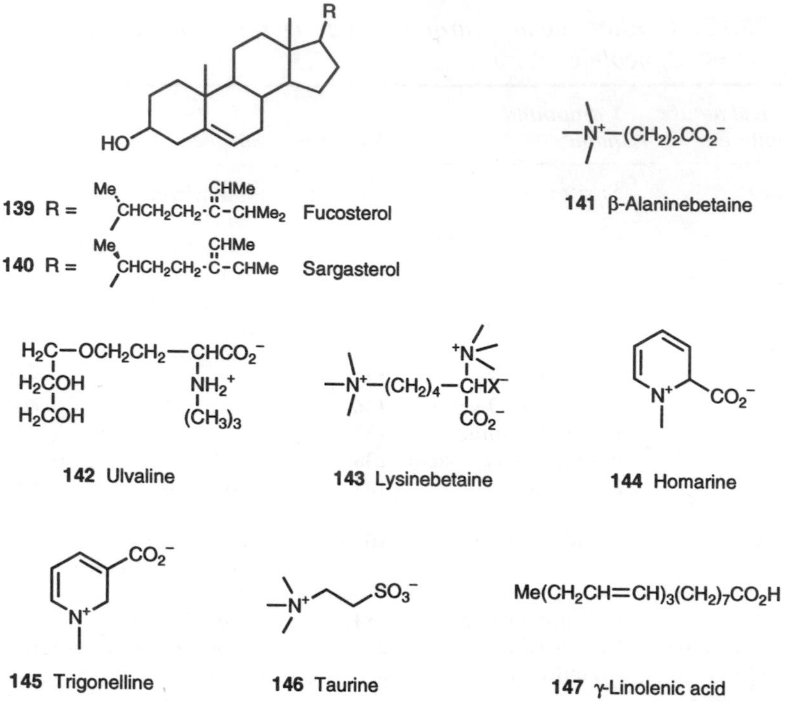

139 R = $\overset{Me}{\diagdown}CHCH_2CH_2\text{-}\overset{CHMe}{\overset{\|}{C}}\text{-}CHMe_2$ Fucosterol

140 R = $\overset{Me}{\diagdown}CHCH_2CH_2\text{-}\overset{CHMe}{\overset{\|}{C}}\text{-}CHMe$ Sargasterol

$—N^+—(CH_2)_2CO_2^-$

141 β-Alaninebetaine

$H_2C—OCH_2CH_2—CHCO_2^-$; H_2COH; H_2COH; NH_2^+; $(CH_3)_3$

142 Ulvaline

$—N^+—(CH_2)_4—CHX^-$; ^+N; CO_2^-

143 Lysinebetaine

N^+ CO_2^-

144 Homarine

CO_2^- N^+

145 Trigonelline

N^+ SO_3^-

146 Taurine

$Me(CH_2CH{=}CH)_3(CH_2)_7CO_2H$

147 γ-Linolenic acid

Figure 5.30 *Hypocholesterolaemic agents*

incorporation of fucosterol and sargasterol, respectively, into their feed.[435] Other sterols from a range of algae have also been reported to possess hypocholesterolaemic properties.[436] Extracts of *Monostroma nitidum*, *Ulva pertusa*, *Enteromorpha compressa*, and *E. prolifera* (Chlorophyta) and *Porphyra tenera* (Rhodophyta) were found to significantly lower plasma cholesterol levels in rats.[437] The betaines β-alaninebetaine (**141**)[438] and ulvaline (**142**)[439] have been identified as the active hypocholesterolaemic compounds in *M. nitidum*. Other betaines, for example lysinebetaine (**143**), homarine (**144**), trigonelline (**145**), and taurine (**146**), have also been found to be active,[439] as has the fatty acid λ-linolenic acid (GLA) (**147**), a constituent of algal lipids.[440]

Hypotensive agents were first reported from the marine algae *Laminaria angustata*, *Heterochordaria abetina*, and *Chondria armata*.[313,441–444] The choline-like basic amino acid, laminine (**107**) was subsequently isolated from *Laminaria angustata* and other marine algae and attributed with the hypotensive activity.[441–443] Laminine also possesses cardiotonic properties (Table 5.15, Section 5.7.2.11). The antihypertensive activity of laminine is thought to be the result of ganglion blockade, but it only occurs at high concentrations.[445]

5.5.7.2.13 *Algae as a Source of Prostaglandins*

Prostaglandins (Figure 5.31) are a group of hormone-like substances which have a number of therapeutic uses, including the induction of labour and

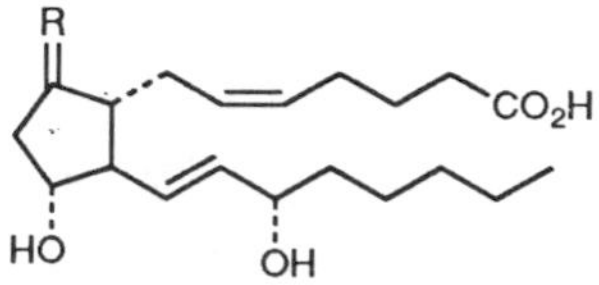

148 Arachadonic acid

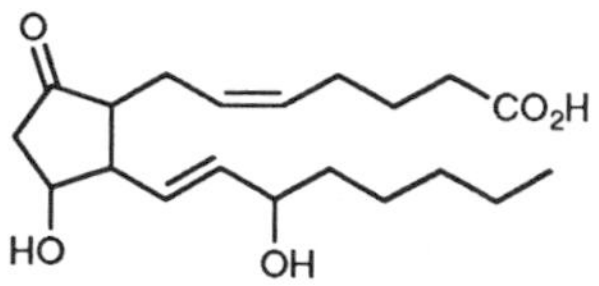

149 Prostaglandin E_2

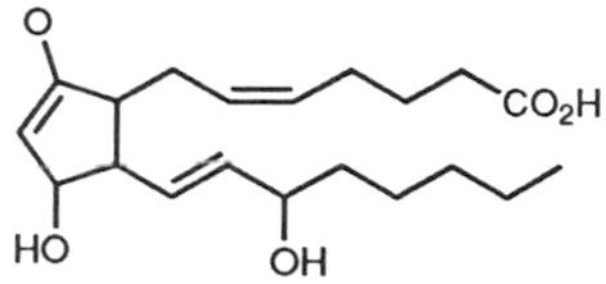

150 Prostaglandin F_2

Figure 5.31 *Algae as a source of prostaglandins*

abortion, contraception, and the treatment of high blood pressure, asthma, bronchitis, ulcers, and thrombosis. The red microalga *Porphyridium cruentum* is reputed to be the richest source of the prostaglandin precursor arachadonic acid (**148**), which constitutes up to 50% of the algal fatty acids at low temperatures.[446] Prostaglandins E2 (**149**) and F2 (**150**) have also been isolated from the red alga *Gracilaria lichenoides*, where they constituted 0.07–0.1% of the algal dry weight. The antihypertensive action of extracts of this alga administered intravenously to rats is presumed to be due to the presence of these prostaglandins.[447]

5.7.2.14 *Compounds Affecting Enzyme Activity*

With the advances in automated enzyme assay technology, extensive screening programmes have been carried out on algal supernatants and organic solvent cell extracts for the presence of enzyme activators and inhibitors[121–128,448–450] (Figure 5.32). The enzymes chosen were either known to be involved in or implicated in human or animal disease conditions.

The novel sphingosine derivative, symbioramide (**151**), isolated from the marine dinoflagellate *Symbiodinium* sp., has been reported to activate the rabbit sarcoplasmic reticulum Ca^{2+}-ATPase by approximately 30%.[448] This enzyme energizes the pumping of calcium from the cytoplasm into the lumen of the sarcoplasmic reticulum, allowing muscle relaxation to occur.[451] The diterpene, phytol (**88**), isolated from the brown alga *Hizikia fusiformis* (hiziki), was found to be an activator of a lipase enzyme which hydrolyses fats to fatty acids and glycerol or monoacylglycerols. Another two activators, A-3 and A-9, from the same alga are also activators of lipase and are thought to belong to terpenes and sterols, respectively.[452]

Other compounds of algal origin have been reported to be inhibitors of ATPase enzymes. The glycolipids, mono- and di-galactosyldiacylglycerol

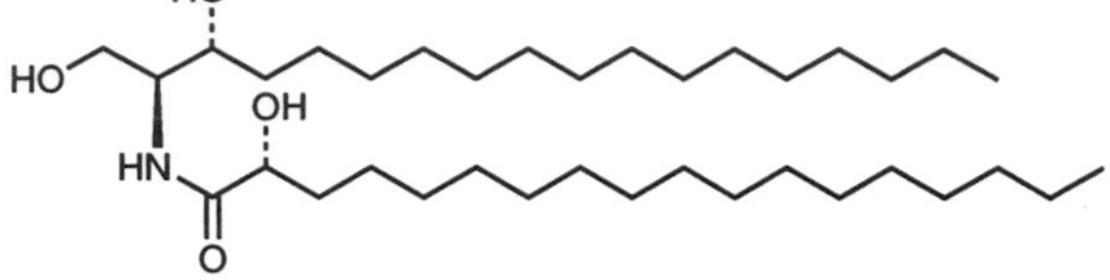

HO HO OH HN O

151 Symbioramide

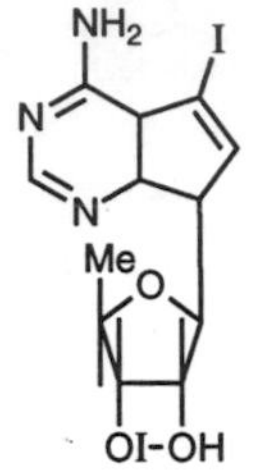

NH_2 I N N Me O OI-OH

152 5-Deoxy-5-iodotubercidin

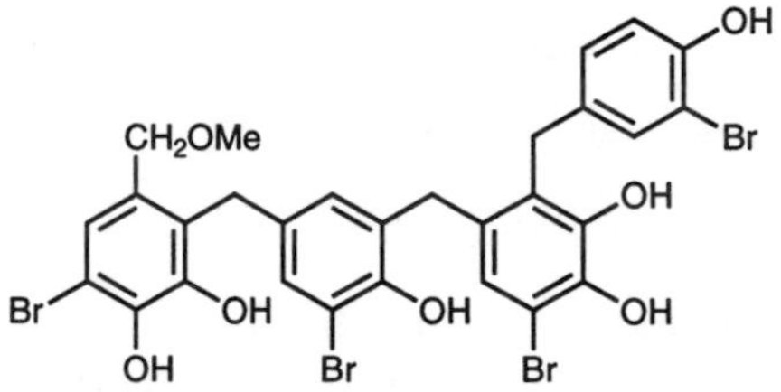

OH Br CH_2OMe OH Br OH OH OH OH Br Br

153 Rawsonol

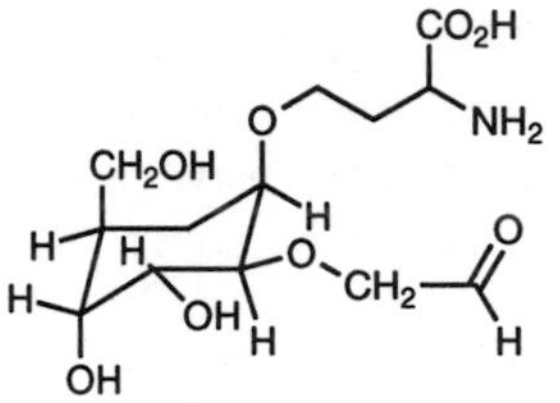

CO_2H O NH_2 CH_2OH H H H O CH_2 O H OH H H OH

154 Amino sugar

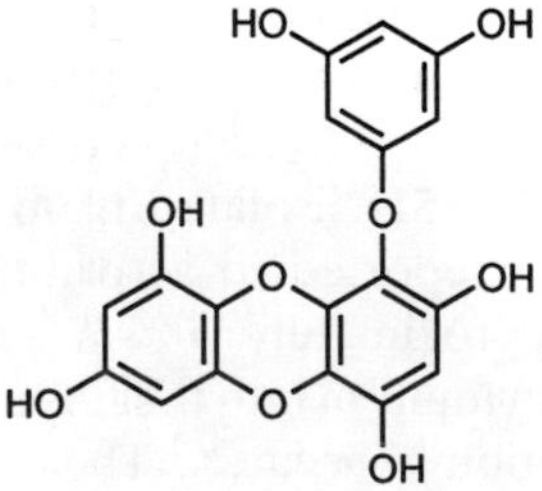

HO OH O OH OH HO O O OH

155 Eckol

HO OH OH O O OH O O OH OH HO O OH OH

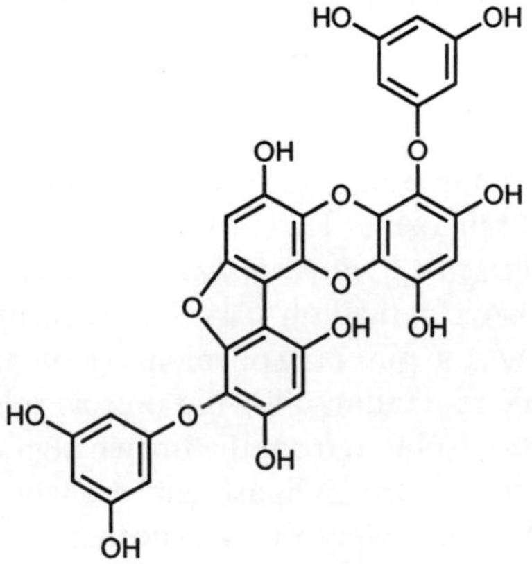

156 Phlorofuco-furoeckol A

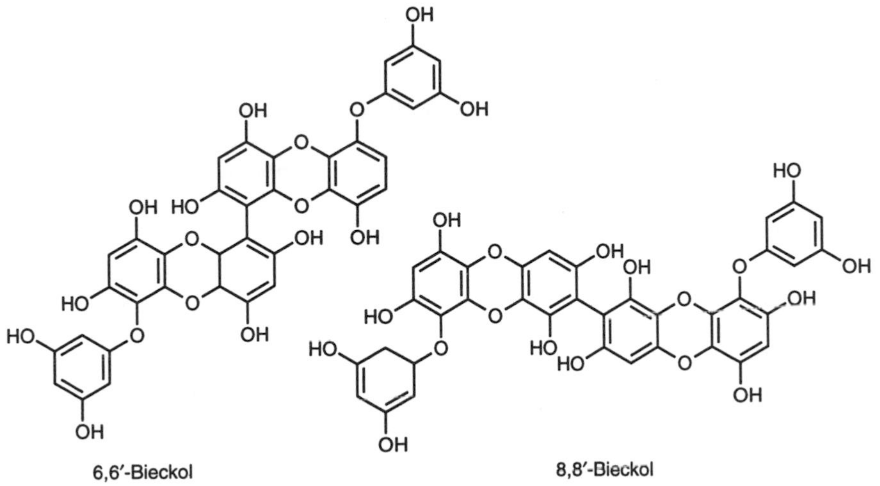

157 Dimeric eckols

Figure 5.32 *Compounds affecting enzyme activity*

(similar in structure to **131**), isolated from the prymnesiophyte *Hymenomonas* sp., have been found to inhibit the Na^+,K^+-ATPase[449] (IC_{50} values of 2×10^{-5} M in both cases), the enzyme which facilitates the energy-dependent transport of sodium and potassium across cell membranes against their concentration gradients. The Na^+ and K^+ gradients across the cell membrane are necessary for the excitation response of muscle cells and for signal transmission by nerve cells. The dinoflagellate toxin, maitotoxin (Section 5.7.1.2.3), from *Gambierdiscus toxicus*, has also been reported to strongly inhibit the Na^+,K^+-ATPase.[450] The halogenated pyrrolopyrimidine analogue of adenosine 5′-deoxy-5-iodotubercidin (**152**), isolated from the red alga *Hypnea valentiae*, has been reported to be a very potent specific inhibitor of the enzyme adenosine kinase.[453] This enzyme is important because it catalyses the first stage [adenosine monophosphate (AMP) into adenosine diphosphate (ADP)] of the reconversion of AMP into adenosine triphosphate (ATP). Inhibitors of adenosine kinase show bronchodilatory activity *in vivo* and are useful in the treatment of asthmatics. The pentagalloylglucose compound, 3-*O*-digalloyl-1,2,6-trigalloylglucose (**55**), isolated from the green alga *Spirogyra varians*, is reported to inhibit irreversibly the enzyme α-glucosidase from yeast, with an IC_{50} value of 3.4×10^{-7} M. Two other α-gluocosidase inhibitors have been reported from the green algae *Mesotaenium caldariorum* and *Mougeotia* sp.;[454] although their structures have not been determined, these are thought to be tannins. By hindering digestion and absorption of carbohydrates, inhibitors of this enzyme have been suggested to be of value in the prevention and control of conditions such as diabetes, obesity, hyperlipoproteinaemia, and hyperlipidaemia.[455] In the search for inhibitors of cholesterol biosynthesis, a novel brominated diphenylmethane derivative, rawsonol (**153**), an inhibitor of HMG-CoA

(3-hydroxy-3-methylglutaryl coenzyme A reductase), was isolated from the green alga *Aurainvillea rawsoni*.[456] Rawsonol had an IC_{50} value of 5 μM against this enzyme in humans which, although not as potent as the fungal metabolite mevinolin (IC_{50} of 2 nM), is of interest because of its structural dissimilarity to other known inhibitors. Other enzyme inhibitors isolated from algae include an amino sugar (**154**), from *Anabaena flos-aquae* (Cyanophyceae), which inhibits α-amylase,[454] anatoxin *a*(s) from *A. flos-aquae* (Section 5.7.1.3), which is reported to be an irreversible inhibitor of acetylcholinesterase,[259] and a series of novel phlorotannins, with a dibenzo-1,4-dioxin skeleton, isolated from the brown alga *Ecklonia kurome* Okamura, which act as specific antiplasmin inhibitors. These include eckol[457–459] (**155**), phlorofucofuroeckol A[460] (**156**), and two dimeric eckols[458] (**157**), and are active as antiplasmin inhibitors by inhibiting the action of α_2-macroglobulin and α_2-plasmin inhibitor, the main plasmin inhibitors in plasma.[460] Plasmin is responsible for fibrinolysis, which is

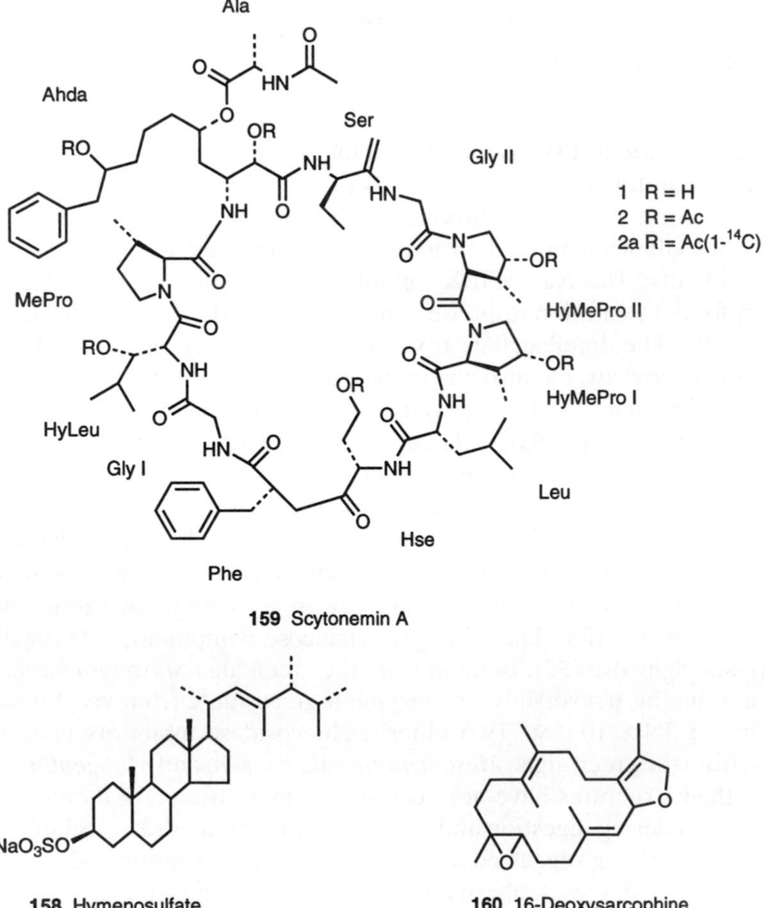

Figure 5.33 *Compounds affecting calcium homeostasis*

the dissolution of blood clots by the proteolytic degradation of fibrin into soluble peptides. These compounds are, therefore, of interest for the prevention and treatment of thrombosis. Inhibitors of various proteases, for example trypsin, elastase, papain, carboxypeptidase A, α-chymotrypsin, and collagenase, have been detected in a number of algae.[128] These may have clinical applications in the treatment of various disease states showing, among others, anti-inflammatory, antihypertensive, and antineoplastic activities.

5.7.2.15 *Compounds Affecting Calcium Homeostasis*

The divalent cation calcium is ubiquitous in nature and has a number of vital functions in organisms, which include being a co-factor for enzymes, stabilizer for proteins, lipids in cell membranes, cytoplasm, organelles, and chromosomes, and second messenger. Because of its importance, calcium has been implicated in various disease states (*e.g.*, cardiovascular diseases, and respiratory and neurological disorders) and compounds affecting its cellular homeostasis have been of great clinical interest. Several compounds of algal origin have been reported (Figure 5.33).

Hymenosulfate (**158**), a novel sterol sulfate isolated from the prymnesiophyte *Hymenomonas* sp.,[449] was found to effect the release of calcium from the muscle sarcoplasmic reticulum, with ten times the potency of caffeine.[461] The novel cyclic peptide, scytonemin A (**159**), isolated from the cyanobacteria *Scytonema* sp. (strain U-3-3), has been reported to be a potent calcium channel antagonist. However, it is 10 times less active than the benzothiazepine calcium channel antagonist diltiazem.[462] Another calcium channel antagonist, 16-deoxysarcophine (**160**), has been detected in extracts of the dinoflagellate *Symbiodinium* sp., which is a symbiont of soft corals[463] of the genus Sarcophyton, from which the compound was first isolated.[464]

5.7.3 Compounds with Agricultural Potential

Algae, in particular the seaweeds, have been collected and used as fertilizers and soil conditioning agents for many years. Seaweed extracts and suspensions are marketed for use in agriculture and horticulture. This subject has been reviewed by Blunden and Gordon[465] and Blunden[466] but is outside the scope of this chapter.

This section deals with 'biologically active' compounds, which can be used to increase yields and improve the quality of agricultural crops. These compounds can influence the crop directly, by regulating plant growth, or indirectly, by reducing damage from pests, pathogens, and weeds. Very few reports are available, however, which provide insecticidal, fungicidal, herbicidal, or plant growth regulatory data, although work is in progress, as illustrated by the occasional patent covering algal-based products for agricultural use.

5.7.3.1 *Insecticides*

Several algal products have been shown to be active against a variety of insect pests (Figure 5.34). The amino acid compounds, domoic acid (**21**) and α-kainic acid (**22**) (Section 5.7.1.6.1), isolated from the red alga *Chondria armata*, have been shown to exhibit insecticidal activity against American cockroaches (*Periplaneta americana*) when administered intraperitoneally[467] (minimum effective dose $0.8\,\mu g\,g^{-1}$ for domoic acid) and to be effective on topical

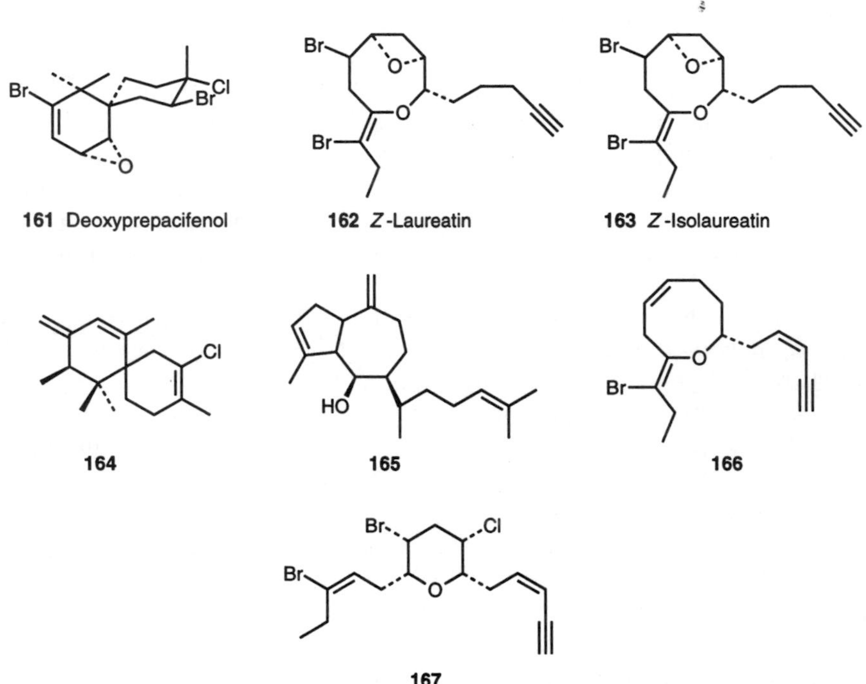

Figure 5.34 *Insecticides*

application to house flies (*Musca domestica*) and German cockroaches (*Blattella germanica*).[468,469] The precise mechanism of insecticidal activity of this class of compounds is uncertain, although they have been shown to be excitory transmitters at the neuromuscular junction of insects.[470] Three insecticidal metabolites have been isolated from the red alga *Laurencia napponica* which are active against mosquito larvae (*Culex pipiens pallens*), with IC_{50} values of 0.48, 0.06, and $0.5\,\mu g\,ml^{-1}$ for deoxyprepacifenol (**161**), *Z*-laureatin (**162**), and *Z*-isolaureatin (**163**), respectively.[471] Methanol extracts of two species of green algae, the filamentous *Rhizoclonium hieroglyphicum* Kütz and the phytoplankton *Chlorella ellipsoidea* Gerneck, were also found to show insecticidal activity against three species of mosquito [*Aedes aegypti* L., *Culex quinquefasciatus* Say, and *Culiseta incidens* (Thomson)]. The extracts exhibited juvenile hormone-like activity and also caused morphogenetic changes in emerging adults.[472] Other compounds reported to possess insecticidal properties include two terpenoid compounds, isolated from *Laurencia* sp. (**164**) and *Pachydictyon coriaceum* (**165**), which are active against spider mites and nematodes, respectively. Two halogen-containing secondary metabolites, isolated from *Laurencia* sp. (**166**) and *L. obtusa* (**167**), are reported to be active against the black bean aphid (*Aphis fabae*) and southern armyworm (*Spodoptera exempta*), respectively, while a phenolic compound from *Odonthalia floccosa* is also active against the black bean aphid.[465] Reduced incidence of nematode infestation has been reported on seaweed-extract treated plants;[609–611] however, this seems to be due to the presence of plant growth regulators, in particular cytokinins, which improve the growth of the plant, rather than a nematicidal effect of the extracts.[611,612]

5.7.3.2 *Herbicides*

The diacyl-substituted lactone cyanobacterin (**44**) (Table 5.10, Section 5.7.2.1.1), isolated from the cyanobacterium *Scytonema hofmanni*, was patented for use as a herbicide in 1986.[473] Effective against both monocotyledonous and dicotyledonous angiosperms, cyanobacterin is known to act by inhibiting photosynthetic electron transport in the oxygen-evolving system of photosynthesis (photosystem II).[474] The toxic principle from the freshwater green alga *Pandorina morum* has also been shown to inhibit the activity of photosystem II,[475] as well as to inhibit the oxidation of succinate or malate/pyruvate in isolated plant mitochondria.[476]

Other compounds reported to exhibit herbicidal activity include a terpenoid (**165**) from *Pachydictyon coriaceum*, which gave preplant control of curlydock (*Rumex crispus*), and two phenolics isolated from *Odonthalia floccosa* (**168**) and *Anacystis marina* (**169**), which control crabgrass (*Digitaria sanguinalis* L.) and bindweed (*Convolvulus arvensis* L.), respectively.[465]

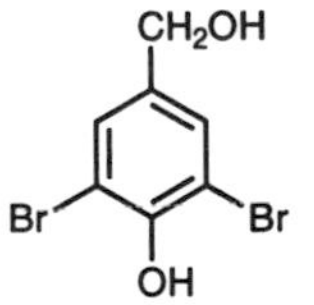

CH_2OH Br Br OH

168

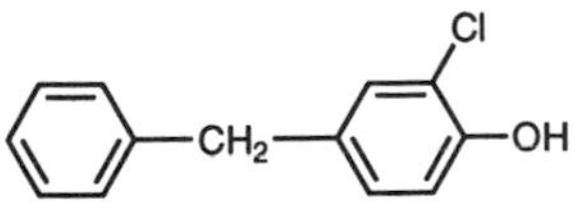

CH_2 Cl OH

169

Herbicides

5.7.3.3 *Plant Growth Regulators*

Compounds exhibiting auxin-like activity (growth promotion) have been detected in extracts of various members of the Chlorophyta, Phaeophyta, and Rhodophyta.[477–485] The chemical nature of many of these compounds is similar to indoleacetic acid (**170**) from higher plants, and examples include indolyl-3-carboxylate, indolyl-3-acetic acid, hydroxyphenylacetic acid, and phenylacetic acid isolated from the brown alga *Undaria pinnatifida.*[486,487] Several non-indolic growth-promoting substances have been isolated from the brown alga *Sargassum polycystum*,[485] which appear to belong to the pyrolle carboxylic acid, polyhydroxyphenol, and phenoxyacetic acid classes of compounds.

Cytokinins (Figure 5.35), compounds which promote growth and cell division and elongation and numerous physiological responses in higher plants, also exist in algae.[488] A commercial preparation of the brown alga *Ecklonia maxima* has been found to contain the cytokinins *cis*- and *trans*-zeatin riboside (**171**), *trans*-zeatin (**172**), dihydrozeatin (**174**) and $N^6(\Delta^2$-isopentenyl)adenosine (**173**).[489] Other compounds with cytokinin-like activities are also found in algae,[490] and include various betaines; for example,

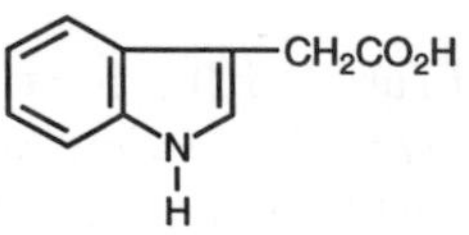

170 Indoleacetic acid

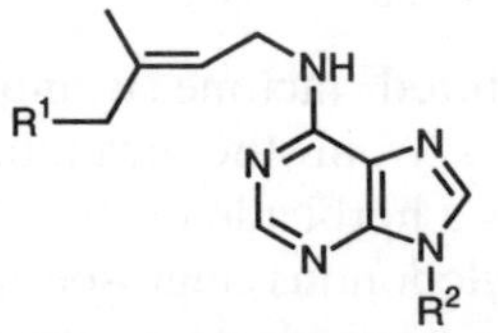

171 R^1 = OH, R^2 = β-D-ribosyl *trans*-zeatin riboside
172 R^1 = OH, R^2 = H *trans*-zeatin
173 R^1 = H, R^2 = H $N^6(\Delta^2$-isopentenyl) adenosine

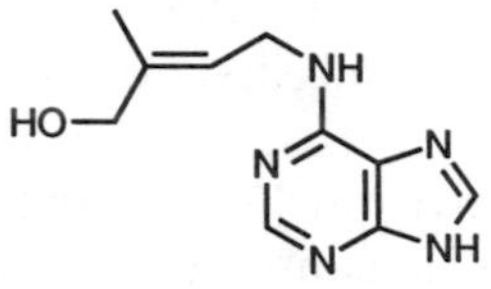

174 Dihydrozeatin

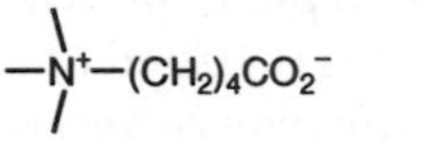

175 δ-Aminovaleric acid betaine

glycinebetaine (**132**), is reported to cause chlorophyll retention in oat leaves,[491] play a role in resistance to frost,[492,493] enhance the efficiency of water utilization and grain yield in winter wheat,[494] and alleviate salt stress in seedlings.[420] Other betaines produced by algae include γ-aminobutyric acid (**23**), δ-aminovaleric acid betaine (**175**), lysinebetaine (**143**), and laminine (**107**).[418,419]

5.7.3.4 *Fungicides*

Although there have been many instances of antifungal activity against human pathogens (Tables 5.10 and 5.11, Sections 5.7.2.1 and 5.7.2.2), there are very few reports involving the screening of algal broths and extracts *versus* economically important phytopathogenic fungi. Of the few cases reported, the most notable activity is that of the cyclic depsipeptide, majusculamide C (**50**), isolated from a deep water variety of the cyanophyte *Lyngbya majuscula*. This

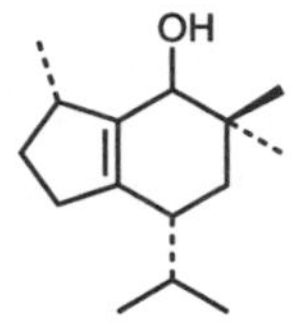

OH

176

OAc

CHO

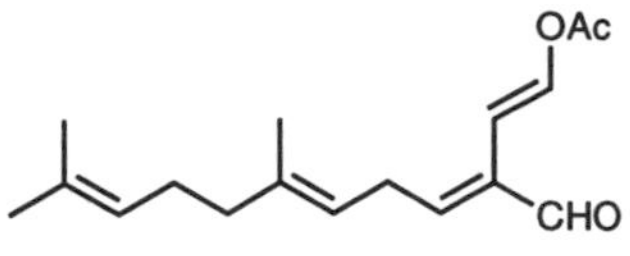

177

has been observed to control the growth of a number of fungal plant pathogens, including the Oomycetes *Phytophthora infestans* (potato late blight) and *Plasmopara viticola* (grape downy mildew).[495] Another compound reported to control *P. infestans* is a terpenoid (**176**) isolated from the brown alga *Laurencia obtusa*.[496] Finally, a terpenoid (**177**) isolated from brown algae of the family *Udoteaceae* has been observed to inhibit the growth of *Leptosphaeria* sp. and *Alternaria* sp. *in vitro*.[366]

5.8 Conclusions and Future Prospects

An increasing variety of interesting compounds, in terms of both structural diversity and biological activity, have been obtained from algae. Many of these have been partially or fully purified and characterized. However, despite the wealth of compounds with useful activities, comparatively few products have reached the market place and, therefore, algae remain a commercially untapped source of pharmaceuticals and/or agrochemicals. This may be attributed to several reasons.

First, there are problems associated with culturing algae to obtain sufficient quantities of the compound of interest. Macroalgae are difficult to obtain on a sufficient scale, culturing in particular posing practical problems. Thus, any product obtained from macroalgae would need to be of high commercial value to cover the cost of large-scale culturing or harvesting of natural populations. Microalgae are probably of more interest in this respect, although availability of axenic cultures, difficulties in culturing and growth of some species, and low yield of biologically active compounds have limited their use. Technologies for the mass production of microalgae in large ponds, photobioreactors, or industrial fermentations do exist, but are still in their infancy when compared with the large-scale production of other micro-organisms. The challenge and costs of developing economically competitive systems for the production of biologically active compounds from algae is a consideration inherent to the future use of these organisms.

Where the production of the active metabolite is under the control of a single gene, yields can be increased substantially by using genetic engineering. This might involve inserting the appropriate algal gene into a heterologous host, such as *Escherichia coli*. The bacterium, more amenable to industrial-scale fermentation procedures, would then produce the algal product of interest in large quantities. With further manipulations the bacterium could be made to secrete the product into the culture medium to make purification relatively easy. If it were not possible to transform a heterologous host, then it may be feasible to engineer the algal producer such that it will overproduce the required substance. An entirely different situation exists, however, when biochemical pathways (such as secondary metabolite production) involving multistep pathways with complex biosynthetic and regulatory processes need to be manipulated. Having said this, recombinant DNA techniques combined with a fuller understanding of biosynthetic processes are likely to have a central role for the future exploitation of rare and novel pharmaceuticals and/or agrochemicals from algae.

Second, a well-designed screening programme to detect these novel compounds, with novel mechanisms of action, is a crucial factor in the exploitation of natural products. Many traditional screens in the pharmaceutical and agrochemical industries are set up to detect compounds with known mechanisms of action and are often inadequate to ensure the detection of novel natural products. For example, many secondary metabolites are insoluble in water or are difficult to place in homogeneous suspension. The screening procedure needs to be flexible enough to accommodate this and other technical problems. As our understanding of disease processes has increased, more direct methods of testing have been utilized by the pharmaceutical industry. For example, enzyme assays, receptor binding and displacement studies, and effects on tissue culture preparations may provide a more reliable evaluation of potential natural product drugs. One has to recognize, however, that such *in vitro* tests do not take into consideration potential problems with the uptake or metabolism of an introduced substance, which can markedly affect the performance of a compound *in vivo*. In spite of this, *in vitro* screening programmes have been incorporated into the natural-product drug discovery programmes of many pharmaceutical companies. This is a relatively new concept for most agrochemical companies for a number of reasons, which include the smaller total market and lower profit margins for agrochemicals compared with pharmaceuticals, and the limited knowledge currently available on plant, fungal, and insect biology at the biochemical and/or molecular level. This state of affairs is reflected in the dearth of natural products reported with herbicidal, insecticidal, fungicidal, or plant growth regulatory activity.

Finally, many biologically active molecules show insufficient activity, poor efficacy profiles, or are just too toxic to be of use as products *per se*. These compounds, however, are of great value as unique templates that could serve as a starting point for synthetic chemists, as experimental tools that can be used to gain a better understanding of biological processes, or as indicators of potentially novel modes of action, which could be subsequently exploited in the search for drugs or agrochemicals. It is my personal opinion that, with improvements in mass culture and/or fermentation technology, improvements in yields of secondary metabolites through genetic engineering and/or molecular biology, the implementation of carefully designed screening programmes, dedication, and a touch of luck, algae will become a valuable source of biologically active compounds for commercial exploitation in the future.

5.9 Acknowledgements

Special thanks to Dr Hilda Canter-Lund FRPS of the Freshwater Biology Association, Windermere for kindly supplying the algal photographs.

5.10 References

1. E. J. Schantz, in 'Toxic Constitutents of Animal Foodstuffs', ed. I. E. Liener, Academic Press, London, 1974, p. 123.
2. K. A. Steidinger and D. G. Baden, in 'Dinoflagellates', ed. D. L. Spector, Academic Press, London, 1984, p. 201.
3. W. W. Carmichael, C. L. A. Jones, N. A. Mahmood, and W. C. Theiss, *CRC Crit. Rev. Environ. Control.*, 1985, **15**, 275.
4. G. A. Codd and G. K. Poon, in 'Biochemistry of the Algae and Cyanobacteria', ed. J. R. Gallon and L. J. Rogers, Proc. Phytochem. Soc. Europe, Clarendon Press, Oxford, 1988, p. 283.
5. R. E. Lee, 'Phycology', Cambridge University Press, Cambridge, 2nd edn, 1989.
6. A. K. Jones, in 'Natural Antimicrobial Systems', ed. G. W. Gould, M. E. Rhodes-Roberts, A. K. Chamley, R. M. Cooper, and R. G. Board, Bath University Press, Bath, 1986, p. 232.
7. A. K. Jones and R. C. Cannon, *Br. Phycol J.*, 1986, **21**, 341.
8. G. E. Fogg, in 'Physiology and Biochemistry of Algae', ed. R. A. Lewin, Academic Press, London, 1962, p. 475.
9. G. E. Fogg, *Oceanogr. Mar. Biol. A. Rev.*, 1966, **4**, 195.
10. G. E. Fogg, *Arch. Hydrobiol.*, 1971, **5**, 1.
11. G. E. Fogg, *Bot. Mar.*, 1983, **26**, 3.
12. J. A. Hellebust, in 'Algal Physiology and Biochemistry', ed. W. D. P. Steward, Blackwell Scientific Publications, Oxford, 1974, p. 838.
13. S. Aaronson, T. Berner and Z. Dubinsky, in 'Algal Biomass', ed. G. Shelef and C. J. Soeder, Elsevier, Amsterdam, 1980, p. 575.
14. A. Jensen, in 'Lecture Notes on Coastal and Estuarine Studies 8. Marine Phytoplankton and Productivity', ed. O. Holm-Hansen, L. Bolis, and R. Gilles, Springer-Verlag, Berlin, 1984, p. 61.
15. A. K. Jones, in 'Biochemistry of the Algae and Cyanobacteria', ed. J. R. Gallon and L. J. Rogers, Proc. Phytochem. Soc. Europe, Clarendon Press, Oxford, 1988, p. 257.
16. R. F. Nigrelli, M. F. Stempien Jr, G. D. Ruggieri, V. R. Liguori, and J. T. Cecil, *Fed. Proc. Fed. Amer. Soc. Exp. Biol.*, 1967, **26**, 1197.
17. M. H. Baslow, 'Marine Pharmacology', Williams & Wilkins, Maryland, 1969.
18. R. J. Chrost, *Acta Microbiol.*, 1975, **7**, 125.
19. J. J. Sasner, in 'Marine Pharmacognosy', ed. D. F. Martin and G. M. Padilla, Academic Press, New York, 1973, p. 127.
20. A. K. Jones, in 'Microbial Interactions and Communities', ed. A. T. Bull and J. H. Slater, Academic Press, London, 1982, p. 189.
21. W. Harvey, *Biotechnol.*, 1988, **6**, 488.
22. R. A. Lincoln, K. Strupinski, and J. M. Walker, *Life Chem. Rep.*, 1991, **8**, 97.

23. D. Mangold, in 'Recent Developments in the Field of Pesticides and their Application to Pest Control', ed. K. Holly, L. G. Copping, and G. T. Brooks, UNDO Publications, 1992, p. 27.
24. J. R. Benemann, D. M. Tillett, and J. C. Weissman, *Tibtech.*, 1987, **5**, 47.
25. H. E. Glover, M. D. Keller, and R. R. L. Guillard, *Nature*, 1986, **319**, 142.
26. R. Margalef, M. Estrada, and D. Blasco, in 'Toxic Dinoflagellate Blooms', ed. D. L. Taylor and H. H. Seliger, Elsevier/North-Holland, New York, 1979, p. 89.
27. J. J. McCarthy, in 'The Physiological Ecology of Phytoplankton', ed. I. Morris, Blackwell Scientific Publications, Oxford, 1980, p. 191.
28. G. E. Fogg, *Phil. Trans. R. Soc. London*, 1982, **B296**, 511.
29. P. M. Holligan, *Rapp. P.-v. R un. Cons. Int. Explor. Mer.*, 1987, **187**, 9.
30. H. Kayers, *Mar. Biol.*, 1979, **52**, 357.
31. I. J. Pintner and V. L. Altmeyer, *J. Phycol.*, 1979, **15**, 391.
32. C. G. Trick, P. J. Harrison, and R. J. Andersen, *Can. J. Fish. Aquat. Sci.*, 1981, **38**, 864.
33. N. G. Carr and B. A. Whitton, 'The Biology of the Blue–Green Algae', University of California Press, Berkeley, 1973.
34. H. Hoecht, H. H. Martin, and O. Kandler, *Z. Pflanzenphysiol.*, 1965, **53**, 39.
35. A. E. Walsby, in 'The Biology of the Blue–Green Algae', ed. N. G. Carr and B. A. Whitton, University of California Press, Berkeley, 1973, p. 340.
36. M. S. Swaminathan, *Sci. Am.*, 1984, **250**, 80.
37. L. A. Kapustaka and J. D. DuBois, *Am. J. Bot.*, 1987, **74**, 107.
38. W. W. Carmichael, *'Proc. 9th World Congr. Animal, Plant and Microbial Toxins'*, 1989, p. 3.
39. G. M. Smith, 'The Freshwater Algae of the United States', McGraw-Hill, New York, 1950.
40. S. W. Jeffery, M. Sielicke, and F. T. Haxo, *J. Phycol.*, 1975, **11**, 374.
41. F. T. Haxo, *J. Phycol.*, 1985, **21**, 282.
42. K. Vogel and B. J. D. Meeuse, *J. Phycol.*, 1968, **4**, 317.
43. J. D. Dodge, *Prog. Protozool.*, 1965, **2**, 264.
44. J. R. Grindley and N. Sapeika, *S. Afr. Med. J.*, 1969, **43**, 275.
45. D. J. Hibberd, *Bot. J. Linn. Soc.*, 1976, **72**, 55.
46. M. Shilo, in 'The Water Environment: Algal Toxins and Health', ed. W. W. Carmichael, Plenum Press, New York, 1981, p. 37.
47. P. M. Holligan, M. Viollier, D. S. Harbour, P. Camus, and M. Champagne-Phillipe, *Nature*, 1983, **303**, 339.
48. R. E. Savage, *Fish. Invest., London, Ser. II*, 1930, **12**, 5.
49. M. Shilo, *Env. Sci. Res.*, 1979, **20**, 37.
50. R. R. L. Guillard and J. A. Hellebust, *J. Phycol.*, 1971, **7**, 330.
51. R. Schmaljohann and R. Roettger, *J. Mar. Biol. Assoc. UK*, 1978, **58**, 227.
52. R. A. Lewin, *Plant Physiol.*, 1955, **30**, 129.
53. J. W. G. Lund, *J. Ecol.*, 1949, **37**, 389.
54. J. W. G. Lund, *J. Ecol.*, 1950, **38**, 1, 15.
55. P. M. Holligan and D. S. Harbour, *J. Mar. Biol. Assoc. UK*, 1977, **57**, 1075.
56. P. M. Holligan, R. P. Harris, R. C. Newell, D. S. Harbour, R. N. Head, E. A. S. Linley, M. I. Lucas, P. R. G. Tranter, and C. M. Weekley, *Mar. Ecol. Prog. Ser.*, 1984, **14**, 111.
57. R. A. Lincoln, K. Strupinski, and J. M. Walker, *Diatom Res.*, 1990, **5**, 337.
58. R. A. Lewin, *Rev. Algologique*, 1959, **3**, 181.
59. M. R. Droop, *J. Mar. Biol. Assoc. UK*, 1954, **33**, 511.

60. M. R. Droop, in 'Methods in Microbiology', ed. R. J. Norris and D. W. Ribbons, Academic Press, New York, 1969, vol. 1, p. 269.
61. J. R. Stein, 'Handbook of Phycological Methods: Culture Methods and Growth Measurements', Cambridge University Press, London, 1973.
62. R. R. L. Guillard, in 'Culture of Marine Invertebrate Animals', ed. W. L. Smith and M. H. Chanley, Plenum, New York, 1975, p. 29.
63. R. R. L. Guillard, in 'Handbook of Phycological Methods: Culture Methods and Growth Measurements', ed. J. R. Stein, Cambridge University Press, London, 1973, p. 69.
64. R. W. Hoshaw and J. R. Rosowski, in 'Handbook of Phycological Methods: Culture Methods and Growth Measurements, ed. J. R. Stein, Cambridge University Press, London, 1973, p. 53.
65. M. D. Kraus, *Nature*, 1966, **211**, 310.
66. L. M. Brown, *Phycologia*, 1982, **21**, 408.
67. M. R. Droop, *Br. Phycol. Bull.*, 1967, **3**, 295.
68. A. R. Loeblich III, *J. Phycol.*, 1975, **11**, 80.
69. A. R. Loeblich III and J. L. Sherley, *J. Mar. Biol. Assoc. UK*, 1979, **59**, 195.
70. E. G. Pringsheim, 'Pure Cultures of Algae', Cambridge University Press, London, 1949.
71. R. E. Moore, C. Cheuk, X. G. Yang, and G. M. L. Patterson, *J. Org. Chem.*, 1987, **52**, 1036.
72. S. Carmeli, R. E. Moore, and G. M. L. Patterson, *J. Nat. Prod.*, 1990, **53**, 1533.
73. S. P. Chu, *J. Ecol.*, 1942, **30**, 284.
74. J. McLaughlin and P. R. Gorham, *Can. J. Microbiol.*, 1961, **7**, 869.
75. J. T. Lehman, *Limnol. Oceanogr.*, 1976, **21**, 646.
76. S. F. Bruno and J. J. A. McLaughlin, *J. Protozool.*, 1977, **24**, 548.
77. F. M. M. Morel, J. G. Rueter, D. M. Anderson, and R. R. L. Guillard, *J. Phycol.*, 1979, **15**, 135.
78. P. J. Harrison, R. E. Waters, and F. J. R. Taylor, *J. Phycol.*, 1980, **16**, 28.
79. E. Booth, *Fish. News Int.*, 1964, **3**, 229.
80. A. Jensen, *Proc. Int. Seaweed Symp.*, 1978, **9**, 17.
81. X. Briand, in 'Seaweed Resources in Europe: Uses and Potential', ed. M. D. Guiry and G. Blunden, John Wiley, New York, 1991, p. 259.
82. T. Levering, in 'The Marine Plant Biomass of the Pacific Northwest Coast', ed. R. W. Krauss, Oregon State University Press, 1977, p. 251.
83. A. C. Mathieson, in 'Proceedings of the Sixth US–Japan Meeting on Aquaculture', Santa Barbara, California, ed. C. J. Sindermann, US Dep. Commerc., Rep. NMFS Circ. 442, 1982, p. 25.
84. A. C. Mathieson, in 'Realism in Aquaculture: Achievements, Constraints, Perspectives', ed. M. Bilio, H. Rosenthal, and C. J. Sindermann, European Aquaculture Society, Belgium, 1986, p. 107.
85. J. M. Kain, in 'Seaweed Resources in Europe: Uses and Potential', ed. M. D. Guiry and G. Blunden, John Wiley, New York, 1991, p. 309.
86. W. Schramm, in 'Seaweed Resources in Europe: Uses and Potential', ed. M. D. Guiry and G. Blunden, John Wiley, New York, 1991, p. 379.
87. A. H. Banner, in 'Biology and Geology of Coral Reefs', ed. O. A. Jones and R. Endean, Academic Press, New York, 1975, p. 177.
88. Y. Hashimoto, in 'Marine Toxins and Other Bioactive Marine Metabolites', Japan Sci. Soc. Press, Tokyo, 1979, p. 205.
89. H. N. Chou and Y. Shimizu, *Tetrahedron Lett.*, 1982, **23**, 5521.

90. E. J. Schantz, E. F. McFarren, M. C. Shafer, and K. H. Lewis, *J. Assoc. Off. Agric. Chem.*, 1958, **41**, 160.
91. K. A. Steidinger, M. A. Burklew, and R. M. Ingle, in 'Marine Pharmacognosy', ed. D. F. Martin and G. M. Padilla, Academic Press, New York, 1973, p. 179.
92. M. R. Ross, A. Siger, and B. C. Abbott, in 'Toxic Dinoflagellates', ed. D. M. Anderson, A. W. White, and D. G. Baden, Elsevier, New York, 1985, p. 433.
93. M. J. Turrell and J. L. Middlebrook, *Toxicon*, 1988, **26**, 1089.
94. D. L. Park, I. Aguirre-Flores, W. F. Scott, and E. Alterman, *J. Tox. Env. Health*, 1986, **18**, 589.
95. G. R. Leather and F. A. Einhellig, in 'The Chemistry of Allelopathy', ed. A. C. Thompson, American Chemical Society, Washington, DC, 1985, p. 197.
96. M. L. Thakur, *J. Exp. Bot.*, 1977, **28**, 795.
97. A. L. Anaya and S. D. Amo, *J. Chem. Ecol.*, 1978, **4**, 289.
98. P. C. Pande, P. K. Dublish, and D. K. Jain, *Bangladesh J. Bot.*, 1980, **9**, 67.
99. A. D. Muir and W. Majak, *Can. J. Plant Sci.*, 1983, **63**, 989.
100. G. F. Nicollier, D. F. Pope, and A. C. Thompson, *J. Agric. Food Chem.*, 1983, **31**, 744.
101. K. L. Stevens and G. B. Merrill, *J. Agric. Food Chem.*, 1980, **28**, 644.
102. H. G. Cutler and R. J. Cole, *J. Nat. Prod.*, 1983, **46**, 609.
103. U. Blum and B. R. Dalton, *J. Chem. Ecol.*, 1985, **11**, 279.
104. A. C. Alfenas, M. Hubbes, and L. Couto, *Can. J. Bot.*, 1982, **60**, 2535.
105. V. K. Tyagi and S. K. Chauhan, *Plant Soil*, 1982, **65**, 249.
106. A. E. Star, *Bull. Torrey Bot. Club*, 1980, **107**, 146.
107. A. T. Chan, R. J. Andersen, M. J. L. Blanc, and P. J. Harrison, *Mar. Biol.*, 1980, **59**, 7.
108. T. Aune and K. Berg, *J. Tox. Environ. Health*, 1986, **19**, 325.
109. T. Aune, *Scan. Cell. Tox. Congress*, 1987, **11**, 172.
110. K. Berg, J. Wyman, W. W. Carmichael, and A. Dabholkar, *Toxicon*, 1988, **26**, 827.
111. R. A. Lincoln, K. Strupinski, and J. M. Walker, *J. Appl. Phycol.*, 1990, **2**, 83.
112. W. W. Carmichael and P. E. Bent, *App. Environ. Microbiol.*, 1981, **41**, 1383.
113. J. Gressel, *Adv. Cell Culture*, 1984, **3**, 93.
114. K. Kogure, M. L. Tamplin, U. Simidu, and R. R. Colwell, *Toxicon*, 1988, **26**, 191.
115. K. Shinohara, Y. Okura, T. Koyano, H. Murakami, and H. Omura, *In Vitro Cell. Develop. Biol.*, 1988, **24**, 1057.
116. M. Fadli, J. Aracil, G. Jeanty, B. Banaigs, and C. Francisco, *J. Nat. Prod.*, 1991, **54**, 261.
117. T. J. Starr, E. F. Deig, K. K. Church, and M. B. Allen, *Texas Rept. Biol. Med.*, 1962, **20**, 271.
118. B. N. Pande and A. B. Gupta, *Phycologia*, 1977, **16**, 439.
119. L. H. Debro and H. B. Ward, *Planta Med.*, 1979, **36**, 375.
120. S. Caccamese, R. M. Toscano, G. Furnari, and M. Cormaci, *Bot. Mar.*, 1985, **28**, 505.
121. R. J. P. Cannell, A. M. Owsianka, and J. M. Walker, *Br. Phycol. J.*, 1988, **23**, 41.
122. R. J. P. Cannell, M. Knowles, and J. M. Walker, *Lett. Appl. Microbiol.*, 1988, **7**, 13.
123. S. J. Kellam, R. J. P. Cannell, A. M. Owsianka, and J. M. Walker, *Br. Phycol. J.*, 1988, **23**, 45.
124. S. J. Kellam and J. M. Walker, *Br. Phycol. J.*, 1989, **24**, 191.

125. H. Umezawa, 'Enzyme Inhibitors of Microbial Origin', Univ. Tokyo Press, Tokyo, 1972.
126. R. J. P. Cannell, S. J. Kellam, A. M. Owsianka, and J. M. Walker, *J. Gen. Microbiol.*, 1987, **133**, 1701.
127. R. J. P. Cannell and J. M. Walker, *Biochem. Soc. Trans.*, 1987, **15**, 521.
128. R. J. P. Cannell, S. J. Kellam, A. M. Owsianka, and J. M. Walker, *Planta Med.*, 1988, **54**, 10.
129. H. A. Bates, R. Kostriken, and H. Rapoport, *J. Agric. Food Chem.*, 1978, **26**, 252.
130. L. J. Buckley, M. Ikawa, and J. J. Sasner, *J. Agric. Food Chem.*, 1976, **24**, 107.
131. L. J. Buckley, Y. Oshima, and Y. Shimizu, *Anal. Biochem.*, 1978, **85**, 157.
132. J. J. Sullivan and M. M. Wekell, in 'Seafood Toxins', ed. E. P. Ragelis, American Chemical Society, Washington, DC, 1984, p. 196.
133. W. E. Fallon and Y. Shimizu, *J. Environ. Sci. Health*, 1977, **A12**, 455.
134. S. R. Davio, J. F. Hewetson, and J. E. Beheler, in 'Toxic Dinoflagellates', ed. D. M. Anderson, A. W. White, and D. G. Baden, Elsevier, New York, 1985, p. 343.
135. Y. Hokama, A. H. Banner, and D. B. Boylan, *Toxicon*, 1977, **15**, 317.
136. Y. Hokama, L. H. Kimuara, M. A. Abad, L. Yokochi, P. J. Scheuer, M. Nukina, T. Yasumoto, D. G. Baden, and Y. Shimizu, in 'Seafood Toxins', ed. E. P. Ragelis, American Chemical Society, Washington, DC, 1984, p. 307.
137. L. H. Kimura, Y. Hokama, M. A. Abad, M. Oyama, and J. T. Miyahara, *Toxicon*, 1982, **20**, 907.
138. R. F. Carlson, M. L. Lever, B. W. Lee, and P. E. Guire, in 'Seafood Toxins', ed. E. P. Ragelis, American Chemical Society, Washington, DC, 1984, p. 181.
139. J. M. Ritchie and R. B. Rogart, *Rev. Physiol. Biochem. Pharmacol.*, 1977, **79**, 1.
140. W. A. Catterall, C. S. Morrow, and R. P. Hartshorne, *J. Biol. Chem.*, 1979, **254**, 11379.
141. W. A. Catterall, *Ann. Rev. Pharmacol. Toxicol.*, 1980, **20**, 15.
142. S. R. Davio and R. P. Pickering, *Fed. Proc., Fed. Am. Soc. Exp. Biol.*, 1983, **42**, 1954 (abstr.).
143. D. J. Faulkner and W. H. Fenical, 'Marine Natural Products Chemistry', Plenum Press, New York, 1977.
144. L. S. Shield and K. L. Rinehart, *J. Chrom. Library*, 1978, **15**, 309.
145. H. A. Hoppe, T. Levring, and Y. Tanaka, 'Marine Algae in Pharmaceutical Science', De Gruyter, Berlin, 1979, vol. 1.
146. J. Berdy, A. Aszalos, M. Bostian, and K. L. McNitt, 'Antibiotics from Higher Forms of Life: Lichens, Algae and Animal Organisms', CRC Press, Florida, 1982.
147. H. A. Hoppe and T. Levring, 'Marine Algae in Pharmaceutical Science', De Gruyter, Berlin, 1982, vol. 2.
148. D. J. Faulkner, *Nat. Prod. Rep.*, 1984, **1**, 251.
149. J. R. Stein and C. A. Borden, *Phycologia*, 1984, **23**, 485.
150. G. Blunden and S. M. Gordon, *Pharm. Int.*, 1986, **7**, 162.
151. D. J. Faulkner, *Nat. Prod. Rep.*, 1986, **3**, 1.
152. D. J. Faulkner, *Nat. Prod. Rep.*, 1987. **4**, 539.
153. D. J. Faulkner, *Nat. Prod. Rep.*, 1988, **5**, 613.
154. D. J. Faulkner, *Nat. Prod. Rep.*, 1990, **7**, 269.
155. D. J. Faulkner, *Nat. Prod. Rep.*, 1991, **8**, 97.
156. W. W. Carmichael, 'The Water Environment – Algal Toxins and Health', Plenum Press, New York, 1981.

157. F. E. Russell, 'Advances in Marine Biology', ed. J. H. S. Blaxter, F. S. Russell, and M. Younge, Academic Press, London, 1984, vol. 21, p. 59.
158. E. J. Schantz, J. D. Mold, D. W. Stanger, J. Shavel, F. J. Reil, J. P. Bowden, J. M. Lynch, R. S. Wyler, B. Reigel, and H. Sommer, *J. Am. Chem. Soc.*, 1957, **79**, 5230.
159. Y. Shimizu, C. Hsu, W. E. Fallon, Y. Oshima, I. Miurra, and K. Nakanishi, *J. Am. Chem. Soc.*, 1978, **100**, 6781.
160. S. Hall, 'Toxins and Toxicity of Protogonyaulax from the North East Pacific', PhD Thesis, University of Alaska, Fairbanks, 1982.
161. N. H. Proctor, S. L. Chan, and A. S. Trevor, *Toxicon*, 1975, **13**, 1.
162. F. E. Koehn, S. Hall, C. Fix Winchmann, H. K. Schnoes, and P. B. Reichardt, *Tetrahedron Lett.*, 1982, **23**, 2247.
163. C. Fix Winchmann, G. L. Boyer, C. L. Divan, E. J. Schantz, and H. K., Schnoes, *Tetrahedron Lett.*, 1981, **22**, 1941.
164. Y. Shimizu and M. Yoshioka, *Science*, 1981, **212**, 547.
165. M. H. Evans, *Br. J. Pharmacol.*, 1964, **22**, 478.
166. M. H. Evans, *Br. J. Pharmacol.*, 1970, **40**, 847.
167. C. Y. Kao and A. Nishiyama, *J. Physiol.*, 1965, **180**, 50.
168. B. Hille, *J. Gen. Physiol.*, 1968, **51**, 199.
169. W. A. Catterall, *Adv. Cytopharmacol.*, 1979, **3**, 305.
170. C. Y. Kao and S. E. Walker, *J. Physiol.*, 1982, **323**, 619.
171. Y. Shimizu, *Pure Appl. Chem.*, 1982, **54**, 1973.
172. S. A. Cohen and R. L. Barchi, *Biochem. Biophys. Acta.*, 1981, **645**, 253.
173. J. M. Cummins and A. A. Stevens, 'Public Health Service Bulletin', US Dept. Health, Education & Welfare, Washington, DC, vol. 3, 1970.
174. D. F. Martin and A. B. Charterjee, *US Fish Wildlife Serv. Fish Bull.*, 1970, **68**, 433.
175. N. M. Trieff, V. M. S. Ramanujam, M. Alam, and S. M. Ray, in 'Proceedings International Conference on Toxic Dinoflagellate Blooms, 1st', ed. V. R. LoCicero, Science & Technology Foundation, Wakefield, Massachusetts, 1975, p. 309.
176. G. M. Padilla, Y. S. Kim, M. Westerfield, E. J. Rauckman, and J. W. Moore, in 'Marine Natural Products Chemistry', ed. D. J. Faulkner and W. H. Fenical, Plenum Press, New York, 1977, p. 271.
177. G. M. Padilla, Y. S. Kim, E. J. Rauckman, and G. M. Rosen, in 'Toxic Dinoflagellate Blooms', ed. D. L. Taylor and H. H. Seliger, Elsevier/North-Holland, New York, 1979, p. 351.
178. M. Risk, Y. Y. Lin, R. D. MacFarlane, V. M. S. Ramanujam, L. L. Smith, and N. M. Trieff, in 'Toxic Dinoflagellate Blooms', ed. D. L. Taylor and H. H. Seliger, Elsevier/North-Holland, New York, 1979, p. 335.
179. Y. Shimizu, H. N. Chou, H. Bando, G. Van Duyne, and J. Clardy, *J. Am. Chem. Soc.*, 1986, **108**, 514.
180. M. A. Poli, T. J. Mende, and D. G. Baden, *Molec. Pharmac.*, 1986, **30**, 129.
181. R. H. Pierce, *Toxicon*, 1986, **24**, 955.
182. B. C. Abbott, A. Siger, and M. Spiegelstein, in 'Proceedings International Conference on Toxic Dinoflagellate Blooms', 1st, ed. V. R. LoCicero, Science & Technology Foundation, Wakefield, Massachusetts, 1975, p. 355.
183. J. J. Sasner, M. Ikawa, F. Thurberg, and M. Alam, *Toxicon*, 1982, **10**, 163.
184. J. L. Parmentier, T. W. Narahashi, W. A. Wilson, N. M. Trieff, V. M. S. Ramanujam, M. Risk, and S. M. Ray, *Toxicon*, 1978, **16**, 235.

185. M. Westerfield, J. M. Moore, Y. S. Kim, and G. M. Padilla, *Am. J. Physiol.*, 1977, **232**, 23.
186. J. J. Shoukimas, A. Siger, and B. C. Abbott, in 'Toxic Dinoflagellate Blooms', ed. D. L. Taylor and H. H. Seliger, Elsevier/North-Holland, New York, 1979, p. 425.
187. P. Shinnick-Gallagher, *Br. J. Pharmacol.*, 1980, **69**, 373.
188. J. M. Huang, C. H. Wu, and D. G. Baden, *J. Pharmacol. Exper. Ther.*, 1984, **229**, 615.
189. D. G. Baden, *FASEB J.*, 1989, **3**, 1807.
190. A. Lombert, J. N. Bidard, and M. Lazdunski, *FEBS Lett.*, 1987, **219**, 355.
191. C. H. Wu, J. M. C. Huang, S. M. Vogel, V. S. Luke, W. D. Atchison, and T. Narahashi, *Toxicon*, 1985, **23**, 481.
192. V. D. Atchison, V. S. Luke, T. Narahashi, and S. M. Vogel, *Br. J. Pharmacol.*, 1986, **89**, 731.
193. Y. S. Kim, G. M. Padilla, and D. F. Martin, *Toxicon*, 1978, **16**, 495.
194. P. J. Scheuer, W. Takahashi, J. Tsutsumi, and T. Yoshida, *Science*, 1967, **155**, 1267.
195. M. Nukina, L. M. Kuroyanagi, and J. P. Scheuer, *Toxicon*, 1984, **22**, 169.
196. K. Tachibana, M. Nukina, Y. Joh, and P. J. Scheuer, *Biol. Bull.*, 1987, **172**, 122.
197. A. M. Legrand, M. Litaudon, J. N. Genthon, R. Bagnis, and T. Yasumoto, *J. Appl. Phycol.*, 1989, **1**, 183.
198. M. Murata, A. M. Legrand, Y. Ishibashi, and T. Yasumoto, *J. Am. Chem. Soc.*, 1989, **11**, 8929.
199. D. N. Lawrence, M. B. Enrique, R. M. Lumish, and A. Maceo, *J. Am. Med. Assoc.*, 1980, **244**, 254.
200. T. Yasumoto, I. Nakajima, Y. Oshima, and R. Bagnis, in 'Toxic Dinoflagellate Blooms', ed. D. L. Taylor and H. H. Seliger, Elsevier/North-Holland, New York, 1979, p. 65.
201. J. N. Bidard, H. P. M. Vijverberg, C. Frelin, E. Chungue, A. M. Legrand, R. Bagnis, and M. Lazdunski, *J. Biol. Chem.*, 1984, **259**, 8353.
202. M. D. Rayner and T. I. Kosaki, *Fed. Proc., Fed. Am. Soc. Exp. Biol.*, 1970, **29**, 548.
203. R. J. Lewis, in 'Toxic Dinoflagellates', ed. D. M. Anderson, A. W. White, and D. G. Baden, Elsevier, New York, 1985, p. 379.
204. R. J. Lewis, *Toxicon*, 1988, **27**, 639.
205. E. Benoit, A. M. Legrand, and J. M. Dubois, *Toxicon*, 1986, **24**, 357.
206. J. A. Setliff, M. D. Rayner, and S. K. Hong, *Tox. App. Pharm.*, 1971, **18**, 676.
207. M. D. Raynor and T. I. Kosaki, *Fed. Proc.*, 1970, **29**, 548.
208. M. D. Raynor, *Fed. Proc.*, 1972, **31**, 1139.
209. Y. Ogura, J. Nara, and T. Yoshida, *Toxicon*, 1968, **6**, 131.
210. J. T. Miyahara and S. Shibata, *Fed. Proc.*, 1976, **35**, 842.
211. Y. Ohizumi, S. Shibata, and K. Tachibana, *J. Pharm. Exp. Ther.*, 1981, **217**, 475.
212. A. M. Legrand and R. Bagnis, *Toxicon*, 1984, **22**, 471.
213. J. T. Miyahara, M. M. Oyama, and Y. Hokama, 'Proceedings Fifth International Coral Reef Conference, Tahiti', ed. C. Sabne and B. Salvat, 1985, p. 467.
214. D. M. Miller, R. W. Dickey, and D. R. Tindall, in 'Seafood Toxins', ed. E. P. Ragelis, American Chemical Society, Washington, DC, 1984, p. 241.
215. T. Yasumoto, N. Seino, Y. Murakami, and M. Murara, *Biol. Bull.*, 1987, **172**, 128.
216. R. W. Dickey, D. M. Miller, and D. R. Tindall, in 'Seafood Toxins', ed. E. P. Ragelis, American Chemical Society, Washington, DC, 1984, p. 257.

217. T. Yasumoto, A. Inoue, R. Bagnis, and M. Garcon, *Bull. Jpn. Soc. Sci. Fish.*, 1979, **42**, 359.
218. A. Yokayama, M. Murata, Y. Oshima, T. Iwashita, and T. Yasumoto, *J. Biochem.*, 1988, **104**, 184.
219. A. M. Legrand, M. Galonnier, and R. Bagnis, *Toxicon*, 1982, **20**, 311.
220. M. Takahashi, Y. Ohizumi, and T. Yasumoto, *J. Biol. Chem.*, 1982, **257**, 7287.
221. M. Kobayashi, Y. Ohizumi, and T. Yasumoto, *Br. J. Pharm.*, 1985, **86**, 385.
222. Y. I. Kim, I. S. Login, and T. Yasumoto, *Brain Research*, 1985, **346**, 357.
223. J. T. Miyahara, C. K. Akau, and T. Yasumoto, *Res. Commun. Chem. Path. Pharm.*, 1979, **25**, 177.
224. Y. Ohizumi, A. Kajiwara, and T. Yasumoto, *J. Pharm. Exp. Ther.*, 1983, **227**, 199.
225. Y. Ohizumi and T. Yasumoto, *J. Physiol.*, 1983, **337**, 711.
226. Y. Ohizumi and T. Yasumoto, *Br. J. Pharmacol.*, 1983, **79**, 3.
227. S. Comi, S. Chaen, and H. Sugi, *Proc. Jpn. Acad.*, 1984, **60**, 28.
228. I. S. Login, A. M. Judd, M. J. Cornin, K. Koike, G. Schettini, T. Yasumoto, and R. M. MacLeod, *Endocrinology*, 1985, **116**, 622.
229. A. M. Legrand and R. Bagnis, *J. Mol. Cell. Cardiol.*, 1984, **16**, 663.
230. M. Takahashi, M. Tarsumi, Y. Ohizumi, and T. Yasumoto, *J. Biol. Chem.*, 1983, **258**, 10944.
231. S. B. Freedman, R. J. Miller, D. M. Miller, and D. R. Tindall, *Proc. Nat. Acad. Sci. USA*, 1984, **81**, 4582.
232. F. Gusovsky and J. W. Daly, *Biochem. Pharmacol.*, 1990, **39**, 1633.
233. F. Gusovsky, O. Choi, J. W. Daly, and T. Yasumoto, *Adv. Second Messenger Phosphoprotein Res.*, 1990, **24**, 723.
234. F. Gusovsky, J. A. Bitran, T. Yasumoto, and J. W. Daly, *J. Pharmacol. Exp. Ther.*, 1990, **252**, 466.
235. T. Yasumoto and M. Murata, in 'Marine Toxins: Origin, Structure and Molecular Pharmacology', ed. S. Hall and G. Strichartz, ACS Symposium Series, 1990, vol. 418, p. 120.
236. Y. Murakami, Y. Oshima, and Y. Yasumoto, *Bull. Jpn. Soc. Fish.*, 1982, **48**, 69.
237. S. Shibata, Y. Ishida, H. Kitano, Y. Ohizumi, J. Harbon, Y. Tsukifani, and H. Kikuchi, *J. Pharmac. Exp. Ther.*, 1982, **223**, 135.
238. K. Terao, E. Ito, T. Yanagi, and T. Yasumoto, *Toxicon*, 1986, **24**, 1141.
239. M. Suganuma, H. Fujiki, H. Suguri, S. Yoshizana, M. Hirota, M. Nakayatsu, M. Ojika, K. Wakamatsu, K. Yamada, and T. Sugimura, *Proc. Nat. Acad. Sci. USA*, 1988, **85**, 1768.
240. W. W. Carmichael and P. R. Gorham, *J. Phycol.*, 1978, **13**, 97.
241. C. S. Huber, *Acta Crystallogr.*, 1972, **B78**, 2577.
242. J. P. Devlin, O. E. Edwards, P. R. Gorham, N. R. Hunter, P. K. Pike, and B. Stavric, *Can. J. Chem.*, 1977, **55**, 1367.
243. H. F. Campbell, O. E. Edwards, and R. J. Kolt, *Can. J. Chem.*, 1977, **55**, 1372.
244. H. F. Campbell, O. E. Edwards, J. W. Elder, and R. J. Kolt, *Pol. J. Chem.*, 1979, **53**, 27.
245. H. A. Bates and H. Rapoport, *J. Am. Chem. Soc.*, 1979, **101**, 1259.
246. M. P. Koskinen and H. Rapoport, *J. Med. Chem.*, 1985, **28**, 1301.
247. J. J. Tufariello, H. Meckler, and K. P. A. Senaratne, *J. Am. Chem. Soc.*, 1984, **106**, 7979.
248. J. J. Tufariello, H. Meckler, and K. P. A. Senaratne, *Tetrahedron*, 1985, **41**, 3447.

249. R. L. Danheiser, J. M. Morin, and E. J. Salaski, *J. Am. Chem. Soc.*, 1985, **107**, 8066.
250. W. W. Carmichael, D. F. Biggs, and P. R. Gorham, *Science*, 1975, **187**, 542.
251. W. W. Carmichael, D. F. Biggs, and M. A. Peterson, *Toxicon*, 1979, **17**, 299.
252. R. S. Aronstam, *The Pharmacologist*, 1980, **22**, 300 (abstr.)
253. C. E. Spivak, B. Witkob, and E. X. Albuquerque, *Mol. Pharmacol.*, 1980, **18**, 384.
254. R. S. Aranstam and B. Witkop, *Proc. Natl. Acad. Sci. USA*, 1981, **78**, 4639.
255. C. E. Spivak, J. Waters, B. Witkop, and E. X. Albuquerque, *Mol. Pharmacol*, 1983, **23**, 337.
256. K. L. Swanson, C. N. Allen, R. S. Aronstam, H. Rapoport, and E. X. Albuquerque, *Mol. Pharmacol.*, 1985, **29**, 250.
257. X. Zhang, P. Stjernlof, A. Adem, and A. Nordberg, *Eur. J. Pharm.*, 1987, **135**, 457.
258. N. A. Mahmood and W. W. Carmichael, *Toxicon*, 1987, **25**, 1221.
259. N. A. Mahmood, W. W. Carmichael, and M. S. Pfaheer, *Am. J. Vet. Res.*, 1988, **49**, 500.
260. T. Krishnamurthy, W. W. Carmichael, and E. W. Sarver, *Toxicon*, 1986, **24**, 865.
261. M. Schwimmer and D. Schwimmer, in 'Algae, Man and the Environment', ed. D. F. Jackson, Syracuse University Press, New York, 1968, p. 279.
262. W. W. Carmichael, in 'Advances in Botanical Research', ed. E. A. Callow, Academic Press, London, 1986, vol. 12, p. 47.
263. P. R. Gorham and W. W. Carmichael, in 'Algae and Human Affairs', ed. C. A. Lambi and J. R. Waaland, Cambridge University Press, Cambridge, 1988, p. 403.
264. W. W. Carmichael, *Environ. Sci. Res.*, 1981, **20**, 491.
265. W. W. Carmichael, *S. Afr. J. Sci.*, 1982, **78**, 367.
266. O. M. Skulberg, G. A. Cood, and W. W. Carmichael, *Ambio*, 1984, **13**, 244.
267. D. P. Botes, H. Kruger, and C. C. Viljoen, *Toxicon*, 1982, **20**, 945.
268. G. K. Poon, I. M. Priestly, S. M. Hunt, J. K. Fawell, and G. A. Codd, *J. Chromatogr.*, 1987, **387**, 551.
269. D. P. Botes, A. A. Tuinman, P. L. Wessels, C. C. Viljoen, H. Kruger, D. H. Williams, S. Santikarn, R. J. Smith, and S. J. Hammond. *J. Chem. Soc., Perkin Trans. 1*, 1984, 2311.
270. D. P. Botes, H. Wessels, H. Kruger, M. T. C. Runnegar, S. Santikarn, R. J. Smith, J. C. J. Barna, and D. H. Williams, *J. Chem. Soc., Perkin Trans. 1*, 1985, 2747.
271. C. T. Bishop, E. F. L. J. Anet, and P. R. Gorham, *Can. J. Biochem. Physiol.*, 1959, **37**, 453.
272. H. Konst, P. D. McKercher, P. R. Gorham, A. Robertson, and J. Howell, *Can. J. Comp. Med. Vet. Sci.*, 1965, **29**, 221.
273. D. P. Botes, in 'Mycotoxins and Phycotoxins. Bioactive Molecules', ed. P. S. Steyn and R. Vleggaar, Elsevier, Amsterdam, 1986, vol. 1, p. 161.
274. T. Kusumi, T. Ooi, M. M. Watanabe, H. Takahagh, and H. Kakisawa, *Tetrahedron Lett.*, 1987, **28**, 4695.
275. P. Painuly, R. Perez, T. Fukai, and Y. Shimizu, *Tetrahedron Lett.*, 1988, **29**, 11.
276. W. W. Carmichael, V. Beasley, D. L. Bunner, N. J. Eloff, I. Falconer, P. R. Gorham, K. Harada, T. Krishnamurthy, Y. Min-Juan, R. E. Moore, K. Rinehart, M. Runnegar, O. M. Skulberg, and M. Watanabe, *Toxicon*, 1988, **11**, 971.
277. G. A. Codd and W. W. Carmichael, *FEMS Microbiol. Lett.*, 1982, **13**, 409.

278. D. S. Richard, K. A. Beattie and G. A. Codd, *Environ. Technol. Lett.*, 1983, **4**, 377.
279. K. Berg and N. E. Soli, *Acta Vet. Scand.*, 1985, **26**, 363.
280. K. Berg and N. E. Soli, *Acta Vet. Scand.*, 1985, **26**, 374.
281. J. E. Eirksson, J. A. O. Meriluoto, H. P. Kujara, and O. M. Skulberg, *Comp. Biochem. Physiol., C: Comp. Pharmacol. Toxicol.*, 1987, **89**, 207.
282. P. R. Hawkins, M. T. C. Runnegar, A. R. B. Jackson, and I. R. Falconer, *Appl. Environ. Microbiol.*, 1985, **50**, 1292.
283. A. R. B. Jackson, A. McInners, I. R. Falconer, and M. T. C. Runnegar, *Vet. Pathol.*, 1984, **21**, 102.
284. M. T. C. Runnegar and I. R. Falconer, in 'The Water Environment – Algal Toxins and Health', ed. W. W. Carmichael, Plenum Press, New York, 1981, p. 225.
285. O. Ostensvik, O. M. Skulberg, and N. E. Soli, *Env. Sci. Res.*, 1981, **20**, 315.
286. I. R. Falconer, A. M. Beresford, and M. T. C. Runnegar, *Med. J. Aust.*, 1983, **1**, 511.
287. A. R. B. Jackson, A. McInnes, I. R. Falconer, and M. T. C. Runnegar, *Toxicon*, 1983, **3**, 191 (suppl.).
288. W. P. Brooks and G. A. Codd, *Pharmacol. Toxicol.*, 1987, **60**, 187.
289. M. T. C. Runnegar and I. R. Falconer, *S. Afr. J. Sci.*, 1982, **78**, 363.
290. M. T. C. Runnegar and I. R. Falconer, *Toxicon*, 1986, **24**, 109.
291. M. T. C. Runnegar, J. Andrews, R. Gerdes, and I. R. Falconer, *Toxicon*, 1987, **25**, 1235.
292. E. Berg, J. Wyman, W. W. Carmichael, and A. S. Dabholkar, *Toxicon*, 1988, **26**, 827.
293. A. S. Dabholkar and W. W. Carmichael, *Toxicon*, 1987, **25**, 285.
294. I. R. Falconer, T. Buckley, and M. T. C. Runnegar, *Aust. J. Biol. Sci.*, 1986, **39**, 17.
295. M. T. C. Runnegar, I. R. Falconer, T. Buckley, and A. R. B. Jackson, *Toxicon*, 1986, **24**, 506.
296. M. T. C. Runnegar, I. R. Falconer, and J. Silver, *Arch. Pharmacol.*, 1981, **317**, 268.
297. I. R. Falconer, A. R. B. Jackson, J. Langley, and M. T. C. Runnegar, *Aust. J. Biol. Sci.*, 1981, **34**, 179.
298. W. C. Theiss and W. W. Carmichael, in 'Mycotoxins and Phycotoxins. Bioactive Molecules', ed. P. S. Steyn and R. Vleggar, Elsevier, Amsterdam, 1986, vol. 1, p. 353.
299. D. N. Slatkin, R. D. Stoner, W. H. Adams, J. H. Kylia, and H. W. Siegelman, *Science*, 1983, **220**, 1383.
300. K. W. Glombitza, in 'Marine Algae in Pharmaceutical Science', ed. H. A. Hoppe, T. Levring, and Y. Tanaka, Walter de Gruyter, Berlin, 1979, p. 303.
301. J. A. Findlay and A. D. Patil, *Phytochemistry*, 1986, **25**, 548.
302. M. D. Higgs, *Tetrahedron*, 1981, **37**, 4255.
303. L. Codomier, Y. Bruneau, G. Combaut, and J. Teste, *C.R. Acad. Sci., Ser. D.*, 1977, 1163.
304. O. McConnell and W. Fenical, *Phytochemistry*, 1977, **16**, 367.
305. J. S. Mynderse and R. E. Moore, *J. Org. Chem.*, 1978, **43**, 2301.
306. R. E. Moore, in 'The Water Environment – Algal Toxins and Health', ed. W. W. Carmichael, Plenum Press, New York, 1981, p. 15.
307. W. W. Carmichael, in 'Handbook of Natural Toxins', ed. A. T. Tu, Marcel Dekker, New York, 1988, vol. 3, p. 121.

308. J. H. Cardellina, R. E. Moore, E. V. Arnold, and J. Clardy, *J. Org. Chem.*, 1979, **44**, 4039.
309. J. H. Cardellina, H. D. Dalietos, F. J. Marner, J. S. Mynderse, and R. E. Moore, *Phytochemistry*, 1978, **17**, 2091.
310. J. S. Mynderse and R. E. Moore, *J. Org. Chem.*, 1978, **43**, 4359.
311. H. Fujiki, M. Mori, M. Nakayasu, M. Terada, T. Sugimura, and R. E. Moore, *Proc. Natl. Acad. Sci. USA*, 1981, **78**, 3872.
312. G. Impellizzeri, S. Mangiafico, G. Oriente, M. Piattelli, S. Sciuto, E. Fattorusso, S. Magno, C. Santacroce, and D. Sica, *Phytochemistry*, 1975, **14**, 1549.
313. T. Takemoto and K. Daigo, *Arch. Pharmacol.*, 1960, **293**, 627.
314. T. Takemoto and K. Daigo, *Archiv. Pharmazie*, 1958, **203**, 627.
315. C. J. Bird, R. K. Boyd, D. Brewer, C. A. Craft, A. S. W. DeFreitas, E. W. Dyer, D. J. Embree, M. Falk, M. G. Flack, R. A. Foxall, M. Gillis, M. Greenwell, M. Hardstaff, W. R. Jamieson, W. D. Laycock, G. K. Leblanc, P. Lewis, A. W. McCulloch, G. K. McCullery, M. McInerney-Northcott, A. G. McInnes, J. L. McLachlan, P. Odense, D. O'Neil, V. P. Pathak, M. Q. Quilliam, M. A. Ragan, P. F. Seto, P. G. Sim, S. D. Tappen, P. Thibault, J. A. Walter, J. C. Wright, A. M. Backman, A. M. Taylor, D. Dewar, M. Gilgan, and D. J. A. Richard, 'Atlantic Research Laboratory Tech. Rep.', NRCC 29083, 1988, vol. 56, p. 86.
316. Y. Shimizu, S. Gupta, K. Masuda, L. Maranda, K. C. Walker, and R. Wang, *Pure Appl. Chem.*, 1989, **61**, 513.
317. S. S. Bates, C. J. Bird, A. S. W. DeFreitas, R. Foxall, M. Gilgan, L. A. Haic, G. R. Johnson, A. W. McCulloch, P. Odense, R. Pocklington, M. A. Quilliam, P. G. Sim, J. C. Smith, D. V. Subba Rao, E. C. D. Todd, J. A. Walter, and L. J. C. Wright, *Can. J. Fish. Aquat. Sci.*, 1989, **46**, 1203.
318. D. V. Subba Rao, M. A. Quilliam, and R. Pocklington, *Can. J. Fish. Aquat. Sci.*, 1988, **45**, 2076.
319. F. Castle, R. H. Evans, and J. N. P. Kirkpatrick, *Comp. Biochem. Physiol.*, 1984, **77**, 399.
320. H. Shinozaki, M. Ishida, and T. Okamoto, *Brain Res.*, 1986, **399**, 395.
321. T. J. Biscoe, R. H. Evans, M. R. Headley, M. R. Martins, and J. C. Watkins, *Nature*, 1975, **255**, 166.
322. T. J. Biscoe, R. H. Evans, M. R. Headley, M. R. Martins, and J. C. Watkins, *Br. J. Pharmacol.*, 1976, **58**, 373.
323. G. Debonnel, L. Beauchesne, and C. DeMontigny, *Can. J. Physiol. Pharmacol.*, 1988, **67**, 29.
324. D. R. Hampson and R. J. Wenthold, *J. Biol. Chem.*, 1988, **263**, 2500.
325. J. Yariv and S. Helstrin, *J. Gen. Microbiol.*, 1961, **24**, 165.
326. S. Ultzur and M. Shilo, *J. Gen. Microbiol.*, 1964, **36**, 161.
327. S. Ultzur, PhD Thesis, Hebrew University, Jerusalem, 1965.
328. K. A. Steidinger, *CRC Rev. Microbiol*, 1973, **3**, 49.
329. M. Imai and K. Inoue, *Biochim. Biophys. Acta*, 1974, **352**, 344.
330. Z. Paster, in 'Marine Pharmocognosy', ed. D. F. Martin and G. M. Padilla, Academic Press, New York, 1973, p. 241.
331. R. Y. Stanier and G. Cohen-Bazire, *Rev. Microbiol.*, 1977, **31**, 225.
332. G. Keleti, J. L. Sykora, L. A. Mailoie, D. L. Doerfler, and I. M. Cambell, in 'The Water Environment – Algal Toxins and Health', ed. W. W. Carmichael, Plenum Press, New York, 1981, p. 447.
333. E. C. Lippy and J. Erb, *Pa. J. Am. Water Works Assoc.*, 1976, **68**, 606.

334. W. H. Billings, in 'The Water Environment – Algal Toxins and Health', ed. W. W. Carmichael, Plenum Press, New York, 1981, p. 243.
335. G. Keleti and J. L. Sykora, *Appl. Environ. Microbiol.*, 1982, **43**, 104.
336. S. Raziuddin, H. W. Siegelman, and T. G. Tornabene, *Eur. J. Biochem.*, 1983, **137**, 333.
337. O. Lûderitz, C. Galanos, M. N. Lehmann, E. T. Rietschel, G. Rosenfelder, M. Simon, O. Westphal, and M. Nurminen, *J. Infect. Dis.*, 1973, **128**, 17 (Suppl.).
338. B. C. Maiti and R. H. Thomson, *J. Chem. Res.*, 1978, **4**, 1669.
339. M. Mahendran, S. Samasundarun, and R. M. Thomson, *Phytochemistry*, 1979, **18**, 1085.
340. P. G. Nielsen, J. S. Carle, and C. Christophersenc, *Phytochemistry*, 1982, **21**, 1643.
341. N. Fusetani, C. Ozawa, and Y. Hashimoto, *Bull. Jpn. Soc. Scient. Fish.*, 1976, **42**, 941.
342. N. Fusetani and Y. Hashimoto, in 'Animal, Plant and Microbial Toxins', ed. A. Ohsaka, K. Hayashi, and Y. Sawai, Plenum Press, New York, 1976, p. 325.
343. K. L. Reinhart, P. D. Shaw, L. S. Shield, J. B. Gloer, G. C. Harbor, M. E. S. Koker, D. Samain, R. E. Schwartz, A. A. Tymiak, D. L. Weller, G. T. Carter, M. H. G. Munro, R. G. Hughes, H. E. Renis, E. B. Swynenberg, D. A. Stringfellow, J. J. Vavra, J. H. Coates, G. E. Zurenko, S. L. Kuentzel, G. J. Briakus, R. C. Brusca, L. L. Craft, D. N. Young, and J. L. Connor, *Pure Appl. Chem.*, 1981, 795.
344. M. Aubert, J. Aubert, and M. Gauthier, in 'Marine Algae in Pharmaceutical Science', ed. H. A. Hoppe, T. Levering, and Y. Tanaka, DeGruyter, Berlin, 1979, p. 267.
345. P. R. Berkholder, L. M. Berkholder, and L. R. Almodovar, *Bot. Mar.*, 1960, **2**, 149.
346. D. C. B. Duff, D. L. Bruce, and N. J. Antia, *Can. J. Microbial.*, 1966, **12**, 877.
347. D. Pesando and M. Gnassia-Barelli, in 'Marine Algae in Pharmaceutical Science', ed. H. A. Hoppe, T. Levring, and Y. Tanaka, De Gruyter, Berlin, 1979, p. 447.
348. J. L. Reichelt and M. A. Borowitzka, *Hydrobiologia*, 1984, **116/117**, 158.
349. A. M. Welch, *J. Bacteriol.*, 1961, **83**, 97.
350. M. E. Espeche, E. R. Fraice, and A. M. S. Mayer, *Hydrobiologia*, 1984, **116/117**, 525.
351. I. S. Hornsey and D. Hide, *Br. J. Phycol.*, 1974, **9**, 353.
352. A. F. Khaleafa, M. A. M. Kharboush, A. Metwalli, A. F. Mohsen, and A. Serwi, *Bot. Bar.*, 1985, **18**, 163.
353. P. Henriquez, A. Candida, R. Norambuena, M. Silva, and R. Zemelman, *Bot. Mar.*, 1979, **12**, 451.
354. J. Moreau, D. Pesandro, and B. Caram, *Hydrobiologia*, 1984, **116/117**, 521.
355. R. Pratt, T. C. Daniels, J. J. Eiler, J. B. Gunnison, W. D. Kummler, J. R. Oneto, H. A. Spoehr, G. J. Hardin, H. W. Milner, J. H. C. Smith, and H. H. Strain, *Science*, 1944, **49**, 351.
356. H. A. Spoehr, J. H. C. Smith, H. H. Strain, H. W. Milner, and G. J. Hardin, *Carnegie Inst. Wash. Publ.*, 1949, **586**, 1.
357. J. E. Scutt, *Am. J. Bot.*, 1964, **51**, 581.
358. K. R. Gustafsson, J. H. Cardellina, R. W. Fuller, O. S. Weislow, R. F. Kiser, K. M. Snader, G. M. L. Patterson, and M. R. Boyd, *J. Nat. Cancer Inst.*, 1989, **81**, 1254.
359. R. Pratt, R. H. Mautner, G. M. Gardner, Y. Sha, and J. Dufrenoy, *J. Am. Pharm. Assoc. Sci. Ed.*, 1951, **40**, 575.

360. K. Kamimoto, *Jap. J. Bact.*, 1955, **10**, 897.
361. C. G. C. Chesters and J. A. Stott, in 'The Production of Antibiotic Substances by Seaweeds, Proc. 2nd Int. Seaweed Symposium', ed. T. Braarudad and N. A. Sørenson, Pergamon Press, New York, 1956, p. 49.
362. H. Roos, *Kiel Meeresforsch.*, 1957, **13**, 41.
363. M. B. Allen and E. Y. Dawson, *J. Bact.*, 1960, **79**, 459.
364. J. McN. Sieburth, *Dev. Ind. Microbiol.*, 1964, **5**, 124.
365. J. McN. Sieburth, in 'Advances in Microbiology of the Sea', ed. M. R. Droop and J. F. Ferguson-Wood, Academic Press, London, 1968, p. 63.
366. W. Fenical and V. J. Paul, *Hydrobiologia*, 1984, **116/117**, 135.
367. V. J. Paul and W. Fenical, *Science*, 1983, **221**, 747.
368. V. J. Paul and W. Fenical, *Tetrahedron*, 1984, **40**, 3053.
369. A. J. Blackman and R. J. Wells, *Tetrahedron Lett.*, 1978, **19**, 3063.
370. V. J. Paul and W. Fenical, *Tetrahedron*, 1984, **40**, 2913.
371. H. Nakashima, Y. Kido, N. Kobayashi, Y. Motoki, M. Neushul, and N. Yamamoto, *Antimicrob. Agents Chemother.*, 1987, **31**, 1524.
372. H. Nakashima, Y. Kido, N. Kobayashi, Y. Motoki, M. Neushul, and N. Yamamoto, *J. Cancer Res. Clin. Oncol.*, 1987, **113**, 413.
373. H. J. H. Cardellina, F. J. Marner, and R. E. Moore, *Science*, 1979, **204**, 193.
374. N. Sakabe, H. Harada, Y. Hirata, Y. Tomiic, and I. Nitta, *Tetrahedron Lett.*, 1966, **23**, 2523.
375. H. Fujiki, M. Suganuma, H. Suguri, S. Yoshizawa, K. Takagi, M. Nakayasu, M. Ojika, K. Yamada, T. Yasumoto, R. E. Moore, and T. Sugimura, in 'Marine Toxins: Origin, Structure and Molecular Biology', ed. S Hall and G. Strichartz, American Chemical Society, Washington, 1990, p. 232.
376. I. R. Falconer and T. H. Buckley, *Med. J. Austr.*, 1989, **150**, 351.
377. M. Kashiwaga, J. S. Mynderse, R. E. Moore, and T. R. Norton, *J. Pharm. Sci.*, 1980, **69**, 735.
378. J. S. Mynderse, R. E. Moore, M. Kashiwaga, and T. R. Norton, *Science*, 1977, **196**, 538.
379. G. M. L. Patterson, T. R. Norton, E. Furusawa, S. Furusawa, M. Kashiwaga, and R. E. Moore, *Bot. Mar.*, 1984, **37**, 485.
380. I. Yamamoto, M. Takahashi, E. Tamura, and H. Maruyama, *Bot. Mar.*, 1982, **25**, 455.
381. H. Noda, H. Amano, K. Arashima, S. Hashimoto, and K. Nisizawa, *Nippon Suisan Gakkaishi*, 1989, **55**, 1259.
382. H. Noda, H. Amano, K. Arashima, S. Hashimoto, and K. Nisizawa, *Nippon Suisan Gakkaishi*, 1989, **55**, 1265.
383. J. L. Schwartz and G. Shklar, *Phytotherapy Res.*, 1989, **3**, 243.
384. I. Yamamoto, T. Nagumo, K. Yagi, H. Tominaga, and M. Aoki, *Jpn. J. Exp. Med.*, 1974, **44**, 543.
385. E. Furusawa and S. Furusawa, *Oncology*, 1985, **42**, 364.
386. J. Schwartz, D. Suda, and G. Light, *Biochem. Biophys. Res. Commun.*, 1986, **136**, 1130.
387. G. Shklar and J. Schwartz, *Eur. J. Cancer Clin. Oncol.*, 1988, **24**, 839.
388. J. L. Urban, H. M., Shepard, J. L. Rothstein, B. J. Sugarman, and H. Schreiber, *Proc. Natl. Acad. Sci. USA*, 1986, **83**, 5233.
389. A. Bendich and S. S. Shapiro, *J. Nutr.*, 1986, **116**, 2254.
390. Y. Tomita, K. Himeno, K. Nomoto, H. Endo, and T. Hirohata, *J. Natl. Cancer Inst.*, 1987, **78**, 679.

391. R. S. Jacobs, S. White, and L. Wilson, *Fed. Proc.*, 1981, **40**, 26.
392. M. Murakami, K. Makabe, S. Okada, K. Yamaguchi, and S. Konoso, *Nippon Suisan Gakkaishi*, 1988, **54**, 1035.
393. W. H. Gerwick, W. Fenical, D. Van Engen, and J. Clardy, *J. Am. Chem. Soc.*, 1980, **102**, 7991.
394. R. L. Margolsi and L. Wilson, *Proc. Natl. Acad. Sci. USA*, 1977, **74**, 3466.
395. R. S. Jacobs, P. Culver, R. Langdon, T. O'Brian, and S. White, *Tetrahedron*, 1985, **41**, 981.
396. M. Lefevere, in 'Algae and Man', ed. D. F. Jackson, Plenum Press, New York, 1964, p. 337.
397. K. Shinohara, Y. H. Zaho, Y. Okazaki, Y. Okura, H. Murakami, H. Omuri, and G. H. Sato, in 'Growth and Differentiation of Cells in Defined Environment', ed. H. Murakami, I. Yamama, D. W. Barns, I. Mather, I. Hayashi, G. H. Sato, and K. Kodansha, Springer Verlag, Berlin, 1985, p. 363.
398. K. Shinohara, Y. Okura, T. Koyana, H. Murakami, E. H. Kim, and H. Omura, *Agric. Biol. Chem.*, 1986, **50**, 2225.
399. R. Kazlauskas, J. Mulder, P. T. Murphy, and R. J. Wells, *Aust. J. Chem.*, 1980, **33**, 2097.
400. J. T. Baker, *Hydrobiologia*, 1984, **116/117**, 29.
401. T. Shiba and T. Wakamiya, Jpn. Kokai Tokkyo Koho, JP 60246360, 1985.
402. W. Hawkins and V. Leonard, *Can. J. Biochem. Physiol.*, 1963, **41**, 1325.
403. W. Anderson and J. G. C. Duncan, *J. Pharm. Pharmacol.*, 1965, **17**, 647.
404. V. Grauffel, B. Kloareg, S. Mabeau, P. Durand, and J. Jozefonvicz, *Biomaterials*, 1989, **10**, 363.
405. H. Mori, H. Kamei, E. Nishide, and K. Nisizawa, in 'Marine Algae in Pharmaceutical Science', ed. H. A. Hoppe and T. Levring, De Gruyter, Berlin, 1982, vol. 2, p. 109.
406. E. T. Dewar, in 'Second International Seaweed Symposium', ed. T. Braarud and N. A. Sorrensen, Pergamon Press, New York, 1956, p. 55.
407. A. E. Myers, *Can. Pharm. J.*, 1965, **98**, 28.
408. H. Kozaki, Y. Oshima, and Y. Yasumoto, *Agric. Biol. Chem.*, 1982, **46**, 233.
409. E. L. Bass, J. P. Pinion, and M. E. Sharif, *Aquat. Toxicol.*, 1983, **3**, 15.
410. W. O. K. Grabow, W. C. Durandt, W. O. Prozesky, and W. E. Scott, *Appl. Environ. Micro.*, 1982, **43**, 1425.
411. H. Sprince, C. M. Parker, and J. A. Josephs, *Agents Actions*, 1969, **1**, 9.
412. H. Sprince, C. M. Parker, J. A. Josephs, and J. Magazino, *Ann. NY Acad. Sci.*, 1969, 323.
413. W. J. Freed, J. C. Gillin and R. J. Wyatt, *Epilepsia*, 1979, **20**, 209.
414. O. Roch, T. A. Crabb, G. Blunden, and S. M. Gordon, *Br. Phycol. J., 1985,* **20**, 190.
415. S. Abe and T. Kaneda, *Bull. Jpn. Soc. Sci. Fish.*, 1973, **39**, 383.
416. K. T. Hori, K. Yamamoto, K. Miyazawa, and K. Ito, *J. Frac. Appl. Biol. Sci. Hiroshima Univ.*, 1979, **18**, 65.
417. G. Blunden, M. M. El Barouni, S. M. Gordon, W. F. H. McLean, and D. J. Rogers, *Bot. Mar.*, 1981, **24**, 451.
418. G. Blunden, S. M. Gordon, W. F. H. McLean, and M. D. Guiry, *Bot. Mar.*, 1982, **25**, 563.
419. G. Blunden, S. M. Gordon, B. E. Smith, and R. L. Fletcher, *Br. Phycol. J.*, 1985, **20**, 105.
420. A. Pollard and R. G. Wyn Jones, *Planta*, 1979, **144**, 291.

421. M. K. Gaitonde, in 'Handbook of Neurochemistry', ed. A. Lajtha, Plenum Press, New York, 1970, vol. 3, p. 225.
422. L. A. Smolin, N. J. Benevenga, and S. Berlow, *J. Pediatr.*, 1981, **99**, 467.
423. D. E. L. Wilken, B. Wilken, N. P. B. Dudman, and P. A. Tyrrell, *New Engl. J. Med.*, 1983, **309**, 448.
424. U. Wendel and H. J. Bremer, *Eur. J. Pediatr.*, 1984, **142**, 147.
425. Y. Sakagami, T. Watanabe, A. Hisamitsu, T. Kamibayashi, K. Honma, and H. Manabe, in 'Marine Algae in Pharmaceutical Science', ed. H. A. Hoppe and T. Levring, De Gruyter, Berlin, 1982, vol. 2, p. 99.
426. J. Houck, J. Bhayana, and T. Lee, *Gasteroenterology*, 1960, **39**, 196.
427. W. Anderson and J. Watt, *J. Pharm. Pharmacol.*, 1959, **11**, 318.
428. W. Anderson and P. D. Soman, *Nature*, 1965, **206**, 101.
429. W. Anderson and J. Watt, *J. Physiol.*, 1959, **147**, 52.
430. W. Anderson, R. Marcus, and J. Watt, *J. Pharm. Pharmacol.*, 1962, **14**, 119T.
431. R. E. Moore, V. Bornemann, W. P. Nieczura, J. M. Gregson, J. L. Chen, T. R. Norton, G. M. L. Patterson, and G. L. Helms, *J. Am. Chem. Soc.*, 1989, **111**, 6128.
432. M. Kobayashi, S. Kondo, T. Yasumoto, and Y. Ohizumi, *J. Pharm. Exp. Ther.*, 1986, **238**, 1077.
433. I. E. Akopov, V. A. Konovabva, and M. M. Mansoura, *Farmakol. I. Toksikol.*, 1958, **21**, 44.
434. E. A. Hosein and H. McLennan, *Nature*, 1959, **183**, 328.
435. E. Reiner, J. Topliff, and J. D. Wood, *Can. J. Biochem. Physiol.*, 1962, **40**, 1401.
436. Y. Tsuchiya, *Proc. Int. Seaweed Symp.*, 1969, **6**, 747.
437. S. Abe and T. Kaneda, 'Proceedings 7th International Seaweed Symposium', ed. K. Nisizawa, University of Tokyo Press, Tokyo, 1972, p. 562.
438. T. Kaneda and S. Abe, in 'Developments in Hydrobiology, Proceedings 11th International Seaweed Symposium', ed. C. J. Bird and M. A. Ragan, Dr. W. Junk Publishers, Boston, 1968, p. 149.
439. S. Abe and T. Kaneda, *Bull. Jpn. Soc. Scient. Fish.*, 1975, **41**, 567.
440. Y. S. Huang and M. S. Manku, *J. Am. Oil. Chem. Soc.*, 1983, **60**, 748.
441. T. Takemoto, K.Daigo, and N. Takagi, *Yakugaku Zasshi*, 1964, **84**, 1176.
442. T. Takemoto, K. Daigo, and N. Takagi, *Yakugaku Zasshi*, 1964, **84**, 1180.
443. T. Takemoto, K. Daigo, and N. Takagi, *Yakugaku Zasshi*, 1965, **85**, 37.
444. T. Takemoto, N. Takagi, and K. Daigo, *Yakugaku Zasshi*, 1965, **85**, 843.
445. P. Girard, C. Marion, M. Liutkus, M. Boucard, E. Rechencq, J. P. Vidal, and J. C. Rossi, *Pl. Med.*, 1988, 193.
446. Z. Cohen, in 'CRC Handbook of Microalgal Mass Culture', ed. A. Richmond, CRC Press Inc., Florida, 1986, p. 421.
447. R. P. Gregson, J. F. Marwood, and R. J. Quinn, *Tetrahedron Lett.*, 1979, **20**, 4505.
448. J. Kobayashi, M. Ishibashi, N. Nakamura, Y. Hirata, T, Yamasu, and Y. Ohizumi, *Experimentia*, 1988, **44**, 800.
449. J. Kobayashi, M. Ishibashi, N. Nakamura, Y. Ohizumi, and Y. Hirata, *J. Chem. Soc., Perkin Trans. 1*, 1989, 101.
450. J. S. Bergmann and B. R. Nechay, *Fed. Proc.*, 1982, **41**, 1562.
451. N. Ikemoto, *Annu. Rev. Physiol.*, 1982, **44**, 297.
452. T. Komura, H. Nagayama, and S. Wada, *Nippon Nogeikagoko Kaishi*, 1974, **48**, 459.

453. L. P. Davies, D. D. Jamieson, J. A. Baird-Lambert, and R. Kazlauskas, *Biochem. Pharmacol.*, 1984, **33**, 347.
454. J. Winder, R. J. P. Cannell, J. M. Walker, S. Delbarre, C. Francisco, and P. Farmer, *Biochem. Soc. Trans.*, 1989, **17**, 1030.
455. W. Puls, U. Keup, H. P. Krause, G. Thomas, and F. Hoffeister, *Naturwissenschaften*, 1977, **64**, 536.
456. B. K. Carte, N. Troupe, J. A. Chan, J. W. Westley, and D. J. Faulkner, *Phytochemistry*, 1989, **28**, 2917.
457. Y. Fukuyama, M. Kodama, I. Miura, Z. Kinzyo, M. Kido, H. Mori, Y. Nakayama, and M. Takahashi, *Chem. Pharm. Bull.*, 1989, **37**, 349.
458. Y. Fukuyama, M. Kodama, I. Miura, Z. Kinzyo, H. Mori, Y. Nakayama, and M. Takahashi, *Chem. Pharm. Bull.*, 1989, **37**, 2438.
459. Y. Fukuyama, I. Miura, Z. Kinzyo, H. Mori, M. Kido, Y. Nakayama, M. Takahashi, and M. Ochi, *Chem. Lett.*, 1989, **14**, 739.
460. Y. Fukuyama, M. Kodama, I. Miura, Z. Kinzyo, H. Mori, Y. Nakayama, and M. Takahashi, *Chem. Pharm. Bull.*, 1990, **38**, 133.
461. Y. Nakamura, J. Kobayashi, J. Gilmore, M. Mascal, K. L. Rinehart, H. Nakamura, and Y. Ohizumi, *J. Biol. Chem.*, 1986, **261**, 4139.
462. G. L. Helms, R. E. Moore, W. P. Niemczura, and G. M. L. Patterson, *J. Org. Chem.*, 1988, **53**, 1298.
463. J. Kobayashi, *J. Nat. Prod.*, 1989, **52**, 225.
464. J. Kobayashi, Y. Ohizumi, N. Nakamura, T. Yamakodo, T. Matsuzaki, and Y. Hirata, *Experimentia*, 1983, **39**, 67.
465. G. Blunden and S. M. Gordon, *J. Agric. Soc. of Univ. Coll. Wales*, 1989, **69**, 184.
466. G. Blunden, in 'Seaweed Resoures in Europe: Uses & Potential', ed. M. D. Guiry and G. Blunden, John Wiley & Sons, New York, 1991, p. 65.
467. M. Maeda, T. Kodama, T. Tanaka, H. Yoshizumi, T. Takemoto, K. Nomoto, and T. Fujita, *Chem. Pharm. Bull.*, 1986, **34**, 4892.
468. Suntory Ltd., Jpn. Kokai Tokkyo Koho, JP 58.222.004 (83 222.004), Ref. C.A. 1984, 100: 116498f, 1984.
469. M. Maeda, T. Kodama, T. Tanaka, Y. Ohfune, K. Nomoto, K. Nishimura, and T. Fujita, *J. Pestic. Sci.*, 1984, **9**, 27.
470. B. J. Cook and G. M. Holman, *Comp. Biochem. Physiol.*, 1979, **64**, 183.
471. K. Watanabe, K. Umeda, and M. Miyakado, *Agric. Biol. Chem.*, 1989, **53**, 2513.
472. M. S. Dhillon, M. S. Mulla, and Y. S. Hwang, *J. Chem. Ecol.*, 1982, **8**, 557.
473. F. K. Gleason, US4.626.271, 1986, Appl. 776.842, 1985, Ref. C.A. 1987, 106:45736t, 1986.
474. F. K. Gleason and J. L. Paulson, *Arch. Microbiol.*, 1984, **138**, 273.
475. G. M. L. Patterson and D. O. Harris, *Br. Phycol. J.*, 1983, **18**, 259.
476. G. M. L. Patterson, D. O. Harris, and W. S. Cohen, *Plant Sci. Lett.*, 1979, **15**, 293.
477. H. G. Van der Weij, *Proc. K. Ned. Ada. Wet.*, 1937, **36**, 759.
478. H. G. DuBuy and R. A. Olson, *Am. J. Bot.*, 1937, **24**, 609.
479. J. A. Bentley, *Nature*, 1958, **181**, 1499.
480. J. A. Bentley-Mowat, *Proc. Int. Seaweed Symp.*, 1963, **4**, 352.
481. B. Moss, *New Phytol.*, 1965, **64**, 387.
482. U. Schiewer, *Planta (Berl.)*, 1967, **74**, 313.
483. R. G. Buggeln and J. S. Craigie, *Planta*, 1971, **97**, 173.
484. A. Ballester, *J. Exp. Mar. Biol. Ecol.*, 1975, **20**, 179.
485. F. C. Sumera and G. B. J. Cajipe, *Bot. Mar.*, 1981, **24**, 157.
486. H. Abe, M. Uchiyama, and R. Sato, *Agric. Biol. Chem.*, 1972, **36**, 2259.

487. H. Abe, M. Uchiyama, and R. Sato, *Agric. Biol. Chem.*, 1974, **38**, 897.
488. E. M. Jones, M. Phil. Thesis, CNAA, Portsmouth Polytechnic, 1979, p. 144.
489. B. C. Featonby-Smith and J. Van Staden, *Bot. Mar.*, 1984, **27**, 527.
490. G. Blunden, D. J. Rodgers, and C. J. Barwell, in 'Natural Products and Drug Development', ed. P. Krogsgaard-Larsen, C. S. Brogger, and K. Kofod, Alfred Benzon Symp. 20, Munksgaard, Copenhagen, 1984, p. 179.
491. A. W. Wheeler, 'Report Rothamsted Experimental Station, Part 1, 1973, p. 101.
492. K. S. Bokarev and R. P. Ivanova, *Sov. Pl. Physiol.*, 1971, **18**, 302.
493. A. Sakai and S. Yoshida, *Teion Kagaka, Siebutsu-Hen*, 1968, **26**, 13.
494. H. Bergman and H. Eckert, *Biologia Pl.*, 1984, **26**, 384.
495. D. C. Carter, R. E. Moore, J. S. Mynderse, W. P. Niemczura, and J. S. Todd, *J. Org. Chem.*, 1984, **49**, 236.
496. W. Fenical, in 'Proceedings Joint China–US Phycology Symposium, ed. C. D. Tseng, Science Press, Beijing, 1983, p. 497.
497. E. O. Hughes, P. R. Gorham, and A. Zehnder, *Can. J. Microbiol.*, 1958, **4**, 225.
498. J. R. Murthy and J. B. Capindale, *Can. J. Biochem.*, 1970, **48**, 508.
499. D. F. Toerien, W. E. Stott, and M. J. Pitout, *Water S.A. (Pretoria)*, 1976, **2**, 160.
500. T. C. Ellman, I. R. Falconer, A. R. B. Jackson, and M. T. Runnegar, *Aust. J. Biol. Sci.*, 1978, **31**, 209.
501. D. P. Botes, C. C. Viljoen, H. Kruger, P. L. Wessels, and D. H. Williams, *Toxicon*, 1982, **20**, 1037.
502. S. Santikarn, D. H. Williams, R. J. Smith, S. H. J. Hammond, D. P. Botes, A. Tuinman, P. L. Wessels, C. C. Viljoen, and H. Kruger, *J. Chem. Soc., Chem. Commun.*, 1983, **275**, 652.
503. W. W. Carmichael, J. T. Eschedor, G. M. L. Patterson, and R. E. Moore, *Appl. Environ. Micro.*, 1988, **54**, 2257.
504. J. E. Eriksson, J. A. O. Meriluoto, H. P. Kujari, K. Osterlund, K. Fagerlund, and L. Hallbom, *Toxicon*, 1988, **26**, 161.
505. M. T. C. Runnegar, A. R. B. Jackson, and I. R. Falconer, *Toxicon*, 1988, **26**, 143.
506. K. C. Jurgens, *Texas Game & Fish.*, 1953, **11**, 8.
507. F. P. Thurberg and J. J. Sasner, *Chesapeake Sci.*, 1973, **14**, 48.
508. W. H. Gerwick, C. H. Mrozek, M. F. Moghadoam, and S. K. Agarwal, *Experimentia*, 1989, **45**, 115.
509. W. H. Gerwick, W. Fenical, N. Fritsch, and J. Clardy, *Tetrahedron Lett.*, 1979, **20**, 145.
510. P. Crews and E. Kho, *J. Org. Chem.*, 1975, **40**, 2568.
511. V. J. Paul, M. M. Littler, D. S. Littler, and W. Fenical, *J. Chem. Ecol.*, 1987, **13**, 1171.
512. V. J. Paul and W. Fenical, *Tetrahedron Lett.*, 1982, **23**, 5017.
513. H. H. Sun and W. Fenical, *Tetrahedron Lett.*, 1979, **20**, 685.
514. H. H. Sun, F. J. McEnroe, and W. Fenical, *J. Org. Chem.*, 1983, **48**, 1903.
515. S. de Rosa, S. de Stefano, S. Marcura, E. Trivellone, and N. Zavodnik, *Tetrahedron*, 1984, **40**, 4991.
516. D. Schlenk and W. H. Gerwick, *Phytochemistry*, 1987, **26**, 1081.
517. M. L. Bittner, M. Silva, V. J. Paul, and W. Fenical, *Phytochemistry*, 1985, **24**, 987.
518. J. J. A. McLaughlin and L. Provasoli, *J. Protozool.*, 1957, **4**, 7.
519. M. Spiegelstein, K. Reich, and F. Bergamann, *Verhandl. Intern. ver. Limnol.*, 1969, **17**, 778.
520. R. E. Moore, *J. Org. Chem.*, 1987, **52**, 1036.

521. C. P. Mason, K. R. Edwards, R. E. Carlson, J. Pignatello, F. K. Gleason, and J. M. Wood, *Science*, 1982, **215**, 400.
522. T. T. Jong, P. G. Williard, and J. P. Porwell, *J. Org. Chem.*, 1984, **49**, 735.
523. F. K. Gleason and J. Porwoll, *J. Org. Chem.*, 1986, **51**, 1615.
524. J. B. Steward, V. Bonremann, J. L. Chen, R. E. Moore, F. R. Chaplan, H. Karuso, L. K. Larsen, and G. M. L. Patterson, *J. Antibiot.*, 1988, **61**, 1048.
525. J. J. Barachi, T. R. Norton, E. Furusawa, G. M. L. Patterson, and R. E. Moore, *Phytochemistry*, 1983, **22**, 2851.
526. M. Ishibashi, R. E. Moore, G. M. L. Patterson, C. Xu, and J. Clardy, *J. Org. Chem.*, 1986, **51**, 5300.
527. V. D. Ramamurthy, *Mar. Biol.*, 1970, **6**, 74.
528. S. Cooper, A. Battat, P. Marsot, and M. Sylvestre, *Can. J. Microbiol.*, 1983, **29**, 338.
529. T. M. Aubert, M. J. Aubert, M. Gauthier, D. Pesando, and S. Daniel, *Rev. Oceanogr. Med.*, 1966, **4**, 23.
530. M. J. Aubert and M. Gauthier, *Rev. Int. Oceanogr. Med.*, 1967, **5**, 63.
531. D. Pesando, *Rev. Int. Oceanogr. Med.*, 1972, **25**, 49.
532. M. Gauthier, P. Bernard, and M. J. Aubert, *J. Exp. Mar. Biol. Ecol.*, 1978, **33**, 37.
533. M. J. Aubert and D. Pesando, *Rev. Int. Oceanogr. Med.*, 1968, **10**, 259.
534. W. Proctor, *Limnol. Oceanogr.*, 1956, **1**, 125.
535. R. Olfers-Weber and U. Mihm, *Hyg. Erste Abt. Orig. Reiche Hyg. Krankenhaushyg. Betriebshyg. Praev. Med.*, 1978, **169**, 287.
536. R. Harder and A. Opperman, *Arch. Mikrobiol*, 1953, **19**, 398.
537. A. Misra and R. Sinha, in 'Marine Algae in Pharmaceutical Science', ed. H. A. Hoppe, T. Levring, and Y. Tanaka, De Gruyter, Berlin, 1979, p. 237.
538. R. J. P. Cannell, P. Farmer, and J. M. Walker, *J. Biochem.*, 1988, **225**, 937.
539. G. Blaauw-Jansen, *Nature*, 1954, **174**, 312.
540. E. G. Jorgensen, *Physiol. Plant.*, 1962, **15**, 530.
541. R. T. Levina, in 'The Purification of Waste-Waters in Biological Ponds', Minsk Akam., Nauk BSSR: 1961, vol. 136 [*Biol. Abstr.*, 1963, **42**, 18828].
542. M. M. Telitchenko, N. V. Davydova, and V. D. Fedorov, *Nauchn Dokl. Vysshei. Shkoly. Biol. Nauki. Moscow.*, 1962, **4**, 157.
543. V. J. Paul, H. H. Sun, and W. Fenical, *Phytochemistry*, 1982, **21**, 428.
544. T. B. Nakatsu, N. Ravi, and D. J. Faulkner, *J. Org. Chem.*, 1981, **46**, 2435.
545. J. M. Sieburth, *J. Bacteriol.*, 1961, **82**, 72.
546. S. Aaronson and B. Bensky, *J. Protozool.*, 1967, **14**, 76.
547. B. R. Berland, D. J. Bonin, A. L. Cornu, S. Y. Maestrini, and J. P. Marino, *J. Phycol.*, 1972, **8**, 383.
548. L. S. Cierszko, *Trans. NY Acad. Sci. Ser. II.*, 1962, **24**, 502.
549. J. M. Sieburth, *Science*, 1960, **132**, 676.
550. K. W. Glombitza, in 'Marine Natural Products Chemistry', ed. D. J. Faulkner and W. H. Fenical, NATO Conference Series IV; Marine Science, Plenum Press, New York, 1977.
551. B. J. Burreson, R. E. Moore, and P. Roller, *Tetrahedron Lett.*, 1975, **7**, 473.
552. B. J. Burreson, R. E. Moore, and P. Roller, *J. Agric. Food Chem.*, 1976, **24**, 856.
553. W. Fenical, *Tetrahedron Lett.*, 1974, **44**, 4463.
554. J. F. Siuda, G. R. Vanblaricom, P. D. Shaw, R. D. Johnson, R. H. White, L. P. Hager, and K. L. Reinhart, *J. Am. Chem. Soc.*, 1975, **97**, 937.
555. R. Kazaluskas, D. T. Murphy, R. J. Quinn, and R. J. Wells, *Tetrahedron Lett.*, 1977, **7**, 37.

556. K. Ohta, in 'Proceedings International Seaweed Symposium 9', ed. A. Jensen and J. R. Stein, Science Press, Princeton, 1979, p. 401.
557. S. J. Wratten and D. J. Faulkner, *J. Org. Chem.*, 1976, **41**, 2465.
558. G. T. Carter, K. L. Reinhart, L. H. Li, S. L. Kuentzel, and J. L. Connor, *Tetrahedron Lett.*, 1978, **46**, 4479.
559. M. P. Kirkup and R. E. Moore, *Tetrahedron Lett.*, 1983, **24**, 2087.
560. W. Fenical and O. J. McConnell, *Phytochemistry*, 1976, **15**, 313.
561. K. Ohta and M. Takagi, *Phytochemistry*, 1977, **16**, 1085.
562. T. Katayama, *Bull. Jpn. Soc. Sci. Fish.*, 1961, **27**, 75.
563. T. Irie, M. Izawa, E. Kurosawa, and T. Masamune, *Tetrahedron*, 1970, **26**, 3271.
564. E. Kurosawa and M. Suzuki, *Kagaku to Seibutsu*, 1983, **21**, 23.
565. J. J. Simms, M. S. Donnell, J. V. Leary, and G. H. Lacy, *Antimicrob. Agents Chemother.*, 1975, **7**, 320.
566. S. Naylor, F. J. Hacke, L. V. Manes, and P. Crews, *Fortschr. Chem. Org. Naturst.*, 1983, **44**, 189.
567. J. A. Pettus, R. M. Wing, and J. J. Simms, *Tetrahedron Lett.*, 1977, **1**, 42.
568. A. G. Gonzalez, V. Darias, and E. Estevez, *Planta Med.*, 1982, **44**, 44.
569. W. Fenical and J. J. Simms, *Tetrahedron Lett.*, 1974, **15**, 1137.
570. J. S. Cragie, A. G. McInnes, M. A. Ragan, and J. A. Walter, *Can. J. Chem.*, 1977, **55**, 1575.
571. K. W. Glombitza and K. Klapperich, *Bot. Mar.*, 1985, **28**, 139.
572. T. Higa, in 'Marine Natural Products, Chemical and Biological Perspectives', ed. P. J. Scheuer, Academic Press, London, 1981, p. 93.
573. J. Tanaka and T. Higa, *Chem. Lett.*, 1984, **3**, 231.
574. J. Finer, J. Clardy, W. Fenical, L. Minale, R. Riccio, J. Battaile, M. Kirkup, and R. E. Moore, *J. Org. Chem.*, 1979, **44**, 2044.
575. N. Enoki, K. Tsuzuki, S. Omura, R. Ishida, and T. Matsumoto, *Chem. Lett.*, 1983, **12**, 1627.
576. A. B. Alvarado and W. H. Gerwick, *J. Nat. Prod.*, 1985, **48**, 132.
577. K. W. Glombitza, *Planta Med.*, 1970, **18**, 210.
578. H. H. Sun, V. J. Paul, and W. Fenical, *Phytochemistry*, 1983, **22**, 743.
579. D. Menzel, R. Kazlauskas, and J. Reichelt, *Bot. Mar.*, 1983, **26**, 23.
580. M. Kuniyoshi, K. Yamada, and T. Higa, *Experimentia*, 1985, **41**, 533.
581. R. E. Moore, *Pure Appl. Chem.*, 1982, **54**, 1919.
582. J. J. Barachi, R. E. Moore, and G. M. L. Patterson, *J. Am. Chem. Soc.*, 1984, **106**, 8193.
583. I. Yamamoto, T. Nagumo, M. Takahashi, M. Fujihara, Y. Suzuki, and N. Iizima, *Jpn. J. Exp. Med.*, 1981, **51**, 187.
584. Y. Suzuki, I. Yamamoto, and I. Umexawa, *Chemotherapy*, 1980, **28**, 165.
585. I. Yamamoto, T. Nagumo, M. Fujihara, M. Takahashi, Y. Ando, M. Okada, and K. Kawaki, *Jpn. J. Exp. Med.*, 1977, **47**, 133.
586. J. Kobayashi, M. Ishibashi, H. Nakamura, and Y. Ohizumi, *Tetrahedron Lett.*, 1986, **27**, 5755.
587. M. Ishibashi, Y. Ohizumi, M. Hamashima, H. Nakamura, Y. Hirata, T. Sasaki, and J. Kobayashi, *J. Chem. Soc., Chem. Commun.*, 1977, 1127.
588. J. Kobayashi, M. Ishibashi, M. R. Walchli, H. Nakamura, Y. Hirata, T. Sasaki, and Y. Ohizumi, *J. Am. Chem. Soc.*, 1988, **110**, 490.
589. W. H. Gerwick, A. Lopez, G. D. Van Duyner, J. Clardy, W. Ortiz, and A. Baez, *Tetrahedron Lett.*, 1986, **27**, 1979.
590. J. F. Biard and J. F. Verbist, *Plant Med. Phytothr.*, 1981, **15**, 167.

591. F. Asari, T. Kusumi, and H. Kakisawa, *J. Nat. Prod.*, 1989, **52**, 1167.
592. K. C. Guven and E. Aktin, *Bot. Mar.*, 1964, **7**, 1.
593. J. Houck, R. Morris, and E. Lazaro, *Proc. Soc. Exp. Biol. Med.*, 1957, **96**, 528.
594. G. Bernadi and G. F. Springer, *J. Biol. Chem.*, 1962, **237**, 75.
595. T. Usui, K. Asari, and T. Mizuno, *Agric. Biol. Chem.*, 1980, **44**, 1965.
596. K. Dobashi, T. Nishino, M. Fujihara, and T. Nagumo, *Carbohydrate Res.*, 1989, **194**, 315.
597. A. Abdel-Fattah, H. Magdel-Din, and H. M. Salem, *Carbohydrate Res.*, 1974, **33**, 1.
598. N. V. S. A. V. Prasado Rao, K. V. Sastry and E. Venkata Rao, *Phytochemistry*, 1984, **23**, 2531.
599. M. Hussein, A. Abdel-Aziz, and H. M. Salem, *Phytochemistry*, 1980, **19**, 2131.
600. A. Abdel-Fattah, M. Magdel-Din Hussein, and S. T. Faouad, *Phytochemistry*, 1978, **17**, 741.
601. D. J. Rodgers, K. Jurd, G. Blunden, F. Zanteei, and S. Paoletti, 'Oral Communication, 38th Meeting British Phycological Society', 1990.
602. E. Percival, in 'Comparative Phytochemistry', ed. T. Swain, Academic Press, New York, 1966, p. 139.
603. P. J. Sawyer, J. H. Gentile, and J. J. Sasner, *Can. J. Microbiol.*, 1968, **14**, 1199.
604. R. F. Taylor, M. Ikawa, J. J. Sasner, F. P. Thurberg, and K. K. Anderson, *J. Phycol.*, 1974, **10**, 279.
605. H. Ozawa, Y. Gomi, and I. Otsuki, *Yakugaku Zasshi*, 1967, **87**, 935.
606. G. L. Johnson, J. J. Spikes, and S. Ellis, *Toxicon*, 1985, **23**, 505.
607. M. Kobayashi, S. Kondo, T. Yasumoto, and Y. Ohizumi, *J. Pharm. Exp. Ther.*, 1986, **238**, 1077.
608. M. Entzeroth, R. E. Moore, W. P. Niemczura, and G. M. L. Patterson, *J. Org. Chem.*, 1986, **51**, 5307.
609. A. C. Tarjan, *J. Nematol*, 1977, **9**, 287.
610. A. C. Tarjan and J. J. Fredrick, *Nematropica*, 1983, **13**, 55.
611. B. C. Featonby-Smith and J. Van Staden, *Sci. Hort.*, 1983, **20**, 137.
612. I. J. Crouch and J. Van Staden, *J. Appl. Phycol.*, 1993, **5**, 37.

CHAPTER 6

Pesticides from Nature

*Part I: Crop Protection Agents from Higher Plants – An Overview**

JILL P. BENNER

6.I.1 Introduction

Over the centuries, man has battled to protect crops against invasion by insects, microbial pathogens, and other pests. Many of the earliest pesticides were extracts of plants, which were used on a local basis to protect crops both in the field and after harvest. As the years progressed, several plants were exploited more widely as sources of commercial insecticides, but from the 1940s onwards, synthetic chemicals largely replaced 'botanicals' as the key commercial products. Research into plant-derived natural products for agriculture went into decline for a number of years, but this trend is now being reversed as it becomes evident that plant natural products still have enormous potential to inspire and influence modern agrochemical research.

From a scientific viewpoint it is not difficult to explain why plants should represent a valuable source of new compounds for agricultural applications. It is estimated that there are at least 250 000 different species of plants in the world today,[1] but the figures could be as high as 500 000. It is also estimated that only about 10% of plant species have been examined chemically, so there is enormous scope for further work. Plants are known to produce a very diverse range of secondary metabolites, such as terpenoids, alkaloids, polyacetylenes, flavonoids, and unusual amino acids and sugars. Some classes of compound are produced by both micro-organisms and higher plants but, to a large extent, higher plants exhibit their own specialized areas of chemistry and therefore research programmes on these complement, rather than overlap with, parallel programmes of research based on micro-organisms.

*Part I previously published in *Pesticide Science,* 1993, **39**(2), 95.

Although it is difficult to define the ecological significance of most of the commercial botanical insecticides, there is good reason to suppose that the secondary metabolism of plants has evolved so as to protect the plant from attack by insects and microbial pathogens. It is also possible to apply folklore to assist targeting of particular plant genera for study. Plants which are reported to have effects against insects, or even which can be used to kill body lice or rats, may represent a good starting point for research.

Few plant natural products will ever reach the market as products *per se,* but others will provide lead structures for programmes of synthetic chemistry and hopefully follow the success story of the synthetic pyrethroids. Structurally complex compounds, which are not amenable to synthetic chemistry programmes, may also have a role to play by validating new modes of action for pesticides.

Examples are presented of compounds that exhibit insecticidal, fungicidal, and herbicidal effects. Consideration is also given to the development of screening programmes to detect new compounds with interesting biological properties. Careful experimental design and thorough recording of procedures and data are crucial to success. Badly designed programmes afford only weakly active compounds or show effects which cannot be reproduced at a later date.

Natural product chemistry, whether based on higher plants, micro-organisms, or other sources, is a very difficult science, but there is little doubt that dedicated research will eventually be rewarded with exciting new lead structures for industrial application.

6.I.2 Development of Commercial Products

Once a plant natural product has been found to have good agrochemical activity, it is necessary to consider how this property can be applied. In very exceptional cases a compound may perform sufficiently well to be a product *per se.* The source of the compound then becomes a key issue. Although it may be practical to grow sufficient plant material for local use in less developed countries, it is highly unlikely that a major agrochemical company would launch a significant commercial product which had to be extracted from whole plant material. The preferred option would be to carry out a total synthesis of the required compound, provided, of course, that the process was economically viable.

Few natural products have all the necessary characteristics to compete with the best synthetic agrochemicals. It is much more likely that a plant natural product will be used as a lead for synthesis, rather than as a product *per se.* The toxophore is used as a basis for the synthesis of analogues which will hopefully show improvements over the original compound. In many cases, improvements in both the potency and physical properties are necessary to generate a commercially viable product. If a compound is very complex and it is not possible to identify a key toxophore, then synthetic chemistry is unlikely to be a realistic option. However, if the mode of action of the compound is novel, it may provide a source of inspiration to biochemists and result in the development of new bioassays capable of detecting other, structurally simpler, compounds with the same mode of action.

6.I.3 Insecticides

There are numerous examples in the literature of plant natural products with interesting agrochemical properties; compounds of this type have played a major role in the development of commercial insecticides. The most important and significant actual application of a plant natural product centres on the insecticidal properties of pyrethrum, which is obtained from *Chrysanthemum cinerariaefolium* Vis. The powdered dry flower has been used as an insecticide from ancient times[2] and the original home of the flower is said to have been the Middle and Near East, although it is now grown mainly in East Africa.[3] Natural pyrethrum, which consists of six closely related compounds, is still used commercially, but it has been superseded largely by the synthetic pyrethroids, which have greatly improved insecticidal properties and photostability. The development of this class of compounds by the Rothamsted group, led by Michael Elliott,[4] represents an important milestone in agrochemical research (see Chapter 7). The relationship between the natural product prototype, pyrethrin I (**1**), and the synthetic pyrethroid cypermethrin (**2**), is shown in Figure 6.1.

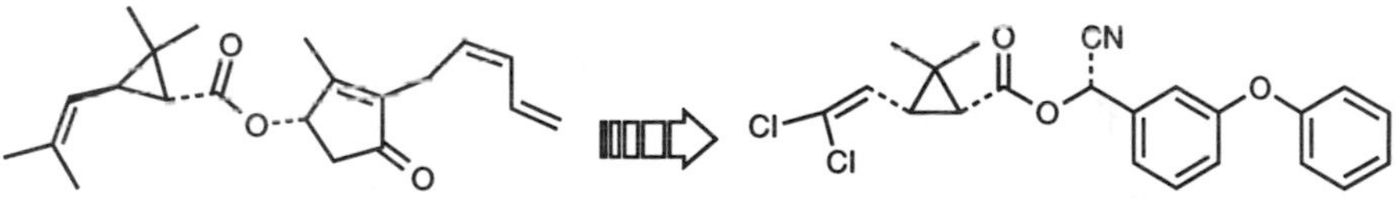

1 Pyrethrin I, Activity = 1 (*Musca*)

2 *cis*-1(*R,S*)-Cypermethrin, Activity = 34

Figure 6.1 *Structures of the natural product Pyrethrin* I *and the synthetic pyrethroid* cis-*1*(R,S)-*cypermethrin*

Rotenone (**3**, Figure 6.2) is a well-known botanical insecticide which still retains popularity with gardeners. Rotenone and related rotenoids are obtained from *Derris, Lonchocarpus,* and *Tephrosia* sp. The original use of these plants in Asia and South America was as a fish poison;[5] the plants were dragged through lakes and streams to stupefy fish, which could then be caught easily. Rotenone is active against a range of insect pests and its mode of action is as a Site I respiration inhibitor.

A very well-known compound which has been used as a commercial insecticide is nicotine (**4**). Reference was made to the use of a tobacco extract as a plant spray in parts of Europe 300 years ago.[6] Nicotine is found in many species of *Nicotiana,* although *Nicotiana rustica* L. is a much better source of this compound than the more familiar *N. tabacum* L. (tobacco), and it has been

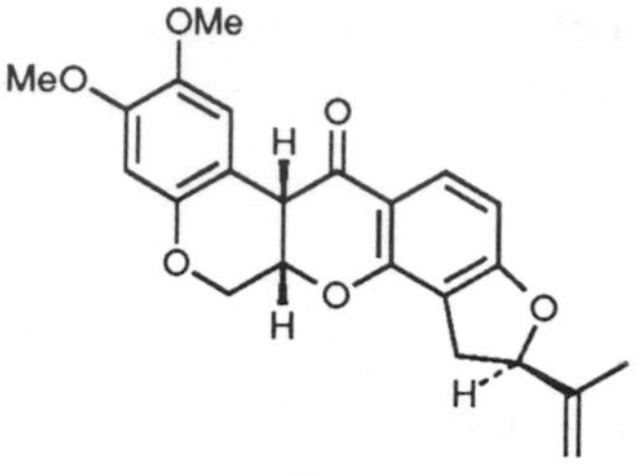

3

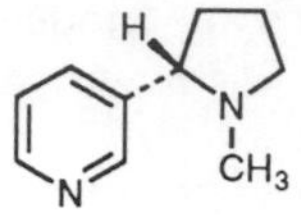

4

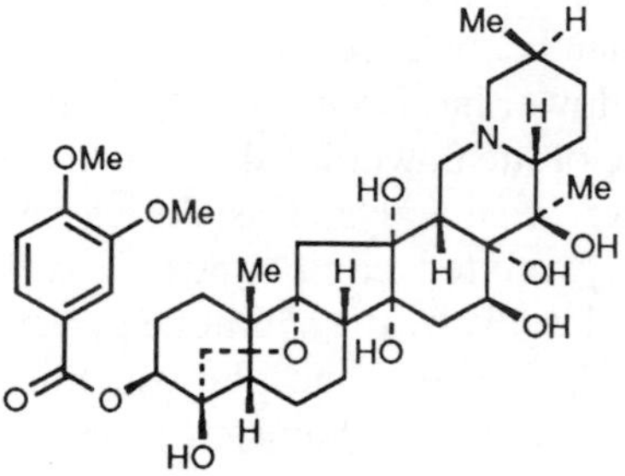

5

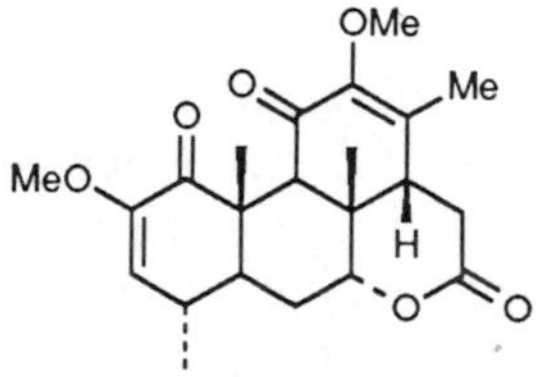

6

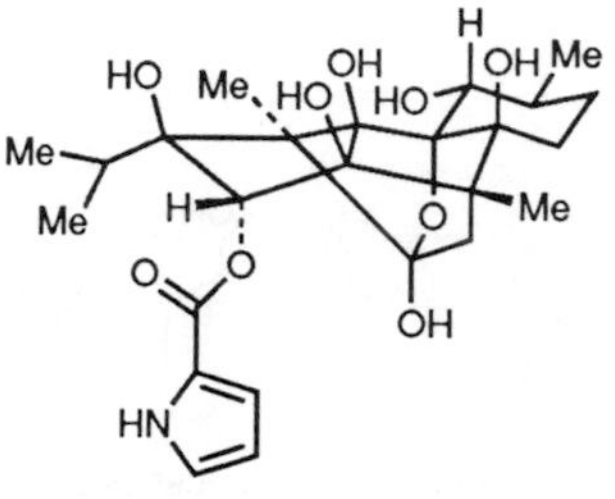

7

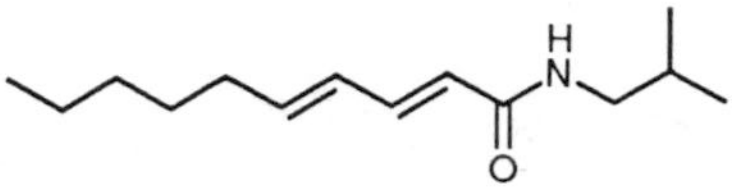

8

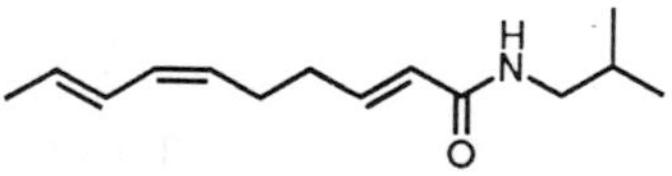

9

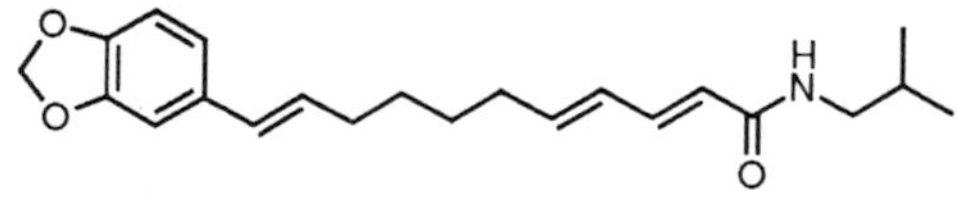

10

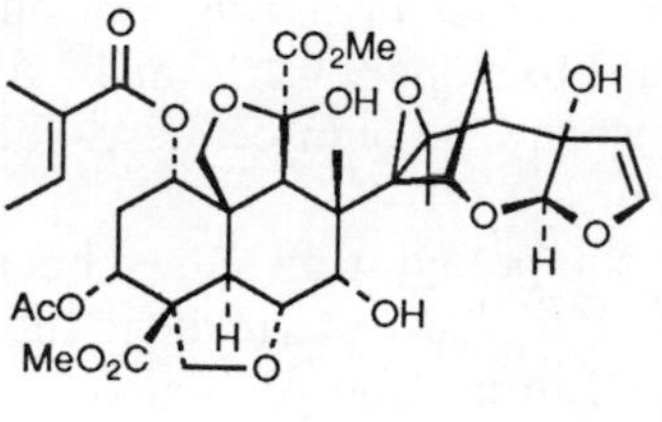

11

Figure 6.2 *Examples of natural products with pesticidal properties*

cultivated specifically for nicotine extraction. Nicotine is a compound which is highly toxic to man as well as to insects; it affects the nervous system by binding to the acetylcholine receptor.[7]

Sabadilla and *Veratrum* species were used for many years as a source of insecticides, although their commercial application is now extremely limited. The active ingredients are a range of ceveratrum alkaloids, *e.g.*, veratridine (**5**). Use of *Sabadilla* in the Americas dates back to the 1500s, and the product was used extensively in Europe and the USA from the late nineteenth to mid-twentieth century.[8] The ceveratrum alkaloids affect the sodium channel[7] and show rather high levels of toxicity to mammals as well as to insects.

Quassin (**6**) was originally derived from the wood of *Quassia amara* L., a small tropical tree which is a member of the Simarubaceae family. A related shrub from the West Indies (*Aeschrion excelsa,* also known as *Picrasma excelsa* Swartz.) eventually became the principal source of the insecticide. Aqueous extracts of *Quassia* chips were used as an insecticide from the late eighteenth to mid-twentieth century; however, it is also claimed that boxes made from *Quassia* wood protect the contents against insects. Quassin is rather a weak insecticide, but it is still used commercially as one of the ingredients in certain types of dog and cat repellent.

Ryanodine (**7**), which is derived from *Ryania* species (Flacourtiaceae), especially *R. speciosa* Vahl, represents the first example of a commercially successful natural insecticide discovered by screening plant extracts for activity. The discovery was made as part of a collaboration between Merck and Rutgers University in the early 1940s.[9] *Ryania* species have been used in South America for euthanasia and as rat poisons, although there was no detailed knowledge of insecticidal properties. Ryanodine poisons muscles by binding to calcium channels in the sarcoplasmic reticulum. This causes calcium ions to flow into the cell and death occurs rapidly.[10] Interestingly, compounds which are closely related to ryanodine are also found in the outer bark of Ceylon cinnamon[11,12] (*Cinnamomum zeylanicum* Nees) and Canary Island laurel[13,14] (*Persea indica* L. Sprengel). Fortunately, the well known spice, cinnamon is free of these compounds since the outer bark is removed prior to processing. There are, however, reports that rats feeding on *P. indica* become severely intoxicated and the effects are believed to be due to the ryanodine analogues.

A group of compounds which have not yet been commercialized, but which have inspired extensive research synthesis, are the unsaturated isobutylamides. One of the first structures of this type to be elucidated was compound (**8**), the principal constituent of the crystalline mixture pellitorine derived from *Anacyclus pyrethrum* DC,[15] but many other examples of the same structural type, including affinin (**9**) from *Heliopsis longipes* (A. Gray) Blake[16] and pipercide (**10**) from *Piper nigrum* L. (black pepper)[17] have been identified subsequently. The isobutylamides cause rapid knockdown and kill of flying insects, but are too unstable for use as products *per se.* They are also very pungent and act as a local anaesthetic when applied to the tongue. These compounds have been found to act as voltage-dependent blockers of the sodium channel.

One of the few botanical insecticides to attract current interest as a product *per se* is neem oil obtained from *Azadirachta indica* A. Juss. Neem oil contains many compounds, but the best known, and one of the most active, is azadirachtin (**11**). This compound was first isolated in 1968, but the structure was not elucidated fully until 1985.[18] Neem oil has been used locally in India for many years, but is now being developed as a significant commercial product. Azadirachtin is an insect antifeedant, but also shows growth inhibitory and endocrine disrupting effects.

6.I.4 Fungicides

Although there are many more examples of plant natural products which show potent activity as insecticides, it is also worth considering the potential of higher plants to yield other types of agrochemicals. The situation regarding production of fungicides by higher plants is more complex than for insecticides. There are a number of conventional secondary metabolites produced by higher plants which show antifungal properties. When investigating the literature there are, however, many more examples of antifungal compounds which are stress metabolites or phytoalexins. Phytoalexins are compounds which are both synthesized by and accumulate in plants after exposure to microorganisms.

Although there is no reason why a higher plant should not yield a commercial fungicide, the following points are worthy of note:

- Many fungicidal compounds show only *in vitro* effects.
- Plant natural products are often fungistatic rather than fungicidal, *i.e.*, they inhibit growth, but do not actually kill the fungus.
- Phytoalexins are frequently phytotoxic – an undesirable effect for a commercial fungicide.
- Phytoalexin responses are difficult to reproduce: this may cause particular difficulties when there is a need to produce reasonable amounts of material for biological testing.

Capillin (Figure 6.3, **12**)[19] is a conventional secondary metabolite obtained from *Artemisia capillaris* Thunb. It has *in vitro* activity against a range of plant pathogens and is also effective against human pathogens. Synthetic programmes based on capillin have been carried out, but have failed to give a useful product. In the medicinal area, capillin itself is too irritant for topical use and the synthetic analogues with lower irritancy generally show reduced antifungal activity.[20]

One of the few higher plant secondary metabolites which is claimed to show *in vivo* control of plant pathogens is sclareol (**13**) which is produced by *Salvia sclarea* L. and *Nicotiana glutinosa* L. The literature[21] reports good control of rust (*Uromyces fabae* de Bary) on beans (*Vicia faba* L.), but the spectrum and potency of the compound are inadequate for commercial application.

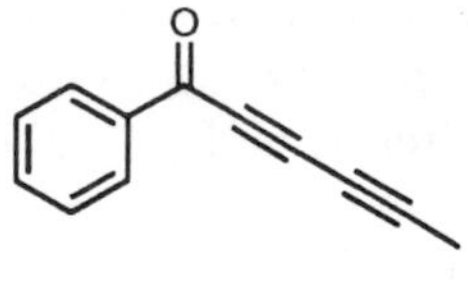

12 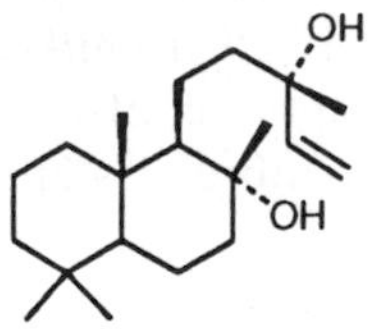 13 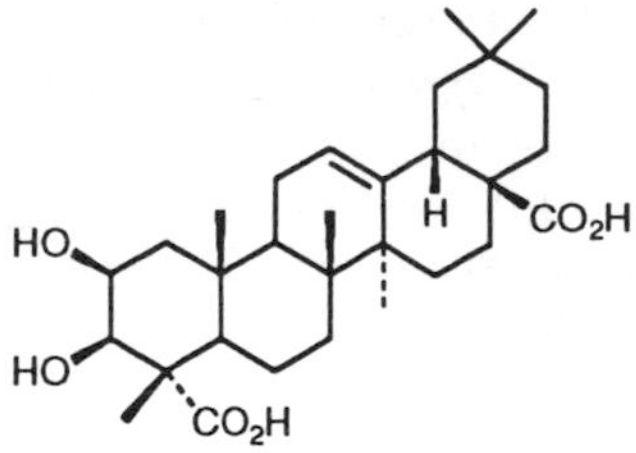 14

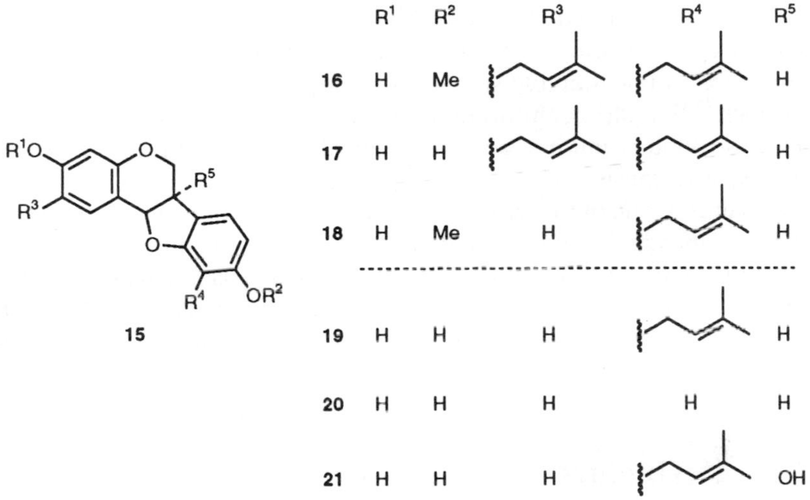

15

	R^1	R^2	R^3	R^4	R^5
16	H	Me			H
17	H	H			H
18	H	Me	H		H
19	H	H	H		H
20	H	H	H	H	H
21	H	H	H		OH

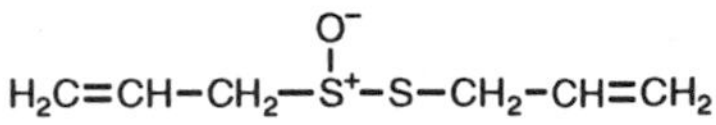

$H_2C{=}CH{-}CH_2{-}S^+(O^-){-}S{-}CH_2{-}CH{=}CH_2$

22

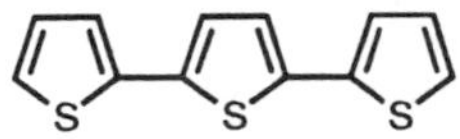

23

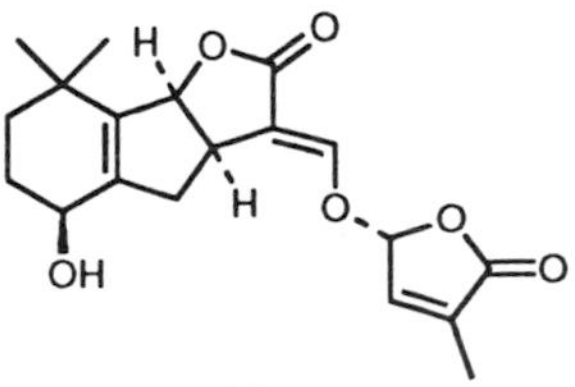

24

Figure 6.3 *Examples of natural products with fungicidal activity*

Plant saponins show a diverse range of biological effects. Medicagenic acid (**14**) and several corresponding glycosides are present in the roots of alfalfa (*Medicago sativa* L.) and exhibit fungicidal properties. Alfalfa meal has been shown to control avocado root rot (*Phytophthora cinnamomi* Rands) and several other plant pathogens.[22] Saponins of this type are believed to exert their effects by interacting with membrane sterols, proteins, and phospholipids.

Another class of compounds with claimed antimicrobial effects are the pterocarpans (**15–21**). The distribution of pterocarpans in *Erythrina crista galli* L., also known as the cockspur coral tree, is particularly interesting. Some of the compounds (**16–18**) are always present in the plant, but other analogues (**19–21**) are produced only when the plant is infected with a pathogen and are, therefore, induced phytoalexins.[23,24] All these compounds have rather weak biological effects when tested *in vivo* and show poor metabolic stability. It has been predicted that suitably stabilized analogues may be much more potent fungicides,[25] but, although this may be true, a natural product must normally show reasonable promise as an agrochemical before resources can be applied to analogue synthesis.

The source of one of the earliest reported fungicides is the well known plant, garlic (*Allium sativum* L). Its juice exhibits activity against a wide range of fungi. In 1944 Sterling Winthrop Chemical Company isolated allicin (**22**) from garlic and showed it to be bactericidal as well as fungicidal.[26,27] Allicin is the subject of a patent (US 2554088), but its use was abandoned because of the substance's odour.

6.I.5 Herbicides

Herbicides are yet another major group of commercial agrochemicals. It is true that a significant number of natural products show very interesting herbicidal properties, but relatively few of these originate from higher plant sources. There are several reasons why plants may be expected to be a poorer source of herbicides than of insecticides or fungicides. First, in the natural environment, selection pressure from other plants is often less important than the serious damage which may be caused by pathogens and insects. Consequently, the production of herbicidal compounds does not necessarily confer major benefits to the producer. A more convincing explanation is that organisms with different physiology and biochemistry from the producing organisms are an easier target for chemical defence. Any plant producing a herbicide must itself be immune to the effects of the compound, otherwise there is no evolutionary advantage. It is also worth noting that many compounds produced by plants are phytotoxins, rather than genuine herbicides. Phytotoxins may cause a range of effects, such as chlorosis or stunting, but if the plant subsequently recovers, the compound cannot be classified as a true herbicide and it is unlikely to be of commercial interest.

One of the best examples of a plant natural product with genuine herbicidal effects is α-terthienyl (**23**).[28] This compound is produced by *Tagetes* species, especially *Tagetes erecta* L., which is more commonly known as the African marigold. In addition to its herbicidal properties, the compound exhibits nematicidal effects. The mode of action of α-terthienyl is not fully understood, but there is considerable evidence that singlet oxygen is generated by α-terthienyl on irradiation with UV light.[29] This, in turn, causes damage in organisms via enzyme deactivation, nucleic acid oxidation, and membrane disruption. Several organizations have carried out synthetic chemistry based on α-terthienyl, but no commercial products have yet arisen from these programmes.

The final example cannot really be classified as a herbicide, but it has attracted significant interest as an alternative means of weed control. Strigol (**24**) is produced by the roots of cotton plants (*Gossypium hirsutum* L.) and is a germination stimulant of the pathogenic weed *Striga* sp. It has been suggested that strigol could be used to induce germination of *Striga* seeds at a time when the seedlings would fail to survive.[30]

6.I.6 Discussion

The examples given above represent only a small proportion of the plant natural products with known agrochemical properties. Considering that only 10% of plant species have been studied chemically, it is apparent that there is enormous scope for further work. Hopefully, the successes achieved to date will offer sufficient incentive to inspire future programmes of research within both academia and industry.

In principle, it is extremely easy to set up a screening programme based on higher plants. In reality, a considerable amount of knowledge, skill, and experimentation is essential to make the programme a success. Careful experimental design and thorough recording of procedures and data are crucial, since badly designed programmes will afford only weakly active compounds or show effects which cannot be reproduced at a later date.

One of the most common mistakes is failure to characterize fully the original plant material. If a plant sample is found to produce a compound with interesting agrochemical effects, then it is invariably necessary to repeat the fractionation process to obtain sufficient material for full biological testing. For this to be achievable, the following data should be recorded:

- Full identification of plant – scientific name and voucher specimen
- Location where collected – as precisely as possible
- Date collected, especially the season
- Part of plant which is of interest, *e.g.*, leaves, stem, bark, roots, fruit, flowers

In addition, it is common sense to avoid diseased or insect-infested tissue and to

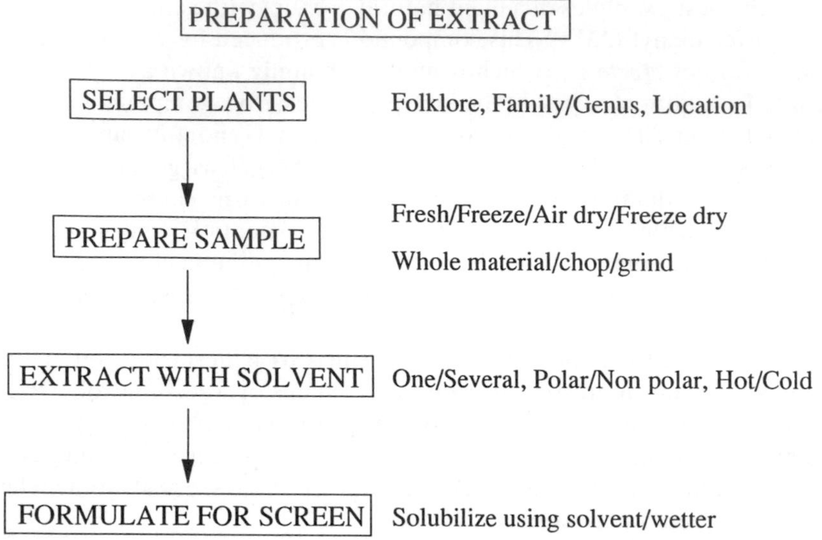

Figure 6.4 *Approaches to the preparation of plant extracts for biological screening*

ensure that the plant material collected has not been treated with an agrochemical, the residues of which could cause a positive screen response.

Once the plant has been collected, it needs to be processed carefully and finally converted into an extract which can be applied to an appropriate screen. There are many different approaches which can be taken (Figure 6.4), and there is no right or wrong method, but it is crucial to keep a record of all the steps involved. For example, a photolabile compound obtained from fresh plant material may not be present in a sample which has been dried in the sun.

Once suitable extracts have been prepared, they need to be submitted to an appropriate screening cascade (Figure 6.5). It is normal to carry out both *in vivo* and *in vitro* testing. The *in vivo* tests show whether there is genuine activity against target pests, while the *in vitro* tests may give information relating to a possible mode of action and offer a simple bioassay to support later fractionation. Selection of an appropriate application rate is crucial to the success of a programme. If the rate is too high, non-specific effects can give rise to positive results; for example, an extract containing a high sugar concentration causes spider mites to stick to the leaves of plants and to die from physical, rather than chemical, causes. High screen rates result in the detection of many weakly active compounds and, although this may be of some academic interest, it is not a profitable approach to seeking new agrochemical leads. All active extracts should be re-tested to confirm the biological effect and, at this stage, it may be appropriate to introduce additional tests or to reduce the screen rate. Confirmed actives are then nominated for full fractionation, hopefully leading to identification of the active ingredient(s). At this stage a detailed survey of the

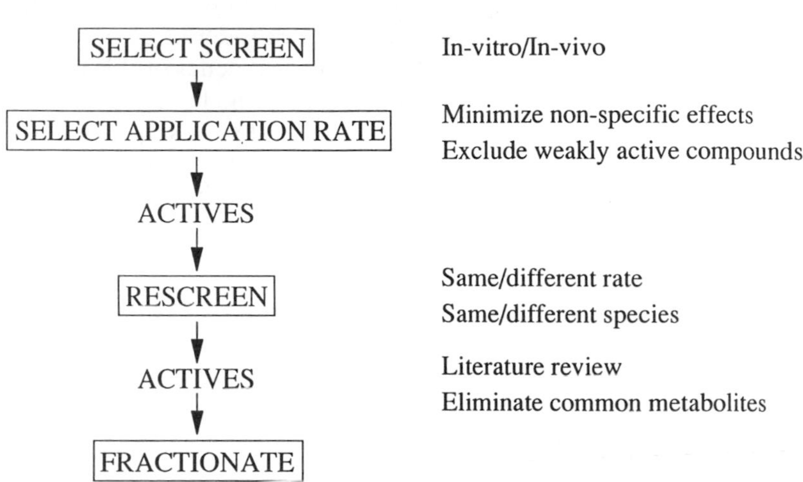

Figure 6.5 *Screening cascade for the selection of plant extracts with interesting biological effects*

literature to establish the known phytochemistry of the genus or family is appropriate, and it is also sensible to check for the presence of 'common metabolites'.

It is always useful to review recent literature on a regular basis to check for new reports of interesting biological effects. Serious difficulties with interpretation of the literature are caused by the nature of the bioassays. Some assays are too sensitive and most use species chosen for experimental convenience rather than relevance. Modern standard compounds are rarely used to calibrate the assays and most authors report results on *in vitro* assays only.

If a literature compound does appear to be of genuine interest, then it is advisable to try to repeat the work and to submit the sample to one's own screens prior to investing effort in a programme of analogue synthesis.

Natural product chemistry is a very difficult science and there are numerous examples of incorrect structural assignments in the literature. One example, experienced in our own laboratories, concerned an insecticide from *Artemisia monosperma* Delile. The paper[31] claimed a compound (**25**) with activity

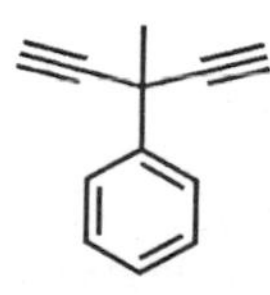

25

comparable to that of DDT (dichlorodiphenyltrichloroethane) on some insect species. Our initial approach was to try to obtain a sample of the plant and to isolate the compound for testing, but it proved extremely difficult to source the plant material. Total synthesis was therefore considered as an alternative option, but it is remarkably difficult to prepare acetylenes of this type. While investigating possible approaches to synthesis, the likely mechanism for biosynthesis of the compound was considered and this led us to question whether the assigned structure was correct. The spectroscopic data for the compound was reviewed carefully, together with literature for other *Artemisia* species, and it was concluded that the compound was almost certainly capillene (**26**). A sample of capillene was synthesized and screened, but its insecticidal effects were far too weak for any further interest.

26

In summary, plants offer an excellent source of biologically active natural products. It requires a considerable amount of time and effort to find new compounds which have the right characteristics for use as agrochemical leads, but the chances of success are greatly improved if appropriate care and consideration is given to the design of the plant screening programmes. There is little doubt that dedicated research will eventually be rewarded with exciting new lead structures for industrial application and it is to be hoped that a few of these compounds will be as successful as the pyrethroids.

6.I.7 Acknowledgements

The work relating to the insecticidal compound from *Artemisia monosperma* was carried out by M. J. Bushell and N. C. Sillars.

6.I.8 References

1. N. R. Farnsworth, in 'Bioactive Compounds From Plants', Ciba Foundation Symposium No 154, John Wiley & Sons, 1990, p. 2.
2. M. Matsui and I. Yamamoto, in 'Naturally Occurring Pesticides', ed. M. Jacobson and D.G. Crosby, Marcel Dekker Inc, New York, 1971, p. 153.
3. G. A. McLaughlin, in 'Pyrethrum, The Natural Insecticide', ed. J. E. Casida, Academic Press, New York & London, 1973, p. 3.
4. M. Elliott and N. F. Janes, *Chem. Soc. Rev.*, 1978, **7**, 473.
5. H. Fukami and M. Nakajima, in 'Naturally Occurring Pesticides', ed. M. Jacobson and D. G. Crosby, Marcel Dekker Inc, New York, 1971, p. 71.
6. I. Schmeltz, in 'Naturally Occurring Pesticides', ed. M. Jacobson and D. G. Crosby, Marcel Dekker Inc, New York, 1971, p. 99.

7. J. A. Benson, in 'Neurotox '91, Molecular Basis of Drug and Pesticide Action', ed. I. R. Duce, Elsevier Applied Science, London & New York, 1992, p. 57.
8. D. G. Crosby, in 'Naturally Occurring Pesticides', ed. M. Jacobson and D. G. Crosby, Marcel Dekker Inc, New York, 1971, p. 186.
9. Merck & Co., Inc., US Patent 2 400 295, 1946.
10. J. E. Casida, I. N. Pessah, J. Seifert, and A. L. Waterhouse, in 'Pesticide Science and Biotechnology', ed. R. Greenlaugh and T.R. Roberts, Blackwell Scientific, Oxford, 1987, p. 177.
11. A. Isogai, S. Murakoshi, A. Suzuki, and S. Tamura, *Agric. Biol. Chem.*, 1977, **41**, 1779.
12. A. Isogai, A. Suzuki, S. Tamura, Y. Ohashi, and Y. Sasada, *Acta Crystallogr. Sect B. Struct. Sci.*, 1977, **B33**, 623.
13. A. Gonzalez-Coloma, M. G. Hernandez, A. Perales, and B. M. Fraga, *J. Chem. Ecol.*, 1990, **16**, 2723.
14. A. Gonzalez-Coloma, R. Cabrera, P. Castanera, C. Guttierez, and B. M. Fraga, *Phytochemistry*, 1992, **31**, 1549.
15. L. Crombie, *J. Chem. Soc.*, 1955, 999.
16. M. Jacobson, in 'Naturally Occurring Pesticides', ed. M. Jacobson and D. G. Crosby, Marcel Dekker Inc., New York, 1971, p. 153.
17. M. Miyakado, I. Nakayama, H. Yoshioka, and N. Nakatani, *Agric. Biol. Chem.*, 1979, **43**, 1609.
18. S. V. Ley, in 'Recent Advances in the Chemistry of Insect Control II', ed. L. Crombie, Royal Society of Chemistry, 1990, p. 90.
19. K. Imai, *J. Pharm. Soc. Jpn*, 1956, **76**, 405.
20. B. W. Nash, D. A. Thomas, W. K. Warburton, and T. D. Williams, *J. Chem. Soc.*, 1965, 2983.
21. J. A. Bailey, G. A. Carter, R. S. Burden, and R. L. Wain, *Nature* (*London*), 1975, **255**, 328.
22. W. Oleszek, M. Jurzysta, and P. M. Gorski, in 'Allelopathy, Basic and Applied Aspects', ed. S. J. H. Rizvi, and V. Rizvi, Chapman and Hall, London, 1992, p. 151.
23. J. L. Ingham and K. L. Markham, *Phytochemistry*, 1980, **19**, 1203.
24. L. A. Mitscher, S. R. Gollapudi, D. C. Gerlach, S. D. Drake, E. A. Veliz, and J. A. Ward, *Phytochemistry*, 1988, **27**, 381.
25. W. Barz, W. Bless, G. Borger-Papendorf, W. Gunia, U. Mackenbrock, D. Meier, Ch. Otto, and E. Super, in 'Bioactive Compounds from Plants', Ciba Foundation Symposium No 154, John Wiley & Sons, 1990, p. 140.
26. C. J. Cavallito and J. H. Bailey, *J. Am. Chem. Soc.*, 1944, **66**, 1950.
27. L. D. Small, J. H. Bailey, and C. J. Cavallito, *J. Am. Chem. Soc.*, 1947, **69**, 1710.
28. J. D. H. Lambert, G. Campbell, J. T. Arnason, and W. Majak, *Can. J. Plant Sci.*, 1991, **71**, 215.
29. J. Bakker, F.J. Gommers, I. Nieuwenhuis, and H. Wynberg, *J. Biol. Chem.*, 1979, **254**, 1841.
30. A. I. Hsiao, A. D. Worsham, and D. E. Moreland, *Weed Sci.*, 1981, **29**, 101.
31. M. A. Saleh, *Phytochemistry* 1984, **23**, 2497.

Part II: Studies from China on Plants as Sources and Models of Insect Control Agents

ZHI-ZHEN SHANG

6.II.1 Introduction

Since the discovery of nicotine, rotenoids, and pyrethrums as insecticides, several thousand species of higher plants have been screened for insecticidal activity by many researchers. In many instances, the plants have a history of use as folk remedies and are still in local use by different societies throughout the world. Today over 2000 species of plant have been shown to possess some insecticidal activity.[1]

Large-scale investigations of insecticidal plants have received considerable attention in China in recent years, because of their abundance and the lack of detailed research on native species. The most common way in which chemical research is done in the area of plants is to synthesize analogues of already known lead compounds. An original lead structure can sometimes open the door to the development of a completely new class of active compounds. The success of synthetic pyrethroids, derived from natural pyrethrum, is a good example of this approach to produce better pesticides that enjoy expanding commercial use (see Chapter 7).[2]

The number of literature references concerned with the subject of natural products as insecticides exceeds 2000.[1,3] In this chapter, the interest in plants as sources and models for insecticides is reviewed, with emphasis given to papers from China. The aim is not only to introduce the progress in research, but also to attempt to discuss the future potential of naturally occurring insecticides for insect control, stimulate interest in the search for new insecticidal natural products, and suggest more satisfactory synthetic compounds modelled upon natural agents.

6.II.2 Current Status

The heavy use of broad spectrum, neurotoxic insecticides, including the 'big three' (chlorinated hydrocarbons, organophosphates, and carbamates), has led to the well-known serious problems of insect resistance, pest resurgence, and residues that affect health and environmental safety.

Many natural plant products are biorational compounds, with increased specificity for target organisms and safety for non-target organisms when compared to the neurotoxin action of conventional agents.[4] They are not persistent in the environment and the rare occurrence of insect resistance has received considerable attention that might be useful in pest control.

Copping[5] has discussed three reasons for looking for new natural products:

- It is possible that the new product can be used itself as the toxiphore.
- The natural product may represent a new class of compounds which, by either minor or major (in the case of synthetic pyrethroids) chemical modification, can lead to a commercially viable product.
- The mode of action of the natural product may be novel and this may lead to a search for new synthetic compounds with a similar mode of action.

Studies on botanical insecticides have been carried out in various universities and institutes in China, and this chapter focuses on the current discovery process for agricultural crop and vegetable protection.

6.II.3 Search for New Insecticides with Plant Origins in China

In China the higher plants have been used as insecticides for 2000 years. A famous Chinese book, 'The Quintessence of the Qi Dynasty' was written in 540AD. In 1639 the book 'Cyclopaedia of Agriculture Management' was produced and described in detail some of the applications of insecticidal plants. More recently the 'Handbook of Chinese Native Pesticides' was published in 1959; it included 220 species of insecticidal plants which were in use in some areas of the country.[6,7]

6.II.3.1 Studies on Meliaceous Plants

Azadirachta indica Juss. (neem) has long been used in India and Africa. Some entomologists now conclude that neem has such remarkable powers for controlling insects that it will usher in a new era of safe, natural pesticides. Neem seems likely to provide non-toxic and long-lived replacements for some of today's synthetic pesticides.[8]

Melia toosendan and *M. azedarach* L. are widely distributed in China and have been studied at the South China Agricultural University for a number of years. It was found that *M. toosendan* contained the triterpenoid toosendanin (**27**) as the main active agent; it exhibited marked stomach poison activity

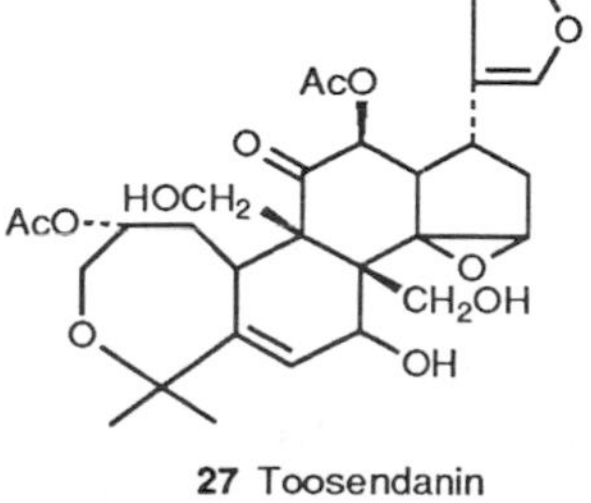

AcO AcO HOCH$_2$ O O O CH$_2$OH OH O

27 Toosendanin

against cabbage worms (*Pieris rapae*) and the rice brown hopper (*Nilaparvata lugens*), showing its potential as a strong antifeedant. During the past few years, analysis of the toosendanin content and its bioactivity against *P. rapae* has shown that there are distinct ecotypes in the different regions of China. However, the toxicity to insects does not correlate closely with the toosendanin content. It may be that other active principles, in addition to toosendanin, play an important role in its bioactivity. It is interesting to note that a series of laboratory and field experiments on the toxicology, formulation, and application of toosendanin, isolated from the bark of the chinaberry tree (*M. azedarach*), showed it to possess antifeedant and growth inhibition of *P. rapae*. The results also showed that the crude products were more effective than the pure sample against cabbage worms.[8,9]

The technology for extracting toosendanin from the bark of chinaberry trees has been studied. A product of '0.5% toosendanin EC' was launched as a new botanical insecticide and has been given a temporary registration in China. This new insecticide could control many species of vegetable and orchard pests, such as three-spotted plusia (*Plusia agnata*), cabbage armyworm (*Barathra brassicae*), turnip sawfly (*Athalia flacca*), potato lady beetle (*Henosepilachna vigintioctomaculata*), and oriental tobacco budworm (*Heliothis assulta*). The acute oral LD_{50} to mice was more than 10 000 mg kg^{-1} for crude technical powder and 3160 mg kg^{-1} for 0.5% toosendanin EC, thereby showing safety to warm-blooded animals. This demonstrated toosendanin as the first novel and potent compound to offer possibilities for commercial exploitation in China. Industrial production and large-scale application in the field may be worthy of further study.[9]

The mode of action of toosendanin is unknown. It seems to be a presynaptic blocking agent which acts on the neuromuscular system. The effect of toosendanin on the lateral and medial maxillary sensilla styloconica of the army worm larvae, using electrophysiological techniques has been reported. Its effect was completely inhibited by treatments in which an equal volume of 2 mg ml^{-1} toosendanin and 1 mM dithiothreital were used.[10]

6.II.3.2 Studies on Yellow Azalea (*Rhododendron molle*)

Since 1986 The Institute of Elemento-Organic Chemistry, Nankai University, has co-operated with The South China Agricultural University in studying the isolation, purification, and use of new bioassay methods for the identification of compounds from *Rhododendron molle*. The flowers and leaves of *R. molle* were collected from Northern Guangdong Province and the presence of bioactive components has been detected in this plant. Six species of insect were used to follow sequentially the various steps of the isolation procedures. The test insects included rice stem borer (*Chilo suppressalis*), brown plant hopper (*Nilaparvata lugens*), corn borer (*Ostrinia furnaculis*), cabbage worm (*Pieris rapae*), and cotton leaf worm (*Prodenia litura*).

It was found that residues from the petrol ether fraction of flowers and from the methanol extracts of leaves exhibited high mortality to rice stem borer and

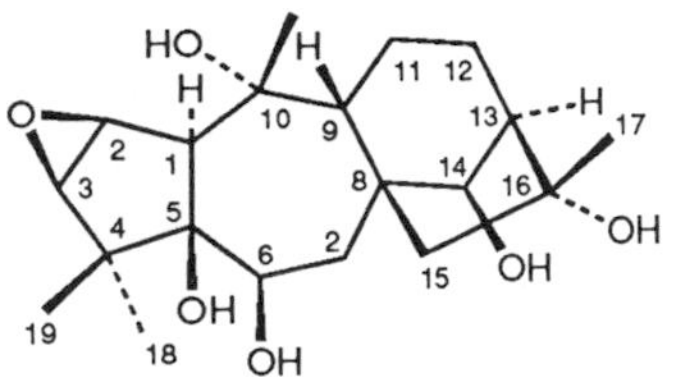

28 Rhodojaponin III

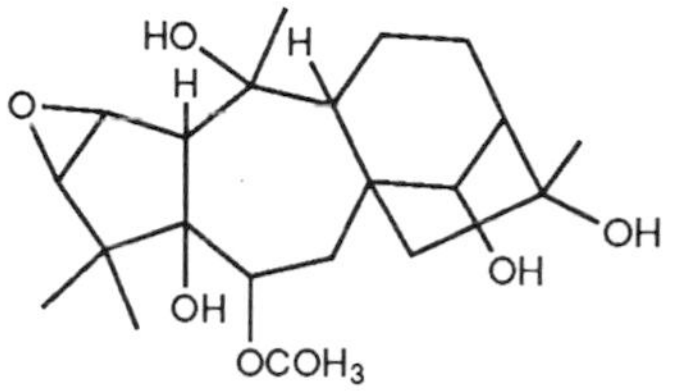

29 Rhodojaponin II

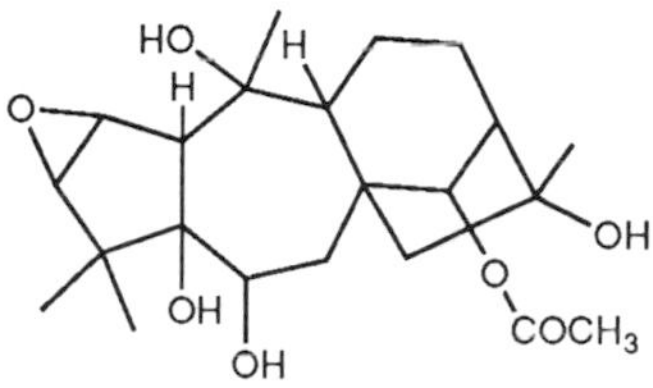

30 Rhodojaponin V

Figure 6.6 *Chemical structure of rhodojaponins*

reduced feeding of army worms. Also, three constituents of flowers and leaves were discovered (Figure 6.6), all chemically similar to the rhodojaponins.[12,13] Later, Klocke *et al.*[14] also found three constituents with different structures which were isolated from the same species of plant. The major compound is rhodojaponin III (**28**), which occurs at a concentration of 0.2% in dried flowers and which exhibits high levels of antifeedant, insect growth inhibition, and insecticidal properties against the larvae of *Spodoptera frugiperda* and *Leptinotarsa decemlineata.* Another two compounds discovered were grayanotoxin II (**31**) and kalmanol (**32**), which were found to be much less effective than **28**,

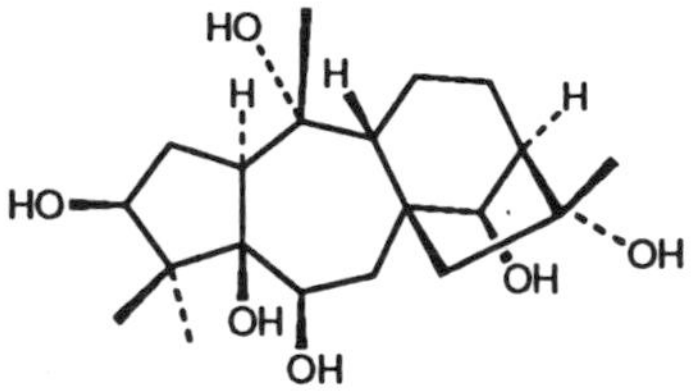

31 Grayanotoxin II

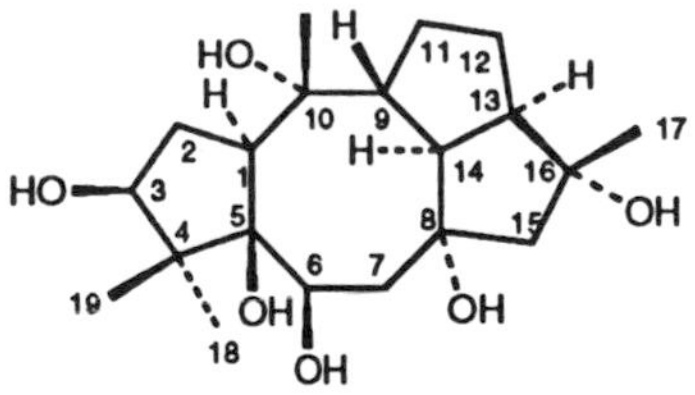

32 Kalmanol

although they are structurally very similar. Rhodojaponin III (**28**) was more active against *Chilo suppressalis*, with a strong contact insecticidal action. The antifeedant effects on *Pieris rapae, Prodenia litura,* and especially *Nilaparvata lugens* were very marked, but the symptoms rapidly changed to knockdown and excitation within a few minutes of topical application to sixth instar larvae of rice stem borer or fifth instar larvae of cabbage worm. Finally, the larvae showed convulsions, temporal paralysis, and death. The bodies of the larvae became black and comma-shaped – symptoms similar to the typical pattern seen with neurotoxins. The action mechanism of the blocking effect of rhodojaponin on neuromuscular junction transmission of fruit fly larvae (*Drosophila melanogaster*) has been studied. Results showed that all three rhodojaponins were found to inhibit and eventually block the neuromuscular excitatory junction potential of the abdominal longitudinal muscle cells of the third instar larvae. **28** was more active than **29** and **30**. Fractionation of *R. molle* should provide more such new compounds for practical exploitation.

6.II.3.3 Studies on *Tephrosia vogelli* and *Amorpha fruticosa*

Tephrosia vogelii grows well in Southern China. Leaves and twigs have been shown to possess strong antifeedant, stomach poison, and growth inhibition effects against cabbage worm. Preliminary field tests with acetone extracts of the plant showed promising results, particularly against diamond back moth (*Plutella xylostella*). Extracts of both *T. vogelii* and *A. fruticosa* were shown to contain compounds that were structurally related to rotenone **33** and its analogues (Figure 6.7).[14] Several groups of compounds have been isolated from various species of *Derris,* the insecticidal efficiency of which has a long history of use in China. The main structural units of all of these 'rotenoids' have been well-documented and several hundred hectares of *Derris* are now cultivated. Two species of plant can be used as a rich source of derris in China, *T. vogelii* in the south and *A. fruticosa* in the north.[15]

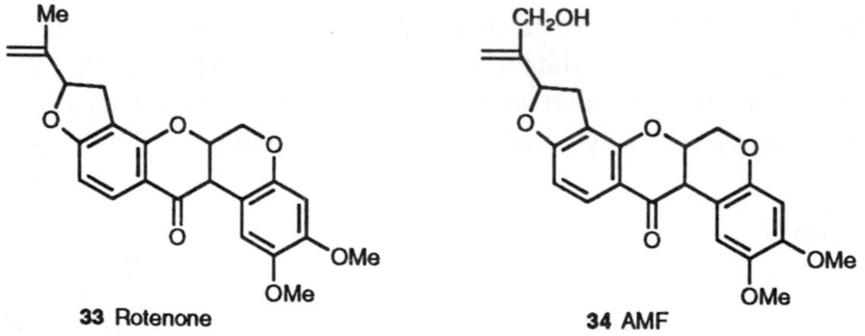

33 Rotenone

34 AMF

Figure 6.7 *Chemical structure of rotenone and its analogue AMF*

6.II.3.4 Studies on *Ajuga nipponensis*

Ajuga nipponensis (Labiatae) is a herbaceous plant distributed widely in Southern China. Preliminary investigations on the insecticidal activity of *A. nipponensis* have been reported by Liu Zhun *et al.*[16] The chloroform extract is very effective against four species of cabbage worm – *Pieris rapae, Plutella xylostella, Spodoptera litura,* and *Mythimna separata.* Courtois *et al.*[17] have isolated several oligosaccharides from *A. nipponensis*. In 1983, four new bitter principles were isolated and identified by Shimomura *et al.*[18] Liu *et al.* are studying the structure of the active chloroform extract from this plant.[19]

6.II.3.5 Studies on the Celastraceae

Plants of the family Celastraceae are moderately well-distributed in tropical and temperate regions of the world. There are about 30 *Celastrus* species widely distributed in central and north-western China. The root bark of Chinese bitter sweet has been used by farmers to control pests of vegetables for many years. Recently, a detailed study has been conducted at several institutes on the bioactive principles from the root barks of *Celastrus angulatus* and *C. laucophyllus*. A new sesquiterpene, celangulin (**35**), thought to be the most

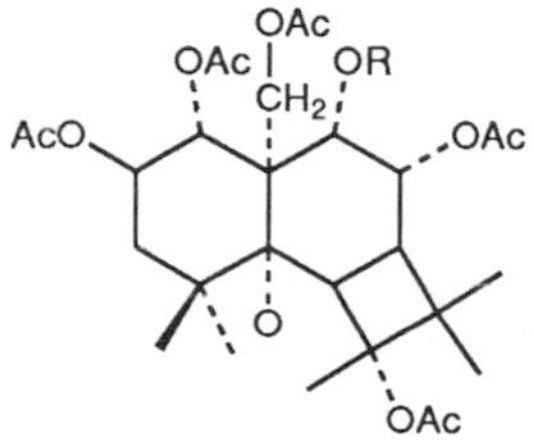

OAc OAc OR CH₂ AcO OAc O OAc

35 Celangulin

active compound from the root bark of *C. angulatus,* was recently isolated by Wakabayashi and co-workers.[20,21] Isolation, guided by fall armyworm (*Spodoptera frugiperda*) bioassay, showed that concentrations in the diet of 5 p.p.m. reduced the average body weight of larvae by 61%. The AF_{50} was 7.08 p.p.m., indicating very strong antifeedant and repellency effects.

Wu[22] established a use for Chinese bitter sweet by developing a 'KPT' dustable powder using root bark dried at 60°C and pulverized in a fodder grinder to pass a 40 mesh. Another formulation, the 'BS' emulsifiable, consists of 20% root bark extract mixed with ethyl acetate (5%) and benzene (65%). The result of field experiments showed that more than 10 species of insect were very well controlled. The acute oral LD_{50} to mice is 6000 mg kg^{-1} for root bark powder.

6.II.3.6 Studies on Other Botanical Insecticides

Extracts from *Plumbago indica* were found to possess growth-inhibiting properties against cabbage worm.[23]

Essential oils from *Ruta graveolens* L., *Feniculum vulgare, Cinnamomum cassia,* and *Clausena dunniana* Level. were tested, using a paper strip method, against adult *Tetranychus castaneum.* It was found that seven constituents of these essential oils are effective against stored product insects,[24] but the bioactive components from *R. graveolens* are still unknown.

The bioactivity of 475 types of potentially insecticidal plants collected from North-western China have been studied by Zhang Xing *et al.*[24] *Tribolium castaneum* was used as the test insect for the screening bioassay. Results have shown that 40 species of plants possess inhibiting activity to the population of

T. castaneum, suggesting that there are many more plant compounds which may be used as insect control agents.

6.II.4 Bioassay Methods for Screening and Searching for Bioactive Substances

When screening higher plants for natural pesticides, bioassay methods are most important. There is a real need for reliable, inexpensive, rapid, and reproducible general bioassays which can detect the broad spectrum of bioactivity present in higher plant extracts.

USDA laboratory assays have used 7-day old larvae of European corn borer (*Ostrinia nubialis*). In China, the Asiatic corn borer (*Ostrinia furnacalus*) was used as the test insect. This insect was chosen because of its economic importance to agriculture and the ease of mass rearing in the laboratory. However, large-scale extraction and fractionation of plants required when monitoring with the corn borer bioassay is laborious. Many laboratories used fruit fly or cabbage worms as the test insects to detect antifeedant activity. For example, in 'choice' bioassays, larvae were presented simultaneously with both treated and untreated leaf discs. In 'no choice' assays, larvae were presented with either treated or control discs. The minimum protective concentration at which 5% of treated leaf discs were consumed while >95% of untreated leaf discs were eaten (PC_{95}) was used to determine the effectiveness. Some laboratories used the brine shrimp (*Artenia salina*) as a bioassay tool.

After considering several possible methods, the electrophysiological technique was selected as a bioassay tool to detect activity and guide fractionation. Isolated heads of lepidopterous larvae were assayed using the tip-recording technique (Figure 6.8). Glass pipettes filled with stimulating chemicals of known concentration were employed as the stimulatory recording electrode. Action

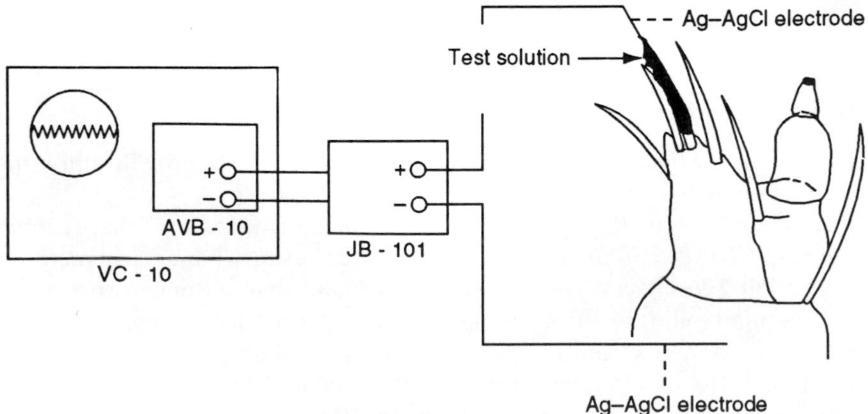

Figure 6.8 *Diagrammatic illustration of the experimental arrangement (tip-recording technique)*

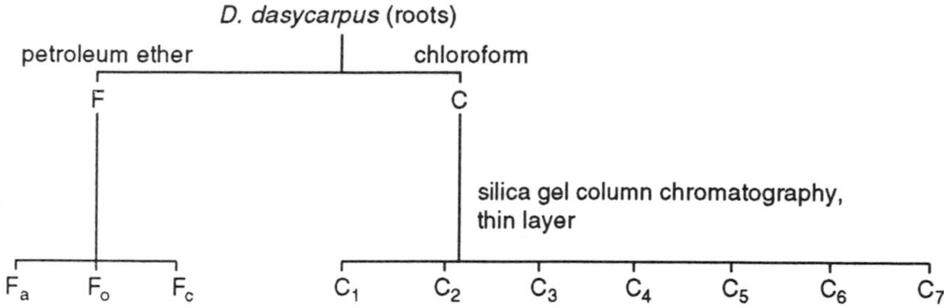

Figure 6.9 *Separation scheme for dividing the extract of* Dictamus dasycarpus

potentials evoked by the test solution were recorded from individual maxillary styloconic sensilla of larvae (army worm, cabbage worm, and corn borer). The electrophysiological responses to different compounds are well correlated with corresponding differences in feeding responses.[25,26] For example, nine active compounds were obtained through compound isolation from roots of *Dictamus dasycarpus* (Figure 6.8). The sugar receptor of *M. separata* larvae showed a considerable reduction in sensitivity when exposed to fractions C_2, C_5, and C_7.[27] This method has improved the bioassay for antifeedants, as it is very convenient, requires limited numbers of test insects, small quantities of chemicals, and the test can be run in a very short time. It contributes to our understanding of the relationship between structure and activity, assisting in the search for highly effective antifeedants with economic potential.

6.II.5 Plant Compounds with Significant Biological Activity for Potential Commercialization

After 44 years of research for new pesticides from several thousand plants from many families, Jacobson encourages expansion in this area.[28,29] He recommends that the most promising botanicals for use as pesticides, both now and in the future, are species from the families Meliaceae, Rutaceae, Asteraceae, Annonaceae, Labiatae, and Canellaceae. He has written excellent reviews on their properties and applicability.[30,31]

The crude alcohol extracts of neem seeds proved effective at very low concentrations against 60 species of insects. In 1985, USEPA, approved a commercial neem-based insecticide for certain non-food uses. Named 'Margosan-O', the product is available at present in the USA. As a result neem is seen as the leading candidate for providing a new generation of pesticides.[32] Although the molecular complexity of azadirachtin probably precludes its synthesis, the crude and partially purified extracts can be used for pest control.

The emulsion formulation of toosendanin 0.5% EC has been produced by the factory of the North-western Agricultural University in China, and both this product and Margosan-O have received registration by the Chinese Ministry of Agriculture. Studies of toosendanin have also increased greatly our understanding of bioactivity, including feeding deterrence, growth disruption, repellency, and toxic properties to many species and stages of insects. The Meliaceae have raised the possibility that crude extracts may eventually be used commercially and seem to have a bright future.

Traditionally, the powdered root bark of Chinese bitter sweet (*Celastrus angulatus*) has been used in China to protect plants from insect damage. Sensitive insect pests include cucumber beetle (*Diabrotica* spp.), cabbage worm (*Pieris rapae*), *Atbalis flacca, Plagiodera versicolora,* and *Calaphellus bouringi.* Insect antifeedant alkaloids, such as wilforine from *Maytenus vigida,* have been isolated.

Plants are the richest source of organic chemicals on Earth, and offer many novel chemical structures. Emphasis will be on studies of the use of botanicals in Integrated Pest Management (IPM).

Sometimes, in the search for the so-called third and fourth generation insecticides, considerable attention has been directed to the research and applications of Insect Growth Regulators (IGR). In a broad sense, Chiu[33] suggests the inclusion of juvenile hormones, antijuvenile (precocenes) compounds and their analogues, and phytoecdysones, in addition to diflubenzuron (Dimilin) and buprofezin (Applaud), as both these synthetic and natural chemicals have specific modes of action.

The goal of biorational approaches is to exploit physiological and biochemical processes, so the information obtained from these basic studies is very important. Attention has been focused on molecular biological studies of IGRs. Plant-derived compounds that regulate insect growth are slow-acting substances that specifically affect the growth and development of insects. This is also true of compounds of insect origin. For example, Mitsui and Matsumoto[34] reported that the non-steroidal ecdysone agonist, RH-5849, was first found during physiological research on the ecdysone sensitive *Drosophila* Kc cell line. The new pyridazinone compounds, NC-170 and NC-184, mimic juvenile hormone activity by inhibiting metamorphosis and suppressing hatchability in the plant and in the leaf hoppers. Identification of functionally critical neuropeptide hormones may provide models for designing highly selective insect control agents.

When we consider the hundreds of thousands of known species of higher plants, the few thousand already examined appear truly insignificant; but the almost untouched plant resources remain as a basis for future research. Few plant compounds have thus far been developed as commercial products. New insect pest species continue to arise and new biologically active compounds continue to be discovered. The past decade, with its revolution in analytical instrumentation, has taught us more about the chemical structures and properties of natural products than has been known throughout history. Natural agents offer a continual source of inspiration and a continual challenge.

Different plant compounds affect different insect targets. These include the nerve axons and synapses (pyrethrums, nicotine), respiration (rotenone), hormonal balance (juvenile hormones, antijuvenile hormones, moulting hormones), and behaviour (antifeedants, attractants, repellents). Some of these compounds have already been exploited in commercial insect control. Others may be the components of host–plant resistance mechanisms, through the possible use of biologically active natural plant compounds in host resistance by the transfer of gene complexes associated with the biosynthesis of active compounds to plants of economic importance.

It is important to stress that for these procedures to work all scientists within the discovery team must communicate and work together. It is also important to recognize that industrial research scientists will need to co-operate more closely with academic centres to generate the basic knowledge in a more efficient discovery process and to provide the industry with the means to combat the challenges that it faces. Also, international co-operation and co-ordination is very important in the implementation of botanical pest control projects. This is an area of continuing technological advance comparable to other branches of modern industry.

6.II.6 References

1. S. Ahmed, M. Grainge, J. W. Hylin, W. C. Mitchel, and J. A. Litsinge, in 'Natural Pesticides from the Neem Tree and other Tropical Plants', ed. H. Schmutterer and K. R. S. Ascher, GTZ, Eschborn, 1984, 565.
2. U. Eder and H. C. von Keyserlingk, 'Approaches to New Leads for Insecticides', Springer-Verlag, Berlin, Heidelberg, 1985.
3. J. A. Klocke, Plant compounds as sources and models of insect control agents, *Econ. Medic. Plant Res.*, 1989, **3**, 103.
4. Chiu Shin-Foon, Recent advances in research in botanical insecticides from China, in 'Insecticides of Plant Origin', ed. J. T. Arnason, ACS Symposium 387, American Chemical Society, Washington, DC, 1989, 69.
5. L. G. Copping, Aspects of pesticide discovery, in 'Proceedings of International Seminar on Pesticides in the Third World', ed. K. Holly, L. G. Copping and G. T. Brooks, United Nations Industrial Development Organization, Vienna, 1990, p. 16.
6. Chiu Shin-Foon, Recent research findings on meliaceae and other promising botanical insecticides in China, *J. Plant Dis. Prot.* 1985, **92**(3), 310.
7. Shang Er-Cae, 'The Application of Natural Insecticides in China'. 6th Japan–China Symposium on Pesticides, 1992, Fukuoka, Japan.
8. F. R. Ruskin, 'Neem – a Tree for Solving Global Problems', Report of an *ad hoc* Panel of the Board of Science and Technology for International Development, National Research Council, National Academy Press, Washington, DC, 1988.
9. Zhang Xing and X. L. Wang, 'Studies on the bioactivities and applications of Chinese botanical insecticide – toosendanin', Proceedings of the XIX International Congress of Entomology', Beijing, 1992.
10. Lao Lin-Er, C. Y. Liao, and P. A. Zhou, Electrophysiological studies of the antifeedant action of toosendanin to the army worm larvae, *Acta Entomol. Sinica,* 1989, **32**(3), 257.
11. Shang Zhi-Zhen, R. Y. Chen, Lio Zhun, Q. L. Zhang, S. F. Chiu, and Fen Xia, 'Preliminary Investigations of Rhodojaponins as Insecticides: Mode of Action and Effects on Rice Pests', Report for the Workshop of Botanical Pest Control in Rice based on Cropping Systems, IRRI, Philippines, 1988.
12. Shang Zhi-Zhen, S. F. Chiu, X. Feng, R. Y. Chen, Z. Liu, and Q. L. A. Zhang, 'A preliminary study on the Mode of Action of Rhodojaponins from *Rhododendron molle* as an Insecticide', IUPAC/China Pesticide Chemistry Workshop, 1988, 67.

13. Shang Zhi-Zhen, Q. L. Zhang, Z. Liu, R. Y. Chen, S. F. Chiu, and X. Feng, Studies on the bioactivities of rhodojaponins, *Ecochemicals*, 1990, **2**, 6.
14. J. Klocke, Mei-Ying, Shin-Foon Chiu, and Isao Kubo, Grayanoid diterpene insect antifeedants and insecticides from *Rhododendron molle, Phytochemistry*, 1991, **30**(6), 1797.
15. Chiu Shin-Foon, Research on insecticides from plants in China, 'Proceedings of XIX International Congress of Entomology', 1992, Beijing.
16. Liu Zhun, Shang Zhi-Zhen, Z. G. Li, R. Y. Chen, Y. G. Zhang, Y. T. Qui, and S. F. Chiu, Preliminary investigations on insecticidal activity of *Ajuga nipponensis, Prog. Nat. Sci.*, 1992, **2**(3), 251.
17. J. E. Courtois, G. Dillemann, and P. L. Dizet, *Ann. Pharm. France*, 1960, **18**, 17.
18. H. Shimomura, Y. Sashida, and K. Ogawa, *Tetrahedron Lett.* 1981, **22**, 1367.
19. Liu Zhun, Li Zhong-Qin, and Chen Ru-Yu, *Prog. Nat. Sci.*, in press.
20. N. Wakabayashi, W. J. Wu, and R. M. Waters, Calengulin: a nonalkaloidal insect antifeedant from Chinese Bittersweet, *Celastrus angulatus, J. Nat. Prod.*, 1988, **51**(3), 537.
21. J. Liu, 'Investigations on the Bioactive Principles of Celastraceae', PhD Thesis, China, 1988.
22. Wu Wen-Jun, Further studies on the insecticidal plant, Chinese bittersweet, *Celastrus angulatus*, 'Proceedings of the International Seminars, Shenyang, China', UNIDO, Vienna, 1990.
23. I. Kubo, J. A. Klocke, T. Matsumoto, and T. Kemikawa, Plumbagin as a model for insect ecdysis inhibitory activity, 'Advances in Pesticide Science', Pergamon Press, Oxford, 1982.
24. Zhang Xing, Wang Xing-Lin, and Hu Zhoa-Nong, Screening studies on insecticidal plants inhibiting population formation of *Tribolium castaneum* Herbst., *Grain Store*, 1992, **21**(3), 3.
25. W. M. Blane, M. S. J. Simonds, S. V. Ley, and R. B. Katz, An electrophysiological and behavioural study of insect antifeedant properties of natural and synthetic drimane-related compounds, *Physiol. Entomol.*, 1987, **12**, 281.
26. W. M. Blancey, Electrophysiological responses of the terminal sensilla on the maxillary palps of *Locusta migratoria* to some electrolytes and nonelectrolytes, *J. Exp. Biol.*, 1974, **60**, 275.
27. Shang Zhi-Zhen, W. Z. Zhoa, Y. S. Zhu, and Q. Li, The use of electrophysiological techniques to explore antifeedants in plants, *Prog. Nat. Sci.*, 1992, **3**, 241.
28. M. Jacobson, 'Insecticides of the Future', Marcel Dekker, New York, 1975.
29. M. Jacobson, Insecticides from plants. A review of the literature 1954–1971, 'USDA Agricultural Handbook', 1975, p. 461.
30. M. Jacobson and D. C. Crosby, 'Naturally Occurring Insecticides', Marcel Dekker, New York, 1971.
31. M. Jacobson, Botanical pesticides, past, present and future, in 'Insecticides of Plant Origin', ed. J. T. Arnason, B. J. R. Philogene, and P. Morand, American Chemical Society, Washington, DC, 1989.
32. Anon, 'Neem: A tree for Solving Global Problems', National Research Council, National Academy Press, Washington, DC, 1992.
33. Chiu Shin-Foon, Studies on plants as a source of insect growth regulators for crop protection, *J. Appl. Entom.*, 1989, **107**, 185.
34. T. Mitsui and S. Matsumoto, New insect growth regulators with insect hormone activities, 'The 6th China–Japan Symposium on Pesticides', Fukuoka, Japan, 1992.

Part III: Pesticides from Nature: Present and Future Perspectives

G. A. MIANA, ATTA-UR-RAHMAN, M. IQBAL CHOUDHARY, G. JILANI, and HAFSA BIBI

6.III.1 Introduction

The observation that leaves and other parts of certain plants are not attacked by pests (insects, bacteria, fungi, *etc.*) led to the belief that these plants may have a well-evolved defence system. During their early evolution many plant species did indeed develop highly sophisticated defence mechanisms, largely based on a complex array of chemicals produced by the plants as secondary metabolites, against insects, mites, pathogenic microbes, and even weeds. These observations formed the foundation of pesticide chemistry from natural sources.

A large number of chemical compounds possessing pesticidal activities have been isolated from a variety of natural sources. Some of these natural products and their synthetic templates are already in use on a variety of crops in different countries. Terrestrial plants are not the only source of these fascinating constituents, as more recently several pesticidal compounds have also been isolated from microbial, marine, and even insects sources.[1] Unfortunately, most of these chemicals are obtained in very limited yields so the necessary extensive testing, field trials, and residue studies often cannot be carried out. Synthetic chemists are generally not attracted to invest their time and funds in the total synthesis of natural pesticides, largely due to the lack of either extensive bioactivity data and/or economic incentives. Presently, there is a considerable interest in the use of biodegradable agrochemicals, particularly pesticides from natural sources, since they are usually friendly and safe to the environment, and may not disturb the delicate eco- and bio-cycles.

Natural products with pesticidal properties can be classified broadly into the following four classes, based on their sources:

- Botanical pesticides (Parts 6.I and 6.II, and Chapter 7).
- Pesticides from micro-organisms (see Chapters 1, 2, and 4).
- Pesticides from insects (see Chapter 9).
- Marine natural products with pesticidal properties (see Chapter 5).

6.III.2 Botanical Pesticides

The most promising botanical pesticides for use at present, and probably in the future, are substances derived from species of the families Meliaceae, Rutaceae, Asteraceae, Annonaceae, Labiatae, and Canellaceae. The single most important botanical source of pesticidal compounds is *Azadirachta indica* Juss. (neem), which is a member of the plant family Meliaceae. Several pure constituents, as well as the crude extracts, of this wonder tree have been shown to possess feeding deterrency, repellency, toxicity, and growth disruptive activities.[2] Neem has also been shown to be safe and abundant, and has a very rich traditional use. Azadirachtin (**36**), a tetranortriterpenoid isolated from the neem tree, is found to be effective as a feeding deterrent, repellent, toxicant, sterilant, and growth disruptant for insects at dosages as low as 0.1 p.p.m.[3] Margosan-O, a neem-based patent formulation, is currently used on non-food crops and in nurseries in the USA and Europe, as approved by EPA (see part 6.II).[4] A neem-based general pesticidal spray for all type of crops is expected to be marketed in the near future.

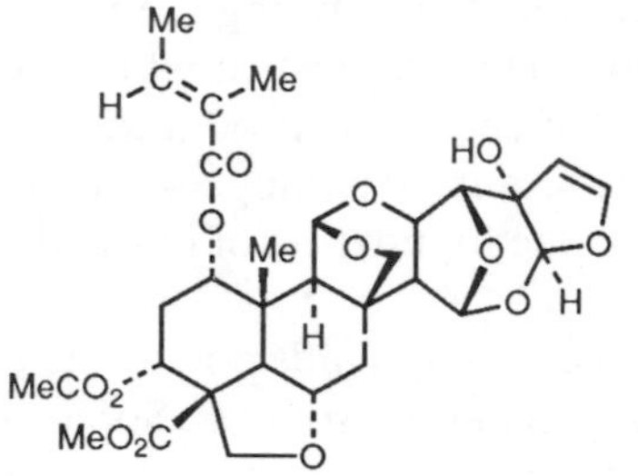

36 Azadirachtin

Limonoids and triterpenoids isolated from various species of *Melia* (Meliaceae) have antifeedant and other related activities. Volkensin (**37**) and salannin (**38**) isolated from *Melia volkensii*, have been reported to be deterrent to larvae of fall army worm. A series of limonoids extracted from *Trichilia* species (Meliaceae) are known to be antifeedants for larvae of *Spodoptera eridania* (Southern army worm) and adult *Epilachna variestis* (Mexican bean beetle).[5–7] The limonoid sendanin (**39**) isolated from fruits of *Trichilia roka* (Meliaceae), is a potent growth inhibitor for *Heliothis virescens, Spodoptera frugiperda,* and *Heliothis zea.*[5] The crude ethanol extract of *Cedrela odorata* L. (Spanish cedar) (Meliaceae) is known to prevent feeding by adult striped cucumber beetles (*Acalymma vittatum*), while extract of *Toona* species (Meliaceae) are also effective antifeedants.[8–11] Toonacilin (**40**) and 6-acetoxytoonacilin (**41**), isolated from the leaves of *Toona ciliata* M. J. Roem,, have shown strong antifeeding and insecticidal activity against *Hypsipyla grandella* Zeller (Lep., Pyralidae) and the Mexican bean beetle.[12]

The plants of the family Rutaceae are also reputed to have pesticidal constituents. Nomilin (**42**), a limonoid isolated from (Rutaceae), is almost as active as azadirachtin (**36**). Limonin (**43**), another limonoid isolated from several citrus species of the family Rutaceae, is ten-fold less active than nomilin as a feeding deterrent for *Heliothis zea.*[13] Limonoids from *Citrus paradisi* Macfad prevent feeding by *Spodoptera litura,* while nomilin prevents feeding

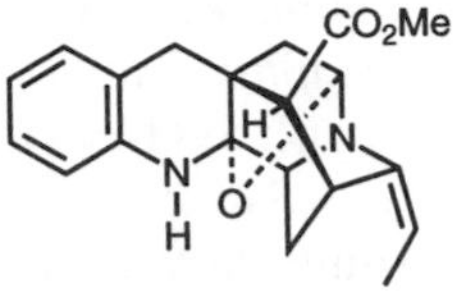

37 Volkensin

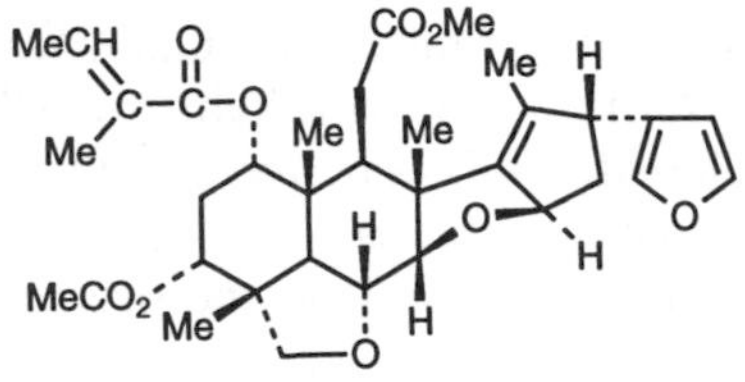

38 Salannin

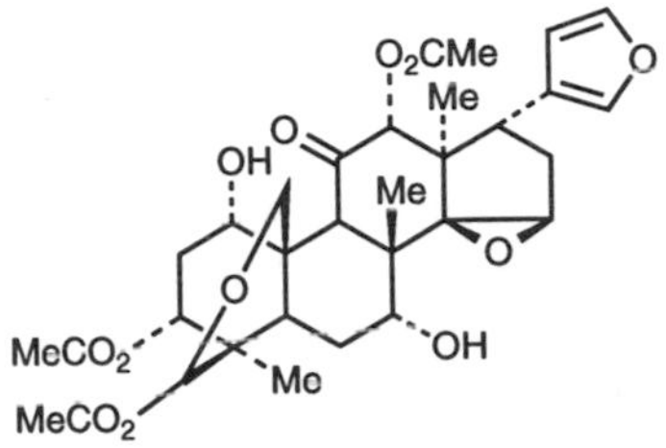

39 Sendanin

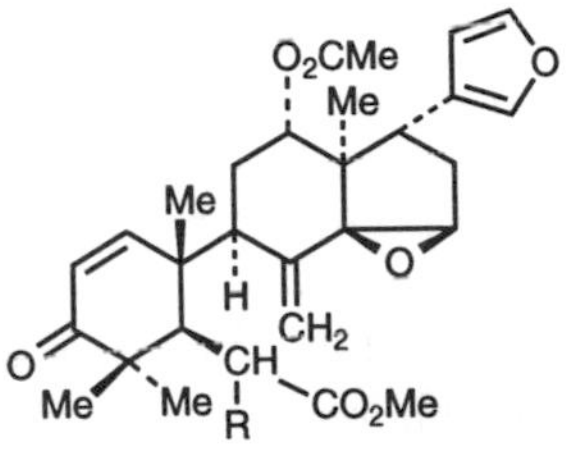

40 R = H Toonacilin
41 R = OAc 6-Acetoxytoonacilin

by *S. frugiperda* and *Trichoplusia ni.*[14] Citrus oil has also proved to be toxic and deterrent to several species of storage pests, such as *Callosobruchus maculatus* (cowpea weevil) and *Sitophilis oryzae* (rice weevil).[15–18] Limonin is an effective antifeedant for *Leptinotarsa decemlineata* (Colorado potato beetle),[18] but not against *Spodoptera exempta* (beet army worm), *Maruca testularis* (bean pod-borer), or *Eldena saccharine.*[19]

Zanthophylline (**44**), an alkaloid isolated from stems and branches of *Zanthoxyllum monophyllum* Lam. (Rutaceae), is a feeding deterrent for

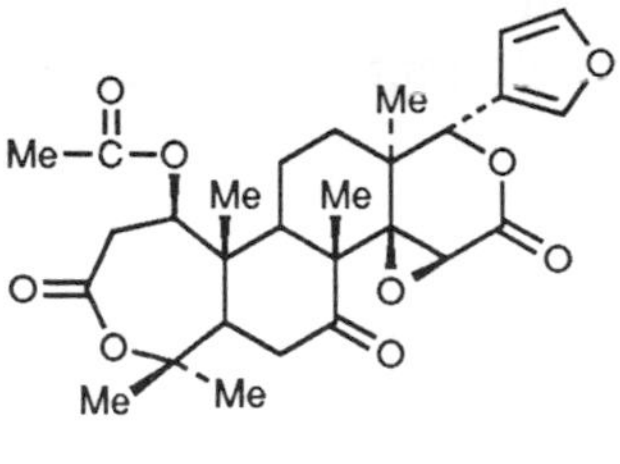

42 Nomilin

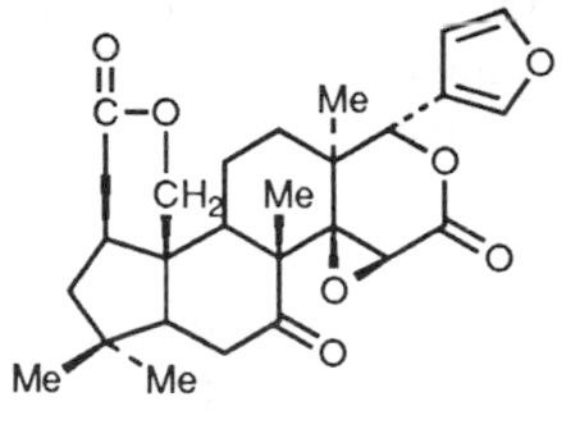

43 Limonin

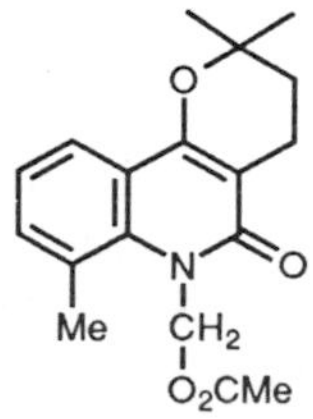

44 Zanthophylline

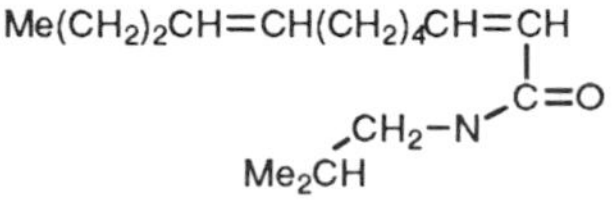

$Me(CH_2)_2CH{=}CH(CH_2)_4CH{=}CH{-}C({=}O){-}N{-}CH_2{-}CHMe_2$

45 Herculin

Hemileuca oliviae (range caterpillar), *Melanopus sanguinipes* (migratory grasshopper), *Hypera postica* (alfalfa weevil), and *Schizaphis graminum* (green bug).[20] Herculin (**45**), a pungent isobutylamide, was isolated from the bark of southern prickly ash *Zanthoxyllum clavaherculis* L. (Rutaceae), and it has been found to be as toxic as the pyrethrins to mosquito larvae, ticks, *Musca domestica* (house fly), and as ovicidal to *Pediculus humans* (body louse).[21] Several benz-(*C*)-phenanthridine alkaloids obtained from the root bark of *Faraga* species (Rutaceae) deterred feeding by beet army worm and the Mexican bean beetle.[22] Indoloquinazoline alkaloids of the fruits of *Evodia rutaecarpa* Hook. F and Thoms (Rutaceae) inhibit the growth of *Bombyx mori* (silk worm) larvae.[23]

The plants of genus *Annona* (Annonaceae) are found to contain a class of chemical compounds, the 'acetogenins', which are cytotoxic and pesticidal. Ethereal extracts of the twigs of *Annona senegalensis* are highly toxic to the large milkweed bug,[24] while the ethanol extracts of the fruit of the custard apple, *Annona reticulata* L., *A. globa,* and *A. purpurea* exhibit a severe juvenilizing effect on the striped cucumber beetle. Several alkaloids of *Annona* species, such as liriodenine (**46**), have also been found to be active,[25]

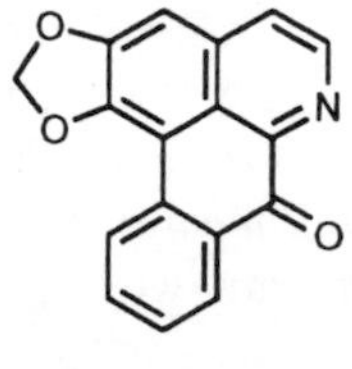

46 Liriodenine

Sesquiterpene lactones, isolated from a number of species of the family Asteraceae, have shown excellent feeding deterrent activity against pest insects. Alantolactone (**47**), a sesquiterpene lactone from *Inula helenium* L., significantly reduces feeding and survival of *Tribolium confusum* (confused flour beetle)[26], while isoalantolactone deters feeding by *Sitophilus granarium* (granary weevil), confused flour beetle, and *Trogoderma granarium* (khapra beetle).[27] Schkuhrin I (**48**) and schkuhrin II (**49**), germacranolides from *Schkuhria pinnata* Thell. (Asteraceae), exhibit antifeedant activity against the beet army worm and the Mexican bean beetle.[28]

Roots of American corn flower *Echinaceae angustifolia* DC. contain components toxic to *Aedes* mosquitoes.[24,29] Precocene II (**51**), a 6,7-dimethoxy-2,2-dimethylchromene isolated from the bedding plant *Ageratum houstonianum* Thill (Asteraceae), has extremely high juvenilizing activity on several species of insects pests.[30] The *Echinaceae* species (Asteraceae) also contain chemical compounds toxic to *Aedes* mosquito larvae and house flies, and they are growth disruptive in the development of yellow meal worm.[31,32]

Plants of the genera *Warburgia* and *Polygonum* (Canellaceae) have shown to be powerful feeding deterrents against larvae of the beet army worm.[33] A series

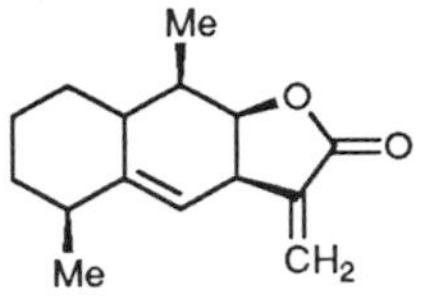

47 Alantolactone

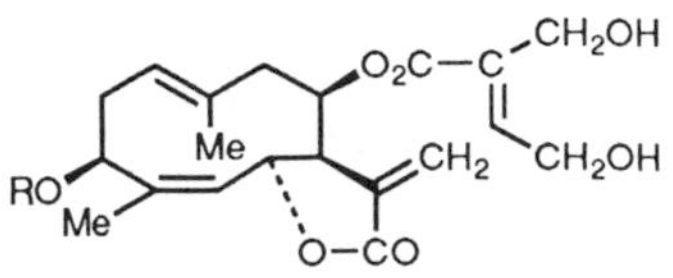

48 R = Ac Schkuhrin-I
49 R = COCH(OH)CHMe$_2$ Schkuhrin-II

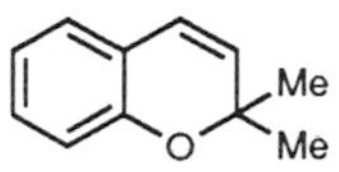

50 Precocene

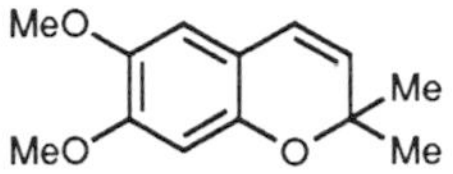

51 6,7-Dimethoxy-2,2-dimethylchromene

of sesquiterpenoid dialdehydes have been isolated from these plants.[34–36] Warburganal (**52**), a compound isolated from the *Warburgia* species, also possesses antiyeast and antifungal activities.[37]

Gossypol (**53**) and other related terpenoids, condensed tannins, and certain monomeric flavonoids isolated from *Gossypium hirsutum* L. (cotton plant) (Malvaceae) are responsible for the resistance of cotton plant to the bollworm *Spodoptera littoralis*. Synthetic gossypol can, therefore, be used as a spray on cotton crops for protection from bollworm attack.[38] Many plants of the family Labiatae and their pure constituents can be used as pesticides. The leaf oil of sweet basil, *Ocimun basilicum* L. (Labiatae), contains clerodanes, known as juvocimene I and juvocimene II, which have a juvenilizing effect on the milkweed bug.[39,40] The oil is also very effective against *Culex* mosquitoes.[41]

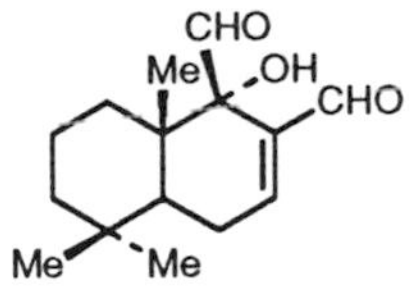

52 Warburganal

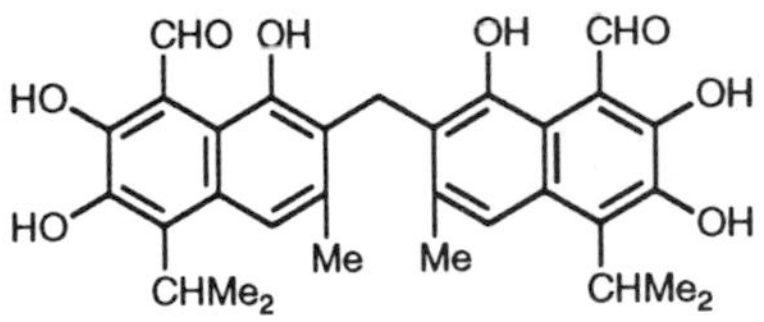

53 Gossypol

More recently, a number of clerodane diterpenoids and phytoecdysteroids with potential insect antifeedant and modulating hormone activities, respectively, have been isolated from *Ajuga* plants (Labiatae).[42] Crude extracts of *Ajuga* plants possess feeding deterrent activity on the Mexican bean beetle and are used in folk medicine. These extracts also prevent feeding by the bollworm and by *Pieris brassicae*,[43] and cause juvenilization of the beetworm.[44] Ajugarin I (**54**), 6-deacetylajugarin (ajugarin II), and ajugarin III (**55**) are believed to be the compounds responsible for the biological activity.[45,46] These compounds were checked by using the host-plant leaf disk method on *Zea mays* (maize).

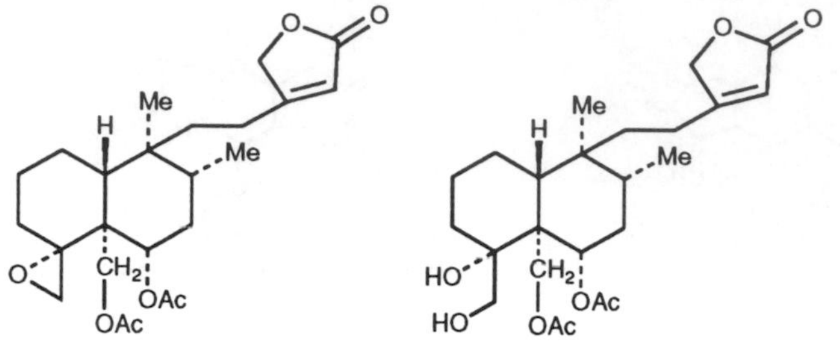

54 Ajugarin I

55 Ajugarin III

Activity levels of 100 p.p.m. against *Spodoptera exempta* and 300 p.p.m. against *S. littoralis* were found.

Many representatives of the genus *Ajuga* contain phytoecdysteroids and polyhydroxysteroids with a 5β-H-7-ene-6-one system, exhibiting well-established physiological activities in insects.[42] Members of the higher plant genus *Artemisia* (Compositae) produce a series of terpenoids, many of which are biologically active as fungicides, herbicides, antimicrobials, insecticides, and insect antifeedents. The antimalarial sesquiterpenoid artemisinin (**56**) from *Artemisia annua* L., is also a potent phytotoxin (see Chapter 4).[47]

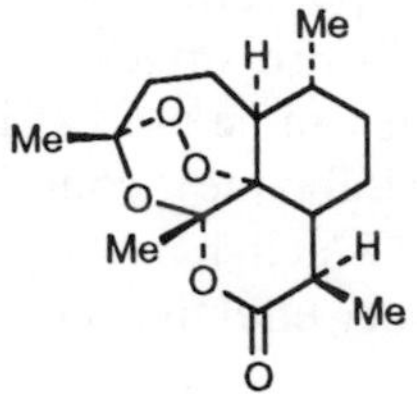

56 Artemisinin

6.III.3 Pesticides from Micro-organisms

The pesticidal compounds from microbial sources can be broadly classified as:

- Compounds with direct activity
- Compounds with indirect activity

6.III.3.1 Direct Activity

A number of micro-organisms produce herbicidal, insecticidal, and nematicidal metabolites. Avermectins, a group of eight macrocyclic lactone metabolites from *Streptomyces avermitilis,* are active against certain nematodes, arthropods and other plant parasitic mites at extremely low doses, but have relatively

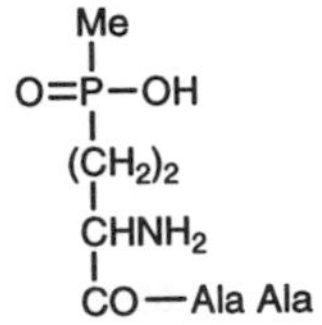

57 Bilanafos

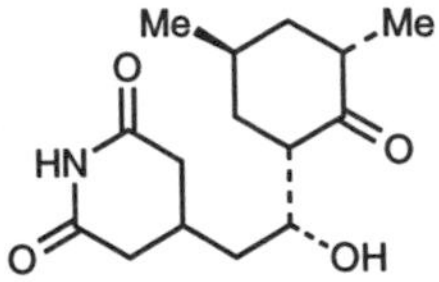

58 Cycloheximide

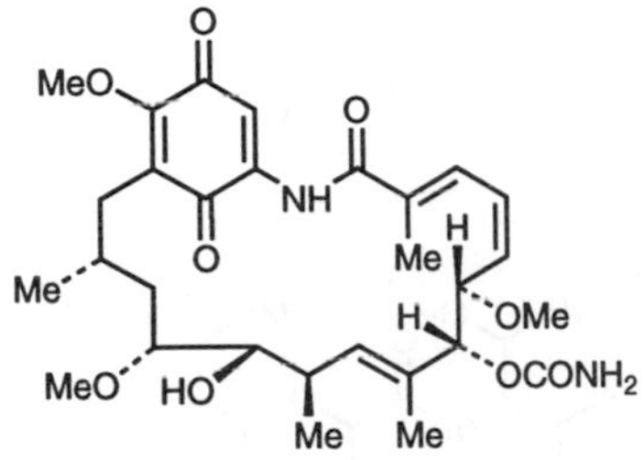

59 Geldanamycin

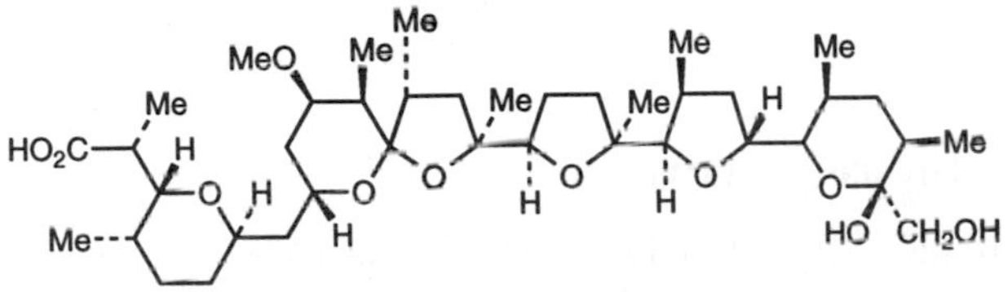

60 Nigericin

low mammalian toxicity.[34,48–50] Avermectin B_1 is currently marketed as 'Avid' and 'Vertimec' for use against plant parasitic mites and insects in agriculture.

Bilanafos (**57**), another microbial metabolite of *S. hygroscopicus* and *S. viridochromogenes,* has very strong post-emergence herbicidal activity against many plants. Actinomycetes, a group of soil micro-organisms, are known to be a source of many bioactive compounds. Cycloheximide (**58**), from a phytotoxic *Streptomyces* isolate, is a potent inhibitor of seed germination and seedling growth, and has been considered for use as a herbicide in Japan.[51] Geldaramycin (**59**) and nigericin (**60**), isolated from a strain of *S. hygroscopicus,* have strong herbicidal activity.

Three phytotoxins, epiepoformin (**61**), 3-hydroxybenzyl alcohol (**62**) and 2-methylhydroquinone (**63**) have been isolated from *Scopulariopsis brumptii* and other fungi. Compounds (**61**) and (**62**) are phytotoxic, while (**63**) has insecticidal and antibiotic activities.[50,52–56] *Streptomyces griseus* produces a strongly mosquitocidal culture broth and has yielded valinomycin (**64**). Valinomycin has been patented for insecticidal, nematicidal, and acaricidal use.[57]

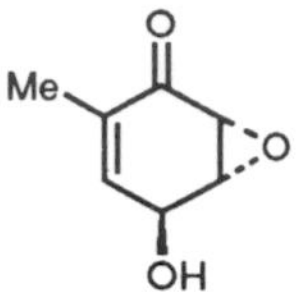

61 Epiepoformin

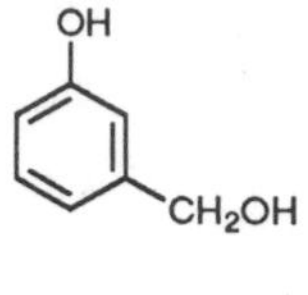

62 3-Hydroxybenzyl alcohol

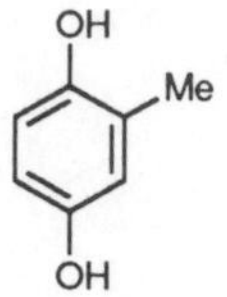

63 2-Methylhydroquinone

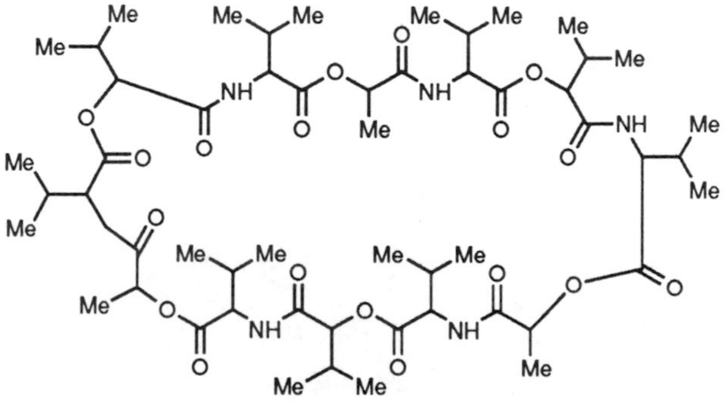

64 Valinomycin

6.III.3.2 Indirect Activity

Several plant pathogenic fungi produce chemicals (elicitors) which are capable of evoking the production and accumulation of phytoalexins (natural plant antibiotics) in plants (see Chapter 8).[58]

Phytoalexins are low molecular weight chemical compounds with antimicrobial activity. These phytoalexins are synthesized and accumulated by plants as a biological response to micro-organism attack. A very large number of phytoalexins from different plants have been characterized chemically. They include isoflavanoid-derived pterocarpan compounds from the Leguminosae, such as pisatin (**65**) from peas, sesquiterpenoid compounds from the Solanaceae family such as rishitin (**66**) from potato, phenanthrene compounds characteristic of the Compositae family such as orchinol (**67**) from *Orchis* spp., and acetylenic compounds from the Compositae, such as safynol (**68**) from safflower.[58] Microbial elicitors were first discovered by Cruickshank, who isolated a protein, monilicon A (M_r = 6 K), from *Monolinia fructicola*, which elicits the phytoalexin phaseolin in garden beans.[59] Another protein isolated recently from *Phytophthora parasitica* var. *nicotianae* is a potent elicitor even at a concentration of 20 μg cotyledon^{-1} (4×10^{-13} mol).[60]

Many of the fungal elicitors are carbohydrate in nature and contain recognizable information. Recognition of micro-organisms is being studied in several plant–micro-organism interactions. Another area of interest is how the elicitor signal is transduced to the plant cell nucleus for the directed processing of nucleic acids and proteins.

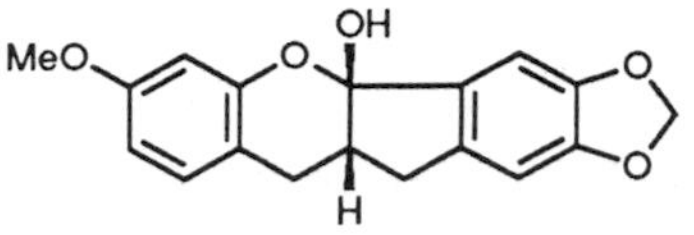

65 Pisatin

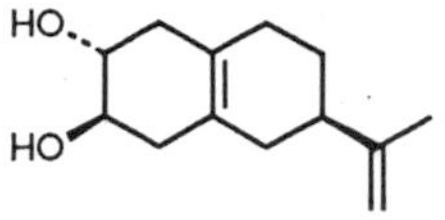

66 Rishitin

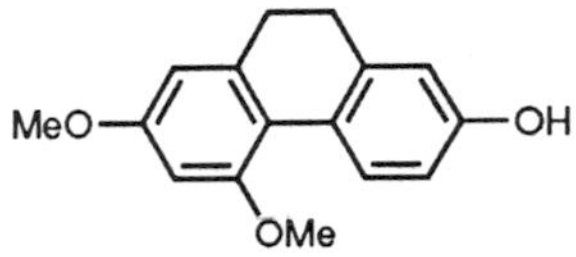

67 Orchinol

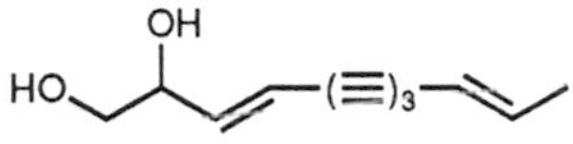

68 Safynol

A glycoprotein elicitor was isolated from *Phytophthora megasperma* f.sp. *glycinea* by Frank and Paxton.[61] It was found that *Colletotichum lindimuthianum,* the causal agent of bean anthracnose, produced a compound in culture which was active at 10 μg ml^{-1}.[62] This fungus also produces a mannose-rich polysaccharide that elicits three enzymes in phytoalexin synthesis.

Chitin, an important component of nematode eggs, insect exoskeletons, and the cell walls of many plant pathogenic fungi, is also an effective elicitor. Chitosan, a deacylated fragment of chitin, elicits phytoalexin accumulation at 10 μg ml^{-1} in peas, and acts as a fungicide against *Fusarium solani,* a pathogen of peas.[63]

The use of elicitors in agriculture offers an exciting promise for the future, since phytoalexins could have an important role in plant disease and pest resistance that does not use harmful synthetic pesticides. Their controlled production (elicitation) could be used to stimulate natural disease and pest resistance. If successful, this technique could replace environmentally damaging compounds.

Phytoalexin production in cotton plants is another example of this class. *Verticillium dahliae,* a pathogenic fungus, evokes anatomical and chemical responses in resistant cotton plants. The anatomical response includes production of tyloses, which block infected sylem vessels and confine the fungal pathogen. The chemical response includes the synthesis of fungitoxic terpenphytoalexins. These phytoalexins have been identified as desoxyhemigossypol (**69**) and hemigossypol (**70**), of which the former is the most active against non-defoliating strains of *V. dahliae.*[64]

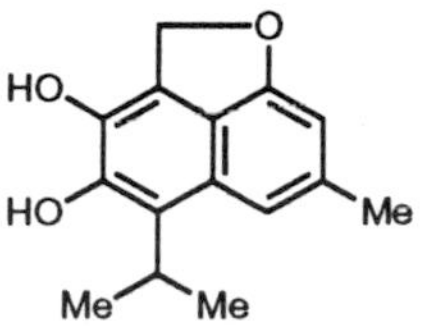

69 Desoxyhemigossypol

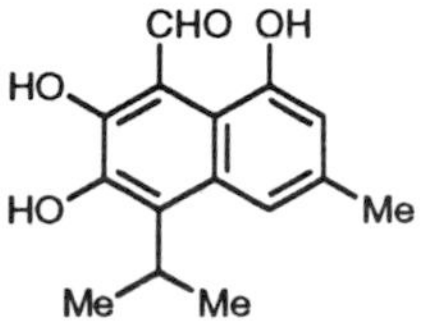

70 Hemigossypol

6.III.4 Pesticides from Insects

Several classes of insects produce or sequester from dietary sources, secondary metabolites which can be used as insecticides (see chapter 9). These metabolites are either a part of the insects' complex chemical defence system or act as sex attractants. For example, chemicals isolated from venomous insects are regarded as promising agents against other types of insects, while the sex pheromones separated from various species of bugs can also be utilized to control insects, particularly to control the opposite sex of the same species.

The glands of poisonous ant species in the genera *Solenopsis* and *Monomorium* are an excellent source of novel nitrogen heterocycles. These compounds have been demonstrated to possess a diversity of biological activities. The pronounced insect repellent properties of these nitrogen heterocycles indicate that they have great potential value as insecticides.

trans-2-Butyl-5-heptylpyrrolidine (**71**), an alkaloid produced in the venom of a thief ant (*Solenopsis* sp.), is an outstanding repellent for other aggressive ant species and, therefore, it can be used as an effective insect deterrent.[48]

The venom of the fire ant *Solenopsis invicta* possess considerable topical insecticidal activity, a property not generally identified with proteinaceous venoms. 2,6-Dialkylpiperidines, such as compound **72**, the toxic constituents of venoms, have been found recently to be termiticidal against species of *Reticulitermes.*[49] 2,5-Dialkylpyrrolidines have topical termiticidal activity against workers of three European species of *Recticulitermes.*[65]

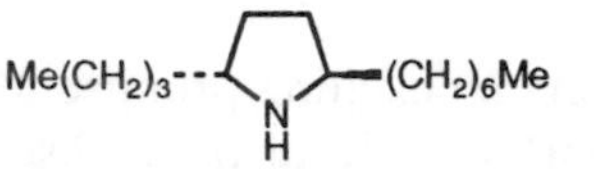

Me(CH$_2$)$_3$ ... N H ... (CH$_2$)$_6$Me

71 *trans*-2-Butyl-5-heptylpyrrolidine

Me ... N ... (CH$_2$)$_n$Me

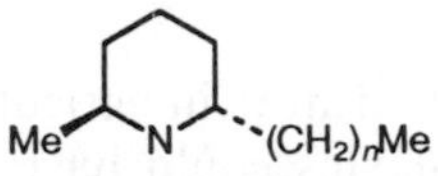

72 2,6-Dialkylpiperidine

The males of many bugs have glands that secrete chemical compounds which act as sex attractants for nearby females. For example, males of an African cotton pest, *Sphaerocoris annulus,* emit a blend of (*Z*)- and (*E*)-4,8-nonadienal, (*Z*)-4-nonenal, and nonanal, which are attractant pheromones. Cottonseed-feeding insect *Tectacorpis diophthalmus* males produce a crystalline deposit of 3,5-dihydroxy-4-pyrone, which may be an aphrodisiac and/or sex attractant.

Pheromones are generally simple acyclic molecules and, if available in large quantities, they can be used as 'biocontrols' for insect pests. Pheromone-baited traps for pest species may also be useful in detecting incipient infestations and insect populations.[66]

6.III.5 Marine Natural Products with Pesticidal Properties

Marine microbe, plant, and animal constituents have proved to be a prodigious source of new and structurally complex secondary metabolites. In some cases, these compounds are promising agents against various agricultural pests. Their unique structural features also provide strong incentives for further work to find new agrochemical agents.

Cardellina[67] isolated a brominated sesquiterpene neomeranol (**73**) from green alga which is phytotoxic to Johnsongrass (*Sorghum halepense* Pers), a

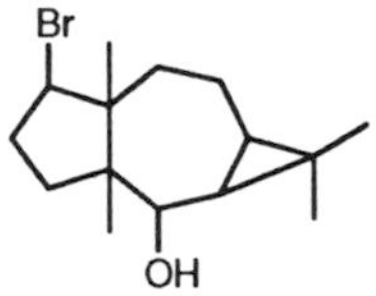

73 Neomeranol

common weed (see chapter 5). They have developed a potent plant-growth inhibitor (lettuce assay) from the water soluble fraction of the sponge *Spheciospongia othella*. These extracts contain exceptionally high levels of nickel, which appear to be bound to peptides of molecular weight *ca*. 1400.[67]

6.III.6 References

1. H. G. Cutler (ed.), 'Biologically Active Natural Products', ACS Symposium 380, American Chemical Society, Washington DC, 1988.
2. M. Jacobson, (ed.), ACS Symposium 296, American Chemical Society, Washington DC, 1986.
3. W. Krauss, M. Bokel, A. Bruhn, R. Cramer, I. Klaiber, G. Nagl, H. Poenhl, H. Sadlo, and B. Vogler, *Tetrahedron*, 1987, **43**, 2817.
4. O. U. Larson, US Patent, 1985, **4**, 556.
5. M. Nakatani, J. C. James, and K. Nakanishi, *J. Am. Chem. Soc.*, 1981, **103**, 1228.
6. I. Kubo and J. A. Klocke, *Experientia*, 1982, **38**, 639.
7. S. D. Jolad, J. J. Hoffman, J. R. Cole, M. S. Tempesta, and R. B. Bates, *J. Org. Chem.*, 1980, **45**, 3132.
8. P. Grijipma and R. Ramalho, *Turrialba*, 1969, **19**, 531.
9. P. Grijipma, *Turrialba*, 1970, **20**, 85.
10. P. Grijipma and S. C. Roberts, *Turrialba*, 1975, **25**, 152.
11. G. C. Allann, R. I. Gara, and S. C. Roberts, *Turrialba*, 1975, **25**, 255.
12. W. Kraus, W. Grimminger, and G. Saitski, *Angew. Chem. (Int. Ed. Engl.)*, 1978, **17**, 452.
13. Anonymous, *Citrus Vegetable Mag.*, 1982, **5**, 32.
14. M. A. Altieri, M. Lippmann, L. L. Schmidt, and I. Kubo, *Prot. Ecol.* 1984, **6**, 91.
15. H. C. F. Su, *J. Georgia Entomol. Soc.*, 1976, **11**, 297.
16. H. C. F. Su, R. D. Speirs, and P. G. Mahany, *J. Econ. Entomol.*, 1972, **65**, 1433.
17. H. C. F. Su, R. D. Speirs, and P. D. Mahany, *J. Econ. Entomol*, 1972, **65**, 1438.
18. A. R. Alford, J. A. Cullen, R. H. Storch, and M. D. Bentley, *J. Econ. Entomol.*, 1987, **80**, 575.
19. A. Hassanalli, M. D. Bentley, E. N. Ole Sitayo, P. E. W. Sjoroge, and M. Yatagai, *Insect Sci. Appl.*, 1986, **7**, 495.
20. J. L. Capinera and F. R. Stermitz, *J. Chem. Ecol.*, 1979, **5**, 767.
21. M. Jacobsen, *J. Am. Chem. Soc.*, 1948, **70**, 4234.

22. F. Y. Chou, K. Hostettmann, I. Kubo, K. Nakanishi, and M. Taniguchi, *Heterocycles*, 1977, **7**, 169.
23. T. Kamidado, S. Murakoshi, and S. Tamura, *Agric. Biol. Chem.*, 1978, **42**, 1515.
24. M. Jacobson, R. E. Redfern, and G. D. Mills, Jr, *Lloydia* 1975, **38**, 455.
25. D. J. Warthen, Jr, E. L. Gooden, and M. Jacobson, *J. Pharm. Sci.*, 1969, **58**, 637.
26. A. K. Pirman, R. H. Elliott, and G. H. N. Towers, *Biochem. Syst. Ecol.*, 1978, **6**, 333.
27. M. Steibl, J. Nawrot, and V. Herout, *Biochem. Syst. Ecol.*, 1983, **11**, 381.
28. M. J. Pettei, I. Miura, I. Kubo, and K. Nakanishi, *Heterocycles*, 1978, **8**, 471.
29. M. Jacobson, *Mitterl. Schweiz Entomol. Ges.*, 1971, **44**, 73.
30. D. C. Deb and S. J. Chakravorty, *J. Insect Physiol.*, 1982, **28**, 703.
31. M. Jacobson, *J. Org. Chem.*, 1967, **32**, 1646.
32. M. Jacobson, R. E. Redfern and G. D. Mills, Jr, *Lloydia*, 1975, **38**, 473.
33. V. Caprioli, G. Cimino, R. Colle, M. Cavagnin, G. Sodano, and A. Spinella, *J. Nat. Prod.* 1987, **30**, 146.
34. W. M. Blaney, M. S. J. Simmonds, S. V. Ley, and R. B. Katz, *Physiol. Entomol.*, 1987, **12**, 281.
35. I. Kubo and I. Ganjian, *Experientia*, 1981, **37**, 1063.
36. M. Taniguchi, T. Adachi, S. Oi, A. Kimura, S. Katsummura, S. Isoe, and I. Kubo, *Agric. Biol. Chem.*, 1984, **48**, 73.
37. I. Kubo, I. Miura, M. J. Pettei, Y. W. Lee, F. Pilkiewicz, and K. Nakanihi, *Tetrahedron Lett.*, 1977, 4553.
38. M. Zur, J. C. James, E. Kabonci, and K. R. S. Ascher, *Phytoparasitica*, 1980, **8**, 189.
39. W. S. Bowers and R. Nishida, *Science*, 1980, **209**, 1030.
40. R. Nishida, W. S. Bowers, and P. H. Evans, *J. Chem. Ecol.*, 1984, **10**, 1435.
41. S. R. Chavan and S. T. Nikam, *Ind. J. Med. Res.*, 1982, **75**, 220.
42. F. Camps and J. Colli, *Phytochemistry*, 1993, **32**, 1361.
43. R.B. M. Geusskens, J. M. Luteijin, and L. M. Schoonohoven, *Experientia*, 1983, **30**, 403.
44. I. Kubo, J.A. Klocke, and S. Asano, *Agric. Biol. Chem.*, 1981, **45**, 1925.
45. I. Kubo, M. Kido, and Y. Fukuyama, *J. Chem. Soc., Chem. Commun.*, 1982, 897.
46. I. Kubo, J. A. Klocke, I. Miura, and Y. Fukuyama, *J. Chem. Soc., Chem. Commun.,* 1982, 618.
47. S. O. Duke, R. N. Paul, and S. M. Lee, in 'Biologically Active Natural Products: Potential Use in Agriculture', ed. H. G. Cutler, ACS Symposium 380, American Chemical Society, Washington DC, 1988.
48. M. S. Blum, T. H. Jones, B. Holldobler, H. M. Fales, and T. Jaouni, *Naturwissenschaften*, 1980, **67**, 144.
49. M. S. Blum, T. H. Jones, and P. Escoubas, unpublished results, 1987.
50. G. T. Bottger, A. P. Yerington, and S. I. Gertler, US Dept. Agric. Bur. Entomol. Plant Quarantine, E-826, 23 pp 1951, *Chem. Abstr.,* **46**, 5778e.
51. Y. Sekizawa and T. Takematsu, in 'Pesticide Chemistry: Human Welfare and the Environment', ed. J. Miyamoto and P. C. Kearncy, Pergamon Press, New York, 1983, vol. 2.
52. H. Nagasawa, A. Suzuki, and S. Tamura, *Agric. Biol. Chem.*, 1978, **42**, 1303.
53. Y. Yoshino, C. Sato, and S. Maeda, Japanese Patent, 73, 41 992, 1973; *Chem. Abstr.*, **81**, 59342 p.
54. D. D. Questel and S. I. Gertler, US Dept. Agric. Bur. Entomol. Quarantine E-840, 12 pp. 1952; *Chem. Abstr.,* **46**, 8311d.

55. Takeda Chemical Company Ltd, Japanese Patent, 1983, **58**, 157; *Chem. Abstr.*, **100**, 3094 m.
56. H. G. Cutler, in 'The Science of Allelopathy', ed. A. R. Putnam and C. S. Tang, Wiley, New York, 1986.
57. E. L. Patterson and P. Wright, US Patent, 1970, **3**, 520.
58. J. D. Paxton, in 'Biologically Active Natural products: Potential Use in Agriculture', ed. H. G. Cutler, ACS Symposium 380, American Chemical Society, Washington DC, 1988.
59. I. A. M. Cruickshank and D. R. Derrin, *Life Sci.*, 1968, **7**, 449.
60. E. E. Farmer and J. P. Helgeson, *Plant Physiol.*, 1994, in press.
61. J. A. Frank and J. D. Paxton, *Phytochemistry*, 1971, **61**, 954.
62. H. M. Griffiths and A. J. Anderson, *Can. J. Bot.*, 1987, **65**, 63.
63. D. F. Kendra and L. A. Hadwiger, *Phytopathology*, 1987, **77**, 100.
64. R. D. Stipanovic, M. E. Mace, D. W. Altman, and A. A. Bell, in 'Biologically Natural Products: Potential Uses in Agriculture', ed. H.G. Cutler, ACS Symposium 380, American Chemical Society, Washington DC, 1988, p. 262.
65. J. L. Clement, M. Lemaire, and C. Lange, *C.R. Acad. Sci. Paris*, 1986, **303**, 669.
66. J. R. Aldrich, in 'Biologically Active Natural products: Potential Uses in Agriculture', ed. H. G. Cutler, ACS Symposium 380, American Chemical Society, Washington DC, 1988, p. 417.
67. J. H. Cardinclla, in 'Biologically Active Natural products: Potential Uses in Agriculture', ed. H. G. Cutler, ACS Symposium 380, American Chemical Society, Washington DC, 1988, p. 305.

CHAPTER 7

*Synthetic Insecticides Related to the Natural Pyrethrins**

MICHAEL ELLIOTT

7.1 Introduction

The natural pyrethrins[1–6] are powerful insecticides that act rapidly against insects of many species, yet they are harmless to mammals under all normal conditions; however, because they lose their activity rapidly on exposure to air and light, the natural compounds and earlier synthetic pyrethroids are generally suitable only for indoor or protected applications. In the past 15 years related synthetic pyrethroids stable enough for agricultural applications have been developed; these retain favourable characteristics like those of the natural compounds and now constitute about one-quarter of all the insecticides used world-wide.

This chapter is a review of the contributions made over some five or six decades at the Rothamsted Experimental Station[7] (RES) to the development of more stable synthetic pyrethroid insecticides.[8–13] Excellent surveys have already dealt with various other aspects of these compounds.[14–16] The main sequence of work at Rothamsted was initiated in the late 1940s by the far-sighted entomologist Charles Potter,[17] who perceived that insecticides to supplement the organochlorine, organophosphorus, and carbamate groups then being intensively studied and developed would eventually be needed. The organic chemist Stanley H. Harper[18] also recognized the merits of natural pyrethrum as an insecticide, partly from experience he had gained at Rothamsted from 1936–1942. In 1942, at the (then) University College, Southampton, he obtained a grant from the Agricultural Research Council to study synthetic compounds related to natural pyrethrum in a project intended to lessen the adverse impact of any failure of plant sources of insecticides in war time.

*Manuscript submitted April 1994.

The visionary decisions that Potter and Harper took then were profoundly to influence insecticide usage world-wide in the latter part of the twentieth century.[13]

7.2 History

Dried pyrethrum flowers were used as insecticides in ancient China and in the middle ages in Persia.[1–3,19] Their introduction to Europe is reputed to have been from both Persia and Dalmatia where, certainly up to the mid 1980s, pyrethrum flowers still graced the countryside (personal observation). The main commercial source of pyrethrum is the mature flower of *Chrysanthemum (Tanacetum) cinerariaefolium,* cultivated principally in Kenya and Tanzania at elevations above 1500 m.[6] Keatings Insect Powder, well known in Europe and the USA, was powdered dried pyrethrum flowers. High-quality flowers, from selective breeding programmes, contain up to 4 mg (1.5% by weight) of pyrethrins. The main supply now is as extracts in hydrocarbon solvents, concentrated to 25%, decolourized or not. Pyrethrum production, *ca* 20 000 tonnes per annum world-wide, has been reviewed recently.[6] Traditionally, harvesting has been by hand, since the optimum yields are obtained from flowers just matured with horizontal disc florets, but alternative locations (*e.g.*, Tasmania) and mechanical harvesting of plants bred to mature over a short period[20] have been explored as the demand for naturally derived insecticides, particularly pyrethrum, increases.

7.3 Early Structure–Activity Correlations

Fujitani[21] and Yamamoto[22,23] in Japan showed that the pyrethrins were esters. Then Staudinger and Ruzicka[24,25] in Zurich between 1910 and 1916, by inspired and innovative degradative and synthetic work, established the main structural features of the insecticides, which were obtained as dark, viscous, heat-, light- and oxygen-sensitive liquids. They isolated and determined the structure of (+)-*trans*-chrysanthemic acid (**1a**),* from which they made 82 synthetic esters with a variety of alcohols; they also esterified naturally derived (+)-pyrethrolone [now recognized to have been a mixture of pyrethrolone (**1f**), cinerolone (**1g**) and jasmololone (**1h**)] with 32 different acids. Earlier investigators[21–23] had established only that cleavage of the esters destroyed insecticidal activity, but Staudinger and Ruzicka, with their synthetic compounds, gained a remarkable insight into the requirements for insecticidal activity. This they assessed by dusting cockroaches ('Schaben'), teased from crevices in their own laboratory, with coffee-spoonfuls of flour impregnated with test compounds, typically at 1:500 dilution rates. By such simple but informative means they showed that very few features in the structures of the

*Compounds are designated by figures (7.1–7.9), followed by letter(s) for the component(s) involved, *e.g.* cinerin I in Figure 7.1 is derived from (+)-*trans*-chrysanthemic acid (**a**) and cyclopentenone (**g**) = (**1ag**). Hydrogens are omitted.

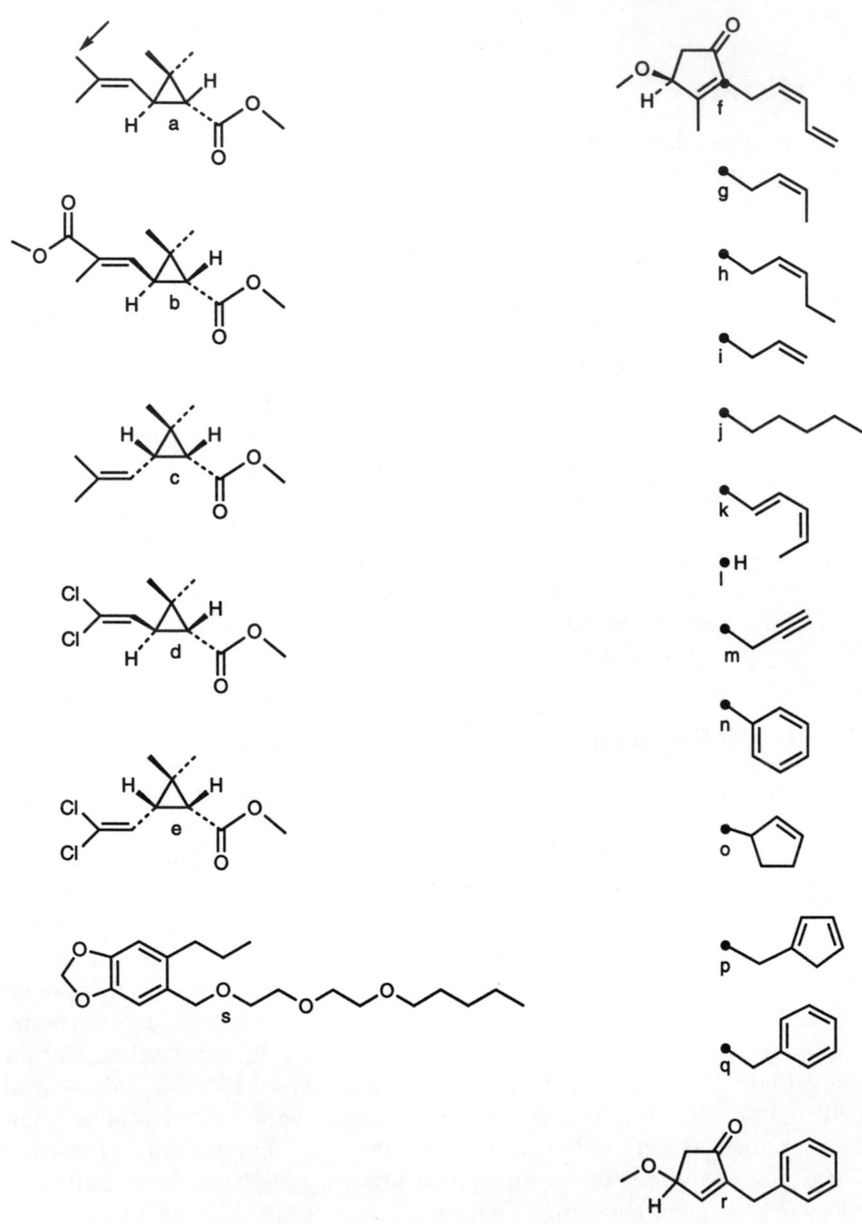

Figure 7.1 *Acidic and alcoholic components of pyrethroids; piperonyl butoxide*

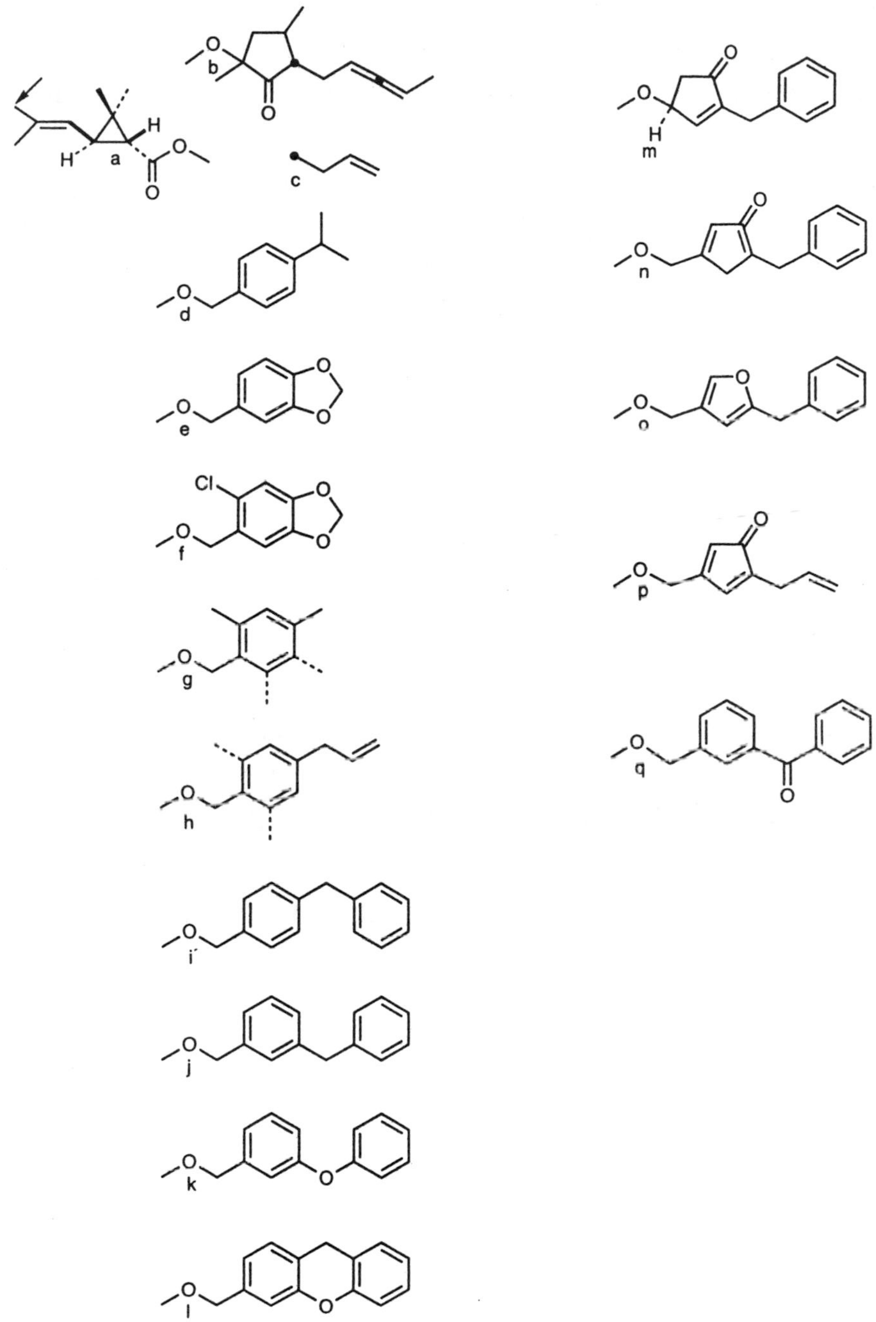

Figure 7.2 *Various chrysanthemates*

esters could be changed or eliminated without loss of insecticidal activity. Unsaturation in the side chains of both acidic and alcoholic components was important, and only acids with structures close to that of chrysanthemic acid gave active esters. Staudinger and Ruzicka formulated pyrethrin I as derived from a 2-hydroxycyclopentanone (**2b**) with cumulated unsaturation in the side chain (**2ab**), to accommodate the confusing evidence from the (by them undetected) presence of cinerolone (**1g**) and jasmololone (**1h**); as an accessible related synthetic ester they prepared a mixture containing a simpler 2-oxocyclopentanyl chrysanthemate (**2ac**) with an allyl side chain. This had some activity, foreshadowing the first synthetic pyrethroid allethrin made by Schechter *et al.*[26] some three decades later. Also significant was the activity of cuminyl chrysanthemate (**2ad**), compared with the inactivity of the parent benzyl ester. A substituent on the alcoholic nucleus remote from the point of attachments of the acid was, therefore, associated with activity, a deduction that was the initial stimulus for the sequence of synthetic compounds which led Rothamsted to the first class of more photostable synthetic pyrethroids. Indeed, most current commercial photostable pyrethroids are substituted benzyl esters. Staudinger and Ruzicka further conceived that an α-isopropyl acid structurally related to chrysanthemic acid by hypothetical ring cleavage might give active esters with appropriate alcohols. Again, their remarkable insight was realized when Sumitomo chemists[27] discovered the activity of some 2-phenylalkanoates, establishing the second major category of photostable pyrethroid insecticides. The influence of the Swiss chemists Staudinger and Ruzicka on the development of pyrethroids has, therefore, been profound.

7.4 Further Progress with Natural and Synthetic Pyrethroids

In the United States Department of Agriculture (USDA), LaForge and his collaborators from 1935–1952 developed the chemistry of the natural esters on the foundations laid by Staudinger and Ruzicka. The difficulty of formulating a structure for pyrethrolone consistent with all the degradative evidence was largely resolved by demonstrating the co-occurrence of esters of cinerolone (**1g**) with those of pyrethrolone (**1f**)[28] [a third alcoholic component, jasmololone (**1h**), was detected by gas–liquid chromatography in 1966[29]]. Spectroscopic evidence by Gillam and West in the UK (1942–1944),[30] cyclopentenone syntheses by Harper, Crombie, and collaborators (1942 onwards) using ring closure of substituted 2,5-diketones, as developed by Hunsdiecker,[31] and some of the first synthetic applications of *N*-bromosuccinimide[32] led to the correct formulation of pyrethrolone (*1f*) as a 4-hydroxy-2-(penta-2,4-dienyl)cyclopent-2-enone, but of undefined side chain stereochemistry. Crombie has recently surveyed these aspects of the chemistry of the pyrethrins in two definitive, personal reviews.[18,32] A route via *N*-bromosuccinimide to 4-bromo-3-methylcyclopent-2-enones,[33] which gave the first totally synthetic pyrethrins, failed when unsaturated side chains were present at position 2, but

Schechter, working with LaForge, developed a general route to 4-hydroxycyclopent-2-enones with saturated or unsaturated side chains involving pyruvaldehyde.[26] Although this was first applied to natural (*Z*)-pyrethrolone itself in 1954,[34] it subsequently enabled the first commercial synthetic pyrethroid, allethrin (still used today), to be developed.[26,35]

7.5 Structure–Activity Studies at Rothamsted Experimental Station

Frederick Tattersfield was appointed the first Head of the Insecticides and Fungicides Department at Rothamsted in 1918 and started to study systematically the variation of insecticidal activity with chemical structure in homologous series of compounds. The precision and value of his results[36] were enhanced by procedures, such as probit analysis, introduced from the Statistics Department at Rothamsted by Fisher and Yates.[37] The chemistry, cultivation, and analysis of the insecticidal principles of pyrethrum were examined using seed produced locally; the quality of this was so high that some of it was used in 1927 to establish the new pyrethrum industry in Kenya. Systematic insecticide tests were restricted to the summer, when suitable species (*e.g.*, aphids) could be collected from the field. One task for a newly appointed entomologist, Charles Potter, was to establish laboratory cultures of insects so that testing could continue throughout the year. Potter also demonstrated in 1935[38] that pyrethrum in heavy white oil formed films that protected dried fruit in darkened warehouses from insect infestation. This application established the practice of using residual insecticide films, destined some years later to be outstandingly successful with DDT (dichlorodiphenyltrichloroethane).

When Potter was appointed Head of the Insecticides and Fungicides Department at Rothamsted in 1948, he decided that a detailed examination of the relationship between the biological properties of insecticides and their chemical constitution would be rewarding. In the case of pyrethroids, a powerful motivation was to explain how these plant-derived compounds could affect insects so rapidly, without adverse influence on mammals. At its inception the project seemed unlikely to overlap or duplicate commercial interests. Natural pyrethrum was expensive and unstable and, as a lead compound, would not attract agrochemical firms impressed with the potential of the accessible and stable organochlorine compounds, the organophosphate insecticides, and the carbamates. Moreover, such organizations were then, unlike Potter, generally unmindful of the future menace of resistance.

In 1948 Potter appointed an organic chemist, myself, and an entomologist, Paul H. Needham, to collaborate in a study of the insecticidal action of the pyrethrins. I had been introduced to and fascinated by the chemistry of the pyrethrins as described in lectures by Stanley H. Harper to undergraduates at the (then) University College, Southampton. Subsequently, as a postgraduate, I studied under him, initially at Southampton and then at King's College, University of London, when he was appointed Reader there. I worked on the

Table 7.1 *Potencies of allethrin and bioallethrin relative to natural pyrethrins (=1)*

Test species	*Source*	*Allethrin*	*Bioallethrin*
Musca domestica (knockdown)	USA	3	6
M. domestica (kill)	USA	1	
Periplanata americana	USA	0.4	
Blatella germanica	USA	0.2	
Oncopeltus fasciatus	USA	0.2	
Plutella xylostella	RES	1.8	4
Macrosiphon euphorbiae	RES	0.06	0.1
Oryzaephilus surinamensis	RES	0.4	0.5
Phaedon cochleariae	RES	0.2	0.4

chemistry of the alcoholic components of the pyrethrins, and the initial results obtained in collaboration with L. Crombie and H.W.B. Reed under Harper (who had just summarized the current state of knowledge of the pyrethrins[39]) were published in 1948.[33]

As Rothamsted's work on the structure–activity relationships of pyrethroids began, Schechter, Green and LaForge (USDA, Section 7.4) reported the first potentially commercial synthetic pyrethroids. Large-scale production of the natural products pyrethrins I (**1af**) and II (**1bf**) was not feasible because no routes to (*Z*)-dienes were then known (and, in fact, the first potentially commercial synthesis of pyrethrin I has only very recently been claimed.[40]). However, with only one terminal double bond in the alcoholic side chain, large-scale synthesis of allethrin (**1ai, 1ci**, ± forms) from accessible precursors was possible.[35] Allethrin excited great interest because it acted more rapidly and effectively than natural pyrethrum[41,42] against the common housefly *Musca domestica* L., an established target used for quantitative evaluations.[43] Through Potter's USDA contacts, Rothamsted received the first samples outside the USA of two allethrolone esters; these were (±)-allethronyl (+)-*trans*-chrysanthemate (**1a** ± **i**) (now called bioallethrin), and (±)-allethronyl (±)-*cis-trans*-chrysanthemate (**1** ± **ac** ± **i**) (allethrin). Needham compared the activities of these samples with those of the natural pyrethrins mixture against insects other than the housefly, and it was soon clear that the exceptional activity reported for allethrin against houseflies was not shown by other species (Table 7.1).[44]

7.5.1 Strategy of Structure–Activity Studies

The initial structure–activity tests at Rothamsted influenced the future direction of the studies there. The most precise results achievable against a broad range of insects were then seen to be desirable. Needham, having initially found that results obtained by spraying or by pick-up from residues on filter paper or glass were unsatisfactory, undertook to develop procedures for rearing consistent cultures of a variety of insect species and to obtain precise relative potency values by topical application of measured drops of insecticide solutions to them. Eventually, over more than three decades, Paul Needham and later Roman Sawicki, Andrew Farnham, and other colleagues produced many very precise bioassays capable of detecting and measuring the influence of small and subtle changes in chemical structures; the progress in developing the synthetic pyrethroid insecticides at Rothamsted depended crucially on these results. The similar or contrasting responses of houseflies (*Musca domestica* L.) and mustard beetles [*Phaedon cochleariae* (Fab.)] to many pyrethroids were found most significant for indicating the directions for syntheses and testing of new compounds. The resources available did not permit sufficiently detailed and extensive testing against other species and, contrary to the general interest at the time, emphasis was placed upon kill of houseflies and other insects, rather than upon speed of action (knockdown), which Needham assessed only subjectively. Any limitation in potency by mixed function oxidases[45,46] (active for detoxification in houseflies, but less so in mustard beetles) was detected by administration without and then with a standard high dose of a synergist (piperonyl butoxide) (**1s**) known itself not to be active as applied. Usually, this adjuvant modified considerably the potency measured against houseflies, but not against mustard beetles. The choice of kill of these two species to establish structure–activity relationships can be judged in retrospect to have been fortunate, because progress was made in directions not followed by other investigators.

A second conclusion from the results in Table 7.1 was that to assess the influence of small structural changes by comparing a mixture of the natural esters (at the time considered to be pyrethrins I and II and cinerins I and II) with the two isomers in (±)-allethronyl (+)-*trans*-chrysanthemate (bioallethrin) and the eight in (±)-allethronyl (±)-*cis-trans*-chysanthemate (allethrin) was difficult. The evaluation would be further complicated by the recognition that jasmolins I (**1ah**) and II (**1bh**) were also present in the natural mixture.[29] This difficulty was resolved when the original chromatographic separation of the natural esters by Ward,[47] as used by Sawicki *et al.*[48] made available pure pyrethrin I, which was then shown to be the most active of the natural esters for kill and was henceforth used as the standard against which other natural and synthetic pyrethroids were assessed. The limited supplies of pyrethrin I isolated by chromatography were supplemented by material regenerated from naturally derived (+)-pyrethrolone, purified as the monohydrate.[49–51]

7.5.2 Results of Structure–Activity Examination

By extended assays at Rothamsted and elsewhere, the outstanding performance of allethrin was confirmed as restricted to the housefly; against other insects (the mustard beetle, for example) pyrethrin I was as much as 40 times more effective.[52] Another early key observation was that isopyrethrin I[53] (**1ak**), obtained when pyrethrin I was thermally isomerized, was inactive. (The thermal isomerization of esters of pyrethrolone was first observed by Brown and Phipers,[54] who generously communicated their discovery to Rothamsted before publication). To obtain pyrethrolone to investigate the phenomenon further, Staudinger and Ruzickas' isolation,[24] achieved by cleaving the semicarbazones of the natural pyrethrin esters, was repeated and thence the previously unreported crystalline monohydrate was discovered.[49] Within a few years of Schechter's 4-hydroxycyclopentenone synthesis,[26] cyclethrin (**1ao**),[55] furethrin (**1ap**),[56,57] a prop-2-ynyl (**1am**) analogue of allethrin,[58] benzylrethrin (**1aq**), phenylrethrin (**1an**),[56] and other compounds[59] had been tested at Rothamsted or elsewhere. In conjunction with the earlier result[33] that dihydrocinerin I and tetrahydropyrethrin I (**1aj**) were very much less active for kill, the conclusion was reached that for the highest activity at least two double bonds in conjugation, appropriately spaced by a methylene group from the cyclopentenone ring (as in pyrethrins I and II), were necessary.

Other investigators (in work reviewed by Barthel[59]) continued testing against houseflies and other organisms, assessing kill and knockdown activity with and without the synergist piperonyl butoxide (**1s**). Extension of Staudinger and Ruzickas' work[24] with substituted benzyl chrysanthemates was emphasized. Activity was shown to depend on the nature and position of substituents on the ring, but no positive leads for further progress were established. The most active compounds were dimethrin (**2ag**) and barthrin (**2af**) (6-chloropiperonyl chrysanthemate), which were even considered for commercial production. At Rothamsted during this phase of synthetic pyrethroid development, the influence of the methylation pattern on the activity of substituted benzyl chrysanthemates (**2ag**) to houseflies (best 2,4,6-) and to mustard beetles (best 2,3,6-) were examined in detail.[60] [Structures (**2ag**) and (**2ah**) represent compounds without and with methyl substituents]. Methyl groups at 2,4, and 6 were more likely to enhance activity than those at positions 3 and 5.[61]

Staudinger and Ruzickas' finding,[24] referred to earlier, that cuminyl, unlike benzyl, chrysanthemate was active indicated that a substituent in the benzyl 4-position was important. Combining this information with developing concepts of the features associated with the activity of allethrin [*e.g.*, the inactivity of the parent 2-methyl-4-oxocyclopent-2-enyl chrysanthemate (**11**)][62] suggested that 4-allylbenzyl chrysanthemate (**2ah**) be examined. The retention or even increase in activity for some species found in assays of this compound by Farnham and Sawicki constituted a significant conceptual and practical advance.[63] The 4-oxocyclopent-2-enyl ring could now be considered a spacer whose function was to hold the alkenyl side chain at an appropriate distance from the esterified hydroxy groups. Since (+)-(**1i**) and (−)-allethrolone gave chrysanthemates with differing activities,[56] the orientation imposed or allowed by the spacer unit, and not just the distance between the two groups, was indicated as important.

The potential for future development of new pyrethroids implied by the activity of 4-allylbenzyl chrysanthemate was recognized by the National Research Development Corporation (NRDC), which provided financial support to Rothamsted and made possible the appointment of Norman F. Janes, also a former student of S. H. Harper; from 1962 onwards, Janes brought a very powerful intellect combined with practical skills to the project.

The results from examination of the most effective methylation pattern in benzyl chrysanthemates were then applied to 4-allylbenzyl chrysanthemate; specifically, the 2,6-dimethyl compound (**2ah**) was found to be as active as the 4-allyl compound to houseflies and 5–10 times more active to mustard beetles.[63]

7.6 Esters with a Benzyl Side Chain in Alcoholic Components: Resmethrin and Bioresmethrin

The decision to simplify interpretation of complex structure–activity relationships by comparing, where possible, the activity of single isomers emphasized that pyrethrin I (**1af**) was more active than the corresponding isomer (+)-allethronyl (+)-*trans*-chrysanthemate (**1ai**) to most species examined. Further compounds with conjugated double bond unsaturation in the alcohol were therefore pursued at Rothamsted. The compounds benzylrethrin (**1aq**) and furethrin (**1ap**) with benzyl and 2-furyl side chains, respectively, had already been synthesized and tested; their activities were not significantly different from that of allethrin.[56] However, the compound benzyl northrin (**1ar**), prepared by a novel route which did not involve pyruvaldehyde,[64] had marked activity, comparable to the best of other 4-oxocyclopent-2-enyl esters. Thus, benzyl was established as a possible and synthetically much more accessible alternative to the (*Z*)-pentadienyl system in pyrethrin I. Further, the 2-methyl group of the natural ring system was thereby demonstrated not to be essential.

Recognition of benzyl (as opposed to allyl) as a side chain potentially conferring high levels of activity rendered accessible a wider range of compounds for testing. Furan, a nucleus close to 4-oxocyclopent-2-enyl in size and shape, was promising and, from results with methyl substituted 3-furylmethyl or furfuryl esters,[65,66] a 5-benzylfurfuryl chrysanthemate (**2an**) was selected for synthesis. The activity of this compound was at a level to stimulate synthesis and examination of closely related analogues, such as the 5-benzyl-3-furylmethyl (5B3F) chrysanthemates (**2ao**), which proved to be even more potent; the development of new syntheses[67] of the system was thereby justified. The potency of the (1*R*)-*trans*-chrysanthemate* (**2ao**), bioresmethrin (NRDC 107), against several insect species was higher than that of any previous synthetic pyrethroid; moreover, its toxicity to mammals (results generously provided by workers at the MRC Toxicology Unit, Carshalton) was very low. The corresponding (1*RS*)-*cis-trans*-chrysanthemate (resmethrin, NRDC 104), using the acid component already available for allethrin, although somewhat less active, was more accessible and the presence of the *cis* isomers conferred a different potency spectrum. Resmethrin and bioresmethrin were developed commercially after a short time by Roussel-Uclaf, France, and S.B. Penick,

*For clarity, compounds are specified by naming, for example, the isomeric forms of chrysanthemic acids as (*1R*)-*trans* [≡(+)-*trans*] and (*1S*)-*cis* [≡(−)-*cis*], and so on (for a fuller explanation, see Elliott *et al.*).[87]

USA under licences from the NRDC. The process was essentially the second method developed at Rothamsted to give supplies larger than the first tedious route could provide easily. A number of 3-furylmethyl, furfuryl, 3-thenyl, and 4-oxazoylylmethyl chrysanthemates with alkyl, alkenyl, benzyl, and substituted benzyl substituents were also synthesized, but none was better than the 5B3F compound.[68] One result, at the time surprising, was the low activity of 5-allylfurfuryl and 5-allyl-3-furylmethyl (**2ap**) compounds in comparison with corresponding benzyl-substituted esters; however, this may now be seen to support the earlier, and possibly naive, concept that the phenyl ring is necessary to stimulate fully the biological function of the (*Z*)-diene in pyrethrins I and II.

7.6.1 Progress from Resmethrin and Bioresmethrin

Curiously, at this stage, the marked progress towards discovery of new insecticides represented by resmethrin and bioresmethrin seemed, when published,[69] to excite little interest from the major agrochemical firms, except Roussel-Uclaf and Sumitomo, already committed to pyrethroid research; Rothamsted workers were left to pursue studies towards the more important compounds now considered.

4-Allyl- and 4-allyl-2,6-dimethylbenzyl chrysanthemates had been conceived as combining the structural features of allethrin and of 2,4-, 3,4-, and 2,4,6-methylbenzyl chrysanthemates. 3-Allylbenzyl chrysanthemate[68] was less active than the 4-allyl compound, but the substitution pattern of 5B3F compounds might be considered equivalent to 1,3- or to 1,4- (**2ai**) aromatic substitution. This concept stimulated synthesis and testing of 3-benzylbenzyl chrysanthemate (**2aj**)[70] which was found to be as much as 25 times more active than the 4-benzyl compound. This was a pivotal observation, because the alcohols of nearly all modern synthetic pyrethroids have two aromatic rings maintained in a comparable mutual *meta* relationship, but a very important difference is that phenyl [*e.g.*, in (**2aj**)] is much more photostable than furyl [*e.g.*, in (**2ao**)].

Pursuing the putative analogy between the (*Z*)-penta-2,4-dienyl and benzyl groups posed the questions as to whether the methylene group functioned merely as a spacer or was itself specifically involved. 3-Phenoxybenzyl (3POB) chrysanthemates, in which oxygen (–O–) served instead of methylene (–CH_2–) to maintain the relative spatial positions of the two rings were therefore synthesized, and were as active or more active than the benzylbenzyl compounds; the spacing function was therefore strongly favoured. Later,[71] CO [*e.g.*, (**2aq**)], C=CH_2, N–R, and even direct bonds[72] (see Section 7.16), in special cases were shown to be adequate substitutes for CH_2. Remarkably, the importance for insecticidal pyrethroids of the *meta* orientation of two aromatic rings was discovered almost simultaneously, but independently, in the laboratories of Rothamsted and in those of the Sumitomo Chemical Company;[73] activity was lost when the two rings were constrained to co-planarity, as in (**2al**). The 5B3F group was, therefore, not only the key intermediate through which the most effective alcoholic radical[74] (*viz.* α-cyano-3-phenoxybenzyl) was eventually discovered, but also led to the acid components of the two main groups of photostable esters (Sections 7.7 and 7.12).

7.7 Development of Acid Components of Pyrethroid Esters

In contrast to the accessibility of (1*R*)-*trans*-chrysanthemic acid to esterify with a range of possibly effective alcoholic components, Staudinger and Ruzicka[24] had only what can now be recognized as a mixture of the natural alcohols pyrethrolone, cinerolone, and jasmolone to examine the influence of changes in acid structure on the insecticidal potency of pyrethroid esters. This mixture was obtained by methanolysis of the pyrethrin semicarbazones to yield the hydroxyketone semicarbazones, which were cleaved by prolonged shaking at room temperatures with potassium hydrogen sulfate. By this extended procedure, Staudinger and Ruzicka overcame the otherwise obstructive base-induced elimination reactions of pyrethroid 4-oxocyclopent-2-enyl esters, which produce cyclopentadienones rather than the required hydroxyketones.[75] Staudinger and Ruzicka tested 32 esters of the derived natural hydroxyketone mixture, but apart from showing that the cyclopropane structure and an unsaturated side chain were associated with activity, made no advances comparable in importance to their work with the substituted benzyl chrysanthemates.

Later, when the constituents of pyrethrum extract were separately available for bioassay,[48] modifications in the acid side chain were shown to influence the nature of the insecticidal activity to houseflies by the greater toxicity and speed of knockdown of pyrethrin II (**1bf**) compared with pyrethrin I (**1af**).[48,76]

Naturally derived pyrethrolone was not sufficiently accessible to induce most research workers after Staudinger and Ruzicka to use it as the constant alcoholic component in a series of esters to determine the influence of acid structure on activity. Allethrolone, synthesized commercially for allethrin,[35] was much more readily available and superficially appeared to be an adequate substitute for pyrethrolone, but subtle enhancements associated with multiple unsaturation in the alcoholic side chains were obscured, leading to conclusions of limited generality. Nevertheless, (1*R*)-*trans*-(**1a**) and (1*R*)-*cis*-(**1c**) chrysanthemates of allethrolone were established to be much more effective than the (1*S*)-isomers,[52,26] and allethrolone was used in other structure–activity studies,[77] but no new useful acidic components for synthetic pyrethroids involving it were discovered. When the effectiveness of the 5B3F chrysanthemate was demonstrated,[69] esters of this same alcohol with a variety of acids clearly defined their relative potential practical value,[78] and for the first time accented the importance of the (1*R*)-*cis* configuration about the cyclopropane ring, in this instance in NRDC 119, cismethrin, 5-benzyl-3-furylmethyl (1*R*)-*cis*-chrysanthemate;[79] this was sufficiently active for both kill and knockdown to be considered for commercial development.

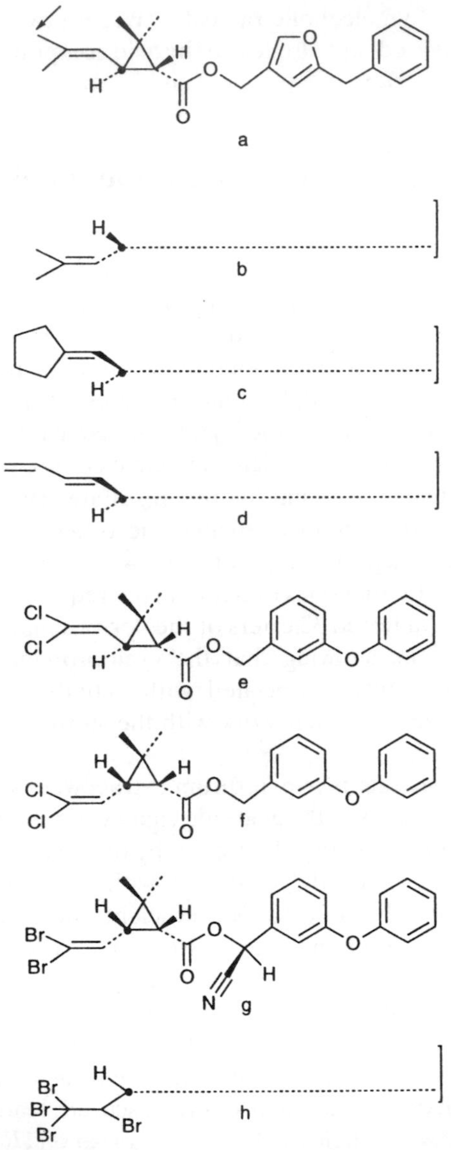

Figure 7.3 *Evolution of more photostable pyrethroids*

In a significant study, Martel and coworkers[80,81] examined 5-benzyl-3-furylmethyl 2,2-dimethylcyclopropanecarboxylates with cycloalkylidene rather than 3-(2-methylprop-1-enyl) substituents. The ethanochrysanthemate (**3c**) was more active than the corresponding chrysanthemate (**3a**). This finding and Casida's demonstration that, both in insects[82] and in mammals,[83] the methyl group in the (*E*)-position [arrowed in (**3a**)] of the chrysanthemate side chain was a site for metabolic attack stimulated a detailed study at Rothamsted[84] of the influence of the structure of the acid side chain on activity. A skilled and innovative organic chemist, David A. Pulman, also supported by NRDC, had joined Rothamsted and in this programme about 50 new esters were synthesized for bioassay,[85–87] and many were found to have high activities. In one, the 3-(2-methylprop-1-enyl) substituent was rearranged to but-1-enyl (*i.e.*, an isomer without a 2-methyl group); another very active example (**3d**) had a buta-1,3-dienyl group. Additional substituents (CO_2R, CN, or Me) mainly diminished activity.[88]

To explore further the influence of unsaturation in the acid side chain, a compound with a 3-alkynyl substituent was envisaged, to be obtained by dehydrochlorination of a 3-(2-chlorovinyl) substituted intermediate. In fact, a 5B3F ester with this acid side chain was more active than the related 3-prop-1-enyl compound, which clearly indicated that a compound with a 3-(2,2-dichlorovinyl) substituent should be synthesized; this was obtained from the corresponding caronaldehyde with triphenylphosphine and carbon tetrachloride. The 5B3F ester was three times more effective than the corresponding chrysanthemate[89] and, yet more significant, in the 3POB esters enhancement was even greater.

Farkas *et al.*[90] had synthesized the racemic *cis*-(**1e**) and *trans*-(**1d**) 2,2-dichlorovinyl analogues of chrysanthemic acid in 1958 and had esterified the *trans*-form (**1d**) (only) with allethrolone (**1i**), the activity of which had been reported with emphasis on knockdown, without recognition of any potential significance. Re-examination later found the allethrolone ester (**1di**) less than one-tenth as active as the 5B3F ester, a striking vindication of the earlier decision to base structure–activity studies of acid structures on 5B3F or 3POB esters.[91]

7.8 Photostable Synthetic Pyrethroids

Chen and Casida[92] identified the (*E*)-methyl group (arrow, **1a**) of the 3-(2-methylprop-1-enyl) side chain of chrysanthemic acid as a site for initial attack during photochemical decomposition of pyrethrin I (**1af**), allethrin (**1ai**), tetramethrin (**7ab**), and dimethrin (**2ag**); in a later study of resmethrin and related pyrethroids, photo-oxidative attack was shown to be on the furan ring, probably via an intermediate peroxide.[93] By 1972, structure–activity studies at Rothamsted had identified acids and alcohols[94] in which such photochemically induced reactions could not take place and which, in combination as esters, were effective insecticides, *viz.* the dichlorovinyl analogue of chrysanthemic acid and 3-phenoxybenzyl alcohol. The ester produced[89] [permethrin (**3ef**) NRDC 143; 3-phenoxybenzyl (1*RS*)-*cis-trans*-3-(2,2-dichlorovinyl)-2,2-dimethylcylopropanecarboxylate] was much more stable under simulated (and later practical) field conditions than any previous synthetic pyrethroid. The combination was particularly favourable because it had greater insecticidal activity than would have been predicted from structure–activity results for the individual components and, moreover, mammalian toxicity was below that of the natural pyrethrins.

The concept and practical demonstration that pyrethroids, with their greater activity and mammalian toxicity lower than that of other classes of insecticide, could be structurally modified to give stability adequate to control agricultural pests was a striking advance that significantly affected the pattern of insecticide development and use world-wide.

7.8.1 Progress in Photostable Pyrethroids

Brown and coworkers reported the insecticidal activity of racemic 5-benzyl-3-furylmethyl 3-(2,2-dibromovinyl)-[95] and 3-(2,2-difluorovinyl)- 2,2-dimethylcyclopropanecarboxylates,[96], but since they did not assay esters from other alcoholic components, failed to discover the full significance of esters of these acids. However, Ohno and his collaborators in the Sumitomo Chemical Company next reported two very important developments. First, in screening as insecticides a range of 5B3F esters derived from acids not necessarily previously considered as possible pyrethroid constituents, activity was detected in the ester (**4c**) from 2-phenylbutyric acid. Then, optimization of potency from this promising precursor led to 3-phenoxybenzyl (*RS*)-2-(4-chlorophenyl)-3-methylbutyrate.[27,97] As noted in Section 7.3, in retrospect this advance can be seen to exploit the concept first proposed by Staudinger and Ruzicka (*loc. cit.*) of breaking one of the bonds of the cyclopropane ring to give an acid (**4a**) with an isopropyl group sterically equivalent to the ring *gem*-dimethyl. Subsequent exploration indicated that a cyclopropyl, but not a larger group, could apparently serve the same function as the *gem*-dimethyl group in chrysanthemic acid.[98]

The second advance by the Sumitomo workers was a definitive observation in a region of structure–activity investigations that had been pursued since the effectiveness of benzyl chrysanthemates had been discovered, *viz.* the influence of substituents at the α-position. Barthel[59] summarized results that showed little improvement in toxicity or synergistic activity in α-substituted piperonyl chrysanthemates. α-Methyl[71] and α-ethynyl[99] substituents did not improve the activity of furylmethyl and benzyl chrysanthemates. However, in a 1970 patent[100] Sumitomo chemists reported α-ethynyl-3-phenoxybenzyl chrysanthemate among other esters, and in 1971,[101] α-cyano-3-benzyl and 3-phenoxybenzyl chrysanthemates. {[About this important discovery, Itaya[74] comments 'We had the luck to prepare and assess the α-ethynyl analogue of furamethrin [5-(prop-2-ynyl)furfuryl chrysanthemate] in the course of the preparation of impurities possibly present in a technical sample of furamethrin', an example of the rewards of meticulous and thorough research.]} α-Cyano was the first substituent actually to increase the activity of esters above the level of potency of unsubstituted compounds. The (*RS*)-α-cyano-3-phenoxybenzyl (*RS*)-2-(4-chlorophenyl)-3-methylbutyrate is the important commercial pyrethroid fenvalerate[102] (**4d**) (see Section 7.12).

As the patent covering the α-cyano modification reached the Patent Office in London, various combinations involving the photostable 3-(2,2-dihalovinyl)-2,2-dimethylcyclopropane acids were being examined at Rothamsted.

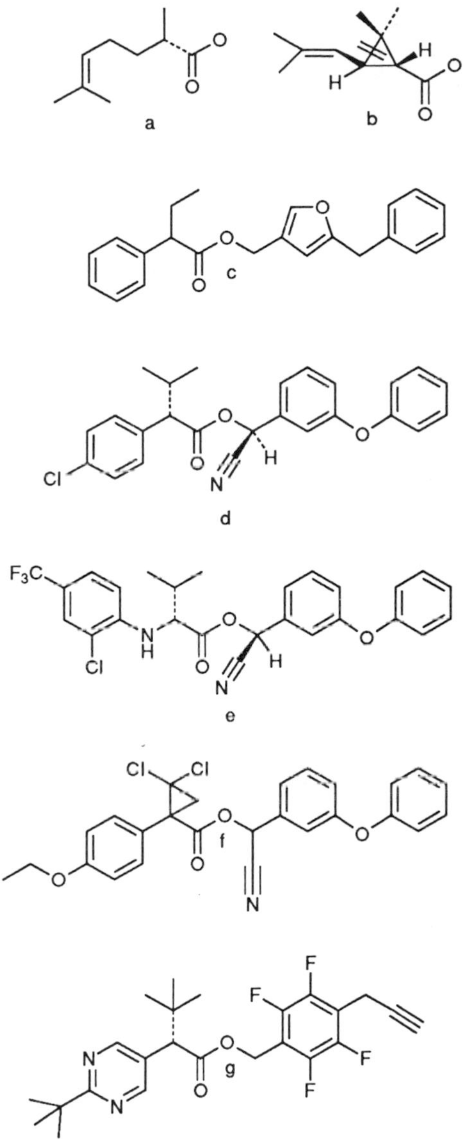

Figure 7.4 *Non-cyclopropane pyrethroids*

3-Benzyl α-cyanobenzyl, and α-cyano-3-phenoxybenzyl esters were, therefore, prepared and found to have valuable properties. Not only were some combinations sufficiently stable to protect agricultural crops in sunlight, but they were also exceptionally active against important crop (particularly cotton) pests. To supplement the 3-phenoxybenzyl ester, NRDC 143, (*RS*)-α-cyano-3-phenoxybenzyl (1*RS*)-*cis-trans*-3-(2,2-dichlorovinyl)-2,2-dimethylcyclopropanecarboxylate (NRDC 149, cypermethrin, **5b, 5d,** and isomers) was found to

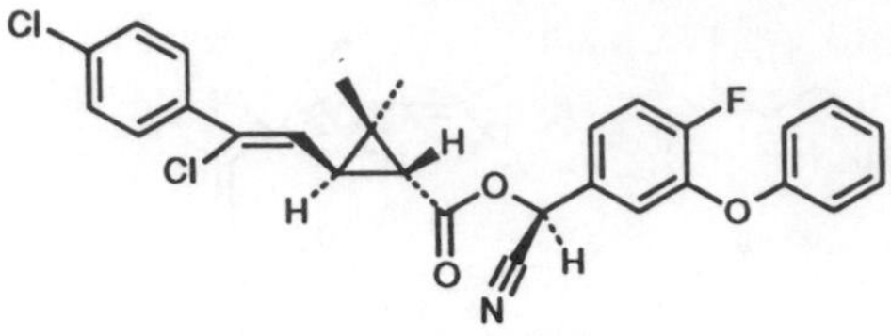

a Flumethrin

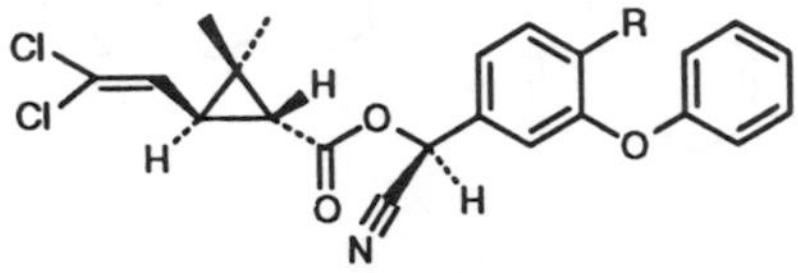

b R = H, Cypermethrin
c R = F, Cyfluthrin

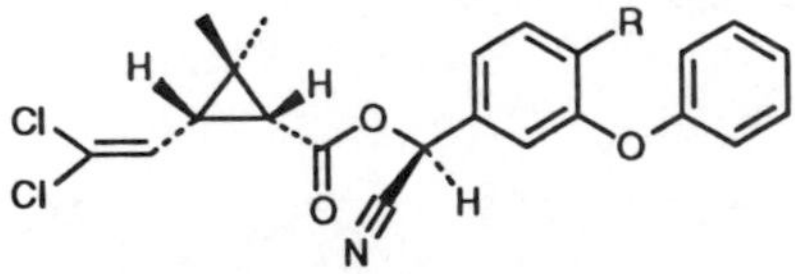

d R = H, Cypermethrin
e R = F, Cyfluthrin

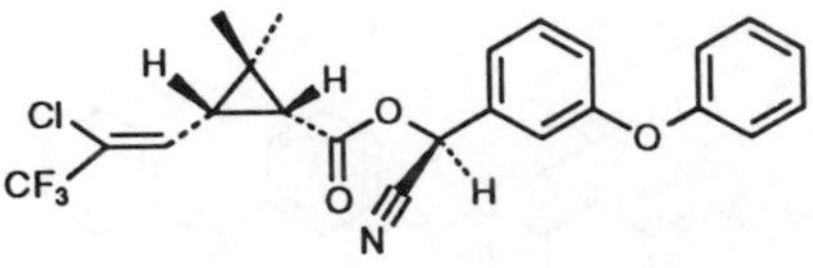

f

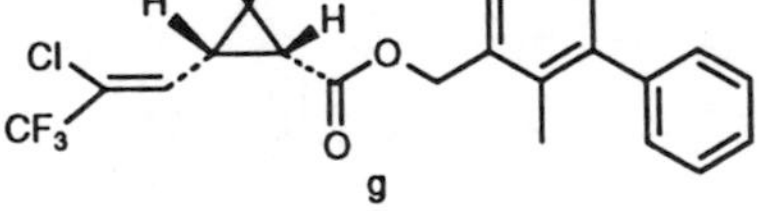

g

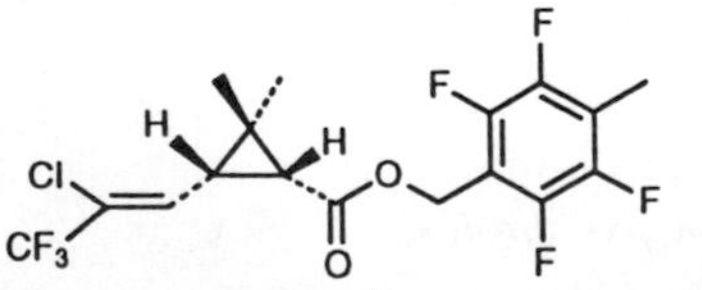

h

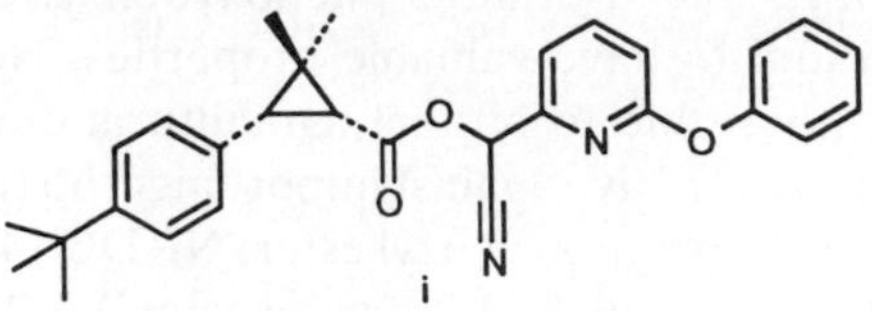

i

Figure 7.5 *Various more photostable pyrethroids*

be an even more active insecticide, with stability adequate to protect crops for significant periods in the field; it also retained low mammalian toxicity. Preliminary results indicated that the (*RS*)-(1*R*)-*cis*-stereoisomers were particularly effective[103]

Cypermethrin is a mixture of eight stereoisomers. To establish a firm base for structure–activity relationships a determination of the relative potencies of the individual isomers was necessary, but at the time no appropriate routes for separating the optical isomers of racemic cyanohydrins for subsequent esterification were known. The next steps were, therefore, taken against the following background.

7.8.2 Influence of Stereochemistry on Potency of Pyrethroids

Much evidence had established that only intact esters of chrysanthemic acids and related compounds and of pyrethrolone, cinerolone, allethrolone, *etc.*, are active insecticides, and therefore the potency of a pyrethroid depends on the stereochemistry of both the acid and the alcohol components. Assay of allethrin isomers, prepared by LaForge *et al.*,[56] showed that (1*R*)-*trans*-(**1a**) and (1*R*)-*cis*-(**1c**) chrysanthemic acids gave much more active esters than the (1*S*)-*trans* and (1*S*)-*cis* forms. Stereochemically related (1*R*)-*trans*-(**1d**) and (1*R*)-*cis*-(**1e**) 3-(2,2-dihalovinyl)-2,2-dimethylcyclopropanecarboxylates were comparably much more active than the (1*S*)-isomers.[94] However, in 1973 when cypermethrin was first prepared.[89,104] the relative potency of esters from the optical isomers of any 4-hydroxycyclopent-2-enones [pyrethrolone (**1f**), cinerolone (**1g**), *etc.*] or, specifically, from those of any α-cyano-3-phenoxybenzyl esters had not been determined.

7.9 Esters of Isomeric Cyanohydrins – Deltamethrin

Initial attempts to isolate (+)- and (−)-forms of α-hydroxy-3-phenoxyphenylacetonitrile following esterification with acids commonly used for resolution failed because the labile alcohol racemized again on attempted cleavage of the separated isomeric esters. In association with Pulman a decision was, therefore, taken to attempt separation of the isomers as esters with an appropriate resolved pyrethroid acid. Earlier, esters of (±)-cinerolone and (±)-allethrolone with naturally derived (+)-*trans*-chrysanthemic acid had been separated via their semicarbazones.[105] Here, a direct approach was explored. Based on experience that *cis*-substituted cyclopropane carboxylic acid derivatives tended to crystallize more readily than the equivalent *trans*-compounds, and knowing that bromo-compounds usually had higher melting points and greater tendency to crystallize than the corresponding chloro-compounds, (1*RS*)-α-cyano-3-phenoxybenzyl (1*R*)-*cis*-3-(2,2-dibromovinyl)-2,2-dimethylcyclopropanecarboxylate was chosen for synthesis. The required acid was obtained using a generous gift of the corresponding (1*R*)-*cis*-chrysanthemic acid (**1c**) by Roussel-Uclaf[106] and, upon esterification and manipulation of the mixture of isomers under solvent, Pulman obtained a crystalline compound (m.p. 100°C, coded NRDC 161), named deltamethrin and subsequently established as (*S*)-α-cyano-3-phenoxybenzyl (1*R*)-*cis*-3-(2,2-dibromovinyl)-2,2-dimethylcyclopropanecarboxylate (**3g**). The absolute configurations of deltamethrin and its related biologically active isomers were confirmed by X-ray analysis.[107–109]

When tested against a wide range of insect species, many of economic importance, deltamethrin was found to have greater insecticidal activity than any previous insecticide, irrespective of class, and at the low application rates made possible by its potency presented a very low hazard to mammals; deltamethrin was stable enough in the environment to protect crops at critical periods of pest infestation.

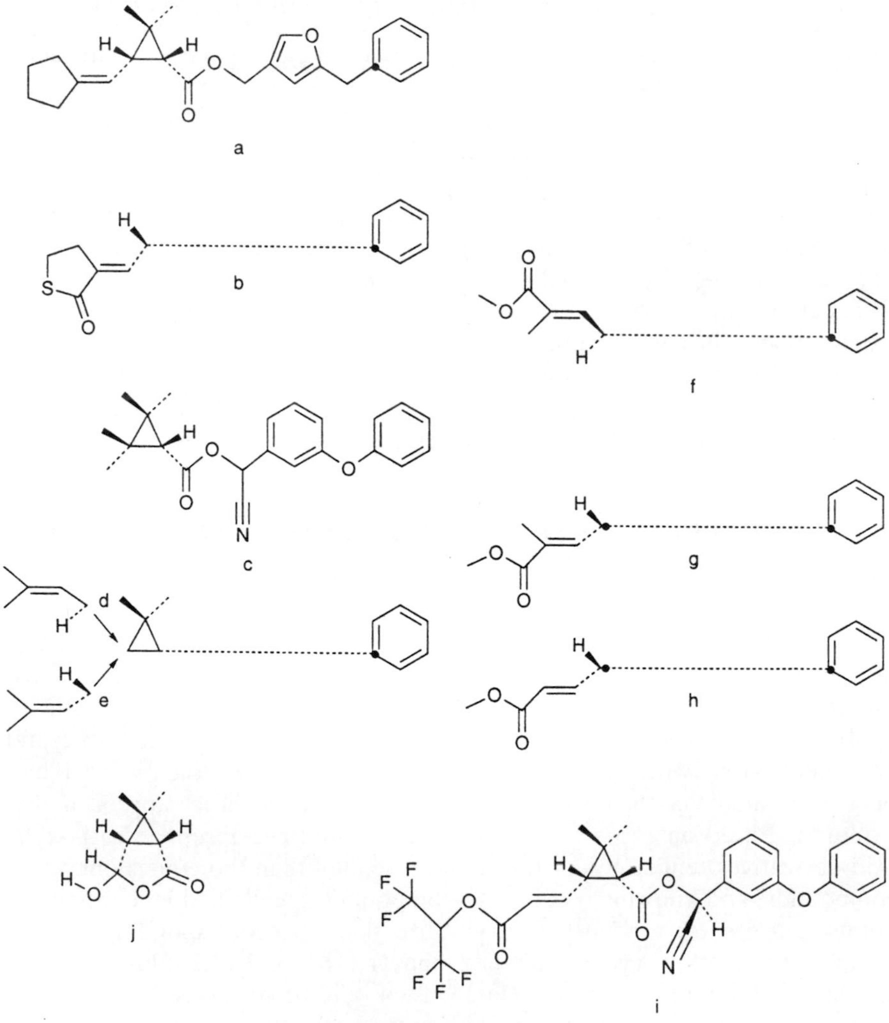

Figure 7.6 *Roussel-Uclaf developments*

7.10 Progress from Permethrin, Cypermethrin, and Deltamethrin

The prospect of insecticides with many of the favourable properties of the earlier photolabile pyrethrins and pyrethroids, yet stable enough to protect agricultural crops, was a powerful incentive to the commercial production of permethrin, cypermethrin, and especially deltamethrin. Thanks to their long-established proficiency in synthesis, especially of chiral compounds, the chemists of Roussel-Uclaf rapidly developed a very efficient manufacturing process for deltamethrin, under licence from NRDC (now the British Technology Group). The high melting point, 100°C, facilitated isolation.

Martel[110] has discussed this commercial synthesis in detail and a monograph on deltamethrin has been published.[111] A new method for the industrial synthesis of biocartol (**6j**), an intermediate in the production of pyrethroids, has been developed, replacing and simplifying more than ten of the production stages for deltamethrin.[112]

The dichloro-analogue (**5d**) (NRDC 182) of deltamethrin is a comparably powerful insecticide, but with a lower melting point, 54°C,[113,114] is somewhat less readily isolated. Commercially, to compensate for this, the product alpha-cypermethrin[115] is manufactured using a synthesis which provides *cis*-rich dichlorovinyl acid. After esterifying this with racemic α-hydroxy-3-phenoxyphenylacetonitrile, a crystalline enantiomer pair, m.p. 80.5°C (pure), is isolated. This has very high insecticidal activity, achieved without a specific resolution step. A related variant, beta-cypermethrin[116] contains *cis*-(**5d**) and *trans*-(**5b**) enantiomeric pairs in a ratio of 2:3.

7.11 Pyrethroids Developed after Cypermethrin – Acaricides, *etc*.

The practical value of cypermethrin (NRDC 149) having been established, other very active pest control agents were developed by modification of the alcohol and acid components.

In the 3-phenoxybenzyl esters studied, only an (S)-α-CN substituent enhanced activity; at the 2-, 6-, or (R)-α-positions it was diminished or eliminated.[113] Similarly, only at position 4 of the phenyl ring of cypermethrin does fluorine enhance insecticidal potency;[117] this effect is exploited in the very active insecticides cyfluthrin (**5c, 5e**) [(RS)-α-cyano-4-fluoro-3-phenoxybenzyl (1RS)-*cis-trans*-3-(2,2-dichlorovinyl)-2,2-dimethylcyclopropanecarboxylate][118] and beta-cyfluthrin,[119] which contains two enantiomeric pairs analogous to beta-cypermethrin.

In structural modifications which might make synthetic pyrethroids more volatile, the influence of replacing Cl by CF_3 in the dihalovinyl side chain was investigated;[120] in cypermethrin this approximately doubled the activity, and established the potential for important new products investigated by ICI,

FMC, and Montedison. In this way were developed cyhalothrin (**5f**), (*RS*)-α-cyano-3-phenoxybenzyl (*Z*)-(1*RS*)-*cis*-3-(2-chloro-3,3,3-trifluoropropenyl)-2,2-dimethylcyclopropanecarboxylate acid and, even more active, the crystalline enantiomeric pair lambda-cyhalothrin, equivalent to alpha-cypermethrin. Lambda-cyhalothrin is more active than cypermethrin, with the additional benefit of greater acaricidal activity. The potentially even more powerful 4-fluoro-3-phenoxy combination is not actually available commercially.

The same acid component esterified with 2,3,5,6-tetrafluoro-4-methylbenzyl alcohol (**5h**) is the first soil-active pyrethroid, tefluthrin, which, with relatively high vapour pressure, can develop a concentration lethal to pests such as the corn root worm in the interstices of the soil.[121] The acaricide bifenthrin (**5g**)[122,123] is also an ester of this acid, but with an original alcohol developed as a result of steric structure–activity analysis.[72] Omission of the oxygen link in 3-phenoxybenzyl alcohol esters usually diminishes potency drastically, but the 2-methyl substituent ensures maintenance of the mutually skew configuration between the two rings necessary for activity which, in this series, is not enhanced by an α-CN substituent.

These acaricidal compounds were based on 3-(2-chloro-3,3,3-trifluoropropenyl)-2,2-dimethylcyclopropanecarboxylate esters. However, the simpler α-cyano-3-phenoxybenzyl 2,2,3,3-tetramethylcarboxylate (**6c**), fenpropathrin has significant acaricidal activity. It was reported in 1981,[124] when its potential for stability in light was recognized from the developments of permethrin, cypermethrin, and fenvalerate.

The addition of bromine to cypermethrin and to deltamethrin gives, respectively, tralocythrin[107] and the commercial product tralomethrin (**3h**).[125] These materials may owe their activity to regeneration of the parent compounds, for the two diastereoisomers, epimeric at C_1' (ratio 3:2), from deltamethrin are, unexpectedly, about equipotent, unusual for chiral isomers in pyrethroids. This result may be best explained by their chemically similar rates of loss of bromine, an identified process of metabolism in insects[126] and mammals,[127] and of photochemical conversion on glass[128] or leaves.[129] Moreover, the tetrachloro analogue, which might be expected to decompose less readily is no more active than would be predicted for a dimethyl cyclopropanecarboxylate.[130]

Replacing the (*Z*)-Cl in (±)-*Z*- [but not in (±)-*E*] cyfluthrin with 4-chlorophenyl gave compound (**5a**), flumethrin, with exceptional activity against cattle ticks.[131] That other isomers are less active[132] is a remarkable example of the specificity of the insecticidal activity of pyrethroids, properties almost without parallel in other classes of pesticides.

7.12 Photostable Pyrethroids: Fenvalerate and Related Compounds

As described in Section 7.8, development of one range of photostable pyrethroids depended upon the discovery that, like 5B3FA, its isostere 3POBA gave esters with enhanced insecticidal activity. Whereas 3-phenyoxybenzyl (1*RS*)-*cis-trans*-2,2-dichlorovinyl)-2,2-dimethylcyclopropanecarboxylate itself had adequate activity to be developed as a commercial insecticide [NRDC 143, permethrin (**3e, 3f**, and isomers)] the potential of the non-cyclopropane pyrethroid acids introduced by Sumitomo was most effectively optimized in (*RS*)-α-cyano-3-phenoxybenzyl (*RS*)-2-(4-chlorophenyl)-3-methylbutyrate (fenvalerate), to be followed some years later by esfenvalerate (**4d**),[133,134] when the separate, most active of the four isomers became available. For a number of years the photostable pyrethroid market was dominated by (% of total, year of commercial introduction): fenvalerate (25%, 1976), permethrin (3%, 1977), deltamethrin (31%, 1977), cypermethrin (21%, 1977), alpha-cypermethrin (6%, 1983), and esfenvalerate (2%, 1986).

Four non-cyclopropane pyrethroids followed the innovative fenvalerate type of structure. The related partly resolved compound flucythrinate (HF_2CO instead of Cl in fenvalerate)[135,136] was proposed in 1981. An important racemic product, fluvalinate, now replaced by the partly resolved tau-fluvalinate (**4e**), was reported by Henrick *et al.*[137,138] who explored the influence of a single atom link introduced between the α-carbon of fenvalerate and the aromatic ring. Earlier, a compound with a methylene group in this position had been found ineffective,[139] but with –NH– and the appropriate aromatic substitution (Cl, CF_3) the product had useful activity, including control of spider mites and non-toxicity and non-repellency to honey bees.

Zeneca Agrochemicals have developed the substituted non-cyclopropane pyrimidinyl compound (**4g**) which, remarkably, has high miticidal, but no insecticidal, activity.[140] This may be considered the culmination of a sequence starting with esters of analogues of chrysanthemic acid in which aryl rings replaced the 2-methylprop-1-enyl side chain[141] and proceeding via insecticidal compounds, such as Nissan NCI 85913[142] (**5i**), with the application of much recent synthetic chemistry and innovative structure–activity exploration.[143]

Holan *et al.*[144,145] exploited a biological and structural link discerned between pyrethroids and DDT in developing cycloprothrin[146] (**4f**), which, although containing a cyclopropane ring, is structurally more related to fenvalerate and fluvalinate than to permethrin, *etc.*

As discussed below (Section 7.15), the generally high toxicity of pyrethroids to fish limits their range of application. However, cycloprothrin has a fish toxicity lower than that of many other pyrethroids and can be used against pests of rice. It illustrates the trend with a number of more specialized compounds having a particular spectrum of biological activities developed after the initial more photostable compounds discussed above.

7.13 Further Developments from the Structure of the Natural Pyrethrins

Consideration of the complex structures of the natural pyrethrins, as with other biologically active compounds, suggests many modifications which might lead to alternative active products. This review indicates some sequences that led eventually to the photostable pyrethroids permethrin, cypermethrin, deltamethrin, fenvalerate, and related products. Earlier refinements of Schechter's synthesis of 4-hydroxycyclopent-2-enones[26] made allethrin (**1ai** with **1ci**, and their optical isomers), then bioallethrin and *S*-bioallethrin (**1ai**) (the two latter compounds having superior properties) became commercially available through the initiative of the chemists of Roussel-Uclaf and of the Sumitomo Chemical Company.[147] Gersdorff and Piquett[58] investigated the activity of a related compound (**1am** and isomer) with prop-2-ynyl in place of allyl, but perhaps due to an inferior sample of the alcohol the favourable properties of the derived chrysanthemate were not fully recognized. An improved recent synthesis of the alcoholic component, via the readily available 2,5-dimethylfuran (sylvan) and enzymatic hydrolysis and chemical transformation of the acetate leads to the resolved alcohol from which the commercial product prallethrin[148,149] is derived. As the (*S*)-2-methyl-4-oxo-3-(prop-2-ynyl)-cyclopent-2-enyl (1*R*)-*cis-trans*-2,2-dimethyl-3-(2-methylprop-1-enyl)cyclopropanecarboxylate (**1am** + **1cm**), this has properties superior to those of allethrin in a number of domestic and public health applications. The commercial availability of this (*S*)-prop-2-ynyl ketone[40,149] has made attainable a practicable synthesis of esters of (*S*)-pyrethrolone, such as pyrethrin I.

As well as manufacturing bioallethrin, *S*-bioallethrin, bioresmethrin, and deltamethrin, the chemists of Roussel-Uclaf alone among pyrethroid producers have pursued structure–activity relationships based on pyrethrin II and other esters of pyrethric, rather than of chrysanthemic, acid.[150] Their key observation was that, whereas esters of pyrethric acid [such as the α-cyano-3-phenoxybenzyl ester (**6f**)] and of its *cis*-isomer (**6g**) had similar insecticidal activities, the ester of the *cis*-nor acid (**6h**) was as much as seven times more active. This finding was exploited in the 2,2,2-trifluoro-1-(trifluormethyl)ethyl ester acrinathrin (**6i**), which is active against a wide range of phytophogous mites as well as sucking insect pests and Lepidoptera.[151]

7.14 Pyrethroids with Rapid Action, Knockdown Activity – Volatility

By tradition, a highly valued characteristic of the natural pyrethrins and of the first synthetic pyrethroids, allethrin and its isomers, was rapidity of action against flying insects. Early synthetic pyrethroids were, therefore, evaluated against the readily accessible test insect, the housefly *Musca domestica* L., and attention was directed to chemical structures and physical properties favouring knockdown as important. Sawicki and Thain,[152] however, showed that pyrethrin II (**1bf**) knocked-down houseflies faster than pyrethrin I (**1af**), and the concept developed that this property was associated with more polar pyrethroids which were, however, generally relatively less effective killing agents.[153,154]

An important application for which compounds with rapid knockdown, heat stability, and volatility are necessary is in mosquito coils, the smoke from which sequentially deters, expels, interferes with host finding, inhibits biting, knocks down, and kills. *S*-Bioallethrin is very active for this purpose and over half the current production of allethrolone chrysanthemates is formulated in coils and vaporizing mats.[155]

After allethrin, the next commercial synthetic pyrethroid was tetramethrin (**7ab**), discovered by chemists of the Sumitomo Chemical Company as a result of empirical screening of commercially accessible esters of chrysanthemic acid.[156] Activity initially detected in phthalimidomethyl chrysanthemate was optimized in 3,4,5,6-tetrahydrophthalimidomethyl (±)-*cis,trans*-chrysanthemate (**7ab**), a contact insecticide with strong knockdown action. It is

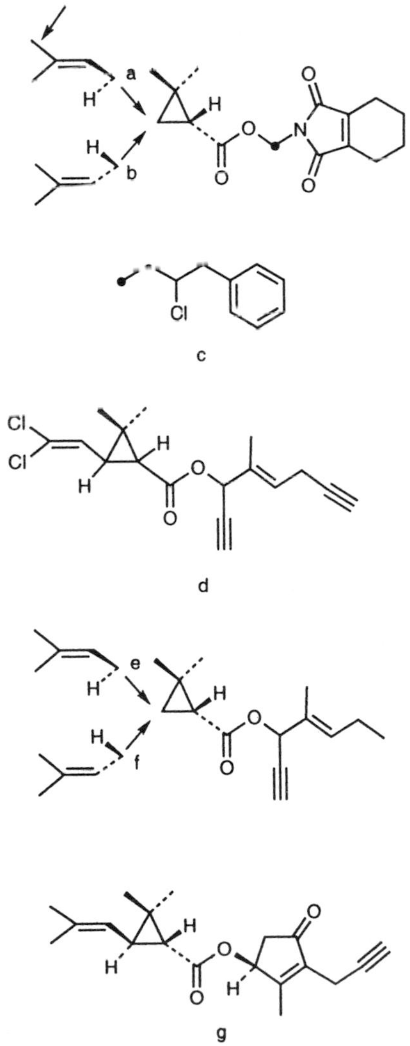

a

b

c

d

e

f

g

Figure 7.7 *Pyrethroids with knockdown activity and volatility*

frequently formulated with the synergist piperonyl butoxide (**1s**), plus a pyrethroid with more powerful killing activity.

The simple 2,2,3,3-tetramethylcyclopropanecarboxylate of allethrolone, terallethrin [*cf.* fenpropathrin (**6c**)] was evaluated for knockdown activity,[157] but further significant conclusions were reached examining 5-benzyl-3-furylmethyl 2,2-dimethylcyclopropanecarboxylates with varied substituents at C-3.[78] The activity of 5-benzyl-3-furylmethyl (1*R*)-*cis*-chrysanthemate (**3b**)[79] for both kill and knockdown led, *via* the corresponding ethanochrysanthemate (**6a**), to a structural analogue, the thiophene derivatives (**6b**), RU 15 525, the most effective commercially available knockdown agent. The superior knockdown properties were realized only in the 3-furylmethyl ester and not in the isosteric 3-phenoxybenzyl ester.[158]

A persistent theme in structure–activity investigations has been the assessment of esters of alcohols in which a double or triple bond replaces the cyclopentenone ring of allethrin or the furan ring of resmethrin. An early example of a compound of this type was butethrin (**7abc**).[159]

In another development, the compound enediynyl 3-(2,2-dichlorovinyl)-2,2-dimethylcyclopropanecarboxylate (**7d**), related to prallethrin (**7g**, mixed stereochemistry) evolved into empenthrin (**7ef**), which is relatively volatile and has been introduced as a domestic insecticide especially effective against the clothes' moth in enclosed volumes, such as wardrobes.[160]

7.15 Recent Developments – Non-Ester Pyrethroids

The cyclopropane ring, for more than five decades considered an important feature of pyrethroids, was shown not to be essential by the high activity of compounds such as fenvalerate (**4d**).[27] The status of the ester link was itself also the subject of comparable speculation, but the potencies of some test compounds reported in 1969 were low, so that activity typical of pyrethroids could not be confirmed.[161] The first definite indication that a non-ester could behave as a pyrethroid was given by a ketoxime *O*-ether (**8a**), with a structure like fenvalerate, but lacking a cyano group.[162]

The most significant practical innovations in non-ester pyrethroids came from the work of the chemists of Mitsui Toatsu from 1980 onwards.[163] From a weakly active compound (**8c**), considered as derived from fenvalerate (**8b**) they produced an ether. Then, omission of the CN group and substitution of $-CH_2-$ for –CO– gave a compound whose activity was optimized in etofenprox (**8d**), MTI 500, an insecticide for a wide range of applications and, most important, with low toxicity to fish.[164] Even a hydrocarbon analogue, MTI 800 (**8e**), and compounds related to it were active.[163]

In other studies, the compound 5-[4-(4-chlorophenyl)-4-cyclopropylbutyl]-2-fluorophenylphenyl ether (**9a**) was found to be more active than MTI 800; the 4-trifluoromethyl (**9b**) and 4-trifluoromethoxy (**9c**) hydrocarbons were also effective.[165] These compounds had low toxicity to fish and to mammals,[166] substitution of cyclopropyl for isopropyl having raised insecticidal potency by as much as 100 times. Analogously, the cyclopropyl isostere of fenvalerate had

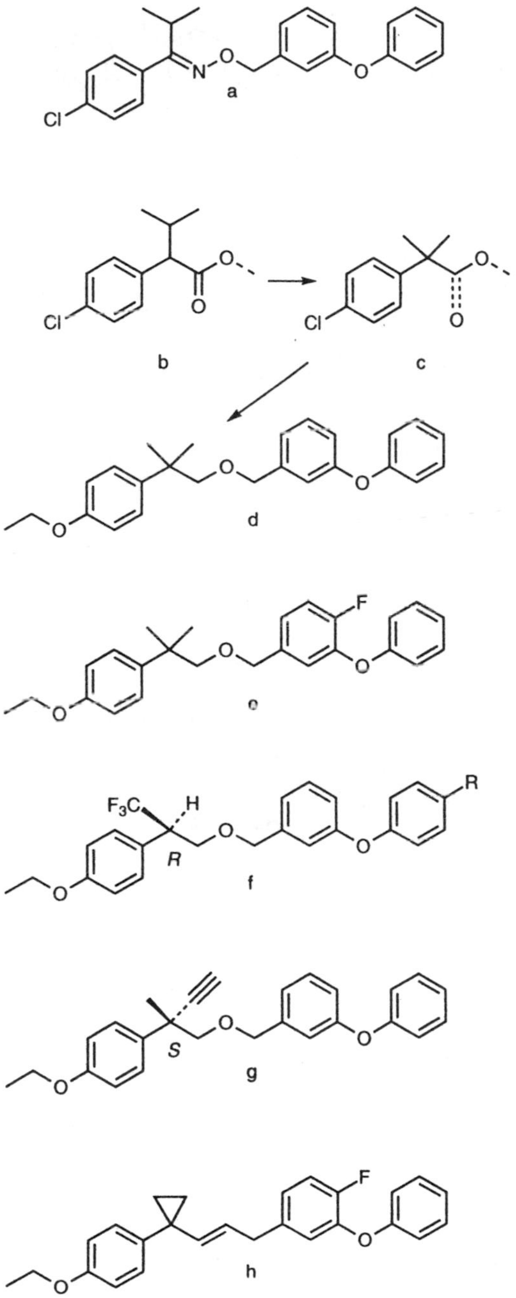

Figure 7.8 *Non-ester pyrethroids*

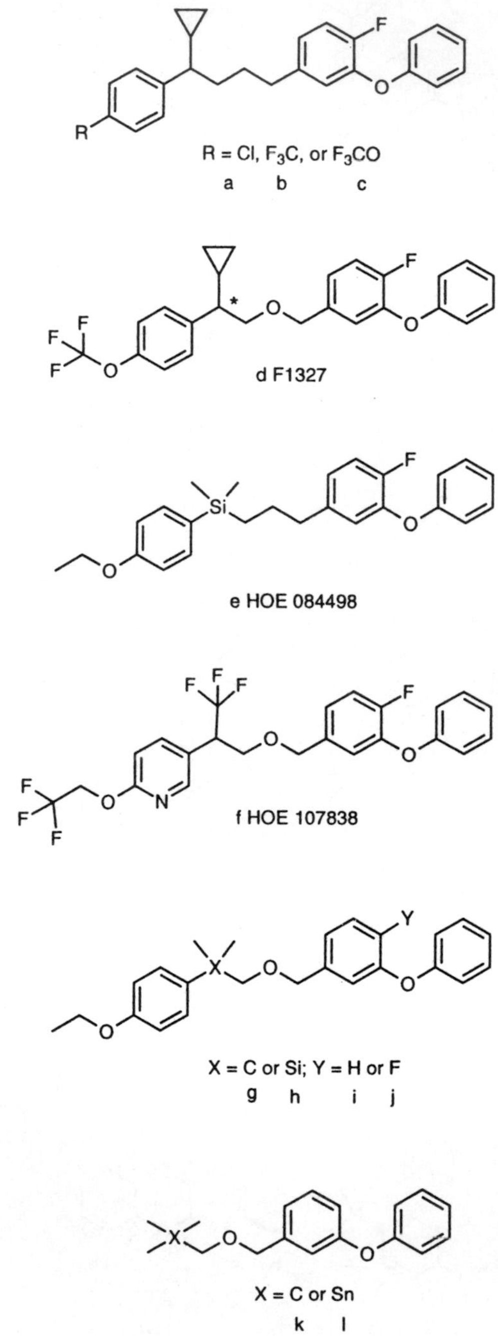

R = Cl, F_3C, or F_3CO
a b c

d F1327

e HOE 084498

f HOE 107838

X = C or Si; Y = H or F
g h i j

X = C or Sn
k l

Figure 7.9 *Non-ester pyrethroids*

undiminished insecticidal activity, but was less toxic to zebra fish and to rats.[98] The single (−)-isomer, F 1327 (**9d**), produced by an efficient enzymic resolution, is reported to be equivalent to bifenthrin, lambda-cyhalothrin, and deltamethrin in insecticidal potency and also to be acaricidal.[167] Comparison of pyrethrin I and F 1327 illustrates the remarkable progress in structure–activity relationships and simplication now achieved in pyrethroids.

Bushell[168] has analysed the stages of development which led to flufenprox (**8f**) (ICI A568), a compound with low toxicity to mammals and fish used for control of rice pests. Replacement of CH_3 by CF_3 in appropriate analogues of MTI 500 led to a greater than ten-fold increase in potency. Like that of F 1327, the structure of flufenprox as an insecticide ultimately evolved by successive modifications of pyrethrin I is noteworthy because it does not contain an ester, a *gem*-dimethyl group, or a cyclopropane ring.

Several non-esters, *e.g.* (**8e**), (**9f**), have significant acaricidal activity;[143] indeed, the insecticidal compound silafluofen (**9e**), HOE 084498, was systematically modified to give the acaricide (**9f**) HOE 107838.[169] The mode of action of two silicon and two carbon analogues (**9g–9j**) was similar and pyrethroidal.[170] Potential replacements for carbon in the chain of these compounds are not restricted to silicon; the simple tin derivative (**9l**) was some ten times more active to houseflies than the corresponding neopentyl compound (**9k**), and both behaved as typical pyrethroids and unlike trimethyltin chloride.[171]

As noted, a favourable property of some non-ester pyrethroids is low toxicity to fish.[172] The toxicity of pyrethroids to fish is associated with their lipophilicity and their strong adsorption by the gills, even from low concentrations in water. However, their impact in practice is less serious than laboratory measurements might suggest, because the concentration of pyrethroids in natural aqueous media is much diminished by adsorption onto river banks, pond sediments, and organic matter with which they are in contact.

7.16 Future Developments

The review of structures in the preceding sections indicates the ever-widening scope for discovering new modifications of the basic structures which retain useful biological activity. Compounds such as bifenthrin, with enhanced acaricidal activity, and others (Section 7.15) with acaricidal activity, but no insecticidal activity, indicate the potential for developing compounds to control particular pests or for specific applications. To develop this theme further, Figure 7.10 shows the structures of some recently patented compounds. Although details of the biological activity of these products are not available, they indicate that further advances are technically feasible. However, economic constraints, including the costs of bringing newly developed products to the market, may seriously limit the introduction of compounds for special applications. Bushell and Salmon (Zeneca)[143] recently reported new compounds with acaricidal activity at levels equivalent to or markedly superior to

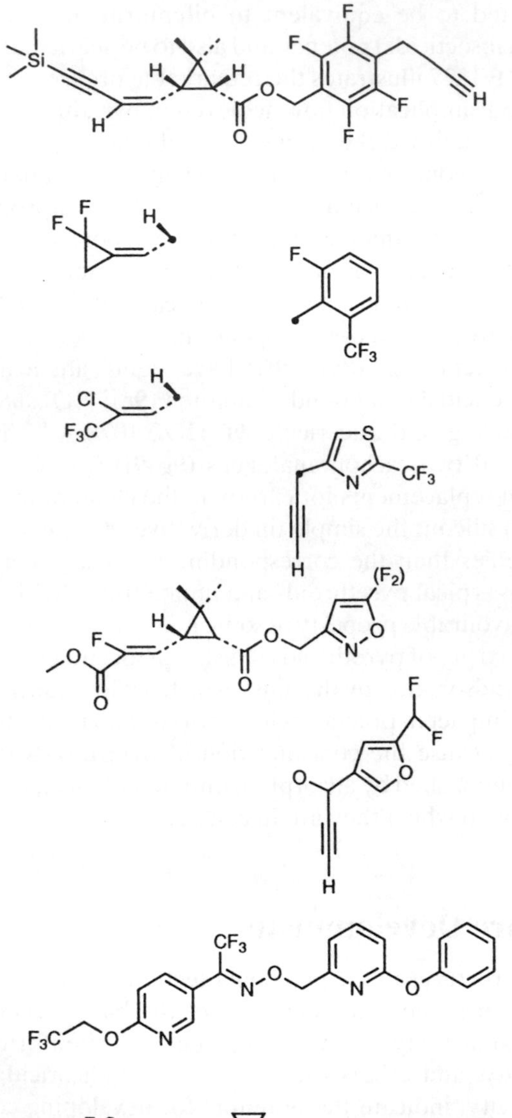

Figure 7.10 *Recent compounds*

commercial standards, but stated that the financial case for the lead compounds had been judged insufficiently attractive for commercialization.

7.17 Mode of Action of Pyrethroids and Resistance

The action of pyrethroids in both insects[173,174] and mammals[175] is necessarily influenced by many factors, such as rate of penetration, susceptibility to metabolism (detoxification) and potency at the site of action, each of which is related to the chemical and physical properties of the compounds. Pyrethroidal activity also depends on the overall shape of the molecules, which may respond to small changes in conformational preference, properties not easily quantified.

Polarities of pyrethroids (expressed as log *P* octanol/water values) range from 4.2 (tetramethrin, **7ab**) through 5.0 (allethrin, **2ai**), 5.9 (resmethrin, **2ao**, and pyrethrin I, **1af**), 6.4 (deltamethrin, **3g**), 6.8 (fenvalerate, **4d**, and permethrin, **3ef**), to 9.5 (MTI 800, **8e**); these values, and also those of other classes of insecticides, relate well with many aspects of behaviour. The values for pyrethroids, in particular, indicate that they will partition preferentially into the lipid rather than into the aqueous tissues of complex organisms.[8]

The manifestations of pyrethroid action (hyperexcitation, tremors, and convulsions followed by paralysis and death) are similar in insects and in mammals, and, with other evidence, indicate that the sodium channels of the nervous system are sites of action common to both types of organisms. Certainly in insects, pyrethroidal interaction with sodium channels to prolong their opening appears to account for most symptoms of pyrethroid poisoning and may explain many of the actions on central and peripheral neurones.[173] The toxicological properties of pyrethroids favourable to mammals depend on the relative levels at which effects occur; for instance, the LD_{50} of deltamethrin applied topically to houseflies is 0.01 mg kg^{-1} implying concentrations in nerve membranes, to which the toxicant has more or less direct access, as low as 1×10^{-12} M.[174] In contrast, after external or oral administration to mammals (typical rat oral LD_{50} for deltamethrin, 13 mg kg^{-1}), pyrethroids are largely converted by hydrolytic or oxidative attack into polar metabolites, which are then eliminated in the faeces or urine, unchanged or as conjugates, before sensitive sites can be reached.[175]

If insect pests did not develop resistance, the current range of insecticides, including the pyrethroids, combined when appropriate with alternative methods of control (semiochemicals, attractants, repellents, *etc.*)[176,177] should be adequate to deal with many insect-related problems. Unfortunately, successive use of various insecticide groups has led to broadly dispersed cross-resistant insects and, by 1990, 504 species had been reported to resist insecticides of one or more groups,[178] and at least 50 economically important pests resisted pyrethroids. Farnham[179] showed that a single gene was responsible for site insensitivity associated with the widely recognized cross-resistance (*kdr*) between DDT and pyrethroids, which delays the onset of knockdown, and which was common to all pyrethroids and DDT.[180]

The extremely grave economic and health consequences[181,182] of widespread resistance to insecticides emphasizes the importance of understanding the modes of action and metabolism of insecticides in general, and especially of pyrethroids, as deeply as possible.

Resistance may arise from differences in access to the site of action (penetration), from increased ability to store or to metabolize (detoxify) the insecticide, or from changes in the action sites. This last may involve modification of the nature or density of the target channels or of the membrane in which they are embedded. Recognizing the implications of recent molecular biological research on the primary structure of sodium channels, Devonshire *et al.*[183] anticipate the ability to characterize regions of the protein in houseflies which differ in susceptible and resistant individuals. Recent work has identified linkage between a mutation at the sodium channel locus and the *kdr* phenotype, providing the first clear genetic evidence for their association and enabling future detailed structural studies of this gene.[184]

7.18 Metabolites of Pyrethroids – Synergism

Pyrethroids, and other xenobiotics, are metabolized in and eliminated from[185] vertebrates and invertebrates by one or more of the following mechanisms:

- Ester cleavage by esterase and oxidase enzymes
- Oxidation
- Hydroxylation
- Conjugation (carboxylic acid glucuronides, sulfate conjugates of phenols)

Since all known and predictable pyrethroids are lipophilic, they will generally be metabolized by one or more of these reactions after persisting long enough to control the pest against which they were applied. The products of metabolism of pyrethroids are polar compounds, which will be bound to soil particles and further degraded and therefore, like the parent compounds, will not be able to persist and contaminate the environment. The behaviour of pyrethroids contrasts with that of chlorinated hydrocarbon insecticides, of which the products of aerobic and non-aerobic metabolism are also non-polar compounds which may persist in the environment for a significant number of years.

The natural pyrethrins and earlier synthetic pyrethroids, such as allethrin, tetramethrin, resmethrin, and bioresmethrin, are readily degraded photochemically and are, therefore, suitable only for applications not involving prolonged exposure to daylight. The more photostable compounds (permethrin, fenvalerate, and later products), like the labile compounds, retain multiple sites at which they are metabolized in biotic systems and, although still susceptible to some photochemical degradation, the reactions by which this proceeds are much slower because light is absorbed and energy is transferred less efficiently between molecules.

The rapid and intense insecticidal activity of the natural pyrethrins, especially against flying insects in light-protected situations, can be usefully extended by application (usually in aerosols) with 5–10 parts of a synergist, such as piperonyl butoxide (**1s**), which acts as a substrate to absorb mixed-function oxidase degradative activity otherwise available to inactivate the esters. Combinations of piperonyl butoxide or other synergists[45] with most synthetic pyrethroids are relatively less effective than with the natural pyrethrins; piperonyl butoxide is itself photolabile and is wasted unless formulated so that toxicant and synergist molecules are present at the same time at the site of action.

Authorities (Lichfield,[175] Demoute,[185] Vijverberg and van den Bercken,[186] and Aldridge[187]) have reviewed various non-insecticidal aspects of the properties of pyrethroids.

7.19 Conformations of Pyrethroids at the Site of Action

The absolute sterochemistries of the natural pyrethrins,[18,32] and of many of the synthetic pyrethroids reviewed here, are known and have stimulated much speculation about the indications their three-dimensional structures give of the nature of the receptors in insects with which they interact so strongly.[188–195] The structure of pyrethrin I and of other very potent esters, such as deltamethrin, can be considered as being divided into seven segments linked by bonds about which free rotation can occur, leading to the possibility of numerous conformations.[9] If the ester group, as is probable,[196] is in an *S-trans* conformation, the number of structures to be considered is significantly diminished. Other evidence suggests that the unsaturated centres in the alcohol side chains [the (*Z*)-diene in pyrethrins I and II, the phenoxy group in 3-phenoxybenzyl esters, and the 3-phenyl group in bifenthrin and its relatives, *etc.*] are not coplanar with the rings to which they are attached.[197] A number of effective alcohol components that can simulate (*S*)-pyrethrolone and also cyclopropane and non-cyclopropane acids with features in common with chrysanthemic and 2-(4-chlorophenyl)-3-methylbutyric acids are known.[197] The activity of non-ester pyrethroids, such as etofenprox (**8d**) show that the ester group is not essential and a number of alternative links are more or less effective.[198–200] The activity of the (*E*)-isomer of compound (**8h**) and the inactivity of the (*Z*)-isomer are significant in defining the requirements for activity in this region of the molecule.[197] Moreover, Ohno and colleagues[201] have shown that the (*R*)-(α)-form of the α-trifluoromethyl ether [flufenprox, ICI A5682, (**8f**)] is related by activity higher than its enantiomer to the (*S*,α*S*) more active isomer of fenvalerate (**4d**), rather than to the superficially similar *S*-(α) form.[202] The (*S*)-α-ethynyl alkane (**8g**), which belongs to the same series as the above compounds, is more active than its (*R*)-enantiomer. A relationship can thereby be discerned by conformation and activity between superficially dissimilar compounds, such as pyrethrin I, deltamethrin (crystalline), and other compounds in Figure 7.8. Naumann[203] gives a more extended and detailed structure–activity analysis on this theme.

Janes[204] has considered the significance of nuclear magnetic resonance (NMR) spectral evidence in relation to the biological action of such compounds. The ^{1}H and ^{13}C spectra show little signs of broadening at room temperature, because rotation about the single bonds referred to above is fast on the NMR time scale and the energy necessary to override any preferred conformation is readily available at room temperature, whether the compound is crystalline [*e.g.*, deltamethrin (**3g**)] or not. When pyrethroids bind to the site of action, even greater energies must be available to restrain the molecule at multiple sites. The interaction probably occurs by a 'zipper' process,[205] with initial attachment first at one position; only when the movement has aligned the next part correctly does a second binding take place, and so on. This concept of sequential action avoids the improbable requirements that the whole molecule presents itself correctly aligned at the site of action, but such considerations suggest that the structure of active, but flexible, insecticides may not indicate with much precision the architecture of the binding site.

7.20 Present Situation

The value of pyrethroids, natural or synthetic, used annually, has risen from about $10 M in 1976 (before the more stable compounds were introduced) to $1360 M (estimated) in 1992. British Technology Group-licensed products, originating from Rothamsted research, were worth about $800 M in 1990, somewhat over 10% of world-wide sales of insecticides of all classes; although pyrethroids intrinsically more potent than permethrin, cypermethrin [*e.g.*, alpha-, beta-, and zeta-cypermethrin,[16] (*S*)-2-cyano-3-phenoxybenzyl (1*RS*)-*cis*-*trans*-3-(2,2-dichlorovinyl)-2,2-dimethylcyclopropanecarboxylate], and deltamethrin have been introduced, the Rothamsted compounds have not been significantly displaced from their market position because later alternatives have not established clear advantages in costs or activity. About 100×10^6 ha are now treated annually with synthetic pyrethroids. Roussel-Uclaf, the world's largest producer of synthetic pyrethroid insecticides, has estimated that the annual world market will grow from below 1500 tonnes of active ingredients in 1990 to more than 2000 tonnes by the year 2000, although this view may be somewhat optimistic in the light of the most recent trends.

About 40% of pyrethroids used are applied to cotton, and almost as much to fruit, vegetables, and other crops. Compounds such as cypermethrin and cyhalothrin are now used in veterinary applications as ectoparasiticides and in the health and household insecticide market. Permethrin, which has the lowest mammalian toxicity of the more stable compounds, is formulated against human head lice and is used increasingly to impregnate clothing because such treatments significantly diminish biting by malaria-carrying mosquitoes.[206]

Table 7.2 shows the pyrethroids commercially available at present, as indicated by their inclusion in the *Pesticide Manual,* 9th edition (1991).[207] The

Table 7.2 *Current pyrethroids**

Compound	*Structure*	*Alcohol type*	*Acute oral toxicity* LD_{50} *(mg kg^{-1}) rat*	*Field rate (g ha^{-1})*
Natural pyrethrins	**1af, 1ag, 1ah, 1bf, 1bg, 1bh**	4-Oxocyclopent-2-enyl	584–900	
Allethrin	**1ai, 1ci**(±)	,,	1100	
Bioallethrin	**1ai**	,,	710	
S-Bioallethrin	**1ai**	,,	780	
Prallethrin	**7g**	,,	640	
Empenthrin	**7ef**	,,	2280	
Tetramethrin	**7ab**	Tetrahydro-phthalimidomethyl	>5000	
Tetramethrin (1*R*)	**7ab**	,,	>5000	
Resmethrin	**3ab**	5B3F	2500	
Bioresmethrin	**7ab**	,,	7070	
RU15,525	**6b**	,,	1320	
Phenothrin	**7ak**	3POB	>10000	
Cyphenothrin	**6de**	α-CN-3POB	320	
Permethrin	**3ef**	3POB	430–4000	25–200
Fenvalerate	**4d**	α-CN-3POB	450	25–250
Esfenvalerate	**4d**	,,	90–330	5–25
Taufluvalinate	**4e**	,,	>3000	60–170
Cypermethrin	**5bd**	,,	250–4120	10–75
Alpha-cypermethrin	**5d**	,,	80–4120	5–30
Beta-cypermethrin	**5d**	,,	170	10–30
Deltamethrin	**3g**	,,	140–>5000	2.5–13
Tralomethrin	**3h**	,,	100–3000	5–20
Fenpropathrin	**6c**	,,	70	20–200
Cyfluthrin	**5ce**(±)	α-CN-4F3POB	900	n.a.
Beta-cyfluthrin	**5e**(±)	α-CN-4F3POB	450	n.a.
Cyhalothrin	**5f**(±)	α-CN-3POB	240	n.a.
Lambda-cyhalothrin	**5f**(±)	α-CN-3POB	55–480	5–30
Cycloprothrin	**4f**	α-CN-3POB	>5000	n.a.
Bifenthrin	**5g**	Biphenyl	55	n.a.
Acrinathrin	**5i**	α-CN-3POB	>5000	30–75
Tefluthrin	**5h**	Benzyl	20–35	n.a.
Etofenprox	**8d**	Non-ester	42880	50–200

*Source, Worthing and Hance[207]; n.a. not given in Worthing and Hance[207]

associated quantitative data indicate that relatively low doses of pyrethroids are necessary; for instance, compounds such as deltamethrin are useful at 10–25 g or even 5 g ha^{-1} for special applications. In contrast, with organochlorine, organophosphorus, or carbamate insecticides, rates of 1 kg ha^{-1} are common, frequently higher if pests are resistant.

Knowledge of how structure affects potency for important practical uses (activity against economically important pest species and resistant strains,[208,209] acaricidal activity, selectivity between pests and beneficial species, persistence on crops, rapidity of action, antifeedant or repellent properties, toxicity to mammals, birds, and fish, *etc.*) is being accumulated. Strategies to guard against resistance developing[210] and to diminish adverse effects to biological systems by the appropriate choice of compounds and of timing and site of application are recognized increasingly as being important. Despite the desirability and greater effectiveness of alternative methods of pest control,[176,211] insecticides will be required in the foreseeable future; because they can still be applied only relatively inefficiently much of the dose does not reach the target and is potentially available to contaminate the environment. Effective compounds with limited persistence, such as pyrethroids, are, therefore, important. In the past decade, synthetic pyrethroids have developed as powerful weapons to benefit mankind by diminishing the adverse influence of insects; although some individual alternative compounds with considerable promise are being exploited, no major group of compounds with scope and potential comparable to that of the pyrethroids has yet been discovered.

7.21 Acknowledgements

I thank many friends and colleagues world-wide most warmly for much help and collaboration over some five decades while this work was in progress: Leslie Crombie (since 1946) and from Rothamsted: (the late) Charles Potter, Alan Lord, Kenneth Jeffs, Norman Janes, David Pulman, Bhupinder Khambay, Paul Needham, (the late) Roman Sawicki, Andrew Farnham, and many others, and, from the University of California, at Berkeley, John Casida and associates. I acknowledge with much gratitude important financial and technical support since 1962 from the National Research Development Corporation and the British Technology Group, which greatly assisted the project to attain its present status. I particularly thank Charles Worthing for a very careful reading of this manuscript, advising on nomenclature and suggesting many improvements, and Mrs Jennie Large for much help in preparing it for publication.

7.22 References

1. C. B. Gnadinger, 'Pyrethrum Flowers', McLaughlin Gormley King Co., Minneapolis, Minnesota, 2nd edn and Supplement, 1936–1945.
2. R. H. Nelson (ed.), 'Pyrethrum Flowers', McLaughlin Gormley King Co., Minneapolis, Minnesota, 3rd edn, 1945–1972.
3. C. C. McDonnell, R. C. Roark, F. B. LaForge and G. L. Keenan, 'Insect Powder', United States Department of Agriculture, Department Bulletin No 824, 1920.
4. L. Crombie and M. Elliott, Chemistry of the natural pyrethrins, *Fortschr. Chem. Org. Naturstoffe*, 1961, **19**, 120.
5. J. E. Casida (ed.), 'Pyrethrum, the Natural Insecticide', Academic Press, New York and London, 1973.
6. J. E. Casida and G. B. Quistad (ed.), 'Pyrethrum Flowers' Production, Chemistry, Toxicology and Uses', Oxford University Press, Oxford, 1994.

7. T. Lewis, The contribution of Rothamsted to British entomology, *Bull. Entomol. Res.*, 1993, **83**, 309.
8. M. Elliott, N. F. Janes, and C. Potter, The future of pyrethroids in insect control, *Ann. Rev. Entomol.*, 1978, **23**, 443.
9. M. Elliott and N. F. Janes, Synthetic pyrethroids – a new class of insecticides, *Chem. Soc. Rev.*, 1978, **7**, 473.
10. J. P. Leahey (ed.), 'The Pyrethroid Insecticides', Taylor and Francis, London and Philadelphia, 1985.
11. C. G. L. Furmidge, G. T. Brooks, and D. W. Gammon (ed.) The pyrethroid insecticides: A scientific advance for human welfare? *Pestic. Sci.*, 1989, **27**, 335.
12. M. Elliott, Pyrethroid insecticides and human welfare, in 'Pesticides and Alternatives', ed. J. E. Casida, Elsevier Science Publishers, Oxford, 1990, p. 345.
13. M. Elliott, Chemicals in Insect Control, in 'Pyrethrum Flowers; Production, Chemistry, Toxicology and Uses', ed. J. E. Casida and G. B. Quistad, Oxford University Press, Oxford, 1994, p. 3.
14. D. Arlt, M. Jautelat, and R. Lantzsch, Synthesis of pyrethroid acids, *Angew. Chem., Int. Ed. Engl.*, 1981, **20**, 703.
15. J. H. Davies, The pyrethroids: an historical introduction, in 'The Pyrethroid Insecticides', ed. J. P. Leahey, Taylor and Francis, London and Philadelphia, 1985, p. 1.
16. K. Naumann, 'Chemistry of Plant Protection, Vol. 4, Synthetic Pyrethroid Insecticides: Structure and Properties', Springer Verlag, Berlin, 1990; 'Vol 5, Synthetic Pyrethroid Insecticides: Chemistry and Patents', Springer Verlag, Berlin, 1990.
17. M. Elliott, Obituary, Charles Potter, 1907–1989, in 'Neurotox '91, Molecular Basis of Drug and Pesticide Action', Elsevier Science Publishers Ltd, London and New York, 1992, p. xxi.
18. L. Crombie, The natural pyrethrins: a chemist's view, in 'Neurotox 88: Molecular Basis of Drug and Pesticide Action', Elsevier Science Publishers Ltd, London and New York, 1988, p. 3.
19. W. D. Gullickson, History of pyrethrum, in 'Pyrethrum Flowers; Production, Chemistry, Toxicology, and Uses', ed. J. E. Casida and G. B. Quistad, Oxford University Press, Oxford, 1994, p. 32.
20. B. K. Bhat and R. C. Menary, Genotypic and phenotypic variation in floral development of different clones of Pyrethrum (*Chrysanthemum cinerariae folium* Vis), *Pyrethrum Post*, 1984, **15**(4), 99.
21. J. Fujitani, Chemistry and pharmacology of insect powder, *Arch. Exp. Pathol. Pharmacol.*, 1909, **61**, 47.
22. R. Yamamoto, The insecticidal principle in *Chrysanthemum cinerariaefolium*, Part I, *J. Tokyo Chem. Soc.*, 1919, **40**, 126.
23. R. Yamamoto and M. Sumi, Studies on the insecticidal principle in *Chrysanthemum Cinerariaefolium*, Parts II and III. On the constitution of pyrethronic acid, *J. Chem. Soc. Jpn.*, 1923, **44**, 311, 1080.
24. H. Staudinger and L. Ruzicka, Insecktentotende Stoffe I–V and X, *Helv. Chim. Acta*, 1924, **7**, 177, 201, 212, 236, 245, 449.
25. H. Staudinger, O. Muntwyler, L. Ruzicka, and S. Seibt, *Helv. Chim. Acta*, 1924, **7**, 390.
26. M. S. Schechter, N. Green, and F. B. LaForge, Constituents of pyrethrum flowers. XXIII Cinerolone and the synthesis of related cyclopentenolones, *J. Am. Chem. Soc.*, 1949, **71**, 3165.

27. N. Ohno, K. Fujimoto, Y. Okuno, T. Mizutani, M. Hirano, N. Itaya, and H. Yoshioka, A new class of pyrethroidal insecticides, alpha substituted phenylacetic acid esters, *Agric. Biol. Chem.*, 1974, **38**, 881.
28. F. L. LaForge and W. F. Barthel, Constituents of pyrethrum flowers XVI. Hetereogeneous nature of pyrethrolone, *J. Org. Chem.* 1944, **9**, 242; *idem., ibid.*, XVII. The isolation of five pyrethrolone semicarbazones, 1945, **10**, 106; *idem., ibid.*, XVIII. The structure and isomerism of pyrethrolone and cinerolone, 1945, **10**, 114; *idem., ibid.*, XIX. The structure of cinerolone, 1945, **10**, 222.
29. P. J. Godin, R. J. Sleeman, M. Snarey, and E. M. Thain, The jasmolins, new insecticidally active constituents of *Chrysanthemum cinerariaefolium* vis., *J. Chem. Soc.*, 1966, 332.
30. L. Crombie and M. Elliott, Chemistry of the natural pyrethrins, *Fortschr. Chem. Org. Naturstoffe*, 1961, **19**, 120.
31. H. Hunsdiecker, On the preparation of γ-diketones. The cyclopentenone ring closure of α-diketones of the type $CH_3COCH_2CH_2COCH_3$, *Chem. Ber.*, 1947, **75**, 460.
32. L. Crombie, Chemistry of the pyrethrins, in 'Pyrethrum Flowers; Production, Chemistry, Toxicology and Uses', ed. J. E. Casida and G. B. Quistad, Oxford University Press, Oxford, 1994, p. 123.
33. L. Crombie, M. Elliott, S. H. Harper, and H. W. B. Reed, Total synthesis of some pyrethrins, *Nature* (*London*), 1948, **162**, 222.
34. L. Crombie, S. H. Harper, and F. C. Newman, Synthesis of *cis*-pyrethrolone and pyrethrin I. Introduction of the *cis*-penta-2,4-dienyl system by selective hydrogenation, *J. Chem. Soc.*, 1954, 3963.
35. J. H. Sanders and H. W. Taft, Allethrin, *Ind. Eng. Chem.*, 1954, **46**, 414.
36. F. Tattersfield, R. P. Hobson, and C. T. Gimingham, Pyrethrin I and Pyrethrin II, their insecticidal value and estimation in pyrethrum (*Chrysanthemum cinerariaefolium*). I, *J. Agri. Sci.*, 1929, **19**, 266.
37. D. J. Finney, 'Probit Analysis', University Press Cambridge, Cambridge, 1947.
38. C. Potter, An account of the constitution and use of an atomised white oil – pyrethrum fluid – to control *Plodea interpunctella* HB and *Ephestia elutella* HB in warehouses, *Ann. Appl. Biol.*, 1935, **22**, 769.
39. S. H. Harper, *Annual Reports of the Chemical Society*, 1948, **45**, 162.
40. N. Matsuo, T. Takagaki, K. Watanobe, and N. Ohno, The first practical synthesis of (*S*)-pyrethrolone, an alcohol moiety of natural pyrethrins I and II, *Biosci. Biotechnol. Biochem.*, 1993, **57**(4), 693.
41. W. A. Gersdorff, Toxicity to houseflies of a new synthetic pyrethroid, *Soap Sanitary Chemic.*, 1949, **25**, 129; 131; 139.
42. W. A. Gersdorff, Toxicity to houseflies of synthetic compounds of the pyrethrin type in relation to chemical structure, *J. Econ. Ent.*, 1949, **42**, 532.
43. W. A. Gersdorff, P. G. Piquett, N. Mitlin, and N. Green. Reproducibility of the toxicity ratio of allethrin to pyrethrins applied to house flies by the turntable method, *J. Econ. Ent.*, 1961, **54**, 580.
44. M. Elliott, P. H. Needham, and C. Potter, The insecticidal activity of substances related to the pyrethrins. I. The toxicities of two synthetic pyrethrin-like esters relative to that of the natural pyrethrins and the significance of the results in the bioassay of closely related compounds, *Ann. Appl. Biol.*, 1950, **37**, 490.
45. J. E. Casida, Mixed-function oxidase involvement in the biochemistry of insecticide synergists, *J. Agric. Food Chem.*, 1970, **18**, 753.
46. I. Yamamoto, Mode of action of synergists in enhancing the insecticidal activity of

pyrethrum and pyrethroids, in 'Pyrethrum, the Natural Insecticide', ed. J. E. Casida, Academic Press, New York and London, 1973, p. 195.

47. J. Ward, Separation of the pyrethrins by displacement chromatography, *Chem. Ind.*, 1953, 586.
48. R. M. Sawicki, M. Elliott, T. C. Gower, M. Snarey, and E. M. Thain, Insecticidal activity of pyrethrum extract and its four insecticidal constituents against houseflies. I. Preparation and relative toxicity of the pure constituents; statistical analysis of the action of mixtures of these components, *J. Sci. Food Agric.*, 1962, **13**, 172.
49. M. Elliott, Purification of (+)-pyrethrolone as the monohydrate and the nature of pyrethrolone C, *J. Chem. Soc.*, 1964, 5225.
50. M. Elliott, Isolation and purification of (+)-pyrethrolone from pyrethrum extract: reconstitution of pyrethrins I and II, *Chem. Ind.*, 1958, 685.
51. M. Elliott, Pyrethrolone and related compounds, *Chem. Ind.*, 1960, 1142.
52. M. Elliott, P. H. Needham, and C. Potter, Insecticidal activity of pyrethrins and related compounds. II. Relative toxicity of esters from optical and geometrical isomers of chrysanthemic, pyrethric and related acids and optical isomers of cinerolone and allethrolone, *J. Sci. Food Agric.*, 1969, **20**, 561.
53. M. Elliott. The pyrethrins and related compounds. III. Thermal isomerization of *cis*-pyrethrolone and its derivatives, *J. Chem. Soc.*, 1964, 888.
54. N. C. Brown, D. T. Hollinshead, R. F. Phipers, and M. C. Wood, New isomers of the pyrethrins formed by the action of heat, *Pyrethrum Post*, 1957, **4**(2), 13.
55. H. L. Haynes, H. R. Guest, H. A. Stansbury, A. A. Sousa, and A. J. Borash, Cyclethrin, a new insecticide of the pyrethrins type, *Contrib. Boyce Thompson Inst.*, 1954, **18**, 1.
56. M. Elliott, Allethrin, *J. Sci. Food Agric.*, 1954, **5**, 505.
57. J. B. Moore, Chemistry of pyrethrinoids, in 'Pyrethrum Flowers', ed. R. H. Nelson, McLaughlin Gormley King Co., Minneapolis, Minnesota, 3rd edn, 1945–1972, p. 28.
58. W. A. Gersdorff and P. G. Piquett, The relative effectiveness of two synthetic pyrethroids more toxic to house flies than pyrethrins in kerosene sprays. *J. Econ. Ent.*, 1961, **54**, 1250.
59. W. F. Barthel, Synthetic pyrethroids, *Adv. Pest Control Res.*, 1961, **4**, 33.
60. M. Elliott, N. F. Janes, and K. A. Jeffs, The pyrethrins and related compounds. X. The methylbenzyl chrysanthemates, *Pestic. Sci.*, 1970, **1**, 49.
61. M. Elliott, A. W. Farnham, M. G. Ford, N. F. Janes, and P. H. Needham, Insecticidal activity of the pyrethrins and related compounds. V, Toxicity of the methylbenzyl chrysanthemates to houseflies (*Musca domestica*) and mustard beetles (*Phaedon cochleariae*), *Pestic. Sci.*, 1972, **3**, 25.
62. M. Elliott, S. H. Harper, and M. A. Kazi, The pyrethrins and related compounds. VIII. Rethrins and the cyclopentadienone related to 3-methylcyclopent-2-enone, *J. Sci. Food Agric.*, 1967, **18**, 167.
63. M. Elliott, N. F. Janes, K. A. Jeffs, P. H. Needham, and R. M. Sawicki, New pyrethrin-like esters with high insecticidal activity, *Nature*, 1965, **207**, 938.
64. M. Elliott, N. F. Janes, and M. C. Payne, The pyrethrins and related compounds. Part XI. Synthesis of insecticidal esters of 4-hydroxycyclopent-2-enones (norrethrins), *J. Chem. Soc. (C)*, 1971, 2548.
65. M. Elliott, N. F. Janes, and B. C. Pearson, The pyrethrins and related compounds. XIII. Insecticidal methyl-, alkenyl and benzyl-substituted furfuryl and furylmethyl chrysanthemates, *Pestic. Sci.*, 1971, **2**, 243.

66. M. Elliott, A. W. Farnham, N. F. Janes, and P. H. Needham, Insecticidal activity of the pyrethrins and related compounds. VI. Methyl-, alkenyl-, and benzyl-furfuryl and -3-furylmethyl chrysanthemates, *Pestic. Sci.*, 1974, **5**, 491.
67. M. Elliott, N. F. Janes, and B. C. Pearson, The pyrethrins and related compounds. XII. 5-Substituted furoates and 3-thenoates, intermediates for synthesis of insecticidal esters, *J. Chem. Soc. (C)*, 1971, 2551.
68. M. Elliott, N. F. Janes, and B. C. Pearson, The pyrethrins and related compounds. IX. Alkenyl- and benzylbenzyl chrysanthemates, *J. Sci. Food Agric.*, 1967, **18**, 325.
69. M. Elliott, A. W. Farnham, N. F. Janes, P. H. Needham, and B. C. Pearson, 5-Benzyl-3-furylmethyl chrysanthemate, a new potent insecticide, *Nature*, 1967, **213**, 493.
70. M. Elliott, The relationship between the structure and the activity of pyrethroids, *Bull. WHO*, 1971, **44**, 315.
71. M. Elliott, N. F. Janes, B. P. S. Khambay, and D. A. Pulman, Insecticidal activity of the pyrethrins and related compounds. XIII. Comparison of the effects of α-substituents in different esters, *Pestic, Sci.*, 1983, **14**, 182.
72. E. L. Plummer and D. S. Pincus, Pyrethroid insecticides derived from [1,1′-biphenyl]-3-methanol, *J. Agric. Food Chem.*, 1981, **29**, 1118.
73. K. Fujimoto, N. Itaya, Y. Okuno, T. Kadota, and T. Yamaguchi, A new insecticidal pyrethroid ester, *Agric. Biol. Chem.*, 1973, **37**, 2681.
74. N. Itaya, Gokilaht, the forerunner of a new era of synthetic pyrethroids is thus invented, *Sumitomo Pyrethroid World*, 1986, No. 7, 2.
75. F. B. LaForge, N. Green, and M. S. Schechter, Dimerized cyclopentadienones from esters of allethrolone, *J. Am. Chem. Soc.*, 1952, **74**, 5392.
76. R. M. Sawicki and E. M. Thain, Insecticidal activity of pyrethrum extract and its four insecticidal constituents against house flies. IV. Knock-down activities of the four constituents, *J. Sci. Food Agric.*, 1962, **13**, 292.
77. F. B. LaForge, W. A. Gersdorff, N. Green, and M. S. Schechter, Allethrin-type esters of cyclopropane carboxylic acids and their relative toxicities to houseflies, *J. Org. Chem.*, 1952, **17**, 381.
78. F. Barlow, M. Elliott, A. W. Farnham, A. B. Hadaway, N. F. Janes, P. H. Needham, and J. C. Wickham, Insecticidal activity of the pyrethrins and related compounds. IV. Essential features for insecticidal activity in chrysanthemates and related cyclopropane esters, *Pestic. Sci.*, 1971, **2**, 115.
79. J. Lhoste, J. Martel, and F. Rauch, Insecticidal activity of 5-benzyl-3-furylmethyl d-*cis*-chrysanthemate, *Med. Fac. Landbouwwet., Rijksuniv. Gent*, 1971, **36**(3), 978.
80. J. Lhoste, J. Martel, and R. Rauch, New insecticide. 5-benzyl-3-furylmethyl d-*trans* ethanochrysanthemate, *Proc. 5th Brit. Insect. Fung. Conf.*, 1969, 554.
81. J. Velluz, J. Martel, and G. Nominée, Synthèse d'analogues de l'acide *trans*-chrysanthémique, *Compt. Rend.* 1969, **268**, 2199.
82. I. Yamamoto, E. C. Kimmel, and J. E. Casida, Oxidative metabolism in houseflies, *J. Agric. Food Chem.*, 1969, **17**, 1227.
83. M. Elliott, N. F. Janes, E. C. Kimmel, and J. E. Casida, Metabolic fate of pyrethrin I, pyrethrin II, and allethrin administered orally to rats, *Agric. Food Chem.*, 1972, **20**, 300.
84. M. Elliott, N. F. Janes, A. W. Farnham, P. H. Needham, and D. A. Pulman, Insecticidal activity of the pyrethrins and related compounds. X. 5-Benzyl-3-

furylmethyl 2,2-dimethylcyclopropanecarboxylates with ethylenic substituents at position 3 on the cyclopropane ring, *Pestic. Sci.*, 1976, **7**, 499.

85. M. Elliott, A. W. Farnham, N. F. Janes, P. H. Needham, and D. A. Pulman, Potent pyrethroid insecticides from modified cyclopropane acids, *Nature*, 1973, **244**, 450.
86. M. Elliott, A. W. Farnham, N. F. Janes, P. H. Needham, and D. A. Pulman, Insecticidal activity of the pyrethrins and related compounds. IX. 5-Benzyl-3-furylmethyl 2,2-dimethylcyclopropanecarboxylates with non-ethylenic substituents at position 3 on the cyclopropane ring, *Pestic. Sci.*, 1976, **7**, 492.
87. M. Elliott, N. F. Janes, and D. A. Pulman, The pyrethrins and related compounds. XVIII. Insecticidal 2,2-dimethyl cyclopropane carboxylates with new unsaturated 3-substituents, *J. Chem. Soc., Perkin Trans 1*, 1947, 2470.
88. L. Crombie, C. F. Doherty, and G. Pattenden, Synthesis of ^{14}C-labelled (+)-*trans*-chrysanthemum mono- and dicarboxylic acids and of related compounds. *J. Chem. Soc.* (*C*), 1970, 1076.
89. M. Elliott, A. W. Farnham, N. F. Janes, P. H. Needham, D. A. Pulman, and J. H. Stevenson, A photostable pyrethroid, *Nature*, 1973, **246**, 169.
90. J. Farkas, P. Kourim, and F. Sorm, Relation between chemical structure and insecticidal activity in pyrethroid compounds. 1. An analogue of chrysanthemic acid containing chlorine in the side chain, *Chem. Listy*, 1958, **52**, 688.
91. M. Elliott, Developments in the chemistry and action of pyrethroids, in 'Natural Products for Innovative Pest Management', ed. D. L. Whitehead and W. S. Bowers, Pergamon Press, Oxford and New York, 1983, p. 127.
92. Y.-L. Chen and J. E. Casida, Photochemistry of pyrethrin I, allethrin, phthalthrin and dimethrin, *J. Agric. Food Chem.*, 1969, **17**, 208.
93. K. Ueda, L. C. Gaughan, and J. E. Casida, Photo-decomposition of resmethrin and related pyrethroids, *J. Agric. Food Chem.*, 1974, **22**, 212.
94. P. E. Burt, M. Elliott, A. W. Farnham, N. F. Janes, P. H. Needham, and D. A. Pulman, The pyrethrins and related compounds. XIX. Geometrical and optical isomers of 2,2-dimethyl-3-(2,2-dichlorovinyl)-cyclopropane carboxylic acid and insecticidal esters with 5-benzyl-3-furylmethyl- and 3-phenoxybenzyl alcohols, *Pestic. Sci.*, 1974, **5**, 791.
95. D. G. Brown, O. F. Bodenstein, and S. J. Norton, New potent pyrethroid, bromethrin, *J. Agric. Food Chem.*, 1973, **21**, 767.
96. D. G. Brown, O. F. Bodenstein, and S. J. Norton, Halo-pyrethroids II. A difluoropyrethroid, *J. Agric. Food Chem.*, 1975, **23**, 115.
97. N. Ohno, K. Fujimoto, N. Okuno, T. Mizutani, M. Hirano, N. Itaya, T. Honda, and H. Yoshioka, 2-Aryl alkanoates, a new group of synthetic pyrethroid esters not containing cyclopropanecarboxylates, *Pestic. Sci.*, 1976, **7**, 241.
98. M. Elliott and N. F. Janes, Recent structure–activity correlations in synthetic pyrethroids, in 'Advances in Pesticide Science, Part 2', ed. H. Geissbühler, Pergamon Press, Oxford and New York, 1979, p. 166.
99. Anon, α-Ethynyl furylmethyl and benzyl esters, Belgian Patent 738 112 to Badische Anilin-Soda-Fabrik-Aktiengellschaft.
100. Anon, JA 7 427 331 to Sumitomo, 1970.
101. Anon, DOS 2 231 312 to Sumitomo, 1971.
102. T. Matsuo, N. Itaya, T. Mizutani, N. Ohno, K. Fujimoto, Y. Okuno, and H. Yoshioka, 3-Phenoxy-alpha-cyanobenzyl esters, the most potent synthetic pyrethroids, *Agric. Biol. Chem.*, 1976, **40**, 247.

103. M. Elliott, A. W. Farnham, N. F. Janes, P. H. Needham, and D. A. Pulman, Insecticidal activity of the pyrethrins and related compounds. VII. Insecticidal dihalovinyl analogues of *cis* and *trans* chrysanthemates, *Pestic. Sci.*, 1975, **6**, 537.
104. UK Patent Applications Nos 20 539/73, 39 539/73 and 4908/73, 1973.
105. F. B. LaForge, N. Green, and M. S. Schechter, Allethrin. Resolution of *dl*-allethrolone and synthesis of the four optical isomers of *trans*-allethrin, *J. Org. Chem.*, 1954, **19**, 457.
106. M. Elliott, A. W. Farnham, N. F. Janes, P. H. Needham, and D. A. Pulman, Synthetic insecticide with a new order of activity, *Nature*, 1974, **248**, 710.
107. P. Ackerman, F. Bourgeois, and J. Drabek, The optical isomers of α-cyano-3-phenoxybenzyl 3-(1,2-dibromo-2,2-dichloroethyl)-2,2-dimethylcyclopropane-carboxylate and their insecticidal activities, *Pestic. Sci.*, 1980, **11**, 169.
108. J. D. Owen, Absolute configuration of the most potent isomer of the pyrethroid insecticide α-cyano-3-phenoxybenzyl *cis*-3-(2,2-dibromovinyl)-2,2-dimethyl-cyclopropanecarboxylate by crystal structure, *J. Chem. Soc., Perkin Trans. 1*, 1975, 1865.
109. J. D. Owen, X-ray crystal structure of two pyrethroid insecticides: *cis*-3-phenoxybenzyl 3-(2,2-dibromovinyl)-2,2-dimethylcyclopropanecarboxylate and the 3-(2,2-dichlorovinyl) analogue, *J. Chem. Soc., Perkin Trans. 1*, 1976, 1231.
110. J. Martel, Deltamethrin – a challenge in process development, in 'Recent Developments in the Field of Pesticides and their Application to Pest Control. Proceedings of the International Seminar, Shenyang, China, 8–12 October 1990', ed. K. Holly, L. G. Copping and G. T. Brooks, 1990, p. 61.
111. *Deltamethrin Monograph*, Roussel-Uclaf, 1982.
112. A. Krief, Ph. Lecomte, J. P. Demoute, and W. Dumont, Straightforward synthesis of isopropylidenediphenyl sulfurane and application to industrially viable stereoselective synthesis of deltamethrin insecticide, *Synthesis*, 1990, 275.
113. M. Elliott, N. F. Janes, A. W. Farnham, and D. M. Soderlund, Insecticidal activity of the pyrethrins and related compounds. XI. Relative potencies of isomeric cyano-substituted 3-phenoxybenzyl esters, *Pestic. Sci.*, 1978, **9**, 112.
114. M. Elliott, N. F. Janes, D. A. Pulman, and D. M. Soderlund, The pyrethrins and related compounds. XXII. Preparation of isomeric cyano-substituted 3-phenoxy-benzyl esters, *Pestic. Sci.*, 1978, **9**, 105.
115. J. P. Fisher, J. Robinson, and P. H. Debray, WL85871 – A new multipurpose insecticide, *10th Int. Congr. Plant. Protection*, 1983, 452.
116. Chinoin Pharmaceutical and Chemical Works Co., 1989.
117. K. Naumann, 'Chemistry of Plant Protection, Vol. 4, Synthetic Pyrethroid Insecticides: Structure and Properties', Springer Verlag, Berlin, 1990, p. 43.
118. I. Hammann and R. A. Fuchs, Baythroid, a new insecticide, *Pflanzenschutz-Nachr.* (*Engl. Ed.*), 1981, **34**(2), 121.
119. W. Behrenz, A. Elbert, and R. Fuchs, Cyfluthrin (FCR 1272), a new pyrethroid with long-lasting activity for the control of public health and stored-products pests, *Pflanzenschutz-Nachr.* (*Engl. Ed.*), 1983, **36**(2), 127.
120. P. D. Bentley, R. Cheetham, R. K. Huff, R. Pascoe, and J. D. Sayle, Fluorinated analogues of chrysanthemic acid, *Pestic. Sci.*, 1980, **11**, 156.
121. A. J. Jutsum, R. F. S. Gordon, and C. N. E. Ruscoe, Tefluthrin – a novel pyrethroid soil insecticide, *Brighton Crop Prot. Conf. – Pests Dis.*, 1986, **1**, 97.
122. H. J. H. Doel, A. R. Crossman, and L. A. Bourdouxhe, FMC54800, a new acaricide–insecticide, *Med. Fac. Landbouwwet., Rijksuniv. Gent*, 1984, **49**(3a), 929.

123. A. R. Crossman, L. A. Bourdouxhe, and H. J. H. Doel, Field experiences in West Europe with FMC 54800 for the control of mites in orchards, vineyards and other crops, *Med. Fac. Landbouwwet., Rijksuniv. Gent*, 1984, **49**(3a), 939.
124. Y. Fujita, Meothrin (fenpropathrin), *Jpn. Pestic. Inf.*, 1981, No. 38, 21.
125. Roussel-Uclaf S.A., German Patent 2742546, 1978.
126. L. O. Ruzo, L. C. Gaughan, and J. E. Casida, Metabolism and degradation of the pyrethroids tralomethin and traolocythrin in insects, *Pestic. Biochem. Physiol.*, 1981, **15**, 137.
127. L. M. Cole, L. O. Ruzo, E. J. Wood, and J. E. Casida, Pyrethroid metabolism: comparative fate in rats of tralomethrin, tralocythrin, deltamethrin and (1*R*,α1*S*)-*cis*-cypermethrin, *J. Agric. Food Chem.*, 1982, **30**, 631.
128. L. O. Ruzo and J. E. Casida, Pyrethroid photochemistry: (*S*)-α-cyano-3-phenoxybenzyl *cis*-(1*R*,3*R*,1′*R* or *S*)-3(1′,2′-Dibromo-2′,2′-dihaloethyl)-2,2-dimethylcyclopropanecarboxylates, *J. Agric. Food Chem.*, 1981, **29**, 702.
129. L. M. Cole, J. E. Casida, and L. O. Ruzo, Comparative degradation of the pyrethroids tralomethrin, tralocythrin, deltamethrin and cypermethrin on cotton and bean foliage, *J. Agric. Food Chem.*, 1982, **30**, 916.
130. M. Elliott, N. F. Janes, and B. P. S. Khambay, Relative insecticidal activity of various 3-polyhalo-ethyl-2,2-dimethylcyclopropane carboxylates, *Brighton Crop Prot. Conf. – Pests Dis.*, 1984, **9A-1**, 849.
131. H. J. Schnitzerling, J. Nolan, and S. Hughes, Toxicology and metabolism of isomers of flumethrin in larvae of pyrethroid-susceptible and resistant strains of the cattle tick *Boophilus microphis* (Acari: Ixodidae), *Exp. Appl. Acarol*, 1989, **6**, 47.
132. K. Naumann, in 'Chemistry of Plant Protection, Vol. 4, Synthetic Pyrethroid Insecticides: Structure and Properties', Springer Verlag, Berlin, 1990, pp. 12, 21, 98, 99.
133. I. Nakayama, N. Ohno, K. Aketa, Y. Suziku, T. Kato, and H. Yoshioka, Chemistry, absolute structures and biological aspects of the most active isomers of fenvalerate and other recent pyrethroids, in 'Advances in Pesticide Science, Symposium papers presented at the 4th International Congress of Pesticide Chemistry, Zauchg, Switzerland', ed. H. Geissbühler, Pergamon Press, Oxford and New York, 1978, Part 2, 174.
134. H. Oouchi, Sumi-alpha (Esfenvalerate), *Jpn. Pestic. Inf.*, 1985, No. 46, 21.
135. K. W. Whitney and K. Wettstein, A new pyrethroid insecticide: performance against crop pests, *Proc. 10th Brit. Insect. Fung. Conf.* (*Brighton*), 1979, **2**, 387.
136. K. Wettstein, AC 222705 pyrethroidal insecticide for use in top fruit and hops, *Proc. 10th Brit. Insect. Fung. Conf.* (*Brighton*), 1981, **2**, 563.
137. C. A. Henrick, B. A. Garcia, G. B. Staal, D. C. Cerf, R. J. Anderson, K. Gill, H. R. Chinn, J. N. Labovitz, M. M. Leippe, S. L. Woo, R. L. Carney, D. C. Gordon, and G. K. Kohn, 2-Anilino-3-methylbutyrates and 2-(isoindolin-2-yl)-3-methylbutyrates, two novel groups of synthetic esters not containing a cyclopropane ring, *Pestic. Sci.*, 1980, **11**, 224.
138. C. A. Henrick, R. J. Anderson, R. L. Carney, B. A. Garcia, and G. B. Staal, Some aspects of structure–activity relationships in pyrethroids, in 'Recent Advances in the Chemistry of Insect Control; Special Publication No. 53', ed. N. F. Janes, Royal Society of Chemistry, London, 1985, 133.
139. M. Elliott, Synthetic pyrethroids in 'Synthetic Pyrethroids', ACS Symposium series No. 42, Amercian Chemical Society, Washington DC.
140. T. R. Perrior, Chemical insecticides for the 21st century, *Chem. Ind.* 1993, 873.

141. J. Farkǎs, P. Kouřim, and F. Sorm, Relation between chemical structure and insecticidal activity in pyrethroid compounds. II. Analogues of chrysanthemic acid containing an aryl group, *Chem. Listy*, 1958, **52**, 695.
142. K. Ozawa, S. Ishii, K. Hirata, and M. Hirose, Phenyl cyclopropanecarboxylic acid esters as new pyrethroid acaricides and insecticides, in 'Abstracts of the 5th International Congress of Pesticide Chemistry', Kyoto, Japan, 1982, Paper 1a–7.
143. M. J. Bushell and R. Salmon, in 'Advances in the Chemistry of Insect Control', III, ed. G. G. Briggs, Royal Society of Chemistry, Cambridge, 1993, p. 103.
144. G. Holan, D. F. O'Keefe, C. Virgona, and R. Walser, Structural and biological link between pyrethroids and DDT in new insecticides, *Nature*, 1978, **272**, 734.
145. G. Holan, W. M. P. Johnson, D. F. O'Keefe, G. L. Quint, K. Rihs, T. H. Sparling, R. Walser, C. T. Virgona, C. Frelin, M. Lazdunski, G. A. R. Johnson, and S. Chen Chow, Multidisciplinary studies in the design of new insecticides, in 'Recent Advances in the Chemistry of Insect Control. Special Publication, No. 53', ed. N. F. Janes, The Royal Society of Chemistry, London, 1985, p. 114.
146. S. Kirihara and Y. Sakurai, Cycloprothrin, a new insecticide, *Jpn. Pestic. Inf.*, 1988, No. 53, 22.
147. K. Naumann, in 'Chemistry of Plant Protection, Vol. 5, Synthetic Pyrethroid Insecticides: Chemistry and patents, Springer Verlag, Berlin, 1990, p. 117.
148. T. Matsunaga, M. Makita, A. Higo, I. Nishibe, K. Dohara, and G. Shinjo, Studies on prallethrin, a new synthetic pyrethroid for indoor applications 1. The insecticidal activities of prallethrin isomers, *Sumitomo Pyrethroid World*, 1988, No. 11, 19.
149. H. Danda, H. Maehara, and T. Umemura, Preparation of (4*S*)-4-hydroxy-3-methyl-2-(2′-propynyl)-2-cyclopentenone by combination of enzymic hydrolysis and chemical transformation, *Tetrahedron Lett.* 1991, **32**, 5119.
150. J. R. Tessier, Stereochemical aspects of pyrethroid chemistry, in 'Recent Advances in the Chemistry of Insect Control. Special Publication No. 53', ed. N. F. Janes, The Royal Society of Chemistry, London, 1985, p. 26.
151. J. R. Tessier, A. P. Tèche, and J. P. Demoute, Synthesis and properties of new pyrethroids, diesters of the nor-pyrethric series, in 'Proceedings of the Fifth International Congress of Pesticide Chemistry, Kyoto, Japan, 29 Aug.–4 Sept. 1982', **1**, p. 95.
152. R. M. Sawicki and E. M. Thain, Insecticidal activity of pyrethrum extract and its four insecticidal constituents against houseflies. IV. Knock-down activities of the four constituents. *J. Sci. Food Agric.*, 1962, **13**, 292.
153. M. Elliott, N. F. Janes, and C. Potter, The future of pyrethroids in insect control, *Ann. Rev. Entomol.*, 1978, **23**, 503.
154. G. G. Briggs, M. Elliott, A. W. Farnham, and N. F. Janes, Structural aspects of the knockdown of pyrethroids, *Pestic. Sci.*, 1974, **5**, 643.
155. M. Elliott, N. F. Janes, and C. Potter, The future of pyrethroids in insect control, *Ann. Rev. Entomol.*, 1978, **23**, 454.
156. T. Kato, K. Ueda, and K. Fujimoto, New insecticidally active chrysanthemates, *Agric. Biol. Chem.*, 1964, **28**, 914.
157. M. Matsui and T. Kitahara, Studies on chrysanthemic acid. Part XVIII. A new biologically active acid component related to chrysanthemic acid, *Agric. Biol. Chem.*, 1967, **31**, 1143.
158. J. Lhoste and F. Rauch, RU-15525, a new pyrethroid with a very strong knockdown effect, *Pestic. Sci.*, 1976, **7**, 247.
159. K. Sota, T. Amano, M. Aida, K. Noda, A. Hyashi, and I. Tanaka, New synthetic

pyrethroids, 4-phenyl-2-buten-1-yl and 4-aryl-2-butyn-1-yl chrysanthemates, *Agric. Biol. Chem.*, 1971, **35**, 968.

160. Y. Fujita, Stages leading to a new pyrethroid – the case of Vaporthrin, *Sumitomo Pyrethroid World*, 1984, No. 3, 2.
161. P. E. Berteau and J. E. Casida, Synthesis and insecticidal activity of some pyrethroid-like compounds lacking cyclopropane or ester groupings, *J. Agric. Food Chem.*, 1969, **17**, 931.
162. M. J. Bull, J. H. Davies, R. J. G. Searle, and A. C. Henry, Alkyl aryl ketone oxime *O*-ethers: a novel group of pyrethroids, *Pestic. Sci.*, 1980, **11**, 249.
163. T. Udagawa, S. Numata, K. Oda, S. Shiraishi, K. Kodaka, and K. Nakatani, A new type of synthetic pyrethroid insecticide, in 'Recent Advances in the Chemistry of Insect Control. Special Publication No. 53', ed. N. F. Janes, The Royal Society of Chemistry, London, 1985, p. 192.
164. T. Udagawa, Trebon (etofenprox), a new insecticide, *Jpn. Pestic. Inf.*, 1988, No. 53, 9.
165. G. A. Meier, T. G. Cullen, S. Sehgel, J. F. Engel, S. E. Burkart, S. M. Sieburth, and C. M. Langevine, Highly efficacious non-ester pyrethroid insecticides with low toxicity to fish, in 'Synthesis and Chemistry of Agrochemicals III, ACS Symposium Series 504', ed. D. R. Baker, J. G. Feynes, and J. J. Staffens, 1992, p. 258.
166. T. G. Cullen, S. M. Sieburth, J. F. Engel, G. A. Meier, A. C. Lew, S. E. Burkart, F. L. Marek, and J. H. Strickland, 1,4-Diaryl-1-cyclopropylbutanes. Highly efficacious insecticides with low fish toxicity, *Jpn. Pestic. Inf.*, 1988, No. 53, 271.
167. J. W. L. Meier, S. Sehgel, T. G. Cullen, K. Burgess, L. D. Jennings, I. R. Silverman, S. F. Ali, J. A. Dybas, and M. A. Walsh, 'Synthesis and Biological Activity of F1327, a Highly Efficacious Non Ester Pyrethroid and Acaricide', Abstract 5, Division of Agrochemicals, 207th ACS Natural Meeting, San Diego, CA, 1994.
168. M. J. Bushell, Synthesis of some fluorinated non-ester pyrethroids, in 'Recent Advances in the Chemistry of Insect Control II. Special Publication No. 79', ed. L. Crombie, The Royal Society of Chemistry, Cambridge, 1990, p. 125.
169. H. Stark, Aspects of the synthesis and structure activity in insecticidal neophane derivatives, '3rd International Symposium on Advances in the Chemistry of Insect Control', 1993.
170. S. McN. Sieburth, C. J. Manlys, and D. W. Gammon, Effects of isosteric replacement of silicon for carbon in etofenprox and MTI-800, *Pestic. Sci.*, 1990, **28**, 289.
171. K. Tsushima, T. Kano, K. Umeda, N. Matsuo, and M. Hirano, Synthesis, insecticidal activity and pyrethroidal mode of action of new tin ether derivatives, *Pestic. Sci.*, 1989, **25**, 17.
172. I. R. Hill, Aquatic organisms and pyrethroids, *Pestic. Sci.*, 1989, **27**, 429.
173. T. A. Miller and V. L. Salgado, The mode of action of pyrethroids on insects, in 'The Pyrethroid Insecticides', ed. J. P. Leahey, Taylor and Francis, London and Philadelphia, 1985, p. 43.
174. D. B. Sattelle and D. Yamamoto, Molecular targets of pyrethroid insecticides, *Adv. Insect Physiol.*, 1988, **20**, 147.
175. M. H. Lichfield, Toxicity to mammals, in 'The Pyrethroid Insecticides', ed. J. P. Leahey, Taylor, and Francis, London and Philadelphia, 1985, p. 99.
176. J. A. Pickett, Chemical pest control – the new philosophy, *Chem. Brit.*, 1988, 137.

177. J. A. Pickett, Potential of novel chemical approaches for overcoming insecticide resistance, in 'Resistance '91: Achievements and Developments in Combatting Pesticide Resistance', ed. I. Denholm, A. L. Devonshire and D. W. Hollomon, Elsevier, London and New York, 1992, p. 354.
178. G. P. Georghiou, Overview of insecticide resistance, in 'Managing Resistance to Agrochemicals, from Fundamental Research to Practical Strategies', ed. M. B. Green, H. M. LeBaron, and W. K. Moberg, Amercian Chemical Society, Washington DC, 1990, p. 18.
179. A. W. Farnham, Genetics of resistance of houseflies (*Musca domestica* L.) to pyrethroids. I. Knock-down resistance, *Pestic. Sci.*, 1977, **8**, 631.
180. R. M. Sawicki, Resistance to pyrethroid insecticides in arthropods, in 'Insecticides', ed. D. H. Hudson and D. R. Roberts, John Wiley, New York, p. 143.
181. J. R. Phillips, J. B. Graves, and R. G. Luttrell, Insecticides resistance management: relationship to integrated pest management, in 'The Pyrethroid Insecticides', ed. J. P. Leahey, Taylor and Francis, London and Philadelphia, 1985, p. 459.
182. I. A. Watkinson, Pyrethroids and the economics of pest management, in 'The Pyrethroid Insecticides', ed. J. P. Leahey, Taylor and Francis, London and Philadelphia, 1985, p. 465.
183. A. L. Devonshire, L. M. Field, and M. S. Williamson, Molecular biology of insecticide resistance, in 'Insect Molecular Science', ed. J. M. Crampton and P. Eggleston, Academic Press, New York and London, 1992, p. 173.
184. M. S. Williamson, I. Denholm, C. A. Bell, and A. L. Devonshire, Knockdown resistance (*kdr*) to DDT and pyrethroid insecticides maps to a sodium channel gene locus in the housefly, *Mol. Gen. Genet.* 1993, **240**, 17.
185. J.-P. Demoute, A brief review of the environmental fate and metabolism of pyrethroids, in 'The Pyrethroid Insecticides', ed. J. P. Leahey, Taylor and Francis, London and Philadelphia, 1985, p. 387.
186. H. P. M. Vijverberg and J. van den Bercken, Neurotoxicological effects and the mode of action of pyrethroid insecticides, *Crit. Rev. Toxicol.*, 1990, **21**(2), 106.
187. W. N. Aldridge, An assessment of the toxicological properties of pyrethroids and their neurotoxicity, *Crit. Rev. Toxicol.*, 1990, **21**(2), 89.
188. M. Elliott and N. F. Janes, Preferred conformations of pyrethroids, in 'Synthetic Pyrethroids, ACS Symposium Series No. 42', American Chemical Society, Washington DC, 1977, p. 29.
189. Y. Kurita, K. Tsushima, and C. Takayama, Conformational analysis of fenvalerate and an ether-type pyrethroid, in 'Probing Bioactive Mechanisms, ACS Symposium Series No. 413', ed. P. S. Magee, D. R. Henry, and J. H. Block, American Chemical Society, Washington DC, p. 183.
190. J. R. Byberg, F. S. Jorgensen, and P. D. Klemmensen, Towards an identification of the pyrethroid pharmacophore. A molecular modelling study of some pyrethroid esters, *J. Comp. Aided Molec. Des.*, 1987, **1**, 181.
191. T. Katagi and Y. Kurita, Similarity in the molecular shapes of the alcohol moieties of pyrethroids, *J. Pestic. Sci.*, 1989, **14**, 93.
192. F. Michelangeli, M. J. Robson, J. M. East, and A. G. Lee, The conformation of pyrethroids bound to lipid bilayers, *Biochem. Biophys. Acta*, 1990, **1028**, 49.
193. A. G. Lee and J. M. East, Interactions of insecticides with biological membranes, *Pestic. Sci.*, 1991, **32**, 317.
194. B. D. Hudson, A. R. George, M. G. Ford, and D. J. Livingstone, Structure–

activity relationships of pyrethroid insecticides. Part 2. The use of molecular dynamics for conformation searching and average parameter calculation, *J. Comp. Aided Molec. Des.* 1992, **6**, 191.

195. H. Yoshioka, Structure evolutions of synthetic pyrethroid insecticides in correlational views guided by template procedure as applied to natural product models with high structural multiplicity, in 'Rational Approaches to Structure, Activity and Ecotoxicology of Agrochemicals', ed. W. Draber and T. Fujita, CRC Press, Boca Raton, FL, ch. 8, p. 185.
196. L. E. Sutton, 'Determination of Organic Structures by Physical Methods', Academic Press, New York, 1955, vol. 1, p. 405.
197. M. Elliott, Lipophilic insect control agents, in 'Recent Advances in the Chemistry of Insect Control, Special Publication No. 53', ed. N. F. Janes, The Royal Society of Chemistry, London, 1985, p. 73.
198. M. Elliott, A. W. Farnham, N. F. Janes, and B. P. S. Khambay, The pyrethrins and related compounds XXXII. Replacement of the central ester link, *Pestic. Sci.*, 1988, **23**, 215.
199. A. E. Baydar, M. Elliott, A. W. Farnham, N. F. Janes, and B. P. S. Khambay, The pyrethrins and related compounds XXXIII. Effective combinations in the gemdimethyl central group region of non-ester pyrethroids, *Pestic. Sci.*, 1988, **23**, 231.
200. A. E. Baydar, M. Elliott, A. W. Farnham, N. F. Janes, and B. P. S. Khambay, The pyrethrins and related compounds XXXIV. Optimisation of insecticidal activity in non-esters, *Pestic. Sci.*, 1988, **23**, 247.
201. K. Tsushima, T. Yano, T. Takagaki, N. Matsuo, M. Hirano, and N. Ohno, Preparation and insecticidal activity of optically active 3-phenoxybenzyl-3,3,3-trifluoro-2-phenylpropyl ethers, *Agric. Biol. Chem.*, 1988, **52**(5), 1323.
202. N. Matsuo, K. Tsushima, T. Takagaki, M. Hirano, and N. Ohno, Syntheses and insecticidal activities of pyrethrins and some new classes of optically active pyrethroid insecticides, in 'Recent Advances in the Chemistry of Insect Control III', ed. G. G. Briggs, Royal Society of Chemistry, Cambridge, 1993, p. 208.
203. K. Naumann, 'Chemistry of Plant Protection, Vol. 4, Synthetic Pyrethroid Insecticides: Structure and Properties', Springer Verlag, Berlin, 1990, p. 64.
204. N. F. Janes, The behaviour of pyrethroids in organic solvents – NMR evidence, and significance in biological action, in 'Tables Rondes Roussel-Uclaf No. 37. Pyrethroid insecticides: chemistry and action', ed. J. E. Casida and M. Elliott, 1980, p. 9.
205. A. S. V. Burgen, J. Feeney, and G. C. K. Roberts, Binding of flexible ligands to macromolecules, *Nature*, 1975, **253**, 753.
206. J. A. Rozendaal, Impregnated mosquito nets and curtains for self-protection and vector control, *Trop. Dis. Bull.*, 1989, **86**(7), R1.
207. C. R. Worthing and R. J. Hance (ed.), 'The Pesticide Manual, A World Compendium', The British Crop Protection Council, 9th edn, 1991.
208. A. W. Farnham and B. P. S. Khambay, The pyrethrins and related compounds, Part XXXIX. Structure–activity relationships of pyrethroidal esters with cyclic side chains in the alcohol component against resistant strains of housefly (*Musca domestica* L.), *Pestic. Sci.*, 1995, **44**, 269.
209. A. W. Farnham and B. P. S. Khambay, The pyrethrins and related compounds. Part XL. Structure–activity relationships of pyrethroidal esters with acyclic side chain in the alcohol component against resistant strains of housefly (*Musca domestica* L.), *Pestic Sci.*, 1995, **44**, 277.

210. M. G. Ford, D. W. Hollomon, B. P. S. Khambay, and R. M. Sawicki, (ed.), 'Combating Resistance to Xenobiotics: Biochemical and Chemical Approaches', Ellis Horwood, Chichester, 1987.

211. J. A. Pickett, L. J. Wadhams, and C. M. Woodcock, New approaches to the development of semiochemicals for insect control, in 'Proceedings of a Conference on Insect Ecology, Tabor', SPB Academic Publishing, The Hague, 1990, p. 333.

CHAPTER 8

Natural Defence Mechanisms

A. J. PARR and M. J. C. RHODES

8.1 Introduction

Sessile organisms, such as plants, are unable to avoid their predators and are exposed in the field to a variety of environmental stresses and to the attacks of viral, bacterial, and fungal pathogens. Plants have evolved elaborate mechanisms to defend themselves against predation and pathogenic attack. Structural features are undoubtedly part of this defence. The presence of thorns, trichomes, and a thick leathery texture may be significant deterrents to predation, while the presence of surface layers, such as the cuticle, may offer important barriers to the ingress of pathogens. However, in addition to these morphological features, plants have important defences based on their chemistry. For instance, they may synthesize chemicals which make them unattractive as food or outrightly poisonous to predators, or play some role in a chemical defence mounted by plants against pathogens. This may be based either on constitutive chemicals or on compounds whose production is stimulated by the attempt of the pathogen to invade the plant tissue. Similarly, chemicals may play a role in protecting the plant from environmental stress resulting from desiccation, halophytic conditions, or from heavy metals, *etc.* An example of this is the accumulation of phenylpropanoid compounds, such as flavonoids, in response to UV-irradiation.[1,2] Such responses to either pathogens or environmental stress are characterized by their physical distribution of the resultant accumulation of chemicals. Typically, reactions to pathogens are concentrated at the plant surface, in the epidermis or on the plant external surface, in either the cuticular layer covering the epidermal layer of cells or in, or on, glands protruding from the surface. The chemicals involved in these defences are termed secondary metabolites. They are considered secondary in the sense that they are not present in all living cells of the plant and are not essential for the processes of basic cell function, cell division, growth, and maintenance. However, there is increasing evidence that secondary metabolites are essential to the survival of the plant in its environment in providing a crucial level of defence against stress and competitive organisms.

Morphology and chemical defences are often deeply intertwined. For instance, in *Mentha* spp. the presence of glandular hairs is combined with the accumulation of high concentrations of antifungal monoterpenes in the extracellular space between exterior walls of the secretory cells within the glandular trichome and the enveloping cuticle. The importance of such structural and morphological features should not be overlooked, but in this chapter we concentrate on the chemical aspects of plant defences. These chemical reactions to infection or stress may be of a rather localized nature, but recent research indicates that plants have signalling systems involving the migration of signal compounds from a site of initial stress or infection to activate responses at distant sites. This opens the possibility that plants have an equivalent of the immunization systems found in animals. Here the chemical nature of these defence compounds and the mechanisms underlying both the local and systemic responses to infection, grazing, and stress are discussed.

8.2 Chemistry of Plant Responses to Infection by Pathogens and to Predation

8.2.1 Endogenous (Constitutive) Chemical Defences

One of the characteristic features of plant metabolism is the synthesis of a wide range of chemically diverse secondary metabolites, which are not normally thought of as being essential to cell survival. The precise function of these compounds is still open to some debate,[3] but it is now clear that many play important roles in the interaction of the plant with its environment. A major part of this is in providing some protection from potential pathogens and predators. The range of secondary products identified from plants is enormous (over 7000 alkaloids alone are known[4]), so it is to be expected that different classes of compounds, or even different individual compounds, may show different patterns of activity. Certain secondary metabolites have a general toxicity against most organisms and, indeed, special compartmentation is needed to prevent toxic effects even in the host plants. Broadly speaking, many compounds that provide protection directed against pathogens tend to have a relatively weak and non-selective activity, but are often present at high concentrations to make up for this. Compounds that provide protection against predators, on the other hand, tend to be more highly targetted in their mode of action and can, in several cases, have low effective doses.

8.2.1.1 General Biocides

A number of secondary metabolites act by interfering with processes that are generally present in all cells, and thus have a potential broad-spectrum protective effect against both pathogens and predators, although under a given set of conditions they may show some degree of selectivity. These are discussed in the following sections.

8.2.1.1.1 Amino Acid Analogues Examples are azetidine-2-carboxylic acid (**1**), an analogue of proline found in Lily of the Valley (*Convallaria majalis* L.) and Solomon's Seal (*Polygonatum multiflorum* All.), and canavanine (**2**), an analogue of arginine found in many legumes within the subfamily Lotoideae. Certain of these analogues function by being incorporated into proteins in place of the normal amino acids, thus producing a product with impaired activity.[5,6] Plants containing these compounds can often discriminate against the analogues during their own protein synthesis, so avoiding self-toxicity.[5] Other amino acid analogues apparently act by being inhibitors of the biosynthesis and transport of the conventional amino acids.[5]

8.2.1.1.2 Sugar Analogues A number of polyhydroxy alkaloids have recently been discovered which are, in many ways, the nitrogenous analogues of monosaccharides (polyhydroxy oxygen-containing heterocycles). Perhaps the best known are swainsonine (**3**) from *Swainsonia* spp. and castanospermine (**4**)

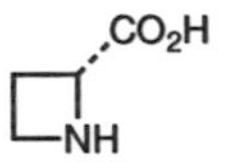

CO_2H NH

1; (*R*)-form 2-Azetidinecarboxlic acid

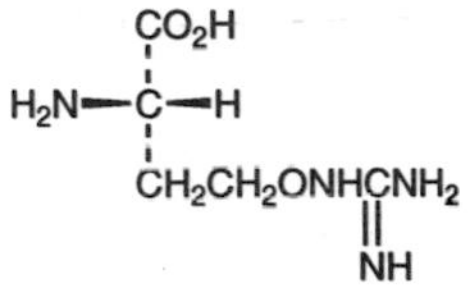

CO_2H
$H_2N—C—H$
$CH_2CH_2ONHCNH_2$
NH

2; (*S*)-form Canavanin

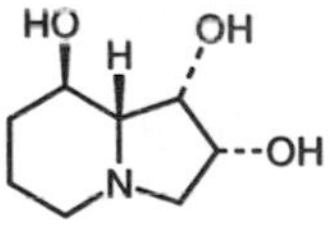

HO H OH OH N

3 Swainsonine

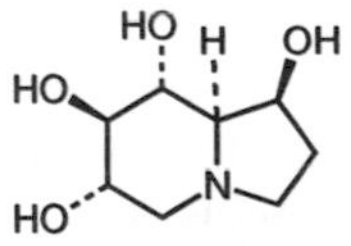

HO H OH HO HO N

4 Castanospermine

from *Castanospermum australe* A. Cunn & Fraser.[7] These compounds interfere with several processes that involve sugars, for instance glycoprotein processing and hydrolysis of sugars during digestion.[7] Castanospermine is currently of note for its anti-HIV activity.

8.2.1.1.3 Cyanogenic Glucosides Examples are amygdalin (**5**) from almonds and linamarin (**6**) from *Trifolium* spp. (for a review, see Conn[8]). These are

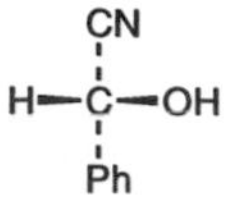

CN
H—C—OH
Ph

5; (*R*)-form 2-Hydroxy-2-phenylacetonitrile (*O*-β-D-glucopyranosyl(1→6)β-D-glucopyranoside: amygdalin)

$Me_2C(OH)CN$

6 2-Hydroxy-2-methylpropanenitrile (*O*-β-D-glucopyranoside: linamarin)

essentially aldehyde- or ketone-cyanide adducts stabilized by being glucosylated. (A few rare cyanogenic lipids are also known, in which the stabilization involves formation of esters with a long fatty acid.[9]) The cyanogenic glucosides are produced from amino acids, via the formation of the corresponding aldoximes, nitriles, and hydroxynitriles.[8] Within the cells of cyanogenic species not only are the glucosides present, but also the glucosidases able to hydrolyse them. Normally, the two are separately compartmentalized, but following the loss of cell integrity subsequent to infection or predation, the enzyme and substrate mix. The released aglycones then decompose by a combination of non-enzymatic and enzymatic routes to release free cyanide, which is the active agent.

8.2.1.1.4 Glucosinolates Examples are sinigrin (allylglucosinolate, **7**) and indol-3-ylmethylglucosinolate (**8**). Glucosinolates are found principally, but not exclusively,[10] in the family Brassicaceae (Cruciferae) and are in many ways similar to the cyanogenic glucosides. They are amino acid-derived thiohydroxamate *O*-sulfonates stabilized by glycosylation (for a review, see Larsen[11]). On

$H_2C{=}CHCH_2C(SGlc){=}NOSO_3H$

7 Glc = glucosyl Sinigrin

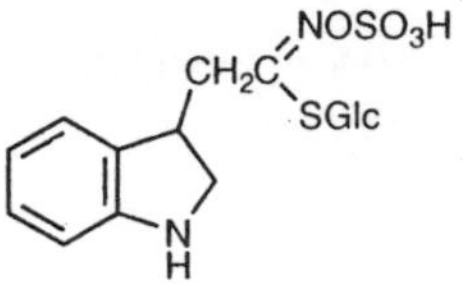

$CH_2C(=NOSO_3H)SGlc$ NH

8 Glc = glucosyl Glucobrassin (3-indolylmethyl glucosinolate)

hydrolysis by the enzyme myrosinase, also present in glucosinolate-bearing species, unstable aglycones are produced. These can undergo spontaneous isomerization and degradation to produce a range of different products, the particular mix being influenced by the precise environmental conditions experienced. Organic isothiocyanates often predominate, but nitriles can be important. Other possible products, depending on the precise nature of the amino acid-derived side chain, include free thiocyanate and oxazolidine-2-thiones.[11] Many of these compounds are highly biologically active.

8.2.1.1.5 Furanocoumarins Examples are xanthotoxin (**9**) and psoralen (**10**). They are typically produced by the plant families Apiaceae (Umbellifera-

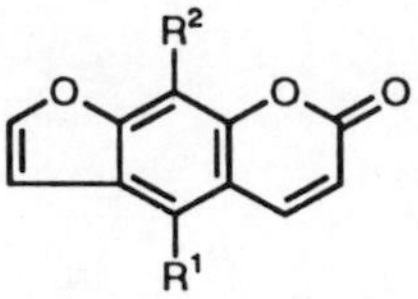

R^2 O O O R^1

Furanocoumarins

9 R^1 = H; R^2 = OMe Xanthotoxin
10 R^1 = R^2 = H Psoralen

ceae), Rutaceae, and certain members of the Fabaceae (Leguminoseae). Many furanocoumarins are able to intercalate into DNA, and then, in the presence of long-wave UV light, become covalently bound.[12] They thus have a photosensitizing and mutagenic capacity. Light-independent toxicity has also been reported,[13] which is presumably equally important in many cases of challenge by pathogens or predators.

It is interesting to note that although the protective effect of furanocoumarins is generally very efficient, certain insects have come to specialize in feeding on furanocoumarin-containing species. In the case of swallowtail butterflies feeding on Apiaceae and/or Rutaceae, the specialization is apparently facilitated by the presence of a specific cytochrome P450 system, which allows the furanocoumarins to be degraded.[14]

8.2.1.1.6 Polyphenolics The polyphenols (or vegetable tannins) represent an extensive group of compounds which are of great significance in many plants' endogenous defensive armoury. Two types are recognized:

- The condensed proanthocyanidins, composed of oligomers and higher polymers of a basic catechin (flavan-3-ol) nucleus.
- Esters (typically with glucose) of gallic acid or an oxidatively derived dimer.

Molecular weights typically range from 500 to 4000.[15]

The key feature of the polyphenols, and the property referred to in their older name of tannins, is their ability to interact with a range of both low and high molecular weight materials (especially proteins)—often causing them to precipitate. In the simplest cases, such interaction is reversible, but secondary irreversible processes are common, particularly in interactions with proteins. From a defensive point of view, polyphenolics have a number of roles. First, when released from the cell vacuole in which they are stored, they function as efficient enzyme inhibitors, so providing a non-specific resistance to invading pathogens. Second, their tanning action accounts for the phenomenon of astringency, which reduces a plant's palatability to herbivores.

8.2.1.1.7 Saponins Saponins are complex glycosides of either pentacyclic triterpenes or, less commonly, steroids. They occur widely throughout the plant kingdom, with the steroidal saponins being more typical of monocotyledons.[10] Their surface-active properties and ability to complex with membrane sterols enable them to interact with, and frequently disrupt, membranes. This gives many saponins a broad-based protective role; their often acrid taste may also serve to deter herbivors (although it should be noted that, in small quantities, certain saponin-bearing species are actually attractive to man—*e.g.*, liquorice, containing glycyrrhizin).

Related to the saponins are the steroidal alkaloid glycosides, such as those found in *Solanum* species. Indeed, many are the nitrogen heterocycle equivalents of the oxygen-containing steroidal saponins [compare solasodine (**11a**) and diosgenin (**11b**)], and not infrequently the two forms occur together in

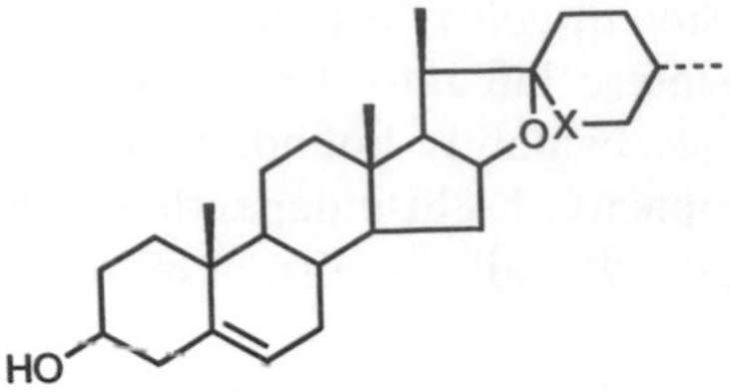

HO X O

11a Solasodine X = NH
11b Diosgenin X = O

glycoalkaloid-containing species. The glycoalkaloids are also able to interact with sterol-containing membranes,[16] and have bitter tastes.

8.2.1.2 Defences against Pathogens

A large number of metabolites, when tested *in vitro*, show antibacterial or antifungal activities (allbeit usually fairly weak), and it has been suggested that these may serve as endogenous defences against attack by pathogens. A recent experiment by Lamb and coworkers has clearly shown that this concept is broadly correct; they used seedlings of tobacco, which had been genetically engineered to contain reduced levels of the enzyme phenylalanine ammonia lyase (PAL), involved in the biosynthesis of phenolic secondary metabolites. At this early stage the principal chemical action was to reduce the levels of chlorogenic acid present. If these seedlings were then challenged with a pathogen, they were found to be more sensitive to infection than were the control plants – so confirming the idea that chlorogenic acid may act as an endogenous defensive agent.

At present, it seems likely that many endogenous plant phenolics may have antipathogen roles. Attention is now also being focussed on the terpenoid essential oils. These are spread quite widely throughout the plant kingdom, and recent work suggests that many of the individual terpenoid components may have antibacterial and/or antifungal activity.[17,18] It is likely that in some instances their mechanism of action is relatively non-specific, but this is not to detract from their efficacy.

8.2.1.3 Defences against Predators

As well as containing chemicals with antipathogen activities, plants also have in their armoury chemicals which deter attack by predators. These are generally highly active and highly specific in their actions, unlike the antipathogen agents, which are more typically of relatively low activity and selectivity. (A plant may, of course, contain high concentrations of a range of antipathogen agents in order to obtain efficient protection.)

8.2.1.3.1 Insects Many plant antipredator agents are active against a range of herbivores, but a number of classes appear specifically targetted against insects, which are, of course, the most successful group in the animal kingdom.

The occurrence in some plants of agents which mimic insect developmental hormones can be of major significance – resulting in abnormal maturation or moulting, and hence ultimately death, of larval insect herbivores. Both phytojuvenile hormones and phytoecdysones are known,[3] as well as certain antihormones, such as the antijuvenile hormone precocenes I and II (both substituted chromenes) from *Ageratum houstonianum* Mill.[19] It is interesting to note that while Butenandt and Karlson[20] isolated a few tens of milligrams of steroidal moulting hormones from 1 ton of silkworms, the same amount of similar compounds could be isolated from just a few grams of the dried rhizome of the fern *Polypodium vulgare* L.![21]

The other main class of defensive compounds targetted against insects are the antifeedants, which act by discouraging insect herbivores from selecting plants containing these compounds as acceptable food material. Chemically, these compounds are relatively diverse, including certain alkaloids, flavonoids and acetogenins, but the terpenoids seem to be particularly well represented.[22] One group of compounds currently attracting considerable attention are the nortriterpenoids found in the family Meliaceae. These are exemplified by azadirachtin, found in the Indian Neem tree (*Azadirachta indica* Juss; see Chapter 6, Parts I and II), but a variety of related compounds from several species are as active.[23] The oxygenated clerudane, ajugarin I, from *Ajuga remota* L., and particularly the sesquiterpene dialdehyde, polygodial, from *Polygonum hydropiper* L., are other compounds which have attracted attention.[24]

8.2.1.3.2 Higher Animals Many plant defensive agents are highly physiologically active in higher animals, a significant number acting either as agonists or antagonists for specific animal receptor systems, while yet others are inhibitors of the enzymes or transport systems involved in receptor-mediated signal transduction. Many of the alkaloids owe their high toxicity in animals to their direct, or indirect, interaction with the nervous system. Nicotine (Figure 8.1) activates one class of acetylcholine receptors, this interaction being so specific that the receptors have come to be known as the nicotinic sub-class.[25] Hyoscyamine and scopolamine are acetylcholine antagonists, while phytostigmine is an acetylcholinesterase inhibitor.[10] Several plant amines also act on the nervous system, or related peripheral receptors. Examples include ephedrine from *Ephedra* spp., mescaline from various Cactaceae, and also histamine and 5-hydroxytryptamine from nettles, *Urtica* spp.[20,25,27]

As well as the alkaloids and amines, the cardenolides are another major group of compounds highly toxic to many higher animals, this time acting on the heart. They are a structurally more homogeneous group than the alkaloids, being essentially C-17 modified 10,13-dimethylcyclopentanoperhydrophenanthrene steroid glycosides. Cardenolides are particularly abundant in the plant families Apocynaceae and Asclepiadaceae, but are also found in several other families.[10] Digitoxin from foxgloves (*Digitalis* spp.) is well-known, but other examples include strophanthin from *Strophanthus* spp. and oleandrin from *Oleander* (*Nerium oleander* L.).

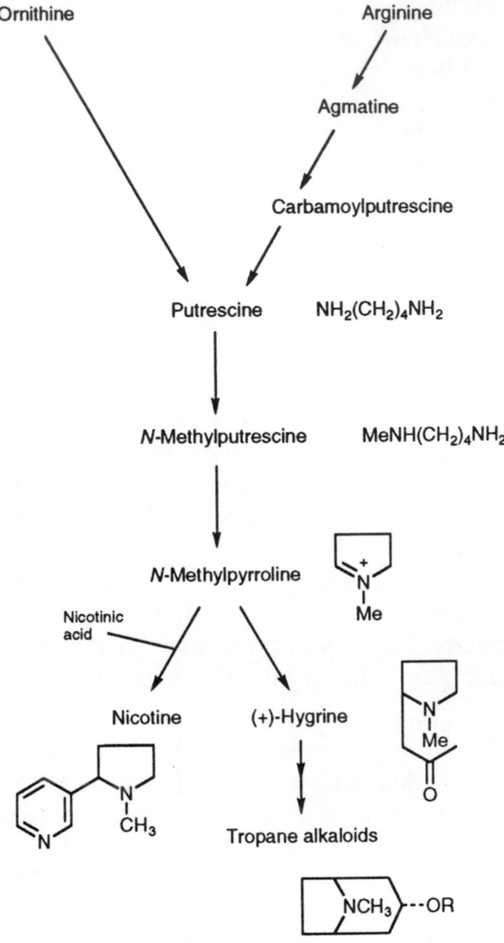

Figure 8.1 *Outline of the biosynthetic routes for the major non-steroidal alkaloids in Solanaceous plant species*

At this point it is worth remembering that to many plants, man and other primates are major predators, and thus it is no surprise to find that many secondary metabolites are highly biologically active in man. Although during the development of cultivation, man has, consciously or otherwise, often selected for foodplant varieties with reduced levels of many defensive compounds, their toxic action cannot always be discounted. This is especially so if infection of the food prior to consumption results in enhanced accumulation of active components (see Section 8.2.2.3.3.). The alkaloids and cardenolides mentioned above represent major hazards to man, although their often extreme toxicity has led to the general avoidance of plants, or plant parts, that contain significant levels of these materials. Perhaps more important, in food terms, are the general biocides described in Section 8.2.1.1. Reports of illness, or even death, resulting from many of the classes mentioned are by no means rare. The unusual amino acids β-cyanoalanine and hypoglycin A, for example, have both caused problems in the developing world.[3,27] Even in Europe and the USA, factors such as the glycoalkaloids of potatoes are monitored during the development of new varieties. Certain glucosinolates also represent a problem, either due to direct goitrogenic activity or, in the case of the indol-3-ylmethyl compound, an ability to react with, and inactivate, vitamin C.[11] Similarly, problems such as contact dermatitis and photosensitization resulting from plant furanocoumarins are well known.

8.2.2 Induced Responses

8.2.2.1 Wound Healing

An obvious response to herbivory is for plants to repair any damage, and to restore their structural integrity. Although this, to some extent, can be considered a passive process, specific chemicals can accumulate during wound healing,[28] and structural barriers do provide some protection against further attack. This aspect is discussed further in Section 8.2.2.3.2.

8.2.2.2 Hypersensitive Responses

A major response during some pathogen–plant interactions involves rapid cell death around the site of infection. As well as the activation of chemical defences around the periphery of the necrotic area, similar to those which occur in other circumstances (see 8.2.2.3), this hypersensitive response carries additional defensive benefits. Intracellular lytic enzymes are released, the local pH falls, and, in these and other ways, the microenvironment is made less suitable for further proliferation of the challenging pathogen.

8.2.2.3 Bioactive Defence Compounds Produced in Response to Challenge

8.2.2.3.1 Proteins Among the range of defence-related genes which become activated following challenge by predators or pathogens, many code for enzymes. Obviously, some of these simply relate to the induction of specific metabolic pathways, such as those for phytoalexin production or for the generation of transmissible signal compounds (see below), which occurs following challenge. A number of enzymes are, however, not of this nature, and clearly have a direct defensive role. A variety of hydrolytic enzymes have been found to be induced, including both acidic and basic chitinases and 1,3-β-glucanases,[29,30] which could act to degrade fungal cell walls. Oxidative enzymes, such as polyphenol oxidases and peroxidases, are also induced, and while these could have specific metabolic functions, their ability to produce highly reactive oxidized chemical species (such as *o*-quinones) may also be important in a broad-spectrum defence against pathogens. Similarly, other oxidases with degradative functions, such as lipoxygenase,[31] may be induced during a defence response.

In addition to enzymes, other proteins may be involved in defence. Proteinase inhibitors seem to play some role.[30,32] The thionines are another group which has received much attention in recent years. These compounds, some constitutively produced and some inducible,[33] are cysteine-rich, low-molecular weight polypeptides which show a wide-ranging toxicity to both procaryotes and eucaryotes.[33]

8.2.2.3.2 Polymeric Compounds Another defence strategy following challenge by pathogens is the development of physical and chemical barriers to the spread of infection. Notably, this involves the increased deposition of polymeric material around the site of infection. The β-1,3-polyglucan, callose, can be involved in this process, as can the lipo-phenolic polymers of suberin. Probably the most general response is an enhanced formation of lignin, which forms a strong, relatively inert, three-dimensional barrier. At the same time as new material is being laid down, there is also a strengthening of pre-existing cell wall material. The incorporation of phenolics is increased,[34] leading to the potential for increased interlinkage between cell wall materials via phenol–phenol coupling. A specific gene coding for the enzyme caffeoyl-CoA SAM *O*-methyl transferase has recently been identified as a key component in the provision of increased amounts of ferulic acid for incorporation into cell walls during a defensive response.[34]

Increased levels of hydroxyproline-rich glycoproteins also accumulate in the cell walls, and it is likely that this has primarily mechanical implications, although the suggestion has also been made that these proteins may be capable of agglutinating, and hence immobilizing, bacteria.[29]

8.2.2.3.3 Monomeric Compounds (Phytoalexins) The production of constitutive defensive compounds may sometimes be stimulated by infection; however, as a major part of the defence reactions occur following challenge by potential pathogens, many plants synthesize specific, biologically active, low molecular weight compounds, referred to as phytoalexins. These are not normally found in unelicited plants. As described below, a very wide range of compounds have been identified as phytoalexins. Different families or genera typically produce different types of phytoalexins and, indeed, a chemical which is a phytoalexin in one species may occur as part of the constitutive defensive armoury of another. Thus, there is nothing special chemically about a phytoalexin, but rather it is the very rapid, and often substantial, localized synthesis of a toxic metabolite in response to challenge which is the key feature of the phytoalexin mode of defensive response.

The induction of phytoalexin production has now been investigated in some detail in a few systems, and is outlined in Section 8.3. It is, however, relevant to note here that in many cases potential pathogens are not the only inducers of phytoalexin biosynthesis. In many instances, abiotic factors, such as heavy metal salts, can also act as elicitors.[35] A number of systems are also sensitive to environmental stresses[36] or, perhaps most significantly for phenolic compounds, to UV light.[1,2]

In a number of instances, the ability to synthesize high levels of specific phytoalexins in response to challenge by a potential pathogen has been shown clearly to be associated with the resistance of a particular plant variety towards that pathogen. A good example is the formation of 6-α-hydroxyphaseollin (**12**)

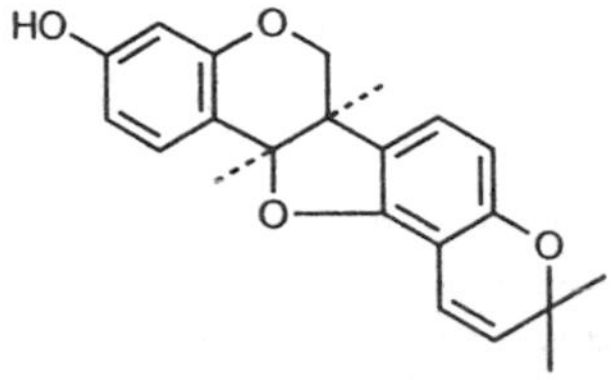

12 Phaseollin

by soybeans in response to challenge by *Phytophthora megasperma* var. *sojae* Drechs. Here, the rate and level of accumulation is 10–100 times greater when the challenge is by an incompatible race, which is resisted, than when the challenge is by a compatible race, which is not.[37,38] It should, however, be noted that in other instances the correlation is less clear cut and other factors must be dominant in controlling resistance in a given plant variety. Given the range of other types of defensive features described above, and the ability of many strains of pathogen to metabolize and degrade phytoalexins, this is perhaps not unexpected.

A very wide range of chemicals have been identified as phytoalexins, some very simple (*e.g.*, 4-hydroxybenzoic acid) and others quite highly elaborated. Perhaps the greatest number are phenolic in nature, generally being derived from the amino acid phenylalanine, although a very few have polyketide origins.[39,40] This presumably relates, in part, to the general enhancement of the deposition of the phenolic polymer, lignin, during a defence response (Section 8.2.2.3.2), and perhaps also to the widespread use of phenolics as constitutive defence metabolites (Section 8.2.1.1, 8.2.1.2). Table 8.1 illustrates some of the metabolites that are currently known to behave as phytoalexins in one or more species. It should be noted here that our knowledge of the area is still very far from complete. The phytoalexins of a number of the major crop species have been investigated in some depth,[55] but there is relatively little work in non-crop species, for which pathogen attack has little economic significance.

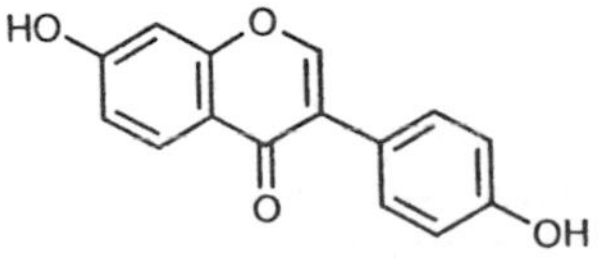

Isoflavone
13 Diadzein

Table 8.1 *Examples of different phytoalexins*

Chemical class/sub-class	*Metabolite with typical species in which it is a phytoalexin*	*Reference*
Phenolic		
Simple	4-Hydroxybenzoic acid – *Daucus carota* L.	41
Acetophenone	2′,6′-Dihydroxy-4′-methoxyacetophenone – *Sanguisorba minor* L.	42
Stilbene	Resveratrol – *Arachis hypogaea* L.	43
9,10-Dihydro-phenanthrene	Hircinol – *Loroglossum hircinum* Rich.	44
Lignan	2,6-Di-(4′-hydroxyphenyl)-3,7-dioxabicyclo[3.3.0]octane – *Vigna angularis* Ohwi & Ohashi	45
Isocoumarin	6-Methoxymellein – *D. carota*	46
Coumarin	Scopolin – *Solanum tuberosum* L.	47
	Scopoletin – *Helianthus annuus* L.	48
	6,7-Methylenedioxycoumarin – *H. annuus*	48
Chromone	Lathodoratin – *Lathryrus odoratus* L.	39
Furanocoumarin	Psoralen, bergapten – *Petroselinum crispum* Nyman	49
	Xanthotoxin – *Pastinaca sativa* L.	50
Benzofuran	Moracin A, moracin B – *Morus* sp.	51
Aurone	Cephalocerone – *Cephalocereus senilis* Pfeiff.	52
Flavanone	Sakuranetin – *Oryza sativa* L.	1
3-Deoxy-anthocyanidin	Apigeninidin, luteolinindin – *Sorghum bicolor* Moench	53
Pterocarpan	Pisatin – *Pisum sativum* L., *Lathyrus* spp. Phaseollin – *Phaseolus vulgaris* L. Medicarpin – *Medicago sativa* L.	38,54,55
Isoflavonoid	Diadzein (**13**) – *V. angularis*	45
	Tectorigenin – *Iris pseudacorus* L.	56
Phenolic biphenyl	Magnolol – *Cercidiphyllum japonicum* Sieb & Zucc.	57
	Aucuparin – *Eriobotrya japonica* Lindl.	58
Terpene		
Furanoterpene	Ipomoeamarone, dehydroipomeamarone – *Ipomoea batatis*	59
Sesquiterpene:		
Secoeudesmane	Phytuberin – *S. tuberosum*	36
Vetispirane	Solavetivone – *S. tuberosum*	36
Eudesmane	Rishitin – *S. tuberosum*, *Lycopersicon esculentum* Mill.	36
Eremophilene	Capsidiol – *Capsicum* sp.	36
Guaianolide	Cichoralexin – *Cichorium intybus* L.	60
Cadalene	Hemigossypol, gossypol – *Gossypium* spp.	61
	2,7-Dihydroxycadalene, lacinilene C – *Gossypium hirisutum* L.	61
Diterpene	Casbene – *Ricinus communis* L.	62

Continued

Table 8.1 *Continued*

Chemical class/sub-class	Metabolite with typical species in which it is a phytoalexin	Reference
Acetylene		
Polyacetylene	Safynol – *Carthamus tinctoria* L.	63
	Falcarindiol – *L. esculentum*	55
	1-Tridecen-3,5,7,9,11-pentayne – *Arctium lappa* L.	64
Furanoacetylene	Wyerone – *Vicia faba* L.	65
Acetylenic spiroketalenol ether	Mycosinol – *Coleostephus myconis* Reich.	66
Anthranilic acid derivative		
	Avenalumin I – *Avena sativa* L.	67
	Dianthramide, dianthalexin – *Dianthus caryophyllus* L.	68
Indole		
Sulfur containing	Brassilexin, brassinin, cyclobrassinin, cyclobrassinin sulfoxide – *Brassica juncea* Coss.	69
	Camalexin, methoxycamalexin – *Camelina sativa* Crantz.	70

As can been seen from Table 8.1, it is not unusual for a given species to produce a range of phytoalexins, which may be either biosynthetically related or belong to very different chemical classes. The ratio of these different phytoalexins is not fixed, but can vary with time, environmental conditions [*e.g.*, low temperatures favour rishitin (**14**) accumulation rather than phytuberin in potatoes infected with *Erwinia carotovora* Holl.[71]] or with the nature

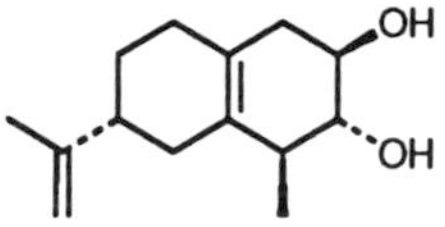

14 Rishitin

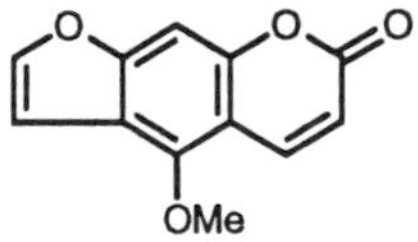

15 Bergapten

of the eliciting agent. A good example of how phytoalexin production is flexible is provided by parsley (*Petroselinum crispum* Nyman) cell cultures. When challenged with the fungal *Alternaria carthami* Chowdhury cell wall elicitor, cells produced mainly the furanocoumarin bergapten (**15**), with some isopimpinellin, while following challenge with *Phytophthora megasperma* Drechs. cell wall elicitor they produced graveolone and psoralen (**10**) as the principal phytoalexins.[49]

8.3 Mechanisms Involved in Plant Responses to Infection and Predation

8.3.1 Localized Metabolic Responses to Infection

Plant responses to infection involve a localized reaction in the vicinity of the point of attack by the pathogen and the transmission of signals from this region to activate other parts of the plant. As we have seen, important features of the localized response are a general mobilization of phenylpropanoid metabolism and the formation of species-specific phytoalexins. Evidence for the importance of phenylpropanoids in such defences is strong. For instance, as discussed above, transgenic tobacco plants with suppressed phenylpropanoid metabolism show increased susceptibility to the pathogen *Cerospora nicotianae*, Ell. & Ev.[72] Such phenylpropanoid compounds are present in uninfected plant tissues, but often their concentration rises significantly following infection. Examples of such compounds are cinnamic acid depsides, such as chlorogenic acid, and coumarins, such as scopoletin and esculin. The accumulation of such compounds apparently results from a general activation of cinnamic acid biosynthesis, much of which is directed initially to the formation of cinnamic acids which are then transported to the cell wall and esterified into the cell structure. There is evidence that an enzyme, caffeoyl CoA *O*-methyltransferase, is activated in tobacco tissue exposed to a microbial pathogen and this provides activated ferulic acid which is transported to the cell wall where it is used to esterify cell wall carbohydrates.[34]

The activation of phenylpropanoid metabolism is also involved in the longer term responses to attack or damage which often entails the deposition of phenolic polymers. These are either suberin, a polymer which consists of long chain fatty acids with cinnamic components,[73] or a form of lignin, 'wound lignin', a polymer of cinnamyl alcohols. Both these materials are laid down at damaged surfaces, where they limit water loss and impede the ingress of pathogens.

The phytoalexins are compounds which are essentially absent from uninfected plant tissues and which are synthesized *de novo* and accumulate on infection. Phytoalexins were first described by Muller and Borger.[74] They include compounds with very different chemical structures representing all of the major classes of secondary products. They tend to have a broad spectrum of activity against micro-organisms. They are generally of relatively low activity, and rely on the generation of very high local concentrations for their effectiveness. The mechanisms involved in phytoalcxin action have been investigated in some detail. Strong evidence for the importance of phytoalexins in defence has come from recent studies of a group of phenylpropanoid-derived phytoalexins, the stilbenes, which accumulate in several plants, including grape (*Vitis vinifera* L.), peanuts (*Arachis hypogaea* L.), and Scots Pine (*Pinus sylvestris* L.) in response to infection.[75] The enzyme stilbene synthase (STS) catalyses the condensation of *p*-coumaroyl CoA and malonyl CoA to form the stilbene

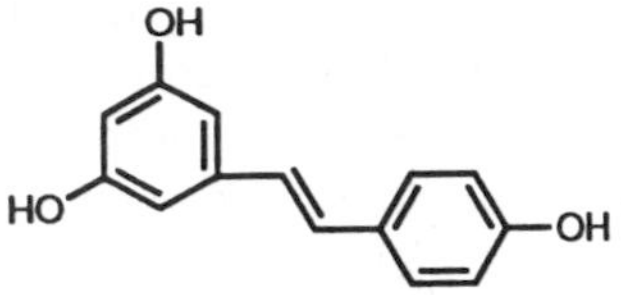

16 Resveratrol

resveratrol (**16**). STS has been cloned from peanut and grape, for which resveratrol is the major phytoalexin.[75] It has been possible to express these genes in tobacco, *Nicotiana tabacum* L., and to establish a pathway in which intermediates in flavonoid biosynthesis are diverted into the formation of a novel phytoalexin.[76,77] With the peanut gene, the concentration of accumulating resveratrol was found to be low and did not lead to any significant changes in resistance of the transgenic tobacco plants to infection.[76] However, in further experiments the STS gene *Vst1* from *V. vinifera* was transformed into tobacco and led to rapid accumulation of high levels of resveratrol. An inverse correlation was found between the levels of resveratrol accumulating in the leaves of transgenic tobacco containing the STS gene and the incidence of infection with the fungal pathogen *Botrytis cinerea* Pars.[77] This work provides strong evidence for the role of stilbenes in plant defence.

8.3.1.1 Pathogen-Derived Elicitors

It was found that the presence of the intact pathogen was not essential for induction of phytoalexin accumulation and that fragments of cell walls and even a range of abiotic factors would trigger the induction of phytoalexin biosynthesis. Such compounds are termed elicitors, and the way in which they trigger phytoalexin biosynthesis has been studied in some detail. Biotic elicitors are isolated from plant as well as from fungal cell walls. They are complex carbohydrates. The best characterized of these is the elicitor obtained from *Phytophthora megasperma* f.sp. *glycinea* T. Kuan & D. C. Erwin, which was shown to be a mixture of branched 3-, 6- and 3,6-linked glucans, typically involving seven glucose residues.[78] Similar β-glucan elicitors have been isolated from brewer's yeast.

These elicitors are thought to be generated by the action of plant or fungal cell wall degrading enzymes which, when attacking either the host or the pathogen cell wall, release fragments that have evolved a role in signalling the presence of the pathogen and invoking the plant's defensive mechanisms. How such a signal transduction pathway is constructed is a subject of much current research.[79] This pathway may involve many components, including protein receptors, reversible covalent modification of key proteins by phosphorylation catalysed by specific protein kinases and phosphorylases, Ca^{2+} flux changes, inositide phosphates, the activation of transcription factors, and hence to changes in the pattern of gene expression in the target tissue leading to phytoalexin biosynthesis.

Our understanding of the mechanisms underlying the induction of phytoalexin accumulation stems from the pioneering work of two groups, both using plant cell cultures as model systems. Hahlbrock and Scheel[80] studied the formation of furanocoumarin phytoalexins, such as marmesin and psoralen, in suspension cultures of *P. crispum* elicited by the fungal elicitor extracted from *Phytophthora* sp., while Dixon *et al.*[81] studied the formation of phytoalexins, such as kievitone, in cell cultures of *Phaseolus* sp. elicited with a preparation from *Colletotrichum* sp. The basic factor in the induction of the biosynthesis of phytoalexins is that elicitation leads to transitory increased transcription of genes coding for the enzymes of the biosynthetic pathway, with increases in the steady-state levels of the relevant mRNAs and a short-lived burst in the activity of the enzyme proteins. As noted above, the factors intervening between the release of the elicitor and the appearance of enhanced gene transcription are unknown, as, indeed, is how the coordinated induction of the whole pathway is achieved.

8.3.2 Systemic Responses to Infection and Damage

As we have seen, plants inoculated with pathogens react by inducing a chemical response at the site of infection. However, the work of Kuć[82] showed that this defence response spread to other parts of the plant, and that distant organs acquired resistance to subsequent attack. This phenomenon was termed systemic acquired resistance (SAR); it can persist for several weeks and often leads to resistance to pathogens other than the one giving the initial induction.[83] There has been considerable interest in the nature of the signal molecules which transmit the stimulus of the invasion of the pathogen to distant organs and induce them to initiate the defence response.

The properties necessary for a compound to act as a signal were defined by Lynn and Chang.[84] It must be capable of being synthesized rapidly, be transported to the target site, carry sufficient structural information to be recognized selectively by a receptor, and then rapidly degraded so that it does not persist in the target site once it has effected its response. One candidate for such a role is salicylic acid [SA,2-hydroxybenzoic acid (**17**) Figure 8.2].

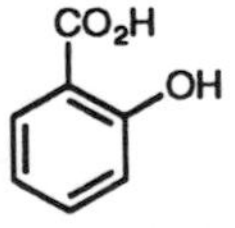

CO_2H OH

17 Salicylic acid

There is evidence that it acts as a messenger in the induction of systemic acquired resistance (SAR) of tobacco and cucumber.[83,85] Untreated tobacco leaves contain levels of SA of about 0.01 $\mu g\ g^{-1}$ fresh weight. Infection of tobacco with TMV leads to a stimulation of the concentration of SA in infected leaves to 1 $\mu g\ g^{-1}$ fresh weight and to about 0.1 $\mu g\ g^{-1}$ fresh weight in

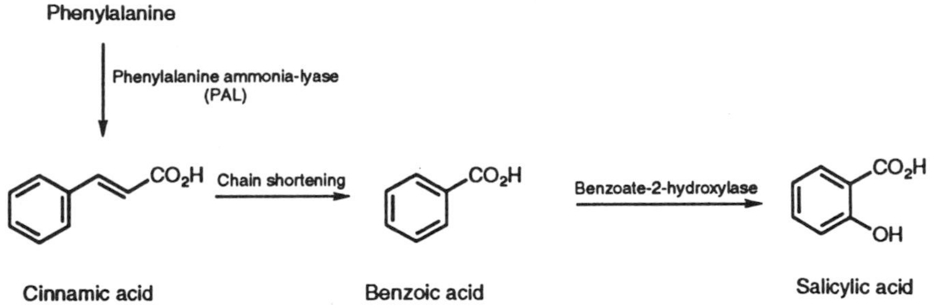

Figure 8.2 *The proposed biosynthetic pathway of salicylic acid in plants*

uninfected leaves in the same plant.[83] SA is transported in the phloem system, and concentrations of SA in the phloem exudate of up to 300 μM have been observed 18 h following inoculation of tobacco leaves with *Pseudomonas syringae* Van Hall.[86] Enyedi *et al.*[87] showed that in TMV-infected tobacco leaves the main product to accumulate in leaves bearing lesions was a conjugate of SA, tentatively identified as *O*-β-D-glucosyl-SA (GSA). GSA, however, was not found either in the phloem system or in uninoculated leaves. An enzyme showing high specificity for the glycosylation of SA, UDP-G: salicylic acid glucosyltransferase (GTase), has been described[88] and this activity is induced in tobacco leaves infected with TMV.[89] Uninfected tobacco leaves have a low basal level of GTase activity, but this level is elevated by about seven-fold within 72 h of infection. However, neither GSA nor GTase activity were found in a healthy leaf immediately above the infected leaves.[89]

In oat leaves treated exogenously with SA, GTase is induced and GSA accumulates. This probably represents a mechanism for controlling the level of SA. It is likely that SA is transported as the free acid, and that this is the active form of the signalling molecule. Glycosylation by the inducible GTase could be a mechanism for controlling SA levels. Evidence for the essential role of SA in SAR has come from experiments in which a bacterial gene which codes for a protein which degrades SA was expressed in transgenic tobacco plants. In these plants, TMV-induced SA accumulation and SAR were both inhibited.[90]

Exogenously supplied SA modifies the pattern of gene expression in treated tissue in a way which parallels that of the presence of a pathogen.[91] Treatment of tobacco with SA leads to the accumulation of pathogenesis-related proteins (PR), such as PR1a, and glycine-rich proteins, such as GPRP8, by a mechanism which involves stimulation of transcription of their genes and the accumulation of the related mRNAs.[83,85]

The route of biosynthesis of SA is unknown. As with other benzoic acid derivatives it is formed by elimination of a two-carbon unit from a cinnamic acid precursor. Two possible routes exit, one *via trans*-cinnamic acid and *o*-coumaric acid, and the other via cinnamic acid and benzoic acid (Figure 8.2). In tobacco, the detection of a benzoic acid 2-hydroxylase activity, which is induced following infection with TMV, suggested that the second of these routes might operate.[92]

This evidence suggests that SA fits the requirements as a signal compound:

- It is readily synthesized *de novo* as a response to the stimulus, in this case the presence of the pathogens, from a branch of an active pathway.
- It is transported in the phloem.
- It influences the pattern of gene expression, both locally and at distant regions of the plant.
- It is readily detoxified as its glycoside.

Whether SA is the only or primary signal in SAR is debatable,[86] but it clearly fits the definition of a signal compound. How a small molecule such as SA can induce changes in the pattern of gene expression leading to induction of the chemical disease response is under intensive research. It is thought that SA must bind to a specific binding protein, probably membrane-associated, and that the act of binding alters the conformation of the protein in such a way as to set off a sequence of signal transduction events, leading to the observed changes in the transcription of genes involved in defence responses. The precise nature of the changes induced by SA are unknown, but SA-binding proteins have been isolated from tobacco leaves.[93,94] SA binding is associated with a large, soluble protein with a molecular weight of 280 kDa, composed of several polypeptides, one of which has a molecular weight of 57 kDa.[94] It shows a high affinity for SA (K_d of 14 μM), within the physiological range of SA.[93] The ability of 23 other phenolic compounds to compete with SA for binding to the protein correlates both quantitatively and qualitatively with their ability to induce defence-related genes. This suggests a role for this binding protein in SA-induced signal propagation, but the nature of this role is still unclear.

A further example of a compound which may play a role in signalling in relation to defence is jasmonic acid {JA, [3-oxo-2-(2′-*cis*-pentenyl)-cyclopentane-1-acetic acid], Figure 8.3}. JA is synthesized, following membrane damage, by a series of reactions initiated by the release of α-linolenic acid [C18:3] by the action of fatty acid acylhydrolase. Linolenic acid then undergoes oxidation by lipoxygenase,[95,96] followed by cyclization of the 13-hydroperoxide[97] to yield 12-oxo-*cis*-*cis*-10,15-phytodienoic acid (12-oxo-PDA).[97,98] 12-Oxo-PDA is then reduced in the C5 side chain to form the *cis* unsaturated derivative, which then undergoes a series of three β-oxidation cycles to form jasmonic acid. Methyl jasmonic acid (MeJA) arises by subsequent esterification of JA. JA formation, unlike that of SA, is induced by wounding as well as by pathogen attack.[99] JA is transported in the phloem.[99]

Exogenous treatment of plant tissues with either JA or its methyl ester, MeJA, induces the information of proteinase inhibitors,[100] which play a role in defence. It is suggested that MeJA may act as a volatile factor in inter-plant communication to stimulate preparedness for defence. Jasmonate has been implicated as a mediator in the wound responses of plants.[101] Wounding of soybean hypocotyl tissue increases jasmonate levels more than five-fold, to about 500 ng g^{-1} fresh weight after 8 h; these elevated levels have been maintained for up to 24 h after wounding.[101] JA has been implicated in the

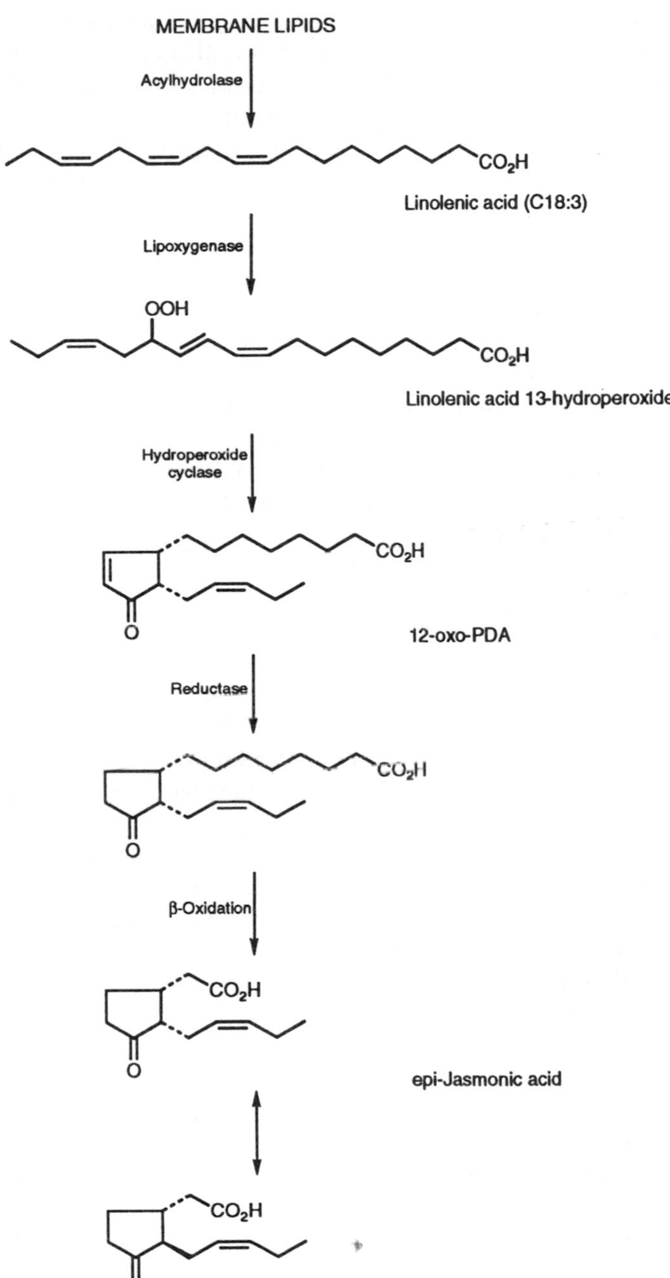

Figure 8.3 *The pathway of jasmonic acid biosynthesis*

responses induced in cell suspension cultures to the presence of biotic elicitors. Elicitation of cell suspension cultures of *Rauvolfia canescens* L. or *Eschscholtzia californica* Cham with a purified yeast cell wall elicitor fraction led to a rise in JA level, from a basal level in untreated cells of about 25 ng g^{-1} dry weight to over 1250 ng g^{-1} dry weight, 45 minutes after elicitor addition.[102] These stimulations of jasmonate lead to average internal concentrations of the order of 1 micromolar, within the range of concentrations of JA and MeJA sufficient to induce physiological changes in plants.[99]

Jasmonates act to alter the pattern of gene expression within plant tissues. In soybean cell suspension cultures, MeJA addition stimulated increases in the steady state levels of mRNA for wound-induced genes, such as chalcone synthase (chs); 20 μM MeJA treatment led to the appearance of chs1 mRNA within 1 h of treatment and to a maximum level of mRNA between 4 and 12 h.[101] JA treatment led to the accumulation of mRNA coding for a Kunitz-type proteinase inhibitor in tissue slices of stored potato tubers.[103] In cell cultures of soybean, treatment with 250 μM MeJA led to the induction of PAL, mRNA accumulation, and subsequent accumulation of isoflavonoid phytoalexins.[102] In *E. californica*, MeJA treatment led to a five-fold increase in mRNA for the berberine bridge enzyme [a key enzyme in benzo[*c*]phenanthridine biosynthesis] 6 h after the start of treatment. This preceded a peak in enzyme activity, 12-fold greater than the control after 17 h, and the accumulation of the benzo[*c*]phenanthridine alkaloid phytoalexins.[102]

JA meets the requirements of a signal compound as it has stringent requirements for activity.[104] It is rapidly synthesized, distributed from its site of synthesis, and influences the defence and wounding response of the plant. However, the nature of the signal transduction pathways initiated by JA is unknown.

A further type of signalling molecule has been demonstrated by the work of Ryan on the systemic induction of the formation of proteinase inhibitors in tobacco following wound damage resulting from mechanical injury or herbivory by insects.[32] Early work showed that wounding not only induced the formation of proteinase inhibitors in damaged leaves of tobacco, but also leaves above the damaged leaf showed the same response, and the presence of a proteinase-inhibitor inducing factor, transported from the site of wounding to other leaves to induce the formation of the inhibitors, was deduced. The transmitted factor was shown to be a small polypeptide consisting of 18 amino acids, which was named systemin.[105] This polypeptide was shown to be transported rapidly in the plant and, when supplied to cut stems of tobacco at femtomolar levels, induced formation of the proteinase inhibitors. Ryan and coworkers developed a model for systemin action in which wounding leads to release of the polypeptide, which is then translocated to other leaves where it interacts at a membrane receptor; this initiates the [α-] linolenic acid cascade, releasing jasmonic acid, which in turn leads to transcription factor formation and to proteinase inhibitor formation.[32] A cDNA has been cloned for the precursor of systemin, prosystemin. This precursor is a peptide of 200 amino acids; systemin is derived from a C-terminal section of this peptide by a process which must involve specific proteinase processing enzymes. Expression of the prosystemin gene in the antisense orientation in transgenic tomato plants abolished the systemic induction of proteinase inhibitors in the leaves following injury.[106] Subsequently, it was shown that such antisense plants showed reduced resistance to feeding by larvae of the tobacco hornmoth (*Manduca sexta* Johannson).[107] If the model of systemin action proposed by Ryan is correct, it provides an interesting interaction between long distance signalling due the very mobile systemin and more localized signalling due to jasmonic acid.

Our understanding of the nature of these signalling events and their contribution to plant defence is at an early stage, but results so far suggest that plants have the ability to signal injury and/or attack at one site and to elicit responses elsewhere in the plant in such a way as to prepare the plant for subsequent attack. In this, it has some similarities to the immune system of animals, but in plants low molecular weight compounds are the principal mediators of the increased resistance. Such signalling may not only be intra-plant and, as we have seen, MeJA may be involved in signalling between plants. A further example of this is that injury to corn caused by insect herbivores leads to the release of relatively large amounts of terpenoid compounds, such as linalool, (3*E*)-4,8-dimethyl-1,3,7,11-tridecatetraene, and (3*E*,7*E*)-4,8,12-trimethyl-1,3,7,11-tridecatetraene. These indicators of damage attract wasps, parasitic on the herbivorous insects, to the wound site.[108] Such interactions between plants and insects are undoubtedly complex, but they provide further evidence of the role of signalling events in the defence of the plant against environmental and pathogenic stresses.

8.4 Conclusion

It is clear that plants have evolved very successful defences which enable them to survive in an ever-changing environment and to resist attack by predators and pathogens. This review describes our current knowledge of the secondary metabolites thought to be involved in the defence of plants and of the biochemical mechanisms employed to facilitate their appropriate production and disposition in the plant.

Plants employ a variety of strategies in their chemical defences. Some compounds, such as the alkaloids and antifeedants involved in restricting predation, are highly targeted to specific receptors in the predator and are extremely potent. However, many phytoalexins involved in resistance to pathogens are non-specific and only weakly active in inhibiting microbial growth when tested *in vitro*, yet *in vivo* they may be effective against organisms ranging from fungi to viruses due to their rapid deposition at high local concentrations at the site of attack. Many secondary products may have many different physiological activities and may play more than one role in plant defence. Two compounds that affect, even weakly, processes within an organism may be potent defensive agents in combination if their targets of action are different. For instance, α-chaconine, the main glycoalkaloid of potato, was found to disrupt membrane structures, while the closely related compound, α-solanine was inactive. However, there was a stimulation of the effect of α-chaconine by α-solanine, which was optimal at roughly the relative proportions of the two compounds found in potato tubers.[109] There are probably many cases of such synergistic effects, which makes evaluation of the effectiveness of the individual secondary metabolites in the defence of the plant very difficult. Understanding the involvement of secondary metabolites in plant defence is confounded by the numbers of different compounds, representing so many different chemical classes, found in all plants, and the sum effectiveness of the chemical defence remains a matter of controversy, although it is clearly substantial.

If the role of an individual compound (or group of related compounds) can be understood and its contribution to overall defence delineated, then there are opportunities to manipulate its production to increase the plant's defence. These considerations have been discussed in detail in a recent review.[72] In the case of certain phenolic phytoalexins the potential of this approach has been demonstrated.[77] Manipulation of endogenous chemical defences has obvious advantages in the development of low-input agricultural systems. However, since these same plant defence compounds may be biologically active in man, and thus could be potential natural toxicants in food crops, the nutritional and health implications of the development of new varieties with improved endogenous defence must be considered. Similar issues may arise over the use of certain natural defensive secondary metabolites as exogenously applied pesticide or antipredator treatments.

8.5 References

1. O. Kodama, J. Miyakawa, T. Akatsuka, and S. Kiyosawa, Sakuranetin, a flavanone phytoalexin from ultraviolet-irradiated rice leaves, *Phytochemistry*, 1992, **31**, 3807.
2. J. Li, T.-M. Ou-Lee, R. Raba, R. G. Amundson, and R. L. Last, *Arabidopsis* flavonoid mutants are hypersensitive to UV-B irradiation, *The Plant Cell*, 1993, **5**, 171.
3. E. A. Bell, The physiological role(s) of secondary (natural) products, in 'The Biochemistry of Plants – A Comprehensive Treatise, Vol. 7, Secondary Plant Products', ed. P. K. Stumpf and E. E. Conn, Academic Press, New York, 1981, p. 1.
4. K. Mothes and M. Luckner, Historical introduction, in 'Biochemistry of Alkaloids', ed. K. Mothes, H. R. Schutte, and M. Luckner, VCH, Weinheim, 1985, p. 15.
5. L. Fouden, Nonprotein amino acids, in 'The Biochemistry of Plants – A Comprehensive Treatise, Vol. 7, Secondary Plant Products', ed. P. E. Stumpf and E. E. Conn, Academic Press, New York, 1981, p. 215.
6. G. A. Rosenthal, The biochemical basis for the deleterious effects of L-canavanine, *Phytochemistry*, 1991, **30**, 1055.

7. L. E. Fellows, G. C. Kite, R. J. Nash, S. J. Simmonds, and A. M. Scofield, Castanospermine, swainsonine and related polyhydroxy alkaloids: structure, distribution and biological activity, *Recent Adv. Phytochem*, 1989, **23**, 395.
8. E. E. Conn, Cyanogenic glycosides, in 'The Biochemistry of Plants – A Comprehensive Treatise, Vol. 7, Secondary Plant Products', ed P. K. Stumpf and E. E. Conn, Academic Press, New York, 1981, p. 479.
9. K. L. Mikolajczak, Cyanolipids, *Prog. Chem. Fats Other Lipids*, 1977, **15**, 97.
10. W. C. Evans, 'Trease and Evans' Pharmacognosy', Baillière Tindall, London, 13th edn. 1989.
11. P. O. Larsen, Glucosinolates, in 'The Biochemistry of Plants – A Comprehensive Treatise, Vol. 7, Secondary Plant Products', ed. P. K. Stumpf and E. E. Conn, Academic Press, New York, 1981, p. 501.
12. G. D. Cimino, H. B. Gamper, S. T. Isaacs, and J. E. Hearst, Psoralens as photoactive probes of nucleic acid structure and function: organic chemistry, photochemistry and biochemistry, *Ann. Rev. Biochem*. 1985, **53**, 1151.
13. A. E. Desjardins, G. F. Spencer, and R. D. Plattner, Tolerance and metabolism of furanocoumarins by the phytopathogenic fungus *Gibberella pulicaris* (*Fusarium sambucinum*), *Phytochemistry*, 1989, **28**, 2963.
14. M. R. Berenbaum, M. B. Cohen, and M. A. Schuler, Cytochrome P450 monooxygenase genes in oligophagous lepidoptera, in 'Molecular Mechanisms of Insecticide Resistance', ed. C. A. Mullin and J. G. Scott, American Chemistry Society, Washington, DC, 1992, p. 114.
15. E. Haslam, Polyphenols – phytochemical chameleons, in 'Phytochemistry and Agriculture (Proceedings of the Phytochemical Society of Europe 34)', ed. T. A. van Beek and H. Breteler, Oxford University Press, Oxford, 1993, p. 214.
16. J. G. Roddick, A. L. Rijnenberg, and M. Weissenberg, Alterations to the permeability of liposome membranes by the solasodine-based glycoalkaloids solasonine and solamargine, *Phytochemistry*, 1992, **31**, 1951.
17. S. G. Deans, Evaluations of antimicrobial activity of essential (volatile) oils, *Mod. Meth. Plant Anal.*, 1991, **12**, 309.
18. C. Bourrel, F. Perineau, G. Michel, and J. M. Bessiere, Catnip (*Nepeta cataria* L.) essential oil: analysis of chemical constituents, bacteriostatic and fungistatic properties, *J. Essent. Oil. Res.*, 1993, **5**, 159.
19. W. S. Bowers, Insecticidal compounds from plants, in 'Phytochemical Resources for Medicine and Agriculture', ed. H. N. Nigg and D. Seigler, Plenum Press, New York, 1992, p. 227.
20. A. Butenandt and P. Karlson, Über die Isolierung eines Metamorphose-Hormons der Insekten in Kristallisierter Form, *Z. Naturforsch*, 1954, **9B**, 389.
21. J. Jizba, V. Herout, and F. Sorm, Isolation of ecdysterone (crustecdysone) from *Polypodium vulgare* L. rhizomes, *Tetrahedron Lett.*, 1967, 1689.
22. D. C. Jain and A. K. Tripathi, Potential of natural products as insect antifeedants, *Phytother. Res.*, 1993, **7**, 327.
23. W. Kraus, M. Bokel, M. Schwinger, B. Vogler, R. Soellner, B. Wendisch, R. Steffens, and U. Wachendorff, The chemistry of azadirachtin and other insecticidal constituents of Meliaceae, in 'Phytochemistry and Agriculture (Proceedings of the Phytochemical Society of Europe 34)', ed. T. A. van Beek and H. Breteler, Oxford University Press, Oxford, 1993, p. 18.
24. J. A. Pickett, Production of behaviour-controlling chemicals by crop plants, *Phil. Trans. R. Soc. Lond. B*, 1985, **310**, 235.

25. E. D. Gundelfinger and N. Hess, Nicotinic acetylcholine receptors of the central nervous system of *Drosophila*, *Biochim. Biophys. Acta*, 1992, **1137**, 299.
26. T. A. Smith, Phenethylamine and related compounds in plants, *Phytochemistry*, 1977, **16**, 9.
27. M. Luckner, 'Secondary Metabolism in Micro-organisms, Plants and Animals', Springler-Verlag, Berlin, 3rd edn, 1990.
28. M. A. Bernards and N. G. Lewis, Alkyl ferulates in wound healing potato tubers, *Phytochemistry*, 1992, **31**, 3409.
29. B. Fritig, S. Kauffman, B. Dumas, P. Geoffrey, M. Kopp, and M. Legrand, Mechanism of the hypersensitive reaction of plants, in 'Plant Resistance to Viruses (CIBA Foundations Symposium 133)', ed. D. Evered and S. Harnett, Wiley, Chichester, 1987, p. 92.
30. Y. Ohashi and M. Oshima, Stress-induced expression of genes for pathogenesis-related protein in plants, *Plant Cell Physiol.*, 1992, **33**, 819.
31. E. Koch, B. M. Meier, H. G. Eiben, and A. Slusarenko, A lipoxygenase from leaves of tomato (*Lycopersicon esculentum* Mill.) is induced by response to plant pathogenic pseudomonads, *Plant Physiol.*, 1992, **99**, 571.
32. C. A. Ryan, The search for the proteinase inhibitor-inducting factor, PIIF, *Plant Molec. Biol.*, 1992, **19**, 123.
33. H. Bohlmann and K. Apel, Thionins, *Ann. Rev. Plant Physiol. Plant Mol. Biol.*, 1991, **42**, 227.
34. D. Schmitt, A.-E. Pakusch, and U. Matern, Molecular cloning, induction, and taxonomic distribution of Caffeoyl-CoA 3-*O*-methyltransferase, an enzyme involved in disease resistance, *J. Biol. Chem.*, 1991, **266**, 1741.
35. D. R. Threlfall and I. M. Whitehead, The use of biotic and abiotic elicitors to induce the formation of secondary plant products in cell suspension cultures of solanaceous plants, *Biochem. Soc. Trans.*, 1988, **16**, 71.
36. A. Stoessl, J. B. Stothers, and E. W. B. Ward, Sesquiterpenoid stress compounds of the Solanaceae, *Phytochemistry*, 1976, **15**, 855.
37. N. Keen, Hydroxyphaseollin produced by soybeans resistant to susceptible to *Phytophthora megasperma* var *sojae*, *Physiol. Plant Pathol.*, 1971, **1**, 265.
38. J. A. Kuć, Phytoalexins, in 'Encyclopedia of Plant Physiology, Vol. 4 – Physiological Plant Pathology', ed. R. Heitefuss and P. H. Williams, Springer Verlag, Berlin, 1976, p. 632.
39. N. A. Al-Douri and P. M. Dewick, Biosynthesis of the 3-ethylchromone phytoalexin lathodoratin in *Lathyrus odoratus*, *Phytochemistry*, 1988, **27**, 775.
40. F. Kurosaki, M. Itoh, and A. Nishi, Interaction between cerulenin and 6-hydroxymellein synthase in carrot cell extracts, *Phytochemistry*, 1994, **35**, 297.
41. V. Harding and J. Heale, Isolation and identification of the antifungal compound accumulating in the induced resistance response of carrot slices to *Botryris cinerea*, *Physiol. Plant Pathol.*, 1980, **17**, 277.
42. T. Kokubun, J. B. Harborne, and J. Eagles, 2′,6′-Dihydroxy-4′-methoxyacetophenone, a phytoalexin from the roots of *Sanguisorba minor*. *Phytochemistry*, 1994, **35**, 331.
43. J. L. Ingham, 3,5,4′-Trihydroxystilbene as a phytoalexin from groundnuts (*Arachis hypogaea*), *Phytochemistry*, 1976, **15**, 179.
44. M. H. Fisch, B. H. Flick, and J. Arditti, Structure and antifungal activity of hircinol, loroglossol and orchinol, *Phytochemistry*, 1973, **12**, 437.
45. M. Kobayashi and Y. Ohta, Induction of stress metabolite formation in suspension cultures of *Vigna angularis*, *Phytochemistry*, 1983, **22**, 1257.

46. B. Herndon, J. Kuć, and E. Williams, The role of 3-methyl-6-methoxy-8-hydroxy-3,4-dihydroisocoumarin in the resistance of carrot root to *Ceratocystis fimbriata*, *Phytopathol.* 1966, **56**, 189.
47. D. Clarke, The accumulation of scopolin in potato tissue in response to infection, *Physiol. Plant Pathol.* 1973, **3**, 347.
48. B. Tal and D. J. Robeson, The induction, by fungal inoculation, of ayapin and scopoletin biosynthesis in *Helianthus annuus*, *Phytochemistry*, 1986, **25**, 77.
49. K. G. Tietjen, D. Hunkler, and U. Matern, Differential response of cultured parsley cells to elicitors from two non-pathogenic strains of fungi, *Eur. J. Biochem.*, 1983, **131**, 401.
50. C. Johnson and D. R. Brannon, Xanthotoxin: a phytoalexin of *Pastinaca sativa root*, *Phytochemistry*, 1973, **12**, 2961.
51. M. Takasugi, S. Nagao, T. Masamune, A. Shirata, and K. Takahashi, Structure of moracin A and B, new phytoalexins from diseased mulberry, *Tetrahedron Lett.*, 1978, 797.
52. P. W. Pare, N. Dmitrieva, and T. J. Mabry, Phytoalexin aurone induced in *Cephalocereus senilis* liquid suspension culture, *Phytochemistry*, 1991, **30**, 1133.
53. Synder B. A. and R. L. Nicholson, Synthesis of phytoalexins in sorghum as a site-specific response to fungal ingress, *Science*, 1990, **248**, 1637.
54. J. L. Ingham, Systematic aspects of phytoalexin formation within the phaseoleae of the Leguminoseae (subfamily Papilionoideae), *Biochem. Syst. Ecol.*, 1990, **18**, 329.
55. J. A. Kuć, Antifungal compounds from plants, in 'Phytochemical Resources for Medicine and Agriculture', ed. N. H. Nigg and D. Seigler, Plenum Press, New York, 1992, p. 159.
56. F. Hanawa, S. Tahara, and J. Mizutani, Isoflavanoids produced by *Iris pseudacorus* leaves treated with cupric chloride, *Phytochemistry*, 1991, **30**, 157.
57. M. Takasugi and N. Katui, A biphenyl phytoalexin from *Cercidiphyllum japonicum*, *Phytochemistry*, 1986, **25**, 2751.
58. K. Watanabe, Y. Ishiguri, F. Nonaka, and A. Morita, Isolation and identification of aucuparin as a phytoalexin from *Eriobotrya japonica* L., *Agric. Biol. Chem.*, 1982, **46**, 567.
59. I. Oguni and I. Uritani, Isolation of dehydro-ipomeamarone, a new sesquiterpenoid from the black-rot fungus infected sweet potato root tissue and its relation to the biosynthesis of ipomoeamarone, *Agric. Biol. Chem.*, 1973, **37**, 244.
60. K. Monde, T. Oya, A. Shirata, and M. Takasugi, A guaianolide phytoalexin, cichoralexin, from *Cichorium intybus*, *Phytochemistry*, 1990, **29**, 3449.
61. A. A. Bell, R. D. Stipanovic, C. R. Howell, and P. A. Fryxell, Antimicrobial terpenoids of *Gossypium*: hemigossypol, 6-methoxyhemigossypol and 6-deoxyhemigossypol, *Phytochemistry*, 1975, **14**, 225.
62. D. Sitton and C. West, Casbene: an antifungal diterpene produced in cell-free extracts of *Ricinus communis* seedlings, *Phytochemistry*, 1975, **14**, 192.
63. E. H. Allen and C. A. Thomas, *Trans-trans*-3,11-tridecadiene-5,7,9-triyne-1,2-diol, an antifungal polyacetylene from diseased Safflower (*Carthamus tinctorius*), *Phytochemistry*, 1971, **10**, 1579.
64. M. Takasugi, S. Kawashima, N. Katsui, and A. Shirata, Two polyacetylenic phytoalexins from *Arctium lappa*, *Phytochemistry*, 1987, **26**, 2957.
65. R. M. Letcher, D. A. Widdowson, B. J. Deverall, and J. W. Mansfield, Identification and activity of wyerone acid as a phytoalexin in broad bean (*Vicia faba*) after infection by *Botrytis*, *Phytochemistry*, 1970, **9**, 249.

66. P. S. Marshall, J. B. Harborne, and G. S. King, A spiroketalenol ether phytoalexin from infected leaves and stems of *Coleostephus myconis*, *Phytochemistry*, 1987, **26**, 2493.
67. S. Mayama, T. Tani, T. Ueno, K. Hirabayashi, T. Nakashima, H. Fukami, Y. Mizuno, and H. Irie, Isolation and structure elucidation of genuine oat phytoalexin, avenalumin I, *Tetrahedron Lett.*, 1981, **33**, 2103.
68. G. J. Niemann, J. Liem, J. B. M. Pureveen, and J. J. Boon, The amide-type phytoalexin activity of carnation extracts is partially due to an artifact, *Phytochemistry*, 1991, **30**, 3923.
69. M. Devys, M. Barbier, A. Kollmann, T. Rouxel, and J. F. Bousquet, Cyclobrassimin sulphoxide, a sulphur-containing phytoalexin from *Brassica juncea*, *Phytochemistry*, 1990, **29**, 1087.
70. L. M. Browne, K. L. Conn, W. A. Ayer, and J. P. Tewari, The camalexins: new phytoalexins produced in the leaves of *Camelina sativa* (Cruciferae), *Tetrahedron*, 1991, **47**, 3909.
71. A. S. Ghanekar, S. R. Padwal-Desai, and G. B. Nadkarni, The involvement of phenolics and phytoalexins in resistance of potato to soft rot, *Potato Res.*, 1981, **27**, 189.
72. C. J. Lamb, J. A. Ryals, E. R. Ward, and R. A. Dixon, Emerging strategies for enhancing crop resistance to microbial pathogens, *Biotechnol.*, 1992, **10**, 1436.
73. P. E. Kolattukudy, Structure, biosynthesis and biodegradation of cutin and suberin, *Ann. Rev. Plant Physiol.*, 1981, **32**, 539.
74. K. O. Muller and H. Borger, Experimentelle Untersuchungen über die Phytophora-Resistenz der Kartoffel, *Arb. Biol. Abt. (Ansl-Reichstanst) Berl.*, 1941, **23**, 189.
75. J. H. Hart, Role of phytostilbenes in decay and disease resistance, *Ann. Rev. Phytopath.*, 1981, **19**, 70.
76. R. Hain, B. Bieseler, H. Kindl, G. Schröder, and R. Stöker, Expression of a stilbene synthase in *Nicotiana tabacum* results in synthesis of the phytoalexin, resveratrol, *Plant Mol. Biol.*, 1990, **15**, 325.
77. R. Hain, H.-J, Reif, E. Krause, R. Langebartels, H. Kindl, B. Vornam, W. Wiese, E. Schmelzer, P. H. Schrerer, R. H. Stöcker, and K. Stenzel, Disease resistance results from foreign phytoalexin expression in a novel plant, *Nature (London)*, 1993, **61**, 153.
78. A. G. Darvill and P. Albersheim, Phytoalexins and their elicitors – a defense against microbial infection in plants, *Ann. Rev. Plant Physiol*, 1984, **35**, 243.
79. J. Ebel, Phytoalexin synthesis. The biochemical analysis of the induction process, *Ann. Rev. Phytopathol.*, 1986, **24**, 235.
80. K. Hahlbrock and D. Scheel, Physiology and molecular biology of phenylpropanoid metabolism, *Ann. Rev. Plant. Physiol. Mol. Biol.*, 1989, **40**, 347.
81. R. A. Dixon, P. M. Dey, M. A. Lawton, and C. J. Lamb, Phytoalexin induction in french bean. Intercellular transmission of elicitation in cell suspension cultures and hypocotyl segments of *Phaseolus vulgaris*, *Plant Physiol*, 1983, **71**, 251.
82. R. A. Dean and J. Kuć, Induced systemic protection in plants, *Trends Biotechnol.*, 1985, **3**, 125.
83. J. Malamy, J. P. Carr, D. F. Klessig, and I. Raskin, Salicylic acid: A likely endogenous signal in the resistance response of tobacco to viral infection, *Science*, 1990, **250**, 1002.
84 D. G. Lynn and M. Chang, Phenolic signals in cohabitation: Implications for plant development, *Ann. Rev. Plant Physiol. Mol. Biol.*, 1990, **41**, 497.

85. J. P. Métraux, H. Signer, J. Ryals, E. Ward, M. Wyss-Benz, J. Gaudin, K. Rashdorf, E. Schmid, W. Blum, and B. Inverardi, Increase in salicylic acid at the time of onset of systemic acquired resistance in cucumber, *Science*, 1990, **250**, 1004.
86. J. B. Rasmussen, R. Hammerschmidt, and M. N. Zook, Systemic induction of salicylic acid accumulation in cucumber after inoculation with *Pseudomonas syringae* pv *syringae*, *Plant Physiol.*, 1991, **97**, 1342.
87. J. J. Enyedi, N. Yalpani, P. Silverman, and I. Raskin, Localisation, conjugation and function of salicylic acid in tobacco during the hypersensitive reaction to tobacco mosaic virus, *Proc. Natl. Acad. Sci. USA*, 1992, **89**, 2480.
88. N. Yalpani, M. Schulz, M. P. Davis, and N. E. Balke, Partial purification and properties of an inducible uridine 5′-diphosphate-glucose: salicylic acid glucosyltransferase from oat roots, *Plant Physiol.*, 1992, **100**, 457.
89. A. J. Enyedi and I. Raskin, Induction of UDP-glucose: salicylic acid glucosyltransferase activity in tobacco mosaic virus-inoculated tobacco (*Nicotiana tabacum*) leaves, *Plant Physiol.*, 1993, **101**, 1375.
90. T. Gaffney, L. Friedrich, B. Vernooij, D. Negrotto, G. Nye, S. Uknes, E. Ward, H. Kessman, and J. Ryals, Requirement for salicylic acid for the induction of systemic acquired resistance, *Science*, 1993, **261**, 754.
91. E. R. Ward, S. L. Uknes, S. C. Williams, S. S. Dincher, D. L. Widerhold, D. C. Alexander, P. Ahl-Goy, J.-P. Métraux, and J. A. Ryals, Coordinate gene activity in response to agents that induce systemic acquired resistance, *The Plant Cell*, 1991, **3**, 1085.
92. J. Leon, N. Yalpani, I. Raskin, and M. A. Lawton, Induction of benzoic acid 2-hydroxylase in virus-inoculated tobacco, *Plant Physiol*, 1993, **103**, 323.
93. Z. Chen and D. F. Klessig, Identification of a soluble salicyclic acid-binding protein that may function in signal transduction in the plant disease response, *Proc. Natl. Acad. Sci., USA*, 1991, **88**, 8179.
94. Z. Chen, J. W. Ricigliano, and D. F. Klessig, Purification and characterisation of a soluble salicyclic acid-binding protein from tobacco, *Proc. Natl. Acad. Sci. USA*, 1993, **90**, 9533.
95. B. A. Vick and D. C. Zimmerman, The biosynthesis of jasmonic acid: a physiological role for lipoxygenase, *Biochem. Biophys. Res. Commun.*, 1983, **111**, 470.
96. B. A. Vick and D. C. Zimmerman, Biosynthesis of jasmonic acid by several plant species, *Plant Physiol*, 1984, **75**, 458.
97. L. Crombie and D. O. Morgan, Synthesis of [14,14-$^2H_2$]-linolenic acid and its use to confirm the pathway to 12-oxophytodienoic acid (12-oxo-PDA) in plants; a conspectus of the epoxycarbonium ion derived family of metabolites from linoleic and linolenic acid hydroperodixes, *J. Chem. Soc., Perkin. Trans. 1*, 1991, 581.
98. D. C. Zimmerman and P. Feng, Characterisation of a prostaglandin-like metabolism of linoleic acid produced by a flaxseed extract, *Lipids*, 1978, **13**, 313.
99. A. J. Enyedi, N. Yalpani, P. Silverman, and I. Raskin, Signal molecules in systemic plant resistance to pathogens and pests, *Cell*, 1992, **70**, 879.
100. E. E. Farmer and R. A. Ryan, Interplant communication; airborne methyl jasmonate induces synthesis of proteinase inhibitors in plant leaves, *Proc. Natl. Acad. Sci. USA*, 1990, **87**, 7713.
101. R. A. Creelman, M. L. Tierney, and J. E. Mullet, Jasmonic acid/methyl jasmonate accumulate in wounded soybean hypocotyls and modulate wound gene expression, *Proc. Natl. Acad. Sci. USA*, 1992, **89**, 4938.

102. H. Gunlach, M. J. Müller, T. M. Kutchan, and M. H. Zenk, Jasmonic acid is a signal transducer in elicitor-induced plant cell cultures, *Proc. Natl. Acad. Sci. USA*, 1992, **89**, 2389.
103. K. Yamagishi, C. Mitosumori, K. Takahashi, K. Fujino, Y. Koda, and Y. Kikuta, Jasmonic acid-inducible gene expression of a Kunitz-type proteinase inhibitor in potato tuber disks, *Plant Mol. Biol.*, 1993, **21**, 539.
104. E. W. Weiler, T. Albrecht, B. Groth, Z.-Q. Xia, M. Luxem, H. Liss, L. Andert and P. Spengler, Evidence for the involvement of jasmonates and their octadecanoid precursors in the tendril coiling response of *Bryonia dioica*, *Phytochemistry*, 1993, **32**, 591.
105. G. Pearce, D. Strydom, S. Johnson, and C. A. Ryan, A polypeptide from tomato leaves activates the expression of proteinase inhibitor genes, *Science*, 1991, **253**, 875.
106. B. McGurl, G. Pearce, M. L. Orozco-Cardenas, and C. A. Ryan, Structure, expression and antisense inhibition of the systemin precursor gene, *Science*, 1992, **255**, 1570.
107. M. L. Orozco-Cardenas, B. McGurl, and C. A. Ryan, Expression of an antisense prosystemin gene in tomato plants reduces resistance toward *Manduca sexta* larvae, *Proc. Natl. Acad. Sci. USA*, 1993, **90**, 8273.
108. T. C. J. Turlings and J. H. Tumlinson, Systemic release of chemical signals by herbivore-injured corn, *Proc. Natl. Acad. Sci. USA*, 1992, **89**, 8399.
109. J. G. Roddick and A. L. Rijnenberg, Synergistic interaction between the potato glycoalkaloids α-solanine and α-chaconine in relation to lysis of phospholipid/sterol liposomes, *Phytochemistry*, 1987, **26**, 1325.

CHAPTER 9

Animal Venoms and Insect Toxins as Lead Compounds in the Design of Agrochemicals – Especially Insecticides

IAN S. BLAGBROUGH* and EDUARDO MOYA

9.1 Introduction

In the search for new biologically active compounds, preferably with novel sites of action, animal venoms and insect toxins, such as argiotoxins and the related spider toxin polyamine amides, are lead structures for new classes of potent compounds with potential as agrochemicals, and especially as insecticides.

In particular, in this chapter, the plethora of published sources of animal toxins are reviewed comprehensively. Vertebrate and invertebrate toxins are reported together with the rich diversity of structures which are produced in the natural world. This includes an appraisal of polyamine amides, derived from certain spiders and wasps, which are potent antagonists of the L-glutamic acid receptors of insect skeletal muscle and of nicotinic acetylcholine receptors. Proteins and polypeptides are possible leads for the production of potent agrochemical agents, so their potential as insecticides, given the new and exciting developments in protein delivery systems, and the possibilities for building protein (*e.g.*, scorpion) toxins into transgenic plants are briefly discussed. In conclusion, a comprehensive, but critical assessment of animal venoms and insect toxins for the design or development of agrochemicals, especially insecticides, is presented.

The gradual, but persistent, development of the resistance of insect and other anthropod pests to chemical-control measures fuels the search for new toxins. In order to overcome this almost insidious problem of pesticide resistance, new control measures should have novel sites of action, provided

*To whom all correspondence should be addressed.

they exist. Although this desirable property may not be designed readily at the present time, it is widely recognized that the natural product library provides the best, in the sense of most comprehensive and potentially selective while structurally diverse, leads for the further design of potent agrochemical agents. It is for these reasons that the discoveries of new, small molecule toxins, larger peptides, and even proteins which are amenable to modern delivery systems, achieved by design of the formulation, continue to excite interest.

Natural toxins do not only include the constituents of animal venoms, but fungal, bacterial, and plant toxins are addressed in other chapters (see Chapters 1–7). In this review, we concentrate on animal toxins which are injected. The electric shocks of manta rays and electric eels are not considered, neither are compounds which are toxic, but usually are encountered only when injested [*Amanita muscaria* and other poisonous fungi toxins from blue–green algae, including microcystin LR (**1**) or anatoxin (**2**)].[1–3] The only exception is a brief mention of the important guanidine-containing neurotoxins, saxitoxin (**3**), and the related shellfish and puffer fish poison, tetrodotoxin (**4**), see Figure 9.1.

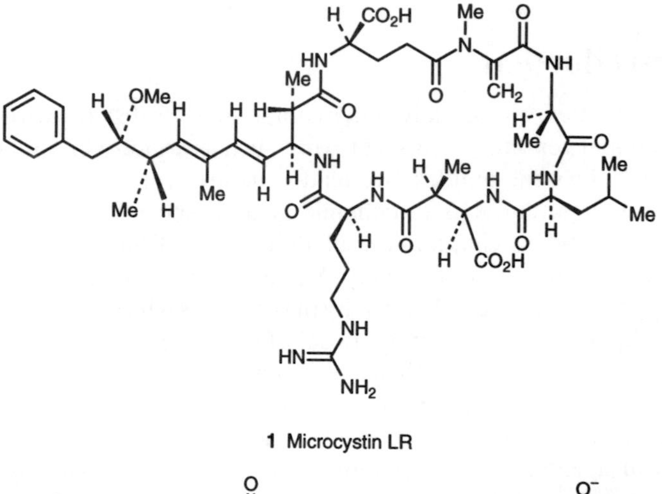

1 Microcystin LR

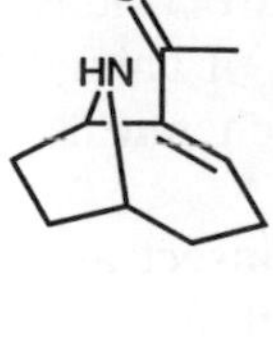

2 Anatoxin A

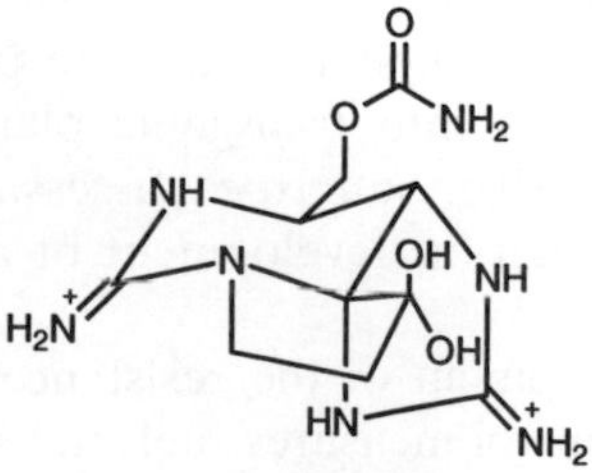

3 Saxitoxin

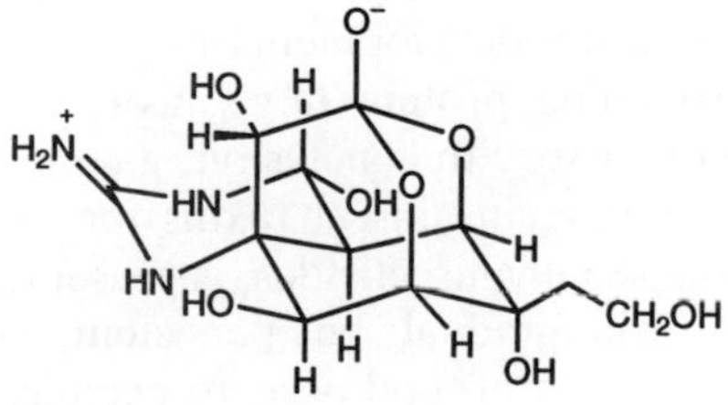

4 Tetrodotoxin

Figure 9.1 *Aquatic toxins from algal blooms [microcystin LR (1), anatoxin A (2), puffer fish [tetrodotoxin (4)], and dinoflagellates, a form of algal bloom [saxitoxin (3)]*

Fish with proteinaceous toxins in spines or glands, such as stonefish, stingrays, and scorpion fish, are reviewed in a book by Habermehl.[4] Not included, in this review, are allomones or pheromones[5,6] of defence or allure of, *e.g.*, the skunk, toad, moth, and beetle. Plant toxins (Figure 9.2), *e.g.*, *β*-erythroidine (**5**), a piperidine alkaloid with curare-like action as a skeletal muscle relaxant antagonized by prostigmine (**6**), and those toxins used by hunters *e.g.*, aconitine[7,8] (**7**) or tubocurarine[9,10] (**8**) to tip arrows and darts, to kill game are outside the scope of this chapter. For recent reviews on the chemistry and mode of action of the insect antifeedant azidarachtin, see Ley and coworkers.[11,12]

Many of the detailed examples come from invertebrates. The literature on insecticidal action of animal toxins is somewhat limited; in particular, scientists have concentrated upon spider, wasp, and ant toxins, for which insects are more obviously the intended prey. A limited discussion of the mammalian pharmacology for the toxins is included where little or no insecticidal data are available. In respect of spider polyamine amide toxins, details of their potential as open-channel blockers of important insect signalling systems are presented. Of course, there can clearly be developments with peptide toxins obtained from or designed on the basis of those used by spiders, (conch) snails, and other animals, but these require a delivery system for a high molecular weight polar compound. In this respect, baculoviruses (see Chapter 11) and transgenic

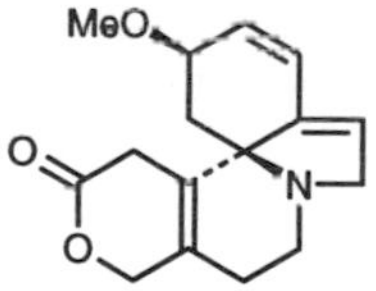

5 β-Erythroidine

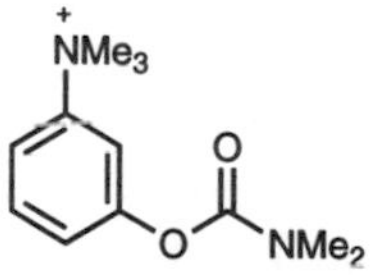

6 Prostigmine (Neostigmine or Prostigmin)

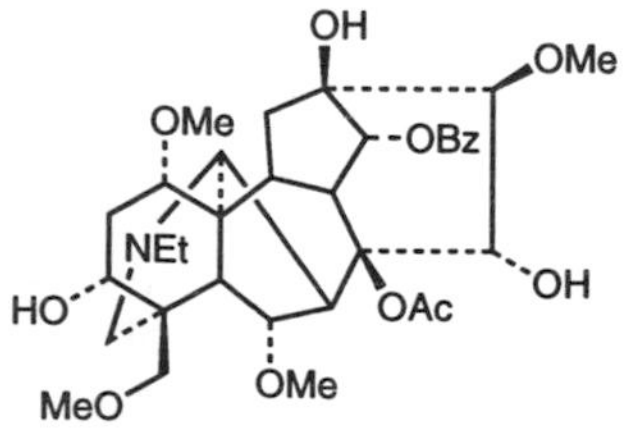

7 Aconitine

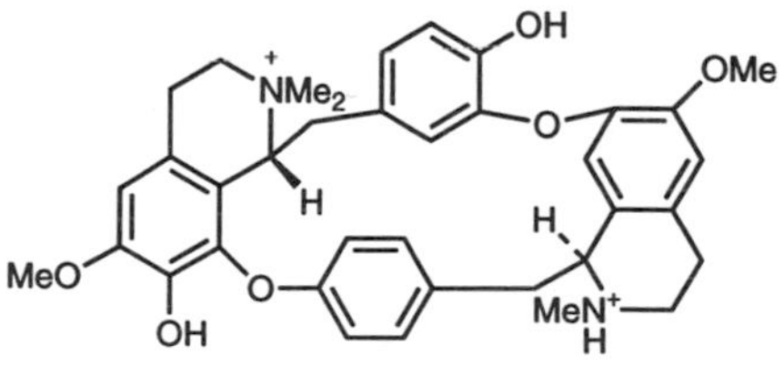

8 (+)-Tubocurarine

Figure 9.2 *Plant alkaloid toxins, β-erythroidine (5) isolated from* Erythrina *spp.,* Leguminosae; *prostigmine (6), aconitine (7) isolated from* Aconitum *spp.; (+)-tubocurarine (8) isolated from* Chondodendron tomentosum

plants, which might incorporate polypeptides whose design is based on animal toxin structures, might provide such a delivery system.

As pests and disease reduce the yields of crops, as well as impair the quality of produce, new chemical entities are required to control these problems in agriculture. However, newer techniques of plant genetic manipulation may allow peptide toxins, animal toxins, or their analogues, to be produced *in situ*. This approach of developing transgenic plants may provide enabling technologies for the efficient use of polypeptides which are genetically encoded and which currently pose significant delivery problems. Another associated problem with this method of pest control is that part of the crop must be consumed before the toxin is delivered. Peptide delivery systems are the subject of current research for pharmaceuticals and for agrochemicals. The potential uses of baculoviruses are covered comprehensively by Leisy and Fuxa in Chapter 11. Some leading references on the use of transformation technology of viruses and bacteria include Bishop and coworkers[13,14] and Winstanley and Rovesti.[15] Key references on transgenic (higher) plants include Hernalsteens *et al.*,[16] Fischoff *et al.*,[17] Vaeck *et al.*,[18] Potrykus,[19,20] and Perlack *et al.*[21]

Toxins are used in hunting and for defensive purposes, for paralysing and for killing prey – two distinctly separate modes of action. Enzymes (proteases) are often used for digestive purposes, before or after ingestion. Hyaluronidase, for instance, facilitates the penetration of toxin by dissolving cementing compounds between cells. Insect poisons, Hymenoptera venoms, and other invertebrate venoms are designed with high specificity for their particular prey. Toxins found in spiders, conch snails, wasps, bees, ants (Hymenoptera), scorpions, Gila monsters (the only species of poisonous lizards), and snakes are assessed. Whereas many mammals outrun, overpower, and then tear up their prey, many slower, weaker, or more vulnerable creatures inject their prey with toxins, although some of these injections are exclusively for defence.

There are many chemical structural variations, a diversity of functional groups, and many pharmacological modes of action. Where there are modifications to a common theme, *e.g.*, within a family or genus, these are in either the peptide or protein backbone (*i.e.*, the three-dimensional shape), but they are generally composed of similar, conservative changes to the side-chain functional groups among constitutive amino acids. The accompanying small molecules may also be variations upon a theme; often, amino acids or terpenes are used as toxins, irritants, or potent pharmacological ion channel openers and/or blockers, if they are not toxins in their own right. Thus, disturbance of metabolism or catabolism is an efficient mechanism for subduing prey, but a more efficient, in the sense of more rapid, method is to disturb the balance of the nervous systems at the same time. Thus, both rapid-acting and longer-lasting toxins can often be identified within one venomous mixture.

In his excellent summary of chemical pest control measures, setting out the new philosophy for the 1990s, Picket[22] highlights the problems of resistance, the potential of semiochemicals, and modifications to the genetics of crop plants, and concludes that insect toxicants based upon 'chemical control will nevertheless be used for the foreseeable future'. It remains true that 'there are still many unexploited natural product leads for new classes of insecticides which could overcome existing resistance mechanisms and give rise to new insecticides with even greater selectivity. In addition, the rapid expansion in work on insect neurohormones and neurotransmitters, *e.g.* the pentapeptide proctolin (RYLPT), a cockroach muscle neurotransmitter, and on toxins produced by insect predators, *e.g.* insecticidal polyamine containing spider toxins, could create new targets for insect control chemistry'.[22]

9.2 Molluscs

Mussels (lamellibronchiata) and other mollusca give rise to shellfish poisoning – a common occurrence. This was recognized as a medical problem as early as in the seventeenth century. Shellfish poisoning occurs throughout the world, wherever shellfish are digested, but the poisoning is not necessarily caused by spoiled shellfish. Typically, toxins are responsible for the symptoms.[23] These toxins have either been assimilated from the food and stored, or were biosynthesized by the shellfish. There has recently been published a total syntheses of the spiropentacyclic cyclitol esters (Figure 9.3) surugatoxin[24] (**9**) and neosurugatoxin[25] (**10**), potent acetylcholine receptor antagonists, isolated from the Suruga bay marine snail *Babylonia japonica*.

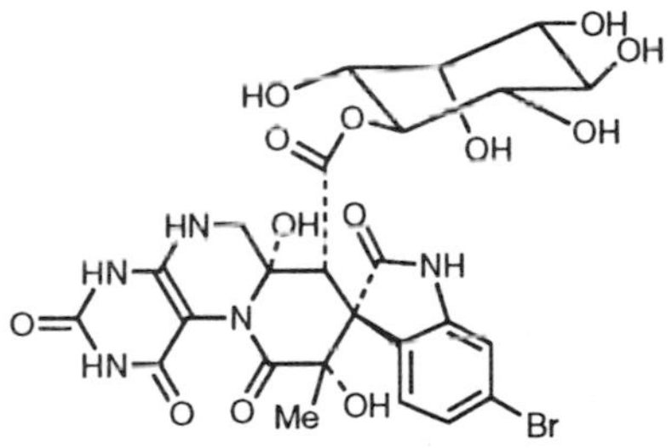

9 Surugatoxin

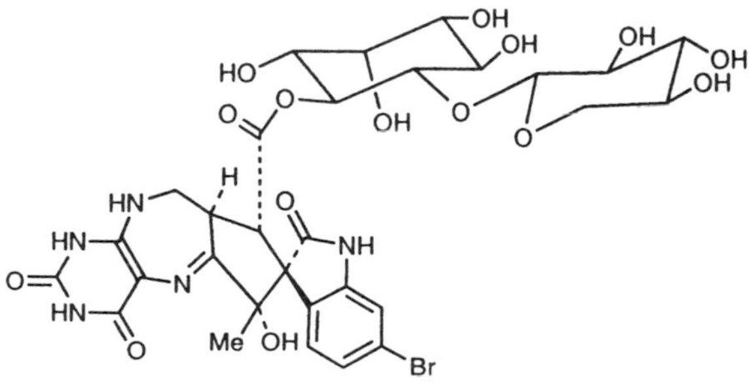

10 Neosurugatoxin

Figure 9.3 *Surugatoxin (9) and neosurugatoxin (10) are algal bloom toxins*

It is not possible to discriminate poisonous shellfish by odour, although toxicity occurs simultaneously with the proliferation of dinoflagellates. The three types of symptomatic poisoning are:

- Gastrointestinal, with a long latency period (10–12 h) and rapid recovery from nausea, diarrhoea, and vomiting.
- Erythematous poisoning, with a short latency period (2–3 h), typically allergenic symptoms, recovery within a few days, and rarely lethal.
- Paralytic shellfish poisoning is lethal, which has no latency period, and the effects are obvious within 30 minutes. In severe cases, ataxia and general motor incoordination are accompanied by a peculiar feeling of lightness, confusion, rapid pulse; death results from respiratory paralysis within 12 h. The lethal dose of the causative agent, saxitoxin (**3**) is 1 mg for an 'average' human.[26–29] This potent neurotoxin is produced by dinoflagellates *Gonyaulax catenella* and *G. excavata*, which are nutrients for shellfish (see Chapter 5).

Cone shells, conidea, and toxoglossa are actively venomous animals. They possess a complicated and well-developed venom-apparatus for use in hunting prey. The conch shells which feed upon fish are particularly toxic to humans, as well as to fish. Most human poisonings happen while collecting the 'beautiful' cone shells. Peptide toxins, from *Conus geographus*, have been studied in detail by Olivera and co-workers.[30] The king-kong peptide has aroused a great deal of interest due to its mammalian pharmacology, and this continues to be a particularly active area of research for Oliveira and many others.[31] Apart from the selective pharmacology of the many peptide toxins in cone shell venom, the tiny harpoon-like tooth, produced and stored in the shell and fired, full of venom, from the proboscis into the prey, is a fascinating area of biology.

9.3 Octopuses

The octopus *Hapalochlaena maculata*, found on the Australian East coast, is venomous. The blue-ringed octopus bite is apparently painless and not often noticed, but it gives rise to powerful systemic symptoms, including vomiting, numbness and loss of coordination, and death occurs by respiratory paralysis. The venom is produced in the anterior salivary glands and the principal toxic component has been called maculotoxin, and has been shown to have the same structure as the puffer fish toxin tetrodotoxin (**4**), a complex guanidinium-containing polycyclic carbonate.[23,26–28]

9.4 Coelenterates and Cridaria

Sea anemonies, jellyfish, corals, sea nettles, sea wasps, and polyps are found in most seas. Poisonings are caused by contact with nettle capsules (nematocysts), venom cells called cridoblasts. These are predominantly found on the tentacles and near the mouth. The sting or venom duct is ejected and the prey injected with the simultaneous extrusion of venom. The venom is a mixture of several low molecular weight peptides which are extremely poisonous to crustaceans and fish. These peptides are rarely toxic to mammals, but they can give rise to hypersensitivity and shock. More generally, contact with the tentacles causes an intense burning, which can be followed by unconsciousness. A marine animal worthy of special mention, in this section, is the sea wasp *Chironex flecheri*, as it is one of the most toxic sea creatures. The venom is cardiotoxic, causing hypertension, lung oedema, acute heart failure, and death within minutes. The venom consists of medium (10 kDa) to high (300 kDa) molecular weight proteins and a cocktail of enzymes, accompanied by, in some species, biogenic amines, histamine, and serotonin (5-HT).

9.5 Amphibians

Frogs, toads, and salamanders are found within the class Amphibia which contains approximately 2000 species. They are divided into Anura (lacking tails, *e.g.*, frogs and toads) and Undela (with a tail, *e.g.*, salamander); these orders are both subdivided into numerous genera.

Amphibians are venomous animals, but they use their toxins for protection and not generally for attack. The venom is produced in skin glands distributed over their bodies and instantly secreted in small quantities. It was once thought that these secretions were used solely for defence against natural predators, but this is not the case. They also protect the host amphibian against some micro-organisms. The skin of amphibians would be a perfect medium for the growth of bacteria and fungi, as it is moist in order to permit the exchange of oxygen and carbon dioxide. It is clearly significant that amphibians live in an environment rich in micro-organisms; in control experiments, detoxified animals died from skin infections within a few days. The antibiotic activity of the toxins is such that they are able to inhibit the growth of bacteria and fungi in concentrations as low as 10 μM. Mammalian poisoning by amphibians is rare. There have been extensive pharmacological studies of the individual toxins from a variety of amphibians and they can be cardiotoxic, neurotoxic, haemotoxic, cholinomimetic, sympathomimetic, vasoconstrictive, hypotensive, hallucinogenic, and local anaesthetic agents. For examples of amine-containing, oxygenated, modified-steroid amphibian toxins, see Figure 9.4.

Amphibians are a source of biologically important, and now well-known, toxins (Figure 9.5). The frog and toad toxins include batrachotoxin (**19**), a steroidal alkaloid from the skin of neotropical poison-dart frog, *Phyllobates aurotaenia*. This substituted indole-3-carboxylate ester of batrachotoxinin A (**18**) was isolated by Daly *et al.*[32] in 1965 and contains an unprecedented 9-α-oxygenation and a 7-membered heterocycle across the *cis*-CD ring junction.[33] Recent studies have focused upon photoaffinity labelling of the voltage-sensitive sodium channel sites that bind this toxin.[34–37] Tritiated batrachotoxinin A has been used as a synaptosome probe of susceptible and *kdr*-type-resistant German cockroaches, *Blatella germanica*,[38] and for

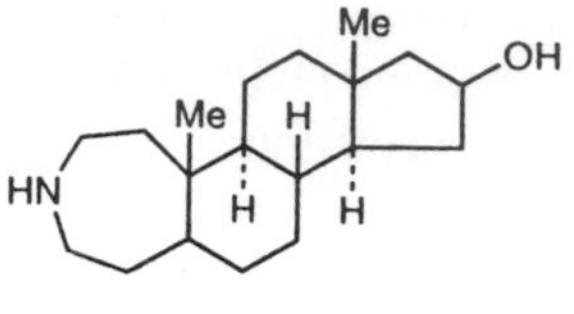

11 Samanine

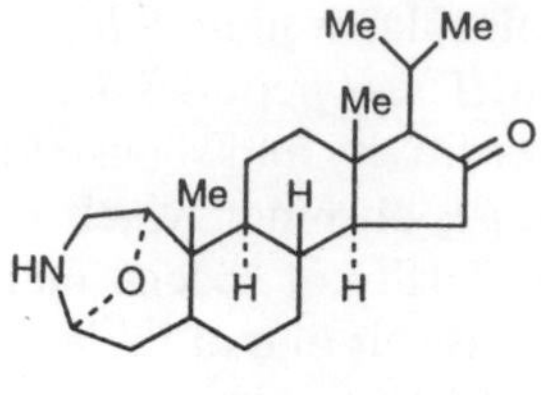

12 Samandenon

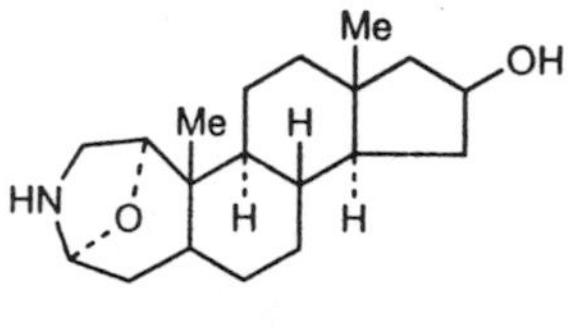

13 Samandarine

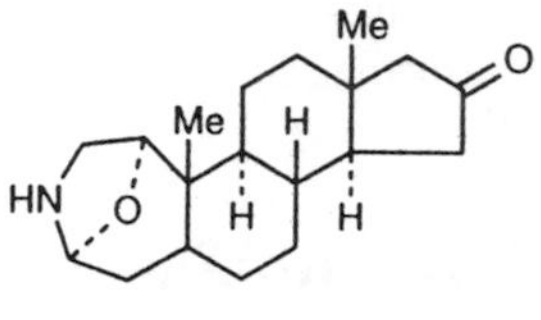

14 Samandarone

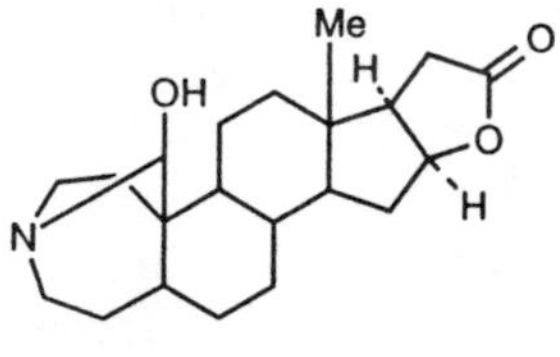

15 Cycloneosamandaridin

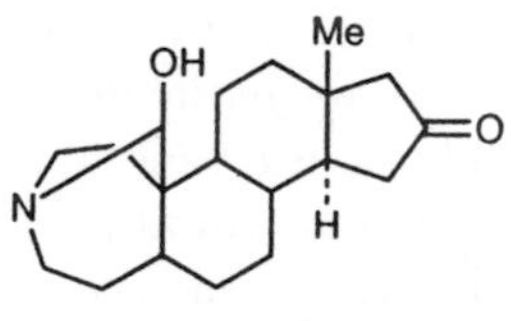

16 Cycloneosamandion

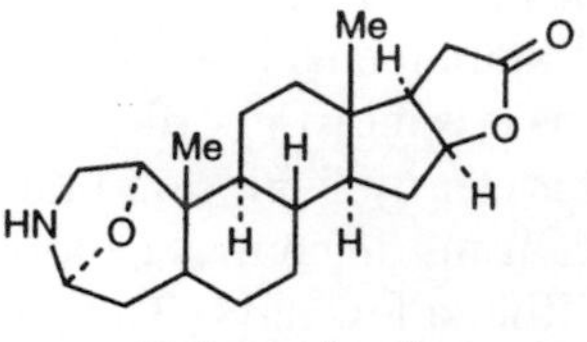

17 Samandaradin

Figure 9.4 *Toxins found in the skin secretions of newts and salamanders (order* Urodela *and genera* Salamandra *and* Triturus*). These highly modified steroidal alkaloids contain either oxazolidine or carbinolamine functional groups*

18 Batrachotoxinin A R = H

19 Batrachotoxin R =

20 Homobatrachotoxin R =

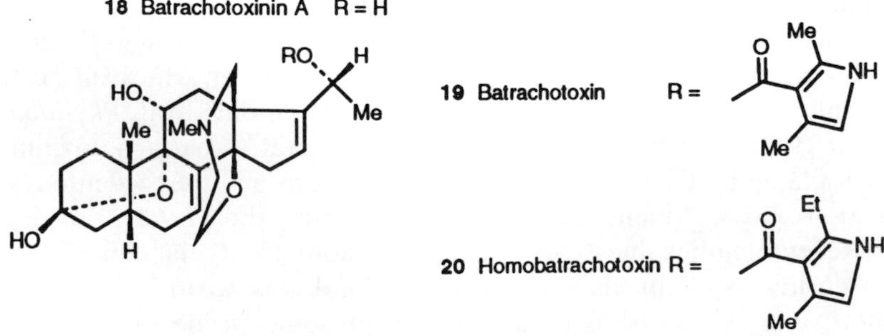

Figure 9.5 *A variety of steroidal alkaloid compounds that are found in certain frogs and toads. The frog* Phyllobates aurolaemia *produces batrachotoxins in its skin secretions*

pyrethroid toxicity modulation in susceptible and resistant tobacco budworm moths, *Heliothis virescens*.[39,40] Grinsteiner and Kishi have recently published their synthetic approaches to batrachotoxin,[41,42] A comprehensive review of the skin alkaloids of neotropical poison frogs (Dendrobatidae) was recently compiled by Daly *et al.*[43]

The most studied and, therefore, best known of the amphibian toxins are the bufotoxins (Figure 9.6), which are steroids. Desacetylbufotalin (bufogenin B) and bufotalin (**33**) are isolated from the Chinese drug Ch'an Su, which are prepared from Chinese toads (*Bufo asiaticus*) and from the venom of the European toad (*Bufa vulgaris*), respectively. They are steroids substituted at C-14 with a tertiary alcohol and an unsaturated δ-lactone at C-17. Bufotoxin (**34**) is the corresponding alkyl (suberoyl) arginine ester at the C-3 β-Hydroxyl functional group.

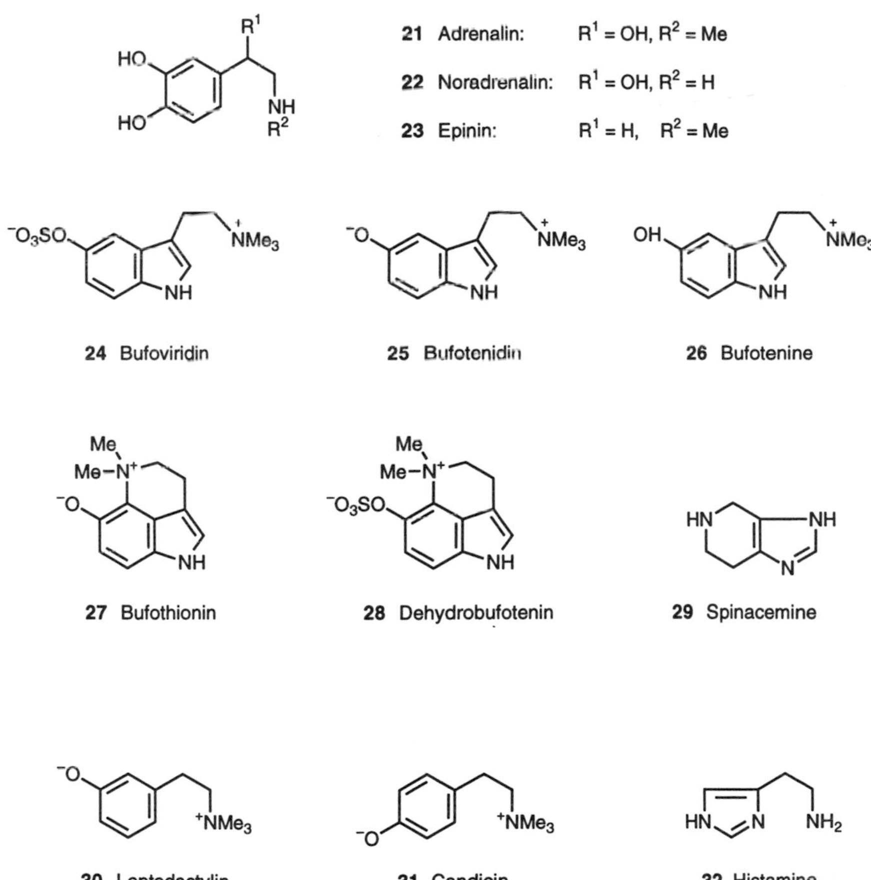

Figure 9.6 *A variety of simple biogenic amines which are found in frogs and toads. These include adrenalin (**21**) and noradrenalin (**22**); functionalized indole amines such as bufoviridin (**24**), bufotenidin (**25**); bufotenine (**26**), bufothionin (**27**), and dehydrobufotenin (**28**), 5-hydroxy tryptophan derivatives (**29**), leptodactylin (**30**) and candiecin (**31**), histamine (**32**).*

(Continued overleaf)

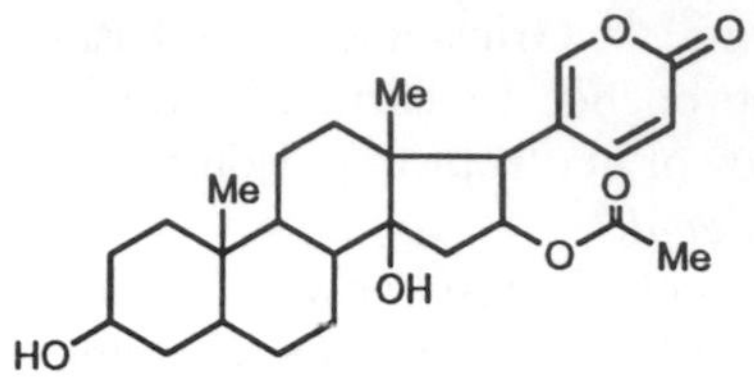

33 Bufotalin

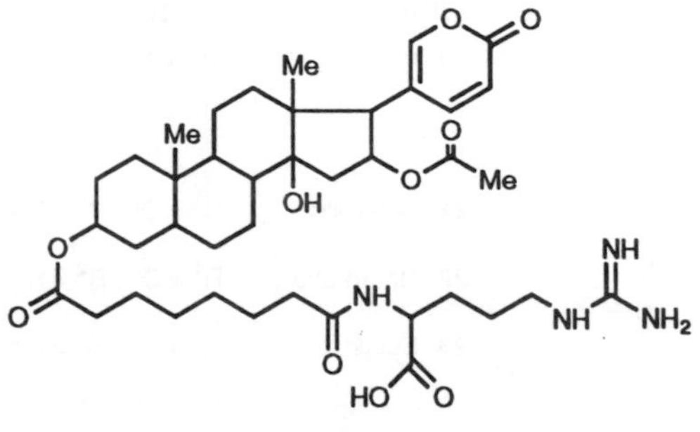

34 Bufotoxin

Figure 9.6 (continued) *Other compounds that are also found include steroidal toxins, bufotalin (**33**) and bufotoxin (**34**) (from* Bufo *species)*

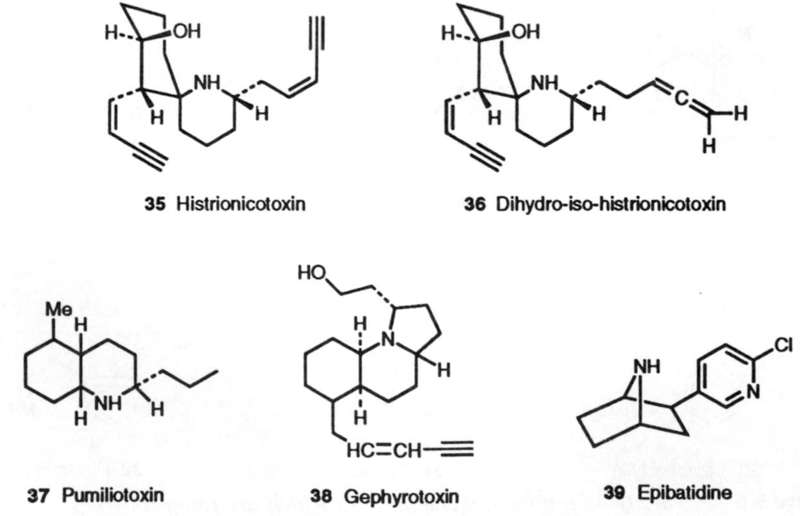

Figure 9.7 *Toxins from frogs. Highly functionalized bicyclic spiropiperidine alkaloids (histrionicotoxins), bicyclic piperidine [pumiliotoxin (**37**), tricyclic piperidine (gephyrotoxin (**38**)], and the pyrrolidine-substituted pyridine, epibatidine (**39**)*

The biologically important reduced alkaloids (perhydro) histrionicotoxins (spiro piperidines),[44–47] pumiliotoxins (quinolizidines),[48,49] and gephyrotoxin (indolizidines)[50,51] (Figure 9.7) from *Dendrobates auratus* (Columbian and South American tree frogs) continue to attract the attention of synthetic chemists.[52]

A chloro-substituted pyridine which is structurally similar to nicotine, epibatidine (**39**), is a newly discovered alkaloid from a Dendrobatidae, *Epipedobates tricolor*. Epibatidine is a potent nicotinic acetylcholine receptor agonist,[53] a powerful ganglionic depolarizing agent,[54] and has been the target of many recent synthetic endeavours, completed 14 times in the past 18 months.

9.6 Snakes

This section on snake venom deals mainly with proteins and polypeptides. Of all the nearly 2000 different species of snakes that exist, about 400 species and sub-species are known to be toxic. These venomous snakes are classified according to morphological characteristics and comprise five families: Crotalidae (crotalids, pit vipers, including rattlesnakes), Viperidae (viperids, vipers), Elapidae (elapids, including cobra, kraits, coral snakes, and mambas), Hydrophiidae (hydrophid, sea snakes), and Colubridae (colubrids or rear fang snakes).

Snakes are among the most widely distributed animals. They inhabit all tropical, subtropical, and most of the temperate zones of the Earth. As a result of this broad distribution and their relatively large numbers, there are more accidents involving snakes than any other kind of venomous animal. Snakes possess a complete venom apparatus, consisting of highly specialized glands (parotid gland, Duvernoys gland) and venom ducts draining into the fangs (except for the aglyphic snakes, which do not posses these fangs), but the venom is kept separate from the CNS of the host. In any event, on ingesting another poisonous snake, it is believed that there are appropriate antivenoms in the snake bloodstream. Snake venom is used to kill or paralyse prey, but it also has other important roles, as it facilitates digestion of the prey, which is swallowed whole.

Contrary to popular perceptions, snakes are more timid than aggressive towards humans; they only bite if they are threatened, frightened, or stepped upon when hidden under grass or leaves. The amount of venom injected by snakes is typically small, but in some cases the results are fatal to humans. Symptoms which are exhibited by the victim result from the combined effects of the complex proteinaceous components present in the venom.

Venom is not composed of a single substance common to all poisonous snakes, although almost all venoms consist of 90% protein. The proportions of the different substances in venom and their specific characteristics vary among the species. Any specific snake venom usually contains more than one toxic principal component, which tend to act in combination in a poisoning. The overall toxicity is due to enzymes and non-enzymatic proteins acting in concert. However, the main lethal action, especially in Elapidae and Hydrophiidae snakes, can be attributed to neurotoxins, most of which are not enzymes (although, presynaptic neurotoxin has phospholipase A_2 activity). This does not mean that enzymes do not participate in the toxic action of venoms – they certainly do. Many enzymes in the venoms are involved in blood coagulation or anti-coagulation, haemorrhage, haemolysis, autopharmacological action, and lysis of cellular and mitochondrial membranes. The homology between sarafotoxins and endothelins illustrates this point.[55] Recent reviews on snake venoms include those which are direct-acting fibrinogenolytic enzymes,[56] potassium channel blockers,[57] and the excellent work of Harvey, Menez, and co-workers[58,59] on green and black mamba and snake α-toxin.

The medicinal uses of snake venoms are well-documented, rather than any agrochemical uses, but parallels may be drawn if the venom can be selectively delivered, incorporated by genetic manipulation, or used as a specific pharmacological probe for detailed receptor characterization (*e.g.*, bungarotoxin).[60] Over the past 90 years, there have been investigations directed towards the utilization of snake venoms in mammalian therapy. Ambiguous and contradicting results, and many failures and disappointments, originate from experiments with inexactly classified snakes or from the use of crude venoms. The composition of snake venom can vary seasonally and locally. Thus, experiments are useful only if performed with purified single components or known, defined preparations. Positive results have been obtained for analgesic action and for influence on blood clotting. The analgesic effect of snake venoms has been used in cases of terminal cancer, because the extreme pain usually associated with this state often requires increasing doses of morphine. Practical levels of analgesia have been obtained with cobratoxin (*Naja naja*), which has been an important and useful alternative to morphine as the analgesic activity of the snake toxin is higher than that of the opioid. In cases of rheumatoid arthritis, toxins from *Vipera aspis* and *V. ammodytes* have produced satisfactory results. Cobratoxin has also been applied in polyarthritides, and Cortalis toxin used for the relief of migraine.

The use of snake venoms as anti-coagulants (defibrinating enzymes) was patented in 1965 and 1972 by NRDC. The venom of *Agkistrodon rhodostoma* (Boie), the Malaysian pit viper, contains a thrombolytic factor which can be isolated and purified. Preparations of this 53.4 kDa protein (known as Ancrod, Arvin, Venacil) produce microclots which are readily dissolved by the endogenous fibrinolytic enzymes of the recipient; thrombophlebitis, and vascular clogging have been treated successfully with this venom. von Klobusitzky isolated clot-promoting fractions from *Bothrops artrox* and from *B. jararaca* venoms and, in 1935, proposed that the protein may find use in the treatment of haemophiliacs.[61] Following from this, a preparation called Reptilase (batroxobin) has proved efficacious as a styptic with no side-effects. Reptilase may be administered intravenously, subcutaneously, or intramuscularly. Prior to 1960, preparations of *Crotalus* venom were used in the treatment of epilepsy with some success, but this therapy has now been replaced with low molecular weight drugs of significantly higher efficacy.

9.7 Gila Monsters

Gila monsters (*Heloderma* sp.) are lizards up to 40 cm long belonging to the order Squamata, the same order to which venomous snakes belong. This is the only venomous reptile (lizard) other than snakes. There are two species of *Heloderma*: *H. suspectum* and *H. horridum*. The Gila monster can be found in North America across the south west (from Mexico, Utah/Nevada, to the Pacific coast), including the Gila river valley in Arizona, from where their common name is derived.

Despite their name, Gila monsters are fearful animals and give warnings by opening their mouth widely and hissing. Typically, they only bite if an attempt is made to seize them. The bite is unpleasant, as they have strong jaws which can clamp on to a limb for 15 minutes. Their muscles are strong – extreme force and a screwdriver are reported to be necessary to prize the jaws open. During this period, sufficient venomous saliva may pass into the wound to cause significant poisoning. Bleeding from the large wound usually begins after 10 seconds and rapidly becomes significant. The wound becomes inflamed and discoloured; the pain level depends upon the size of the wound.

The general symptoms of Gila monster poisoning[62] are not distinct. Nausea, elevated pulse and temperature, and low blood pressure are typical. Although the mouse LD_{50} of the crude venom is 0.5–4.0 $\mu g\ g^{-1}$ (depending upon the route of administration), human fatalities are extremely rare, and those reported have been associated with subsequent events, *e.g.*, heart failure through fear, secondary infections of the untreated wound. The most important consideration is treatment of the wound with a broad-spectrum antibiotic. Beyond this, symptomatic treatment suffices and the victim recovers after 4–14 days. Gila monster venom primarily comprises proteins,[62–64] but is accompanied by serotonin (5-HT).

9.8 Scorpions

Scorpions are a group in the animal kingdom that contains about 650 species; they belong to the class Arachnida and order Scorpinia. They vary in size from 3–25 cm, inhabit large parts of southern USA, the central (Mexico) and southern (Brazil) Americas, and are also native to the Mediterranean. Scorpions are also found in the Middle East, Africa, and Asia (India, Malaysia, China), but are not yet reported from Japan or Taiwan. Unlike snakes, all scorpions are venomous. The venom is injected by means of a stinger found at the tip of the telson, the terminal structure of the scorpion tail.

The symptoms of mammalian (human) scorpion poisoning vary greatly. According to the individual scorpion species, envenomation may be completely harmless or it may fatal. The majority of these toxins are medium sized peptides (57–78 amino acid residues).[59] Different components in the venom of *Androctonus australis* are responsible for the mice mortality and larvae contraction paralysis effects. The insect toxin (*i.e.*, specifically toxic to insects) is also different from the mammalian-specific toxins in amino acid composition.

Pure neurotoxins (separated from *A. australis*), which are highly toxic to mammals, prove inactive when tested on several species of arthropods, and *A. australis* insect-specific toxin is used to transform baculoviruses (Chapter 11). The fly larvae toxin, from the same venom, displays a strong toxicity to insects, but is completely inactive when applied to an arachnid or a crustacean. Within this venom there is also another discrete protein that is specifically active on crustaceans.

The insect toxin and the mammal toxin have different physiological responses. In general, the insect toxin induced an afferent transynaptic response at the sixth abdominal ganglion of the cockroach, but the mammal toxin did not affect synaptic transmission. There have been extensive pharmacological and physiological studies of scorpion toxins. Typical mammalian symptoms from such scorpion poisoning include severe local pain, swelling, occasional discolourment, sweating, pallor, restlessness, anxiety, confusion, salivation, nausea, abdominal cramps, chest pains, and headaches. There is often also a sensation of choking, muscle weakness, and twitching. Initial tachycardia changes to brachycardia and initial hypertension to hypotension. There is respiratory distress and subsequent cyanosis. Death results from cardiovascular collapse and pulmonary oedema. In view of these widespread mammalian effects, only careful targetting of scorpion peptides will result in any agrochemical application. The central and peripheral effects on mammals, together with the attendant risks, probably minimize the agrochemical potential of scorpion toxins. At the fore in scorpion venom research are Zlotkin and his co-workers, who have published prodigiously, especially with respect to the insect neuronal sodium channels as targets for insect-selective neurotoxins.[65–67]

9.9 Centipedes

The class Chilopoda (centipedes) is broadly distributed and contains 3000 species. They have a long (up to 30 cm) and slender (1 cm diameter) body, inhabiting caves or dark places under stones, bark, or leaves. For their predation, they possess a pair of venom glands with furcipules at the first trunk segment.

Bites of the larger species of centipedes may have medical significance, although the many reported poisonings are still the subject of some doubt. Human envenomation occasionally results in irritation, swelling of the bite site, and ulceration, but these could be due to secondary infections of the wound. The venom is highly potent in small mammals. It acts as a neurotoxin and produces corresponding symptoms – tachypnoea, sweating, dizziness, nausea, vomiting, and ultimately death by respiratory paralysis.

In addition to these active venoms, centipedes also possess a variety of defence secretions as weapons against ants, beetles, and other predators. These are produced and secreted in the ventral and axol glands. Some of them are sticky, some luminescent, and others have an intense odour. Centipede toxic secretions are typically peptides.

9.10 Millepedes

Diplopoda (millepedes) are an ancient group of animals comprising about 7500 species. They have developed an effective defence mechanism that consists of secretions excreted from glands (located in the segments) when the millepede is frightened or endangered. Chemically, they consist of simple substituted aromatic (phenolic) compounds and a few simple alkaloids. These low molecular weight benzene derivatives and the alkaloids are frequently secreted with proteins, which cause the compound mixture to harden quickly on exposure to air and to stick to attacking animals. Thus, predators are efficiently repelled by a cocktail of odoriferous components and precursors for hydrocyanic acid. The alkaloids glomerin and homoglomerin have paralytic action on mice as well as spiders. Some of the phenols and quinones are powerful antibiotics.

9.11 Beetles

This section does not deal, in any detail, with semiochemicals or pheromones. For leading recent studies on these and related low molecular weight, volatile natural products (Figure 9.8), see Francke and coworkers[68,69] and Mori,[70] including 'Pheromone synthesis, part 166'.[71] The toxins used by the members

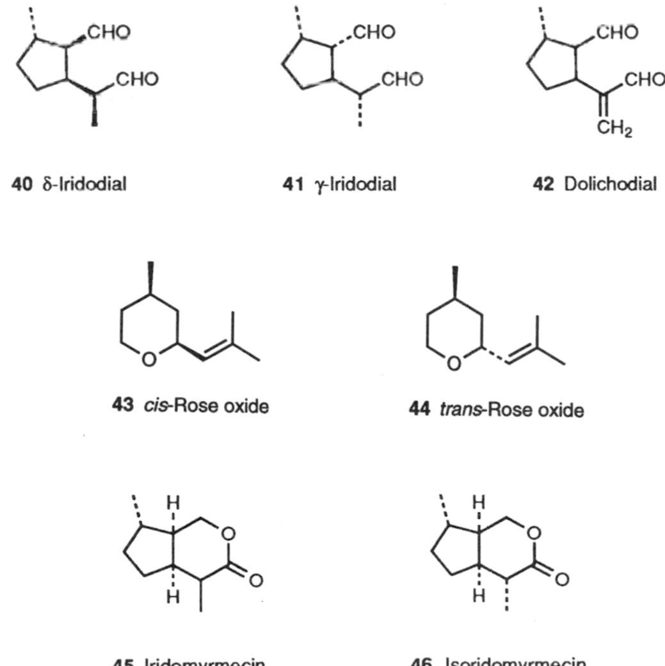

Figure 9.8 *Lactones, pyrans, and cyclic dialdehydes found in the venoms of certain ants and beetles. Rose oxides and iridodials are found in Cerambycidae,* Aromia moschata

of the order Coleoptera, comprising 200 families and over 250,000 species, are of diverse chemical structures and generally are squirted from their scent glands in defence.

9.11.1 Carabidae

The bombarding beetle (*Brachynus creptitans*) has an unusual defence mechanism. The defence chemicals consist of benzoquinone and toluquinone. The secretion consists of a solution of the corresponding hydroquinones in hydrogen peroxide. The reaction between these components is prevented by chemical stabilization of the mixture until the beetle must defend itself. Then, the solution is brought from the pygidial gland into the 'exploding chamber', where the mixture is enzymatically destabilized. The hydrogen peroxide oxidizes the hydroquinones to give quinones which are ejected a few centimetres with a resoundingly loud 'crack'. This can be repeated several times in succession. Other beetle species contain a complex variety of semiochemicals.

9.11.2 Cerambycidae

There are a variety of Cerambycidae toxins, which vary between species, from salicylaldehyde and monoterpenes to functionalized benzaldehydes, phoracanthal, and phoracthol.

9.11.3 Coecinellidae

These animals use alkaloids, *e.g.* concinelline, preconcinelline, hippodamine and adaline, in their defence.

9.11.4 Dytiscidae

The 'diving beetle' contains, in its pygidial gland, a mixture of *p*-hydroxybenzaldehyde, *p*-hydroxybenzoic acid, and benzoic acid, which protects the beetle against other animals. Furthermore, this mixture protects against micro-organisms during hibernation in muddy pools. Also found in the gland are steroids, *e.g.*, cortexan and testosterane.

9.11.5 Grinidae

The pygidial glands of Grinidae contain norsesquiterpenoids, *e.g.* gyrinidal and gyridinane.

9.11.6 Meloidae

The toxic effects of the 'blister beetle' are notorious for its alleged aphrodisiac properties; it is better known as 'Spanish fly'. The toxin responsible is the alkaloid cantharidin.

9.11.7 Silphidae

The carniu beetles use ammonia in their defence.

9.11.8 Staphylinidae

These beetles have a diverse chemical armoury, including alkyl ketones, long chain and cyclic aldehydes, carboxylic acids, lactones, and alkenes.

9.12 Ants

Ants, by weight, are the major predatory animal. These arthropods are all members of the family Formicidae – 15,000 species of formicids are known, they are social, and form colonies. Formicidae comprises about 260 genera with 11 sub-families. Most ants are capable of stinging and have a similar venom delivery system to bees, hornets, and wasps – together these are the Hymenoptera. In some cases, reaction from ant poisoning is severe enough to require medical attention. Substances from the Dufour gland and from venom glands can be discharged *via* the sting into the victim or prey. Compounds formed in the mandibular glands can also be used for defence. Ant toxins in the venom or mandibular glands vary between species. Examples of the alkyl pyrrolidines, substituted piperidines, and related low molecular weight components found in ant venom are given in Figure 9.9.

The majority of ants are carnivorous; their prey is subdued by stinging or by spraying into abrasion wounds made by their mandibles. Although there is sometimes some confusion between ants (Hymenoptera) and termites, the latter, occasionally called white ants, are vegetarian, not carnivorous. Termites are the only social animal not in Hymenoptera – they are members of Igloonoptera. Although it is a common belief that rhubarb is made of oxalic acid and that ants inject their prey with formic acid, the latter is quite limited in its occurrence. The toxins of recognized sub-families of ants have been reviewed: formicinae, myrmeciinae, myrmicinae, ponerinae, dorylinae, and dolichoderinae. There are now many fully characterized ant toxins which cover a wide range of structural types. These toxins have been reviewed in several papers and monographs, but preeminent among the publications of ant natural product chemists are the studies of Blum and his co-workers.[72,73]

9.12.1 Formicinae

Formicinae, scaly ants, are distributed world-wide and form a sub-family whose major toxic constituent is formic acid. First isolated by Wray, in 1670, formic acid is produced by all ants in Formicinae. Ants from other sub-families do not produce formic acid. The venom is a strongly acidic solution, 70% aqueous formic acid. Thus, the venom is highly cytotoxic and admirably suited to defence as a respiratory toxin. Formic acid is accompanied by free common

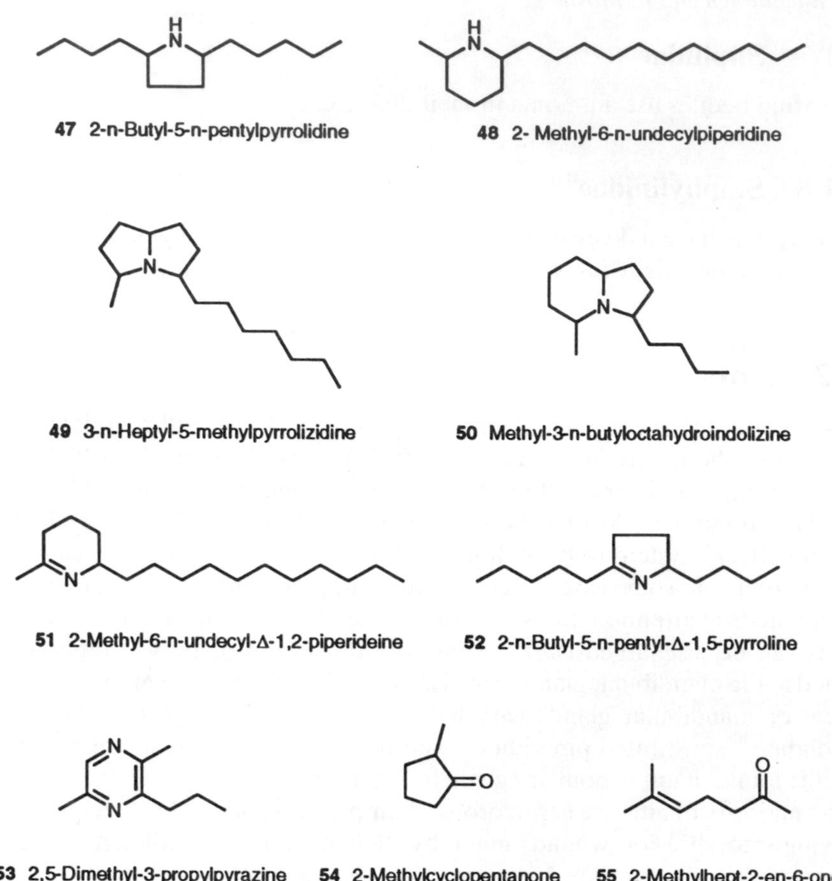

Figure 9.9 *Ant venom components, including alkyl piperidines, alkyl pyrrolidines, bicyclic piperidines and pyrrolidines, alkyl anhydropiperidines and pyrrolidines, alkyl enones, and alkylated cyclopentanones*

amino acids, some small peptides, terpenes, and ketones. The venom is 22% of the individual's total weight, another attribute which contributes to their powerful self-defence mechanism. The sub-family Formicinae are all venom sprayers, direct from the venom gland, not injectors, as they possess no sting. The LD_{50} (i.v.) of formic acid is 145 mg kg^{-1}. Therefore, although formic acid is highly cytotoxic, Formicinae venom is less active than any other ant venom.

9.12.2 Myrmeciinae

The Australian bulldog ant, *Myrmecia pyriformis*, of southern Australia, south eastern USA, and South America, produces a sting similar to, but more painful than, that of a bee, with an LD_{50} of 5.0 mg kg^{-1} in mice (i.p.). Their infamous venom (hence their trivial name) produces a wheal and immediate erythema, followed by oedema, itching, and then tenderness which last for days. The venom contains a smooth muscle stimulating fraction and hyaluronidase. As well as endogenous histamine (3%), a histamine-releasing factor is also present in the venom, accompanied by phospholipase A_2 and a peptide of similar molecular weight to melittin, isolated from honey bees (*vide infra* Section 9.13). There are proteins of 11–23 kDa and, possibly to assist in the delivery of these polar macromolecules, there are lipids (alkanes and alkenes).

9.12.3 Myrmicinae

The desert ant, *Pogonmyrmex barbatus*, has a potent venom with an LD_{50} of 1.3 mg kg^{-1} in mice i.p. Injection of the proteinaceous venom causes stretching and twisting of the abdomen, ataxia, respiratory distress, lethargy, and then death. These ants have a well-developed venom apparatus and use their mandibles for defence. They are widely distributed throughout tropical and temperate zones and are credited with the most painful stings of all North American ant species. In humans, sweating becomes extensive, accompanied by prolonged pain, and then lymph node tenderness is caused. The most enzyme-rich venom of any arthropod is that of *P. badius*. This venom includes four esterases, acid phosphatase, phospholipase A_2, phospholipase B, hyaluronidase, lipase, a powerful direct haemolytic factor, histamine, and 17 free, common amino acids. Accompanying toxins are piperidine and pyrrolidine alkaloids, terpenes, *e.g.* citronellol and geraniol, benzaldehyde, and C7–C10 ketones. Synergistically, *P. badius* has the lowest LD_{50} of any insect species, 0.42 μg kg^{-1} in mice i.p. The venom acts on the mouse CNS, causing colonic convulsions, lachrymation, and death in 4 h. These effects may be accompanied by terror, but are certainly accompanied by hair follicles standing on end (piloerection). *Solenopsis invicata* are fire ants, of Myrmicinae, and not to be walked on.

9.12.4 Myrmica

Painful stings (*e.g.*, of *Myrmica rubra*) comprise protein (12%), hyaluronidase, and histamine as the only biogenic amine in this venom. These toxins are accompanied by 16 free common amino acids, and although there is no phospholipase A_2, there is a 15 kDa convulsive component. Envenomation causes wheal, oedema, pain, sweating and fever, with an LD_{50} of 60 mg kg^{-1} in mice i.v., but the venom is much more toxic to insects than to mammals.

9.12.5 Myrmicaria

This genus is distinguished from all others, as monoterpene hydrocarbons (Figure 9.10) are found in the venom. The primary toxin of *Myrmicaria natalensis* is limonene, accompanied by eight other monoterpenes. α- and β-

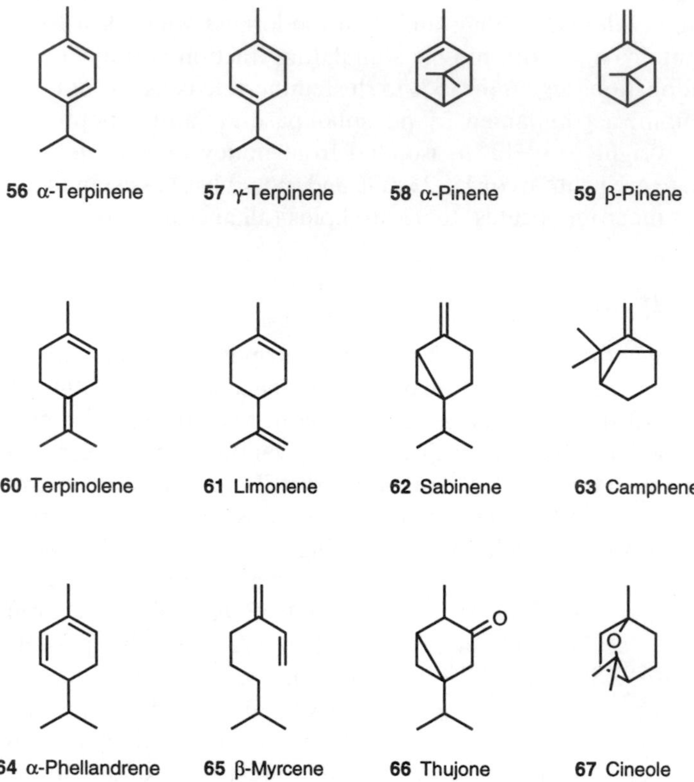

Figure 9.10 *Ant venom monoterpenes*

pinene (**58** and **59**, respectively) and sabinene (**62**), terpinolene (**60**), α-terpinene (**56**), β-myrcene (**65**), α-phellandrene (**64**), and camphene (**63**); such monoterpene hydrocarbons are typical of those found in the venom of termite soldiers.

9.12.6 Ponerinae

Ponerinae, sting ants, possess a highly developed sting. The majority of these species are carnivorous, feeding upon insects. They are found in tropical countries and the Americas, where they have been described as fierce and aggressive. One of the most painful stings comes from *Paraponera clavata*, a native of North America. The venom contains mainly peptides and proteins.

9.12.7 Dorylinae

These are aggressive ants (*e.g.*, driver and migratory ants) whose raids spare no creature. The migratory ant possesses a sting and is distributed throughout the tropical and sub-tropical zones. Their toxins are methylheptanones and methylindole.

9.12.8 Dolichoderinae

Dolichoderinae, gland ants (*e.g.*, *Iridomyrmex detectus*), have retarded stings; interestingly, their defensive action has been taken over by the anal glands. The venom comprises bicyclic lactones, alkyl pyrazines, alkyl ketones, alkyl acids, aldehydes, and alcohols. Such a chemical cocktail of carbonyl functional groups with alcohols would be certain to cross-react, one might speculate. Indeed, the mixture polymerizes, and immobilizes the enemy – certainly an interesting mixture to store in anal glands.

9.13 Bees

Bees, hornets, and wasps are closely related members of the order Hymenoptera. There are several excellent reviews of this area, but 'Venoms of the Hymenoptera' by Piek[74] is the best starting point. In this book, the extensive literature up to 1985 is covered thoroughly. The classification of Hymenoptera is complex and outside the scope of this chapter. However, their venoms and methods of toxin delivery are similar. Their specialized venom equipment consists of gland, canal, and sting which produce, store, and inject the venom; this is predominantly used in defence, but exceptions include the parasitic wasps (*vide infra*).

The venom consists mainly of enzymes, peptides, and biogenic amines. The major bee toxins are apamin and melittin, promelittin, kinin, and mast-cell degranulating (MCD) peptide. The predominant enzymes are types of phospholipase A_2, hyaluronidase, esterase, and acid and alkaline phosphatases. Enzyme inhibitors include a protease inhibitor to interfere with vital protein processing. Non-peptidic components include histamine, free common amino acids, carbohydrates, lipids, biogenic amines, serotonin, noradrenalin, dopamine, and acetylcholine. As can be readily imagined, such a cocktail ensures that the nervous system of the attacker or prey is disrupted efficiently, and many channels are opened for small molecules to gain facile access to the intracellular fluid, accompanied by gross desensitization of receptors.

The local reactions to bee stings are well-documented. Painful inflammation is caused by all of the venom components. Histamine and serotonin are the major mediators of pain and swelling. However, the LD_{50} of bee venom is not particularly low, and it is estimated that over 100 stings would be required to kill an 'average' human. Nevertheless, it is not uncommon for one or two stings to be fatal due to anaphylactic shock, and certain allergic groups are especially vulnerable. More comforting, in a sense, are the reports of people who have been attacked by Brazilian killer bees, have suffered over 400 stings, and yet have survived.

9.14 Wasps

The recent chemical and biochemical manipulations by Nakanishi and his co-workers and by Piek and his co-workers on the polyamine amide toxins (Figure 9.11) from *Philanthus triangulum* have been reviewed.[75,76] *N*-(n-Butanoyl)-L-tyrosinylthermospermine (philanthotoxin-4.3.3; PhTX-4.3.3, **68**) is isolated from the venom of the parasitic, solitary wasp *P. triangulum* and established as a potent glutamate receptor antagonist. PhTX-4.3.3 is an antagonist of insect ligand-gated ion channels, quisqualif acid sensitive glutamate receptors (in the concentration range 10–100 μmol dm^{-3}, when applied to locust nerve–muscle preparations). Pharmacological studies suggest that this antagonism is non-competitive. The analogue PhTX-3.4.3 (**69**), in which the polyamine spermine replaces the thermospermine of the natural toxin, is essentially as potent as the natural product.[77]

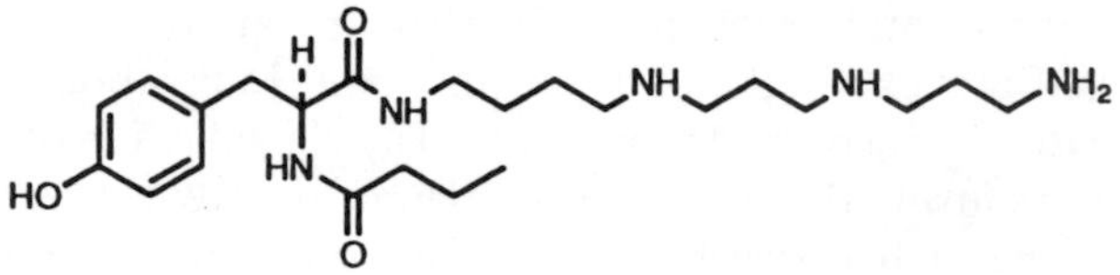

68 PhTX-4.3.3

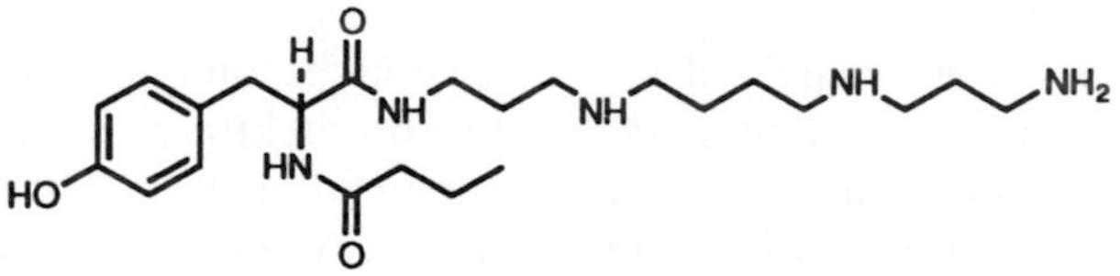

69 PhTX-3.4.3

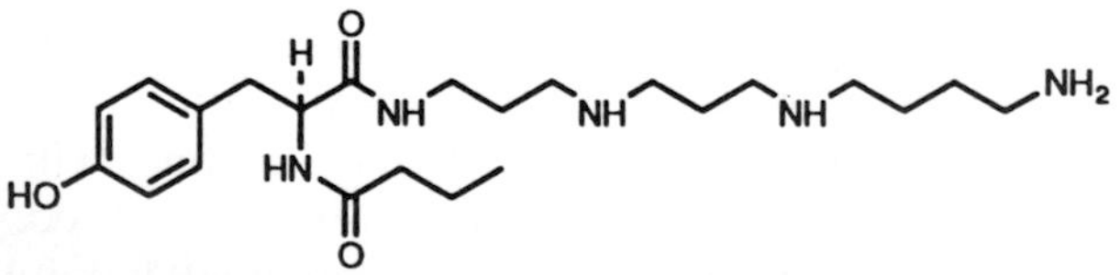

70 PhTX-3.3.4

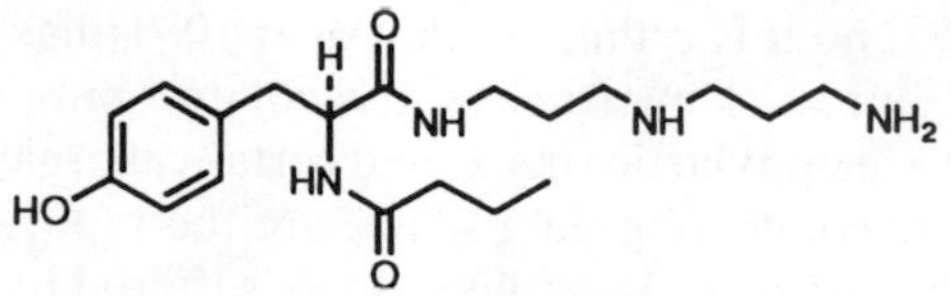

71 PhTX-3.3

Figure 9.11 *PhTX-4.3.3 and analogues of this wasp toxin*

Activity is enhanced in the corresponding thermospermine analogue PhTX-3.3.4 (**70**), but only by 30%. PhTX-3.3 (**71**), an analogue with a shorter chain, and which is capable of protonation at only two nitrogen atoms, is inactive at 10 μmol dm^{-3}. Although the polyamine thermospermine has not been tested, the isomeric spermine has been established as interacting with insect muscle glutamate receptors. Other results from our laboratories confirm this non-competitive antagonism, but only at millimolar concentrations.[77] Similar results with other naturally occurring polyamines and diamines, including spermidine, putrescine, and cadaverine, have been obtained.[77]

9.15 Spiders

Spiders (Araneae) form a fitting end to the venomous animals reviewed in this chapter. Their prey are generally insects, but not always agricultural pests. Systematic classification of various families is often complex, and several different classification schemes have been proposed.

Spiders have a special place in folklore and in many more modern books and films with respect to arachnophobia and the apparent hatred of their enemies. They are portrayed as deadly to humans, cold killers, and we reserve a special place for them in our subconscious. Far be it for us even to attempt to dispel this myth in one chapter, but spiders are more often the victims than the perpetrators. Nevertheless, they possess potent venoms which contain a rich diversity of toxins. In addition to many of the small molecules and polypeptides cited above, spider venoms contain high concentrations of glutamic acid, spermine, and related polyamine amides (Figure 9.12). As these molecules have been studied in more pharmacological detail, it has become clear that they act together to block cation channels in the insect neuromuscular junction, which have been gated open by glutamic acid. The study of spider toxins is one area which may lead to new agrochemical agents with a novel mode of action, blocking the ion channel associated with the quisqualic acid subtype of glutamic acid receptor.

Argiotoxins are isolated from the venoms of orb-web spiders, as are agatoxins and hettoxins, and are rich sources of novel polyamine amides. These polyamine-derived agents target, in particular, those L-glutamate receptors (GluR) of skeletal muscle which gate cation-selective, transmembrane ion channels resulting in paralysis. These compounds (polyamine amides) comprise a new class of bioactive agents; their antagonism of GluR provides the pesticide industry with opportunities to target a hitherto unexploited system in pest insects. The potential of these compounds and their synthetic analogues as pharmaceutical agents is currently under equally intensive development.

The pharmacological properties of the argiotoxins (a general term which includes those compounds discovered in the venoms of Japanese and New Guinean spiders and which are typified by the presence of a polyamine amide moiety) and the philanthotoxins (wasp toxins, *vide supra*) have been extensively reviewed, and synthetic pathways and approaches to these compounds

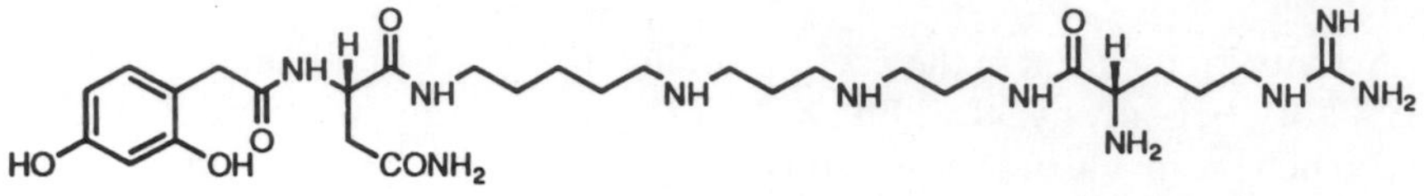

72 ArgTX-636

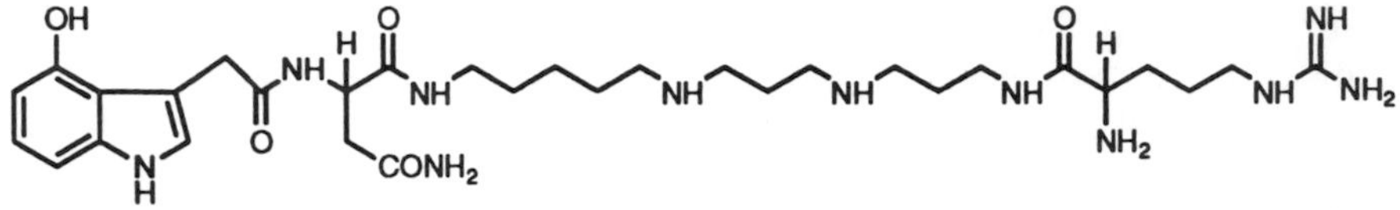

73 ArgTX-659

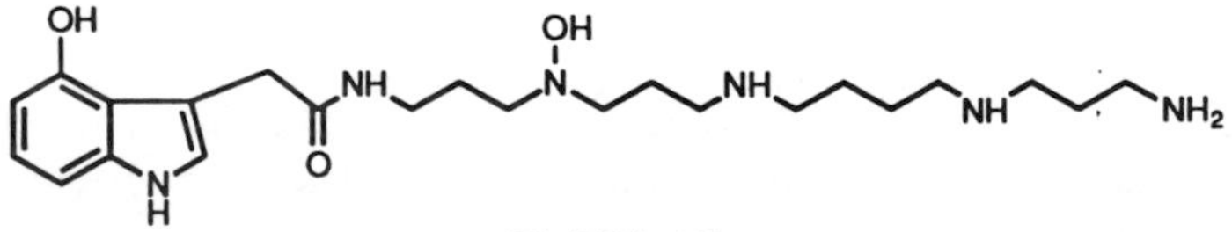

74 AGEL-448

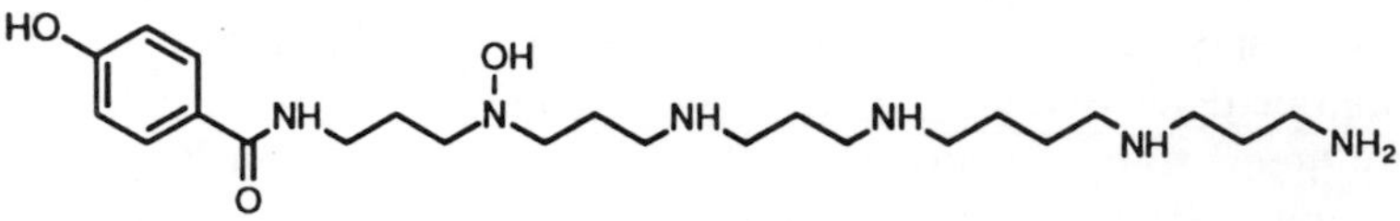

75 AGEL-452

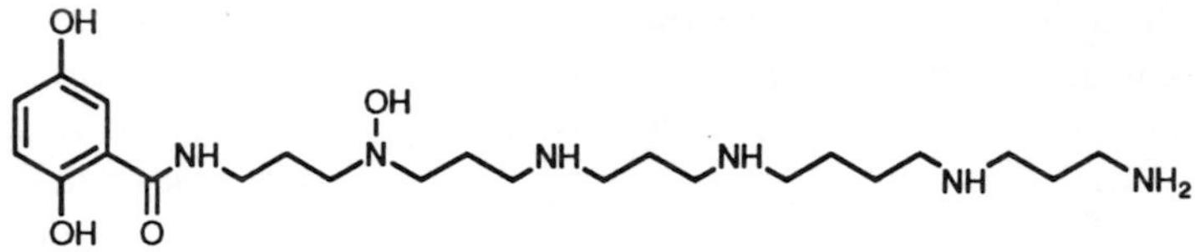

76 AGEL-468

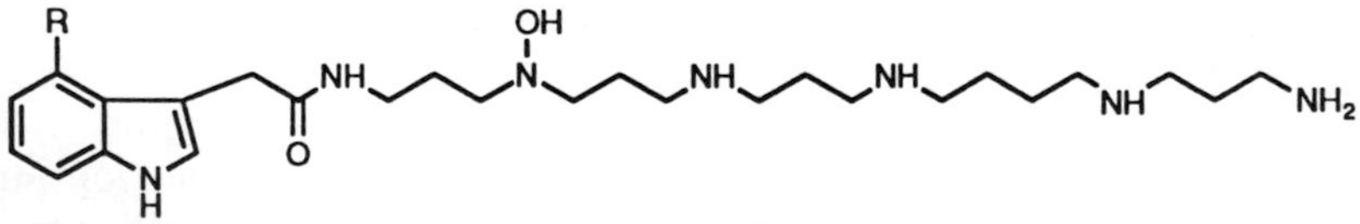

77; R = H AGEL-489
78; R = OH AGEL-505

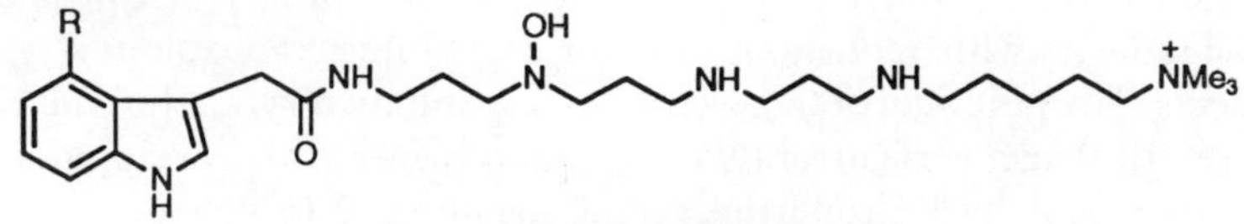

79; R = H AGEL-489a
80; R = OH AGEL-505a

Figure 9.12 *Polyamine amide toxins found in* Argiope *and* Agelenopsis *species of spider*

have been described.[78–80] A number of laboratories have produced synthetic analogues of these natural products, the structures and pharmacological properties of which have also been published.[77,81] Synthetic studies undertaken in our laboratories to produce hybrid molecules based upon the argiotoxins and PhTX-4.3.3 have been reported.[81] These synthetic analogues are less complex structurally than the natural products, but they exhibit similar pharmacological properties.[82]

9.16 Conclusions

Polyamine amides constitute novel, low molecular weight, polar compounds which are lead structures for the development of selective pharmacological tools and thence for agrochemical agents. Their high polarity may present some significant difficulties in selective delivery and targetting, but this should be offset by their high potency. Polyamine amides are potent, non-competitive antagonists of cation channels. However, although they are not selective between quisqualic acid sensitive GluR and nAChR, they are channel blockers and form an exciting new class of compounds – low molecular weight polyamine-containing toxins, isolated from garden spiders (*Araneidae*), with important role(s) to play in new agrochemical control methods and invertebrate neuroscience research. Such toxins may have more use as selective pharmacological tools and probes for receptors. This is of importance in insecticide design, as protein receptor subunit mutations can afford significant protection against bungarotoxin, at least to the mongoose.[83] A molecular rationale for such receptor specificity is emerging.[60]

Broad, high throughput screening programmes will continue to dominate the insecticide and pharmaceutical discovery processes, although the possibility of rational design based upon the structure of a toxin or the target of a toxicant is a worthwhile goal. The future development of insecticides will result in chemicals which are less damaging to the environment than those in current use. Novel modes of action may go some way to alleviate the current problems of rapid resistance to insecticides. New strategies to minimize the build-up of resistance may develop from studies of animal venoms and insect toxins, in particular the genetic incorporation of insect toxicants into crops to form part of an integrated pest management strategy.[22] The structures highlighted in this chapter are only the first fruits to be gleaned from venoms and toxins.

There are four excellent journals as literature sources for animal toxins: *Toxicon* (Elsevier Pergamon Press), *J. Toxicology – Toxin Reviews* (Marcel Dekker Inc.), *J. Natural Toxins* (Alaken Inc.), and *Natural Toxins* (John Wiley). In addition to these primary and secondary sources, there are several recent books (see the Bibliography), but special mention must be made of the eight-volume 'Handbook of Natural Toxins' by Tu,[64,84] which are excellent summaries and provide a thorough introduction to the biology,[85] as well as to the chemistry of this large research area.

It is hoped that this review of animal venoms and insect toxins will help to generate some new ideas for lead structures for scientists searching through natural product libraries for new compounds to screen. Nevertheless, they will still require active collaboration with teams of synthetic chemists to gain an understanding of the mode(s) of action of the novel agents and to refine the structure with respect to the metabolism, pharmacokinetics, and field stability.

Finally, and by no means last, this chapter will hopefully also contribute to our understanding that 'natural is not safe', an idea which still needs to gain hold, even though Ames and co-workers[86,87] have established that 99.99% of the pesticides which enter our diet are natural products of plant origin which would fail the current toxicological testing for synthetic pesticides.

9.17 Acknowledgements

We are grateful to BBSRC for their financial support. This work was also, in part, supported by grants from the Wellcome Trust and the Medical Research Council. ISB is the recipient of a Nuffield Foundation Science Lecturer's award. We also acknowledge gratefully the help of Professor Anthony T. Tu (Colorado State University, Fort Collins) who made many helpful comments on the content and direction of this chapter.

9.18 Bibliography

1. M. S. Blum, 'Chemical Defenses of Arthropods', Academic Press, London, 1981.
2. I. R. Duce (ed.), 'Neurotox '91, Molecular Basis of Drug and Pesticide Action', Elsevier, London, 1992.
3. G. G. Habermehl, 'Venomous Animals and their Toxins', Springer Verlag, Berlin, 1981.
4. G. G. Lunt (ed.), 'Neurotox '88, Molecular Basis of Drug and Pesticide Action', Elsevier, London, 1989.
5. T. Piek (ed.), 'Venoms of the Hymenoptera. Biochemical, Pharmacological and Behavioural aspects', Academic Press, London, 1986.
6. B. D. Roitberg and M. B. Isman (ed.), 'Insect Chemical Ecology. An Evolutionary Approach', Chapman and Hall, London, 1992.
7. H. H. Shorey and J. J. McKelvey, Jr (ed.), 'Chemical Control of Insect Behaviour. Theory and Application', John Wiley and Sons, New York, 1977.
8. A. T. Tu, 'Venoms: Chemistry and Molecular Biology', John Wiley and Sons, New York, 1977.
9. A. T. Tu (ed.), 'Handbook of Natural Toxins', Marcel Dekker, Inc., New York, 1991, vols 1–8.

9.19 References

1. H. C. Tang, D. X. Huang, Y. L. Li, X. B. Du, H. S. Lian, and J. S. Chen, Synthetic study of analogs of anatoxin, *Acta Chim. Sinica*, 1994, **52**, 306.
2. P. Thomas, P. A. Brough, T. Gallagher, and S. Wonnacott, Alkyl modified side-chain variants of anatoxin A – a series of potent nicotinic agonists, *Drug Dev. Res.*, 1994, **31**, 147.
3. A. Hernandez and H. Rapoport, Conformationally constrained analogs of

anatoxin-chirospecific synthesis of *S-trans* carbonyl ring-fused analogs, *J. Org. Chem.*, 1994, **59**, 1058.

4. G. G. Habermehl, 'Venomous Animals and their Toxins', Springer Verlag, Berlin, 1981.
5. B. D. Roitberg and M. B. Isman (ed.), 'Insect Chemical Ecology. An Evolutionary Approach', Chapman and Hall, London, 1992.
6. H. H. Shorey and J. J. McKelvey, Jr (ed.), 'Chemical Control of Insect Behaviour. Theory and Applications', John Wiley and Sons, New York, 1977.
7. T. Y. K. Chan, L. B. Tomlinson, L. K. K. Tse, J. C. N. Chan, W. W. M. Chan, and J. A. J. H. Critchely, Aconitine poisoning due to chinese herbal medicines – A review, *Veter. Human Toxicol.*, 1991, **36**, 452.
8. T. Y. K. Chan, Aconitine poisoning, a global perspective, *Veter. Human Toxicol.*, 1994, **36**, 326.
9. M. E. O'Leary, G. N. Filator, and M. M. White, Characterisation of d-tubocurarine binding site of torpedo acetylcholine receptor. *Am. J. Physiol.*, 1994, **266**, 648.
10. T. Fraenkel, J. M. Gershoni, and G. Navon, NMR analysis reveals a positively charged hydrophobic domain as a common motif to bond acetylcholine and d-tubocurarine, *Biochemistry*, 1994, **33**, 644.
11. S. V. Ley, Synthesis of antifeedants for insects – novel behaviour modifying chemicals from plants, *Ciba Foundation Symp.*, 1990, **154**, 80.
12. S. V. Ley, A. A. Denholm, and A. Wood, The chemistry of azadirachtin, *Nat. Prod. Rep.*, 1993, **10**, 109.
13. D. H. L. Bishop, M. P. G. Harris, M. Hirst, A. T. Merryweather, and R. D. Possee, The control of insect pests by viruses; Opportunities for the future using genetically engineered virus insecticides, in 'Progress and Prospects in Insect Control', British Crop Protection Council Monograph 43, 1989.
14. J. S. Cory, D. H. L. Bishop, *et al.*, Field trial of a genetically improved baculovirus insecticide, *Nature (London)*, 1991, **370**, 138.
15. D. Winstanley and L. Rovesti, Insect viruses as biocontrol agents, in 'Exploitation of Micro-organisms', ed. D. G. Jones, Chapman and Hall, London, 1993.
16. J.-P. Hernalsteens, F. Van Vliet, M. DeBeuckeleer, A. Depicker, G. Engler, M. Lemmers, M. Holsters, M. Van Montagu, and J. Schell, The *Agrobacterium tumefaciens* Ti Plasmid as a host vector for introducing DNA in plant cells, *Nature (London)*, 1980, 654.
17. D. A. Fischhoff, K. S. Bowdish, F. J. Perlack, P. J. Marrone, S. M. McCormick, J. G. Niedermeyer, D. A. Dean, K. Kusano-Kretzmer, E. J. Mayer, D. E. Rochester, S. G. Rogers, and R. T. Fraley, Insect tolerant transgenic tomato plants, *Bio/technol.*, 1987, **5**, 807.
18. M. Vaeck, A. Reynaerts, H. Hofte, S. Jansens, M. DeBeuckeleer, C. Dean, M. Zabeau, M. Van Montagu and J. Leemans, Transgenetic plants protected from insect attack, *Nature (London)*, 1987, **327**, 33.
19. I. Potrykus, Gene transfer to cereals: an assessment, *Bio/technol.*, 1990, **8**, 535.
20. I. Potrykus, Gene transfer to plants: an assessment and perspectives, *Bio/technol.*, 1990, **79**, 125.
21. F. J. Perlack, R. W. Deaton, T. A. Armstrong, R. L. Fuchs, S. R. Simms, J. T. Greenplate, and D. A. Fischhoff, Insect resistant cotton plants, *Bio/technol.*, 1990, **8**, 939.
22. J. A. Pickett, Chemical pest control – the new philosophy, *Chem. Britain*, 1988, **24**, 137.

23. T. Narahashi, M. L. Roy, and K. S. Ginsberg, Recent advances in the study of mechanism of action of marine neurotoxins, *Neurotoxicol.*, 1994, **15**, 545.
24. S. Inoue, K. Okada, H. Tanido, K. Hashizume, and H. Kakoi, Total synthesis of (±)-surugatoxin, *Tetrahedron*, 1994, **50**, 2729.
25. S. Inoue, K. Okada, H. Tanido, and H. Kakoi, Total synthesis of neosurugatoxin, *Tetrahedron*, 1994, **50**, 2753.
26. J. Satin, J. T. Limberis, J. W. Kyle, R. B. Rugart, and H. A. Fozzard, The saxitoxin, tetrodoxin binding site on cloned rat brain IIA sodium channels is in the transmembrane electric field, *Biophys. J.*, 1994, **67**, 1007.
27. G. M. Lipkind and H. A. Fozzard, A structural model of the tetrodotoxin and saxitoxin binding site of the sodium channel, *Biophys. J.*, 1994, **66**, 1.
28. M. Noda, Structure and function of sodium channels, *Ann. NY Acad. Sci.*, 1993, **707**, 20.
29. L. E. Llewelly and E. G. Moczydlowski, Characterization of saxitoxin binding to saxiphilin, a relative of the transferrin family that displays pH dependent ligand binding, *Biochemistry*, 1994, **33**, 12312.
30. B. M. Olivera, L. J. Cruz, R. A. Myers, D. R. Hillyard, J. Rivier, and J. K. Scott, Conus peptides and biotechnology, in 'Neurotox '91, Molecular Basis of Drug and Pesticide Action', ed. I. R. Druce, Elsevier Applied Science, London, 1992, Ch 4, p. 45.
31. B. M. Olivera, G. P. Miljanich, J. Ramachandran, and M. E. Adams, Calcium channel diversity and neurotransmitter release – the ω-conotoxins and ω-agatoxins, *Ann. Rev. Biochem.*, 1994, **63**, 823.
32. J. W. Daly, B. Witkop, P. Bommer, and K. Biemann, Batrachotoxin. The active principle of the Columbian arrow poison frog. *Phyllobates bicolor*, *J. Am. Chem. Soc.*, 1965, **87**, 124.
33. J. W. Daly, Biologically active alkaloids from poison frogs (Dendrobatidae), *J. Toxicol–Toxin Rev.*, 1982, **1**, 33.
34. T. L. Casebolt and G. B. Brown, Batrachotoxinin A-ortho azidobenzoate – A photoaffinity probe of the batrachotoxin binding site of voltage-sensitive sodium channels, *Toxicon*, 1993, **31**, 1113.
35. V. L. Trainer, E. Moreau, D. Guedin, D. G. Baden, and W. A. Catherall, Neurotoxin binding and allosteric modulation at receptor site 2 and site 5 on purified and reconstituted rat-brain sodium-channels. *J. Biol. Chem.*, 1993, **268**, 17114.
36. J. G. Rubin and D. M. Soderlund, Binding of [H-3] batrachotoxinin A 20-alpha-benzoate and [H-3] saxiotoxin to receptor sites associated with sodium channels in trout brain synaptoneurosomes, *Comp. Biochem. Physiol. C: Comp. Pharmacol. Toxicol.*, 1993, **105**, 231.
37. W. Thompson, S. J. Hays, J. L. Hicks, R. D. Schwarz, and W. A. Catterall, Specific binding of the novel sodium channel blocker PD85, 639 to the alpha-subunit of rat brain sodium channels. *Mol. Pharmacol.*, 1993, **43**, 955.
38. K. Dong, J. G. Scott, and G. A. Weiland, [H-3] Batrachotoxinin A-20-alpha-benzoate binding in synaptosomes from susceptible and *kdr*-type resistant German cockroaches, *Blattella germanica* (L), *Pestic. Biochem. Physiol.*, 1993, **46**, 141.
39. C. J. Church and C. O. Knowles, Relationship between pyrethroid enhanced batrachotoxinin A 20-alpha-benzoate binding and pyrethroid toxicity to susceptible and resistant tobacco budworm moths, *Heliothis virescens*, *Comp. Biochem. Physiol., C: Comp. Pharmacol. Toxicol.*, 1993, **104**, 279.

40. J. G. Rubin, G. T. Payne, and D. M. Soderlund, Structure–activity relationships for pyrethroids and DDT analogs as modifiers of [H-3] batrachotoxinin A 20-alpha-benzoate binding to mouse brain sodium channels, *Pestic. Biochem. Physiol.*, 1993, **45**, 130.
41. T. J. Grinsteiner and Y. Kishi, Synthetic studies towards bactrachotoxin 1. A furan-based intramolecular Diels–Alder route to construct the A–D ring system, *Tetrahedron Lett.*, 1994, **35**, 8333.
42. T. J. Grinsteiner and Y. Kishi, Synthetic studies towards batrachotoxin 2. Formation of the oxazepane ring system via a Michael reaction. *Tetrahedron Lett.*, 1994, **35**, 8337.
43. J. W. Daly, C. W. Myers, and N. Whittaker, Further classification of skin alkaloids from neotropical poison frogs (*Dendrobatidae*), with a general survey of toxic noxious substances in the amphibia, *Toxicon.*, 1987, **25**, 1023.
44. P. J. Parsons, R. Angell, A. Naylor, and E. Tyrell, A novel approach to the histrionicotoxin framework, *J. Chem. Soc., Chem. Commun.*, 1993, 366.
45. N. Maezaki, H. Fukuyama, S. Yagi, T. Tanaka, and C. Iwata, Two types of stereocontrol in the formation of spiro skeletons via a carbonyl ene reaction and a palladium-catalyzed carbonyl allylation – A formal synthesis of (+)-perhydrohistrionicotoxin, *J. Chem. Soc., Chem. Commun.*, 1994, 1835.
46. C. M. Thompson, A route to 1-azaspirans related to perhydrohistrionicotoxin, *Heterocycles*, 1992, **34**, 979.
47. G. D. Harris, R. J. Herr, and S. M. Weinreb, A palladium-mediated approach to construction of nitrogen heterocycles. *J. Org. Chem.*, 1992, **57**, 2528.
48. D. L. Comins and H. Dehghani, A short asymmetric synthesis of (−)-pumiliotoxin C, *J. Chem. Soc., Chem. Commun.*, 1993, 1838.
49. D. L. Comins and D. H. Lamunyon, Grignard addition to 1-acyl salts of chiral 4-alkoxypyridines – a new enantioselective preparation of 2-alkyl-2,3-dihydro-4-pyridones, *Tetrahedron Lett.*, 1994, **35**, 7343.
50. G. G. Habermehl and J. H. Wilhelm, Synthesis of *N*-[(3-ethoxycarbonylmethyl) cyclohexyl]-azetidin-2-yl propionic acid hydrochloride, *Z. Naturforsch., B. Chem. Sci.*, 1992, **47**, 1779.
51. A. B. Holmes, A. B. Hughes, and A. L. Smith, *N*-Alkenyl nitrone dipolar cycloaddition routes to piperidines and indolizidines. 5. Preparation of a gephyrotoxin precursor, *J. Chem. Soc., Perkin Trans. 1*, 1993, 633.
52. C. Kibayashi, A new synthetic approach to dendrobatid alkaloids, *Pure Appl. Chem.*, 1994, **66**, 2079.
53. B. Badio and J. W. Daly, Epibatidine, a potent analgesic and nicotinic agonist, *Mol. Pharmacol.*, 1994, **45**, 563.
54. M. Fisher, D. Huangfu, T. Y. Shen, and P. G. Guyenet, Epibatidine, an alkaloid from the poison frog *Epipedobates tricolor*, is a powerful ganglionic depolarizing agent. *J. Pharmacol. Exp. Ther.*, 1994, **270**, 702.
55. E. Kochva, A. Bdolah, and Z. Wollberg, Sarafotoxins and endothelins – evolution, structure and function. *Toxicon*, 1993, **31**, 541.
56. J. Siigur and E. Siigur, The direct acting α-fibrin(ogen)olytic enzymes from snake venoms, *J. Toxicol–Toxin Rev.*, 1992, **11**, 91.
57. J. O. Dolly, Components involved in neurotransmission probed with toxins: voltage dependant K^+ channels, in 'Probes for Neurochemical Target Sites', ed. K. F. Tipton and L. L. Iversen, Royal Irish Academy, Dublin, 1991, p. 127.
58. A. L. Harvey, From venom to toxin to drug?, in 'The Advancement of Drug

Delivery', ed. A. L. Harvey and W. C. Bowman, Proceedings of the Royal Society of Edinburgh, 1992, vol. 99, p. 55.

59. A. Menez, F. Bontems, C. Roumestand, B. Gilquin, and F. Toma, Structural basis for functional diversity of animal toxins, in 'The Advancement of Drug Delivery', ed. A. L. Harvey and W. C. Bowman, Proceedings of the Royal Society of Edinburgh, 1992, vol. 99, p. 83.
60. R. H. Loring, The molecular basis of curaremimetic snake neurotoxin specificity for neuronal nicotinic receptor subtypes, *J. Toxicol.–Toxin Rev.*, 1993, **12**, 105.
61. D. von Klobusitzky, *Arch. Exp. Pathol. Pharmakol.*, 1935, **179**, 204.
62. R. A. Hendon and A. T. Tu, Biochemical characterization of the lizard toxin, Gilatoxin, *Biochemistry*, 1981, **20**, 3517.
63. P. Utaisincharoen, S. P. Mackessy, R. A. Miller, and A. T. Tu, Complete primary structure and biochemical properties of Gilatoxin, a serine protease with kallikrein-like and angiotensin-degrading activities. *J. Biol. Chem. 1993,* **268**, 21975.
64. A. T. Tu (ed.), A lizard venom, Reptile Venoms and Toxins, in 'Handbook of Natural Toxins', Marcel Dekker, New York, 1991, vol. 5, Ch 23, p. 755.
65. D. Gordon, H. Moskowitz, and E. Zlotkin, Biochemical characterization of insect neuronal sodium channels, *Arch, Insect Biochem. Physiol.*, 1993, **22**, 41.
66. E. Zlotkin, M. Gurevitz, E. Fowler, and M. E. Adams, Depressant insect selective neurotoxins from scorpion venom – chemistry, action, and gene cloning, *Arch. Insect Biochem. Physiol.*, 1993, **22**, 55.
67. E. Zlotkin *et al.*, Insect specific neurotoxins from scorpion venom that affect sodium current inactivation, *J. Toxicol.–Toxin Rev.*, 1994, **13**, 25.
68. Z. Wimmer, D. Saman, V. Nemec, and W. Francke, Carbamate series of Juvenoids – variation of the *O*-alkyl substituent, *Helv. Chim. Acta*, **77**, 1994, 561–568.
69. Z. Wimmer, D. Saman, and W. Francke, Novel Juvenoids of the 2-(4-hydroxybenzyl)cyclohexan-1-one series, *Helv. Chim. Acta*, 1994, **77**, 502.
70. K. Mori, Synthetic and stereochemical aspects of pheromone chemistry. *Pure Appl. Chem.*, 1994, **66**, 1991.
71. K. Mori and M. Amaike, Pheromone synthesis. 166. Synthesis of (2*E*, 5*R*, 6*E*, 8*E*)-5,7-dimethyldeca-2,6,8-trien-4-one, the major component of the sex-pheremone of the Israeli pine bast scale, and its antipode. *J. Chem. Soc., Perkin Trans 1*, 1994, 2727.
72. M. S. Blum, Ant venoms: Chemical and pharmacological properties, *J. Toxicol–Toxin Rev.*, 1992, **11**, 115.
73. M. S. Blum, 'Chemical Defenses of Arthropods', Academic Press, London, 1981.
74. T. Piek (ed.), 'Venoms of the Hymenoptera. Biochemical, Pharmacological and Behavioural Aspects', Academic Press, London, 1986.
75. K. Nakanishi, Natural products as sources of pest management agents. *ACS Symp. Ser.*, 1994, **551**, 11.
76. T. Piek, Arthropod-derived neurotoxic insecticides – a lead in pesticide science, *ACS Symp. Ser.*, 1993, **524**, 233.
77. P. N. R. Usherwood and I. S. Blagbrough, Spider toxins affecting glutamate receptors: polyamines in therapeutic neurochemistry, *Pharmacol. Therapeu.*, 1991, **52**, 245.
78. K. D. McCormick and J. Meinwald, Neurotoxic acylpolyamines from spider venoms, *J. Chem. Ecol.*, 1993, **19**, 2411.
79. H. Benz and M. Hesse, Herstellung von Analoga von Polyamin-Spinnentoxinen, *Helv. Chim. Acta*, 1994, **77**, 957.

80. A. Schäfer, H. Benz, W. Fiedler, A. Gruggisberg, S. Bienz and M. Hesse (ed.), Polyamine toxins from spiders and wasps, in 'The Alkaloids', Academic Press, New York, London, 1994, vol. 45, p. 1.
81. I. S. Blagbrough, P. T. H. Brackley, M. Bruce, B. W. Bycroft, A. J. Mather, S. Millington, H. L. Sudan, and P. N. R. Usherwood, Arthropod toxins as leads for novel insecticides: an assessment of polyamine amides as glutamate antagonists, *Toxicon*, 1992, **30**, 303.
82. I. S. Blagbrough and P. N. R. Usherwood, Polyamine amide toxins as pharamcological tools and pharmaceutical agents, in 'The Advancement of Drug Delivery', ed. A. L. Harvey and W. C. Bowman, Proceedings of the Royal Society of Edinburgh, 1992, vol. 99, p. 67.
83. D. Barcham, S. Fuchs, *et al.*, How the mongoose can fight the snake – the binding site of the mongoose acetylcholine receptor, *Proc. Natl. Acad. Sci. USA*, 1992, **89**, 7717.
84. A. T. Tu (ed.), 'Insect Poisons, Allergens and Other Invertebrate Venoms. Handbook of Natural Toxins', Marcel Dekker, New York, 1984, vol. 2.
85. A. T. Tu, 'Venoms: Chemistry and Molecular Biology', John Wiley and Sons, New York, 1977.
86. B. N. Ames and L. S. Gold, Misconceptions on pollution and the causes of cancer, *Angew. Chem., Int. Ed. Engl.*, 1990, **29**, 1197.
87. B. N. Ames, M. Profet, and L. S. Gold, Dietary pesticides (99.99% all natural), *Proc. Natl. Acad. Sci. USA*, 1990, **87**, 7777.

CHAPTER 10

Diversity and Biological Activity of Bacillus thuringiensis

LEE F. ADAMS, CHI-LI LIU, SUSAN C. MacINTOSH
and ROBERT L. STARNES

10.1 Introduction

10.1.1 History and Insecticidal Activity

Bacillus thuringiensis (Bt) is a motile, gram-positive bacterium that is widely distributed in nature, especially in soil and insect-rich environments. During sporulation, Bt produces parasporal crystal inclusion(s) which are insecticidal upon ingestion to susceptible insect larvae of the orders Lepidoptera, Diptera, and Coleoptera. The inclusions may vary in shape (the most common being a bipyramid), number, and composition. They comprise one or more proteins called delta-endotoxins, which may range in size from 27–140 kDa. The insecticidal delta-endotoxins are processed and activated by proteases in the larval gut, causing midgut destruction and, ultimately, death of the insect.

The insecticidal action of Bt was first observed by S. Ishiwata in 1901 as the cause of the sotto disease in *Bombyx mori* (Linnaeus) (silkworm). Later, in 1911, the bacterium was discovered in the pupae of *Anagasta kuehniella* (Zeller) (Mediterranean flour moth), and named by E. Berliner as *Bacillus thuringiensis*.[1] The species name originated from the German province in which it was discovered. In 1950, Steinhaus[2] determined that the bacterium had commercial potential, which culminated in 1957 with the commercialization of Thuricide by Pacific Yeast Products.[3]

Growing recognition that different Bt strains have different insecticidal activities towards a given insect pest prompted recommendations for the use of a standard Bt preparation, prepared from *B. thuringiensis* subsp. *thuringiensis* by the Pasteur Institute in Paris, called E-61. In 1970, Dulmage reported the discovery of a Bt subspecies, which he termed *alesti* (now called *kurstaki*, and, most likely, the same isolate as that discovered by Kurstak in 1962) with 200-fold better insecticidal activity against *Trichoplusia ni* (Hubner) (cabbage looper), *Heliothis virescens* (Fabricius) (tobacco budworm), and *Pectinophora gossypiella* (Saunders) (pink bollworm) than E-61 or any existing Bt product sold.[4,5] In 1970, Abbott Laboratories introduced Dipel™ based on the subspecies *kurstaki*. This strain remained as the focus of most commercial products for caterpillar control in the forestry and agricultural markets throughout the 1970s and 1980s. The commercial Bt market then expanded to public health with the discovery in 1977, by Goldberg and Margalit, of Bt *israelensis* with mosquito and black fly activity.[6] In 1983, Krieg *et al.*[7] discovered Bt *tenebrionis* with activity against beetle larvae.

The discovery of new Bts with different insecticidal spectra provided an impetus for a number of companies to initiate research programmes searching for novel Bts. Today, Abbott, Ciba-Geigy, Ecogen, Mycogen, Novo Nordisk, Plant Genetic Systems, Sandoz, and Toa Gosei are major companies involved in this effort. Several novel Bts have been discovered in recent years, with activity towards nematodes, ants, and fruit flies.

10.1.2 Delivering the Bt Insecticide to the Field

New techniques in biotechnology have greatly expanded the ways in which Bt is employed. Genes from Bt have been used to transform other microbes and plants to improve the way in which delta-endotoxins are delivered to the field. Microbial transformation for agricultural use was first initiated by Monsanto, who introduced the *cryIAc* gene into *Pseudomonas fluorescens*,[8] a bacterium that commonly inhabits the rhizosphere of corn plants. (The *cry* designation refers to the specific gene that actually encodes the crystal protein possessing insecticidal activity.) This new technology faced difficulties with governmental agencies, which prevented commercialization. Mycogen optimized the process further in a patented system called CellCap™, which was the first genetically engineered microbe for agricultural use to receive registration in the USA.[9] In this process, the *Pseudomonas fluorescens* host expressing the Bt gene is treated with cross-linking agents to fortify the bacterial cell wall. The encapsulation process reportedly lengthens the field residual activity of the crystals contained within. Another interesting delivery method is the introduction of the Bt gene into a bacterial endophyte (*Clavibacter xyli*), that can then be inoculated into a wild type corn plant.[10] In this manner, the corn plant is protected from insects in much the same was as if it had been transformed directly with Bt.

Plant transformation has been a much more difficult process, requiring the ability to regenerate whole plants from single cell transformants. This technology was first reported in 1985[11] and is widely used today by a number of companies including Agrigenetics, Calgene, Ciba-Ceigy, Zeneca, Monsanto, Pioneer Seed, Plant Genetics Systems, and others. Expression of the toxin in plants and field efficacy have been greatly improved by tailoring of the bacterial genetic codons to ones resembling those preferred by the plant.[12] Registration requirements will delay commercialization until the late 1990s.

10.1.3 The Market for Bt Products

The global agrochemical market is estimated at $22 billion, of which insecticides constitute 27% or $5.9 billion, the majority being chemical insecticides. Of the biological control agents, Bt has been commercially the most successful. The world-wide market for Bt insecticides was $60 million in 1991, and is estimated to be $375 million by the year 2001. However, the current Bt market is only 1% of the total insecticide market. It is expected the market will grow at a rate of 20% per year as a result of the environmental safety and insect selectivity of Bt, increasing resistance of insects to chemical insecticides, and re-registration requirements that reduce the variety of chemical insecticides available to the agrochemical market.

In this chapter several areas of Bt research that will aid efforts in the future to develop new and improved Bt insecticides are reviewed. These areas include classification, diversity, mode of action, structure–activity relationships, and expression.

10.2 Classification

10.2.1 Bt Nonmenclature and Classification

Adoption of the E-61 standard in the 1960s (see Section 10.1.1) resolved problems in the determination of Bt potencies, as did adoption of Dulmage's HD-1-70 standard.[13] However, the problem of classifying Bt strains with similar crystal morphologies but different activities towards various lepidopteran pests remained. The search for methods to classify Bt strains has led to serological identification schemes based on flagella and crystal antigens, the *cry* gene nomenclature, polymerase chain reaction (PCR) for the detection of specific gene segments within a Bt strain, and random-primed PCR with short oligonucleotides for high-resolution distinction between Bt strains within a variety. Researchers in the commercial Bt arena rely heavily on one or more of the following classifications in order to direct their screening efforts into desired areas.

10.2.2 Flagella Serology

At the Pasteur Institute in Paris, de Barjac and Bonnefoi recognized the taxonomic confusion regarding various Bt strains, and lamented the profusion of the various subspecies or varietal designations, particularly in instances where there were no taxonomic criteria to separate strains bearing different epithets.[14] Therefore, they undertook a detailed study to compare carbohydrate utilization, proteolytic abilities, pigmentation, and other microbiological characteristics. Most importantly, they also raised antisera against the flagella of each Bt strain for serological typing. They included an interesting twist in their serological typing by flagellar, or so-called 'H', antigens; after obtaining a precipitation reaction with antisera raised against the flagella of one strain with the flagella of another strain, they then used the 'preabsorbed' antisera to see if they could still obtain a precipitation reaction with the flagellar antigens of the original strain used to prepare the antisera. The cases where further precipitation occurred permitted establishment of more than one antigenic determinant on the original strain, not shared with the second strain. They designated multiple antigenic determinants within a major serovar with small-type letters. Such 'preabsorption reactions' explain the appearance of serovar subtypes, such as Bt *kurstaki*, which is 3a3b, and distinct from Bt *alesti*, which is 3a, possessing only one flagellar determinant or H-antigen in common with *kurstaki*.

In their paper of 1962[14] and in subsequent work, de Barjac and Bonnefoi have shown a strong correlation between flagellar serology and the more tedious biochemical and microbiological examination of strains.[15,16] Importantly, their original study reduced the 24 isolates to six different varieties; their most recent study has assigned the literally thousands of Bt strains submitted to 34 different varieties (frequently termed subspecies by other researchers). Although de Barjac has admitted that there exists no logical reason for the strong correlation between biochemical characters and flagellar serotype,[15] the robustness of the flagellar or H-antigen classification scheme is evidenced by the submission to the Pasteur Institute of innumerable new Bt strains for flagellar serotyping and by its wide acceptance in the literature. Also, notably, organizations such as the United States Environmental Protection Agency require flagellar serotyping for registration of Bt strains proposed for use as biopesticides.

10.2.3 Crystal Serology

Classification schemes based on crystal rather than on flagellar antigens have more logical appeal. de Barjac herself first explored crystal serology before turning to flagellar serology.[17] Krywienczyk noted in double diffusion tests in agarose that the subspecies *thuringiensis*, *sotto*, and *entomocidus* shared crystal antigenic determinants in common, as well as having important unique determinants.[18,19] Subsequent work established that several different crystal types could be found within a given serotype;[20], *e.g.*, the *thuringiensis* subspecies (serotype H-1) may contain crystal types *thu*, *k-1*, and both *thu* and *k-1*.[21] Krywienczyk's work had particular value in differentiating between two subtypes of Bt *kurstaki* (both 3a3b),[22] which produce identically shaped bipyramidal crystals that nevertheless differ greatly in their species-specific toxicity. Serological determination of these crystal types, designated *k-1* and *k-73* by Krywienczyk *et al.*,[23] reflecting their isolation from the HD-1 and HD-73 strains of Howard Dulmage's Bt collection, was the only means at the time for differentiating between the two strains which both possessed the 3a3b H-antigen. Now, we know that the principal crystal of HD-1 comprises three insecticidal polypeptides, designated CryIAa, CryIAb, and CryIAc, whereas the crystal of HD-73 comprises only the CryIAc polypeptide (see below).

Aside from this success, crystal serology has not enjoyed widespread use and adoption.

Other researchers have also employed crystal serology as a classification aid. Pendleton and Morrison[24] investigated the antigenic determinants of both solubilized and intact crystals. They found that solubilization led to the loss of some antigenic determinants and the creation of new ones, and, perhaps on account of the complexity of their findings, did not relate their results to existing classification schemes. Baumann *et al.*[25] and Lynch and Baumann[26] presented extensive studies that compared a variety of phenotypic characters and antigenic determinants of solubilized protein with flagellar serotyping results. Their results showed that phenotypic differences could not serve for the identification of a serovar, and that no good correlation between crystal type and flagellar serotype could be detected. On the other hand, Höfte *et al.*,[27] in an extensive characterization of crystals from 29 *Bacillus thuringiensis* strains, found a strong correlation between crystal type, as determined by monoclonal antibody analysis and insect specificity.

10.2.4 *cry* Gene Classification Scheme

The gene–toxin classification scheme of Höfte and Whiteley[28] has found utility, not so much in classifying various Bt strains, but rather in categorizing and characterizing the various *cry* (*cry* designating crystal) genes and deduced polypeptides (Cry proteins) that the various subspecies possess. Following the cloning and characterization of the first toxin-encoding gene by Ernest Schnepf,[29] other academic and industrial organizations quickly identified new genes that encoded related insecticidal proteins, and named them with alphabetic or alphanumeric codes reflecting (roughly) the order of their discovery. Indeed, some genes were named according to the size of the HindIII restriction fragment on which they were found. Thus, at one time, the *cryIAa*, *cryIAb*, and *cryIAc* genes were known not only by various permutations of the epithets *tox*, *cry*, *icp* (for insecticidal crystal protein), and *bt*, but also by the 4.5, 5.3, and 6.6 kb designations.

To resolve the disorganization in toxin gene designations, Höfte and Whiteley proposed a nomenclature keeping the *cry* epithet, followed by a Roman numeral indicating the insect class against which the encoded toxin is active, and by two alphabetic designations to indicate relative amino acid homology (see Table 10.1 for an explanation of the Roman numeral classes). This nomenclature was immediately accepted (undoubtedly with great relief) by most Bt researchers. Recent modifications and additions are the inclusion of an Arabic numeral at the end of a toxin designation to indicate minor differences from the holotype (*e.g.*, *cryIAc2*, which differs from the holotype *cryIAc* by only 10 nucleotides),[30] and the addition of new Roman numeral classes to indicate entirely new toxin families. Researchers at Zeneca proposed a new *cryV* designation for their newly isolated gene that encoded a protein with both lepidopteran and coleopteran activity.[31] Mycogen Corporation has proposed the use of Roman numerals V and VI for their nematocidal proteins, whose

Table 10.1 *The* cry *gene classification scheme*

cry *gene class**	*Insect order against which the encoded Cry protein has activity*	*Size range of encoded proteins (kDa)*
I	Lepidoptera	130–140
II	Lepidoptera and Diptera	70–73
III	Coleoptera	67–75
IV	Diptera	70–140
V	Lepidoptera and Diptera (Zeneca)†	81
V and VI	Nematocera (Mycogen)	44–153

*Following common usage, gene designations are italicized. The proteins encoded by those genes are indicated by capitalization (and no italics).
†Zeneca proposes the gene designation *cryV* for the gene encoding their lepidopteran/dipteran-active toxin; Mycogen claims *cryV* for their nematode-active toxin-encoding genes.

sequences have so far appeared only in the patent literature. We propose that Mycogen adopts Roman numerals VI and VII for those gene families, recognizing the priority of Tailor *et al.*[31] in the peer-reviewed literature.

The *cry* gene nomenclature of Höfte and Whiteley serves a very valuable function, and, for the most part, reflects accurately amino acid differences between gene families designated by different Roman numerals. However, weaknesses in the nomenclature appear occasionally. For example, as the dendogram in a recent review indicates,[32] CryIB has lepidopteran activity as do other CryI proteins, but nevertheless shows more amino acid homology with the CryIII family, which possesses coleopteran activity.

10.2.5 Plasmid Profiling

The crystal proteins of Bt are, for the most part, encoded by genes located on one or more of the 6 to 11 plasmids that the Bt cell normally harbours. One might consider that the plasmid profile of a particular strain would therefore correlate well with Bt subspecies/variety designations and/or serotype.[33] It is known, for example, that crystal production in Bt *kurstaki* HD-1 is associated with a 75 Mda plasmid, in Bt *kurstaki* HD-73 with a 50 Mda plasmid,[34] and in Bt *israelensis* with a 75 Mda plasmid.[35] These studies have been refined further by determination of specific plasmid types within a subspecies by Southern-blot hybridization of probes specific to the various origin of replication types.[36] While the plasmid profiles of many Bt subspecies have been published, in particular those of Bt *thuringiensis*, *kurstaki*, *israelensis*, *tenebrionis*, and their variants, the plasmid profile of most non-commercial strains has not been determined. Failure to correlate plasmid profile with a meaningful classification scheme undoubtedly reflects the difficulty in:

- Determining an accurate plasmid profile in a species where the largest plasmids may exceed the nucleotide size of 250 kb (165 MDal).

- Preventing plasmid destruction during isolation and agarose gel electrophoresis.

10.2.6 Molecular Classification

The use of molecular techniques has permitted high resolution classification of various Bt strains. At the species level, 165 rRNA sequencing of *B. cereus*, *B. anthracis*, and Bt has shown that the highly conserved segment is identical for the former two, and differs by only 3–4 nucleotides from the latter.[37] Thus, separation of the species *thuringiensis* from the other two bacilli is barely justified. Arbitrary primer polymerase chain reaction (AP-PCR) studies, however, have shown distinct differences between *B. cereus* and Bt,[38] justifying separation of the two species. In addition, AP-PCR could readily distinguish between 33 different varieties of Bt. However, electrophoretograms of the amplification products also suggested strong similarities between the serovars *aizawa* and *colmeri*, and between *sotto* and *thuringiensis* as well.

Judicious selection of primers has also enabled separation between individual strains of a given serovar. Thus, for example, Brousseau and coworkers[38] were able to distinguish between various HD-1 (serotype 3a3b) strains isolated from the commercial products Bernan, Biodart, Dipel, Condor, Foray, Futura, Javelin, and Thuricide. Miteva *et al.*[39] have also achieved high resolution differentiation between strains of a given serovar by an M13 DNA fingerprinting method. In this method, M13mp8 DNA was hybridized at low stringency to restriction digests of various Bt strains. The distinctions permitted by high resolution methods, such as AP-PCR and M13 fingerprinting, do not, of course, indicate significant differences between the various HD-1 strains isolated from their respective commercial products, or between various isolates of one serovar, but nevertheless represent a powerful tool in the identification of a specific Bt isolate.

The large number of unique *cry* genes that appear in the patent and refereed literature and that have been cloned and sequenced, now totalling over 40, has permitted the design of specific and unique oligonucleotides for the detection of a given *cry* gene. Some reports that describe PCR methods for identifying individual *cry* genes have appeared recently from non-industrial laboratories;[40–42] most companies have extensive in-house capabilities for the PCR-identification of any given *cry* gene. The ability to determine the *cry* gene profile of a particular Bt strain may permit rapid classification of that strain according to the scheme established by the work of de Barjac, as well as rapid prediction of a strain's insecticidal activity.[43,44]

The high resolution of molecular techniques has led to a tendency for the use of molecular criteria in microbial classification.[45] The availability of sequence information for the Bt varieties, particularly of non-*cry* genes, should permit rapid PCR identification of any Bt strain. For example, sequencing of the hypervariable regions of the flagellar genes of the Bt varieties, whose existence may be deduced from the large number of Bt serotypes, could conceivably permit classification of any new Bt strain by PCR according to the scheme originally developed by de Barjac.

10.3 Diversity

10.3.1 Diversity of Serotype

The introduction of flagellar serotyping has brought order to the Bt group.[14] In 1990, the number of serotypes was updated to 27.[16] Subsequently, additional serotypes have appeared, such as serotype H-30, representing Bt *medellin*, discovered in Colombia in 1992 and active against mosquito larvae.[46]

Bt flagellar serotype 3 (H-3) was established in 1962[14] on the basis of the strain isolated in France from diseased silkworm.[47] This serotype was later divided into two subserotypes, 3a (subsp. *alesti*) and 3a3b (subsp. *kurstaki*), by discovery of the 3b H-antigenic subfactor in several isolates.[22] In 1989, two new flagellar antigenic subfactors, 3d and 3e, were discovered.[48] Two new subspecies, Bt *sumiyoshiensis* (H-3a3d) and Bt *fukuokaensis* (H-3a3d3e) were described.[48] The type strains of Bt *alesti* and Bt *kurstaki* were re-described as H-3a3c and H-3a3b3c, respectively. Flagellar serotyping has resolved some important discrepancies that appeared in the literature over the years, and has determined (along with other biochemical criteria), for example, that the Bt *san diego* epithet is not justified.[2]

The flagellar antigens (H-antigens) used to describe the serotypes may show little correlation to the crystal antigens (Cry proteins) possessed by those strains. For example, strain PG-14 (serotype H-8a8b[49]), strains 73-E-10-2 and 73-E-10-16 (both serotype H-10[50]), Bt *kyushuensis* (serotype 11a11c[51]), Bt *israelensis* (serotype H-14[6]), and isolate 163-131 (serotype H-30[46]), are all active against mosquito larvae, and, in many cases, possess identical Cry protein components. On the other hand, Bt *morrisoni* (serotype 8a8b) is active against caterpillars, whereas Bt *tenebrionis*, also serotype 8a8b, is active against *Leptinotarsa decemlineata* (Say) (Colorado potato beetle). Recently, an Italian Bt isolate, NCIMB 40152, serotyped as H-9, has also demonstrated activity against *L. decemlineata*.[52] It is important to recognize the fact that H-antigens may be shared by strains that differ in other properties, such as phage susceptibility, crystal protein composition, and pathogenicity, and that, conversely, strains with different H-antigens may produce identical Cry proteins.

10.3.2 Diversity of Pathogenicity

Since the discovery of Bt *tenebrionis*, large efforts, mostly from industry, have been devoted to discover new Bt strains. With the constantly increasing number of new Bt isolates, more variations of pathogenicity have been found. A small number of research articles have been published about the effects of delta-endotoxins from Bts on the viability of nematode eggs. It has been reported[53] that Bt *kurstaki* and Bt *israelensis* are active *in vitro* against eggs of the nematode *Trichostrongylus colubriformis*. In 1987 Mycogen filed a patent application for five nematode-active Bts.[54] The alkali-soluble proteins of these strains all have very different molecular weights compared to those of other known Bt strains.

More recently, in 1991, Mycogen reported the discovery of a Bt strain, PS81F, which can be used to treat humans and animals hosting parasitic protozoans.[55] Furthermore, several Bt strains were found by Mycogen to have activity against acaride pests. These isolates produce crystals in various shapes – spheric, amorphic, elongated, and long rectangular – with molecular weights of the component proteins in the (wide) range of 35 kDa to 155 kDa.[56] Mycogen's most recent discovery is Bt strains with activity against pests of the order *Hymenoptera*.[57] Strains producing protein crystals of ellipsoidal, long, and amorphic shapes are claimed to be effective in controlling hymenopteran pests in various environments.

Yet another strain of Bt *kurstaki*, WB3S-16, isolated from Australian sheep wool clippings, demonstrated toxicity against biting louse[58] (*Damalinia ovis*, a *Phthiraptera* pest).

Nature seems to be able to supply all kinds of Bt strains which can control any type of pest. One need only look!

10.4 Mode of Action

The mode of action of Bt is multi-step, complex, and not completely understood. The earliest studies on Bt activity focused on the tendency of intoxicated insects to be paralysed and the cessation of feeding.[59,60] Once these actions were determined to be orally mediated, histological studies centred on the insect gut lining. It was common to find swollen midgut cells that eventually lysed, causing exposure of the insect haemolymph and mortal infections.[61–64] In recent years, this area has received extensive study due primarily to the increased profile of and commercial interest in Bt as an insect control agent. Studies of Bt action have been approached from very diverse disciplines: biochemistry of the Bt toxin, pharmacology, and insect and cell physiology. Bt mode of action can be divided into a series of critical steps: crystal solubilization, protoxin proteolysis, peritrophic membrane transport, brush border membrane binding, and pore formation. Each step of Bt *kurstaki*'s mode of action is described in detail below. Also the modes of action of Bt *tenebrionis* and Bt *israelensis* are addressed at the end of this section.

10.4.1 Crystal Solubilization and Protein Processing

The crystals of Bt *kurstaki* are stabilized by a high number of disulfide bridges between the individual proteins within each crystal. Bietlot *et al.*[65] found no evidence of intraprotein disulfide bonds, despite the presence of 10–18 cysteines in the CryIA protoxin. The midgut environment of a lepidopteran larva is both alkaline (pH>9.5) and probably reductive, allowing the dissolution of Bt *kurstaki* crystals.[66–68] Theoretically, a highly reductive environment in the insect midgut breaks the interprotein disulfide bonds easily, liberating the full length protoxins of 130 kDa.[68] Some researchers believe that the high alkalinity, in combination with gut proteases, negates the need for a reductive environment for solubilization.[66] The requirement of alkaline pH and specific larval gut proteases is an important feature in Bt safety, since all mammals and other non-target pests (including most other insects) are unable to dissolve Bt crystals, allowing their passage through the digestive system in an unaltered form. Furthermore, the identification of specific receptor-binding sites in the larval midgut (discussed below) gives additional support for the safety of Bt products.

Once the large soluble protoxins are free of the crystal structure, the larval midgut proteolytic enzymes begin their action. The primary protease components of most lepidopterans are serine proteases, particularly trypsin- and chymotrypsin-like. Information from the structure of Bt *tenebrionis* crystals suggests that the *C*-terminus of the protein is susceptible to proteolytic action, while the *N*-terminus is tightly packaged and protected. For Bt *kurstaki*, the first 29 amino acids of the *N*-terminal are cleaved along with a large portion of the *C*-terminus. The final toxin core, often referred to as the trypsin-resistant fragment, is only half the protoxin size, *ca.* 65–67 kDa. Further cleavages destroy the toxin's bioactivity. The cleavage of the protoxin plays a role in specificity, as evidenced by the variety of proteolytic patterns from different insects.[67] For example, two strains, Bt *aizawai* and Bt *colmeri*, produce a 130 kDa protoxin that can be proteolytically activated by either lepidopteran or dipteran gut juice. Bt treated with lepidopteran gut juice produces a lepidopteran toxin that is not active on dipterans, whereas Bt treated with dipteran gut juice is only active towards dipteran insects. Recently, Lambert *et al.*[69] discovered a new Bt insecticidal protoxin that only displays coleopteran activity after *in vitro* trypsin activation.

10.4.2 Role of the Peritrophic Membrane

The peritrophic membrane, a continuous tube lining the midgut, is an often ignored, but potentially crucial aspect of Bt mode-of-action. The role of the peritrophic membrane is to protect the larval gut epithelium from various bacteria, fungi, and other microbial contamination present on the leaf surface.[70,71] It acts like a molecular sieve, allowing the free flow of digestive enzymes and nutrients up to a molecular mass of about 100 kDa, in the case of *Manduca sexta* (Linnaeus) (tobacco hornworm). Larger molecules are retained by the peritrophic membrane and are eliminated in larval frass. The Bt *kurstaki* toxin easily diffuses across the peritrophic membrane to interact with the midgut epithelium. Therefore, the full length protoxin or intact crystals may be excreted intact. Recently, immunohistological studies have shown that the toxin (and protoxin) bind non-specifically to the peritrophic membrane in both lepidopteran and coleopteran larvae.[72,73] The consequence of this interaction, if any, is unknown. Various techniques to increase the permeability of the peritrophic membrane, thus enhancing the toxicity of Bt, have been pursued, with marginal success.[74,75]

In addition, the role of the spore has been shown to be very important for Bt *kurstaki* activity in specific, 'class III', insects, *e.g.*, *Spodoptera exigua*

(Hubner) (N. DuBois, 1992, personal communication). However, the size of the spore would preclude its movement through the peritrophic membrane. The function of the spore in enhancing Bt activity is not clearly understood.

10.4.3 Specific Midgut Binding

The interaction of the toxin fragment with the larval epithelium has received the most attention in the past 5 years. A family of putative receptor-binding sites has been described that bind Bt *kurstaki* toxins with high affinity. This binding is generally irreversible, perhaps due to integration into the membrane. Hofmann *et al.*[76] were the first to illustrate toxin binding to isolated brush border membrane vesicles (BBMVs) from the larvae of *Pieris brassicae* (Linnaeus) (cabbage white butterfly). Since then, BBMVs from many different insects have been studied using several different Bt toxins.[77–86] These insects represent a wide range of susceptibility to Bt, and include several species that have developed resistance to Bt either through laboratory selection experiments or in the field. It is clear that binding must occur for insecticidal activity, but that receptor binding does not always predict activity nor does it predict the degree of susceptibility. For example, vesicles isolated from *Lymantria dispar* (Linnaeus) (gypsy moth),[81] *H. virescens* (Fabricius) (tobacco budworm),[80] and *Spodoptera frugiperda* (Smith) (fall army worm)[82] bind Bt toxins CryIAb, CryIC, and CryIAc, respectively, which produce minimal or no insecticidal activity. In fact, this result may account for the lowered activity found with strains containing multiple Bt *kurstaki* proteins. If non-active toxin molecules bind to shared binding sites that preclude the efficient binding of toxic molecules, the overall activity of the Bt strain could be reduced.

The nature of the binding sites is largely unknown. Using competitive binding experiments to *H. virescens*, Van Rie *et al.*[79] explained the relationship between binding of the CryIA proteins. Three unique binding sites were identified. CryIAc binds to all three sites, CryIAb recognizes only two of the binding site types, and CryIAa binds to only a single site. CryIIA, CryIC, and CryIE each have separate binding sites, different from that of any of the CryIA proteins.[77]

The binding activity of the Bt *kurstaki* toxins interferes with the K^+-dependent active amino acid symport system located in the larval midgut.[87] The midgut epithelium maintains a high electrical potential and potassium gradient across the membrane.[88] Bt toxins short-circuit this symport system by causing large-pore formation in the apical membrane, destroying the potassium gradient and increasing the water permeability of the membrane. Barium and calcium reverse the action of Bt on this symport system.[89] The results of other studies using planar lipid bilayers, phospholipid vesicles, and cultured insect cells also support the theory that Bt toxins cause the formation of a cation-selective pore.[90–93] A large uptake of water results, causing cell swelling, eventual rupture, and the disintegration of the midgut lining.

10.4.4 Cellular Pore Formation

The nature of pore formation is an area of research that remains largely obscure. Is the pore formed by the Bt toxin or is it a combination of the receptor molecule and the Bt toxin? A third option would suggest that the toxin alters the receptor so that the receptor molecule forms the pore alone. Others have speculated further that the toxin signals a secondary messenger system that is related to pore formation.[93] The isolation and characterization of receptor molecules will clarify some of these questions and is the most intensely studied area of Bt mode of action.[77,82,94]

10.4.5 Bt *tenebrionis* Mode of Action

The discussion above focuses on the Bt *kurstaki* mode of action, since it is the most widely studied of the Bt proteins. Progress is being made also in understanding the Bt *tenebrionis* and Bt *israelensis* modes of action. Bt *tenebrionis* is a much easier system to study, since this strain produces a single insecticidal protein of 73 kDa, which does not require activation for full activity.[95] The midgut environment of the target insect, *Leptinotarsa decemlineata*, is nearer to neutral pH,[96] about 6, and the protease components are not serine-type, as in lepidopteran larvae, but thiol-type proteases.[97] Interestingly, the Bt *tenebrionis* protein is soluble at both high pH (>10) and low pH (<4) extremes and lacks disulfide bonds, despite the content of three cysteine residues.[98] Ultrastructural effects on the insect midgut are quite similar to those found with Bt *kurstaki* toxin on lepidoptera, with a few important exceptions. No membrane lesions or microvillar damage are observed and the first cellular response (*e.g.*, swelling, elongation) is relatively slow.[99] Receptor binding studies have been difficult due to the problem of obtaining:[100]

- High specific radiolabelling of the Bt *tenebrionis* protein
- Saturated binding

Slaney *et al.*[100] compared the mode-of-action of the Bt *tenebrionis* toxin toward two coleopterans, *L. decemlineata* (Say) (Colorado potato beetle) and *Diabrotica undecimpunctata howardi* (Barber) (southern corn rootworm), which are, respectively, susceptible and tolerant to the protein. They found that susceptibility was correlated, at least in part, with increased receptor-binding affinity and pore formation. However, saturated binding was not observed and overall receptor affinity was a log lower than that measured in lepidoptera using Bt *kurstaki* toxins, making interpretation of the results difficult.

10.4.6 Bt *israelensis* Mode of Action

Bt *israelensis* crystals are probably the most insoluble of any Bt crystals, requiring very high pH (>11) for full solubilization. Until recently, the insecticidal activity of the individual proteins was difficult to assess, partially because of the need for a specific particle size for bioactivity in mosquitoes.[101] All of the Bt *israelensis* proteins (27, 70, 135, and 140 kDa in size) have dipteran activity, and the 27 kDa cytolytic toxin (CytA) appears to synergize the others.[102] Ultrastructural effects mimic those found in lepidopteran larvae intoxicated with Bt *kurstaki* – insect paralysis, midgut lesions, destruction of the gut wall, and finally death of the pest.[103] It is presumed that this action is receptor mediated.[104] Additional indirect support has been provided with cultured insect cells and brush border membrane vesicles from various mosquito species, which were used to study toxin interactions from a related strain, *Bacillus sphaericus*, that is also active on diptera.[105–107]

Our understanding of the Bt mode of action has progressed rapidly in the past few years. Characterization of the Bt *kurstaki* receptor molecules will greatly enhance our knowledge in this area. Studies on Bt *tenebrionis* and Bt *israelensis* will, undoubtedly, develop more fully in the near future. In total, our knowledge of Bt toxin mode of action will influence our ability to control insects in a more directed manner.

10.5 Structure–Activity Relationships

The role of structure in specificity and insecticidal activity has been the subject of research in many laboratories. This work is particularly important because knowledge of the structural basis for insecticidal activity will eventually enable the use of protein engineering to increase activity and to broaden the insect activity spectrum.

10.5.1 Activation of Protoxins

As described earlier, the crystal inclusions are solubilized in the midgut of susceptible insect larvae, releasing the protoxins which are then converted into toxin moieties by the action of gut proteases. Choma *et al.*[108] have shown, using trypsin as a model protease, that the insecticidal core of the CryIAc protoxin of Bt *kurstaki* HD73 is generated by an unusual ordered cleavage process, beginning at the *C*-terminus of the protoxin and proceeding towards the *N*-terminal region. The sequential proteolysis ends with a 67 kDa toxin, which is resistant to further proteolysis. This processing is generally true for most Cry protoxins, including those from the subspecies *kurstaki* and *israelensis*. In contrast to other Cry protoxins, the CryIIIA protoxin from Bt *tenebrionis* is 73 kDa in size and is proteolytically processed to a 67 kDa protein.[109]

10.5.2 Secondary Structure of Protoxins

Amino acid sequence alignments of different delta-endotoxins have shown that the *C*-terminal half of the protoxin is highly conserved, while the *N*-terminal half contains many variable regions. The *N*-terminal regions of the protoxin proteins contain two multidomain regions with unique structural and biological properties. Based on secondary structure determinations, Choma *et al.*[110] have shown that the native conformation of the CryIAc protein from Bt *kurstaki* HD73 is highly folded and contains considerable amounts of both alpha-helical and beta-sheet structures.

10.5.3 *N*-Terminal Region

The *N*-terminal half of the protoxin is responsible for insecticidal activity. Location of the toxic domain was shown by deletion analysis of the CryIAa protoxin from Bt *kurstaki* HD-1.[111] This toxic portion contains conserved hydrophobic regions in the amino half and a more variable carboxyl half. There are five distinct regions which are highly conserved and which alternate with highly variable regions.[28] These variable regions have been shown to be responsible, in some cases, for insect specificity. Convents *et al.*[112] demonstrated that the hypervariable fragment of the CryIAb protein from Bt *berliner* 1715 comprises two structural domains. The *N*-terminal half of the fragment contains several alpha-helices, while the *C*-terminal half is rich in beta-strand conformation.[112] They have observed similar results with the CryIC toxic fragment.[113]

The region responsible for activity towards the mosquito *Aedes aegypti* (Linnaeus) (yellow fever mosquito) in the CryIIA delta-endotoxin from Bt *kurstaki* was identified by Widner and Whiteley.[114] The CryIIA protein possesses activity towards *A. aegypti* and *M. sexta*, while the CryIIB protein is only active against *M. sexta*. A series of hybrid protein products prepared by molecular recombination of their encoding genes revealed that the sequence responsible for *A. aegypti* activity resides between amino acids 307–382 in the CryIIA protein. Interestingly, the CryIIA and CryIIB proteins differ by only 18 amino acids in this region. Therefore, few amino acid changes are necessary to produce a change in the insect spectrum. Site-directed mutagenesis of the 27 kDa cytolytic protein from Bt *israelensis*, which is unrelated to the CryIIA protein, yielded several mutants which produced proteins with acidic and basic amino acids replaced with alanine (Ala154, Ala163, Ala164, Ala213, and Ala225). These proteins were significantly less active against *A. aegypti* relative to the wild-type polypeptide.[115]

In another study, Ge *et al*[116] determined that the specificity region for activity towards *B. mori* resides between amino acids 332–450 in the CryIAa protein.[116] In receptor binding studies, the *B. mori* receptor binding region on the CryIAa toxin was located using mutant proteins derived from CryIAa and CryIIA.[117] The binding region included the amino-terminal portion of the hypervariable region and amino acids 332–450, which correspond to the specificity-determining region. Therefore, this study provides evidence that delta-endotoxins possess a sequence of amino acids that comprise a receptor binding region determining specificity.

Localized mutagenesis of two portions of the gene encoding the CryIAc toxin by Wu and Aronson[118] defined regions of the toxin involved in specificity and insecticidal activity. Mutations in the highly conserved hydrophobic region substantially reduce activity against three insects: *H. virescens*, *T. ni*, and *M. sexta*. In general, changes which preserve the hydrophobicity of the region retain toxicity, whereas the introduction of charged residues or other major changes of the amphophilic helical structure result in reduced or no activity. However, all of these mutant toxins compete with the wild type CryIAc toxin for binding to *M. sexta* BBMVs but lose the ability to inhibit K^+-dependent leucine uptake. These results are consistent with the proposal that the pore may comprise amphophilic helices of the Bt toxin. Substitutions at Ala92 and Arg93, which are located in an alpha-helical region, indicate their involvement in the formation of specific binding domains. Almond and Dean[119] were able to show that residues 428–450 do not account completely for the specificity determining region. Site-directed mutagenesis of this region, as well as secondary site mutations, revealed that three regions are of structural importance to the CryIA toxins: amino acid residues 428–450, residues 347–349, and 370–373. Therefore, tertiary structural interactions between non-contiguous regions are very likely important for the preservation of the conserved hydrophobic regions implicated in insecticidal activity.

10.5.4 *C*-Terminal Region

The *C*-terminal half of the protoxin is highly conserved at the amino acid level, compared to the *N*-terminal half. It is rich in proteolytic sites and is removed proteolytically from the protoxin to form the toxin. All of the cysteine amino acids of CryIA toxins are located in this half of the molecule and are oxidized to disulfide bridges in the crystal inclusions. Disulfide bond formation is probably responsible for the crystal form of the protoxin and for the water solubility of the inclusions. Arvidson *et al.*[120] have also proposed that disulfide bonds may be essential for specificity. The reduction of the disulfide bonds in the CryIAc protoxin was shown to result in a significant reduction of selectivity towards the insect *H. virescens*. However, Pfannenstiel *et al.*[121] have shown that when the protein disulfide bonds were cleaved and blocked in the mosquito larval toxin from Bt *israelensis*, there was no loss of toxicity.[121] Further, Choma and Kaplan[122] modified chemically the cysteine and lysine residues of the reduced CryIAc protoxin from Bt *kurstaki* HD-73 and found that all the derivatives retained full insecticidal activity towards *Choristoneura fumiferana* (Clemens) (spruce budworm). This result suggests that these residues are on the surface of the protein, and that derivatization does not alter the conformation of the solubilized protoxin. MacIntosh *et al.*[98] showed that the CryIIIA toxin from Bt *tenebrionis* contains no inter- nor intra-molecular disulfide bonds,[98] despite the presence of three cysteine residues. However, replacement of cysteine 540 with serine caused complete inactivation of the protein's toxicity towards *L. decemlineata* larvae. Disulfide bonds in the full-length CryIA protoxins, may be required for crystal formation and stability, but not for insecticidal activity.

10.5.5 Role of Multiple Toxins

The presence of multiple toxins in a crystal inclusion appears to play a role in broadening insect specificity, as well as in enhancing insecticidal activity. Arvidson *et al.*[120] have shown that the absence of the CryIAb protoxin in a crystal inclusion that normally contains the CryIAa, CryIAb, and CryIAc protoxins increases the specificity towards *T. ni*,[117] providing indirect evidence for this theory. The stability of the protoxins is also important in determining specificity. The differences in the stability of the protoxins is probably attributable to the variable regions of the *N*-terminal half and also to the interaction of the protoxins within the crystal. The interaction of protoxins has also been shown by Aronson *et al.*[123] to affect solubility under alkaline conditions. When Bt *aizawai* (containing inclusions comprising CryIAa, CryIAb, CryIC, and CryID proteins) was cured of a 25 MDa plasmid encoding CryIAb, only 35–40% of the inclusions were solubilized in 10 mM beta-mercaptoethanol, pH 9.2, compared with 63–74% for the parent strain and 65–80% for Bt *kurstaki* HD1.

10.5.6 X-Ray Crystal Structure of CryIIIA

The first X-ray crystal structure of a Bt delta-endotoxin, CryIIIA from Bt *tenebrionis*, has been demonstrated.[124] Three domains comprise the delta-endotoxin:

(I) A seven-helix bundle (residues 1–290).
(II) A three-sheet domain (residues 292–500).
(III) A sandwich of two antiparallel beta sheets (residues 501–644).

The five regions conserved with other Cry proteins, comprising residues 189–218, 239–305, 491–538, 560–569, and 633–644, form the core of the molecule, while the variable regions are on the outside of the molecule. Domain II and, probably, its apical region are likely to be responsible for binding to the receptor, based on studies mapping the specificity regions identified by molecular means to equivalent positions in the structure of the molecule. The long, hydrophobic, and amphipathic helices of domain I have been proposed as suitable for insertion into a membrane to allow pore formation.

The structure of the coleopteran-active delta-endotoxin can be extrapolated to other delta-endotoxins, based on the extensive amino acid sequence homology that nearly all Cry proteins share, which will prove important for modification of toxins to improve stability, increase activity, and broaden insect specificity.

10.6 Expression of Bt Delta-Endotoxins

The molecular details of *cry* gene expression continue to be a fascinating area in the Bt field. Researchers have elucidated some of the transcriptional events in expression of the *cryIA*, *cryII*, and *cryIIIA* toxin genes, but many aspects remain unclear. This section addresses *cry* gene transcription, the relation between *cry* gene expression and sporulation, and the influence of external factors in Cry toxin expression, such as chaperonins and scaffolding proteins.

10.6.1 Transcriptional Regulation of *cry* Gene Expression

Factors involved in the initiation of transcription of *cry* genes have been reviewed previously,[28,125] and only novel features are emphasized in the discussion that follows. Many, if not all, *cryIA* genes are regulated apparently, by a tandem promoter, as shown by transcriptional mapping. The elements of the tandem promoter are also controlled temporally – in the case of *cryIAa*, the downstream Bt I promoter is turned on first at approximately stage II of sporulation, then diminishes in activity. At stage IV of sporulation, the upstream Bt II promoter becomes active and remains so until lysis of the mother cell and release of the crystal and mature endospore.[126]

Casual examination of the *cryIAa* promoter reveals several DNA sequence motifs, implying additional layers of transcriptional regulation. For example, two segments of a dyad surround the Bt I promoter and direct repeats bracket both the Bt I and II promoters. The function of these DNA regions is unknown. Transcription in *Escherichia coli* occurs from a fortuitous promoter near the Bt II promoter, and is greatly enhanced by deletion of the 150 nucleotides lying immediately upstream.[27] The effect of the upstream region in Bt is unknown at this time.

Other factors may influence *cry* gene expression as well. Minnich and Aronson[128] performed plasmid curing experiments on Bt *kurstaki* HD-1 and found that toxin expression from one plasmid required the presence of other plasmids that did not encode *cry* genes. From these results, they inferred that two cryptic plasmids in strain HD-1 encoded factors that acted in *trans* to enhance *cry* gene expression on another plasmid. Plasmid-curing experiments have also shown that *cry* gene expression can be eliminated without elimination of the *cry* gene itself, again indicating the necessity of factors encoded by other plasmids for toxin expression.

Unknown factors also contribute to different expression of individual *cry* genes. The *cryIAa*, *cryIAb*, and *cryIAc* genes in Bt *kurstaki* show different rates of expression depending on the particular isolate,[129] and many researchers have shown that HD-1 crystals themselves contain different proportions of each Cry toxin. Depending on the strain, some toxins may be expressed preferentially to other toxins, and some toxins may not be expressed at all.[130] As an example, Masson *et al.*[131] have shown that the percent composition of the individual CryIAa, CryIAb, and CryIAc toxins of Bt *kurstaki* is 23, 39, and 38% in the NRD-12 strain, and 14, 54, and 32% in the HD-1 strain.[131] Because the promoter regions of the three genes encoding the toxins are identical for at least 150 nucleotides upstream of the ATG start codon, other factors must influence the differential rates of toxin expression seen in these strains.

10.6.2 The Link Between Sporulation and Crystal Formation

Despite the fact that spore formation and crystal production occur in separate compartments of the sporulating cell, researchers have long suspected that the two processes share a common mechanism.[132,133] Recent observations supporting this idea include experiments showing that the cloned *cryIAa* gene was not expressed in *B. subtilis* harbouring mutations in sigma factors required for sporulation[28], such as *spoIIG41* (defective in σ^E expression). More definitive evidence has come from the biochemical isolation of the sigma factors directing transcription from the Bt I and II promoters,[134,135] and cloning and sequencing of the encoding genes[136] The deduced amino acid sequences of the genes encoding σ^{35} and σ^{28}, which direct transcription from the Bt I and Bt II promoters, share 88 and 85% homology, respectively, with the well-known sporulation factors σ^E and σ^K from *B. subtilis*. To demonstrate functional homology as well as amino acid homology, Adams *et al.*[136] used the σ^{35} and σ^{28} genes to restore sporulation in σ^E- and σ^K-defective *B. subtilis* strains. Further, they demonstrated that core polymerase that had been reconstituted with either σ^{35} or σ^{28} could recognize *B. subtilis* promoters normally recognized by σ^E or σ^K-containing polymerase.[136] The sequence and functional homology between σ^{35} and σ^{28}, required for *cry* gene expression in Bt, and their counterparts σ^E and σ^K, required for sporulation in *B. subtilis*, implies that crystal production and sporulation are intimately connected events at the molecular level.

10.6.3 Chaperonins and Scaffolding Proteins Involved in *cry* Gene Expression

In addition to the unidentified *trans*-acting factors required for toxin expression mentioned above, other factors are believed to be involved as well. During efforts to clone and express the 27 kDa cytolytic protein (CytA) from Bt *israelensis* in *E. coli*, McLean and Whiteley[137] found that proper expression of CytA required the presence of a region nearly 4 kb upstream. Adams *et al.*[138] found that the region encoded a 20 kDa protein, which acted after the initiation of translation to enhance CytA production. Visick and Whiteley[139] showed that the same 20 kDa protein also enhanced expression of CryIVD (another Bt *israelensis* toxin), as well as expression of LacZX90, a β-galactosidase enzyme mutant with a very short half-life. However, the 20 kDa protein was not required for CytA expression in *E. coli* cells with defects in *rpoH*, *groEL*, or *dnaK*. Those genes encode proteins (or so-called chaperonins) implicated in the proper folding, secretion, or degradation of other proteins. Thus, it appears that the 20 kDa protein acts as a chaperonin itself, somehow sequestering or protecting Cry proteins from degradation. Recently, Chang *et al.*[140] have shown that the 20 kDa protein leads to more efficient expression of CryIVD and CytA in Bt *israelensis*, and Wu and Federici[141] have overexpressed the 20 kDa protein in a Bt *kurstaki* strain, which also led to substantial overexpression of the CytA toxin. Thus, it appears that the 20 kDa protein enhances toxin expression in Bt as well. Curiously, a related protein, CytB, cloned from Bt *kyushuensis*, did not require an accessory protein, such as the 20 kDa protein, for expression in *E. coli*.[142]

Another interesting protein required for expression of the CryIIA protein is encoded by an upstream open-reading frame termed *orf2*. (CryIIA is encoded by *orf3* in this operon.) The deduced amino acid sequence of *orf2* contains a 15 amino acid segment repeated 11 times, representing two-thirds of the protein.[143] By itself, the protein has no toxicity to insects. In its absence, however, Bt cells do not make a CryIIA crystal, although they do make CryIIA protein.[144] It therefore appears that the *orf2* protein is required, not as a chaperonin, but rather as a scaffolding protein that ensures the proper assembly of the CryIIA-containing crystal.

10.7 Bt as an Insecticide: Effectiveness and Potential for Insect Resistance

10.7.1 Bt as an Insecticide

The Bt delta-endotoxin is potently insecticidal, as demonstrated by effective LC_{50}s (the concentration required to kill 50% of susceptible larvae) that are often as low as a fraction of a μg per ml (less than 1 part per million). Feitelson *et al.*[32] converted typical field application rates (*i.e.*, those that exert acceptable insect control in the field) of various insecticides from pounds per acre to moles per acre, and found that, on a per molecule basis, the Bt delta-endotoxin is 300-fold more toxic than a synthetic pyrethroid, and 80,000-fold more toxic than an organophosphate. However, if used improperly, Bts will not be as effective as chemical insecticides. Generally, less than optimal field efficacy results when a Bt is applied in a manner similar to that used for a chemical. Bts, because of their unique mode of action and biological nature, require greater care in handling and in application than do other insecticides.

Specific areas likely to lead to improved Bt performance include proper storage in a dry, cool environment to prevent degradation of the active insecticidal protein. Bts are most effective against neonates and early larval instars; therefore, spraying must be timed for egg hatch, and not for later stages, to account for the inevitable delay between field scouting and insecticide application. Further, because Bts have no contact activity and must be ingested to be toxic, plants must be thoroughly covered to ensure that an insect larva receives a lethal dose. Finally, Bts exhibit a very short half-life in the field (as short as 4 h), again necessitating careful timing of product application to appearance of susceptible instars. Because of the requirements for maximum effectiveness that are distinct from those of chemicals, some manufacturers are beginning to recognize end-user education as an important component of their Bt sales. Yet, when used properly against susceptible insect pests in an integrated pest management (IPM) programme (see below), Bts exhibit cost-effective control.

10.7.2 Resistance to Bts

Increasing awareness of the possibility for insect resistance to Bts and, in particular, to Bt endotoxins expressed in plants (as well as to other chemical insecticides) has prompted researchers to encourage Bt use within the context of an IPM programme. Many excellent articles reviewing the subject of insect maintenance to Bts have been published in the past few years.[145–149] Several different strategies have been outlined recently by McGaughey and Whalon[150] to introduce Bt transformed plants, as well as Bt microbial products, within a balanced IPM programme to delay or avoid resistance development. An industrial group, the Bt Management Working Group, was formed in 1988 to foster the judicious use of Bt-based products. The management group has disbursed approximately $250,000 since its inception for research to address the potential for development of resistance to Bt and to devise IPM strategies that will minimize or prevent resistance.

10.8 Conclusions

The use of Bt-based products is an excellent means of controlling specific classes of insect pests. Although this method of biological control has been available for many years, it has seen only limited use, as evidenced by low market share (1% of the total insecticide market). However, product improvement and enhanced public awareness is pushing the use of ecologically sound techniques of crop pest control. Bt is positioned at the forefront of biological control measures, particularly within the context of an IPM programme, and is the focus of strong research efforts, which include screening for new strains.

Protein engineering of known delta-endotoxins with improved stability, increased activity, and broader insect specificity will eventually become a reality as the structural basis for the mode of action of Bt delta-endotoxins becomes better understood. In particular, characterization of receptors and information on the interaction of Bt proteins with the midgut binding receptors will aid in improving the biological control of insects.

A better understanding of the sporulation process should lead to a better understanding of crystal production. Furthermore, the discovery and identification of transcriptional factors, chaperonins, and scaffolding proteins offer new avenues for the enhancement of *cry* gene expression by molecular means.

Several key factors will insure the growth of the Bt market. These factors include:

- Discovery of new and improved Bt strains active against caterpillars as well as against other insects.
- Improvement of product performance in application and field efficacy.
- Use of integrated pest management to minimize or delay the development of insecticide resistance.
- Education of insecticide distributors, sales representatives, pest control agents, and growers.
- Unavailability of new, environmentally safe chemical insecticides.

10.9 References

1. E. Berliner, Uber die Schlaffsucht der Mehlmottenraupe (*Ephestia kuhniella*, Zell.) und ihren Errger *Bacillus thuringiensis*, n. sp., *E. Z. Angew. Entomol.*, 1915, **2**, 29.
2. E. Steinhaus, Potentialities for microbial control of insects, *Agric. Food Chem.*, 1956, **4**, 676.
3. J. Briggs, Commercial production of insect pathogens, in 'Insect Pathology: An Advanced Treatise', ed. E. A. Steinhaus, Academic Press, New York, 1963, vol. 2, p. 519.
4. H. Dulmage, Insecticidal activity of HD-1, a new isolate of *Bacillus thuringiensis* subsp. *alesti.*, *J. Invert. Path.*, 1970, **15**, 232.
5. E. Kurstak, 'Microbial and Viral Pesticides', Marcel Dekker, New York, 1982.
6. L. Goldberg and J. Margalit, A bacterial spore demonstrating rapid larvicidal activity against *Anopheles sergenti, Uranataenia unguiculata, Culex univitatius, Aedes aegypti* and *Culex pipiens, Mosq. News*, 1977, **37**, 355.
7. A. Krieg, A. Huger, G. Langenbruch, and W. Schnetter, *Bacillus thuringiensis* var. *tenebrionis*: ein neuer, gegenuber Larven von Coleopteren wirksamer Pathotyp, *Z. Angew. Entomol.*, 1987, **96**, 500.
8. M. Obukowicz, F. Perlak, K. Kusano-Kretzmer, E. Mayer, and L. Watrud, Integration of the delta-endotoxin gene of *Bacillus thuringiensis* into the chromosome of root colonizing strains of pseudomonads using Tn5, *Gene*, 1986, **45**, 327.
9. W. Gelernter, Targeting insecticide-resistant markets, in 'Managing Resistance to Agrochemicals: From Fundamental Research to Practical Strategies', ed. M. B. Green, W. K. Moberg, and H. LeBaron, American Chemical Society, New York, 1990, 109.
10. M. Dimock, R. Beach, and P. Carlson, Endophytic bacteria for the delivery of crop protection agents, in 'Biotechnology, Biological Pesticides and Novel Plant–Pest Resistance for Insect Pest Management, ed. D. Roberts and R. Granados, Cornell University, Ithaca, 1989, p. 88.
11. R. Horsch, J. Fry, N. Hoffmann, D. Eichholtz, S. Rogers, and R. Fraley, A simple and general method for transferring genes into plants, *Science*, 1985, **227**, 1229.
12. F. Perlak, R. Fuchs, D. Dean, S. McPherson, and D. Fischhoff, Modification of the coding sequence enhances plant expression of insect control protein genes, *Proc. Natl. Acad. Sci. USA*, 1991, **88**, 3324.
13. H. Dulmage, O. Boening, C. Rehnborg, and G. Hansen, A proposed standardized bioassay for formulations of *Bacillus thuringiensis* based on the international unit, *J. Invert. Path.*, 1971, **18**, 240.
14. H. de Barjac and A. Bonnefoi, Essai de classification biochimique et serologique de 24 souches de *Bacillus* du type *B. thuringiensis Entomophaga*, 1962, **1**, 5.
15. H. de Barjac and A. Bonnefoi, Mise au point sur la classification des *Bacillus thuringiensis*, *Entomophaga*, 1973, **18**, 5.
16. H. de Barjac and E. Frachon, Classification of *Bacillus thuringiensis* strains, *Entomophaga*, 1990, **35**, 233.
17. H. de Barjac and M. Lecadet, Relations immunologiques des protéines toxiques extraits de trois souches de *Bacillus thuringiensis*, *C. R. Acad. Sci.*, 1961, **252**, 3160.
18. J. Krywienczyk and G. Bergold, Serological relationship between inclusion body proteins studied by means of agar diffusion method, *J. Insect Pathol.*, 1960, **2**, 118.

19. J. Krywienczyk and T. Angus, A serological comparison of the parasporal bodies of three insect pathogens, *J. Insect Pathol.*, 1960, **2**, 411.
20. J. Krywienczyk and T. Angus, A serological comparison of several crystalliferous insect pathogens. *J. Insect. Pathol.*, 1967, **9**, 126.
21. H. Dulmage *et al.*, Insecticidal activity of isolates of *Bacillus thuringiensis* and their potential for pest control, in 'Microbial Control of Pests and Plant Diseases 1970–1980', ed. H. D. Burgess, Academic Press, New York, 1981, p. 193.
22. H. de Barjac and F. Lemille, Presence of flagellar antigenic subfactors in serotype 3 of *Bacillus thuringiensis*, *J. Invert. Path.*, 1970, **15**, 139.
23. J. Krywienczyk, H. Dulmage, and P. Fast, Occurrence of two serologically distinct groups within *Bacillus thuringiensis* serotype 3 ab var. *kurstaki, J. Invert. Pathol.*, 1978, **31**, 372–375.
24. I. Pendleton and R. Morrison, Antigenic analysis of the digests of the crystal toxins of *Bacillus thuringiensis, J. Appl. Bacteriol.*, 1967, **30**, 402.
25. L. Baumann, K. Okamoto, B. Unterman, M. Lynch, and P. Baumann, Phenotypic characterization of *Bacillus thuringiensis* (Berliner) and *B. cereus* (Frankland and Frankland), *J. Invert. Path.*, 1984, **44**, 329.
26. M. Lynch and P. Baumann, Immunological comparisons of the crystal protein from strains of *Bacillus thuringiensis, J. Invert. Path.*, 1985, **46**, 47.
27. H. Höfte, J. Van Rie, S. Jansens, A. Van Houtven, H. Vanderbruggen, and M. Vaeck, Monoclonal antibody analysis and insecticidal spectrum of three types of Lepidopteran-specific insecticidal crystal proteins of *Bacillus thuringiensis, Appl. Environ. Microbiol.*, 1989, **54**, 2010.
28. H. Höfte and H. Whiteley, Insecticidal crystal proteins of *Bacillus thuringiensis, Microbiol. Rev.*, 1989, **53**, 242.
29. H. Schnepf and H. Whiteley, Cloning and expression of the *Bacillus thuringiensis*, crystal protein gene in *Escherichia coli, Proc. Natl. Acad. Sci. USA*, 1981, **78**, 2893.
30. F. Dardenne, J. Seurinck, B. Lambert, and M. Peferoen, Nucleotide sequence and deduced amino acid sequence of a *cryIa(c)* variant from *Bacillus thuringiensis. Nucleic Acids Res.*, 1990, **18**, 5546.
31. R. Tailor, J. Tippett, G. Gibb, S. Pells, D. Pike, L. Jordan, and S. Ely, Identification and characterization of a novel *Bacillus thuringiensis* δ-endotoxin entomocidal to coleopteran and lepidopteran larvae, *Mol. Microbiol.*, 1992, **6**, 1211.
32. J. Feitelson, J. Payne, and L. Kim, *Bacillus thuringiensis*: insects and beyond, *Bio/Technol.*, 1992, **10**, 271.
33. J. Gonzalez and B. Carlton, Patterns of plasmid DNA in crystalliferous and acrystalliferous strains of *Bacillus thuringiensis, Plasmid*, 1980, **3**, 92.
34. J. Gonzalez, H. Dulmage, and B. Carlton, Correlation between specific plasmids and δ-endotoxin production in *Bacillus thuringiensis, Plasmid*, 1981, **5**, 351.
35. J. Gonzalez and B. Carlton, A large transmissible plasmid is required for crystal toxin production in *Bacillus thuringiensis* variety *israelensis, Plasmid*, 1984, **11**, 28.
36. J. Baum and J. Gonzalez, Mode of replication, size, and distribution of naturally occurring plasmids in *Bacillus thuringiensis, FEMS Microbiol. Let.*, 1992, **96**, 143.
37. C. Ash, J. Farrow, M. Dorsch, E. Stackebrandt, and M. Collins, Comparative analysis of *Bacillus anthracis, Bacillus cereus*, and related species on the basis of reverse transcriptase sequencing of 16S rRNA, *Int. J. Syst. Bacteriol.*, 1991, **41**, 343.

38. R. Brousseau, A. Saint-Onge, G. Préfontaine, L. Masson, and J. Cabana, Arbitrary primer polymerase chain reaction, a powerful method to identify *Bacillus thuringiensis* serovars and strains, *Appl. Environ. Microbiol.*, 1993, **59**, 114.
39. V. Miteva, A. Abadjieva, and R. Grigorova, Differentiation among strains and serotypes of *Bacillus thuringiensis* by M13 DNA fingerprinting, *J. Gen. Microbiol.*, 1991, **137**, 593
40. M. Starkey, T. Fenning, M. Davey, and B. Mulligan, Design and use of synthetic oligonucleotide probes in the cloning of δ-endotoxin genes from *Bacillus thuringiensis, Enzyme Microbiol. Technol.*, 1991, **13**, 661.
41. S. Bourque, J. Valero, J. Mercier, M. Lavoie, and R. Levesque, Multiplex polymerase chain reaction for detection and differentiation of the microbial insecticide *Bacillus thuringiensis, Appl. Environ. Microbiol.*, 1993, **59**, 523.
42. G. Préfontaine, P. Fast, P. Lau, M. Heffords, Z. Hanna, and R. Brousseau, Use of oligonucleotide probes to study relatedness of delta-endotoxins genes among *Bacillus thuringiensis* subspecies and strains, *Appl. Environ. Microbiol.*, 1987, **53**, 2808.
43. N. Carozzi, V. Kramer, G. Warren, S. Evola, and M. Koziel, Prediction of insecticidal activity of *Bacillus thuringiensis* strains by polymerase chain reaction product profiles, *Appl. Environ. Microbiol.*, 1991, **57**, 3057.
44. S. Kalman, K. Kiehne, J. Libs, and T. Yamamoto, Cloning of a novel *cryIC*-type gene from a strain of *Bacillus thuringiensis* subsp. *galleriae*, *Appl. Environ. Microbiol.*, 1993, **59**, 1131.
45. C. Ash, J. Farrow, S. Wallbanks, and M. Collins, Phylogenetic heterogeneity of the genus Bacillus revealed by comparative sequence analysis of small-subunit-ribosomal RNA sequences, *Lett. Appl. Microbiol.*, 1991, **13**, 202.
46. S. Orduz, W. Rojas, M. Correa, A. Montoya, and H. de Barjac, A new serotype of *Bacillus thuringiensis* from Colombia toxic to mosquito larvae, *J. Invert. Path.*, 1992, **59**, 99.
47. C. Toumanoff and C. Vago, L'agent pathogène de la flacherie des vers à soie endemique dans la region des cevennes: *Bacillus cereus* var. *alesti* var. nov., *C. R. Acad. Sci.*, 1951, **233**, 1504.
48. M. Ohba and K. Aizawa, New flagella (H) antigenic subfactors in *Bacillus thuringiensis* H Serotype 3 with description of two new subspecies, *Bacillus thuringiensis* subsp. *sumiyoshiensis* (H serotype 3a:3d) and *Bacillus thuringiensis* subsp. *fukuokaensis* (H Serotype 3a:3d:3e), *J. Invert. Path.*, 1989, **54**, 208.
49. L. Padua, M. Ohba, and K. Aizawa, Isolation of a *Bacillus thuringiensis* strain (serotype 8a:8b) highly and selectively toxic against mosquito larvae, *J. Invert. Path.*, 1984, **44**, 12.
50. L. Padua, M. Ohba, and K. Aizawa, The isolation of *Bacillus thuringiensis* serotype 10 with a highly preferential toxicity to mosquito larvae, *J. Invert. Path.*, 1980, **36**, 180.
51. M. Ohba and K. Aizawa, A new subspecies of *Bacillus thuringiensis* possessing 11a:11c flagellar antigenic structure: *Bacillus thuringiensis* subsp. *kyushuensis, J. Invert. Path.*, 1979, **33**, 387.
52. D. Cidaria, A. Cappai, A. Vallesi, V. Caprioli, and G. Pirali, A novel strain of *Bacillus thuringiensis* (NCIMB 40152) active against coleopteran insects, *FEMS Microbiol. Lett.*, 1991, **81**, 129.
53. K. Bottjer, L. Bone, and S. Gill, Nematoda: Susceptibility of the egg to *Bacillus thuringiensis* toxins, *Exp. Parasit.*, 1985, **60**, 239.

54. D. Edwards, J. Payne, and G. Soares, Novel isolates to *Bacillus thuringiensis* having activity against nematodes, Mycogen Corporation, Europe Pat. Appl. No. 0 303 426 B1 8 Aug 1988.
55. M. Thompson and F. Gaertner, Novel *Bacillus thuringiensis* isolate having anti-protozoan activity, Mycogen Corporation, Europe Pat. Appl. No. 0 461 799 A2, 6 Apr 1991.
56. J. Payne, R. Cannon, and A. Bagley, Novel *Bacillus thuringiensis* isolates for controlling acarides, Mycogen Corporation, WO 92/19106, International filing date: 30 Apr 1992.
57. J. Payne, M. Kennedy, J. Randall, H. Meier, and H. Uick, Novel *Bacillus thuringiensis* isolates active against hymenopteran pest and gene(s) encoding hymenopteran-active toxins, Mycogen Corporation, Eur. Pat. Appl. No. 0 516 306 A2, 5 Dec 1992.
58. J. Drummond, D. Miller and D. Pinnock, Toxicity of *Bacillus thuringiensis* against *Damalinia ovis* (Phthiraptera: Mallophaga), *J. Invert. Path.*, 1992, **60**, 102.
59. T. Angus, The reaction of certain lepidopterous and hymenopterous larvae in *Bacillus soto* toxin, *Canad. Entomol.*, 1956, **88**, 280.
60. H. Dulmage, H. Graham, and E. Martinez, Interactions between the tobacco budworm, *Heliothis virescens*, and the delta-endotoxin produced by the HD-1 isolate of *Bacillus thuringiensis* var. *kurstaki*: relationship between length of exposure to the toxin and survival, *J. Invert. Path.*, 1978, **32**, 40.
61. A. Heimpel and T. Angus, The site of action of crystalliferous bacteria in Lepidoptera larvae, *J. Insect Path.*, 1959, **1**, 152.
62. J. Nishiitsutsuji-Uwo and Y. Endo, Mode of action of *Bacillus thuringiensis* delta-endotoxin: general characteristics of intoxicated *Bombyx* larvae, *J. Invert. Path.*, 1980, **36**, 219.
63. J. Percy and P. Fast, *Bacillus thuringiensis* crystal toxin: ultrastructural studies of its effect on silkworm midgut cells, *J. Invert. Path.*, 1983, **41**, 86.
64. E. deLello, W. Hanton, S. Bishoff, and D. Misch, Histopathological effects of *Bacillus thuringiensis* on the midgut of tobacco hornworm larvae (*Manduca sexta*): low doses compared with fasting, *J. Invert. Path.*, 1984, **43**, 169.
65. H. Bietlot, I. Vishnubhatla, P. Carey, M. Pozsgay, and H. Kaplan, Characterization of the cysteine residues and disulfide linkages in the protein crystal of *Bacillus thuringiensis* subsp. *kurstaki* and *entomocidus, Biochem. J.*, 1990, **267**, 309.
66. A. Tojo and K. Aizawa, Dissolution and degradation of *Bacillus thuringiensis* δ-endotoxin by gut juice protease of the silkworm *Bombyx mori., Appl. Environ. Microbiol.*, 1983, **45**, 576.
67. K. Ogiwara, L. Indrasith, S. Asano, and H. Hori, Processing of δ-endotoxin from *Bacillus thuringiensis* subsp. *kurstaki* HD-1 and HD-73 by gut juices of various insect larvae, *J. Invert. Path.*, 1992, **60**, 121.
68. R. Fast, The crystal toxin of *Bacillus thuringiensis*, in 'Microbial Control of Pests and Plant Diseases 1970–1980', ed. H. D. Burgess, Academic Press, New York. 1981, p. 223.
69. B. Lambert, H. Höfte, K. Annys, S. Jensens, P. Soetaert, and M. Peferoen, Novel *Bacillus thuringiensis* insecticidal crystal protein with a silent activity against coleopteran larvae, *Appl. Environ. Microbiol.*, 1992, **58**, 2536.
70. V. Wigglesworth, Digestion and nutrition, in 'The Principles of Insect Physiology', Chapman and Hall, London, 7th edn, 1972, p. 476.
71. W. Terra, Physiology and biochemistry of insect digestion: An evolutionary perspective, *Braz. J. Med Biol. Res.*, 1988, **21**, 675.

72. A. Bravo, S. Jansens, and M. Peferoen, Immunocytochemical localization of *Bacillus thuringiensis* insecticidal crystal proteins in intoxicated insects, *J. Invert. Path.*, 1992, **60**, 237.
73. A. Bravo, K. Hedrickx, S. Jansens, and M. Peferoen, Immunocytochemical analysis of specific binding of *Bacillus thuringiensis* insecticidal crystal proteins to lepidopteran and coleopteran midgut membranes, *J. Invert. Path.*, 1992, **60**, 247.
74. P. Fast, Laboratory bioassays of mixtures of *Bacillus thuringiensis* and chitinase, *Can. Ent.*, 1978, **110**, 201.
75. R. Granados and B. Corsaro, Baculovirus enhancing proteins and their implication for insect control, in 'Proceedings of the Fifth International Colloquium on Invertebrate Pathology and Microbial Control', ed. D. E. Pinnock, Society of Invertebrate Pathology, Adelaide, Australia, 1990, p. 174.
76. C. Hofmann, P. Lüthy, R. Hütter, and V. Pliska, Binding of the delta-endotoxin from *Bacillus thuringiensis* to brush-border membrane vesicles of the cabbage butterfly (*Pieris brassicae*), *Eur. J. Biochem.*, 1988, **173**, 85.
77. B. Knowles and D. Ellar, Characterization and partial purification of a plasma membrane receptor for *Bacillus thuringiensis* var. *kurstaki* lepidopteran-specific δ-endotoxin, *J. Cell Sci.*, 1986, **83**, 89.
78. C. Hofmann, H. Vanderbruggen, H. Höfte, J. Van Rie, S. Jansens, and H. Van Mellaert, Specificity of *Bacillus thuringiensis* δ-endotoxins is correlated with the presence of high-affinity binding sites in the brush border membrane of target insect midguts, *Proc. Natl. Acad. Sci. USA*, 1988, **85**, 7844.
79. J. Van Rie, S. Jansens, H. Höfte, D. Degheele, and H. Van Mellaert, Specificity of *Bacillus thuringiensis* δ-endotoxins, *Eur. J. Biochem.*, 1989, **186**, 239.
80. J. Van Rie, S. Jansens, H. Höfte, D. Degheele, and H. Van Mellaert, Receptors on the brush border membrane of the insect midgut as determinants of the specificity of *Bacillus thuringiensis* δ-endotoxins, *Appl. Environ. Microbiol.*, 1990, **56**, 1378.
81. M. Wolfersberger, The toxicity of two *Bacillus thuringiensis* δ-endotoxins to gypsy moth larvae is inversely related to the affinity of binding sites on midgut brush border membranes for the toxins, *Experientia*, 1990, **46**, 475.
82. S. Garczynski, J. Crim, and M. Adang, Identification of putative insect brush border membrane-binding molecules specific to *Bacillus thuringiensis* δ-endotoxin by protein blot analysis, *Appl. Environ. Microbiol.*, 1991, **57**, 2816.
83. J. Van Rie, W. McGaughey, D. Johnson, B. Barnett, and H. Van Mellaert, Mechanism of insect resistance to the microbial insecticide *Bacillus thuringiensis, Science*, 1990, **247**, 72.
84. S. MacIntosh, T. Stone, R. Jokerst, and R. Fuchs, Binding of *Bacillus thuringiensis* proteins to a laboratory-selected line of *Heliothis virescens, Proc. Natl. Acad. Sci. USA*, 1991, **88**, 8930.
85. J. Ferré, M. D. Real, J. Van Rie, S. Jansens, and M. Peferoen, Resistance to the *Bacillus thuringiensis* bioinsecticide in a field population of *Plutella xylostella* is due to a change in a midgut membrane receptor, *Proc. Natl. Acad. Sci. USA*, 1991, **88**, 5119.
86. F. Gould, A. Martinez-Ramirez, A. Anderson, J. Ferré, F. Silva, and W. Moar, Broad-spectrum resistance to *Bacillus thuringiensis* toxins in *Heliothis virescens, Proc. Natl. Acad. Sci. USA*, 1992, **89**, 7986.
87. V. Sacchi, P. Parenti, G. Hanozet, B. Giordana, P. Lüthy, and M. Wolfersberger, *Bacillus thuringiensis* toxin inhibits potassium-gradient-dependent amino-acid

transport across the brush border membrane of *Pieris brassicae* midgut cells, *FEBS Lett.*, 1986, **204**, 213.

88. B. Giordana, V. Sacchi, P. Parenti, and G. Hanozet, Amino acid transport systems in intestinal brush-border membranes from lepidopteran larvae, *Am. J. Physiol.*, 1989, **257**, R494.
89. D. Crawford and W. Harvey, Barium and calcium block *Bacillus thuringiensis* subsp. *kurstaki* δ-endotoxin inhibition of potassium current across isolated midgut of larval *Manduca sexta, J. Exp. Biol.*, 1988, **137**, 277.
90. S. Slatin, C. Abrams, and L. English, Delta-endotoxin for cation-selective channels in planar lipid bilayers, *Biochem. Biophys. Res. Commun.*, 1990, **169**, 765.
91. L. English, T. Readdy, and A. Bastian, Delta-endotoxin-induced leakage of $^{86}Rb^+$–K^+ and H_2O from phospholipid vesicles is catalyzed by reconstituted midgut membrane, *Insect Biochem.*, 1991, **21**, 177.
92. B. Knowles and D. Ellar, Colloid-osmotic lysis is a general feature of the mechanism of action of *Bacillus thuringiensis* δ-endotoxins with different insect specificity, *Biochim. Biophys. Acta.*, 1987, **924**, 509.
93. J.-L. Schwartz, L. Garneau, L. Masson, and R. Brousseau, Early response of cultured lepidopteran cells to exposure to δ-endotoxin from *Bacillus thuringiensis*: involvement of calcium and anionic channels, *Biochim. Biophys. Acta*, 1991, **1065**, 250.
94. M. Haider and D. Ellar, Analysis of the molecular basis of insecticidal specificity of *Bacillus thuringiensis* crystal δ-endotoxin, *Biochim. J.*, 1987, **248**, 197.
95. S. McPherson, F. Perlak, R. Fuchs, P. Marrone, P. Lavrik, and D. Fischhoff, Characterization of the coleopteran-specific protein gene ot *Bacillus thuringiensis* var. *tenebrionis, Bio/Technol.*, 1988, **6**, 61.
96. C. Koller, L. Bauer, and R. Hollingworth, Characterization of the pH-mediated solubility of *Bacillus thuringiensis* var. *san diego* native δ-endotoxin crystals, *Biochem. Biophys. Res. Commun.*, 1992, **184**, 692.
97. L. Murdock, G. Brookhart, P. Dunn, D. Foard, S. Kelley, L. Kitch, R. Shade, R. Shukle, and J. Wolfson, Cysteine digestive proteinases in coleoptera, *Comp. Biochem. Physiol., B: Comp. Biochem.*, 1987, **87**, 783.
98. S. MacIntosh, S. McPherson, F. Perlak, P. Marrone, and R. Fuchs, Purification and characterization of *Bacillus thuringiensis* subsp. *tenebrionis* insecticidal proteins produced by *E. coli, Biochem. Biophys. Res. Commun.*, 1990, **170**, 665.
99. L. Bauer and H. Pankratz, Ultrastructural effects of *Bacillus thuringiensis* var. *san diego* on midgut cells of the cottonwood leaf beetle, *J. Invert. Path.*, 1992, **60**, 15.
100. A. Slaney, H. Robbins, and L. English, Mode of action of *Bacillus thuringiensis* toxin CryIIIA: An analysis of toxicity in *Leptinotarsa decemlineata* (Say) and *Diabrotica undecimpunctata howardi* Barber, *Insect Biochem. Mol. Biol.*, 1992, **22**, 9.
101. G. Couche, M. Pfannenstiel, and K. Nickerson, Parameters affecting attachment of *Bacillus thuringiensis* var. *israelensis* toxin to latex beads, *Appl. Microbiol. Biotechnol.*, 1986, **24**, 128.
102. A. Delecluse, C. Bourgouin, A. Klier, and G. Rapoport, Specificity of action on mosquito larvae of *Bacillus thuringiensis israelensis* toxins encoded by two different genes. *Mol. Gen. Genet.*, 1988, **214**, 42.
103. G. Singh, L. Schouest, and G. Sill, The toxic action of *Bacillus thuringiensis* var. *israelensis* in *Aedes aegypti in vivo, Pest Biochem. Physiol.*, 1986, **26**, 36.

104. B. Knowles, M. Blatt, M. Tecster, J. Horsnell, J. Carroll, G. Menestrina, and D. Ellar, A cytolytic δ-endotoxin from *Bacillus thuringiensis* var. *israelensis* forms cation-specific channels in planar lipid bilayers, *FEBS Lett.*, 1989, **244**, 259.
105. E. Davidson, Binding of the *Bacillus sphaericus* (Eubacteriales: Bacillaceae) toxin to midgut cells of mosquito (Diptera: Culicidae) larvae: relation to host range, *J. Med. Entomol.*, 1988, **25**, 151.
106. E. Davidson, Variation of binding of *Bacillus sphaericus* toxin and wheat germ agglutinin to larval midgut cells of six species of mosquitoes, *J. Invert. Path.*, 1989, **53**, 251.
107. C. Nielsen-Leroux and J.-F. Charles, Binding of *Bacillus sphaericus* binary toxin to a specific receptor on midgut brush-border membranes from mosquito larvae, *Eur. J. Biochem.*, 1992, **210**, 585.
108. C. Choma, W. Surewicz, P. Carey, M. Pozsgay, T. Ratnor, and H. Kaplan, Unusual properties of the protoxin and toxin from *Bacillus thuringiensis*: structural implications, *Eur. J. Biochem.*, 1990, **189**, 523.
109. J. Carroll, J. Li, and D. Ellar, Proteolytic processing of a Coleopteran-specific delta-endotoxin produced by *Bacillus thuringiensis* subsp. *tenebrionis, Biochem. J.*, 1989, **261**, 99.
110. C. Choma, W. Surewicz, P. Carey, M. Pozsgay, and H. Kaplan, Secondary structure of the entomocidal toxin from *Bacillus thuringiensis* subsp. *kurstaki* HD73, *J. Prot. Chem.*, 1990, **9**, 87.
111. H. Schnepf and H. Whiteley, Delineation of a toxin-encoding segment of a *Bacillus thuringiensis* crystal protein gene, *J. Biol. Chem.*, 1985, **260**, 6273.
112. D. Convents, C. Houssier, I. Lasters, and M. Lauwerey, The *Bacillus thuringiensis* delta-endotoxin: Evidence for a two domain structure of the minimal toxic fragment, *J. Biol. Chem.*, 1990, **265**, 1369.
113. D. Convents, M. Cherlet, J. Van Damme, I. Lasters, and M. Lauwerey, Two structural domains as a general fold of the toxic fragment of the *Bacillus thuringiensis* delta-endotoxins, *Eur. J. Biochem.*, 1991, **195**, 631.
114. W. Widner and H. Whiteley, Location of the Diptera specificity region in a Lepidopteran–Dipteran crystal protein from *Bacillus thuringiensis, J. Biol. Chem.*, 1990, **172**, 2826.
115. E. Ward, D. Ellar, and C. Chilcott, Single amino acid changes in the *Bacillus thuringiensis* subsp. *israelensis* delta-endotoxin affect the toxicity and expression of the protein, *J. Mol. Biol.*, 1988, **202**, 527.
116. A. Ge, N. Shivarova, and D. Dean, Location of the *Bombyx mori* specificity domain on a *Bacillus thuringiensis* delta-endotoxin protein, *Proc. Natl. Acad. Sci. USA*, 1989, **86**, 4037.
117. M. Lee, R. Milner, A. Ge, and D. Dean, Location of a *Bombyx mori* receptor binding region of a *Bacillus thuringiensis* delta-endotoxin, *J. Biol. Chem.*, 1992, **267**, 3115.
118. D. Wu and A. Aronson, Localized mutagenesis defines regions of the *Bacillus thuringiensis* delta-endotoxin involved in toxicity and specificity, *J. Biol. Chem.*, 1992, **267**, 2311.
119. B. Almond and D. Dean, Suppression of protein destabilizing mutations in *Bacillus thuringiensis* delta-endotoxins by second site mutations, *Biochemistry*, 1993, **32**, 1040.
120. H. Arvidson, P. Dunn, S. Strand, and A. Aronson, Specificity of *Bacillus thuringiensis* for Lepidopteran larvae: Factors involved *in vivo* and in the structure of a purified protoxin, *Mol. Microbiol.*, 1989, **3**, 1533.

121. M. Pfannenstiel, G. Couche, G. Muthukumar, and K. Nickerson, Stability of the larvicidal activity of *Bacillus thuringiensis* subsp. *israelensis*: amino acid modification and denaturants, *Appl. Environ. Microbiol.*, 1985, **50**, 1196.
122. C. Choma and H. Kaplan, *Bacillus thuringiensis* crystal protein: Effect of chemical modification of the cysteine and lysine residues, *J. Invert. Path.*, 1992, **59**, 75.
123. A. Aronson, E.-S. Han, W. McGaughey, and D. Johnson, The solubility of inclusion proteins from *Bacillus thuringiensis* is dependent upon protoxin composition and is a factor in toxicity to insects, *Appl. Environ. Microbiol.*, 1991, **57**, 981.
124. J. Li, J. Carroll, and D. Ellar, Crystal structure of insecticidal delta-endotoxin from *Bacillus thuringiensis* at 2.5 Å resolution, *Nature (London)*, 1991, **353**, 815.
125. H. Whiteley and H. Schnepf, The molecular biology of parasporal crystal body formation in *Bacillus thuringiensis, Ann. Rev. Microbiol.*, 1986, **40**, 549.
126. H. Wong, H. Schnepf, and H. Whiteley, Transcriptional and translational start sites for the *Bacillus thuringiensis* crystal protein gene, *J. Biol. Chem.*, 1983, **258**, 1960.
127. H. Schnepf, H. Wong, and H.Whiteley, Expression of a cloned *Bacillus thuringiensis* crystal protein gene in *Escherichia coli, J. Bacteriol.*, 1987, **169**, 4110.
128. S. Minnich and A. Aronson, Regulation of protoxin synthesis in *Bacillus thuringiensis, J. Bacteriol.*, 1984, **158**, 447.
129. H. Rothnie and M. Geiser, Differential expression of the 3 δ-endotoxin genes in *Bt* subsp. *kurstaki* HD1, *UCLA Symp. Mol. Cell. Biol. New Ser.*, 1990, **112**, 599.
130. A. Pang and B. Mathieson, Peptide mapping of different *Bacillus thuringiensis* toxin gene products by CnBr cleavage in SDS-PAGE gels, *J. Invert. Path.*, 1991, **57**, 82.
131. L. Masson, G. Préfontaine, L. Péloquin, P. Lau, and R. Brousseau, Comparative analysis of the individual protoxin components in P1 crystals of *Bt* subsp. *kurstaki* isolates NRD-12 and HD-1, *Biochem J.*, 1989, **269**, 507.
132. M. Lecadet and R. Dedonder, Biogenesis of the crystalline inclusion of *Bacillus thuringiensis* during sporulation, *Eur. J. Biochem.*, 1971, **23**, 282.
133. H. Whiteley, H. Schnepf, J. Kronstad, and H.Wong, Structural and regulatory analysis of a cloned *Bacillus thuringiensis* crystal protein gene, in 'Genetics and Biotechnology of Bacilli', ed. A. Ganeson and J. Hoch, Academic Press, New York, 1984, p. 375.
134. K. Brown and H. Whiteley, Isolation of a *Bacillus thuringiensis* RNA polymerase capable of transcribing crystal protein genes, *Proc. Natl. Acad. Sci. USA*, 1988, **85**, 4166.
135. K. Brown and H. Whiteley, Isolation of the second *Bacillus thuringiensis* RNA polymerase that transcribes from a crystal protein gene promoter, *J. Bacteriol.*, 1990, **172**, 6682.
136. L. Adams, K. Brown, and H. Whiteley. Molecular cloning and characterization of two genes encoding sigma factors that direct transcription from a *Bacillus thuringiensis* crystal protein gene promoter, *J. Bacteriol.*, 1991, **173**, 3846.
137. K. McLean and H. Whiteley, Expression in *Escherichia coli* of a cloned crystal protein gene of *Bacillus thuringiensis* subsp. *israelensis*, *J. Bacteriol.*, 1987, **169**, 1017.
138. L. Adams, J. Visick, and H. Whiteley, A 20 kilodalton protein is required for efficient production of the *Bacillus thuringiensis* subsp. *israelensis* 27 kilodalton crystal protein in *Escherichia coli, J. Bacteriol.*, 1989, **171**, 521.

139. J. Visick and H. Whiteley, Effect of a 20 kilodalton protein from *Bacillus thuringiensis* subsp. *israelensis* on production of the CytA protein by *Escherichia coli, J. Bacteriol.*, 1991, **173**, 1748.
140. C. Chang, Y.-M. Yu, S.-M. Dai, S. Law, and S. Gill, High-level *cryIVD* and *cytA* gene expression in *Bacillus thuringiensis* does not require the 20 kilodalton protein and the coexpressed gene products are synergistic in their toxicity to mosquitoes, *Appl. Environ. Microbiol.*, 1993, **59**, 815.
141. D. Wu and B. Federici, A 20 kDa protein preserves cell viability and promotes CytA crystal formation during sporulation in *Bacillus thuringiensis*, *J. Bacteriol.*, 1993, **175**, 5276.
142. P. Koni and D. Ellar, Cloning and characterization of a novel *Bacillus thuringiensis* cytolytic delta-endotoxin, *J. Mol. Biol.*, 1993, **229**, 319.
143. W. Widner and H. Whiteley, Two highly related crystal proteins of *Bacillus thuringiensis* subsp. *kurstaki* possess different host range specificities, *J. Bacteriol.*, 1989, **171**, 965.
144. N. Crickmore and D. Ellar, Involvement of a possible chaperonin in the efficient expression of a cloned CryIIA δ-endotoxin gene in *Bacillus thuringiensis, Mol. Microbiol.*, 1992, **6**, 1533.
145. F. Gould, Evolutionary biology and genetically engineered crops, *BioScience*, 1988, **38**, 26.
146. B. Tabashnik and R. Roush, Introduction, in 'Pesticide Resistance in Arthropods', ed. R. Roush and B. Tabashnik, Chapman and Hall, New York, 1990.
147. P. Marrone and S. MacIntosh, Resistance to *Bacillus thuringiensis* and resistance management, in 'Bacillus thuringiensis. An Environmental Biopesticide: Theory and Practice', ed. P. Entwistle, J. Cory, M. Bailey, and S. Higgs, John Wiley & Sons, Chichester, 1993, p. 221.
148. P. Marrone and S. MacIntosh, Insect resistance to biotechnology products: An overview of research and possible management strategies, in 'Resistance '91: Achievements and Developments in Combating Pesticide Resistance', ed. I. Denholm, A. Devonshire, and D. Hollomon, Elsevier Applied Science, London, 1992, p. 272.
149. W. Gelertner, Targeting insecticide-resistant markets: New developments in microbial-based products, in 'Managing Resistance to Agrochemicals', ed. M. Green, H. LeBaron, and W. Moberg, American Chemical Society, Washington, DC, 1990, p. 105.
150. W. McGaughey, and M. Whalon, Managing insect resistance to *Bacillus thuringiensis* toxins, *Science*, 1992, **258**, 1451.

CHAPTER 11

*Natural and Engineered Viral Agents for Insect Control**

D. J. LEISY and J. R. FUXA

11.1 Introduction

Our current over-reliance on chemical insecticides causes a threat to ground and surface water, non-target organisms, and human safety, and creates insect resistance and secondary pest problems as well. More and more chemical insecticides are being removed from the market for various reasons, such as environmental concerns or lack of incentive for re-registration.[1] Additionally, certain methods of 'microbial control' of insects have the potential to reduce inputs at the farm or producer level, such as labour costs of pest control agents, and use of petroleum-based products.[2]

Interest in controlling insect populations with viruses, which dates back at least to the early 1890s,[3] is currently greater than ever. The major reason for research and development of viruses, as well as of other types of entomopathogens (bacteria, fungi, protozoa, and nematodes), is their environmental safety. In this current climate of environmental concerns coupled with the need for insect control to maintain human health and well-being, the entomopathogenic viruses have important assets.[4] Their narrow host ranges – baculoviruses, the major group of interest, usually are infective only to one genus of insect – make them safe to all non-target organisms, including humans, animals, plants, and even beneficial insects. They are capable of causing natural epizootics and infecting a large proportion of insects in a population. Baculoviruses replicate rapidly and cause extensive cell and tissue destruction in the host insect, and, thus, are virulent mortality agents.

Many entomopathogenic viruses induce the host cell to produce a proteinaceous crystal in which the virions become embedded. This crystal gives these

*Approved for publication by the Director of the Louisiana Agricultural Experiment Station as manuscript number 93-17-7110. This is Technical Report no. 101196 from the Oregon State University Agricultural Experimental Station.

viruses great persistence in water and other carriers (important for formulation of viral pesticides), as well as in soil (important for long-term approaches to insect control). Other advantages of viruses include their relative compatibility with other control methods (including chemical insecticides and biological control with parasitoids) and the low cost of development and registration compared to chemical insecticides.

Another major feature of viruses is the versatility in the approaches by which they can be used for insect control. There are four such approaches.[2]

- The microbial insecticide approach uses much of the same technology as for chemical insecticides; large quantities of the virus are produced and released for quick suppression of the pest population. Residual effects are not significant; thus, any subsequent increase in the pest population to damaging levels requires another release of the pathogen.
- The seasonal colonization approach is similar to a 'booster shot'; the release results in replication of the virus and suppression of more than one pest generation. Seasonal colonization may or may not be aimed at immediate knockdown of the pest population, and subsequent viral releases are required, usually once in each new growing season.
- The introduction–establishment approach is sometimes called 'classic biocontrol'; a viral 'species' or strain is released in an area where it did not occur previously. The virus becomes a permanent part of the ecosystem in which it is released, resulting in permanent pest population suppression.
- The final approach is environmental manipulation or conservation. This is the only approach in which virions are not actually released into the environment, although this approach can supplement the release-type approaches. Instead, the usual farming or resource management practices are altered to enhance a viral population that is already present, thereby reducing the pest population without significant interference to the management of the resource, whether agricultural or otherwise.

Unfortunately, insect viruses also have numerous shortcomings which have prevented widespread use of these agents in pest control.[4] Several of these problems are targets for improvement by recombinant-DNA and formulation technologies. The dependence of viruses on host-cell biochemistry for replication necessitates their mass production in living insects using current technology, decreasing their cost-competitiveness in the insecticide market. Although often considered an advantage, their relatively narrow host ranges can be viewed as a disadvantage as well, limiting market size and effectiveness in pest complexes. Another major weakness is the time required for viral disease to debilitate or kill the host insect. Before succumbing to an infection, the insects normally continue to feed for several days. Unlike the situation with rapidly acting chemical insecticides, the user of a virus insecticide sees continuing crop damage following application before the pest is finally eradicated. Another shortcoming of viruses can be termed 'maturation immunity' in the host insect. Host insects become increasingly refractive to infection as they age; once the caterpillars, or larvae, grow beyond mid-size, generally very large doses of virus are required to initiate infection. Finally, the proteinaceous crystals do not protect virions from sunlight, resulting in only short viral persistence and a small 'window of opportunity' for the insect to become infected on crop plants.

In the remainder of this chapter, the biology of the entomopathogenic viruses, their uses in the microbial control of insects, and the role of the recombinant-DNA revolution in the development of new viral insecticides are discussed. Some of the safety and regulatory issues relevant to their use is also explored.

11.2 Viral Diseases of Insects

Viral diseases have been characterized in more than 600 species of insects.[5] The most important groups of insect viruses are listed in Table 11.1. In addition to those listed, there are a number of viruses that replicate in insects, but which are considered non-pathogenic for their insect hosts. For example, aphids, whiteflies, and certain beetles carry and spread plant viruses from plant to plant, and in some cases a limited amount of viral replication occurs within the insect vector without causing disease symptoms.[6]

Of all the insect viruses that have been characterized, only the baculoviruses have been given serious consideration for development into microbial insecticides. Some of the characteristics listed in Table 11.1 have been used to determine the suitability of baculoviruses for biocontrol purposes.

For example, in terms of safety it is considered prudent not to select viruses from families that contain members that infect vertebrates. Most of the viruses listed, with the exceptions of the baculoviruses, polydnaviruses, and ascoviruses, have near relatives that are vertebrate viruses, or belong to groups that are suspected of containing members capable of infecting vertebrates. The ascoviruses, however, are usually considered unsuitable for development as microbial insecticides because they show little infectivity *per os* (by mouth).[7]

Table 11.1 *Major classes of insect viruses*

Virus	*Occlusion bodies*	*Genome*	*Shape*	*Infects vertebrates?*
Baculoviruses	+	ds DNA	Rod	No
Polydnaviruses	–	ds DNA	Spherical	No
Ascoviruses	–	ds DNA	Bacilliform	No
Nudalerelia B virus group	–	ss RNA	Icosahedral	?
Noda viruses	–	ss RNA	Icosahedral	Yes
Iridoviruses	–	ds DNA	Icosahedral	Yes
Poxviruses	+	ds DNA	Spherical	Yes
Reoviruses	+	ds RNA	Icosahedral	Yes
Parvoviruses	–	ss DNA	Icosahedral	Yes
Rhabdoviruses	–	ss RNA	Bacilliform	Yes
Many small RNA viruses	–	RNA	Icosahedral	Yes

Likewise, the polydnaviruses have little potential as biocontrol agents since they have been found to replicate only in the ovaries of parasitic wasps, where they cause no apparent pathology. These viruses are injected along with the parasitic egg into a host larvae during oviposition. Hence, they apparently 'infect' cells, but are not replicated; their DNA genomes are transcribed and translated into proteins which may cause physiological changes to the parasitized larvae, such as immunosuppression[8,9] and alteration of hormone levels.[10] These effects are thought to condition the larvae to provide an optimized environment for the development of the parasitic egg. It is unclear how polydnaviruses are transmitted, but the evidence indicates that transmission occurs vertically from mothers to daughters, possibly through a proviral form that is integrated into the wasp genome.[11]

On the other hand, the baculoviruses are infectious *per os* and exhibit efficient horizontal transmission. In addition, many are highly infectious for insects which are important pests of agriculture and forestry. Baculoviruses are also one of three insect virus families that have members which encase their virions in large proteinaceous crystals, known as occlusion bodies. This is a perceived benefit for an insecticidal virus since occlusion body formation is a naturally occurring process of microencapsulation that protects viruses from degradation in the open environment.[12–14]

For the development of a viral insecticide it is desirable to choose a virus with a narrow host range in order to prevent the infection and elimination of beneficial insects along with the targeted species. Historically, the host range of a particular baculovirus species has been thought to be fairly restricted, ranging from a single species to 10 or more families of insects within an insect order.[15] There is limited evidence, however, which suggests that one species of baculovirus, usually thought of as specific for a lepidopteran insect, may be capable of infecting insects from other insect orders.[16,17] While viruses with very narrow host ranges are desirable for safety considerations, because of the high cost of development and registration, viruses with host ranges that include a fairly large number of pest species are often regarded as having more potential for commercialization.

11.2.1 Baculovirus Biology

11.2.1.1 Classification and Nomenclature

The baculoviridae are a family of arthropod-specific viruses which have large rod shaped virions ensheathed within membranous envelope structures. They contain large, circular, double-stranded DNA genomes ranging in size from approximately 80 to 160 kilobase pairs.[18–21]

Taxonomically, baculoviruses are divided into three subgroups based on morphological criteria. Subgroup A, the nuclear polyhedrosis viruses (NPVs) form large occlusion bodies (1–15 μm in diameter), called polyhedra or polyhedral occlusion bodies (PIB), and many virus particles are encased within each crystal. The NPVs are further subdivided into the single nucleocapsid types (SNPVs), containing only one nucleocapsid per viral envelope, and the multi-nucleocapsid types (MNPVs), which contain up to 29 nucleocapsids within each viral envelope.[22] Subgroup B, the granulosis viruses (GVs), form much smaller occlusion bodies (0.1–0.5 μm in diameter), termed granula, and contain only one virus particle per crystal.[23] A small number of baculoviruses are classified into a third subgroup (subgroup C), termed the nonoccluded viruses (NOVs), which do not form occlusion bodies.

The vast majority of baculovirus species infect the larval stages of insects from the order Lepidoptera (butterflies and moths). Some NPVs have also been found in the orders Hymenoptera (ants, bees, and wasps), Diptera (flies, gnats, and midges), Coleoptera (beetles), and Trichoptera (caddis flies). There are also occurrences of NPV infections of non-insect arthropods, including several species of shrimp[24] and mites.[5]

In general, baculoviruses are given names that identify their subgroup and the insect species from which they were first isolated. For instance, *Anagrapha falciphera* nuclear polyhedrosis virus (AfNPV) identifies an NPV first isolated from larvae of the celery looper, *Anagrapha falciphera*. This standard of nomenclature, although widespread, is somewhat unfortunate since many baculoviruses show varying degrees of virulence to a fairly wide variety of insect hosts (for example, AfNPV has been shown to be infectious for more than 30 insect species),[16] and the host species from which a particular baculovirus is initially isolated is not necessarily the one towards which that virus is most virulent.

11.2.1.2 Baculovirus Life Cycle

The major features of the baculovirus life cycle are depicted in Figure 11.1. A susceptible insect larva becomes infected when it consumes foliage or diet that

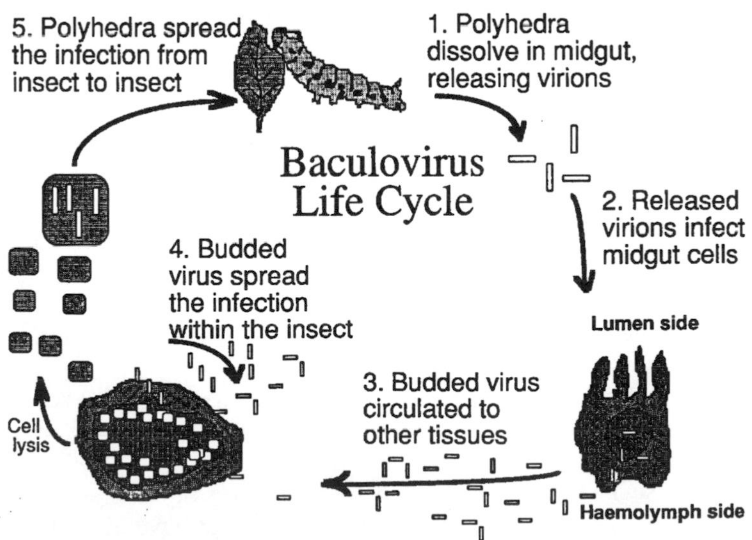

Figure 11.1 *Major features of the baculovirus life cycle*

is contaminated with occlusion bodies. In both NPVs and GVs, the occlusion bodies dissolve in the alkaline environment of the larval midgut. This releases the virus particles, which must then pass through the peritrophic membrane, a rigid barrier lining the lumen of the midgut, before invading midgut cells by fusion with microvilli.[25]

The nucleocapsids are transported to the nucleus, where uncoating followed by virus replication and maturation takes place. For some baculoviruses, such as an NPV infectious for the pine sawfly (*Neodiprion sertifer*)[26] and a GV infectious for the grape leaf skeletinizer (*Harrisina brillians*), [27] the virus infection appears limited to the midgut, and occluded progeny viruses are produced in midgut tissues. With most baculovirus species, however, the infected midgut cells do not produce occluded virions, but rather produce and release free progeny viruses, called extracellular viruses (ECV), into the haemolymph. The haemolymph then transports these virions to the remaining tissues of the insect.

In secondarily infected tissues, additional ECV are first produced and budded, and then later in infection another form of the virus is made that stays within the nucleus and becomes occluded. These virions are often referred to as polyhedra derived viruses (PDV). PDV are spread from insect to insect as the occlusion bodies become released into the environment by cell lysis and the subsequent decay of the cadaverous insect. An infected insect may produce more than 10^9 occlusion bodies before succumbing to the viral disease.

As previously mentioned, the infection process is relatively slow. Infected insect larvae often survive for periods ranging from 3 days to several weeks, depending on the virus and host species, and additional factors such as the dose, temperature, and developmental stage of the insects. This is a major drawback against the use of these viruses for controlling insects which damage crops with low economic thresholds for damage, such as cotton and apples.

11.2.1.3 Structure and Molecular Biology

There has been an explosion in our understanding of the molecular biology and biochemistry of insect baculoviruses during the past decade. Detailed knowledge has been gained about the structural components of baculoviruses, how their genomes are arranged and expressed, their evolutionary relationships, and the manner in which they interact with their hosts.[28] Our research efforts on baculoviruses have resulted in several practical benefits, including the development of an efficient and widely used system for the high level expression of genes, and improvements, via genetic engineering, of the insecticidal properties of these viruses.[29]

Most of the detailed genetic and biochemical characterizations of baculoviruses have been performed with subgroup A viruses; in particular, the *Autographa californica* multicapsid nuclear polyhedrosis virus (AcMNPV) has been a favourite for such studies. Most of the basic features of other subgroup A and subgroup B viruses are probably similar to those that have been described for AcMNPV, although many important aspects may vary, even between closely related baculovirus species. A number of insect cell lines are established that support the replication of AcMNPV and several other NPVs, and plaque assay systems are available which allow for the selection of mutant and recombinant viruses.[30] Further development of insect cell lines that support the growth of baculoviruses will greatly enhance our understanding of the diversity of these viruses. Winstanley reported recently on the development of insect cell lines that support the growth of GVs,[31] an important accomplishment that proved recalcitrant for many years.

Detailed restriction site maps have been published for AcMNPV, as well as for a number of other baculoviruses.[32] The entire genomes of AcMNPV and *Bombyx mori* MNPV have been sequenced,[33,34] although not all of the sequence information for the BmMNPV virus has become available to the public at the time of writing.

Baculovirus occlusion bodies consist of a matrix material surrounding the PDV virions that is produced mostly from the crystallization of a single virally encoded protein with molecular weights in the approximate range 28–31 kDa in different species. In NPVs, the matrix protein is termed polyhedrin, and in the case of GVs this protein is given the name granulin. Polyhedrins and granulins are structurally and functionally homologous proteins.[35] This polyhedrin or granulin matrix material, in turn, is surrounded by a mucopolysaccharide coat composed of protein and carbohydrates.[36,37]

Baculovirus genes have been divided into several categories based on the timing of their expression following infection.[32,38] The early genes are the first to be transcribed; their promoters are recognized by the host RNA polymerase II. This group includes genes that function to transactivate other viral genes, as well as genes involved in viral DNA replication. After the onset of viral DNA replication the late genes are transcribed. Included among the late genes are those which encode the structural components of the virions. Late genes are governed by promoters which are apparently recognized by a distinct viral specific RNA polymerase.[39–41]

Yet another class are the hyper-expressed late genes. Like the late genes, the hyper-expressed late genes are transcribed by a viral specific RNA polymerase. During infection the hyper-expressed late genes are the last genes to be transcribed – their expression occurring concurrently with the production of occlusion bodies. These genes are expressed at extremely high levels. In NPVs the hyper-expressed late genes include the polyhedrin gene, which encodes the major structural component of polyhedra, and the p10 gene, whose function has not yet been ascribed. The promoters for these hyper-expressed late genes are exploited for the high-level expression of proteins in the baculovirus expression vector system.

The occluded baculoviruses are unique in that during the viral infection two phenotypes of the virus are produced; the ECV, which are responsible for tissue-to-tissue spread within an infected insect, and the occluded forms, responsible for the spread of viruses between insects. Although the two forms of virus are genetically identical[19] and appear to have nearly identical nucleocapsid structures, there are considerable differences in their antigenicity and infectivity, the timing of their development, and the composition of their viral envelopes.

In cell types that produce both phenotypes, the ECV appear early in infection, whereas the PDV are produced only late. In cell culture, ECVs are highly infectious, but PDVs have very little infectivity, even when liberated from the occlusion bodies by alkaline treatment.[42] This difference in infectivity of the two virion phenotypes is likely to result from different compositions of the viral membranes that surround the nucleocapsids. Whereas the membranes for PDV appear to be formed *de novo* in the nucleus, the membranes of the ECV are obtained from the cytoplasmic membrane as the virions are budded out of the cell. Analysis of the budding process in two NPVs (AcMNPV and a virus that infects the Douglas fir tussock moth, *Orgyia pseudotsugata*) has shown that shortly after infection a virus-encoded glycoprotein, termed gp64 (or gp67), appears in the host cell membrane.[43,44] ECV nucleocapsids are assembled in the nucleus, then transported through the cytoplasm to the cell periphery, where they bud out through patches of cell membrane containing gp64. In the process of budding, they become enveloped by the gp64 modified membrane.

During infection of target cells, the ECV virions enter into an intracellular vesicle, called an endosome, by endocytosis.[45] The nucleocapsids are then released into the cytoplasm after a pH-dependent fusion of the virus envelope with the endosomal membrane.[45,46] This event is probably mediated by gp64, since antibodies against gp64 block fusion of the virus with the endosome, but not the initial endocytic event.[45]

The infection process with PDV is less well-characterized. PDV envelopes do not contain gp64, although several virally encoded proteins are reportedly found in these structures.[46] The PDVs are specific for midgut epithelial cells and show little infectivity by injection.[47] The PDV envelope may have an important role in determining the tissue specificity of these virions. Deletion of a gene for a protein termed p74 renders occlusion bodies produced in cell culture to be non-infectious when fed to insects, suggesting that this protein may have a role to play in the specificity of PDV.[48]

Evidence has accumulated to indicate that a minor component of the occlusion bodies may enhance the infectivity of the viruses by disrupting the peritrophic membrane.[49] The peritrophic membrane is a barrier structure formed from secretions of the midgut cells which harden into a tube-like structure composed principally of glycosaminoglycans, protein, and hyaluronic acid embedded in chitin fibrils. It serves several functions, including protection of the midgut from insect pathogens, prevention of abrasion of midgut cells, and the maintenance of a pH gradient of approximately 9.5 to 7.5 across the peritrophic membrane (the lumen side being most alkaline). The peritrophic membrane contains pores generally estimated at less than 30 μm in size; too small to allow passage of the baculovirus virions. Derksen and Granados[49] found that inoculation of fifth instar *Tricoplusia ni* larvae with either 10^6 AcMNPV polyhedra or 10^8 *T. ni* GV granula caused the peritrophic membranes to become very fragile for several hours following infection. Inoculation with 2–4 × 10^6 *T. ni* SNPV occlusion bodies caused a similar, but less severe, disruption of the peritrophic membrane. SDS-PAGE analysis showed that both AcMNPV and TnSNPV infections resulted in the loss of a 68 kDa glycoprotein (gp68) from the peritrophic membrane, whereas infections with TnGV caused degradation of three peritrophic membrane proteins (gps 123, 194, and 253). Presumably the degradation of these proteins results in the observed gross disruption of the peritrophic membrane.

Fractionation of AcMNPV and TnGV occlusion bodies indicated there is a factor associated with the occlusion bodies from each of these viruses which is responsible for the disappearance of the specific peritrophic membrane glycoproteins. These factors have been termed viral enhancing factors (VEFs). The gene for the TnGV-VEF has been cloned and sequenced, and encodes a protein with a predicted molecular weight of 104 kDa.[50] When *T. ni* larvae are challenged with a constant dose of AcMNPV in the presence of increasing amounts of purified TnGV-VEF, there is a significant rise in both peritrophic membrane damage and larval mortality.

The size and amino acid composition of *T. ni* VEF is very similar to a factor from *Pseudaletia unipuncta* granulosis virus Hawaiian strain (PuGV-H) occlusion bodies, which has been termed synergistic factor (SF).[51] Work from Tanada's laboratory has shown that SF increases the susceptibility of *P. unipuncta* larvae to infection by a *P. unipuncta* NPV.[51] The SF is released from the PuGV-H occlusion bodies and becomes localized to the midgut microvilli, where it apparently increases nucleocapsid entry into the midgut cells. It is entirely possible that VEF, like SF, may have an activity that increases viral attachment to midgut microvilli, in addition to its peritrophic membrane-disrupting activity.

11.2.1.4 Epizootiology

The goal of microbial control of insects is simply to increase levels of disease in pest populations, thereby reducing pest population density and harmfulness. Thus, microbial control can be considered as applied epizootiology. The population dynamics of entomopathogenic viruses have many characteristics that lend themselves as viral epizootics in pest populations; in fact, the general capability of baculoviruses to cause natural disease epizootics in insect populations is a major advantage for their use in microbial control.

Persistence in the environment is a characteristic that affects microbial control regardless of the approach. The two most important aspects of persistence are, first, the capability of many viruses for long-term survival in soil, and, second, their relatively brief survival on foliage and other surfaces exposed to sunlight. Probably due to the protection provided by inclusion bodies, NPVs and GVs survive for years in soil, even in relatively unstable ecosystems such as row crops.[52] In one case, an NPV survived for 41 years in forest soil.[53] This persistence in soil greatly increases the potential of these viruses to be used in the seasonal colonization, introduction–establishment, and environmental manipulation approaches. On the other hand, the baculoviruses generally are inactivated on aerial plant surfaces within days or even hours,[52] primarily due to sunlight. The polyhedral inclusion bodies of NPVs and granules of GVs protect the virions to a large degree in the environment, but they do not protect the virions from sunlight.[54,55] The short life span of baculoviruses on plant surfaces is detrimental to all approaches of microbial control, because it reduces the probability that the pest insect will ingest active virus.

The other epizootiological factor that has a major impact on microbial control is transmission or transport of the virus in the environment. In the microbial insecticide approach, human manipulation largely replaces natural transmission; therefore, it must do so in a manner that deposits active virus in a location where the target insect will ingest it before the virus is inactivated or the insect grows too old to become infected easily. In the seasonal colonization and introduction–establishment approaches, natural transmission is essential for the virus to infect pest generations subsequent to the one against which the virus is originally released.

Epizootics of entomopathogenic viruses often are dependent on short-range transport, particularly from a soil reservoir onto the host plant where the insect feeds. Abiotic agents, in particular, such as rainfall, air currents, and gravity, have been implicated in the movement of viruses onto the insect's host plant and within the plant.[56] Many of the data concerning such transport are correlational or circumstantial,[56] although there have been experimental studies on transport by water movement[57–59] and air currents.[60,61]

Biotic agents can transport viruses over short or long distances. Larvae of the host insect can transport viruses short distances of 2–8 m or more before they become incapacitated.[62,63] Adult hosts can transport viruses over greater distances, because this stage of the insect generally is specialized for movement and reproduction. Adult transport has been implicated or demonstrated experimentally several times, particularly for Lepidoptera and one beetle.[64–70] One mechanism of viral transport by adult insects is by means of vertical transmission or, in other words, parent-to-offspring passage of the virus. Evidence is increasing that vertical transmission may be common in baculoviruses and that it is important to their epizootiology.[71] For example, it has been hypothesized that the reason *Spodoptera frugiperda* NPV is indigenous to the USA is that it is vertically transmitted, whereas the NPV of a similar migratory pest, *Anticarsia gemmatalis*, is not indigenous because it is not vertically transmitted.[72] Avian and mammalian predators of host insects have been implicated in viral transport, although this body of research consists primarily of sampling viable viruses from the gut or faeces of those predators.[73–78] Research has indicated that parasitic, and especially predatory, arthropods might be a major transport agent for entomopathogenic viruses.[58,79–93] For example, predatory arthropods caused the nuclear polyhedrosis virus of *A. gemmatalis* to spread from a release site in a soybean field at a rate of approximately 1 m per day during the growing season (J. R. Fuxa and A. R. Richter, unpublished data). Saprozoites, such as birds and sarcophagid flies, also have been implicated in viral transport,[64,74,94] as have grazing mammals.[56]

In addition to viral persistence and transmission, viral and host population parameters affect epizootiology and microbial control. The most obvious of these is population density. A large body of microbial control research supports the idea that a greater viral population density increases disease prevalence in the insect population, probably by increasing the chance of contact between the virus and uninfected host.[95–97] The pest insect population density primarily affects the seasonal colonization and introduction–establishment approaches to control. Viral epizootics, like those of most mortality agents, are host-density dependent, although the viruses are unusual in that they can act in a host-density independent fashion for long periods due to viral persistence.[98]

The quality of the host and virus populations can affect any of the approaches to microbial control. There is substantial evidence that insect populations can develop resistance to viruses, although the early research indicates that such resistance will have features that lend themselves to resistance management.[99] Similarly, different viral populations can differ in virulence,[100–102] and baculoviruses in the laboratory have been artificially selected for increased virulence, resistance to ultraviolet radiation, and vertical transmission.[71,103–105]

11.3 Natural Strains of Viruses as Commercial Products

Two of the approaches to microbial control with viruses lend themselves well to commercial development: the microbial insecticide and seasonal colonization approaches. In both cases, virions must be mass produced and released repeatedly for insect control. The technology is very similar for these two approaches, but they differ greatly in the way the virus performs and in the way the strengths and weaknesses of viruses contribute to their success or failure.

11.3.1 Viruses as Microbial Insecticides

The microbial insecticide approach is the one of primary interest to companies involved in the commercial development of viruses, because the multiple applications create the best opportunity for product sales. In this approach, the viruses are mass produced in reared, living insects. The costs of this technology have been greatly reduced, as in the case of the *Heliothis* (cotton bollworm) and *Lymantria dispar* (gypsy moth) NPVs,[106,107] but it is still too expensive for viral production to compete on an equal basis with the production of chemical insecticides.[108] Production of baculoviruses in cell culture may be approaching commercial competitiveness, but this has not yet been tested at the pilot plant scale.[109] The viruses generally are formulated and applied in much the same way as are chemical insecticides. The formulations must be standardized in their activity, must contribute to viral persistence in storage and on the substrate where the target insect feeds, and must remain low cost to facilitate competition in the insecticide market. The viruses generally have been formulated as aqueous concentrates or wettable powders,[110,111] and usually applied as sprays using conventional equipment.

The microbial insecticide approach has been the one that has received the most research attention, but viruses developed for this approach have not been very successful in terms of competitiveness in the insecticide market. The closest thing to a virus being used successfully as a microbial insecticidc is the *Heliothis armigera* NPV on cotton, tobacco, and tomato in China.[3,112,113] This virus was used on 8094 ha in 1991,[114] but it may have been used partially in a seasonal colonization approach,[3,112] and probably cannot be considered a truly commercial product. The *Heliothis* NPV in the USA which was developed for control of *Heliothis* in cotton in the 1970s, has been considered technically, but not economically, successful.[3] This NPV failed as a commercial product because it was too expensive to compete with chemicals, too slow, and too host-specific to be useful in a pest complex.[3,113] Additionally, *Heliothis* in cotton presented a difficult pest-control situation in that the target insect burrows into the fruiting structure and thus is not in a position to ingest the virus for a large portion of its life cycle. The *Cydia pomonella* (codling moth) GV has been under development for many years as a commercial product in Europe and the USA. This virus seems unlikely ever to capture a major share of the market, except perhaps in the production of 'organically grown' fruit, because it is too slow, too host-specific, and less persistent than the chemical insecticides.[114] In all the other cases of viruses as microbial insecticides, primarily in developing nations, it is not clear whether the product has been successful or whether it was used as a microbial insecticide or as a seasonal colonization agent. These examples include the *Adoxophyes orana* GV in Switzerland,[115] *Autographa californica* NPV against *Trichoplusia ni* in Guatemala,[116] (E. A. Alvarado, 1988, personal communication), *Pieris rapae* GV in China[3], *Spodoptera exigua* NPV in Thailand,[3] *Spodoptera sunia* and *S. exigua* NPVs in Guatemala,[117] (E. A. Alvarado, 1988, personal communication), and *Trichoplusia orichalcea* NPV in Zimbabwe.[117]

11.3.2 Viruses in Seasonal Colonization

Seasonal colonization with viruses is usually, although not always, a commercial venture. The methods of production, formulation, application, and commercial development are similar to those for the microbial insecticide approach. However, the seasonal colonization approach at least partially negates some of the disadvantageous characteristics of viruses as insecticides. Since more than one pest generation is suppressed by one viral application, viruses used in this approach are more cost-competitive than most chemicals, partially offsetting the expensive *in vivo* production. Due to the host specificity and the reduced number of applications, seasonal colonization can be attractive for commercial development by the cottage industry; the resultant 'niche' markets usually are unattractive to major companies with large overheads.[114] Quick mortality in the target insect is not as critical as with the microbial insecticide approach, since the virus has more time to act against at least some of the pest generations, thus partially alleviating one of the major disadvantages of viruses. On the other hand, the seasonal colonization approach demands some capability for causing natural epizootics, which in turn requires some degree of persistence and efficient transmission in the target insect and ecosystem. Also, viruses used in seasonal colonization are still at a disadvantage to chemicals in pest complexes, unless the insect is a key pest. A major stumbling block to this approach in the USA is that, even though registration costs for a viral insecticide (approximately $1.25 million) are only a fraction of those for chemicals, this usually is a prohibitive amount for cottage industry or venture-capital type companies.

The success of seasonal colonizations of viruses has varied considerably, although some of the most notable examples of microbial control have been through this approach. For example, the *Anticarsia gemmatalis* NPV currently is used on approximately one million ha of soybean each year in Brazil,[118,119] where one application gives season-long control. *Anticarsia gemmatalis* is a key pest in Brazil, and the virus is produced commercially in field populations of the insect, eliminating insect-rearing costs[120] (F. Moscardi, 1988, personal communication). Another notable success is the *Oryctes* baculovirus against coconut palm rhinoceros beetle in the South Pacific islands. This is primarily a seasonal colonization, although successive releases are at less than yearly intervals and occasionally are not necessary at all.[70,96,121,122] Its success has been attributed to relatively stable habitats and host populations, as well as to good horizontal and vertical transmission of the virus.[96] Another success is the *Dendrolimus* CPV, which is used on approximately 28 000 ha for control of pine caterpillar in China.[123] In North America, viruses to date have had limited success through seasonal colonization. The examples include NPVs of *Lymantria dispar* (gypsy moth), *Orgyia pseudotsugata* (Douglas-fir tussock moth), *Neodiprion lecontii* (redheaded pine sawfly), and *Neodiprion sertifer* (European pine sawfly).[3,117,124–127] Other examples of seasonal colonization include the *Dendrolimus spectabilis* CPV in Japan[128] and *N. sertifer* NPV in England, Finland, and Russia.[3,117,126,127]

11.4 Non-Commercial Uses of Viruses for Insect Control

Two of the approaches to microbial control have virtually no appeal for commercial development, and therefore must be advanced primarily by university and government researchers. Introduction–establishment, by definition, entails only a one-time release of a virus in a given area, and usually is characterized by replication and spread of the virus in the environment. Thus, there is limited opportunity for commercial-scale production and sales, although a bacterium is sold commercially for insect control by introduction–establishment.[103,129] Environmental manipulation involves no viral release at all. Due to their low long-term cost, these two approaches can be advantageous for low-value crops and commodities.

11.4.1 Introduction–Establishment of Viruses

Introduction–establishment is attractive because its primary feature, permanent pest suppression after a one-time viral release, utilizes many of the assets of viruses, while counteracting several of their shortcomings. This approach takes advantage of one of the great assets of the viruses: of all the insect pathogens, the baculoviruses probably produce the most effective disease epizootics in terms of host mortality coupled with high prevalence, particularly in terrestrial ecosystems. Their persistence in soil, usual lack of short-term host-density dependence, rapid generation time, and relatively efficient transmission give them potential as classical biocontrol agents, even in row crops, where biological control has been notoriously unsuccessful.[2] On the other hand, three major problems with viruses are only minor disadvantages with the introduction–establishment approach. The high cost of production in living host insects is largely negated due to the one-time application and the capability of these viruses to replicate and spread after release. Host specificity becomes a minor problem only because there is no concern over a limited market and because the producer does not have to worry about applying an insecticide that controls only one insect. The producer simply concentrates on the insects reaching pest status and ignores any that are not, due to suppression by a 'natural' control agent. Likewise, slow kill is only a minor problem because the virus has more time to function and because a user does not need to see quick kill after making an expensive application.

Another major advantage of introduction–establishment, one that is usually ignored, is that the established virus does not have to kill a large portion of the pest insects, as long as the farmer or resource manager adheres to pest population thresholds for treatment with insecticides. For example, if an introduced virus kills only 25% of the insects, and if this level of control only eliminates the necessity of one pesticide application out of every five, this can still be economically and environmentally worthwhile, because that 25% level of control is completely cost-free after the initial viral release.

There also are major questions about viral introductions, particularly concerning how widely applicable they might be. It is possible that the number of opportunities to release a viral 'species' where it does not already occur may be somewhat limited. This will be particularly true if introductions in row crops do not prove to be generally successful, although the few such attempts that have been made have been encouraging.[130] Similarly, if an imported or artificially improved 'strain' of a virus is released in an area where a similar 'strain' of the same 'species' is indigenous, it is not known whether the introduced strain would be able to compete for a niche in that particular ecosystem. If not, opportunities for viral introductions would be further limited. One possible such example of a strain being successfully introduced where another was indigenous is being investigated currently.[131]

According to the scientific literature, most of the 'successes' of viruses in insect control have been by introduction–establishment. However, this literature can be difficult to evaluate. First, such literature usually reports research results, not implementation of control attempts on a wide scale. Second, classical biological control is inherently difficult to evaluate, and criteria to help one judge its success, such as reduction in crop damage or elimination of a proportion of treatments with chemical insecticides, often are not reported. Third, introduction sometimes cannot be evaluated properly because there were no zero-baseline data to show that the released virus was not indigenous to the target ecosystem.

With these caveats in mind, there have been at least 16 'successful' viral introductions reported.[2] Five were in row crops and the other 11 in more stable ecosystems, primarily forests. Perhaps the most cited example is that of the *Gilpinia hercyniae* (=*Diprion hercyniae*) NPV to suppress populations of European spruce sawfly in forests.[121] The success of this virus, which was introduced accidentally from Europe into North America and has reduced the sawfly to the status of a minor pest, was attributed to the relative stability of the forest ecosystem and host populations as well as to efficient horizontal and vertical transmission of the virus. The examples in row crops are not as clear-cut. In the only case in which complete control was claimed,[3,132] an NPV introduced to control *Trichoplusia ni* on cotton in Colombia, the data have not been published. The release of a Guatemalan NPV in Louisiana soybean for control of *Pseudoplusia includens* almost certainly was still resulting in substantial host mortality after 12–15 years, but questions remain concerning its status.[131] This release, which is currently being investigated (J. R. Fuxa and A. R. Richter, unpublished data), was intended originally as a large, microbial insecticide trial, and there is a lack of zero-baseline data.

11.4.2 Environmental Manipulation and Conservation

Environmental manipulation is a logical approach; viral epizootics can be dependent on environmental variables, so it makes sense to use inexpensive cultural manipulations to make that environment more favourable for the viral population. This approach requires a good understanding of, and some ability to predict, viral epizootics in insects. The cultural manipulations, which can include the lack of an action that normally would be taken, usually are aimed at aiding viral persistence or viral transport from the soil reservoir to the insect's feeding substrate.

The environmental manipulation approach receives relatively little research attention; there are few examples, and it is not clear whether these are in use. Changes in sowing, grazing, cultivation, and chemical use were recommended to increase natural control by NPV of *Wiseana* species of Lepidoptera in New Zealand pastures.[133] Movement of cattle similarly enhanced NPV transport and natural control of *Spodoptera frugiperda* in pastures in Louisiana.[56,134] The only other example is interesting, because environmental manipulation was combined with seasonal colonization in the case of the *Oryctes* baculovirus, discussed in Section 11.3.2. Viral spread and control of the beetle populations are enhanced if some of the dead palms are left standing and the others are piled and overgrown with crops rather than left lying around the plantation.[122]

11.5 Recent Advances in Formulation, Genetic Engineering, and Production of Baculoviruses

There is a concerted world-wide effort in various industry, government, and academic laboratories to overcome the many shortcomings that baculoviruses have as commercial products. Research on new ways of formulating baculoviruses suggests that methods to improve pathogenicity and field stability may be within our grasp. Recent advances in the genetic engineering of baculoviruses point towards ways of reducing the survival time of infected insects, and the crop damage which occurs after insecticide application. Also, the development of high volume, low cost insect cell culture technology may give us a new method of insect virus production that is more economical and which results in a better quality-controlled product.

11.5.1 Improved Formulation

Baculovirus insecticide formulations usually contain a number of adjuvants which are added to improve their performance.[111] Normally included are agents, such as sticker-spreaders, to increase coverage and retention on plant surfaces, and some form of sunscreen to reduce UV inactivation of the virus. Gustatory stimulants are also sometimes added. Of course, when contemplating the addition of any adjuvant to a baculovirus formulation it is imperative that the effect on the overall cost of the formulation is considered, as well as the environmental impact that the added ingredients may impose.

Recent findings indicate that it may be possible to improve further formulations by the addition of constituents that increase the infectivity of the active viral ingredient. Many baculoviruses are already extremely virulent towards their specific insect hosts; in several virus–host interactions, ingestion of only a single occlusion body is sufficient to initiate a lethal infection (N. van Beek, 1982, personal communication). In such instances, formulation technology can be used to increase the stability of the virus, both in the field and in storage, but there is no possibility of using formulation technology to enhance further the inherent infectivity of the viruses. There are, however, several economically important species of insects for which only marginally infective baculovirus strains have been characterized. In such cases, there is an enormous potential for the enhancement of virulence via formulation technology.

Shapiro and Robertson[135] recently reported dramatic increases in the activity of the gypsy moth NPV (LdMNPV) against its host, *Lymantria dispar*, by formulation with several of a series of compounds known as optical or fluorescent brighteners. These agents absorb UV radiation and re-emit the energy as visible light. They are commonly used in commercial laundry detergents to make washed clothes look brighter. Originally tested as UV protectants in baculovirus formulations,[136] they have subsequently been found to increase greatly the virulence of at least two different viruses that show marginal infectivity against their hosts, even when the viruses were not subjected to UV irradiation.[135,137] For example, in one laboratory test the LC_{50}

of LdMNPV against 4th instar gypsy moth larvae was reduced 400–1000-fold by the addition of selected optical brighteners to the formulation at levels of 1%. The mechanism of this enhancement is not known, but it has been suggested that the fluorescent brighteners may enhance infectivity by somehow causing disruption of the peritrophic membrane.[135] Although field tests of baculovirus formulations containing fluorescent brighteners have not yet been reported, the laboratory tests suggest that these compounds may be effective in greatly reducing the amount of active material that needs to be incorporated into a viral insecticide, thereby significantly lessening the cost of production.

Another possibility for increasing infectivity is to incorporate VEF (see Section 11.2.1.3) into baculovirus formulations. Since VEF is heat and UV stable, it has been suggested that it could be added to baculovirus formulations to increase the effective potency of the virus in field applications.[138] Of course, this approach to improve biopesticide action would be practical only if a method were devised to produce large quantities of VEF in a cost-effective manner. It has also been suggested that VEF may be effective if expressed by transgenic plants.[50] Insects feeding on such plants would experience a continual disruption of their peritrophic membranes, causing their midguts to disfunction and increasing their susceptibility to insect pathogens, ultimately leading to death of the insect.

11.5.2 Genetic Engineering of Baculoviruses

In recent years baculoviruses have become more widely known for their use in the pharmaceutical and biotechnology industries as eukaryotic expression systems, than for their application as biological insecticides. Many of the techniques that have been developed for the baculovirus expression vector system, however, are also being utilized for the development of viruses with enhanced biopesticidal activities.

In general, the baculovirus expression vector system makes use of a promoter from one of the very late hyper-expressed genes (p10 or polyhedrin) to drive high-level expression of a foreign gene in infected insect cells. With rare exceptions, the foreign gene products are made to high levels, become correctly processed, and are biologically active.[139]

In its most common form, the baculovirus expression system utilizes recombinant AcMNPV baculoviruses in which the polyhedrin gene has been replaced by the foreign gene of interest. Because of the large size of the baculovirus DNA it is not possible to ligate a foreign gene DNA segment directly into the baculovirus DNA. Instead a transfer vector is first constructed. The transfer vector consists of a segment of baculovirus DNA containing the polyhedrin gene and its flanking sequences inserted into a standard plasmid cloning vector. All or part of the polyhedrin gene contained within the transfer vector is then replaced by the foreign gene of interest, so that this gene comes under the control of the polyhedrin promoter. After cotransfection of insect cells with the transfer vector and wild type viral DNA, at low frequency (0.1–1%), homologous recombination between the transfer vector and the viral DNA results in the production of progeny viruses carrying the foreign gene.[29] These are usually selected by microscopic examination; since the recombinant viruses no longer contain an intact polyhedrin gene they form plaques that lack the highly refractile occlusion bodies characteristic of wild-type plaques.[29]

Unfortunately, recombinant viruses which fail to make occlusion bodies cannot be used in biopesticide applications because they are environmentally unstable and have such poor oral activity. To overcome this problem, several transfer vector systems have been invented that allow construction of occluded viruses. Among them are those that utilize the p10 site.[140,141] In these systems, the parental DNA contains a lacZ gene at the p10 site and the background plaques formed from the parental DNA have a characteristic blue colour when the chromogenic substrate X-gal is added to the culture vessels. Recombination occurs in insect cells cotransfected with a p10 transfer vector and p10-LacZ parental DNA, resulting in the production of recombinant viruses containing the gene of interest at the p10 site. The recombinant viruses are selected by their $LacZ^-$ (colourless) phenotype. Potentially, systems such as this could be developed to introduce foreign DNA at any non-essential locus in the baculovirus genome.

Another class of transfer vector used to produce occluded recombinant baculoviruses has a p10 or polyhedrin promoter duplicated and placed at a site upstream from an intact polyhedrin promoter.[142] Foreign genes are cloned into the transfer vector in such a fashion that they become controlled by the duplicated promoter. In these systems, the parental DNA is derived from polyhedrin viruses and the recombinant viruses are easily selected by their ability to form plaques containing occlusion bodies.

A major drawback to using the baculovirus expression system has been that it is often difficult to identify the recombinant viruses from among the background of the parental type viruses. Furthermore, as many as four sequential plaque assays are usually required before a pure recombinant virus is obtained, taking as long as 6 weeks. In the past few years, however, several systems have been developed which greatly improve the efficiency of recombinant virus selection.

Kitts *et al.*[143] developed one of the first systems used to increase the frequency by which recombinant virus are produced after cotransfection. They introduced a unique Bsu36I restriction site into the viral DNA at the polyhedrin locus, which is used to linearize the viral DNA. Upon transfection with linearized viral DNA and standard transfer vectors, up to 30% of the progeny viruses were recombinants. Similar results were obtained using p10 transfer vectors with viral DNA linearized at the p10 locus. An improvement of the system developed by Kitts *et al.*[143] contains a Bsu36I restriction site at the polyhedrin locus, another in a LacZ gene, which has been inserted at the polyhedrin locus, and yet another which was engineered into an essential gene located adjacent to the polyhedrin gene. In this improved system, recombinant viruses are produced at a frequency approaching 100%. This system is currently available as a commercial product from both Invitrogen (San Diego, CA), and Clonetech (Palo Alto, CA).

Another system, described by Peakman *et al.*[144] utilizes the bacteriophage P1 encoded Cre recombinase and its substrate loxP to allow recombination to occur *in vitro*. Recombinant viruses are produced at frequencies of up to 50%. To obtain pure stocks of recombinant virus, sequential plaque purification is still required.

Perhaps the most innovative system for making recombinant baculoviruses was developed by Patel *et al.*[145] Yeast sequences were introduced into the baculovirus DNA that allow stable replication of the entire baculovirus genome in transformed *Saccharomyces cerevisiae*. Transfer vectors were developed which allow recombination of foreign genes into the baculovirus genome being carried by the yeast cells. The transfer vector also provides markers which allow the selection of yeast cells containing recombinant baculovirus genomes after plating on appropriate medium. The recombinant baculovirus DNA is then isolated from the yeast cells and transfected into insect cells to give rise to recombinant progeny viruses. With this system stocks of pure recombinant virus are obtained in as little as 10–12 days.

The yeast system of Patel *et al.*[145] offers the most rapid selection of recombinant viruses to date. There is likely to be some resistance to the use of this system, since many laboratories that exploit the baculovirus expression vector system lack experience with yeast. Owing to the widespread use and interest in the baculovirus expression system, other improved methods for recombinant virus construction and selection will undoubtedly be developed.

11.5.2.1 Recombinant Baculovirus with Faster Action

Currently, baculoviruses cannot compete with chemical insecticides in the pesticide market place because they are too slow in their action. Whereas chemical insecticides are usually effective in a matter of hours, viral infections can take from several days to several weeks to kill, depending on the virus species and insect host. During most of this period the infected insects continue to feed and can cause significant damage to crops before they perish.

A great deal of research is being conducted on limiting the feeding damage that occurs after infection. Most strategies involve the engineering of viruses to express genes that are expected to be detrimental to the insects. Examples of such genes are those encoding insect neurohormones or hormone regulators, and insect-specific toxins. One strategy, developed by O'Reilly and Miller,[146] involves the deletion of an existing baculovirus gene, rather than the addition of a foreign gene. These researchers characterized a baculovirus gene which encodes an enzyme, ecdysteroid glucosyl transferase (EGT), which acts to modify ecdysteroid hormones in the haemolymph of infected insects.

Ecdysteroid hormones, in particular 20-hydroxyecdysone, are key elements in triggering larval moults from one instar to the next, or from larval to pupal stages. By producing an enzyme that effectively removes biologically active ecdysones from the haemolymph, the virus can prevent the insect from moulting. Infected insects stuck in the intermoult phase continue to feed and grow larger until they finally succumb to the viral infection. Inactivation of

EGT by gene deletion allows the infected insects to moult normally, thereby preventing some of the feeding that usually occurs after infection.

In one experiment, diet consumption by fifth instar larvae injected with 2×10^5 p.f.u. of wild-type virus was some 40% higher than it was by insects infected with the same amount of EGT^- virus.[146] The EGT gene deletion was also found to reduce the survival time of neonate larvae feeding on virus-laced media (an approximate 20% reduction). The timing of infection in these experiments may be critical, however. Insects infected with EGT^- viruses at late times in the intermoult period are often not prevented from moulting (N. van Beek, 1992, personal communication). Perhaps in late intermoult infections, insufficient EGT is produced before the moult is initiated. The effectiveness of EGT^- viruses in preventing field damage by infected insects has not yet been determined.

Maeda[147] was the first to report the achievement of a significant reduction in survival time of infected larvae by the addition of an insect hormone gene to a baculovirus. The hormone gene, in this case, encoded a 41 amino acid diuretic hormone from the tobacco hornworm, *Manduca sexta*. Diuretic and anti-diuretic hormones are used by insects to control water excretion. In their experiments, haemolymph volume and survival time were compared in fifth instar silkworm larvae injected with 3×10^5 p.f.u. of wild-type *Bombyx mori* nuclear polyhedrosis virus (BmNPV), or an equivalent amount of BmNPV expressing either the diuretic hormone gene or an interferon gene under the control of the polyhedrin promoter. After 2 days, haemolymph volumes in larvae expressing the DH gene were significantly lower than in insects infected with either of the two other viruses, and the ST_{50} occurred about 25% faster with the DH recombinant virus. Many attempts to produce a lower ST_{50} in *T. ni* larvae infected with a DH-recombinant AcMNPV have not succeeded, however (N. van Beek, 1992, personal communication). Another approach to controlling larval feeding after a baculovirus infection is to engineer into the virus a gene encoding an enzyme involved in the regulation of an insect neurohormone. Juvenile hormone esterase (JHE) is an example of one enzyme that has been tried. Juvenile hormone (JH) plays a critical role in the control of insect larval development and reproduction. JHE causes a reduction in haemolymph JH titres by hydrolysing the chemically stable, conjugated methyl ester into the biologically inactive JH acid. During the development of lepidopteran larvae, reductions in the level of JH, caused by the production of high levels of JHE, lead to the cessation of feeding, melanization of the insects, and initiation of moulting or metamorphosis. Hanzlik *et al.*[148] isolated and characterized several JHE clones from a cDNA library made from fat bodies of *Heliothis virescens*. *Manduca sexta* larvae turned black after injection with JHE purified from a baculovirus expression system, indicating that the *in vitro* produced enzyme was biologically active.[149] Surprisingly, although *T. ni* larvae fed on a diet laced with JHE-expressing baculoviruses exhibited an increased level of JHE, a reduction in larval growth and feeding occurred only with insects from the earliest instar.[149] Recent experiments from Hammock's laboratory, however, have indicated that expression of a gene encoding a form of JHE designed to be more stable *in vivo* results in a significant reduction in survival time, even in later instars (B. Hammock, 1992, personal communication).

One of the more promising developments in improving baculoviruses has come from the introduction of genes for insect-specific toxins, derived from insect predators and parasites, into the baculoviruses genome. A gene encoding a 70 amino acid peptide toxin derived from the North African scorpion, *Androctonus australis* Hector, has been used to shorten the survival time after infection in at least three species of insects.[150–152] This toxin, termed AaIT, acts by causing specific modifications to the Na^+ conductance of neurons, resulting in a presynaptic excitatory effect leading to paralysis and death. Stewart *et al.*[150] reported that neonate *T. ni* larvae fed droplets containing 2×10^6 occlusion bodies ml^{-1} from a recombinant AcMNPV virus, in which the AaIT gene was expressed from the p10 promoter, had survival times 25–30% shorter than those infected similarly with wild-type viruses. Also, under laboratory conditions, feeding damage by third instar larvae infected with the toxin-expressing virus was reduced by approximately 50%, compared with larvae infected with wild-type viruses. McCutchen *et al.*[151] described a similar reduction in survival time after infection of the cotton bollworm, *H. virescens*, using a comparable scorpion toxin expressing AcMNPV. Maeda *et al.*[152] showed that second instar *Bombyx mori* injected with 10^5 occlusion bodies of a scorpion toxin expressing BmNPV exhibited toxicological effects after about 40 h post infection. The survival time of these insects was reduced to a little over 60 h, compared to an ST_{50} for wild-type BmNPV of between 90 and 100 h.

Another toxin which has been used successfully to shorten survival time is derived from a predatory mite, *Pyemotes tritici*.[153,154] This toxin has a molecular weight of approx. 27 kDa, which is much larger than that of the scorpion toxin.[153] Like the scorpion toxin, the mite toxin elicits a contractile paralysis after injection into insects.[153] Tomalski and Miller[155] demonstrated that expression of the mite toxin gene from a modified polyhedrin promoter elicits a significant reduction in the survival time (an approximately 40% decrease) after injection of 400 000 p.f.u. of AcMNPV budded virus into fifth instar *T. ni* larvae, or after feeding of neonate larvae with occluded virus.

In a separate publication, Tomalski and Miller[156] explored the effects that different promoters can have on the rate at which paralysis occurs after infection. They found that viruses which expressed the mite neurotoxin from a hybrid promoter, containing promoter elements from a hyper-expressed late gene (polyhedrin) promoter and a later gene (viral capsid protein) promoter linked in tandem, caused paralysis of insects about a day sooner than viruses expressing the toxin from a standard polyhedrin promoter. Presumably, this quicker paralysis results from the more rapid accumulation of toxin to effective threshold levels in the insects infected with the viruses containing the capsid-polyhedrin hybrid promoters.

In the future, additional reduction in survival time may be realized by using genes for toxins with even greater inherent activities than either the scorpion toxin or the straw itch mite toxin. Coupled with super-hyper-expressing dual promoters, such genes may turn baculoviruses into viable competitors in the insecticide market.

11.5.2.2 Recombinant Baculoviruses with Expanded Host Specificity

The host specificity of a baculovirus being considered for use as an insecticide is an issue of utmost importance. In order to keep the costs of development and registration as low as possible, it is desirable to choose a virus that infects a broad spectrum of pest insects. This must be balanced, however, by the need to choose viruses that do not cause significant harm to non-targeted species.

The AfNPV virus has the broadest host range of any known baculovirus; it infects at least 39 agronomically important species from 10 families of Lepidoptera.[15] AfNPV is genetically closely related to AcMNPV and grows well in insect cell culture. It is amenable to genetic engineering using the same transfer vectors that have been developed for AcMNPV (Leisy, unpublished data). This virus is a prime candidate for development into virus insecticides, both as an unmodified natural product and as a genetically engineered virus.

The molecular mechanisms that control the host range of a given baculovirus are not known, making it difficult to manipulate host range directly by genetic engineering. Kondo and Maeda[157] reported that replication of *Bombyx mori* NPV could be induced in several non-permissive cell lines by coinfection with AcMNPV, which is capable of growing on these lines. This suggested the existence of a helper function of AcMNPV for BmNPV in the non-permissive lines. Progeny viruses from the coinfected cells were plaque purified on BmN cells (a *Bombyx mori* cell line). Most of the isolates were incapable of replicating in the cell lines in which they were produced, but some isolates were capable of growing on cell lines non-permissive for BmNPV. DNA restriction analysis showed that the viral isolates with the wider host range were recombinant viruses resulting from crossovers between the two parent baculoviruses, AcMNPV and BmNPV. These isolates were also characterized as having a broader host range in insect larvae than either of the parental viruses. Thus, it may be possible to create a baculovirus with an optimal host range by the selection of hybrid viruses produced during coinfection of cultured insect cells.

It is also possible that viruses that express potent insect toxin genes may exhibit broadened host ranges, since the expressed toxin may elicit killing in insects to which the unmodified virus is only marginally infective. Host range testing of such recombinant viruses will be an important aspect of their safety evaluation.

11.5.3 Production Technology

The practice of producing baculoviruses for insecticidal applications in mass-reared insect larvae (*in vivo* production) has several problems that may be overcome by the development of efficient *in vitro* production methods. *In vivo* production is relatively labour intensive, limiting the possibility of achieving a significant scale-up advantage in the cost per unit produced. Many insect larvae are cannibalistic, requiring that they be reared in individual containers. Quality control is difficult and the final product is often contaminated with micro-organisms other than the virus. The *in vivo* production of genetically engineered viruses that kill their hosts faster may also poses a problem; since the larvae die sooner, the number of polyhedra produced may be expected to be lower than with wild-type viruses.

There are many problems still associated with *in vitro* production as well, but significant progress is being made in this area. Insect cells typically have been grown in expensive media containing 10% fetal calf serum. The recent development of serum-free media, such as SF-900 (Gibco) and EX-CELL 400 (JR Scientific), is encouraging, although these media are still quite expensive. Large capacity airlift bioreactors have been used for growing recombinant baculoviruses for the production of overexpressed products.[158] Continuous passage of baculovirus in cell culture has been shown to lead to decreased virulence, because of spontaneously occurring deletion mutations in the virus, indicating that batch production may be required.[159] All of these difficulties must be overcome before *in vitro* production will become a viable alternative to *in vivo* production.

11.6 Safety and Regulatory Aspects

The inherent safety of entomopathogenic viruses, due to their host specificity, is one of the major reasons for interest in their development for insect control. The baculoviruses have undergone extensive safety testing in the laboratory[160] and have caused almost no harmful effect when released into the environment.[52,160] The only detrimental effect of environmental releases has been indirect harm to populations of invertebrate parasitoids and predators due to removal of their food source, the pest insect; however, even in this respect the viruses do not have as severe an effect as do chemical insecticides.[52]

Despite their inherent safety, regulatory issues are a major bottleneck in the development of viruses and other micro-organisms for microbial control. Much of this is due to the advent of recombinant-DNA technology in the improvement of micro-organisms. Concerns about releasing recombinant organisms into the environment have even caused a re-evaluation of releasing natural, non-indigenous strains of entomopathogens.

In the USA, the Environmental Protection Agency (EPA) has been the primary regulatory body involved in releases of entomopathogens. Their initial case involved the *Heliothis* NPV in the early 1970s. The EPA had no policy for evaluating risks posed by this virus, and the ensuing difficulties became an unreasonable impediment to developing microbial agents.[161,162] After the *Heliothis* NPV was registered finally in 1975, registration of other 'biorational' pesticides proceeded on a case-by-case basis.[163] Other viruses were registered,[164] but the requirements were so inconsistent that this *ad hoc* approach was no longer considered appropriate by the late 1970s.[165] The EPA finally developed a set of guidelines, commonly called the 'Subdivision M' guidelines, for the evaluation of hazards and environmental impact.[166] These guidelines have two important features, 'tier testing' and a 'maximum challenge' approach, that help keep the entire process relatively inexpensive. These guidelines became considered by microbial control researchers as striking an excellent balance between assuring the public of safety and environmental protection, while at the same time allowing rapid commercial development.[167] The major problem with the guidelines today is that, even though the cost of research required to register a virus is only a fraction of that for a chemical, that cost is still prohibitive to cottage industry and venture-capital companies or any company developing viruses for narrow, 'niche' markets.

The EPA guidelines for microbial pesticides had barely evolved when, in the mid 1980s, the recombinant-DNA revolution impacted microbial control and confused the entire registration issue. The concern generally was not over the usual types of safety tests, such as laboratory tests of the modified micro-organism against non-target organisms in the laboratory. Rather, concern was expressed that modified organisms might have unexpected, hazardous characteristics after they had been released into the environment. The controversy settled down to three major issues.[52,167,168]

- First, some scientists believe that recombinant organisms might have unexpected properties, such as unusual evolutionary potential and pathogenicity for the wrong host.
- Second, the recombinant organisms might themselves become pest species. The history of agriculture is based almost entirely on biological introductions, but a number of introductions, accidental and intentional, have had adverse consequences.
- Third, there could be unintended transfer of genetic material from the recombinant organism to other organisms in the environment, causing the first two areas of concern to resurface.

There are good arguments both for and against all three of these issues. However, it is generally believed that, if care is exercised, introductions will have a low risk of environmental damage.[169,170] Even the severest critics of releasing recombinant-DNA organisms acknowledge that there will be few or no problems in the vast majority of cases.[171]

Two factors make these questions very difficult for regulatory agencies, the same two factors that caused problems with registering the *Heliothis* NPV in the early 1970s. The first is that one can never prove the negative hypothesis that harm will not result from a release. The best one can do is simply accumulate a large amount of negative data. However, this raises difficult questions: what types of data are most pertinent, and at what point do we have sufficient negative data to approve a release? The second factor is a classic 'Catch 22'; ultimately, one cannot obtain the best data to indicate that a release will be safe without actually releasing the organism, and one should not release the organism, in even a small-plot test (biological entities always have the potential to reproduce and spread), without having the 'best' data indicating that the release will be safe.

Thus, before an actual field trial, risk assessment must be based on three other types of data:

- First, safety tests in the laboratory with recombinant organisms, according to the Subdivision M guidelines.
- Second, a thorough knowledge of the biology and ecology of the 'parental' organisms from which the recombinant is derived.
- Third, 'microcosm' studies, consisting of contained systems, ranging from test tubes to greenhouses, in which a portion of the ecosystem is duplicated to test the 'environmental' effects of a release.[168]

The arguments about releasing recombinant organisms had an interesting side effect. Prestigious groups of scientists concluded that the organism to be released, not the method by which it was modified, should be the basis for risk assessment.[169,172] That, together with the fact that many biological introductions have caused ecological disruptions,[168] has led many to conclude that the releases of natural strains, including entomopathogens, should be regulated more closely. This is in spite of the fact that, up to the present time, releases of entomopathogens have not caused any appreciable environmental harm.[167,168] All types of entomopathogen releases have been adversely affected by these developments, particularly introduction–establishment research. The United States Department of Agriculture (USDA) Animal and Plant Health Inspection Service (APHIS) now primarily regulates the introduction of natural strains of non-indigenous entomopathogens, although EPA also still has a role.

Thus, regulation of microbial control agents in the USA is proceeding at a mixed pace. Registrations of strains not produced by recombinant-DNA techniques are again progressing; 18 entomopathogen species or strains are now registered, including four viruses.[166,173,174] One of these pathogens is a recombinant bacterium, but this recombinant is not released alive. Other recombinant entomopathogens are being considered for registration on a case-by-case basis. The only small-plot releases allowed to date in the USA have been those with a low probability of subsequent spread in the environment.[138,175] Introduction–establishment research is progressing, but has been hampered. Work underway in Europe is summarized by Vlak[176] and Possee *et al.*[177]. Significant progress has been made in testing the persistence in the environment and the biological effects of baculoviruses transformed with toxin genes (see Chapter 9).

11.7 Summary

Insect baculoviruses have become prime candidates in our search for safe and effective replacements for chemical insecticide products. During the past few decades we have attained a detailed understanding of baculovirus biology, ranging from the molecular mechanisms governing gene expression to population dynamics. This knowledge is being used to overcome the many problems associated with the use of baculoviruses, such as slowness of action, host-range restrictions, low infectivity in some cases, and lack of field persistence. Recently, we have witnessed the development of genetically engineered viruses that act more quickly. New formulation adjuvants that overcome the problems of persistence and infectivity on a laboratory scale also look promising. The next step is to test these new technologies in field experiments.

From a commercial point of view, the prospects for microbial control of insects with natural insect viruses are mixed. In general, natural strains are too slow and costly to compete with chemical insecticides produced by industrial nations, particularly in major markets. Exceptions could arise rapidly if chemical insecticides become unavailable for particular crops due to safety and regulatory considerations. Perhaps the greatest potential of natural insect viruses is their use in the seasonal colonization and introduction–establishment approaches. The suppression of several host generations could make the seasonal colonization approach cost-competitive. Although there has been insufficient research to evaluate how widely the introduction–establishment approach might be used, it has the potential to impact insect control in many situations, including row crops. In many cases, however, it might result only in partial pest suppression and elimination of only a portion of the chemical pesticide eliminations. Since this approach does not afford the opportunity for repeated sales, it is unlikely to gain the attention of private industry. The environmental manipulation approach receives so little research attention that it will probably be used rarely in insect control. In the short term, natural strains will probably be used only in niche markets and production will be restricted primarily to cottage industry or companies attempting to enter the market while they develop recombinant viruses.

Recombinant viruses that kill more quickly than natural strains could eventually prove successful as microbial insecticides. Owing to the high cost of development and registration, however, such genetically engineered viruses are likely to be used only in large markets for which chemical insecticides have become unavailable. A major unknown with respect to such viruses is the effect of public opinion on regulatory agencies; even a minor environmental problem with recombinant viruses could delay their use for many years.

The safety of baculoviruses to humans and other mammals is, for the most part, undisputed. With reasonable risk-assessment guidelines, any risks from releases of natural and recombinant baculoviruses are likely to be much less than those from the continued over-reliance on chemicals. Of course, zero risk will not be possible, just as it has not been possible in traditional agriculture. Nevertheless, if environmental problems associated with the release of baculoviruses are kept to a minimum and used as experience to avoid further mistakes, then society undoubtedly will benefit, just as it has from many other introductions during the history of agriculture.

11.8 References

1. G. G. Kennedy and F. Gould, Future challenges for entomology and the Entomological Society of America, *Bull. Entomol. Soc. Am.*, 1989, **35**, 190.
2. J. R. Fuxa, Microbial control of insects: status and prospects for IPM, in 'IPM Systems in Agriculture', ed. R. L. Rajak, R. K. Upadhyay, and K. G. Mukerji, Aditya, India, in press.
3. J. Huber, Use of baculoviruses in pest management programs, in 'The Biology of Baculoviruses. Volume II. Practical Application for Insect Control', ed. R. R. Granados and B. A. Federici, CRC Press, Boca Raton, 1986, p. 181.
4. J. R. Fuxa, Insect control with baculoviruses, *Biotech. Adv.*, 1991, **9**, 425.
5. M. E. Martignoni and P. J. Iwai, 'A Catalog of Viral diseases of Insects, Mites, and Ticks', USDA Forest Service PNW-195, Washington DC: USGPO, 1986, 50 pp.
6. L. Bos, 'Introduction to Plant Virology', Center for Agricultural Publishing and Documentation, Wageningen, The Netherlands, 1983.
7. R. Govindarajan and B. A. Federici, Ascovirus infectivity and the effects of infection on the growth and development of Noctuid larvae, *J. Invert. Pathol.*, 1991, **56**, 291.
8. K. M. Edson, S. B. Vinson, D. B. Stoltz, and M. D. Summers, Virus in a parasitoid wasp *Compoletis sonorensis*: suppression of the cellular immune response in the parasitoid's host *Heliothis virescens*, *Science*, 1981, **211**, 582.
9. D. B. Stoltz and D. Guzo, Apparent haemocytic transformations associated with parasoitoid-induced inhibition of immunity in *Malacosoma disstria* larvae, *J. Insect Physiol.*, 1986, **32**, 377.
10. N. E. Beckage and T. J. Templeton, Physiological effects of parasitism by *Appanteles congregatus* in terminal-state tobacco hornworm larvae, *J. Insect Physiol.*, 1986, **32**, 299.
11. J. A. G. W. Fleming and M. D. Summers, *Compoletis sonorensis* endoparasitic wasps contain forms of *Compoletis sonorensis* virus DNA suggestive of integrated and extrachromosomal polydnavirus DNAs, *J. Virol.*, 1986, **57**, 552.
12. H. F. Evans and K. A. Harrap, Persistence of insect viruses, in 'Virus persistence', ed. A. C. Minson and G. K. Darby, SGM Symposium, Cambridge University Press, Cambridge, 1982, p. 57.
13. R. P. Jaques, Persistence, accumulation, and denaturation of nuclear polyhedrosis and granulosis viruses, in 'Baculoviruses for Insect Pest Control: Safety Considerations', ed. M. Summers, American Society for Microbiology, Washington, DC, 1975, p. 90.
14. P. F. Entwistle and H. F. Evans, Viral control, in 'Comprehensive Insect Physiology, Biochemistry and Pharmacology', ed. G. A. Kerkut and L. I. Gilbert, Pergamon Press, Oxford, 1986, vol. 13, p. 347.
15. D. L. Hostetter and B. Puttler, A new broad host spectrum nuclear polyhedrosis virus isolated from a celery looper, *Anagrapha falcifera* (Kirby), (Lepidoptera: Noctuidae), *Env. Entomol.*, 1991, **20**, 1480.
16. A. Bensimon, S. Zinger, E. Gerassi, A. Hauschner, I. Harpaz, and I. Sela, 'Dark Cheeks', a lethal disease of locusts provoked by a lepidopterous baculovirus, *J. Invert. Pathol.*, 1987, **50**, 254.
17. A. A. Faizairy and F. A. Hassan, Infection of termites by *Spodoptera littoralis* nuclear polyhedrosis virus, *Insect Sci. Applic.*, 1988, **9**, 37.
18. S. Burgess, Molecular weights of lepidopteran baculovirus DNAs: derivation by electron microscopy, *J. Gen. Virol.*, 1977, **37**, 501.

19. G. E. Smith and M. D. Summers, Analysis of baculovirus genomes with restriction endonucleases, *Virology*, 1978, **89**, 517.
20. M. P. Schafer, G. Rohrmann, U. Heine, and G. S. Beaudreau, DNA from two *Orgyia pseudotsugata* baculoviruses: Molecular weight determination by means of electron microscopy and restriction endonuclease analysis, *Virology*, 1979, **95**, 176.
21. C. Goto, Y. Minobe, and T. Iizuka, Restriction endonuclease analysis and mapping of the genomes of granulosis viruses isolated from *Xestia c-nigrum* and five other noctuid species, *J. Gen. Virol.*, 1992, **73**, 1491.
22. J. R. Adams and J. T. McClintock, Baculoviridae. Nuclear polyhedrosis viruses of insects, in 'Atlas of Invertebrate Viruses', ed. J. R. Adams and J. R. Bonami, CRC Press, Boca Raton, 1991, Ch. 6, pt 1.
23. R. R. Granados, Infectivity and mode of action of baculoviruses, *Biochem. Bioengineer*, 1980, **22**, 1370.
24. L. D. Bruce, B. B. Trumper, and D. V. Lightner, Methods for viral isolation and DNA extraction for a penaid shrimp baculovirus, *J. Virolog. Meth.*, 1991, **34**, 245.
25. R. R. Granados and K. A. Lawler, *In vivo* pathway of *Autographa californica* baculovirus invasion and infection, *Virology*, 1981, **108**, 297.
26. F. T. Bird and M. M. Whalen, A virus disease of the European pine sawfly, *Neodiprion sertifer* (Geoffr.), *Can. Entomol.*, 1953, **85**, 433.
27. B. A. Federici and V. M. Stern, Replication and occlusion of a granulosis virus in larval and adult midgut epithelium of the western grapeleaf skeletonizer, *Harrisina brillians*, *J. Invert. Pathol.*, 1990, **56**, 401.
28. G. W. Blissard and G. F. Rohrmann, Baculovirus diversity and molecular biology, *Ann. Rev. Entomol.*, 1990, **35**, 127.
29. B. R. Jasny, Insect viruses invade biotechnology, *Science*, 1988, **238**, 1653.
30. M. D. Summers and G. E. Smith, 'A Manual of Methods for Baculovirus Vectors and Insect Cell Culture Procedures', Tex. Agric. Exp. Stn. Bull. No. 155, 1987, 56 pp.
31. D. Winstanley and N. E. Crook, Replication of *Cydia pomonella* granulosis virus in cell cultures, *J. Gen. Virol.*, 1993, **74**, 1599.
32. M. A. Cochran, S. E. Brown, and D. L. Knudson, Organization and expression of the baculovirus genome, in 'The Biology of Baculoviruses, Volume 1. Biological Properties and Molecular Biology', ed. R. R. Granados and B. A. Federici, CRC Press, Boca Raton, 1986, p. 239.
33. M. D. Ayres, S. C. Howard, J. Kuzio, M. Lopez-Ferber, and R. D. Possee, The complete DNA sequence of *Autographa californica* nuclear polyhedrosis virus, *Virology*, 1994, **202**, 586.
34. S. Maeda and K. Majima, The American Society for Virology 1992 Annual Meeting, Cornel University, Ithaca, NY, July 11–15, 1992.
35. G. F. Rohrmann, Polyhedrin structure, *J. Gen. Virol.*, 1986, **67**, 1499.
36. F. C. Minion, L. B. Coons, and J. R. Broome, Characterization of the polyhedral envelope of the nuclear polyhedrosis virus of *Heliothis virescens*, *J. Invert. Pathol.*, 1979, **34**, 303.
37. M. A. Whitt and J. S. Manning, A phosphorylated 34 kDa protein and a subpopulation of polyhedrin are thiol linked to the carbohydrate layer surrounding a baculovirus occlusion body, *Virology*, 1988, **163**, 33.
38. P. D. Friesen and L. K. Miller, The regulation of baculovirus gene expression, *Curr. Top. Microbiol. Immunol.*, 1986, **131**, 31.

39. C. L. Yang, D. A. Stetler, and R. F. Weaver, Structural comparison of the *Autographa californica* nuclear polyhedrosis virus-induced RNA polymerase and the three nuclear RNA polymerases from the host, *Spodoptera frugiperda*, *Virus Res.*, 1991, **20**, 251.
40. C. Rankin, B. G. Ooi, and L. K. Miller, Eight base pairs encompassing the transcriptional start point are the major determinant for baculovirus polyhedrin gene expression, *Gene*, 1988, **70**, 39.
41. U. Weyer and R. D. Possee, Analysis of the promoter of the *Autographa californica* nuclear polyhedrosis virus p10 gene, *J. Gen. Virol.*, 1989, **70**, 203.
42. L. E. Volkman and M. D. Summers, *Autographa californica* nuclear polyhedrosis virus: Comparative infectivity of the occluded, alkali-liberated, and nonoccluded forms, *J. Invert. Pathol.*, 1977, **30**, 820.
43. L. E. Volkman, P. A. Goldsmith, R. T. Hess, and P. Faulkner, Neutralization of budded *Autographa californica* NPV by a monoclonal antibody: Identification of the target antigen, *Virology*, 1984, **133**, 354.
44. G. W. Blissard and G. F. Rohrmann, Location, sequence, transcriptional mapping, and temporal expression of the gene encoding p39, a major structural protein of the *Orgyia pseudotsugata* multicapsid nuclear polyhedrosis virus, *Virology*, 1989, **170**, 537.
45. L. E. Volkman and P. A. Goldsmith, Mechanism of neutralization of budded *Autographa californica* nuclear polyhedrosis virus by a monoclonal antibody: Inhibition of entry by adsorptive endocytosis, *Virology*, 1985, **143**, 185.
46. G. W. Blissard and J. R. Wenz, The American Society for Virology 1992 Annual Meeting, Cornel University, Ithaca, NY, July 11–15, 1992.
47. G. F. Rohrmann, Baculovirus structural proteins, *J. Virol.*, 1992, **73**, 749.
48. J. D. Kuzio, R. Jaques, and P. Faulkner, Identification of p74, a gene essential for virulence of baculovirus occlusion bodies, *Virology*, 1989, **173**, 759.
49. A. C. G. Derksen and R. R. Granados, Alteration of a lepidopteran peritrophic membrane by baculoviruses and enhancement of viral infectivity, *Virology*, 1988, **167**, 242.
50. Y. Hashimoto, B. G. Corsaro, and R. R. Granados, Location and nucleotide sequence of the gene encoding the viral enhancing factor of the *Tricoplusia ni* granulosis virus, *J. Gen. Virol.*, 1991, **72**, 2654.
51. K. Uchima, D. E. Egerter, and Y. Tanada, Synergistic factor of a granulosis virus of the army worm, *Pseudaletia unipuncta*: Its uptake and enhancement of virus infection *in vitro*, *J. Invert. Pathol.*, 1989, **54**, 156.
52. J. R. Fuxa, Fate of released entomopathogens with reference to risk assessment of genetically engineered microorganisms, *Bull. Entomol. Soc. Am.*, 1989, **35**, 12.
53. C. G. Thompson, D. W. Scott, and B. E. Wickman, Long-term persistence of the nuclear polyhedrosis virus of the Douglas-fir tussock moth, *Orgyia pseudotsugata* (Lepidoptera: Lymantriidae), in forest soil, *Environ. Entomol.*, 1981, **10**, 254.
54. G. Benz, Environment, in 'Epizootiology of Insect Diseases', ed. J. R. Fuxa and Y. Tanada, Wiley, New York, 1987, p. 177.
55. C. M. Ignoffo, W. C. Rice, and A. H. McIntosh, Inactivation of nonoccluded and occluded baculoviruses and baculovirus-DNA exposed to simulated sunlight, *Environ. Entomol.*, 1989, **18**, 177.
56. J. R. Fuxa, Release and transport of entomopathogenic microorganisms, in 'Risk Assessment in Genetic Engineering', ed. M. Levin and H. Strauss, McGraw-Hill, New York, 1991, p. 83.

57. W. A. L. David and B. O. C. Gardiner, The persistence of a granulosis virus of *Pieris brassicae* in soil and in sand, *J. Invert. Pathol.*, 1967, **9**, p. 342.
58. R. P. Jaques, Leaching of the nuclear-polyhedrosis virus of *Trichoplusia ni* from soil, *J. Invert. Pathol.*, 1969, **13**, 256.
59. S. Y. Young, Influence of sprinkler irrigation on dispersal of nuclear polyhedrosis virus from host cadavers on soybean, *Environ. Entomol.*, 1990, **19**, p. 717.
60. C. G. Thompson and E. A. Steinhaus, Further tests using a polyhedrosis virus to control the alfalfa caterpillar, *Hilgardia*, 1950, **19**, 411.
61. E. Olofsson, Dispersal of the nuclear polyhedrosis virus of *Neodiprion sertifer* from soil to pine foliage with dust, *Entomol. Exp. Appl.*, 1988, **46**, 181.
62. H. F. Evans and G. P. Allaway, Dynamics of baculovirus growth and dispersal in *Mamestra brassicae* L. (Lepidoptera: Noctuidae) larval populations introduced into small cabbage plots, *Appl. Environ. Microbiol.*, 1983, **45**, 493.
63. P. H. Sterling, P. M. Kelly, M. R. Speight, and P. F. Entwistle, The generation of secondary infection cycles following the introduction of nuclear polyhedrosis virus to a population of the brown-tail moth, *Euproctis chrysorrhoea* L. (Lep., Lymantriidae), *J. Appl. Entomol.*, 1988, **106**, 302.
64. G. R. Stairs, Artificial initiation of viral epizootics in forest tent caterpillar populations. *Can. Entomol.*, 1965, **97**, 1059.
65. E. C. Young, The epizootiology of two pathogens of the coconut palm rhinoceros beetle, *J. Invert. Pathol.*, 1974, **24**, 82.
66. I. E. Gard, 'Utilization of Light Traps to Disseminate Insect Viruses for Pest Control', PhD Dissertation, University of California, Berkeley, 1975.
67. B. Zelazny, Transmission of a baculovirus in populations of *Oryctes rhinoceros*, *J. Invert. Pathol.*, 1976, **27**, 221.
68. B. D. Gorick, Release and establishment of the baculovirus disease of *Oryctes rhinoceros* (L.) (Coleoptera: Scarabaeidae) in Papua New Guinea, *Bull. Entomol. Res.*, 1980, **70**, 445.
69. N. Suzuki and Y. Kunimi, Dispersal and survival rate of adult females of the fall webworm, *Hyphantria cunea* Drury (Lepidoptera: Arctiidae), using the nuclear polyhedrosis virus as a marker, *Appl. Entomol. Zool.*, 1981, **16**, 374.
70. G. O. Bedford, Biological control of the rhinoceros beetle (*Oryctes rhinoceros*) in the South Pacific by baculovirus, *Agric. Ecosyst. Environ.*, 1986, **15**, 141.
71. J. R. Fuxa and A. R. Richter, Selection for an increased rate of vertical transmission of *Spodoptera frugiperda* (Lepidoptera: Noctuidae) nuclear polyhedrosis virus, *Environ. Entomol.*, 1991, **20**, 603.
72. J. R. Fuxa and A. R. Richter, Lack of vertical transmission in *Anticarsia gemmatalis* (Lepidoptera: Noctuidae) nuclear polyhedrosis virus, a pathogen not indigenous to Louisiana, *Environ. Entomol.*, 1993, **22**, 425.
73. P. F. Entwistle, P. H. W. Adams, and H. F. Evans, Epizootiology of a nuclear polyhedrosis virus in European spruce sawfly (*Gilpinia hercyniae*): the status of birds as dispersal agents of the virus during the larval season, *J. Invert. Pathol.*, 1977, **29**, 354.
74. P. F. Entwistle, P. H. W. Adams, and H. F. Evans, Epizootiology of a nuclear polyhedrosis virus in European spruce sawfly, *Gilpinia hercyniae*: birds as dispersal agents of the virus during winter, *J. Invert. Pathol.*, 1977, **30**, 15.
75. P. F. Entwistle, P. H. W. Adams, and H. F. Evans, Epizootiology of nuclear polyhedrosis virus in European spruce sawfly (*Gilpinia hercyniae*): the rate of passage of infective virus through the gut of birds during cage tests. *J. Invert. Pathol.*, 1978, **31**, 307.

76. P. F. Entwistle, P. H. W. Adams, H. F. Evans, and C. F. Rivers, Epizootiology of a nuclear polyhedrosis virus (Baculoviridae) in European spruce sawfly (*Gilpinia hercyniae*): spread of disease from small epicentres in comparison with spread of baculovirus diseases in other hosts, *J. Appl. Ecol.*, 1983, **20**, 473.
77. R. A. Lautenschlager and J. D. Podgwaite, Passage of nucleopolyhedrosis virus by avian and mammalian predators of the gypsy moth, *Lymantria dispar*, *Environ. Entomol.*, 1979, **8**, 210.
78. R. A. Lautenschlager, J. D. Podgwaite, and D. E. Watson, Natural occurrence of the nucleopolyhedrosis virus of the gypsy moth, *Lymantria dispar* [Lep.: Lymantriidae] in wild birds and mammals, *Entomophaga*, 1980, **25**, 261.
79. F. T. Bird, Transmission of some insect viruses with particular reference to ovarial transmission and its importance in the development of epizootics, *J. Insect Pathol.*, 1961, **3**, 352.
80. F. M. Laigo and M. Tamashiro, Virus and insect parasite interaction in the lawn army worm, *Spodoptera maurita acronyctoides* (Guenée), *Proc. Hawaii Entomol. Soc.*, 1966, **19**, 233.
81. T. A. Irabagon and W. M. Brooks, Interaction of *Campoletis sonorensis* and a nuclear polyhedrosis virus in larvae of *Heliothis virescens*, *J. Econ. Entomol.*, 1974, **67**, 229.
82. C. C. Beegle and E. R. Oatman, Effect of a nuclear polyhedrosis virus on the relationship between *Trichoplusia ni* (Lepidoptera: Noctuidae) and the parasite, *Hyposter exiguae* (Hymenoptera: Ichneumonidae), *J. Invert. Pathol.*, 1975, **25**, 59.
83. J. L. Capinera and P. Barbosa, Transmission of a nuclear-polyhedrosis virus to gypsy moth larvae by *Calosoma sycophanta*, *Ann. Entomol. Soc. Am.*, 1975, **68**, 593.
84. B. Raimo, R. C. Reardon, and J. D. Podgwaite, Vectoring gypsy moth nuclear polyhedrosis virus by *Apanteles melanoscelus* (Hym.: Braconidae), *Entomophaga*, 1977, **22**, 207.
85. D. B. Levin, J. E. Laing, and R. P. Jaques, Transmission of granulosis virus by *Apanteles glomeratus* to its host *Pieris rapae*, *J. Invert. Pathol.*, 1979, **34**, 317.
86. A. G. Beekman, The infectivity of polyhedra of nuclear polyhedrosis virus (NPV) after passage through gut of an insect-predator, *Experientia*, 1980, **36**, 858.
87. D. J. Cooper, The role of predatory Hemiptera in disseminating a nuclear polyhedrosis virus of *Heliothis punctigera*, *J. Aust. Entomol. Soc.*, 1981, **20**, 145.
88. M. A. Mohamed, H. C. Coppel, D. J. Hall, and J. D. Podgwaite, Field release of virus-sprayed adult parasitoids of the European pine sawfly (Hymenoptera: Diprionidae) in Wisconsin, *Great Lakes Entomol.*, 1981, **14**, 177.
89. M. S. T. Abbas and D. G. Boucias, Interaction between nuclear polyhedrosis virus-infected *Anticarsia gemmatalis* (Lepidoptera: Noctuidae) larvae and predator *Podisus maculiventris* (Say) (Hemiptera: Pentatomidae), *Environ. Entomol.*, 1984, **13**, 599.
90. J. J. Hamm, D. A. Nordlund, and O. G. Marti, Effects of a nonoccluded virus of *Spodoptera frugiperda* (Lepidoptera: Noctuidae) on the development of a parasitoid, *Cotesia marginiventris* (Hymenoptera: Braconidae), *Environ. Entomol.*, 1985, **14**, 258.
91. S. Y. Young and W. C. Yearian, *Nabis roseipennis* adults (Hemiptera: Nabidae) as disseminators of nuclear polyhedrosis virus to *Anticarsia gemmatalis* (Lepidoptera: Noctuidae) larvae, *Environ. Entomol.*, 1987, **16**, 1330.

92. T. J. Kring, S. Y. Young, and W. C. Yearian, The striped lynx spider, *Oxyopes salticus* Hentz (Araneae: Oxyopidae), as a vector of a nuclear polyhedrosis virus in *Anticarsia gemmatalis* Hübner (Lepidoptera: Noctuidae), *J. Entomol. Sci.*, 1988, **23**, 394.
93. J. R. Fuxa, A. R. Richter, and M. S. Strother, Detection of *Anticarsia gemmatalis* nuclear polyhedrosis virus in predatory arthropods and parasitoids after viral release in Louisiana soybean, *J. Entomol. Sci.*, 1993, **28**, 51.
94. D. L. Hostetter, A virulent nuclear polyhedrosis virus of the cabbage looper, *Trichoplusia ni*, recovered from the abdomens of sarcophagid flies, *J. Invert. Pathol.*, 1971, **17**, 130.
95. H. D. Burges, (ed.), 'Microbial Control of Pests and Plant Diseases 1970–1980', Academic Press, London, 1981.
96. J. C. Cunningham, Field trials with baculoviruses: control of forest insect pests, in 'Microbial and Viral Pesticides', ed. E. Kurstak, Marcel Dekker, New York, 1982, p. 335.
97. W. C. Yearian and S. Y. Young, Control of insect pests of agricultural importance by viral insecticides, in 'Microbial and Viral Pesticides', ed. E. Kurstak, Marcel Dekker, New York, 1982, p. 387.
98. J. R. Fuxa, Ecological considerations for the use of entomopathogens in IPM, *Annu. Rev. Entomol.*, 1987, **32**, 225.
99. J. R. Fuxa, Insect resistance to viruses, in 'Parasites and Pathogens of Insects', ed. N. E. Beckage, S. N. Thompson and B. A. Federici, Academic Press, San Diego, 1993, 197.
100. L. L. J. Ossowski, Variation in virulence of a wattle bagworm virus, *J. Insect Pathol.*, 1960, **2**, 35.
101. R. G. Harrison, S. F. Wintermeyer, and T. M. Odell, Patterns of genetic variation within and among gypsy moth, *Lymantria dispar* (Lepidoptera: Lymantriidae), populations, *Ann. Entomol. Soc. Am.*, 1983, **76**, 652.
102. J. R. Fuxa, *Spodoptera frugiperda* susceptibility to nuclear polyhedrosis virus isolates with reference to insect migration, *Environ. Entomol.*, 1987, **16**, 218.
103. K. Aizawa, Strain improvement and preservation of virulence of pathogens, in 'Microbial Control of Insects and Mites', ed. H. D. Burges and N. W. Hussey, Academic Press, London, 1971, p. 655.
104. J. Brassel and G. Benz, Selection of a strain of the granulosis virus of the codling moth with improved resistance against artificial ultraviolet radiation and sunlight, *J. Invert. Pathol.*, 1979, **33**, 358.
105. M. Shapiro and R. A. Bell, Selection of a UV-tolerant strain of the gypsy moth, *Lymantria dispar* (L.) (Lepidoptera: Lymantriidae), nucleopolyhedrosis virus, *Environ. Entomol.*, 1984, **13**, 1522.
106. C. M. Ignoffo and T. L. Couch, The nucleopolyhedrosis virus of *Heliothis* species as a microbial insecticide, in 'Microbial Control of Pests and Plant Diseases 1970–1980', ed. H. D. Burges, Academic Press, London, 1981, p. 330.
107. M. Shapiro, *In vivo* production of baculoviruses, in 'The Biology of Baculoviruses. Volume II. Practical Application for Insect Control', ed. R. R. Granados and B. Federici, CRC Press, Boca Raton, 1986, p. 31.
108. M. Shapiro, *In vivo* mass production of insect viruses for use as pesticides, in 'Microbial and Viral Pesticides', ed. E. Kurstak, Marcel Dekker, New York, 1982, p. 463.
109. S. A. Weiss and J. L. Vaughn, Cell culture methods for large-scale propagation of baculoviruses, in 'The Biology of Baculoviruses. Volume II. Practical Application

for Insect Control', ed. R. R. Granados and B. A. Federici, CRC Press, Boca Raton, 1986, p. 63.

110. T. L. Couch and C. M. Ignoffo, Formulation of insect pathogens, in 'Microbial Control of Pests and Plant Diseases 1970–1980', ed. H. D. Burges, Academic Press, London, 1981, 621.
111. S. Y. Young, III and W. C. Yearian, Formulation and application of baculoviruses, in 'The Biology of Baculoviruses. Volume II. Practical Application for Insect Control', ed. R. R. Granados and B. A. Federici, CRC Press, Boca Raton, 1986, p. 157.
112. C. T. Chin, Insect toxicology and insect pathology in the People's Republic of China, *Acta Entomol. Sinica*, 1979, **22**, 249 (in Chinese).
113. G.-Y. Zhang and C.-Z. Bai, Research and development of first commercial viral pesticide – *Heliothis* nuclear polyhedrosis virus pesticide in China, in 'Proceedings XIX International Congress of Entomology', Beijing, China, 1992, p. 264.
114. A. R. Jutsum, Commercial application of biological control: status and prospects, *Phil. Trans. R. Soc. Lond. B*, 1988, **318**, 357.
115. R. P. Jaques, J. E. Laing, D. R. Laing, and D. S. K. Yu, Effectiveness and persistence of the granulosis virus of the codling moth *Cydia pomonella* (L.) (Lepidoptera: Olethreutidae) on apple, *Can. Entomol.*, 1987, **119**, 1063.
116. M. S. Goettel, 'Directory of Industries Involved in the Development of Microbial Control Products', Society for Invertebrate Pathology, 1991.
117. J. C. Cunningham, Baculoviruses: their status compared to *Bacillus thuringiensis* as microbial insecticides, *Outlook Agric.*, 1988, **17**, 10.
118. F. Moscardi, Use of viruses for pest control in Brazil: the case of the nuclear polyhedrosis virus of the soybean caterpillar, *Anticarsia gemmatalis*, *Mem. Inst. Oswaldo Cruz*, 1989, **84** (Suppl. III), 51.
119. F. Moscardi and D. R. Sosa-Gomez, Use of viruses for insect control in Brazil, in 'Proceedings XIX International Congress of Entomology', Beijing, China, 1992, p. 301.
120. F. Moscardi and B. S. C. Ferreira, Biological control of soybean caterpillars, in 'World Soybean Research Conference III: Proceedings', ed. R. Shibles, Westview Press, London, 1985, p. 703.
121. J. D. Harper, Applied epizootiology: microbial control of insects, in 'Epizootiology of Insect Diseases', ed. J. R. Fuxa and Y. Tanada, Wiley, New York, 1987, p. 473.
122. B. Zelazny, A. Lolong, and B. Pattang, *Oryctes rhinoceros* (Coleoptera: Scarabeidae) populations suppressed by a baculovirus, *J. Invert. Pathol.*, 1992, **59**, 61.
123. C.-J. Chen, Pine caterpillar control by *Dendrolimus* cytoplasmic polyhedrosis viruses in China, 'Proceedings XIX International Congress of Entomology', Beijing, China, 1992, p. 266.
124. H. D. Burges and N. W. Hussey, Past achievements and future prospects, in 'Microbial Control of Insects and Mites', ed. H. D. Burges and N. W. Hussey, Academic Press, London, 1971, p. 687.
125. H. D. Burges, Strategy for the microbial control of pests in 1980 and beyond, in 'Microbial Control of Pests and Plant Diseases 1970–1980', ed. H. D. Burges, Academic Press, London, 1981, p. 797.
126. O. N. Morris, J. C. Cunningham, J. R. Finney-Crawley, R. P. Jaques, and G. Kinoshita, Microbial insecticides in Canada: their registration and use in agriculture, forestry and public and animal health, *Bull. Entomol. Soc. Can. (Suppl.)*, 1986, **18**, 1.

127. J. C. Cunningham and K. van Frankenhuyzen, Microbial insecticides in forestry, *Forest. Chron.*, 1991, **67**, 473.
128. K. Katagiri, Pest control by cytoplasmic polyhedrosis viruses, in 'Microbial Control of Pests and Plant Diseases 1970–1980', ed. H. D. Burges, Academic Press, London, 1981, p. 433.
129. P. O. Hutton and P. B. Burbutis, Milky disease and Japanese beetle in Delaware, *J. Econ. Entomol.*, 1974, **67**, 247.
130. J. R. Fuxa, New directions for insect control with baculoviruses, in 'New Directions in Biological Control. Alternatives for Suppressing Agricultural Pests and Diseases', ed. R. R. Baker and P. E. Dunn, Liss, New York, 1990, p. 97.
131. J. R. Fuxa, A. R. Richter, and P. J. McLeod, Virus kills soybean looper years after its introduction into Louisiana, *Louisiana Agric.*, 1992, **35**(3), 20.
132. A. C. Bellotti and J. A. Reyes, South and Central America, in 'Proceedings of a Workshop on Insect Pest Management with Microbial Agents: Recent Achievements, Deficiencies, and Innovations', Boyce Thompson Institute, Ithaca, NY, 1980, p. 20.
133. J. Kalmakoff and A. M. Crawford, Enzootic virus control of *Wiseana* spp. in the pasture environment, in 'Microbial and Viral Pesticides', ed. E. Kurstak, Marcel Dekker, New York, 1982, p. 435.
134. J. R. Fuxa and J. P. Geaghan, Multiple-regression analysis of factors affecting prevalence of nuclear polyhedrosis virus in *Spodoptera frugiperda* (Lepidoptera: Noctuidae) populations, *Environ. Entomol.*, 1983, **20**, 603.
135. M. Shapiro and J. L. Robertson, Enhancement of gypsy moth (Lepidoptera: Lymantriidae) baculovirus activity by optical brighteners, *J. Econ. Entomol.*, 1992, **85**, 1120.
136. M. Shapiro, Use of optical brighteners as radiation protectants for gypsy moth (Lepidoptera: Lymantriidae) nuclear polyhedrosis virus, *J. Econ. Entomol.*, 1992, **85**, 1682.
137. J. J. Hamm and M. Shapiro, Infectivity of fall army worm (Lepidoptera: noctuidae) nuclear polyhedrosis virus enhanced by a fluorescent brightener, *J. Econ. Entomol.*, 1992, **85**, 2149.
138. H. A. Wood and R. R. Granados, Genetically engineered baculoviruses as agents for pest control, *Ann. Rev. Microbiol.*, 1991, **45**, 69.
139. S. Maeda, Expression of foreign genes in insects using baculovirus vectors, *Ann. Rev. Entomol.*, 1989, **34**, 351.
140. J. M. Vlak, A. Schouten, M. Usmany, G. J. Belsham, E. C. Klingeroode, A. J. Maule, J. W. M. van Lent, and D. Zuidema, Expression of cauliflower mosaic virus gene I using a baculovirus vector based upon the p10 gene and a novel selection method, *Virology*, 1990, **179**, 312.
141. U. Weyer, S. Knight, and R. D. Possee, Analysis of the very late gene expression by *Autographa californica* nuclear polyhedrosis virus and the further development of multiple expression vectors, *J. Gen. Virol.*, 1990, **71**, 1525.
142. X. Wang, B. G. Ooi, and L. K. Miller, Baculovirus expression vectors for multiple gene expression and for occluded virus production, *Gene*, 1991, **100**, 131.
143. P. Kitts, M. D. Ayres, and R. D. Possee, Linearization of baculovirus DNA enhances the recovery of recombinant virus expression vectors, *Nucleic Acids Res.*, 1990, **18**, 5667.
144. T. C. Peakman, R. A. Harris, and D. R. Gewert, Highly efficient generation of recombinant baculoviruses by enzymatically mediated site-specific *in vitro* recombination, *Nucleic Acids Res.*, 1992, **20**, 495.

145. G. Patel, K. Nasmyth, and N. Jones, A new method for the isolation of recombinant baculovirus, *Nucleic Acids Res.*, 1992, **20**, 97.
146. D. R. O'Reilly and L. K. Miller, Improvement of a baculovirus pesticide by deletion of the EGT gene, *Bio/Technology*, 1991, **9**, 1086.
147. S. Maeda, Increased insecticidal effect by a recombinant baculovirus carrying a synthetic diuretic hormone gene, *Biochem. Biophys. Res. Commun.*, 1989, **165**, 1177.
148. T. N. Hanzlik, Y. A. I. Abdel-Aal, L. G. Harshman, and B. D. Hammock, Isolation and characterization of cDNA clones coding for juvenile hormone esterase from *Heliothis virescens*. Evidence for a catalytic mechanism for the serine carboxylesterase different from that of the serine proteases, *J. Biol. Chem.*, 1989, **264**, 12419.
149. B. D. Hammock, B. C. Bonning, R. D. Possee, T. N. Hanzlik, and S. Maeda, Expression and effects of the juvenile hormone esterase in a baculovirus vector, *Nature (London)*, 1990, **344**, 458.
150. L. M. D. Stewart, M. Hirst, M. L. Ferber, A. T. Merryweather, P. J. Cayley, and R. D. Possee, Construction of an improved baculovirus insecticide containing an insect-specific toxin gene, *Nature (London)*, 1991, **352**, 85.
151. B. F. McCutchen, P. V. Choudary, R. Crenshaw, D. Maddox, S. G. Kamita, N. Palekar, S. Volrath, E. Fowler, B. D. Hammock, and S. Maeda, Development of a recombinant baculovirus expressing an insect-selective neurotoxin: Potential for pest control, *Bio/Technol.*, 1991, **9**, 848.
152. S. Maeda, S. L. Volrath, T. N. Hanzlik, S. A. Harper, K. Majima, D. W. Maddoz, B. D. Hammock, and E. Fowler, Insecticidal effects of an insect-specific neurotoxin expressed by a recombinant baculovirus, *Virology*, 1991, **184**, 777.
153. M. D. Tomalski, R. Kutney, W. A. Bruce, M. R. Brown, M. S. Blum, and J. Travis, Purification and characterization of insect toxins derived from the mite, *Pyemotes tritici*, *Toxicon*, 1989, **27**, 1151.
154. M. D. Tomalski, W. A. Bruce, J. Travis, and M. S. Blum, Preliminary characterization of toxins from the straw itch mite, *Pyemotes tritici*, which induce paralysis in the larvae of a moth, *Toxicon*, 1988, **26**, 127.
155. M. D. Tomalski and L. K. Miller, Insect paralysis by baculovirus mediated expression of a mite neurotoxin gene, *Nature (London)*, 1991, **352**, 82.
156. M. D. Tomalski and L. K. Miller, Expression of a paralytic neurotoxin gene to improve insect baculoviruses as biopesticides, *Bio/Technol.*, 1992, **10**, 545.
157. A. Kondo and S. Maeda, Host range expansion by recombination of the baculoviruses *Bombyx mori* nuclear polyhedrosis virus and *Autographa californica* nuclear polyhedrosis virus, *J. Virol.*, 1991, **65**, 3625.
158. B. Maiorella, D. Inlow, A. Shaughen, and D. Harano, Large-scale insect cell culture for recombinant protein production, *Bio/Technol.*, 1988, **6**, 1406.
159. F. L. J. van Lier, M. Kool, E. J. van den End, C. D. de Gooijer, M. Usmany, J. M. Vlak, and J. Tramper, Production of baculovirus or recombinant derivatives in continuous insect-cell bioreactors, *Ann. NY Acad. Sci.*, 1990, **613**, 183.
160. M. Laird, L. A. Lacey, and E. W. Davidson, (ed.), 'The Safety of Entomopathogens', CRC Press, Boca Raton, 1990.
161. L. D. Newsom, The role of entomopathogens in pest management systems. Opening remarks, in 'Microbial Control of Insect Pests: Future Strategies in Pest Management Systems', ed. G. E. Allen, C. M. Ignoffo, and R. P. Jaques, NSF-USDA-Univ. Florida Workshop, Gainesville, 1978, p. 131.
162. C. M. Ignoffo, Tests used in the United States to evaluate the safety of insect

viruses, in 'Characterization, Production and Utilization of Entomopathogenic Viruses', ed. C. M. Ignoffo, M. E. Martignoni, and J. L. Vaughn, American Society of Microbiology – National Science Foundation, 1980, p. 162.

163. A. K. Chock and M. J. Dover, Registration, in 'Proceedings of a Workshop on Insect Pest Management with Microbial Agents: Recent Achievements, Deficiencies, and Innovations', Insect pathology Resource Center, Boyce Thompson Institute, Ithaca, NY, 1980, p. 44.
164. F. Betz, A. Rispin, and W. Schneider, Biotechnology products related to agriculture. Overview of regulatory decisions at the US Environmental Protection Agency, in 'Biotechnology in Agricultural Chemistry', ed. H. M. LeBaron, R. O. Mumma, R. C. Honeycutt, and J. H. Duesing, ACS Symposium Series 334, American Chemical Society, Washington, DC, 1987, p. 316.
165. M. H. Rogoff, Regulatory safety data requirements for registration of microbial pesticides, in 'Microbial and Viral Pesticides', ed. E. Kurstak, Marcel Dekker, New York, 1982, p. 645.
166. F. S. Betz, Registration of baculoviruses as pesticides, in 'The Biology of Baculoviruses. Volume II. Practical Application for Insect Control', ed. R. R. Granados and B. A. Federici, CRC Press, Boca Raton, 1986, p. 203.
167. J. R. Fuxa, Risk assessment of genetically engineered entomopathogens: effects of microbial control agents on the environment including their persistence and dispersal, in 'Environmental Science and Engineering Fellows Report', American Association Advisory Scientists/US Environmental Protection Agency, Washington, DC, 1987.
168. J. R. Fuxa, Environmental risks of genetically engineered entomopathogens, in 'The Safety of Entomopathogens', ed. M. Laird, L. A. Lacey, and E. W. Davidson, CRC Press, Boca Raton, 1990, p. 203.
169. National Academy of Sciences, 'Introduction of Recombinant DNA-Engineered Organisms into the Environment: Key Issues', National Academy Press, Washington, DC, 1987.
170. P. J. Regal, Models of genetically engineered organisms and their ecological impact, in 'Ecology of Biological Invasions of North America and Hawaii', ed. H. A. Mooney and J. A. Drake, Springer Verlag, New York, 1986, p. 111.
171. S. A. Levin and M. A. Harwell, Environmental risks and genetically engineered organisms, in 'Biotechnology. Implications for Public Policy', ed. S. Panem, The Brookings Institution, Washington, DC, 1985, p. 56.
172. J. M. Tiedje, R. K. Colwell, Y. L. Grossman, R. E. Hodson, R. E. Lenski, R. N. Mack, and P. J. Regal, The planned introduction of genetically engineered organisms: ecological considerations and recommendations, *Ecology*, 1989, **70**, 298.
173. Interregional Research Project-4, Biopesticides in the 1990s, *IR-4 Newsletter*, 1990, **21**(3), 6.
174. J. L. Kerwin, EPA registers *Lagenidium giganteum* for mosquito control, *SIP Newsletter*, *Soc. Invert. Pathol.*, 1992, **24**(2), 8.
175. M. J. Adang, *Bacillus thuringiensis* insecticidal crystal proteins: gene structure, action, and utilization, in 'Biotechnology for Biocontrol of Pests and Vectors', ed. K. Maramorosch, CRC Press, Boca Raton, 1991, p. 3.
176. J. M. Vlak, Genetic engineering of baculoviruses, in 'Opportunities for Molecular Biology in Crop Production', ed. D. J. Beadle, D. H. Bishop, L. C. Copping, G. K. Dixon, and D. W. Hollomon, BCPC Monograph 55, BCPC, Farnham, 1993, p. 11.

177. R. D. Possee, D. Hurst, L. O. Jones, D. H. L. Bishop, and P. J. Cayley, Field tests of genetically engineered baculoviruses, in 'Opportunities for Molecular Biology in Crop production', ed. D. J. Beadle, D. H. L. Bishop, L. G. Copping, G. K. Dixon, and D. W. Hollomon, BCPC Monograph 55, BCPC, Farnham, 1993, p. 23.
178. D. H. Bishop, R. A. Harris, D. Hirst, A. T. Merryweather, and R. D. Possee, BCPC Monograph 43, 1989.
179. Cory, D. H. Bishop *et al.*, Field trials of a genetically improved baculovirus, *Nature*, 1994, **370**, 138.

CHAPTER 12

Micro-organisms as Agents in Plant Disease Control

P. LAWRENCE PUSEY

12.1 Historical Background

For more than 100 years, it has been known that micro-organisms interact with each other.[1] Roberts[2] was the first to note the inhibition of one micro-organism by another. This phenomenon was subsequently observed by many other workers, but was not exploited until the discovery of the antibiotic penicillin by Fleming in 1928, and its purification and use in medicine by Florey and co-workers in 1939.[3] The advent of antibiotics in medicine is thought to be the single greatest stimulus to studies of micro-organisms antagonistic to plant pathogens.[4]

The first attempts at biological control of plant pathogens with introduced microbial antagonists were made in the 1920s and 1930s. Hartley[5] introduced antagonistic fungi, isolated from soil, to control damping-off of pine seedlings (caused by *Pythium debaryanum*) in partially sterilized soil. Millard and Taylor[6] were able to control potato scab (caused by *Streptomyces scabies*) by the addition of green grass cuttings plus the antagonist *S. praecox*, an obligate saprophyte, to sterilized soil in pot tests. No antibiosis was exhibited by *S. praecox* in culture, but numbers of *S. scabies* were decreased in the presence of *S. praecox* in soil. It was speculated that competition for nutrients was involved. Henry[7] reported that isolates of actinomycetes, bacteria, and fungi from soil, when applied in various combinations to sterilized soil, slightly reduced the disease of wheat caused by *Helminthosporium sativum*.

In the ensuing years there was a continuous, although slow, turnout of reports relating to biocontrol of plant diseases through the introduction of antagonistic micro-organisms. Many of the early experiments were unsuccessful and, consequently, plant pathologists developed negative attitudes towards this approach to disease control. The mood began to change in the 1960s when the first international conference on biocontrol of plant pathogens was held.[8] It was also during this period that Rishbeth[9] reported on the control of *Fomes* (now *Heterobasidion*) *annosus* by *Peniophora* (now *Phlebia*) *gigantea*, which became the first commercially available biocontrol agent for the control of a plant pathogen. This fungal antagonist, which colonizes tree stumps and competitively excludes *Fomes*, has been marketed on a small scale. In 1974, the first book devoted wholly to the subject of biocontrol of plant pathogens was published.[10] Since then the volume of information on biocontrol has increased dramatically.

The success of chemical fungicides has, in general, worked against the development of biocontrol technology. While chemicals have offered a relatively cheap and effective means of controlling plant diseases, there has been little or no incentive to develop alternative methods. However, recent concerns about pesticides in relation to the environment and food safety have led to increased support for research on biocontrol.

There are other commercial reasons for the current interest in biocontrol agents. Fungal pathogens often develop resistance to fungicides, so new chemicals are constantly needed. At the same time, it is becoming more difficult, and thus more expensive, to find new antifungal compounds.[11–13] A further stumulus is the prospect that development and registration will cost less for biological agents than for chemicals.[14] However, biocontrol agents must be subjected to toxicological testing. It is important that they are proved safe for humans and the environment before they are put into commercial use.

12.2 Antagonists in the Soil Environment

Biological control of soil pathogens has been studied far more extensively than biocontrol of pathogens in other environments.[4,15–18] Campbell,[14] who presented reasons for this, pointed out that plant breeders and agrochemical companies have been successful in improving disease control on the aerial parts of plants, but have given relatively little attention to root diseases. Often, diseases that occur below ground have escaped attention, while the more noticeable leaf and stem diseases have been perceived as the major limitations on production. Many of these aerial diseases are now controlled through improved host resistance or with fungicides. Consequently, root diseases have become recognized as the limiting factors in many crop production systems.

Soil pathogens have traditionally been managed by cultural practices,[19] such as crop rotation, tillage methods, or the addition of organic substances. It is thought that natural biological control has been involved in these systems. Tillage not only facilitates a rapid breakdown of crop residues, which serve as a food base for pathogens, but it also exposes pathogens to antagonists. Crop rotations are effective because the pathogen must survive without its host for a prolonged period, and inoculum potential is reduced by the activity of soil microbes.

Despite extensive research on the use of microbial antagonists to suppress soil pathogens, few of these organisms have been developed commercially as biocontrol agents. An overwhelming problem has been the lack of consistency in results from year to year and with different climatic and soil types. The variability is attributed to many factors, including:

- Clays in the soil that adsorb the antagonist or its metabolites.
- Unfavourable weather leading to antagonist death.
- Instability of antagonist cultures.
- Other soil microbes that inhibit or kill the antagonist.

In addition, it has been a major challenge to distribute the antagonist uniformly through the inert, virtually immobile soil system.[20] Hence, it must be applied in relatively large and unwieldy quantities to be effective. Application rates of 6000 kg ha^{-1} are not uncommon.[21]

Although variability is encountered with soil systems, the composition of microbial populations in soil of a given type and location is generally quite stable. It is, therefore, difficult to introduce a 'foreign' organism into soil where it does not already exist. Conversely, incorporation may be accomplished by altering resident populations through cultivation practices or by more drastic measures, such as partial sterilization by fumigation.

It is usually desirable that the introduced organism be able to survive for a very long time. However, an antagonist that persists indefinitely in the soil may raise greater concerns from environmental protection agencies than one that survives only for a short period. Also, commercial development and production of some long-time survivors (especially those sold to small markets) may not be profitable because of the lack of repeat sales.

12.2.1 Seed Treatments

Conventional seed treatments involve coating the seed with fungicides to protect them from decay during germination or to eliminate seed-borne fungal pathogens. This has been especially important for crops with large seeds that are vulnerable to decay by fungi, such as *Fusarium* spp., *Pythium* spp., and *Rhizoctonia solani.* Although chemical seed treatments are relatively inexpensive and safe compared with other pesticides, advantages of biological seed treatment as compared to chemicals may include:

- Less disruption to natural microflora in the soil.
- Ability of organisms to increase and provide subsequent protection of root systems.
- Ability of organisms to induce growth responses.
- The lower risk of using an excess of seed to feed poultry and livestock.

Fungal antagonists that have been most effective as biological seed treatments are species of *Chaetomium*, *Penicillium*, and *Trichoderma*.[4] Treatment with *C. globosum* was as effective as thiram or captan on corn seed[22] and *P. oxalicum* equalled captan on pea seed.[23] Strains of *T. harzianum* derived by protoplast fusion were consistently as effective as thiram in seed treatment trials that involved a number of pathogen and host combinations.[24]

Among the most effective bacterial antagonists used on seeds or seed-species have been *Bacillus subtilis*, *Streptomyces* spp., and strains of the *Pseudomonas fluorescens-putida* group. Evidence for the potential of *B. subtilis* was provided by researchers in Australia,[25] who reported dramatic yield increases with strain A13 tested on cereals and carrots, and by others in Canada,[26] who showed that four strains protected onions from *Sclerotium cepivorum*. In the People's Republic of China, *Streptomyces* strain 5406 has been used with success for several decades in the treatment of cotton seed.[27,28]

Species of *Pseudomonas*, particularly *P. fluorescens* and *P. putida*, are now considered by many workers to be prime candidates for biocontrol in soil environments. Various strains and their secondary metabolites have been patented and some are marketed commercially. These organisms are:

- Normal inhibitants of the soil and root surfaces.
- Relatively easy to isolate and identify.
- Easy to culture.
- Nutritionally versatile.
- Produce a variety of antibiotics and siderophores.
- Have been implicated in connection with plant growth promotion.

Examples of biocontrol with pseudomonads applied by seed or seed-piece treatment include studies with potato,[29,30] onion,[31] sugar beet,[32] and wheat.[33] In studies with *P. fluorescens* applied to wheat seed,[34] yield increased 10–27% in fields where take-all (caused by *Gaeumannomyces graminis*) had been serious. There is now considerable evidence that strains of *P. fluorescens* control take-all through both antibiotic and siderophore production.[35,36] Siderophores are low molecular weight, high-affinity iron(III) chelators which, under low-iron conditions, can supply iron to bacterial cells, but limit iron availability to plant pathogens.[37]

12.2.2 Preplant Treatment of Roots

Microbial antagonists may be established on cuttings or transplants of ornamentals, vegetable seedlings, young fruit trees, or other plants that are propagated initially in a nursery and moved to the field. The antagonist can be applied by incorporation in the nursery medium or by dipping the cutting or transplant. Biocontrol of crown gall (caused by *Agrobacterium radiobacter* pv. *tumefaciens*) with *A. radiobacter* K84 involves dipping the cuttings, transplants, or sometimes seeds of the susceptible plants into a cell suspension or commercial preparation of the biocontrol bacterium and then planting or sowing immediately.[38] Treatment with this antagonist, which produces a bacteriocin, has proved effective over much of the world. It is the most widely used biocontrol of plant pathogens by an introduced antagonist.

The inoculation of plants with mycorrhizal fungi is most commonly done during production of the seedling plants or at the time of transplanting. This practice can result in biocontrol, either through the protection of roots from pathogens or by the enhancement of host resistance. The effects of vesicular-arbuscular mycorrhizae on disease are complex,[39] but usually beneficial.[40]

12.2.3 Soil Treatments

Direct application of antagonists to soil, either as a broadcast incorporation or as an application in the seed furrow, has greatest potential for use in plant production systems that involve a limited amount of soil or land (*e.g.*, glasshouse operations). However, this approach could also become economically feasible in some systems involving large fields. One reason for adding antagonists to soil is to destroy pathogen inoculum. This has been attempted with hyperparasites, including the widely tested *Trichoderma* spp.[41,42] The efficacy of these hyperparasites has been established; but, in most cases the approach is not economically comparable to conventional methods, or the technology is inadequate for delivering the antagonist. A particularly promising hyperparasite is *Sporidesmium sclerotivorum*, which invades and destroys sclerotia of *Sclerotina* spp., *Sclerotium cepivorum*, *Botrytis* spp., and *Claviceps purpurea* in field soil.[43] This fungus controlled lettuce-drop caused by *Sclerotinia minor* in the field[44] and increased its population once established in soil.[45] Currently, it is being developed commercially as a biocontrol agent. The major disadvantage with *S. sclerotivorum* is that it is highly specific to sclerotia and cannot be mass produced with conventional substrates.

Another reason why antagonists might be applied to soil is to prevent pathogen recolonization of soil that has been treated with steam or fumigants. Elad *et al.* reported[46] that the introduction of *T. harzianum* into strawberry nursery beds following soil fumigation gave significant protection of plants from black root rot caused by *Rhizoctonia solani*. Application of this antagonist to fumigated soil in field plots resulted in an 88% reduction in reinfestation by *Sclerotium rolfsii* and *R. solani*.[47]

In addition, antagonists may be incorporated in soil to protect germinating seeds and roots from infection. When fluorescent pseudomonads were added to conducive soil planted with seeds of flax, cucumber, or radish, the soil became suppressive to Fusarium wilt pathogens that normally infect these plants.[48,49] Evidence was presented to indicate that the biocontrol involved siderophore production by the antagonists.

12.3 Antagonists in the Aerial Environment

12.3.1 Treatment of Aerial Surfaces of Plants

The methodology of applying biocontrol agents to aerial surfaces of plants has developed slowly, as compared with the dramatic successes in early work with chemical controls. However, research in this area, particularly as it relates to the phylloplane (leaf surface), has intensified in recent years.[50–54]

The environment surrounding the above-ground portion of plants is very different from that in the soil. On aerial surfaces, the environment is dynamic with cyclic and non-cyclic variables, including temperature, relatively humidity, dew, rain, wind, and radiation. Although variable itself, the soil medium removes some environmental factors (*e.g.*, light and wind) and attenuates variation in others (*e.g.*, moisture and temperature).

During the growing period in temperate regions, bacteria, yeasts, and filamentous fungi grow on the leaves, stems, flowers, and fruit of plants. These micro-organisms live on nutrients from plant exudates or exogenous deposits, such as dust, pollen, and insect excreta. Blakeman and Brodie[55] recognized that chemical or nutritional variables can be manipulated more easily than the physical microenvironment to encourage epiphytic bacteria. Morris and Rouse[56] determined the differential ability of epiphytic bacteria from snap bean leaves to utilize single carbon and nitrogen sources. By applying selective nutrients to foliage in field plots, they modified the composition of the bacterial community, altered the population size of fluorescent pseudomonads, and, in some cases, reduced disease caused by *Pseudomonas syringae*.

Microbial populations on plant surfaces fluctuate widely, depending on environmental factors, and are negatively affected by low moisture conditions. Introduced organisms most often die out quickly or are not maintained in numbers sufficient to be effective as biocontrol agents. Early attempts by Leben and Daft[57] to control foliar diseases of cucumber, tomato, and corn with an epiphytic bacterium were encouraging in greenhouse studies, but were a total failure in field trials because of poor antagonist survival.[58] Spurr[59] demonstrated that strains of *Pseudomonas cepacia* and *Bacillus thuringiensis* provided limited, but significant, control of peanut Cercospora leaf spot in field trials. These results were presented by Spurr and Knudsen[60] as an example of the 'silver bullet' approach. It has been their thesis that such results from trial and error demonstrate the potential of biocontrol of foliar diseases in the field, but real progress will be realized only when the complex interactions that make up the microecology of leaf surfaces are better understood. They proposed the use of systems analysis and modelling as research tools in this effort. Andrews[50,51] later echoed these expressed views and conceptualized further with regard to biocontrol on the phylloplane.

An aerial disease that has been the target of biocontrol research from as early as the 1930s[61,62] is fireblight (caused by *Erwinia amylovora*), which develops primarily from infections initiated in the flowers of rosaceous plants, such as apple and pear. *Erwinia herbicola* has received the most attention as a potential biocontrol agent against this disease.[61,63] Although antibiotic production by strains of *E. herbicola* has been correlated with control of fireblight in the orchard,[64,65] trials with antibiotic-deficient mutants[66] indicate that factors other than antibiosis are involved. Based on electron microscopy studies,[67] prior colonization by the antagonist on the stigma effectively prevents the pathogen from gaining access to this site. More recently, *Pseudomonas fluorescens* strain A506 was also shown to be a promising candidate for biocontrol of fire blight.[62,68,69] This bacterium does not exhibit antibiosis against *E. amylovora* in culture[62] and probably prevents fire blight in the field by pre-emptively excluding the pathogen.[69] Antagonist populations on flowers were higher when *P. fluorescens* A506 and *E. herbicola* C9-1 were applied in combination than when either was applied alone.[70] Recently, an interesting system involving the employment of honey bees has been used to deliver bacterial antagonists to apple and pear flowers.[68] Johnson *et al.*[71] demonstrated control of fire blight of pear in the field with A506 and C9-1 applied by the bee-dispersal method.

Biocontrol agents have also been used to reduce airborne inoculum of pathogens affecting aerial parts of plants. Heye and Andrews[72] showed that the fungi *Chaetomium globosum* and *Athelia bombacina* inhibited or prevented the overwinter development of sexual fruiting bodies of *Venturia inaequalis*, the cause of apple scab, when the antagonists were applied to leaves on the orchard floor. Later trials in the field with improved formulations resulted in the reduction of airborne spores by 60–70% with *C. globosum* and by 100% with *A. bombacina.*[73]

12.3.2 Treatment of Wounds and Internal Tissues

Much work has been done on the use of biocontrol agents to protect tree wounds, particularly those resulting from pruning. The silver leaf disease of fruit trees caused by *Chondrostereum purpureum* can be prevented or cured by inoculation of the tree with a strain of *Trichoderma viride* antagonistic to the pathogen.[74] The antagonist has been introduced by using modified pruning shears which permit automatic inoculation at the time of pruning[74] or by implanting commercially available wood dowels or pellets impregnated with *Trichoderma*.[4,14,75] In Australia, application of *Fusarium lateritium* to pruning wounds on apricot protected trees against the pathogen *Eutypa armeniacae*.[76] Wound protection was further improved by applying benomyl along with *Fusarium*, which is much more tolerant than *Eutypa* to benzimidazole fungicides. Swinburne and co-workers[77] were successful in using isolates of *Bacillus subtilis* on apple to protect leaf scar wounds from the canker fungus *Nectria galligena.*

Biocontrol of chestnut blight caused by *Endothia parasitica* (now *Cryphonectria parasitica*) became a possibility when it was noticed in Italy in the early 1950s[78] that cankers were healing or growing very slowly. By the 1970s[79] the disease had declined to tolerable levels in some regions of Europe. Strains of the pathogen with low virulence were isolated from these areas.[80] This hypovirulence is transmitted cytoplasmically and virulent strains are transformed into hypovirulent types. Most of the hypovirulent strains contain virus-like particles consisting of one or more species of double stranded ribose nucleic acid (dsRNA) within membrane-bound vesicles.[81] Transmission depends largely on the fusion of hyphae of compatible fungi; but, unfortunately, there are many vegetative compatibility groups.[82] However, in Europe, chestnut trees in large areas are being treated commercially by introducing mixtures of hypovirulent strains that match local vegetative compatibility types.[14] This form of biocontrol also appears feasible in North America, where problems were encountered initially because of incompatibility among fungal strains.[83]

Dutch elm disease is another destructive disease of trees that has been the subject of much research during this century. The pathogen is the fungus *Ceratocystis ulmi*, which is spread by bark beetles (*Scolytus* spp.) and invades the vascular system of trees to cause wilt symptoms. Attempts at biocontrol as a possible solution have been made by a number of researchers, but little success has been realized thus far. Nevertheless, work during the past 15 years has raised hopes that biocontrol agents might be used in the future to limit *C. ulmi*. Weber[84] discovered that *Phomopsis oblongata* colonizes infected trees in northern England and Scotland and is detrimental to the beetle larvae. Several saprophytic fungi, notably *Botryosphaeria stevensii* and *B. ribis*, were found to compete with *C. ulmi* and suppress inoculum formation.[85] It has also been noted that dsRNA may be related to the virulence of *C. ulmi*,[86] as with the chestnut blight fungus, although transmission of this factor between fungal strains has never been demonstrated. Strobel and co-workers[87] injected trees in the greenhouse and field with the bacterium *Pseudomonas syringae* and were moderately successful in protecting trees from *C. ulmi*.

One success story involving biocontrol of a pathogen that infects internal tissues is the commercial use of mild strains of citrus tristeza virus (CTV) to control severe strains of this virus in Brazil. Studies begun in 1951[88] led to commercial testing in 1968 with CTV-protected budwood on a tristeza-tolerant rootstock.[89] By 1980, eight million trees in Brazil had been cross-protected by this method.[90]

12.4 Antagonists in the Postharvest Environment

The best esimates place losses of harvested crops from spoilage at around 24% in the USA[91] and 50% in underdeveloped, tropical countries.[92] Current losses world-wide are probably in the range 10–30% on most crops and as high as 30–50% for some perishable crops.[93] During the past several decades there has been a growing dependency by the fruit and vegetable industry on the use of fungicides to control storage rots.[94,95] However, recent concerns about the health risks associated with fungicides[96] have led to the removal or voluntary withdrawal of a number of compounds from the market for some or all postharvest use. In 1991 the European Parliament voted in favour of a total ban on the postharvest treatment of fruits and vegetables with pesticides as soon as this practice becomes feasible.[97] There is clearly an urgent need for alternative methods of postharvest disease control that are effective, pose minimal or no risk to human health, and are perceived as safe by the general public. Biological control with microbial antagonists is one alternative being considered.

12.4.1 Preharvest Treatment

Many of the storage rot diseases of fruits and vegetables actually begin with infections that occur before harvest. In such cases, application of protective agents must start in the field. Biocontrol agents have been tried in place of conventional fungicides, but few successes can be cited. In Norway, Tronsmo and Dennis[98] applied spore suspensions of *Trichoderma viride* and *T. polysporum* to strawberry plants at flowering, and every 14 days thereafter. These treatments were as effective as dichlofluanid against subsequent storage rot caused by *Botrytis cinerea* and *Mucor mucedo*. However, neither the chemical nor the biological treatment provided the level of control desired.

12.4.2 Postharvest Treatment

The postharvest environment offers a unique opportunity for researchers because, unlike the preharvest environment, conditions can be controlled and maintained to a great extent. During the past 10 years a number of micro-organisms have been shown experimentally to inhibit postharvest pathogens on fruits and vegetables, and a few show promise of being developed commercially as biocontrol agents.[99–103]

Early examples of postharvest disease control with antagonists include work with a strain of the bacterium *Bacillus subtilis* used to control *Monilinia fructicola* on stone fruits.[104] The primary mode of action appears to be the production of iturin antibiotics, which are inhibitory to several important pathogens of fruits and vegetables.[105,106] *Pseudomonas cepacia* also controls a number of postharvest pathogens and produces the antibiotic pyrrolnitrin.[107] The fact that these antagonists utilize antibiotic production may be of special concern when assessing their potential for commercial use on produce that will be consumed a short time later. Stringent toxicological testing could be required for approval.

The possible effects on man and animals by candidate organisms to be used on harvested crops certainly must be considered. On the other hand, micro-organisms have been used since ancient times to pickle and ferment foods to preserve them.[108] Among the wide array of microbial antagonists available, it seems probable that safe and effective biocontrol agents can be found.

Many researchers are now focusing attention on microbial antagonists which do not produce detectable antibiotics and apparently control postharvest diseases through other modes of antagonism. Of particular interest are yeast organisms, such as *Pichia guilliermondii*, which was isolated by Wilson and Chalutz[109] and subsequently studied by co-workers[110–112] for control of post-harvest rots of several commodities. Other promising yeast and yeast-like antagonists are *Cryptococcus* spp.[113,114] and *Acremonium breve*,[115] used against postharvest rots of apple and pear. Yeasts may be preferred candidates as antagonists for reasons discussed elsewhere.[116]

Possible modes of antagonism other than antibiotic production include competition with the pathogen for nutrients, induced resistance in the host, and direct interactions with the pathogen. Investigations with *Pichia guilliermondii* (strain US-7) indicate that a combination of these mechanisms may be involved in the control of postharvest fungal rots. The antagonist showed less activity against *Penicillium digitatum* on citrus fruit when nutrients were added to the fruit, and it effectively competed with the pathogen in a synthetic medium and in wound leachate solutions.[110] Yeast US-7 may also induce resistance in citrus, as evidenced by an increase in ethylene production and activity by phenylalanine ammonia lyase.[102] From light and electron microscopy studies, it was found that US-7 attaches to the mycelium of the pathogen *Botrytis cinerea* and possibly produces cell wall degrading enzymes.[112]

12.5 Commercialization of Antagonists

Despite thousands of reports of potential biocontrol agents against plant diseases, very few have been developed and marketed as commercial products.[117,118] Biocontrol systems that involve introduced micro-organisms are often successful in the laboratory and greenhouse, but fail or are inconsistent when tested in the field. It is hoped in many cases that difficulties can eventually be overcome with an increased understanding of the mechanisms and interactions involved. In addition to the ecological barriers encountered, other problems may limit the implementation of biocontrol.

A major concern in commercial production systems is achieving adequate growth of the antagonist. Sometimes large-scale production is difficult or not economically feasible because of specific nutritional and environmental conditions required for growth of the organism.[119] It is also essential that the inoculants, whether produced by liquid or semi-solid fermentation, are viable as colonizers and antagonists where they are needed.[120] Selective pressures that exist in laboratory cultures or in commercial production plants could result in the loss of characteristics important to the success of biocontrol. For example, structural features of bacteria that function in nature as attachment devices, permeability barriers, ion exchange resins, or as protection against osmotic stress may be lost under conditions that prevail in laboratory cultures.[121] Concepts used to deal with the mass production of micro-organisms are modelled on pharmaceutical products, rather than on microbial agents for the control of plant diseases. Therefore, biocontrol organisms present a unique challenge to industry because of the need for mass quantities of 'competent' inoculum compatible with existing storage and application technology.[50,122]

Maintaining viability is also fundamental to formulating antagonists as commercial products. Biocontrol agents must have a storage life of at least 6 months and preferably 1–2 years.[123–127] Perhaps recent advances in encapsulation technology using cross-linked matrix organic polymers[118,123,126,128,129] will increase the shelf-life of products. Other marketing aspects of formulations relate to their attractiveness as a product and their compatibility with the soil, crop, application equipment, and prevailing pest control practices. To accommodate these frequently conflicting demands, additional adjuvants may include nutrients, binders, or stickers, and inert carriers or bulking agents. Formulations may be in the form of dusts, wettable powders, granules, pellets, gels, or emulsifiable liquids.[126,129] Delivery systems vary widely depending on the crop and disease being targetted.[118,126–129]

In addition to potential problems with the production and formulation of biocontrol agents, other factors that might impede commercialization are:

- Narrow market potential.
- Low efficiency of preparations compared with cost.
- Unknown health and environmental risks and lack of clear guidelines by regulatory agencies.
- Apprehension by users about safety and efficacy.

12.6 Future Outlook

Although the use of micro-organisms as an approach to plant disease control has been fraught with difficulties and the reality of commercialization has been slow in coming, a changing socio-economic climate is likely to encourage a continued emphasis on research and development in this area. Concerns about the adverse effects of pesticides on human health and the environment are greater than ever, pesticide usage is being restricted, and many chemicals are being voluntarily or forcibly withdrawn from the market. Microbial biocontrol is still one of the most viable alternatives we have. However, much perseverance may be necessary before success is achieved. As pointed out by Andrews,[51] the two classic biocontrol successes in plant pathology – that of crown gall and Fomes (Heterobasidion) root rot – each involved a persistent research effort extending for more than two decades.

Fortunately, the prospects for biocontrol have never been better. Knowledge relating to plant–microbe interactions has increased dramatically in recent years and the development of biotechnology has opened up new possibilities. Conceivably, it may be possible through genetic manipulation to combine the most desirable characteristics from different organisms into one agent. Organisms generated could be produced more efficiently in fermentation systems, possess greater stability, have several modes of antagonism, and have enhanced survival and colonization capabilities.

Finally, some workers now suggest that we lower our expectations of biocontrol agents and avoid making comparisons with chemical fungicides. For many plant diseases, it is doubtful that biocontrol agents by themselves will ever provide the efficacy or consistency associated with synthetic fungicides. Nonetheless, the probability is increasing that they will eventually constitute a significant part of integrated systems that can provide adequate disease control.

12.7 References

1. K. F. Baker, Evolving concepts of biological control of plant pathogens, *Ann. Rev. Phytopathol.*, 1987, **25**, 67.
2. W. Roberts, Studies on biogenesis, *Phil. Trans. Roy. Soc. London*, 1874, **164**, 457.
3. R. Hare, 'The Birth of Penicillin and the Disarming of Microbes', Allen and Unwin, London, 1970.

4. R. J. Cook and K. F. Baker, 'The Nature and Practice of Biological Control of Plant Pathogens', American Phytopathology Society, St. Paul, 1983.
5. C. Hartley, Damping-off in forest nurseries, *US Dept. Agric. Bull.*, 1921, **934**, 1.
6. W. A. Millard and C. B. Taylor, Antagonism of micro-organisms as the controlling factor in the inhibition of scab by green-manuring, *Ann. Appl. Biol.*, 1927, **14**, 202.
7. A. W. Henry, The natural microflora of the soil in relation to the foot-rot problem of wheat, *Can. J. Res.*, 1931, **4**, 69.
8. K. F. Baker and W. C. Snyder, 'Ecology of Soil-Borne Plant Pathogens', University of California Press, Berkeley, 1965.
9. J. Rishbeth, Stump protection against *Fomes annosus.* III. Inoculation with *Peniophora gigantea, Ann. Appl. Biol.*, 1963, **52**, 63.
10. K. F. Baker and R. J. Cook, 'Biological Control of Plant Pathogens', W. H. Freeman, San Francisco, 1974 (Reprinted edn, American Phytopathology Society, St. Paul, 1982).
11. C. J. Delp, Privately supported disease management activities, in 'Plant Disease', ed. J. G. Horsfall and E. B. Cowlingin, Academic Press, New York, 1977, vol. 1, p. 381.
12. C. J. Lewis, The economics of pesticide research, in 'Origins of Pest, Parasite, Disease and Weed Problems', ed. J. M. Cherrett and C. R. Sagar, Blackwell Scientific Publishers, Oxford, 1977, p. 237.
13. R. Campbell, The search for biological control agents: a pragmatic approach, *Biol. Agric. Hort.*, 1986, **3**, 317.
14. R. Campbell, 'Biological Control of Microbial Plant Pathogens', Cambridge University Press, Cambridge, 1989.
15. I. Chet (ed.), 'Innovative Approaches to Plant Disease Control', John Wiley, New York, 1987.
16. D. Hornby (ed.), 'Biological Control of Soil Borne Plant Pathogens', C.A.B. International, Wallingford, 1990.
17. J. M. Lynch, Biological control within microbial communities of the rhizosphere, in 'Ecology of Microbial Communities', ed. M. Fletcher, T. R. G. Gray, and J. G. Jones, Society of General Microbiology, Symposium 41, Cambridge University Press, Cambridge, 1987, 55.
18. C. A. Parker, A. D. Rovira, K. J. Moore, P. T. W. Wong, and J. F. Kollmorgan (ed.), 'Ecology and Management of Soilborne Plant Pathogens', American Phytopathology Society, St. Paul, 1985.
19. J. Palti, 'Cultural Practice and Infectious Crop Diseases', Springer Verlag, New York, 1981.
20. U. Gisi, R. Schenker, R. Schulin, F. Stadelman, and H. Sticher, 'Bodenokologie', Thieme Verlag, Stuttgart, 1990.
21. P. B. Adams, The potential of mycoparasites for biological control of plant diseases, *Ann. Rev. Phytopathol.*, 1990, **28**, 59.
22. I.-P. Chang and T. Kommedahl, Biological control of seedling blight of corn by coating kernels with antagonistic microorganisms, *Phytopathology*, 1968, **58**, 1395.
23. T. Kommedahl, C. E. Windels, G. Sarbini, and H. B. Wiley, Variability in performance of biological and fungicidal seed treatments in corn, peas, and soybeans, *Prot. Ecol.*, 1981, **3**, 55.
24. G. E. Harman, A. G. Taylor, and T. E. Stasz, Combining effective strains of *Trichoderma harzianum* and solid matrix priming to improve biological seed treatments, *Plant Dis.*, 1989, **73**, 631.

25. P. R. Merriman, R. D. Price, F. Kollmorgen, T. Piggott, and E. H. Ridge, Effect of seed inoculation with *Bacillus subtilis* and *Streptomyces griseus* on the growth of cereals and carrots, *Austr. J. Agric. Res.*, 1974, **25**, 219.
26. R. S. Utkhede and J. E. Rahe, Biological control of onion white rot, *Soil Biol. Biochem.*, 1980, **12**, 101.
27. S. Y. Yin, J. K. Chang, and P. C. Xun, Studies in the mechanisms of antagonistic fertilizer '5406' IV. The distribution of the antagonist in soil and its influence on the rhizosphere, *Acta Microbiol. Sinica*, 1965, **11**, 259.
28. R. J. Cook, Biological control of plant pathogens: overview, in 'Biological Control in Crop Production', ed. G. C. Papavizas, Beltsville Symposium in Agricultural Research 5, Allanheld, Osmun, London, 1981.
29. P. D. Colyer and M. S. Mount, Bacterization of potatoes with *Pseudomonas putida* and its influence on postharvest soft rot diseases, *Plant Dis.*, 1984, **68**, 703.
30. G. W. Xu and D. C. Gross, Field evaluation of interactions among fluorescent pseudomonads, *Erwinia carotovora*, and potato yields, *Phytopathology*, 1986, **76**, 423.
31. S. O. Kawamoto and J. W. Lorbeer, Protection of onion seedlings from *Fusarium oxysporum* f. sp. *cepae* by seed and soil infestation with *Pseudomonas cepacia*, *Plant Dis. Reptr.*, 1976, **60**, 189.
32. T. V. Suslow and M. N. Schroth, Rhizobacteria of subar beets: effects of seed application and root colonization on yield, *Phytopathology*, 1982, **72**, 199.
33. H. D. Weller and R. J. Cook, Suppression of take-all of wheat by seed-treatment with fluorescent pseudomonads, *Phytopathology*, 1983, **73**, 463.
34. D. M. Weller, Application of fluorescent pseudomonads to control root diseases, in 'Ecology and Management of Soilborne Plant Pathogens', ed. C. A. Parker, A. D. Rovira, K. J. Moore, P. T. W. Wong, and J. F. Kollmorgan, American Phytopathology Society, St. Paul, Minnesota, 1985, p. 137.
35. B. H. Ownley, D. M. Weller, and L. S. Thomashow, Influence of *in situ* and *in vitro* pH on suppression of *Gaeumannomyces graminis* var. *tritici* by *Pseudomonas fluorescens* 2-79, *Phytopathology*, 1992, **82**, 178.
36. L. S. Thomashow, D. M. Weller, R. F. Bonsall, and L. S. Pierson, Production of the antibiotic phenazine-1-carboxylic acid by fluorescent *Pseudomonas* species in the rhizosphere of wheat, *Appl. Environ. Microbiol.*, 1990, **56**, 908.
37. J. Leong, Siderophores: their biochemistry and possible role in the biocontrol of plant pathogens, *Annu. Rev. Phytopathol.*, 1986, **24**, 187.
38. A. Kerr, Biological control of crown gall through production of agrocin 84, *Plant Dis.*, 1980, **64**, 25.
39. D. Bagyaraj, Biological interactions with VA mycorrhiza, in 'VA Mycorrhiza', ed. C. L. P. Powell and D. Bagyaraj, CRC Press, Boca Raton, 1984, p. 131.
40. N. C. Schenck and M. K. Kellam, 'The influence of vesicular arbuscular mycorrhizae on disease development', Fla. Agric. Exp. Stn. Tech. Bull. 798, 1978.
41. L. Sundheim and A. Tronsmo, Hyperparasites in biological control, in 'Biocontrol of Plant Diseases', ed. K. G. Mukerji and K. L. Garg, CRC Press, Boca Raton, 1988, vol. 1, 53.
42. H. D. Wells, *Trichoderma* as a biocontrol agent, in 'Biocontrol of Plant Diseases', ed. K. G. Muderji and K. L. Garg, CRC Press, Boca Raton, 1988, vol. 1, 71.
43. W. A. Ayers and P. B. Adams, Mycoparasitism of sclerotia of *Sclerotinia* and *Sclerotium* species by *Sporidesmium sclerotivorum, Can. J. Microbiol.*, 1979, **25**, 17.

44. P. B. Adams and W. A. Ayers, Biological control of *Sclerotinia* lettuce drop in the field by *Sporidesmium sclerotivorum, Phytopathology*, 1982, **72**, 485.
45. P. B. Adams, J. J. Marios, and W. A. Ayers, Population dynamics of the mycoparasite, *Sporidesmium sclerotivorum*, and its host, *Sclerotinia minor*, in soil, *Soil Biol. Biochem.*, 1984, **16**, 627.
46. Y. Elad, I. Chet, and Y. Henis, Biological control of *Rhizoctonia solani* in strawberry fields by *Trichoderma harzianum*, *Plant Soil*, 1981, **60**, 245.
47. Y. Elad, A. Kalfon, and I. Chet, Control of *Rhizoctonia solani* in cotton by seed-coating with *Trichoderma* spp., *Plant Soil*, 1982, **66**, 279.
48. M. Scher and R. Baker, Mechanism of biological control in a Fusarium-suppressive soil, *Phytopathology*, 1980, **70**, 412.
49. F. M. Scher and R. Baker, Effect of *Pseudomonas putida* and a synthetic iron chelator on induction of soil suppressiveness to Fusarium wilt pathogens, *Phytopathology*, 1982, **72**, 1567.
50. J. H. Andrews, Biological control in the phyllosphere: realistic goal or false hope? *Can. J. Plant Pathol.*, 1990, **12**, 300.
51. J. H. Andrews, Biological control in the phyllosphere, *Annu. Rev. Phytopathol.*, 1992, **30**, 603.
52. J. P. Blakeman, Competitive antagonism of air-borne fungal pathogens, in 'Fungi in Biological Control Systems', ed. M. N. Burge, Manchester University Press, Manchester, 1988, p. 141.
53. N. J. Fokkema, The phyllosphere as an ecologically neglected milieu: a plant pathologist's perspective, in 'Microbial Ecology of Leaves', ed. J. H. Andrews and S. S. Hirano, Springer Verlag, New York, 1991, p. 3.
54. C. D. Upper, Manipulation of microbial communities in the phyllosphere, in 'Microbial Ecology of Leaves', ed. J. H. Andrews and S. S. Hirano, Springer Verlag, New York, 1991, p. 451.
55. J. P. Blakeman and I. D. S. Brodie, Inhibition of pathogens by epiphytic bacteria on aerial plant surfaces, in 'Microbiology of Aerial Plant Surfaces', ed. C. H. Dickinson and T. F. Preece, Academic Press, New York, 1976, p. 529.
56. C. E. Morris and D. I. Rouse, Role of nutrients in regulating epiphytic bacterial populations, in 'Biological Control on the Phylloplane', ed. C. E. Windels and S. E. Lindow, American Phytopathology Society, St. Paul, 1985, p. 63.
57. C. Leben and G. C. Daft, Influence of an epiphytic bacterium on cucumber anthracnose, early blight of tomato, and northern leaf blight of corn, *Phytopathology*, 1965, **55**, 760.
58. C. Leben, G. C. Daft, J. D. Wilson, and H. F. Winter, Field tests for disease control by an epiphytic bacterium, *Phytopathology*, 1965, **55**, 1375.
59. H. W. Spurr, Experiments on foliar disease control using bacterial antagonists, in 'Microbial Ecology of the Phylloplane', ed. J. P. Blakeman, Academic Press, London, 1981, p. 369.
60. H. W. Spurr and G. R. Knudsen, Biological control of leaf diseases with bacteria, in 'Biological Control on the Phylloplane', ed. C. E. Windels and S. E. Lindow, American Phytopathology Society, St. Paul, 1985, p. 45.
61. S. V. Beer, J. R. Rundle, and J. L. Norielli, Recent progress in the development of biological control for fire blight, *Acta Hortic.*, 1984, **151**, 195.
62. S. E. Lindow, Integrated control and the role of antibiosis in biological control of fire blight and frost injury, in 'Biological Control on the Phylloplane', ed. C. E. Windels and S. E. Lindow, American Phytopathology Society, St. Paul, 1985, p. 83.

63. J. L. Vanneste, J. Yu, and S. V. Beer, Role of antibiotic production by *Erwinia herbicola* Eh 252 in biological control of *Erwinia amylovora*, *J. Bacteriol.*, 1992, **174**, 2785.
64. R. S. Wodzinski, S. J. Coval, C. H. Zumoff, J. C. Clardy, and S. V. Beer, Antibiotics produced by strains of *Erwinia herbicola* that are highly effective in suppressing fire blight, *Acta Hortic.*, 1990, **273**, 411.
65. R. S. Wodzinski, T. E. Umholtz, K. Garrett, and S. V. Beer, Attempts to find the mechanism by which *Erwinia herbicola* inhibits *Erwinia amylovora*, *Acta Hortic.*, 1987, **217**, 223.
66. J. L. Vanneste, J. Yu, and S. V. Beer, Role of antibiotic production by *Erwinia herbicola* Eh252 in biological control of *Erwinia amylovora*, *J. Bacteriol.*, 1992, **174**, 2785.
67. M. J. Hattingh, S. V. Beer, and E. W. Lawson, Scanning electron microscopy of apple blossoms colonized by *Erwinia amylovora* and *E. herbicola, Phytopathology*, 1986, **76**, 900.
68. S. V. Thomson, D. R. Hansen, and K. M. Flint, Dissemination of bacteria antagonistic to *Erwinia amylovora* by honey bees, *Plant Dis.*, 1992, **76**, 1052.
69. M. Wilson and S. E. Lindow, Interactions between the biological control agent *Pseudomonas fluorescens* A506 and *Erwinia amylovora* in pear blossoms, *Phytopathology*, 1993, **83**, 117.
70. V. O. Stockwell, K. B. Johnson, and J. E. Loper, Establishment of bacterial antagonists on blossoms of pear, *Phytopathology*, 1992, **82**, 1128.
71. K. B. Johnson, V. O. Stockwell, D. Sugar, and J. E. Loper, Effect of antagonistic bacteria on establishment of honey bee-dispersed *Erwinia amylovora* in pear blossoms and on fire blight control, *Phytopathology*, 1992, **82**, 1130.
72. C. C. Heye and J. H. Andrews, Antagonism of *Athelia bombacina* and *Chaetomium globosum* to the apple scab pathogen, *Venturia inaequalis*, *Phytopathology*, 1983, **73**, 650.
73. U. Miedtke and W. Kennel, *Athelia bombacina* and *Chaetomium globosum* as antagonists of the perfect stage of the apple scab pathogen *Venturia inaequalis* under field conditions, *J. Plant Dis. Prot.*, 1990, **97**, 24.
74. A. T. K. Corke and J. Rishbeth, Use of microorganisms to control plant diseases, in 'Microbial Control of Pests and Plant Diseases 1970–1980', ed. H. D. Burges, Academic Press, New York, 1981, p. 717.
75. A. T. K. Corke, Interactions between microorganisms, *Annu. Appl. Biol.*, 1978, **89**, 89.
76. M. V. Carter, Biological control of *Eutypa armeniacae*, *Austr. J. Exp. Agric. Anim. Husb.*, 1971, **11**, 687.
77. T. R. Swinburne, Post-infection antifungal compounds in quiescent or latent infections, *Annu. Appl. Biol.*, 1978, **89**, 322.
78. A. Biraghi, Possible active resistance to *Endothia parasitica* in *Castanea sativa*, 'Report of Congress of International Union of Forest Research Organizations 11th', Rome, 1953, p. 643.
79. J. Grente and S. Berthelay-Sauret, Biological control of chestnut blight in France, in 'Proceedings of the American Chestnut Symposium', Morgantown, WV, 1978, West Virginia University Books, Morgantown, 1978, p. 30.
80. J. Grente and S. Sauret, L'hypovirulence exclusive phenomene original en pathologie vegetale, *C. R. Acad. Sci., Ser. D*, 1969, **268**, 2347.
81. J. A. Dodds, Association of type 1 viral-like dsRNA with club-shaped particles in hypovirulent strains of *Endothia parasitica*, *Virology*, 1980, **107**, 1.

82. S. L. Anagnostakis and P. E. Waggoner, Hypovirulence, vegetative incompatibility, and the growth of cankers of chestnut blight, *Phytopathology*, 1981, **71**, 1198.
83. R. A. Jaynes and J. E. Elliston, Pathogenicity and canker control by mixtures of hypovirulent strains of *Endothia parasitica* in American chestnut, *Phytopathology*, 1980, **70**, 453.
84. J. Weber, A natural biological control of Dutch elm disease, *Nature*, 1981, **292**, 449.
85. J. N. Gibbs and M. E. Smith, Antagonism during the saprophytic phase of the life cycle of two pathogens of woody hosts – *Heterobasidion annosum* and *Ceratocystis ulmi*, *Annu. Appl. Biol.*, 1978, **89**, 125.
86. P. L. Pusey and C. L. Wilson, Detection of double-stranded RNA in *Ceratocystis ulmi*, *Phytopathology*, 1982, **72**, 423.
87. D. F. Myers and G. A. Strobel, *Pseudomonas syringae* as a microbial antagonist of *Ceratocystis ulmi* in the apoplast of American elm, *Trans. Br. Mycol. Soc.*, 1983, **80**, 389.
88. T. J. Grant and A. S. Costa, A mild strain of the tristeza virus of citrus, *Phytopathology*, 1951, **41**, 114.
89. L. R. Fraser, K. Long, and J. Cox, Stem pitting of grapefruit – field protection by the use of mild virus strains, 'Proceedings 4th Conference of the International Organization of Citrus Virologists', University of Florida Press, Gainesville, 1968, p. 27.
90. A. S. Costa and G. W. Muller, Tristeza control by cross protection: a US–Brazil cooperative success, *Plant Dis.*, 1980, **64**, 538.
91. USDA, 'Losses in Agriculture', United States Department of Agriculture, Agricultural Research Service, Washington, DC, Handbook 291, 1965.
92. D. G. Coursey and R. H. Booth, The post-harvest phytopathology of perishable tropical produce, *Rev. Plant Pathol.*, 1972, **51**, 751.
93. A. Kelman, Introduction: The importance of research on the control of postharvest diseases of perishable food crops, *Phytopathology*, 1989, **79**, 1374.
94. J. W. Eckert and J. M. Ogawa, The chemical control of postharvest diseases: deciduous fruits, berries, vegetable, and root/tuber crops, *Annu. Rev. Phytopathol.*, 1988, **26**, 433.
95. J. W. Eckert and J. M. Ogawa, The chemical control of postharvest diseases: subtropical and tropical fruits, *Annu. Rev. Phytopathol.*, 1985, **23**, 421.
96. National Resource Council, Board of Agriculture, 'Regulating Pesticides in Food – The Delaney Paradox', National Academy Press, Washington, DC, 1987.
97. Anon, General news, *Postharvest News Inf.*, 1991, **2**, 3.
98. A. Tronsmo and C. Dennis, The use of Trichoderma species to control strawberry fruit rots, *Neth. J. Plant Pathol.*, 1977, **83**, 449.
99. P. Jeffries and M. J. Jeger, The biological control of postharvest diseases of fruit, *Postharvest News Inf.*, 1990, **1**, 365.
100. P. L. Pusey, C. L. Wilson, and M. E. Wisniewski, Management of postharvest diseases of fruits and vegetables: strategies to replace vanishing fungicides, in 'Pesticide Interactions in Crop Production: Beneficial and Deleterious Effects', ed. J. Altman, CRC Press, Boca Raton, 1993, p. 477.
101. C. L. Wilson and M. E. Wisniewski, Biological control of postharvest diseases of fruits and vegetables: an emerging technology, *Annu. Rev. Phytopathol.*, 1989, **27**, 425.
102. C. L. Wilson, M. E. Wisniewski, C. L. Biles, R. McLaughlin, E. Chalutz, and S.

Droby, Biological control of post-harvest diseases of fruits and vegetables: alternatives to synthetic fungicides, *Crop Prot.*, 1991, **10**, 172.

103. M. E. Wisniewski and C. L. Wilson, Biological control of postharvest diseases of fruits and vegetables: recent advances, *Hort. Sci.*, 1992, **27**, 94.
104. P. L. Pusey and C. L. Wilson, Postharvest biological control of stone fruit brown rot by *Bacillus subtilis*, *Plant Dis.*, 1984, **68**, 753.
105. R. C. Gueldner, C. C. Reilly, P. L. Pusey, C. E. Costello, R. F. Arrendale, R. H. Cox, D. S. Himmelsbach, F. G. Crumley, and H. G. Cutler, Isolation and identification of iturins as antifungal peptides in biological control of peach brown rot with *Bacillus subtilis*, *J. Agric. Food Chem*, 1988, **36**, 366.
106. P. L. Pusey, Use of *Bacillus subtilis* and related organisms as biofungicides, *Pestic. Sci.*, 1989, **27**, 133.
107. W. J. Janisiewicz and J. Roitman, Biological control of blue-mold and gray-mold on apple and pear with *Pseudomonas cepacia*, *Phytopathology*, 1988, **78**, 1697.
108. S. E. Gilliland, Role of starter culture bacteria in food preservation, in 'Bacterial Starter Cultures for Foods', ed. S. E. Gilliland, CRC Press, Boca Raton, 1985, 175.
109. C. L. Wilson and E. Chalutz, Postharvest biological control of Penicillium rots of citrus with antagonistic yeasts and bacteria, *Sci. Hortic.*, 1989, **40**, 105.
110. S. E. Droby, E. Chalutz, C. L. Wilson, and M. E. Wisniewski, Characterization of biocontrol activity of *Debaryomyces hansenii* in the control of *Penicillium digitatum* on grapefruit, *Can. J. Microbiol.*, 1989, **35**, 794.
111. R. J. McLaughlin, C. L. Wilson, E. Chalutz, C. P. Kurtzman, W. F. Fett, and S. F. Osman, Characterization and reclassification of yeasts used for biological control of postharvest diseases of fruits and vegetables, *Appl. Environ. Microbiol.*, 1990, **56**, 3583.
112. M. Wisniewski, C. Biles, S. Droby, R. McLaughlin, C. Wilson, and E. Chalutz, Mode of action of the postharvest biocontrol yeast, *Pichia guilliermondii*. I. Characterization of attachment to *Botrytis cinerea*, *Physiol. Molec. Plant Pathol.*, 1991, **39**, 245.
113. R. G. Roberts. Postharvest biological control of gray mold of apple by *Cryptococcus laurentii*, *Phytopathology*, 1990, **80**, 526.
114. R. G. Roberts, Biological control of Mucor rot of pear by *Cryptococcus laurentii*, *C. flavus* and *C. albidus*, *Phytopathology*, 1990, **80**, 1051.
115. W. J. Janisiewicz, Biocontrol of postharvest diseases of apples with antagonist mixtures, *Phytopathology*, 1988, **78**, 194.
116. W. J. Janisiewicz, Biological control of diseases of fruit, in 'Biocontrol of Plant Diseases', ed. K. G. Mukergi and K. L. Garg, CRC Press, Boca Raton, 1988, p. 153.
117. K. A. Powell, J. L. Faull, and A. Renwick, The commercial and regulatory challenge, in 'Biological Control of Soil-borne Plant Pathogens', ed. D. Hornby, CAB International, Wallingford, 1990, p. 445.
118. J. A. Lewis and G. C. Papavizas, Biocontrol of plant diseases: the approach for tomorrow, *Crop Prot.*, 1991, **10**, 95.
119. S. G. Lisansky, Production and commercialization of pathogens, in 'Biological Pest Control', ed. N. W. Hussey and N. Scopes, Blandford Press, Poole, 1985, p. 210.
120. G. Lethbridge, An industrial view of microbial inoculants for crop plants, in 'Microbial Inoculation of Crop Plants', ed. R. Campbell and R. M. Macdonald, IRL Press, Oxford, 1989, p. 11.

121. G. R. Knudsen and H. W. Spurr, Management of bacterial populations for foliar disease biocontrol, in 'Biocontrol of Plant Diseases', ed. K. G. Mukerji and K. L. Garg, CRC Press, Boca Raton, 1988, p. 83.
122. S. G. Lisansky and R. A. Hall, Fungal control of insects, in 'The Filamentous Fungi, Vol. 4, Fungal Technology', ed. J. E. Smith, D. R. Berry, and B. Kristiansen, Edward Arnold, London, 1983, 327.
123. C. A. Baker and J. M. S. Henis, Comercial production and formuation of microbial biocontrol agents, in 'New Directions in Biological Control: Alternatives for Suppressing Agricultural Pests and Diseases', ed. R. R. Baker and P. E. Dunn, Alan R. Liss, New York, 1990, p. 333.
124. A. J. Caesar and T. J. Burr, Effect of conditioning, betaine, and sucrose on survival of rhizobacteria in powder formulations, *Appl. Environ. Microbiol.*, 1991, **1**, p. 23.
125. B. C. Carlton, Economic considerations in marketing and application of biocontrol agents, in 'New Directions in Biological Control: Alternatives for Suppressing Agricultural Pests and Diseases', Alan R. Liss, New York, 1990, p. 419.
126. W. J. Connick Jr, J. A. Lewis, and P. C. Quimby Jr, Formuation of biocontrol agents for use in plant pathology, in 'New Directions in Biological Control: Alternatives for Suppressing Agricultural Pests and Diseases', ed. R. R. Baker and P. E. Dunn, Alan R. Liss, New York, 1990, p. 345.
127. J. L. McIntyre and L. S. Press, Formulation, delivery systems and marketing of biocontrol agents and plant growth promoting rhizobacteria, in 'The Rhizosphere and Plant Growth', ed. D. L. Keister and P. B. Cregan, Kluwer, Dordrecht, 1991, p. 289.
128. G. R. Harman, Deployment tactics for biocontrol agents in plant pathology, in 'New Directions in Biological Control: Alternatives for Suppressing Agricultural Pests and Diseases', ed. R. R. Baker and P. E. Dunn, Alan R. Liss, New York, 1990, p. 779.
129. J. A. Lewis, Formulation and delivery systems of biocontrol agents with emphasis on fungi, in 'The Rhizosphere and Plant Growth', ed. D. L. Keister and P. B. Cregan, Kluwer, Dordrecht, 1991, p. 279.

CHAPTER 13

Microbial Herbicides – Factors in Development

M. P. GREAVES

13.1 Introduction

Following the introduction of synthetic herbicides in the 1940s, there was a rapid increase in the discovery, development, and marketing of herbicides. Within ten years scientists were starting to warn of the possibility of side-effects from the use of these agrochemicals on flora, fauna, and the environment as a whole. Although these warnings were heeded to some extent, and research in a number of disciplines aimed at developing an integrated approach to pest management was initiated, both the development of new herbicides and the concerns about their use continued to increase to the late 1980s. Since then, increasing stringency of pesticide regulations and, in a sense fortuitously, world recession, have reduced the number of herbicides available for use. In particular, those herbicides perceived as offering serious risk of contamination of ground-water supplies have been withdrawn from use.

Despite this, concerns about environmental damage are still prevalent. Undoubtedly, many of the concerns are exaggerated, either deliberately for political reasons or through ignorance. Nonetheless, the concerns are generally based on a genuine desire to avoid damage to the environment and, as such, deserve a genuine response in attempts to establish alternative approaches to weed management in modern agriculture. This response is clearly evident in the expanding interest in developing so-called 'sustainable agricultural systems', which depend heavily on integrated pest management.

It would be foolish to imagine that agriculture can dispense with chemical herbicides in the foreseeable future. Projected increases in world population dictate that the losses of crops caused by weeds be reduced, if not eliminated, in order to maintain the increase in food supplies which are required. As yet, the only effective ways of achieving this aim depend on chemical herbicides. The agrochemical industry has made significant advances in developing safer, more effective products, especially in the past ten years. At the same time, it has shown interest in the development of alternative weed management strategies, including biological control. Even so, progress in this area has been slow. In reviewing this area of research and development, some of the reasons for lack of progress are addressed.

13.2 The Microbial Herbicide Strategy

The first demonstration that an endemic, indigenous, plant pathogenic micro-organism might be exploited as a biological control agent in agriculture (*i.e.*, as a microbial herbicide) was by Daniel *et al.*[1] They showed that the natural constraints operating to restrict disease development in the field could be overcome by inundating the target weed, at an appropriate time, with an inoculum of a phytopathogenic fungus. This approach allowed exploitation of any conditions required to optimize disease development. Daniel *et al.*[1] summarized the requirements for a successful microbial herbicide ('mycoherbicide' if the agent is a fungus) as:

- Being capable of producing abundant and durable inoculum in artificial culture.
- Being genetically stable and specific to the target weed.
- Being infective and able to kill the target weed in a wide range of environments.

Subsequently, the microbial herbicide concept has been expanded and refined and has been described comprehensively in many excellent reviews.[2–24]

In the course of this expansion the original definitions have been redefined variously. Thus, the term 'mycoherbicide', coined at the start of the research in this area, was defined as 'plant pathogenic fungi developed and used in the inundative strategy to control weeds in the way chemical herbicides are used'.[17] Later,[22] the definition was rephrased as 'living products that control weeds in agriculture as effectively as chemicals'. This rephrasing is welcome in that it expands the concept to include organisms other than fungi. Certainly, there is every reason to expect that many organisms, including bacteria and viruses, can be exploited as weed control agents to be used in the inundative strategy. For this reason, the term 'mycoherbicide' is slowly being replaced by more generic terms, such as 'bioherbicide' or, as preferred for this review, 'microbial herbicide'.

As the first microbial herbicides to reach the market were plant pathogenic fungi, it is understandable that the resulting expansion in research focused on this group. Following the early work of Daniel *et al.*,[1] which established the microbial herbicide strategy, interest developed in two other forms of biological weed control for use in agriculture, as opposed to stable ecosystems where the classical biological control strategy is appropriate.[25] These two approaches are the augmentative strategy and the use of necrogenic micro-organisms.

In the former, microbial inoculum is introduced in localized areas within larger areas that are already infected at a low level by the disease, thus producing an augmentation effect. This differs from the microbial herbicide approach in that the inoculum is not necessarily mass-produced, a few grams of inoculum per hectare is usually sufficient, and it is not inoculated inundatively over the whole of the weed-infested crop. To date, there has been relatively little development of this approach, probably because it, generally, involves biotrophic fungi that are difficult or even impossible to mass produce on artificial media. However, one potential product has been developed. This is based on *Puccinia canaliculata* Schw., a rust fungus that is native to the USA and that is effective against yellow nut-sedge, *Cyperus esculentus* L.[26,27] This product, to be called Dr Biosedge, is a preparation of the uredospores which are obtained from infected plants especially grown to produce the fungus. Such fungi, used in the augmentative approach, are usually host-density dependent for dissemination from inoculum sites to infect the weed throughout the infested area.

The use of necrogenic organisms for weed control relies on their ability to produce phytotoxins in the soil environment. They are, generally, not pathogenic organisms. Several fungi have been shown to offer potential as weed control agents in this approach. *Gliocladium virens* Miller, Giddens & Foster[28–30] is, perhaps, the best known example of a phytotoxin-producing non-pathogenic fungus with potential for weed control. Provided the inoculum is prepared in a suitable nutrient base, the fungus can produce sufficient viridiol to control redroot pigweed, *Amaranthus retroflexus* L., in cotton.[30] Control was achieved by causing root necrosis and was effective against several annual composites and some monocots. Crop phytotoxicity was avoided by placement of the fungus away from the crop root zone. As well as providing an alternative strategy for the biological control of weeds, necrogenic organisms are potentially a rich source of herbicidal chemicals which may be of value to the conventional herbicide industry (see Chapter 4).

13.3 Development of Microbial Herbicides

The early work of Daniel *et al.*[1] was the start of an intensive study of the potential for *Colletotrichum gloeosporioides* (Penz) & Sacc. f.sp. *aeschynomene* to control the legume weed northern joint vetch, *Aeschynomene virginica* (L.) B.S.P. This work, which has been published in detail[31–37] resulted in the formulation and marketing of the fungus as COLLEGO, which is used widely against its weed target in rice and soybean in Arkansas and adjacent states in the USA. During the development of COLLEGO, a second microbial herbicide, ultimately marketed as DeVine, was being developed.[38–42] This, the first microbial herbicide registered with the US Environmental Protection Agency, is a preparation of a suspension of chlamydospores of a pathotype of *Phytophthora palmivora* (Butl.) Butl., which is native to Florida and infects stranglervine, *Morrenia odorata* (HBA) Lindl., a vine weed of citrus in Florida. Although not entirely specific to stranglervine, selective site-specific application avoids compromising safety of non-target species. This product is unique in the sense that it has no appreciable shelf-life and must, therefore, be custom-prepared for each purchaser, supplied and stored in refrigerated containers, and used soon after receipt. These restrictions are now seen as inimicable to the development of a commercially viable microbial herbicide which should, ideally, have a shelf-life in normal ambient conditions (5–50°C) of 18 months or more.

Since these commercial products were introduced in the early 1980s, few other products have reached the market. Indeed, the only other registered product, in the commercial sense, is BioMal, a preparation of *Colletotrichum gloeosporioides* f.sp. *malvae* for the control of round-leaved mallow, *Malva pusilla* Sm., in Canada and the northern USA. This was registered for practical use in 1992, ten years after the first products. Registration was achieved only after exhaustive tests to ensure that, even though safflower is also a host for the fungus, use of the microbial herbicide posted no unacceptable risk for the safflower crop. Several other microbial herbicides were put into use in the intervening period, although not on a truly commercial basis. For example, a wilt-causing fungus, *Cephalosporium diospyri* Crandall, is one in use in Oklahoma against persimmon, *Diospyros virginiana* L.[43–46] This fungus, which is supplied by a charitable foundation, is wound-inoculated into the trunks of the persimmon trees, which it then kills quite rapidly. *Colletotrichum gloeosporioides* f.sp. *cuscutae* is used in the People's Republic of China to control dodders, *Cuscuta* spp., under the name of LUBOA II.[47] *Colletotrichum gloeosporioides* f.sp. *clidemiae* is used to control Kosters curse, *Clidemia hirta* Don., in Hawaii, the inoculum being distributed by hikers.[48]

As well as the few microbial herbicides in practical use there are several which are at an advanced stage of commercial development. Of these, CASST, a wettable powder preparation of *Alternaria cassiae* Jurair & Khan for the control of sicklepod, *Cassia obtusifolia* L., is, perhaps, the best known example.[49–54] This organism appears to have an advantage over other microbial herbicides in that, rather than being totally specific to one weed, it can control three economically important leguminous species. In addition to sicklepod, it is effective against coffee senna (*Cassia occidentalis* L.) and showy crotalaria (*Crotalaria spectabilis* Roth.).[55] Despite this advantage, CASST has yet to appear on the market, for reasons discussed later in this chapter. Another promising microbial herbicide is *Colletotrichum coccodes* (Wallr.) Hughes, proposed trade name VELGO, for the control of velvetleaf, *Abutilon theophrasti* Medic.[56,57] Like CASST, this agent has been under commercial development for some time, but has not yet reached the market.

In Holland, Scheepens and his co-workers[58,59] have developed a somewhat revolutionary herbicide for the control of black cherry (*Prunus serotina* Enrl.), a serious woody weed of forestry plantations. They have elected to use *Chondrostereum purpureum* (Pers.) Pouzar as the control agent, a fungus which is the causal agent of serious disease in commercial fruit tree orchards. Through exhaustive epidemiological studies they have been able to show that the organism does not disseminate widely from the treated areas. Thus, if its use is restricted to infestations of black cherry which are distant (by 1 km) from *Prunus* orchards there is no risk of disease spreading to the crop. All infected *P. serotina* material must be burned in the treatment area and not removed. On the basis of these restrictions, the agent was given clearance for experimental field use by the Dutch regulatory authorities and it is now under commercial development.

The most recent example of a microbial herbicide undergoing exhaustive development for commercial use is that of *Colletotrichum orbiculare* (Berk. & Mort) v. Arx for the control of Bathurst burr or spiny cockleburr, *Xanthium spinosum* L., in Australia.[60–63] Field trials in both dryland pasture and irrigated soybean crops gave levels of kill ranging from 50 to 100%, with the best results (98–100%) occurring in the dryland pasture.[63] Field efficacy was dependent on adequate periods of dew or high humidity. Formulation of a spore suspension to reduce evaporation is required to ensure reliable efficacy with *C. orbiculare* throughout the weed's range.

Many other fungi have been the subjects of exhaustive research, as pre-requirements for commercial development. These include *Cercospora rodmanii* Conway for the control of water hyacinth, *Eichhornia crassipes* (Mart.) Solms;[64] *Colletotrichum coccodes* (Wallr.) Hughes for velvetleaf, *A. theophrasti*;[57] *Colletotrichum gloeosporioides* f.sp. *jussiaeae* for winged water primrose, *Jussiaea decurrens* (Walt.) Dc.;[65] *Colletotrichum malvarum* (Braun & Caspary) Southworth for prickly sida, *Sida spinosa* L.;[66] *Fusarium solani* f.sp. *cucurbitae* Snyd. & Hans. for Texas gourd, *Cucurbita texana* Gray;[67–70] and *Phomopsis convolvulus* Ormeno for field bindweed, *Convolvus arvensis* L.[71]

The literature contains references to many other fungi which have been examined for potential as microbial herbicides. Charudattan[8] has provided a comprehensive world-wide listing of 100 projects, dealing with 69 weed species and 41 genera of fungi. As there is also a considerable amount of work that is commercially funded and, therefore, confidential, there are, undoubtedly, many more candidate pathogens being studied. Even so, the intensity of microbial herbicide research is impressive. Projects identified by Charudattan,[8] in literature published up to 1989, were being conducted in 16 countries and 44 locations. In the USA alone, 18 research groups are involved. The 5 years since that survey have seen some changes. In particular, there appears to have been some decrease in interest shown by major pesticide companies and, due to the world financial recession, some decrease in funding, especially from the public sector. Nonetheless, the scope and scale of research is still significant.

In view of the extent of research over the past two decades, it is, perhaps, surprising that the number of products developed is so small. It appears that the identification of potential value as a microbial herbicide may only be in the eye of the researchers and be exaggerated by a desire to stay with a fascinating area of research. While this may be so in some cases, objective analysis of the literature shows clearly that more tangible reasons are operating.

13.4 Microbial Herbicide Patents

A convenient means of reviewing progress in microbial herbicide research is to examine the disclosures found in patent specifications. Much of the information presented in patents has a unique nature in that it is, often, not published elsewhere. A recent monograph published by the UK Patent Office[72] presents a valuable synthesis of the state-of-the-art in microbial herbicides. While not claiming to be totally comprehensive it does, through the judicious selection of 43 sets of patent publications (patent families of published patent applications and/or granted patents) or related disclosures, present a valuable focus on the strengths and weaknesses of microbial herbicides and of the associated research and development. Obviously, there are many patents which do not specifically mention microbial herbicides, but which describe highly relevant technology. An example is European Patent Application No. 88305535.2 (1988) which refers to maintenance of the viability of micro-organisms for use in microbial inoculants and describes the use of soluble, non-crosslinked polysaccharides (*e.g.*, alginate) as stabilizing agents. A search of this area of technology patents could be most rewarding.

The two earliest patents identified originate, interestingly, from Russia in 1970 and 1971. Thus, they pre-date the most significant patents which relate to COLLEGO and other mycoherbicide products in the USA. Surprisingly, this patent list does not include any reference to *Phytophthora palmivora* (DeVine). The only agent for *Morrenia odorata* which is referred to is the *Arauijio* mosaic virus. The patents refer to one other virus (TMV) for control of *Solanum carolinense* L. in Russia (Patent No. SU663357, 1977), and to three bacteria. Otherwise, all the control agents disclosed are 32 species of 14 genera of fungi. The three genera represented most frequently are *Alternaria*, *Colletotrichum*, and *Fusarium*, which are mentioned in 15, 13, and 10 patents, respectively.

Many of the patents refer to methods of formulation and delivery but only five refer to these specifically. These cover alginate granules (two patents), synergistic formulations with crop oils and surfactant, encapsulation in polymers, and oil-in-water invert emulsion. A further six patents refer to combinations of fungi with chemical herbicides and six others refer to combinations of two fungal pathogens to broaden the spectrum of weed control. Broad-spectrum control is the subject of two patent applications which introduce the important concept of pathogens, namely *Alternaria cassiae* and *Fusarium lateritium* Nees, being able to control several species of weed.

The pathogens, weed targets, and relevant patents are listed in Table 13.1. As mentioned in the preceding section, this evidence of considerable research effort and progress in the development of microbial herbicides contrasts with the lack of practical achievement, as demonstrated by products on the market. While the patent disclosures would suggest that effective means of formulating and delivering the large numbers of candidate microbial herbicides are available (Table 13.2), it is clear that there are significant factors that inhibit the translation of promise into practice. In many cases these must be matters of speculation, as companies are often reluctant to disclose reasons for failures. In the following, some possible reasons for failure are discussed.

Table 13.1 *Microbial herbicide patents (from Bridges[72])*

Pathogen	*Target weed*	*Patent No.*; Date†*
Albugo tragopognis (Pers.) S. F. Gray	*Ambrosia artemisiifolia* L.	SU 343671; 1970
Alternaria alternata (Fr.) Keissler	*Xanthium* spp.	JP 62278978; 1987
Alternaria cassiae Jurair & Khan	*Cassia occidentalis* L. *Cassia obtusifolia* L. *Crotalaria spectabilis* Roth.	US 4390360; 1983
Alternaria crassa (Sacc.) Rands	*Datura stramonium* L.	US 7092100; 1987
Alternaria euphorbiicola Simmons & Engelhard	*Euphorbia* spp.	US 4755208; 1988 US 4871386; 1989
Alternaria (*zinniae*?)	*Carduus tenuiflorus* Curt.	US 4636386; 1987
Amphobotrys ricini (Buchwald) Hennebert	*Caperonia palustris* St.-Hil	US 4909826; 1990
Arauijio mosaic virus	*Morrenia odorata* (H&A) Lindl.	US 4162912; 1979
Ascochyta hyalospora Lib.	*Chenopodium* spp.	EP 296057; 1988
Bipolaris sorghicola (Lefebvre Sherwin) Alcorn.	*Sorghum halepense* (L.) Pers.	US 4606751; 1986
Cercospora rodmanii Conway	*Eichhornia crassipes* Solms.	US 4097261; 1978
Colletotrichum coccodes (Wallr.) Hughes	*Abutilon theophrasti* Medic.	CA 1224055; 1987
Colletotrichum coccodes (Wallr.) Hughes	*Solanum ptycanthum* L.	US 4715881; 1987
Colletotrichum gloeosporioides (Penz.) Sacc. f.sp. *aeschynomene*	*Aeschynomene virginica* (L.) B.S.P.	US 3849104; 1974
Colletotrichum gloeosporioides (Penz.) Sacc. f.sp. *malvae*	*Malva pusilla* Sm. *Abutilon theophrasti*	EP 218386; 1987
Colletotrichum malvarum (Braun & Caspary) Southworth	*Sida spinosa* L.	US 3999973; 1976
Colletotrichum orbiculare (Berk. & Mont.) Arx.	*Xanthium spinosum* L.	AU 8818454; 1989
Colletotrichum truncatum (Schw.) Andrus & Moore	*Desmodium tortuosum* DC. *Sesbania exaltata* (Raf.) Rydb.	US 4643756; 1987 US (NTIS) 7338680; 1989
Colletotrichum sp.	*Cyperus rotundus* L.	WO 90/06056; 1990
Drechslera spp.	*Echinochloa crus-galli* Beauv.	EP 374499; 1990
Fusarium lateritium Nees	*Abutilon theophrasti* Medic. *Sida spinosa* L. *Anoda cristata* Schlecht	US 4419120; 1983
Fusarium orobanches fom.	*Orobanche* spp.	SU 387689; 1973
Fusarium oxysporum Schl.	*Echinochloa* sp.	UP 02013367; 1990
Fusarium roseum Lk.	*Hydrilla verticillata* Presl.	US 4263036; 1981
Fusarium tricinctum (Corda) Sacc.	*Cuscuta* spp.	US 4915726; 1990
Fusarium sp.	'Arrowroot'	JP 53099321; 1978

Continued

Table 13.1 *Continued*

Pathogen	*Target weed*	*Patent No.*; Date†*
Hyphomycetes sp.	*Eleocharis kurogauwai* L.	JP 01238507; 1989
Lactic acid bacteria	General weeds	JP 01100106; 1989
Phomopsis cirsii (Sacc.) Sacc.	Compositae	EP 136850; 1985
Phomopsis convolvulus (Sacc.) Sacc.	*Convolvulus arvensis* L.	EP 2777736; 1988
Pseudomonas spp.	*Bromus tectorum* L.	WO 89/12691; 1989
Puccinia canaliculata Schw.	*Cyperus esculentus* L.	US 4731104; 1988
Septoria cirsii	Compositae	EP 136850; 1985
Tobacco mosaid virus	*Solanum carolinense* L.	SU 663357; 1979
Xanthomonas campestris	*Poa annua* L.	WO 88/01172; 1988

*Letters prefacing patent numbers: AU, Australia; CA, Canada; JP, Japan; SU, Russia; US, United States of America; EP, European Patent Office; WO, World Patent (Patent Co operation Treaty).
†Dates refer to date of publication of the patent application.

Table 13.2 *Microbial herbicide technology patents (from Bridges[72])*

Technology	*Patent*; Date†*
Microbial herbicides containing chemical herbicides and plant growth regulators	EP 207653; 1987
Alginate granule formulation	US 4718935; 1988 US 4767441; 1988
Synergistic formulation with crop oil	US 4755207; 1988
Encapsulation with non-ionic polymer	EP 320483; 1989
Invert emulsion (oil-in-water) formulation	US 4902333; 1990
Mixed microbial herbicide inoculum	WO 90/06056; 1990
Chemical herbicide mixed with pathogenic bacterium	WO 91/03161; 1991

*Letters prefacing patent numbers: US, United States of America; EP, European Patent Office; WO, World Patent (Patent Co-operation Treaty).
†Dates refer to publication of the patent application.

13.5 Commercial Factors

Auld[73] has pointed out that many of the considerations made before developing a microbial herbicide are identical to those for a chemical herbicide. He identified three main areas to be considered, summarized in Table 13.3.

Of these considerations, one which seems to be ignored in many research projects is that of market size. Obviously, where the research is commercially funded, market size has been carefully researched before the project commences. In contrast, many publicly funded projects address weeds that appear

Table 13.3 *Commercial considerations in development of microbial herbicides (from Auld[73])*

1. Market size and stability:
   (a) Does the weed cause a similar level of problem each year?
   (b) What area of use is likely?
   (c) What is the value per unit area of affected crops?
   (d) What is the cost of alternative effective controls?
2. Is the technology protected by patent or secrecy?
3. Product development costs.

to offer little or no attraction for commercial development investment. Understandably, public funders, especially those at a local level, may be anxious to provide control strategies for weeds that are a significant problem at a local level. A case in point is the development of a microbial herbicide to control *A. virginica.* This leguminous weed is a serious problem in rice and soybean in Arkansas and a few neighbouring states in the USA. However, this weed does not appear in a listing of 224 species of weeds selected by US weed scientists as prevalent in croplands, grazing lands, non-croplands, and aquatic sites in the USA.[74] In practice, this microbial agent did achieve commercial product status, as did COLLEGO, and is used successfully to control the weed. This is despite the fact that the market is small, being a maximum of 500,000 ha or *ca.* 6 million dollars at 1990 prices. Indeed, only a part of this market has been penetrated and, although farmers are making repeat purchases, it would not be considered as commercially viable by many companies. The fact that the principal chemical herbicide to give satisfactory control of the weed was 2,4,5-T, which was withdrawn, obviously helped sales, but, even so, they remain small by pesticide market standards. The product DeVine is another example of an agent developed for a small, local, specialist market, which is not commercially viable in the sense used by pesticide manufacturers.

These small local markets might, in modern terms, be defined as niche markets. However, even this is doubtful as most companies see niche markets as having wide, often multinational distribution. Thus, weed controls in aquatic sites, forestry, and 'lawn care' (urban, sports, and domestic sites) are now regarded as potentially viable niche markets. These are seen as providing the financial return on sales of approximately £10 million per annum quoted as the minimum required by larger pesticide companies. It is worth noting that COLLEGO sales do not reach this level. The financial requirements of smaller biotechnology-based companies may be less stringent, but, even so, there must be guaranteed minimum sales on an annual basis. These must be achieved in a market dominated by and, therefore, in competition with, chemical herbicides.

It would be foolish to imagine that chemical herbicides are going to be anything other than the dominant means of weed control in the foreseeable future. Therefore, microbial herbicides must be capable of withstanding the competition these chemicals present. Equally, it is obvious that, if microbial herbicides are to be used in practical agriculture, forestry, horticulture, *etc.*, they must be developed commercially. That being so, they must compete with chemicals in the market. A major factor in establishing a viable competitive position is to target an appropriate market before initiating the research. Even a cursory examination of the literature shows that, in many instances, such targetting has not been addressed or that it has been attempted but without adequate guidance from the appropriate commercial organizations. Public sector funding is, and will remain, a critically important contribution to microbial herbicide research. It will be the power behind the initial discovery of suitable micro-organisms and their characterization and development to the point of proving feasibility of practical use in small-scale field trials. Thus, it is vital that those benefitting from such funding ensure that appropriate target weeds (markets) are selected to maximize the opportunity for public investment to result in public good *via* practical use of the discoveries.

13.6 Product Registration

It is beyond the scope of this chapter to provide a detailed account of registration procedures for microbial herbicides in different countries. However, the subject is of such importance to successful commercialization of the products that it must be mentioned.

At first glance, it seems that the registration of biological agents offers a major advantage compared to that of chemicals. Essentially, it is very much cheaper. However, the picture may not be so simple. At present there are major differences in national attitudes to registration requirements, and these are exaggerated when the question of genetic modification of the agent is raised. In extreme cases, some countries have refused to consider biological agents through fears of mutation leading to epidemics of disease that affect crop species. Such a fear, indeed, is commonly expressed in virtually all countries by certain sections of the population. While there is no evidence to show that mutation is likely in conditions of practical use, neither is there evidence to prove that it will not and, so, allay the concerns.

Despite some efforts to harmonize registration requirements, notably in the European Community, differences are bound to persist. These will provide hindrance, perhaps seriously, to widespread use of a product – a prerequisite, in many instances, for commercial development. For example, any country proposing to permit widespread use of a microbial herbicide may face objections from countries with whom they share a frontier and who may anticipate spread into their territory of what they view as an alien species. Countering such objections would, at least, involve considerable expense in, for example, proving that the exact strain in question was already endemic in these neighbouring countries, that it would not spread from the use sites, or that it was sufficiently selective not to affect native species. Such expense, of course, erodes the cost advantage that the microbial herbicide may have over chemical herbicides in the country of registration. Until this matter is resolved, there is bound to be some reluctance by potential developers to make the necessary financial investment.

13.7 Product Selectivity

The great majority of candidate microbial herbicides and microbial herbicide products are claimed to be totally specific to their selected host weed. Indeed, this is a major factor in their environmental safety. The absolute specificity claimed may, in some cases, be an artifact of limited host-range testing. This was so for COLLEGO which, since its introduction to the market, has been shown to infect, but not kill, some leguminous crops.[36] Under field use there is unlikely to be spread to such crops, even if they are grown close to crops treated with COLLEGO, as the agent's spores are sticky and it spreads poorly. Similarly, *Colletotrichum gloeosporioides* f.sp. *malvae*, being developed for *Malva pusilla* Sm control, infects safflower and an extensive risk–benefit analysis has been necessitated.[8] A similar analysis, based on detailed epidemiological studies, was made before commercial development of *Chondrostereum purpureum*, a serious pathogen of fruit trees, as a control agent for *P. serotina* was initiated in Holland.[58,59] In this case, also, release of the agent was approved as its spread by natural means was limited.

These few examples of lack of absolute specificity do not seriously impair the perceived environmental safety of the agents. However, specificity to one weed target does seriously impair marketability in most instances. It is axiomatic that farmers and growers rarely face a problem from a monoculture of a weed and they are, at present at least, supplied with a formidable array of broad-spectrum chemical herbicides. Simple economics dictates that a single application of a herbicide will always be preferred to multiple applications. Thus, host-specific microbial herbicides face a considerable disadvantage.

There is already considerable evidence that this problem may, in part at least, be resolved. Table 13.2 includes two relevant patents, one concerned with mixtures of microbial herbicides and one with a mixture of microbial and chemical herbicides. Both these approaches can broaden the spectrum of weeds controlled. While the opportunity to use mixed micro-organisms may be limited to two or three species, due to adverse inter- and intra-specific interactions in the inoculum resulting from dense spore concentration, there is less limitation on mixing microbial and chemical herbicides. Clearly, some herbicides have microbicidal activity, but many do not, and may give advantages of synergistic interactions with the microbial herbicide.[56,57] This approach has much potential for exploitation as demonstrated, indirectly, in the integrated control of weeds in rice and soybean in parts of the USA.[33,34,75] In addition to these approaches, there is the possibility of using pathogens which attack more than one weed species.[55,76]

13.8 Product Reliability

The great strength of chemical herbicides is that they give reliable weed control in a very wide range of edaphic and climatic conditions, thus opening up world-wide markets. The truly successful microbial herbicide, even if intended for use in niche markets, such as aquatic weed control and forestry, must do no less.

Table 13.4 *Variable performance of* Alternaria cassiae *on sicklepod* (Cassia obtusifolia) *in field trials in the USA (from Charudatton* et al.[54]*)*

State	*Weed control (%)*	
	1982	*1983*
Arkansas	71	78
Florida	94	98
Mississippi	74	80
North Carolina	100	100
South Carolina	0	82

This presents a difficult challenge for using organisms which have evolved to have quite specific requirements for successful infection. There are many examples of the lack of the required reliability of microbial herbicide efficacy. This is, indeed, why most companies will not become involved in development research until efficacy has been proven in field trials. A good example of the difficulty of moving from the laboratory to the field is given by work at Long Ashton. Strains of *Phoma exigua* Desm. were isolated that gave rapid, total kill of cleavers (*Galium aparine* L.) in laboratory and greenhouse studies. In the field in two seasons, however, only very limited kill was achieved and, overall, only a 40% reduction in weed biomass occurred (Greaves, McQueen, York, and Williamson, unpublished data). This failure was attributed to lack of sufficient available water at the target surface to promote effective germination of the spore inoculum. Hence disease expression was restricted. In particular, disease failed to develop at or below the cotyledonary node. Thus, even though all parts of the plant above this node were often killed, regrowth from the cotyledonary node occurred and, at best, the competitive effect of the weed would have been delayed by a few weeks only.

The data in Table 13.4 illustrate the problem of poor field performance of *Alternaria cassiae*, with efficacy ranging from zero to complete control of *C. obtusifolia*. The failure in South Carolina in 1982 was attributed to very dry weather and the development of a second flush of sicklepod seedlings. Similarly, inoculum desiccation was blamed for poor performance at other sites in other years. The same reason has been cited for poor performance in field trials by other mycoherbicides, notably CASST and BIOMAL, although in the latter case the problem was overcome sufficiently well to allow introduction of the product to the market in 1992 to be contemplated.

It must be emphasized that those mycoherbicides which have suffered from poor efficacy as a result of inoculum desiccation are, almost without exception, foliage-applied preparations. The leaf surface is a hostile environment for fungal spores in many ways. It may be protected against retention of spores by a waxy coat, germination is constrained by the dry atmosphere commonly present, especially on the upper leaf surface, and penetration of the fungus is resisted by the cuticle and chemical defence reactions in the epidermal cells. The challenge facing microbial herbicide research is to develop strategies to overcome these constraints, a challenge which has not been met fully as yet.

13.9 Formulation to Improve Field Reliability

It is clear from the accumulated data on microbial herbicides that the most serious constraint on efficacy arises from the lack of sufficient free-water to enable efficient spore germination and pre-infection growth of the microbial inoculum on the leaf surface. An obvious strategy to reduce, or eliminate, the problem is to inoculate the soil in which the weed target is growing. This takes advantage of the buffering capacity of the soil against large-scale fluctuations in both moisture and temperature, so extending, considerably, the application window for the agent. Furthermore, soil inoculation allows the use of granular formulations which can be designed to promote germination and proliferation of the micro-organisms by the incorporation of suitable substrates. Another considerable advantage is that there is a wealth of information available about the physiology, biology, and ecology of soil-borne plant pathogenic micro-organisms.

In view of this, it is more than surprising that there has been relatively little research into soil-borne pathogens as microbial herbicides. This being so despite the first registered microbial herbicide, DeVine, being a soil-borne pathogen, *Phytophthora palmivora*. To date, the main genus studied in this context is *Fusarium* (Table 13.5). The soil-borne pathogen *Rhizoctonia solani*

Table 13.5 *Soil-borne plant pathogenic fungi as microbial herbicides*[77]

Fungus	*Target weed*
Phytophthora palmivora (Butl.) Butl.	*Morrenia odorata* (H&A) Lindl. (strangler vine)
Sclerotinia sclerotiorum (Lib.) de Bary	*Cirsium arvense* (L.) Scop. (creeping thistle)
Rhizoctonia solani Kuhn	*Eichhornia crassipes* (Mart.) Solms (water hyacinth)
Fusarium lateritium (Nees)	*Sida spinosa* L. (prickly sida) *Anoda cristata* Schlecht (spurred anoda) *Abutilon theophrasti* Medic. (velvet leaf)
F. oxysporum Schl. f.sp. *cannabis* Novello & Snyder	*Cannabis sativa* L. (marijuana)
F. oxysporum Schl. f.sp. *carthami* Klisiewicz & Houston	*Centaurea solstitialis* L. (yellow starthistle)
F. roseum Lk.	*Hydrilla verticillata* Presl. (water hydrilla)
F. roseum Lk.	*Cirsium arvense* (L.) Scop. (creeping thistle)
F. solani Appl. & Wr. f.sp. *cucurbitae* Snyd & Hans	*Cucurbita texana* Gray (Texas gourd)

Kuhn has been studied only as an agent against an aquatic weed species (*Eichhornia crassipes*).

Although having advantages over foliage-applied pathogens, soil inoculated microbial herbicides still suffer from disadvantages such as specificity (see earlier). In addition, many of the pathogens used so far are closely related to serious crop pathogens. Thus, there may be farmer resistance to using them, stemming from the widely held concern that the applied organism may mutate, or change in some other way, and so become virulent on the treated, or neighbouring, crops. The accumulated evidence of many years of field trials and/or practical use of a number of microbial herbicides shows that such changes have not occurred as yet and suggests they are unlikely to in the future. Nonetheless, such evidence is not convincing to many in the non-scientific community and the concern persists. This provides yet another challenge to the researchers and industrial developers of microbial herbicides. In common with those in many other areas in crop protection research, they have been relatively ineffectual in countering the bad publicity which is so readily generated by detractors of modern agriculture. The constructive use of publicity must be learned, to highlight the positive aspects of our research, and the negative aspects should be aired in a balanced, realistic way. All this must be done in attractive, readable prose which is easily understood by the non-scientific reader. This is patently not the forte of many in the scientific community.

Despite the advantages offered by soil application, it seems clear that most effort will be directed towards foliage-applied microbial herbicides in the foreseeable future. That being so, it is essential that considerable effort is given to developing appropriate formulations to overcome the constraints mentioned earlier. Much has been done already, but this has served to identify the even greater amount of research that is still needed, particularly as there is a growing awareness that many, if not all, of the pathogens being studied at present are likely to need individual formulations designed to meet their particular needs. The subject of formulation has been reviewed[78–81] and no more than a superficial consideration is given here.

The literature demonstrates clearly that there are many options available to overcome the principal constraint on reliability – moisture shortage. These include formulation as invert emulsions,[82] standard emulsions with vegetable oils (Potyka, quoted in Greaves[81]), granules of alginate gels, inert solid carriers, or as colonized seeds. The liquid formulations, whether they are simple aqueous suspensions of spores or complex emulsions, may be variously mixed with surfactants, humectants, stickers, *etc.*, to suit particular applications. Whatever the composition, the objective remains the same, to reduce the inoculum dependency on the minimum 12–36 h of free water commonly essential for spore germination and disease establishment with many fungal pathogens. The progress towards meeting this objective, as evidenced by the literature reviewed by Daigle and Connick[78] and Boyette *et al.*,[79] is promising.

It is interesting and important to note, and exploit, other reported benefits of formulation to reduce dew dependency. In some ways, many of these incidental benefits may prove to be as or more important than protecting against desiccation. For example, formulation as an invert emulsion appears to reduce the inoculum threshold for pathogen infection of leaves.[83] Thus, when formulated in this way only one spore of *Alternaria crassa* in a 2 μl droplet effectively infects the target weeds (*Cassia obtusifolia* and *Datura stramonium*). In contrast, in aqueous suspension 1000 spores per droplet are required to ensure that every droplet produced a successful infection. The implications of such an effect are very significant. Stowell,[84] in his detailed and comprehensive review of production of biological herbicides, has calculated that, in order to ensure satisfactory economics, a potential microbial herbicide product should yield, in liquid fermentation, 2.7×10^{11} propagules per litre and the virulence should be high enough to allow 5.5×10^{11} propagules per hectare to control the weed. Obviously then, a 1000-fold increase in infectivity ($\equiv$ virulence) obtained by formulation has great economic significance. Other, simpler, approaches may also achieve similar results. Sucrose, for example, has been shown to enhance disease development.[85] Sorbitol has been associated with a 20-fold increase in survival of viable spores of *Colletotrichum coccodes* on inoculated leaves[86] and, more importantly, has been shown to render three consecutive short dew periods as effective as one long one. In many areas, especially in the UK, appropriate spray windows for microbial herbicides coincide with short periods of dews.

The production of appressoria is a critical factor in successful infection for many microbial herbicides. Observations at Long Ashton Research Station (McQueen, Martin, Potyka, and Greaves, unpublished data) have shown that, although many potential adjuvants for microbial herbicides may inhibit spore germination, they can simultaneously enhance the numbers of appressoria formed and, so, increase disease development. If such a phenomenon can be induced reliably on the leaf it should reduce inoculum thresholds and significantly improve the economics of microbial herbicide production.

Perhaps one of the most promising formulation approaches for microbial herbicides, which offers many opportunities for practical exploitation, is a mixture with chemical herbicides. On the one hand, such mixtures can be used to provide control of a broad spectrum of weeds, using the microbial herbicide to control an important weed not controlled by the chemical. On the other hand, the known synergism between some chemical herbicides and their biological partners can be exploited usefully to control weeds normally resistant to each component of the mixture when used separately. This synergism often allows very low doses of herbicide to be used effectively. For example, *Colletotrichum coccodes*, an effective microbial herbicide for velvetleaf (*Abutilon theophrasti*), is unreliable if environmental conditions are unfavourable, killing only infected leaves and not preventing continued growth of the weed. Mixing with the growth regulator, thidiazuron, produces a synergistic interaction and weed mortality is enhanced.[56,57] Other similar interactions have been reported.[78,87] In a collaborative programme between the International Institute of Biological Control and Long Ashton Research Station (Evans, Ellison, Martin, and Greaves, unpublished data), synergism between a *Colletotrichum* spp. and a very low dose (*ca.* 5% of recommended field rate) of an experimental sulfonyl urea herbicide has given effective control of itch grass [*Rottboellia cochinchinensis* (Lour.) W. D. Clayton]. At the low dose used, the herbicide had only slight effects on sprayed leaves. Similarly, when applied alone the pathogen infected and killed only the leaves that received inoculum. In both cases, the meristem was unaffected and grew away from the treated tissues rapidly. The mixture of the two, however, killed plants within 7–14 days after treatment. It is thought that the sub-lethal dose of herbicide acts as a growth retardant, so giving the fungus time to develop and grow on sprayed leaves and move to and infect the plant meristem.

The foregoing superficial consideration of formulation of microbial herbicides serves to show that the progress towards improving reliability of field performance is constrained only by limitations of time invested in appropriate research. Inevitably, much of what has been done is redolent of the 'pick and mix' approach common in pesticide formulation. However, it is clear that formulation of living agents demands a more rational approach which should exploit appropriate physiological and biochemical attributes of the microbial agent. This will only be achieved by intensive, focused research

13.10 Genetic Manipulation of Microbial Herbicides

Inevitably, any consideration of the possibility of improving reliability of microbial herbicides includes reference to the need to increase pathogenicity. While this can be achieved, in some instances, by traditional methods, such as selection, induced mutation, or optimization of production and stability of the agent, increasingly the opportunities offered by genetic means are discussed. Such approaches include sexual and parasexual crossing, somatic hybridization, and the use of recombinant DNA, *i.e.*, genetic engineering. The last offers unique opportunities, although public and political pressures may restrict the release of genetically engineered organisms into the environment.

The technology of transferring, and gaining expression of, specific genes in filamentous fungi has developed rapidly in recent years and gene transformation has been achieved by many different groups. This subject is reviewed by Greaves *et al.*[88] and Kistler.[89]

The first stage in the consideration of how to improve the performance of a wild strain of plant pathogen, so that it can be developed as a commercial microbial herbicide, is to identify the factors that limit efficacy. In some instances there is inadequate information about the genetic basis of characteristics which could be targets for manipulation, including, importantly, specificity and pathogenicity. There is an urgent need to further our understanding of these. However, other characteristics are open to genetic manipulation. These were identified by Greaves *et al.*[88] as:

- Genes involved in the production of specialized structures essential for colonization and reproduction. The targets here include recognition of the host-plant surface, production of sexual and asexual spores, spore germination, production of appressoria and haustoria, and adherence of pathogens to host surfaces. All of these, if improved, would contribute significantly to reducing the necessary inoculum levels. However, they are all under multigenic control and, thus, advances in the short term are most likely to come from strain selection rather than genetic manipulation.

- Genes involved in pathogenesis. Genetic control of the production of enzymes that destroy plant tissue components or inactivate plant defence chemicals and of toxins that destroy tissue are the principal targets for genetic manipulation. There is much knowledge of the enzymes implicated in pathogenesis, but only a few of the genes which encode for these enzymes have been cloned and characterized. Similarly, restricted progress has been made with genes controlling enzymes that inactivate plant defence mechanisms. Although much more research into such genes is required, the availability of some genes raises important possibilities and questions.[88] The possibility to enhance pathogenesis by improving fungal ability to colonize tissue or to overcome natural defence mechanisms in plants clearly exists. However, we need answers to several questions before this ability can be exploited safely and effectively. Will multiple copies of a naturally existing gene lead to increased enzyme production and, hence, to increased infection and weed control? What will be the effect of introducing a foreign gene, such as a wood-degrading cellulase gene, into a plant pathogen? Will this upset or enhance the naturally regulated pathogenesis process? Will such manipulations be stable and remain confined to the microbial herbicide or will they be transferred to other closely related species?
- Genes that regulate pathogen specificity. One of the most frequently cited 'disadvantages' of microbial herbicides is that they are too specific. Conversely, some of the most powerful plant pathogens have too broad a host range to be suitable as microbial herbicides. Obviously, knowledge of, and the ability to manipulate, genes controlling specificity would be of considerable value, allowing defined construction of the host range or any microbial herbicide. Currently, much research is aimed at cloning genes that determine specificity, but success with fungi is still awaited.
- Genes determining fitness for survival in the environment. It was mentioned earlier that the factor which contributes most to the unreliable field performance of microbial herbicides is the constraint of unfavourable environmental conditions, particularly inadequate moisture. Unfortunately, resistance to desiccation and temperature extremes appears to be controlled by polygenic mechanisms, which are too complex to be readily amenable to genetic manipulation. Thus, at present, strain selection is likely to be the most effective method of obtaining tolerance to these factors. Formulation will, undoubtedly, be needed to enhance this. It is essential, for environmental and commercial reasons, that a microbial herbicide does not persist in the environment after it has killed its target weed. In practice, many pathogens have limited saprophytic ability and do not survive well in the absence of their host. Nevertheless, genetic alterations, such as spore-killer factors, double stranded RNA and viruses, and specific metabolite control of introduced genes suggest possible approaches. Sands and Miller[90] have

exploited induced mutation to limit the survival of the environment of *Sclerotinia sclerotiorum*, used to control thistle. They also suggest other innovative and, perhaps, controversial strategies. Microbial herbicides will be only one part of an integrated crop protection programme. Thus, it will be advantageous for them to be resistant to a range of pesticides that are likely to be encountered in the crop. This will greatly extend the window of opportunity for applying the microbial product. Such pesticide resistance is often readily obtained by strain selection or induced mutation, but it should also be relatively easily achieved by gene transformation.

The general public are, to some degree, conditioned by the media to perceive that biological control may pose a threat to the environment. Headline phrases, such as 'Biological Warfare Against Pests' are used frequently and serve to perpetuate the perceived threat. The addition of 'Genetically Engineered' to such media hype is usually the final straw in breaking the camel's back. Such exaggerated concern does nothing to ease the strictness of regulatory control of the release of genetically manipulated organisms which, clearly, constrains development of genetically manipulated microbial herbicides.

Most recent studies tend to lead to the conclusion that genetically modified organisms offer no more risks than unmodified organisms. The report by Kelman *et al.*[91] is typical of many other studies in concluding that there is no evidence of unique risks in the use of recombinant techniques or gene transfer between unrelated organisms, that the risks associated with organism release into the environment are the same for both genetically modified and unmodified organisms, and that the assessment of these latter risks should be based on the nature of the organism and the receiving environment, not on the method by which the organism was modified.

Assuming that these conclusions can be widely promulgated and accepted, the genetic manipulation of plant pathogenic micro-organisms selected as potential microbial herbicides offers significant potential for improvement of effective, reliable, and environmentally safe products.

13.11 Conclusions

The broad scope of this chapter has necessitated a selective consideration of factors that are of importance in the future development of reliable microbial herbicides. Research in the past two decades has demonstrated clearly that it is relatively easy to isolate, identify, and characterize large numbers of plant pathogenic micro-organisms which offer real potential as microbial herbicides. Every plant species studied so far is susceptible to control in this way. Equally, however, the same research programmes have demonstrated, only too clearly, that it is very difficult to translate successful laboratory and glasshouse experiments into effective, reliable weed control in the field. It is essential, therefore, that researchers learn from the mistakes of the past and resist the temptation to add to the long and impressive list of laboratory winners that will never enter the commercial development race. Market identification and, thus, weed target selection must be much more controlled in future and, perhaps, local needs made subservient to the greater needs of the science as a whole. After all, if commercial successes are not forthcoming, the strategy will lose credibility and funding sources, which are already limited, may dry up altogether. In this context, it is essential that there is greater cohesion between public and private sector funding. The present, commonly held, concept that public sector funding supports research only to the point of proving feasibility, mainly through glasshouse or small field plot experiments, and that private sector funding is available for development once reliable performance in the field is established, must be challenged as it leaves a large grey area of field evaluations at a range of sites in the unfundable category. Thus, a massive barrier exists between research and development, acting in effect as a technology transfer blockage. It is to be hoped that this blockage can be removed, possibly through more cooperative or club research involving both sectors. The Ministry of Agriculture, Fisheries and Food LINK scheme in the UK is a good example of one way forward, in which joint private and public funding is used to promote research liable to lead to technology transfer in the short-to-medium term.

This chapter, in describing the many constraints on microbial herbicides, may give the impression that the author is doubtful of the future of microbial herbicides. Nothing could be further from the truth. The conclusion I wish to leave with the reader is that, given a properly focused, logical approach, combining scientific rigour and commercial realities, microbial herbicides have an important, albeit niche, role to play in world agriculture, horticulture, forestry, and urban development. I look forward to the arrival of that reality eagerly.

Added in proof. Since this chapter was written, COLLEGO and DeVine were withdrawn from the market for commercial reasons, principally limited market size. Happily, the decision has been revised for DeVine and the University of Arkansas are close to making an agreement for COLLEGO to re-enter production. BioMal, CASST, VELGO and the Bathurst burr agent all appear unlikely to be commercialized.

13.12 References

1. J. T. Daniel, G. E. Templeton, R. J. Smith, and W. T. Fox, Biological control of northern jointvetch in rice with an endemic fungal disease, *Weed Sci.*, 1973, **21**, 303.
2. J. Altman, S. Neate, and A. D. Rovira, Herbicide–pathogen interactions and mycoherbicides as alternative strategies for weed control, in 'Microbes and Microbial Products ad Herbicides', ed. R. E. Hoagland, ACS Symposium Series 429, American Chemical Society, Washington DC, 1990, p. 240.
3. B. Auld, 'Potential for Mycoherbicides in Australia', Workshop Proceedings, Agricultural Research and Veterinary Centre, Orange, NSW, Australia, 1986.
4. W. Bernmann, Fungi isolated from weeds and their capability for weed control – a review of publications, *Zentralbl. Mikrobiol.*, 1985, **140**, 111.

5. R. Charudattan and C. J. DeLoach, Jr, Management of pathogens and insects for weed control in agroecosystems, in 'Weed Management in Agroecosystems: Ecological Approaches', ed. M. A. Altieri and M. Liebman, CRC Press, Inc., Boca Raton, 1980, p. 246.
6. R. Charudattan, Inundative control of weeds with indigenous fungal pathogens, in 'Fungi in Biological Control Systems', ed. M. N. Burge, Manchester University Press, Manchester, 1988, p. 88.
7. R. Charudattan, Pathogens and potential for weed control, in 'Microbes and Microbial Products as Herbicides', ed. R. E. Hoagland, ACS Symposium Series 439, American Chemical Society, Washington DC, 1990, p. 132.
8. R. Charudattan, The mycoherbicide approach with plant pathogens, in 'Microbial Control of Weeds', ed. D. O. TeBeest, Chapman and Hall, New York, 1991, p. 24.
9. T. E. Freeman and R. Charudattan, Conflicts in the use of plant pathogens as biological control agents for weeds, in 'Proceedings of the VI International Symposium, Biological Control of Weeds', 1984, p. 351.
10. S. Hasan, Industrial potential of plant pathogens as biocontrol agents of weeds, *Symbiosis*, 1986, **2**, 151.
11. S. Hasan, Biocontrol of weeds with microbes, in 'Biocontrol of Plant Diseases', ed. K. G. Mukerji and K. L. Carg, CRC Press, Inc., Boca Raton, 1988, p. 129.
12. R. E. Hoagland, Microbes and microbial products as herbicides: an overview, in 'Microbes and Microbial Products as Herbicides', ed. R. E. Hoagland, ACS Symposium Series 439, American Chemical Society, Washington DC, 1990, p. 2.
13. R. W. Jones and J. G. Hancock, Soilborne fungi for biological control of weeds, in 'Microbes and Microbial Products as Herbicides', ed. R. E. Hoagland, ACS Symposium Series 439, American Chemical Society, Washington DC, 1990, p. 276.
14. G. F. Joye, Biological control of aquatic weeds with plant pathogens, in 'Microbes and Microbial Products as Herbicides', ed. R. E. Hoagland, ACS Symposium Series 439, American Chemical Society, Washington DC, 1990, p. 155.
15. K. Mortensen, Biological control of weeds with plant pathogens, *Can. J. Plant Pathol.*, 1986, **8**, 229.
16. P. C. Quimby and H. L. Walker, Pathogens as mechanisms for integrated weed management, *Weed Sci.*, 1982, **30**(Suppl.), 30.
17. D. O. TeBeest and G. E. Templeton, Mycoherbicides: progress in the biological control of weeds, *Plant Dis.*, 1985, **69**, 6.
18. G. E. Templeton, D. O. TeBeest, and R. J. Smith, Jr, Biological weed control with mycoherbicides, *Annu. Rev. Phytopath.*, 1979, **17**, 301.
19. G. E. Templeton, R. J. Smith, and W. Klomparens, Commercialization of fungi and bacteria for biological control, *Biocontrol News Inf.*, 1980, **1**, 291.
20. G. E. Templeton, Status of weed control with plant pathogens, in 'Biological Control of Weeds with Plant Pathogens', ed. R. Charudattan and H. L. Walker, Wiley, New York, 1982, p. 29.
21. G. E. Templeton and M. P. Greaves, Biological control of weeds with fungal pathogens, *Trop. Pest Manag.*, 1984, **30**, 333.
22. G. E. Templeton, R. J. Smith, Jr, and D. O. TeBeest, Progress and potential of weed control with mycoherbicides, *Rev. Weed Sci.*, 1986, **2**, 1.
23. G. E. Templeton, Biological control of weeds, *Am. J. Alt. Agric.*, 1988, **3**, 69.
24. G. E. Templeton, Weed control with pathogens: future needs and directions, in 'Microbes and Microbial Products as Herbicides', ed. R. E. Hoagland, ACS Symposium Series 439, American Chemical Society, Washington DC, 1990, p. 320.

25. A. K. Watson, The classical approach with plant pathogens, in 'Microbial Control of Weeds', ed. D. O. TeBeest, Chapman and Hall, New York, 1991, p. 3.
26. S. C. Phatak, D. R. Sumner, H. D. Wells, D. K. Bell, and N. C. Glaze, Biological control of yellow nutsedge with the indigenous rust fungus *Puccinia canaliculata*, *Science*, 1983, **219**, 1446.
27. S. C. Phatak, M. B. Callaway, and C. S. Vavrina, Biological control and its integration in weed management systems for purple and yellow nutsedge (*Cyperus rotundus* and *C. esculentus*), *Weed Technol.*, 1987, **1**, 84.
28. C. R. Howell, Effect of *Gliocladium virens* on *Pythium ultimum*, *Rhizoctonia solani*, and damping off of cotton seedlings, *Phytopathology*, 1982, **72**, 496.
29. C. R. Howell and R. D. Stipanovic, Phytotoxicity to crop plants and herbicidal effects on weeds of viridiol produced by *Gliocladium virens*, *Phytopathology*, 1984, **74**, 1346.
30. R. W. Jones, W. T. Lanini, and J. G. Hancock, Plant growth response to the phytotoxin viridiol produced by the fungus *Gliocladium virens*, *Weed Sci.*, 1988, **36**, 683.
31. R. C. Bowers, Commercialization of microbial biological control agents, in 'Biological Control of Weeds with Plant Pathogens', ed. R. Charudattan and H. L. Walker, Wiley, New York, 1982, p. 157.
32. R. C. Bowers, Commercialization of Collego – an industrialist's view, *Weed Sci.*, 1986, **34**(Suppl.), 24.
33. R. J. Smith, Jr, Integration of microbial herbicides with existing pest management programs, in 'Biological Control of Weeds with Plant Pathogens', ed. R. Charudattan and H. L. Walker, Wiley, New York, 1982, p. 189.
34. R. J. Smith, Jr, Biological control of northern jointvetch in rice and soybeans – a researcher's view, *Weed Sci.*, 1986, **34**(Suppl.), 17.
35. D. O. TeBeest, Survival of *Colletotrichum gloeosporioides* f. sp. *aeschynomene* in rice irrigation water and soil, *Plant Dis.*, 1982, **66**, 469.
36. D. O. TeBeest, Additions to the host range of *Colletotrichum gloeosporioides* f. sp. *aeschynomene*, *Plant Dis.*, 1988, **72**, 16.
37. G. E. Templeton, D. O, TeBeest, and R. J. Smith, Jr, Biological weed control in rice with a strain of *Colletotrichum gloeosporioides* (Penz.) Sacc. used as a mycoherbicide, *Crop Prot.*, 1984, **3**, 409.
38. H. C. Burnett, D. P. H. Tucker, and W. H. Ridings, *Phytophthora* root and stem rot of milkweed vine, *Plant Dis. Rep.*, 1974, **58**, 355.
39. D. S. Kenny, DeVine – the way it was developed – an industrialist's view, *Weed Sci.*, 1986, **34**(Suppl.), 15.
40. W. H. Ridings, Biological control of stranglervine in citrus – a researcher's view, *Weed Sci.*, 1986, **34**(Suppl.), 31.
41. W. H. Ridings, D. J. Mitchell, C. L. Schoulties, and N. E. El-Gholl, Biological control of milkweed vine in Florida citrus groves with a pathotype of *Phytophthora citrophthora*, in 'Proceedings IV International Symposium Biological Control of Weeds', 1976, p. 224.
42. S. H. Woodhead, Field efficacy of *Phytophthora palmivora* for control of milkweed vine, *Phytopathology*, 1981, **71**, 913.
43. B. S. Crandall and W. L. Baker, The wilt disease of American persimmon caused by *Cephalosporium diospyri*, *Phytopathology*, 1950, **40**, 307.
44. C. A. Griffith, 'Persimmon wilt research', Annual Report 1969–1970, Noble Foundation Agricultural Division, Ardmore, Oklahoma, 1970.

45. C. L. Wilson, Wilting of persimmon caused by *Cephalosporium diospyri*, *Phytopathology*, 1963, **53**, 1402.
46. C. L. Wilson, Consideration of the use of persimmon wilt as a silvicide for weed persimmon, *Plant Dis. Rep.*, 1965, **49**, 789.
47. G. E. Templeton and D. K. Heiny, Improvement of fungi to enhance mycoherbicide potential, in 'New Directions in Biological Control. Alternatives for Suppressing Agricultural Pests and Diseases', ed. R. Baker and P. Dunn, UCLA Symposia on Molecular and Cellular Biology, New Series, A. R. Liss, New York, vol. 112, p. 279.
48. E. E. Trujillo, F. M. Latterell, and A. E. Rossi, *Colletotrichum gloeosporioides*, a possible biological control agent for *Clidemia hirta* in Hawaiian forests, *Plant Dis.*, 1986, **70**, 974.
49. J. S. Bannon, CASST™ herbicide (*Alternaria cassiae*): a case history of a mycoherbicide, *Am. J. Alt. Agric.*, 1988, **3**, 73.
50. H. L. Walker and J. A. Riley, Evaluation of *Alternaria cassiae* for the biocontrol of sicklepod (*Cassia obtusifolia*), *Weed Sci.*, 1982, **30**, 651.
51. H. L. Walker, A seedling blight of sicklepod caused by *Alternaria cassiae*, *Plant Dis.*, 1982, **66**, 426.
52. H. L. Walker and C. D. Boyette, Biological control of sicklepod (*Cassia obtusifolia*) in soybeans (*Glycine maxima*) with *Alternaria cassiae*, *Weed Sci.*, 1985, **33**, 212.
53. H. L. Walker and C. D. Boyette, Influence of sequential dew periods on biocontrol of sicklepod (*Cassia obtusifolia*) by *Alternaria cassiae*, *Plant Dis.*, 1986, **70**, 962.
54. R. Charudattan, H. L. Walker, C. D. Boyette, W. H. Ridings, D. O. TeBeest, C. G. Van Dyke, and A. D. Worsham, 'Evaluation of *Alternaria cassiae* as a mycoherbicide for sicklepod (*Cassia obtusifolia*) in regional field tests', Southern Coop. Ser. Bulletin 317, Alabama Agricultural Experimental Station, Auburn University, Alabama, 1986.
55. C. D. Boyette, Biocontrol of three leguminous weed species with *Alternaria cassiae*, *Weed Technol.*, 1988, **2**, 414.
56. L. A. Wymore and A. K. Watson, Interaction between velvetleaf isolate of *Colletotrichum coccodes* and thidiazuron for velvetleaf (*Abutilon theophrasti*) control in the field, *Weed Sci.*, 1989, **37**, 478.
57. L. A. Wymore, A. K. Watson, and A. R. Gotlieb, Interaction between *Colletotrichum coccodes* and thidiazuron for the control of velvetleaf (*Abutilon theophrasti*), *Weed Sci.*, 1987, **35**, 377.
58. P. C. Scheepens and A. Hoogerbrugge, Control of *Prunus serotina* in forests with the endemic fungus *Chondrostereum purpureum*, in 'Abstracts VII International Symposium Biological Control of Weeds', Biological Control of Weeds Laboratory, USDA-ARS, Rome, 1988.
59. M. D. de Jong, Risk to fruit trees and native trees due to control of black cherry (*Prunus serotina*) by silverleaf fungus (*Chondrostereum purpureum*), PhD Thesis, Landbouwuniversiteit te Wageningen, The Netherlands, 1988.
60. C. F. McRae and B. A. Auld, The influence of environmental factors on anthracnose of *Xanthium spinosum*, *Phytopathology*, 1988, **78**, 1182.
61. C. F. McRae, H. I. Ridings, and B. A. Auld, Anthracnose of *Xanthium spinosum* – quantitative disease assessment and analysis, *Austral. Plant Path.*, 1988, **17**, 11.
62. B. A. Auld, M. M. Say, and G. D. Millar, Influence of potential stress factors on anthracnose development on *Xanthium spinosum*, *J. Appl. Ecol.*, 1990, **27**, 513.

63. B. A. Auld, M. M. Say, H. I. Ridings, and J. Andrews, Field applications of *Colletotrichum orbiculare* to control *Xanthium spinosum*, *Agric., Ecosyst. Env.*, 1990, **32**, 315.
64. R. Charudattan, Role of *Cercospora rodmanii* and other pathogens in the biological and integrated controls of water hyacinth, in 'Proceedings International Conference on Water Hyacinth', UN Env. Prog., Nairobi, 1984, p. 834.
65. C. D. Boyette, G. E. Templeton, and R. J. Smith, Jr. Control of winged waterprimrose (*Jussiaea decurrens*) and northern jointvetch (*Aeschynomene virginica*) with fungal pathogens, *Weed Sci.*, 1979, **27**, 497.
66. T. L. Kirkpatrick, G. E. Templeton, D. O. TeBeest, and R. J. Smith, Jr, Potential of *Colletotrichum malvarum* for biological control of prickly sida, *Plant Dis.*, 1982, **66**, 323.
67. C. D. Boyette, G. E. Templeton, and L. R. Oliver, Texas gourd (*Cucurbita texana*) control with *Fusarium solani* f. sp. *cucurbitae*, *Weed Sci.*, 1984, **32**, 649.
68. G. J. Weidemann, Effects of nutritional amendments on conidial production of *Fusarium solani* f. sp. *cucurbitae* on sodium alginate granules and on control of Texas gourd, *Plant Dis.*, 1988, **72**, 757.
69. G. J. Weidemann and G. E. Templeton, Efficacy and soil persistence of *Fusarium solani* f. sp. *cucurbitae* for control of Texas gourd (*Cucurbita texana*), *Plant Dis.*, 1988, **72**, 36.
70. G. J. Weidemann and G. E. Templeton, Control of Texas gourd, *Cucurbita texana*, with *Fusarium solani* f. sp. *cucurbitae*, *Weed Technol.*, 1988, **2**, 271.
71. J. Ormeno-Nunez, R. D. Reeleder, and A. K. Watson, A foliar disease of field bindweed (*Convolvulus arvensis*) caused by *Phomopsis convolvulus*, *Plant Dis.*, 1988, **72**, 338.
72. G. M. Bridges, Mycoherbicides, in 'Patent Information Monograph 3(c)', ed. T. H. Lemon and M. J. R. Blackman, The Patent Office, London, 1992, p. 127.
73. B. A. Auld, Economic aspects of biological weed control with plant pathogens, in 'Microbial Control of Weeds', ed. D. O. TeBeest, Chapman and Hall, New York, 1991, p. 262.
74. USDA-ARS, 'Common Weeds of the United States', Dover Publications, New York, 1971, p. 463.
75. K. Khodayari and R. J. Smith, Jr, A mycoherbicide integrated with fungicides in rice, *Oryza sativa*, *Weed Technol.*, 1988, **2**, 282.
76. H. L. Walker, *Fusarium lateritium*: a pathogen of spurred anoda (*Anoda cristata*), prickly sida (*Sida spinosa*) and velvetleaf (*Abutilon theophrasti*), *Weed Sci.*, 1981, **29**, 629.
77. R. W. Jones and J. G. Hancock, Soilborne fungi for biological control of weeds, in 'Microbes and Microbial Products as Herbicides', ed. R. E. Hoagland, ACS Symposium Series 439, American Chemical Society, Washington DC, 1990, p. 276.
78. D. J. Daigle and W. J. Connick, Jr, Formulation and application technology for microbial weed control, in 'Microbes and Microbial Products as Herbicides', ed. R. E. Hoagland, ACS Symposium Series 439, American Chemical Society, Washington, DC, 1990, p. 488.
79. C. D. Boyette, P. C. Quimby, Jr, W. J. Connick, Jr, D. J. Daigle, and F. E. Fulgham, Progress in the production, formulation and application of mycoherbicides, in 'Microbial Control of Weeds', ed. D. O. TeBeest, Chapman and Hall, New York, 1991, p. 209.
80. J. S. Bannon, J. C. White, D. Long, J. A. Riley, J. Baragona, M. Atkins, and R. H.

Crowley, Bioherbicide technology: An industrial perspective, in 'Microbes and Microbial Products as Herbicides', ed. R. E. Hoagland, ACS Symposium Series 439, American Chemical Society, Washington, DC, 1990, p. 305.

81. M. P. Greaves, Formulation of microbial herbicides to improve performance in the field, in 'Proceedings 8th EWRS Symposium, Quantitative Approaches in Weed and Herbicide Research and their Practical Application, Braunschweig', 1993, p. 219.
82. P. C. Quimby, Jr, F. E. Fulgham, C. D. Boyette, and W. J. Connick, Jr, An invert emulsion replaces dew on biocontrol of sicklepod – a preliminary study', in 'Pesticide Formulation and Application Systems', ed. D. A. Houde and G. B. Beestman, ASTM-STP 980, American Society for Testing and Materials, Philadelphia, 1988, vol. 8, p. 264.
83. W. J. Connick, Jr, D. J. Daigle, and P. C. Quimby, Jr, An improved invert emulsion with high water retention for mycoherbicide delivery, *Weed Technol.*, 1991, **5**, 442.
84. L. J. Stowell, Submerged fermentation of biological herbicides, in 'Microbial Control of Weeds', ed. D. O. TeBeest, Chapman and Hall, New York, 1991, p. 225.
85. H. L. Walker, Granular formulations of *Alternaria macrospora* for control of spurred anoda (*Anoda cristata*), *Weed Sci.*, 1981, **29**, 342.
86. L. A. Wymore and A. K. Watson, An adjuvant increases survival and efficacy of *Colletotrichum coccodes*, a mycoherbicide for velvetleaf (*Abutilon theophrasti*), *Phytopathology*, 1986, **76**, 1115.
87. J. D. Caulder, A. R. Gotlcib, L. Stowell, and A. K. Watson, Herbicidal compositions comprising microbial herbicides and chemical herbicides or plant growth regulators, European Patent Office, Application No. EP 0 207 753 A1, 1987.
88. M. P. Greaves, J. A. Bailey, and J. A. Hargreaves, Mycoherbicides, opportunities for genetic manipulation, *Pestic. Sci.*, 1989, **26**, 93.
89. H. C. Kistler, Genetic manipulation of plant pathogenic fungi, in 'Microbial Control of Weeds', ed. D. O. TeBeest, Chapman and Hall, New York, 1991, p. 152.
90. D. C. Sands and R. V. Miller, Evolving strategies for biological control of weeds with plant pathogens, *Pestic. Sci.*, 1993, **37**, 399.
91. A. Kelman, W. Anderson, S. Falkow, N. V. Federoff, and S. Levin, 'Introduction of Recombinant DNA-Engineered Organisms in the Environment, Key Issues', National Academy Press, USA, 1987.

CHAPTER 14

The Registration of New Natural Pesticides

JACK R. PLIMMER

14.1 Introduction

Although the use of conventional chemical pesticides is predominant among pest management practices, there is strong continuing interest in natural products as sources of new pesticides and novel pest management technology. For registration, the US Environmental Protection Agency (EPA) separates pesticides into two general categories:

- Conventional chemical pesticides.
- Biochemical and microbial pesticides.

Natural products generally fall into the latter category. The EPA has specified the test requirements for registration in the USA in guidelines for registration of biorational pesticides (Subdivision M of CFR 158).

The guidelines prescribe a tier testing system. Toxicity tests must be conducted using the maximum hazard concept in terms of dose, concentration, and route of administration. Regulatory issues are discussed in relation to the accompanying scientific developments. The potential application of chemicals that affected insect behaviour and development appeared an extremely promising approach to the reduction of adverse environmental and health impacts associated with many conventional pesticides. In addition to these benefits, the new biopesticides generally affected a very specific range of target species. However, although juvenile hormones and semiochemicals offer exciting new approaches to pest control and are potentially valuable components of pest management strategies, commercialization of these techniques has been slow.

Profitable development has been difficult for several reasons:

- Regulatory requirements may appear burdensome.
- Unconventional technologies may call for special formulations or application equipment.
- It may not be easy to demonstrate the efficacy of non-lethal techniques.

- Investigators who derive funds from the public sector may not have the resources to carry the technology beyond the experimental stage.
- As most activities in this field have been heavily research-oriented, few workers possess the additional resources to obtain safety data to satisfy regulatory requirements.

Difficulties in preparing and administering guidelines for registration of biopesticides lie in the novelty of the pest control agents and the diversity of the techniques involved in their effective use. The expertise of regulators who are responsible for making safety assessments is constantly challenged by the rapid progress in molecular biology and other disciplines. Although regulatory authorities are familiar with conventional pesticides, safety assessment of biopesticides may demand totally different approaches.

14.2 Natural versus Synthetic Products

Regulatory policies and actions play a critical role in the development of pest control technologies. In the USA, the EPA has recognized 'biorational' pesticides as meriting special consideration in the registration process. Some critics maintain that the regulatory actions taken have not sufficiently stimulated the development of alternatives to conventional insecticides. This attitude is reflected in some drafts of new legislation.

The final chapter of Rachel Carson's book 'Silent Spring', published in 1962, is entitled 'The Other Road'.[1] In it are described some alternative approaches to the control of insect pests that showed promise as replacements for the synthetic insecticides that had been used extensively in the preceding two decades. Chemicals derived from the insects' own biochemical processes were being identified and their physiological roles were becoming recognized. The function of hormones and semiochemicals in insect behaviour, reproduction, and development offered vistas of new, specific, control chemicals that would differ radically from their predecessors.

There are a variety of reasons why the vision of such natural products as insecticides of the future has not been realized in practice. Synthetic chemicals continue to dominate the market. The example of the synthetic pyrethroids shows how a natural insecticidal class of compounds can be modified structurally to afford extremely valuable commercial insecticides. At the same time, synthesis chemists, in elaborating new biologically active molecules, have taken pains to observe constraints imposed by the increasing knowledge of effects on the environment and on non-target organisms.

Many currently used pesticides were discovered by random screening of thousands of organic compounds of synthetic origin. Although this approach originally proved extremely fruitful, successes became rarer and more costly as increasing numbers of candidate pesticides were examined to recognize potential leads. This increase was necessitated, in part, by more stringent requirements for registration, particularly considerations of effects on the environment and non-target organisms.

Improved understanding of the biochemical basis of pesticide action made it feasible to design molecules to target a specific enzyme system or biological process, and suggested new approaches to discovering new biologically active molecules. When the study of structure–activity relationships was linked with biological understanding, the outcome provided a rational approach to synthesis. New molecular structures derived by this process were termed 'biorational pesticides', and their synthesis offered a new approach to the production of biologically active molecules. An advantage of such compounds is that their effects may be limited to the target species.

As the knowledge of the biological role of natural products increased and new, rapid methods for elucidation of their structures became available, they also began to receive new attention as leads for pesticide structures.[2,3] Because the biochemical bases of pesticidal action are now better understood, more sophisticated screening procedures are being devised. The identification of receptors associated with particular biochemical sites may afford screening procedures that reveal leads to selected modes of action and may provide rapid and effective approaches to the recognition of pesticidal and pharmacological activity.

Plants have long been a source of medicine and insecticides. Their medicinal use dates from the early history of human society – the processes by which medicinal principles of plants were recognized originally by mankind have been obscured by the passage of time. Little is known of what must have been an empirical process of testing, trial, and observation.

Similarly, the ability of certain plants to withstand the attack of insects and other pests has long been part of the traditional body of knowledge that has contributed to the development of agriculture and determined many of its practices. However, the evolution of scientific thought and the emergence of scientific disciplines brought new, more rational, approaches to the description of natural phenomena.

The components of plants that possessed toxicity to mammals were a primary focus for scientific investigation. The recognition that plants contained insecticidal and other biological principles of potential economic value came at a time when structural organic chemistry was at an early stage of development. Many years of painstaking effort were required to elucidate and confirm the structures of complex plant chemicals. For example, determination of the structure of strychnine took many years and required the resources of major academic laboratories.

The industrial revolution stimulated the development of chemical and physical sciences. Progress in these sciences was dominated by nations in which institutions of higher learning were closely associated with the needs of industrial production and chemical industry. Western industrialized society favoured manufacturing systems that were based on concentrations of expertise, capital, raw materials, and proximity to markets. Under these conditions, early successes in the treatment of disease by synthetic drugs strongly supported an approach to the discovery of biological activity by screening available compounds that could be synthesized with relative ease (and subsequently manufactured on a large scale), in preference to seeking exotic plants or other sources of new molecules for the same purpose.

From the industrial standpoint, empirical screening of synthetic organic compounds to discover new medicines or pesticides has the merit that successful leads can be integrated readily into the process of manufacture. If their synthesis or discovery takes place in the manufacturer's own laboratories, patent protection and subsequent development can be facilitated. The raw materials are generally petrochemicals that can be converted into more complex molecules or intermediates, so drug or pesticide production becomes part of a well-established technology.

14.3 Regulatory Aspects

Increasing public awareness of the implications of pesticide use has stimulated legislative activities; these concerns have been reflected in the evolution of current regulation. Public pressure for action has progressed from concerns over problems of potential acute intoxication to less defined, but equally troubling, questions of subtle effects on health and the environment.

The increasing quantity and variety of synthetic organic chemicals used for pest control and the growth of the industry were responsible for the introduction of new legislation to control the movement of pesticides in interstate commerce. The authority for regulation of pesticides lies in the *Federal Insecticide, Fungicide, and Rodenticide Act* (FIFRA),[4] which states that a pesticide must be registered with the Administrator of the EPA before it can be sold or distributed in the USA, and a pesticide is defined as 'any substance or mixture of substances intended for preventing, destroying, repelling or mitigating any pest'. Attractants are identified as a class of pesticide-active ingredients, and pheromones are included in this class under the definition of attractant 'as any substance or mixture of substances which through their property of attracting certain animals are intended to mitigate a population of pest species'.

The history of pesticide regulation in the USA has been reviewed by Upholt.[5] The *Federal Insecticide Act* was approved by the US Congress in 1910. Primary concerns were proper ingredient statements, adulteration, and misbranding. The Food and Drug Administration (then part of the US Department of Agriculture) set tolerances for lead and arsenic (used as insecticides) for shipments in interstate commerce, because in 1926 Great Britain had refused to import American apples contining excessive residues.

Special requirements exist for pesticides that will be used on food crops. The requirement that a tolerance must be established for all pesticides registered for use on food crops was introduced by modification of the *Federal Food Drug and Cosmetic Act* in 1954.[6] Section 408 sets tolerances for pesticides on raw agricultural commodities. Section 409 of the FFDC Act concerns residues that concentrate in processed food above the level authorized to be present in or on the raw commodities. This clause states that when the additive is put to its intended use there must be 'reasonable certainty' that there will be 'no harm' to consumers. The 'Delaney clause' is included in this section; it prohibits the approval of a food additive that has been shown to 'induce cancer'. While Section 408 recognizes that both risks and benefits are involved in pesticide use, Section 409 does not authorize the consideration of benefits. Strong controversy remains over this complex issue. A report commissioned by the National Academy in 1987 contains the background and a discussion of the process of tolerance setting with special emphasis on the effect of the Delaney clause.[7]

The responsibility for pesticide registration resided in the Secretary of Agriculture and, from 1964, the Secretaries of the Interior and of the Department of Health, Education and Welfare have provided advice on the questions of undue hazards to fish and wildlife and to human health. There have been many changes in the provision of this *Act* as it has evolved from the 1947 version. The responsibility for implementing the regulations was transferred from the Secretary of Agriculture to the Administrator of the Environmental Protection Agency under the *Federal Environment Pest Control Act* of 1972 (FEPCA).[8] This *Act* was basically an amendment of *FIFRA* and there have been subsequent amendments to this act.

14.4 Pesticides and Biorationals: Policy Considerations

The popularity of 'Silent Spring' and many other publications that are concerned with pesticides and environmental issues made it clear to legislators that these public issues demand serious consideration. Techniques for controlling pests have continued to receive considerable legislative scrutiny and oversight. Some difficulties that confront the policy maker or the regulator have been outlined by Perkins,[9] who has discussed insect control and entomologists and reviewed the changing approaches to control of insect pests.

He describes what he terms 'the three paradigms of control'. The first of these, chemical control, appeared to be dramatically successful and conferred, in fact, outstanding benefits in terms of public health and crop production. These benefits were offset to some extent by adverse effects as problems of environmental contamination, effects on non-target organisms, the emergence of resistance, and the potential for more profound, long-term harmful effects of man were recognized. Two successor paradigms were developed during the late 1950s and each has been advocated strongly by its protagonists. The concepts were those of integrated pest management and total population management. The latter was exemplified in the successful, large-scale USDA programme to eradicate the screw worm fly in the USA by releasing sterile male insects.[10]

Each approach has been marked by outstanding successes and also by strong partisanship. There have been major differences and divisions among adherents who strongly uphold their claim to an infallible solution to the pest problem. However, from the regulatory standpoint, there is a strong commitment on the part of the Federal government to the concept of integrated pest management (IPM), so compatibility of new pest control chemicals or technologies with IPM will be an important factor in approval of their use.

Regulatory policy makers, whose jurisdiction relates only to certain components of the technology, may find the complexity of modern pest control technology perplexing. The manufacture, distribution, and use of synthetic organic chemical pesticides has become a paradigm that dominated the formative stages of regulatory philosophies. Considerable knowledge of the consequences of widespread use of chemical pesticides has accumulated and this information is disseminated widely. It is often the subject of public debate and the position of the regulatory agencies reflects their accommodation of both technical needs and public opinion. The agencies have promulgated methods of assessment of risks and benefits and their guidelines have, for the most part, been recognized as capable of providing valuable data for acceptance or denial of registration of the product. Among the major criticisms of the guidelines by the pesticide industry are the costs of such test programmes and the delays that a company must face in bringing a new product to market.

These costs and delays are passed on to the producer and the consumer. Additional costs entailed in obtaining data to support use on specialty crops and the slow pace of approval of new pesticides means that some agricultural producers have difficulty in obtaining pesticides to meet specialized needs.

Important reasons why many chemical pesticides have become unacceptable are their potential long-term, low-level toxicity to mammals, the potential for adverse effects on the environment or wildlife, and the development of pesticide resistance by target species.

These factors are recognized in the process of discovery and are important in designing new active ingredients. At the same time, regulatory agencies have developed reliable methods of predicting many adverse effects that might be associated with new chemicals. However, the assessments can never be totally complete because scientific knowledge changes continuously and many questions, particularly those relating to long-term ecological implications of pesticide use, remain incompletely answered. The complexity of the factors influencing large-scale, long-term ecological changes and the quantitative and qualitative nature of such effects generally limits investigations to the few institutions that have the required combination of intellectual skills, experimental techniques, and financial resources. Additionally, the interpretation of the accumulated data presents a major challenge.

The widespread introduction of synthetic organic pesticide use in the 1950s was followed by the recognition of the ubiquitous presence of environmentally persistent chemicals. After monitoring capabilities were strengthened by the availability of gas chromatography and other instrumental techniques at the end of the decade, environmental monitoring became a rapid, routine procedure and a flow of scientific publications resulted. Popular awareness of the issues followed dissemination of this information and demands for government action were generated by the public perception that the environmental burden of pesticides was increasing and that adverse effects were being observed.

The regulatory agencies reacted by requesting more data from registrants. The first notice requesting data on the environmental behaviour of pesticides was sent to registrants (PR 70-15) by the USDA in 1970.[11] Environmental issues were becoming more prominent on the national agenda and, in 1970, the EPA was formed. Responsibility for registration of pesticides was transferred from the Secretary of Agriculture to the Administrator of the EPA.

Uncertainties associated with assessment of the risks posed by conventional pesticides generated pressures to develop alternative methods of pest control. Such concerns were reflected in the wording of the legislation establishing the EPA, and the Administrator of the EPA was charged with the responsibility to conduct research in alternative methods of pest control.

14.5 Scientific Progress and Response to Environmental Concerns

Among the new pest control strategies, the concept of IPM found strong advocates. Although this term signified many different things to many different people, in practice its adoption opened the way for combination and rationalization of pest control techniques to minimize harmful environmental effects and, at the same time, meet the needs of economically sound agricultural production systems.

Conventional chemicals continue to play an important role, but a variety of techniques were recognized as potential components of IPM. Among these were the use of semiochemicals (*i.e.*, pheromones and other compounds that affect behaviour), juvenile hormones (JH) and their analogues, and biological controls (predators, parasites, and pathogenic organisms). Some natural products, such as pyrethrum, nicotine, and other natural insecticides, already had well-established niches in insect control.

During the 1960s and 1970s progress in biological science generated new approaches to pest control that appeared to offer preferable alternatives to the application of conventional chemical pesticides and it became apparent that natural products might merit special regulatory consideration.

14.6 Testing Requirements

To register a pesticide in the USA requires that the registrant submit a formal application and provide the EPA with data concerning the ingredients of the pesticide, its intended use, and data in support of the registration. Much of the data required relates to the toxicology of the pesticide and its potential for adverse effects on man and the environment. For conventional pesticides, a comprehensive testing programme is usually necessary and a considerable amount of data must be developed at substantial cost. The tests must be conducted in compliance with 'Good Laboratory Practices' (GLPs).[12] The conduct of tests and reporting of data for the purposes of *FIFRA* are described in the 'Pesticide Assessment Guidelines' prepared by EPA, and GLP guidelines are mandatory for such studies.

For a tolerance to be granted before the pesticide is used on crops or feed requires that the registrant conduct extensive mammalian toxicity testing and demonstrate that the pesticide and its metabolites or degradation products are not carcinogens. To accomplish this usually requires studies of metabolism in plants and animals and studies of environmental fate. The levels of residues in treated products and the environment must be determined. Tests may also be necessary to show whether there is any impact on wildlife, beneficial species, and non-target organisms. From the test data described above and the intended use patterns of the pesticide, EPA makes a safety assessment and determines whether or not the pesticide should be registered.

Not only has the cost of tests to meet registration requirements grown enormously, but the time required for review has grown longer as more tests are included in the process and the personnel of regulatory agencies become more heavily burdened. As a consequence, the registrant may have only a limited period to exploit a patented product. These factors have had a negative impact on the rate and number of new products entering the market and, in view of the costs entailed, industry can only recoup its development costs by designing products intended for major crop markets. Gaps in the pest control armoury are being filled by maintaining the use of older compounds through re-registration or through government-sponsored programmes to meet minor-use needs. There are needs that are not met adequately by the existing arrangements, and the categorization and registration of certain pest control active ingredients as 'biorational pesticides' was intended to improve the situation by providing a route to registration that recognized their special nature.

To accommodate 'third generation' pesticides and chemicals derived from studies of biochemical pathways, EPA issued guidelines in 1983 to establish data requirements for registration. 'Subdivision M: Guidelines for Testing Biorational Pesticides' describes protocols which may be used for testing biochemical and microbial pest control agents.[13] EPA drew several distinctions between conventional pesticides and biorationals. The latter were characterized by 'unique non-toxic mode of action, low use volume, target species specificity and natural occurrence'. They comprised two major categories: the 'biochemical pest control agents (*e.g.*, pheromones, natural insect and plant growth regulators and enzymes) and the microbial pest control agents (*e.g.*, micro-organisms).' The caveat that the chemical must be naturally occurring or structurally identical with the naturally occurring material if it were a product of synthesis appeared to eliminate JH analogues and analogues of semiochemicals from consideration. Minor differences in stereochemical isomer ratios would, however, not normally eliminate a classification as a biorational unless the isomers showed significantly different toxicological properties.

Although analogues would be excluded by the above criteria, EPA stated that these would be considerd on a case-by-case basis. Chemicals similar to biorationals would be classified as biorationals or conventional pesticides based on their chemical structure and mode of action in the target species. EPA stated that it will base its decision on the chemical and toxicological significance of differences in chemical structure, mode of action of synthetic analogues compared with those of the naturally occurring compounds, and differences in toxicity (in at least Tier I screening tests for biorational pesticides). Direct toxicity to the non-target organism based on Tier I testing may or may not exclude classification as a biorational pesticide.

EPA uses a maximum hazard testing concept in designing its toxicology and non-target risk assessment strategy. The first tier of test requirements embodies the maximum hazard challenge in terms of dose, concentration, and route of administration. If the results of the first tier tests indicate no adverse effects, data from the second and third tiers are not required.

Data requirements include:

- Product analysis.[14]
- Residue chemistry.
- Toxicology.
- Non-target organism hazard and environmental fate and expression.

Product information must describe the identification and quantitation of active ingredients and of the impurities likely to occur as part of the manufacturing process. The manufacturing process and purification steps must be described fully and there should be a demonstration of the chemical structure of each active ingredient.

A major advantage of pest control chemicals categorized under Subdivision M is that their environmental degradation will generally be facile and they will be unlikely to leach and contaminate ground water.

14.7 Provisions of Subdivision M

EPA assumes that the potential for exposure of human or non-target organisms (hazard) to biochemicals is often very limited. EPA's criteria for reduced data requirements are:

- Low exposure pesticide formulation (traps, controlled release formulations, *etc.*)
- Low rates of application [20 g or less active ingredient (a.i.) per acre (4000 m^2)].
- Non-aquatic use sites (applied directly to land).
- High volatility (reduces likelihood of residues on food or feed crops).

Biochemicals are subjected to a tiered sequence of tests. The first tier requires acute toxicity, irritation and hypersensitivity tests, short-term mutagenicity, and cellular immune response studies. The toxicity tests are summarized below.[13,15,16]

14.7.1 Tier I Tests

14.7.1.1 Acute Toxicity – Oral

This test (rat) is needed to support the registration of every biochemical (each manufacturing-use product and each end-use formulated product), except for highly volatile materials that cannot be administered by intubation. If the LD_{50} is less than $5\ g\ kg^{-1}$ body weight, additional tests must be conducted to establish an LD_{50} with 95% confidence intervals.

14.7.1.2 Acute Toxicity – Dermal

In this test (rat or mouse), single-dose LD_{50} is required for each manufacturing-use product and each end-use formulated product, except for highly volatile materials and gases that cannot be administered dermally.

14.7.1.3 Acute Toxicity – Inhalation

This test (mouse, rabbit, or guinea pig) is required if the manufacturing-use product is a gas, produces a respirable vapour, or 20% or more of the aerodynamic equivalent of the product is composed of particulates not larger than $10\ \mu m$ in diameter. These requirements also apply to end-use formulated products.

14.7.1.4 Irritation

The primary ocular (albino rabbit) and primary dermal (guinea pig or albino rabbit) tests are required for each manufacturing-use product and each end-use product. If the dermal irritation tests show that the material causes severe dermal irritation, it will be assumed that the material will also be a severe eye irritant.

14.7.1.5 Hypersensitivity

Data on any incidence of hypersensitivity, immediate or non-immediate, to humans or domestic animals that occurs during production or testing of the technical chemical, the manufacturing-use product, or end-use products must be reported.

14.7.1.6 Genotoxicity

A battery of tests is used. Data derived from short-term microbial mutagenicity tests are required to support the registration of each end-use product that requires a tolerance; or is likely to result in significant human exposure by inhalation or dermal routes, before or during the normal reproductive portion of the human lifespan; or if the active ingredient, or any metabolite thereof, is structurally related to a known oncogen, or belongs to any chemical classes containing known mutagens or oncogens. The test systems should address gene mutations, structural chromosomal aberrations, and other genotoxic effects, *e.g.* direct DNA damage and repair.

14.7.1.7 Cellular Immune Response

A battery of tests (mouse) is used. The data are required to support registration for each manufacturing-use product and each end-use formulated product. Studies of effects on immunocompetence include blood cell counts, leukocyte number and classes, including T and B cells, functional activity of blood leukocytes, macrophage number and function, and serum protein determination.

14.7.2 Tier II Tests

The observation of adverse effect(s) in Tier I studies may trigger the requirement for studies in the following areas:

- Mutagenicity.
- Toxicity – subchronic oral (mouse, rat or dog), dermal (rabbit or guinea pig), and subchronic inhalation (rat).
- Cellular Immune Response.
- Teratogenicity.

14.7.3 Tier III Tests

If testing indicates potential for adverse chronic effects based on subchronic effect levels in subchronic oral, dermal, or inhalation toxicity studies, pesticide use pattern (rate, frequency, site of application) or frequency level of repeated human exposure, data on a chronic exposure study are required to support registration for each end-use formulated product and each manufacturing-use product that may be used legally to formulate such an end-use product.

14.7.4 Oncogenicity

Oncogenicity studies may be required when active ingredients, or any of their metabolites, degradation products, or impurities produce a morphologic effect in any organ that could potentially lead to neoplastic change; or if adverse cellular effects suggesting oncogenic potential are observed in cellular immune response studies or in mammalian mutagenicity assays.

14.7.5 Non-Target Organisms

Data on hazards to non-target organisms proposed in Subdivision M cover the four areas of terrestrial wildlife, aquatic animals, plants, and beneficial insects.

The testing follows a tier scheme analogous to that for toxicological tests. Tier I represents the maximum hazard test and Tier II consists of environmental fate studies that will be required only if indicated by the outcome of Tier I tests.

14.7.6 Efficacy

Field tests carried out to determine efficacy trigger the requirement for an EPA Experimental Use Permit (EUP) if they are to be conducted on an area $\geqslant 10$ cumulative acres (40 000 m^2) of land or $\geqslant 1$ surface acre (4000 m^2) of water. Any crops grown on these sites must be destroyed or used for feeding experimental animals unless an exemption from tolerance has been granted by EPA. Data that must be submitted for an EUP include product chemistry and toxicological testing.

In the case of semiochemicals, this acreage restriction has made it difficult for investigators to conduct exploratory studies of pheromone effectiveness in the field. Determination of the efficacy of pheromones depends on assessments of population reduction or reduction of damage to crops or plants, and it has proved difficult to obtain statistically satisfactory data within small plots, although these may be useful for screening chemicals, blends, and concentrations.[17] The data requirements for an EUP are substantially the same as the data that would be required for full registration as a biochemical.

14.8 Biochemicals

There are four major classes of biochemical agents – semiochemicals, hormones, natural plant regulators, and enzymes.

14.8.1 Hormones

Hormones are defined as 'biochemical agents that are synthesized in one part of an organism and are translocated to another where they have controlling, behavioural, or regulating effect.'

14.8.2 Natural Plant Regulators

'Natural plant regulators are chemicals produced by plants that have toxic, inhibitory, stimulatory, or other modifying effects on the same or other species of plants' (some termed 'plant hormones' or 'phytohormones').

14.8.3 Enzymes

These are defined as 'protein molecules that are instruments for expression of gene action and that catalyse chemical reactions.'

14.8.4 Semiochemicals

Semiochemicals may be naturally occurring compounds or synthetics. They may be defined as substances or mixtures of substances emitted by one species that modify the behaviour of receptor organisms of other individuals of like or different species ('EPA Subdivision M'). Several types of semiochemicals may be useful for pest management. Although interest focused mainly on sex attractant pheromones during the 1970s and 1980s, many types of behavioural activity may be elicited by chemical stimuli – the investigation of the potential application of various types of semiochemicals to pest management has continued to be an active research field.[18,19] A number of pheromones have been registered and there is considerable field experience of pest control techniques that employ pheromones.

Some definitions of other types of behaviour-affecting substances have been given ('EPA Subdivision M'):

- Kairomones – chemicals emitted by one species that modify the behaviour of a different species to the benefit of the receptor species.
- Allomones – chemicals emitted by one species that modify the behaviour of a different species to the benefit of the emitting species.
- Synomones – chemicals emitted by one species that modify the behaviour of a different species to the benefit of both the emitting and receptor species.

Pheromones may be used in pest management as follows:[20]

- In insect traps for monitoring or survey purposes.
- In a mass trapping programme to reduce insect populations.
- In combination with insecticides to attract insects to an area treated with insecticides or to a device containing insecticide.
- To permeate the air and suppress insect population by disrupting mating or aggregation.

The above uses of pheromones require extremely small amounts of material and the targets are quite specific. Many of the semiochemicals are identical to, or closely resemble, naturally occurring materials in terms of chemical composition. They are generally readily degraded in the environment and show low toxicity to non-target species.

Registration of pheromones became an important issue in the USA and in Europe.[21] This class of compounds appeared to have considerable promise as pest management chemicals. They are species specific. Sex attractant pheromones of many major lepidopteran pests are blends of aldehydes, alcohols, or acetates derived from long chain aliphatic hydrocarbons containing one or two double bonds. Their mammalian toxicity is very low and they are rapidly degraded in the environment. The major challenge to the users was to formulate these compounds and prolong their activity to extend throughout the mating season of the target insect. They are volatile and readily oxidized and controlled-release formulations are necessary to maintain their efficacy.[22] Suppression of population by disruption of mating communication appeared to be a promising strategy for control.[23] It was essential that the technique be applied to a low population because mating frequency could be reduced most effectively as natural sources of pheromone (calling females) were widely dispersed.

Semiochemicals were recognized as receiving special consideration by regulators because they were generally non-toxic to man, rapidly biodegradable, highly specific in their activity, and the amounts used were extremely small. The regulatory acceptance of semiochemicals in pest management was the topic of a North Atlantic Treaty Organization (NATO) study under the auspices of the NATO Committee on the Challenges of Modern Society. The recommendations of a NATO Advanced Research Institute on Chemical Ecology held in 1978 were that regulatory agencies of all countries should publish guidelines for registering behaviour-modifying chemicals for pest control.[21] Chemicals that provided insectistasis without insecticidal action should be distinguished from conventional insecticides. At the time that the NATO study was planned, there was uncertainty within regulatory agencies as to the special consideration, if any, that should be given to semiochemicals, as distinct from conventional pesticides. Minks has summarized the position in a number of countries world-wide.[24] Some countries make no distinction and the progress of registration is slow. In Europe, in 1987, one pheromone was registered for mass trapping in two countries (France and Norway), three pheromones had been granted registration for mating disruption in two countries (Germany and the USSR), and two provisional registrations had been granted in Switzerland.

The EPA issued a guidance document and, in May 1979, the EPA published its proposed policy on the regulation of biorational pesticides.[25,26] In the document, attention was drawn to the active ingredients that functioned by modes of action other than innate toxicity. These included living or replicable biological entities, such as viruses, bacteria, protozoans, and fungi. Also included were the naturally occurring biochemicals, such as plant growth hormones, insect semiochemicals, and hormones.

The stated policy of EPA was that, in its data requirements, it would take into account the fundamental differences between biorational pesticides and conventional pesticides and the lower risks of adverse effects. It also committed the EPA to develop programmes to resolve safety issues and monitor the effects of biorational pesticides on man and the environment. Guidelines were to be issued within 2 years of the issuance of the policy statement. Registration of environmentally acceptable biorational pesticides would be accorded some priority. Submissions would be reviewed expeditiously and the data requirements would not be unduly burdensome. Furthermore, EPA would encourage demonstration of the practical value and safety of biorational pesticides and would rely heavily on the expertise of other Federal agencies concerned with biological pesticides for development and implementation of the biorational pesticides regulations and programmes.

The policy document recognizes specifically the low risks associated with the use of pheromones:

Of all the biorational pesticides, the naturally occurring biochemicals, namely insect hormones and pheromones, present the fewest problems to the Agency in hazard assessment. Their chemical nature, patterns of use and extremely small amounts, which must be applied for effect in the field (at least for pheromones) may, in fact, indicate risk levels of so low a magnitude that much formal safety testing for mammalian effects prediction may not be necessary. For example, when pheromones currently registered are used under actual agricultural (or silvicultural) field conditions, no significant changes in the environmental background of these naturally occurring biochemicals has been observed. This indicates little or no potential for ecosystem effects since levels of the compounds found in nature are not significantly altered.

The EPA has excluded the use of pheromones and substantially similar compounds from regulation if they are used in insect traps for detection or survey purposes.[27] This includes their use for pest control by attraction and removal of target organisms from the treated area, as long as this does not result in increased levels of the pheromone (or similar compound) in the treated area.

The commitment of the Federal government to IPM remains an additional stimulus for the development of biorational pesticides.

14.9 Microbials

Microbial pest control agents include bacteria, fungi, viruses, and protozoans. Each variety of subspecies of a microbial pest control agent should be tested. Biotechnology has progressed rapidly in recent years and many strategies for modifying organisms are becoming available. This has raised many major issues of policy that affect many areas of biological sciences, so questions relating to the products of biotechnology have been addressed by a number of advisory groups who have provided guidelines and regulations to supplement existing legislation. One aspect of these developments is their application to pest control, and the USDA, the EPA, and the FDA are individually responsible for the administration of regulations concerning the products of biotechnology. The EPA is responsible for registration and issued an updated version of the guidelines for the registration of microbial pesticides in 1989.[28]

Although the topic of registration of living organisms (microbials), whether genetically modified or not, as biopesticides falls generally outside the scope of this review, experience suggests that indigenous micro-organisms will be easier to register, whereas introduced species will require more safety, environmental, and non-target impact studies, and genetically modified organisms will be treated on a case-by-case basis with more caution. The regulations are complex and must respond to a rapidly developing field of science; therefore, the following section is only a brief outline of procedures and data requirements.

Biocontrol agents include a variety of organisms and, in terms of sales, the most important class is represented by the bioinsecticides, of which *Bacillus thuringiensis* Berliner (Bt) dominates the market (Chapter 10). Sales in 1990 were $120 million and the sales of Bt accounted for $110 million of this total. By contrast, in 1991, the total sales of agrochemicals were $26,800 million.[29] Microbial insecticides have the largest share of the market for biorational pesticides, a share which will probably expand as technical problems are overcome.

The major commercial products and the most widely used are based on Bt. This organism exerts its activity after ingestion by the larvae when it releases a toxin that degrades the midgut epithelium. The insect ceases to feed and death occurs within 2–6 days.

There are many different strains of this soil-borne bacterium, which has been marketed since the late 1950s. It is primarily used for the control of lepidopteran species, although a wide variety of Bt strains exist which show varying degrees of insecticidal activity against specific target species. For example, the strain Bt *israelensis* shows activity against dipterans, including mosquitoes and biting flies.[30]

The introduction of engineered or nonindigenous organisms is regulated by the Animal and Plant Health Inspection Service of the USDA under the *Plant Pest Act*. The introduction and movement of plants, animals, organisms, *etc.*, into areas where they are not indigenous is controlled by the Plant Protection and Quarantine Division of APHIS. *FIFRA* represents a licensing programme with data requirements as specified under CFR 158 Subpart M, as outlined below, but commonly some data requirements may be waived.

APHIS regulations serve as a permitting programme (the authorities are the *Federal Plant Pest Act* and the *Plant Quarantine Act*). The scope of the regulations includes interstate movement of any plant pest and field release of genetically engineered organisms. For commercialization no license is needed, but delisting is required.

Procedurally, APHIS and EPA each require applications using specific formats and each Agency has a different format. Under the regulations a period is allowed for public comment. After review by appropriate organizations, a permit is issued stipulating requirements to be observed during the test programme and data are then obtained to show activity and efficacy.

The guidelines for registration of microbials were based largely on experience with Bt and some insect viruses, but microbial pest control agents also include algae, fungi, and protozoa. Both naturally occurring organisms and those that are strain improved by natural selection or by genetic manipulation are included. The guidelines apply to both types, but additional data may be required for the latter as determined in each individual case.

The data requirements for toxicology, ecological effects, and environmental expression are based on a tiered testing scheme. Product analysis requires the provision of sufficient data to identify the active ingredient(s) and any inert additives. Contaminants and impurities must also be identified. In the case of genetically altered organisms, the method of alteration must be described, identity of inserted or deleted genes must be stated, information on the gene control region and on genetic stability (reversion tendency or rate of exchange with other organisms), and descriptions of the phenotypic traits to be gained or lost must be provided. Microbes are included under the definition of new chemicals under the *Toxic Substances Control Act.* Therefore, before testing of intergeneric organisms can proceed, a Premanufacturing Notice must be submitted and EPA approval must be given before production commences.

Toxicology data are arranged in three tiers. Tier I contains a battery of short term tests to evaluate potential for toxicity, infectivity, and pathogenicity. If, in Tier I testing, either toxicity or infectivity is observed, and there is no evidence of pathogenicity, Tier II testing data are required. The tests in Tier III are required to resolve issues of known or suspected human pathogenicity or to test for particular adverse effects of intracellular parasites of mammalian cells.

Residue data may be required to show the amount of a microbial (or associated toxins) on a food or feed crop and are only required if there are significant human health concerns that arise from toxicology testing.

The tests for ecological effects, and environmental expression, are also organized on a tiered basis. Tier I consists of maximum dose single species hazard testing on non-target organisms. If adverse effects are observed in Tier I, Tier II tests for population dynamics (fate and expression) in the environment must be carried out to assess potential exposure to the microbial agent. If there appears to be any significant exposure to the organism as shown by the data from Tier II tests, Tier III test data are required. Tier III studies are designed to show dose–response effects or to examine chronic effects. These will be performed to determine whether the minimum infective dose is less than the exposure or whether there are factors in the environment that might decrease the observed effects. Tier IV tests, under environmental or simulated environmental conditions, are required on a case-by-case basis if there are factors that are not resolved in lower Tiers.

The manipulation of genetic material and its insertion into the gene sequences of micro-organisms provide a valuable technique for the selective delivery of toxins or bioregulators to pest species. The potential for commercial development of bioinsecticides and the utilization of baculoviruses involving gene insertion technology is being explored by industry.[31] Examples of approaches that may abe exploited are the inhibition of insect development by interference with a critical biochemical pathway, such as the synthesis of ecdysone glucotransferase, the enzyme responsible for the inactivation of ecdysone.

There are difficulties in the production of such biopesticides. There are also major technical problems in effectively delivering organisms to ensure rapid onset of biological activity and to ensure that the duration of such activity is sufficiently long to combat pest infestation. Baculoviruses are also deactivated rapidly in sunlight – the problem of devising satisfactory formulations has challenged scientists in industry and the public sector since their introduction. Attempts are continuing to exploit insect viruses, but these do not yet match conventional insecticides in terms of cost and performance. However, as pressures to provide alternatives to conventional chemical pesticides grow stronger, there is greater incentive to develop microbially based technologies.

The exploitation of biotechnology is taking many experimental forms, such as the expression of insect toxins by the introduction of genetic sequences into plants or organisms endogenous to plants. The potential effects of the introduction of novel genetic material into the environment are unknown; therefore, such experiments must initially be conducted under conditions that prevent escape of material to the environment. They have attracted considerable attention from regulatory agencies and the public. As these developments are now reaching the field evaluation stage, they represent new challenges for the regulatory authorities, who are adopting a case-by-case basis until there is more widespread understanding of the consequences of their large scale use in agriculture.

14.10 Commercial Development and Current Legislative Activity

Conventional chemical pest control has generally served agriculture well, but the appearance of pesticide resistance in the target species is of concern to both users and producers of pesticides. New chemicals may provide a temporary remedy, but the phenomenon of resistance will necessitate the introduction of new approaches to pest control; such changes will require legislative recognition if they are to be adopted rapidly. The EPA has stated that it wishes to encourage the development of newer pesticides that would present less risk and could be registered with reduced data requirements.

There has been an effort to accelerate the registration of microbials. The factors considered are low acute and chronic toxicity, low persistence, high K_d, low toxicity to wildlife and aquatic organisms, formulations that minimize exposure, and the absence of effects such as skin sensitization, mutagenesis, carcinogenesis, *etc.* A number of microbial pesticides have been registered in recent years and the registration process appears to have dealt expeditiously with this class of pesticides.

Unfortunately, few of the pest-control techniques that offer promising alternatives to chemical control are highly profitable for a variety of reasons. Many are abandoned or neglected because neither industry nor the public sector have adequate resources for their investigation and development. The market is a dominant force and insect pheromones have been discussed at some length in this review because they exemplify the vicissitudes of a highly sophisticated technology during the struggle to achieve and maintain a commercial position among available pest management techniques.

Although the EPA has attempted to facilitate registration of biorationals the comment has been made that the registration process remains a disincentive to the development of new pheromones for pest management.[32] The reasons presented are the limited market, the cost of generating data, the time required for registration, and the practical constraints on the use of pheromones.

Gossyplure was the first pheromone to be registered for protection of a field crop. Potential residues in cottonseed oil were one of the EPA's major concerns and the cost of registration was reported to be close to \$1 million.[33,34]

Siddall estimated the costs of developing a crop protection pheromone chemical product at about $1.3 million.[35] Present costs would be considerably greater than these 1979 figures, which represent primarily the cost of development of information for a government registration petition. Currently, the expressed requirements of Subdivision M indicate that some costs could be eliminated if the compound showed no questionable effects in the course of the first tier of tests. Although costs of registration are considerably lower than those for conventional pesticides, the specificity of individual pheromones considerably limits their market potential. In addition, they must generally be used in combination with other pest control measures, as efficacy for population suppression is population dependent and their use may be restricted to controlling incipient infestations when population levels are low.

The requirements for the development of residue analytical methods for semiochemicals also appear to impose an additional constraint on the registrant unless a waiver is granted by the EPA.[32] The structure of many pheromone components would make it technically very difficult to devise methods that would have adequate sensitivity and be applicable to complex biological matrices.

Additional pressures are building to provide legislative force to encourage the use of alternative pest control agents. O'Connor[32] states that there is no clear rationale for the imposition of excessive data requirements because pheromones are not persistent and their volatility implies that exposure will be of very short duration. The addition of teratogenicity and subchronic toxicity tests of Subdivision M has substantially increased the costs of testing.

A further added cost is that of test chemicals. The testing requirements for pheromones call for larger quantities of expensive material than would be used in practice in an experimental programme. More material may be required to provide the toxicity data for an Experimental Use Permit than is needed to evaluate the efficacy of the material in the field. As an example, acute oral, dermal, and inhalation toxicity tests required 100 g of purified *Heliothis* sp. pheromone components. This amount far surpassed that recommended for application to several acres when intended for use in mating disruption programmes.

Adoption of the proposal that the EUP trigger for pheromones should be increased to an area of 500 acres (2×10^5 m^2) for non-food applications and 200 acres (8×10^4 m^2) for food use would create a more reasonable situation for experimental evaluation of efficacy. Relatively large areas are essential for adequate, statistically based demonstrations of efficacy, so this change may remove a major obstacle to the evaluation of this technique.

In 1982, pheromones, attractants, and parapheromones were available for 86 species of insects in the USA and formulations for insect control by mass trapping or mating disruption were available for 12 species.[18] On an international basis, pheromones and parapheromones were listed as being available for 138 insect species in 1986.[18]

During 1990 and 1991, 50% of the new compounds registered by the EPA were biological. The average time required for EPA registration was 30 months for chemical and 10 months for biological compounds.

Updated guidelines for the registration of biochemicals (Part B of the 'EPA Pesticide Assessment Guidelines') are awaited (1993). Future prospects for biochemicals could be improved if new legislation were to offer some relief from the burden of data required to support registration.

Added in proof. Since this chapter was prepared, the USEPA relaxed the requirements for experimental use permits and, in August 1995, expanded the minimum acreage from 10 to 250 acres. This policy includes the majority of lepidopteran pheromones, regardless of formulation, if applied at a maximum use rate of 150 grams a.i. per acre per year. Other arthropod pheromone products on food crops entering commerce would still require an EUP and a temporary tolerance (or exemption from the requirement of a temporary tolerance). Tests conducted on areas greater than 250 acres for all pheromones still require an EUP (Federal Register, August 30, 1995 Volume 60, Number 168 pp. 45156–45157).

On the same date, residues of certain lepidopteran pheromones were exempted from the requirements of a food tolerance on all raw agricultural commodities when the annual application did not exceed 150 grams per acre. The exemption refers only to the active ingredient and encapsulating materials must be cleared as inerts for use on food crops (Federal Register, August 30, 1995 Volume 60, Number 168 pp. 45060–45062).

14.11 References

1. R. Carson, 'Silent Spring', Houghton Mifflin, Boston, 1962, 368 pp.
2. J. J. Menn, Present insecticides and approaches to discovery of environmentally acceptable pesticides for pest management, in 'Natural Products for Innovative Pest Management', ed. D. L. Whitehead and W. S. Bowers, Pergamon, New York, 1983, p. 5.
3. M. Jacobson, Botanical pesticides, in 'Insecticides of Plant Origin', ed. J. T. Arnason, B. J. R. Philogene, and P. Morand, ACS Symposium Series 387, American Chemical Society, Washington, DC, 1989, p. 1.
4. Anon, 'Federal Insecticide, Fungicide, and Rodenticide Act', Public Law 152, US Statutes at Large 1, p. 331, April 26, 1910.
5. W. M. Upholt, Regulation of pesticides, in 'Handbook of Natural Pesticides: Methods', ed. N. B. Mandava, CRC Press, Boca Raton, 1985, vol. 1, 273.
6. Anon, 'Federal Food, Drug, and Cosmetic Act', Public Law 95-532, USCA Title 21, 321 A,B,C.
7. Committee on Scientific and Regulatory Issues, Board on Agriculture, National Research Council, 'Regulating Pesticides in Food: The Delaney Paradox', National Academy Press, Washington, DC, 1987, 272 pp.
8. Anon, 'Federal Insecticide, Fungicide, and Rodenticide Act', Public Law 92-516; US Code Title 7, Pt. 136 *et seq.*, October 21, 1972.
9. J. H. Perkins, Naturally occurring pesticides and the pesticide crisis, 1945 to 1980, in 'Handbook of Natural Pesticides: Methods', ed. N. B. Mandava, CRC Press, Boca Raton, 1985, vol. 1, p. 297.
10. E. F. Knipling, Use of insects for self destruction, in 'The Basic Principles of Insect Population Suppression and Management', US Dept. of Agriculture Handbook 512, Washington, DC, 1979, p. 318.

11. H. Alford, 'Notice to Registrants', PR 70-15, USDA, Washington, DC, 1970.
12. Anon, 'Title 40 Code of Federal Regulations Part 160', Good Laboratory Practice Standards; Final Rule.
13. US Environmental Protection Agency, 'Pesticide Assessment Guidelines Subdivision M: Microbial and Biochemical Pest Control Agents – Guidelines for Testing Biorational Pesticides', October 1981 (EPA No. 540/09-82-028), USEPA, Washington, DC, 1982.
14. Anon, 'Code of Federal Regulations', Title 40 Section 158.690 (a).
15. E. F. Tinsworth, Regulation of pheromones and other semiochemicals in the United States, in 'Behavior-Modifying Chemicals for Insect Management', ed. R. L. Ridgway, R. M. Silverstein, and M. N. Inscoe, M. Dekker, New York, 1990, p. 569.
16. R. J. Hodosh, E. M. Keough, and Y. Luthra, Toxicological evaluation and registration requirements for biorational pesticides, in 'Handbook of Natural Pesticides: Methods', ed. N. B. Mandava, CRC Press, Boca Raton, 1985, vol. 1, p. 231.
17. W. L. Roelofs and M. A. Novak, Small plot disorientation tests for screening potential mating disruptants, in 'Management of Insect Pests with Semiochemicals', ed. E. R. Mitchell, Plenum Press, New York, 1981, p. 229.
18. M. N. Inscoe, B. A. Leonhardt, and R. L. Ridgway, Commercial availability of insect pheromones and other attractants, in 'Behavior-Modifying Chemicals for Insect Management', ed. R. L. Ridgway, R. M. Silverstein, and M. N. Inscoe, M. Dekker, New York, 1990, p. 631.
19. 'Insect Pheromones and Other Behaviour-Modifying Chemicals: Application and Regulation', BCPC Monograph No. 51, BCPC, Farnham, 1990.
20. R. M. Silverstein, Pheromones: background and potential for use in insect control, *Science*, 1981, **213**, 1326.
21. F. Ritter, Conclusions and recommendations of the NATO Advanced Research Institute on Chemical Ecology, in 'Chemical Ecology: Odour Communication in Animals', ed. F. Ritter, Elsevier, Amsterdam, 1979, p. 403.
22. J. R. Plimmer, Formulation and regulation: Constraints on the development of semiochemicals for pest management, in 'Management of Insect Pests with Semiochemicals', ed. E. R. Mitchell, Plenum Press, New York, 1981, p. 403.
23. E. F. Knipling, Insect suppression by the use of attractants, in 'The Basic Principles of Insect Population Suppression and Management', US Dept. of Agriculture Handbook 512, Washington, DC, 1979, p. 421.
24. A. Minks, Registration requirements and status for pheromones in Europe and other countries, in 'Behavior-Modifying Chemicals for Insect Management', ed. R. L. Ridgway, R. M. Silverstein, and M. N. Inscoe, M. Dekker, New York, 1990, p. 557.
25. W. G. Phillips, EPA's registration requirements for insect behavior controlling chemicals – philosophy and mandates, in 'Pest Control with Sex Attractants and Other Behavior Controlling Chemicals', ed. M. Beroza, ACS Symposium Series 23, American Chemical Society, Washington, DC, 1976, p. 135.
26. S. D. Jellinek, 'Regulation of "biorational pesticides"; Policy statement and notice of availability of background document', Fed. Reg, **44**(94): 28093, 1979.
27. Anon, 'Code of Federal Regulations', Title 40 Section 162.3 (ff)(3).
28. Anon, 'Pesticide Assessment Guidelines Subdivision M: (Part A Microbial)', EPA No. 540/09-89-056, USEPA, Washington, DC, 1989.
29. K. A. Powell and A. R. Jutsum, Technical and Commercial Aspects of Biocontrol Products, *Pestic. Sci.*, 1993, **37**, 315–321.

30. B. C. Carlton and J. M. Gonzalez, Jr, Biocontrol of insects – *Bacillus thuringiensis*, in 'Biotechnology for Solving Agricultural Problems', ed. P. C. Augustine, H. D. Danforth, and M. R. Bakst, Beltsville Symposia on Agricultural Research 10, Martinus Nijhoff, Dordrecht, 1986, p. 253.
31. B. C. Black, Commercialization and potential of baculoviruses, *Agrochemical Division: Abstracts*, 206th ACS National Meeting, Chicago, 1993, Abstract 89.
32. C. A. O'Connor, III, Registration of pheromones in practice, in 'Behavior-Modifying Chemicals for Insect Management', ed. R. L. Ridgway, R. M. Silverstein, and M. N. Inscoe, M. Dekker, New York, 1990, p. 605.
33. E. L. Johnson, Approval of application to register pesticide product containing a new active ingredient, *Fed. Reg.*, **43**(82), 18018, 1978.
34. W. Tucker, Of mites and men, *Harper's*, 1978, **257**(1539), 43.
35. J. Siddall, Commercial production of insect pheromones – problems and prospects, in 'Chemical Ecology: Odour Communication in Animals', ed. F. Ritter, Elsevier, Amsterdam, 1979, p. 389.

Index